Description	Formula	Equation number	Page
Pooled variance for equal or unequal sample sizes	$s^2_{\text{pooled}} = \dfrac{(n_1 - 1)s_1^2 + (n_2 - 1)s_2^2}{n_1 + n_2 - 2}$	(11.5a)	220
Standard error of the difference between two means when sample sizes are equal	$s_{\bar{X}_1 - \bar{X}_2} = \sqrt{s^2_{\text{pooled}} \dfrac{2}{n}}$ [only when $n_1 = n_2 = n$]	(11.6)	220
Effect size index d, two-independent-samples test for means	$d = \dfrac{\lvert(\bar{X}_1 - \bar{X}_2)_{\text{obs}}\rvert}{s_{\text{pooled}}}$ (nondirectional test) or $d = \dfrac{(\bar{X}_1 - \bar{X}_2)_{\text{obs}}}{s_{\text{pooled}}}$ (directional test)	(11.7)	224
Test statistic (t), two-dependent-samples test for means	$t = \dfrac{\bar{D}}{s_{\bar{D}}}$ [$df = n - 1$]	(12.3)	249
Standard error of the mean differences	$s_{\bar{D}} = \dfrac{s_D}{\sqrt{n}}$ [where n is number of *pairs*]	(12.4)	249
Standard deviation of the differences, mean squared deviation and computational forms	$s_D = \sqrt{\dfrac{\sum(D_i - \bar{D})^2}{n-1}} = \sqrt{\dfrac{\sum D_i^2 - \dfrac{(\sum D_i)^2}{n}}{n-1}}$	(12.5)	249
Effect size index d, two-dependent-samples test for means	$d = \dfrac{\lvert\bar{D}_{\text{obs}}\rvert}{s_D}$ (nondirectional test) $d = \dfrac{\bar{D}_{\text{obs}}}{s_D}$ (directional test)	(12.6)	253
Power	Power $= 1 - \beta$	(12.7)	258
Population effect size index	$d_{\text{population}} = \dfrac{\mu_{\text{real}} - \mu_{H_0}}{\sigma}$ (one-sample test) $d_{\text{population}} = \dfrac{\mu_1 - \mu_2}{\sigma}$ (two-independent-samples test)	(12.8) (12.9)	266 267
	$d_{\text{population}} = \dfrac{\mu_{D_{\text{real}}}}{\sigma_D}$ (two-dependent-samples test)	(12.10)	267
Variance of the sample means, definitional formula	$s^2_{\bar{X}} = \dfrac{\sum(\bar{X}_j - \bar{X}_G)^2}{k-1}$	(13.2)	288
Mean square between groups, definitional formula	$\text{MS}_B = s^2_{\bar{X}}(n)$ [only if $n_1 = n_2 = \cdots = n_k = n$]	(13.3)	289
Pooled variance, k equal-sized groups	$s^2_{\text{pooled}} = \dfrac{s_1^2 + s_2^2 + s_3^2 + \cdots + s_k^2}{k}$ [only if equal n_j]	(13.4)	289
Mean square within groups, definitional formula	$\text{MS}_W = s^2_{\text{pooled}}$	(13.5)	290
Test statistic (F), analysis of variance	$F = \dfrac{\text{between-group point-estimate of } \sigma^2}{\text{within-group point-estimate of } \sigma^2} = \dfrac{\text{MS}_B}{\text{MS}_W}$	(13.6)	290
Degrees of freedom for the analysis of variance	$df_B = k - 1$ (between groups) $df_W = n_G - k$ (within groups) $df_T = n_G - 1$ (total)	(13.7) (13.8c) (13.13)	291 291 295
Total sum of squares	$\text{SS}_T = \sum(X_{ij} - \bar{X}_G)^2$	(13.9)	294
Sum of squares within groups	$\text{SS}_W = \sum(X_{ij} - \bar{X}_j)^2$	(13.11)	295
Sum of squares between groups	$\text{SS}_B = \sum(\bar{X}_j - \bar{X}_G)^2$	(13.12)	295

SECOND EDITION

Comprehending Behavioral Statistics

SECOND EDITION

Comprehending Behavioral Statistics

Russell T. Hurlburt
University of Nevada, Las Vegas

Brooks/Cole Publishing Company
I(T)P® An International Thomson Publishing Company

Pacific Grove • Albany • Belmont • Bonn • Boston • Cincinnati • Detroit • Johannesburg • London
Madrid • Melbourne • Mexico City • New York • Paris • Singapore • Tokyo • Toronto • Washington

Sponsoring Editors: *Jim Brace-Thompson, Denis Ralling*
Marketing Representative: *Hazel Dunlap*
Marketing Team: *Michael Campbell, Christine Davis, Deborah Petit*
Editorial Assistants: *Terry Thomas, Stephanie Andersen*
Production Editor: *Marlene Thom*
Manuscript Editor: *Carol Reitz*
Permissions Editor: *Cathleen C. Morrison*
Design Editor: *E. Kelly Shoemaker*
Interior & Cover Design: *Lisa Thompson*
Cover Photo: *Lee Hocker*
Interior Illustration: *Integre Technical Publishing Co., Inc.*
Typesetting: *Integre Technical Publishing Co., Inc.*
Cover Printing: *Phoenix Color Corporation*
Printing and Binding: *World Color*

For more information, contact:

BROOKS/COLE PUBLISHING COMPANY
511 Forest Lodge Road
Pacific Grove, CA 93950
USA

International Thomson Publishing Europe
Berkshire House 168–173
High Holborn
London WC1V 7AA
England

Thomas Nelson Australia
102 Dodds Street
South Melbourne, 3205
Victoria, Australia

Nelson Canada
1120 Birchmount Road
Scarborough, Ontario
Canada M1K 5G4

International Thomson Editores
Seneca 53
Col. Polanco
11560 México, D.F., México

International Thomson Publishing GmbH
Königswinterer Strasse 418
53227 Bonn
Germany

International Thomson Publishing Asia
221 Henderson Road
#05–10 Henderson Building
Singapore 0315

International Thomson Publishing Japan
Hirakawacho Kyowa Building, 3F
2-2-1 Hirakawacho
Chiyoda-ku, Tokyo 102
Japan

Printed in the United States of America

10 9 8 7 6 5 4 3

Library of Congress Cataloging in Publication Data
Hurlburt, Russell T.
Comprehending behavioral statistics / Russell T. Hurlburt.—2nd ed.
p. cm.
Includes bibliographical references and index.
ISBN 0-534-34889-0
1. Psychology—Statistical methods. 2. Social sciences—Statistical methods. I. Title.
BF39.H87 1997
150′.7′27—dc21 97–26026
CIP

Credits continue on page 548.

To My Wife

BRIEF CONTENTS

CONTENTS

PREFACE

Modern learning theory holds that constructive visualization, active involvement, immediate feedback, and multiple discrimination trials are essential for effective learning. Statistics textbooks typically fail to incorporate these components. As a result, students commonly master computational skills but gain little conceptual understanding of the statistics they compute.

Comprehending Behavioral Statistics remedies these omissions with three innovations: *eyeball-estimation*, a technique for assessing a distribution and quickly predicting the approximate magnitude of a statistic; *eyeball-calibration*, massed practice in estimating the characteristics of a distribution; and *progressive cumulative review*, chapter sections that develop students' ability to discriminate between new concepts and those presented in previous chapters.

These techniques are the results of more than 20 years spent helping my own students to visualize statistical concepts. They were made widely available for the first time in the first edition of *Comprehending Behavioral Statistics*. Students find these techniques actively engaging, easy, and fun. I am gratified by responses from students and teachers around the world who have written comments such as "I expected to hate statistics, but I loved your book—especially the eyeball-estimation" and "I've been using your text for the last three years and I think it's a terrific book. It really helps to dispel some of the anxiety that students have about statistics, which is one of my goals for the course."

To reap these benefits, the busy teacher need not rethink the structure of the statistics course; the core structure of *Comprehending Behavioral Statistics* is consistent with traditional texts. Eyeball-estimation, eyeball-calibration, and progressive cumulative review techniques are formatted in the text for the instructor's flexibility and ease of use.

The second edition of *Comprehending Behavioral Statistics* has benefitted from feedback offered to me by numerous instructors who have used the first edition. In addition, I have implemented a process I call the "student sieve": a line-by-line evaluation designed to screen out minute stumbling blocks in the text. While using the book, 50 students marked lines they had to reread to understand. By collating the 50 sets of markings, I have gained a focused view of the text from the student perspective, which enabled me to clarify the precise spots that students found troublesome.

Eyeball-estimation

eyeball-estimation: predicting the approximate magnitude of a statistic

Eyeball-estimation techniques enable students to predict, without the use of a calculator or statistical tables, the approximate magnitude of statistics. Sections of the text that present eyeball-estimation skills are flagged with the eyeglass symbol shown here. Eyeball-estimation is not a substitute for accurate computation; *Comprehending Behavioral Statistics* is thorough in its treatment of computation skills. Students benefit from eyeball-estimation, however, for these reasons:

- Students who eyeball-estimate are actively involved. They inspect the data and decide for themselves the approximate magnitude of a reasonable answer.
- Eyeball-estimation cultivates genuine understanding of statistical concepts. The ability to make an educated guess is better evidence of comprehension than is the ability to compute a result.
- Following eyeball-estimation, computation has an element of excitement because it provides immediate feedback on the accuracy of the estimate.
- Eyeball-estimation is quick. A beginning student can eyeball-estimate a standard deviation in about *15 seconds*. Computation would take the same student about *15 minutes*. A class hour is ample time for eyeball-estimation and discussion of more than a dozen standard deviations. Students remain actively involved throughout the discussion because they provide the eyeball-estimations in each case.
- Eyeball-estimation engages students regardless of their level of mathematical sophistication. With eyeball-estimation, a classroom discussion of the standard deviation is *comprehensible* to inexperienced students, who can practice fundamentals such as locating an inflection point and estimating its distance from the mean. The same discussion is *challenging* for mathematically sophisticated students, who can practice refining such awarenesses as the effects of outliers or skew on the standard deviation.
- Eyeball-estimation is a valuable skill. It enables students to spot mistakes immediately. Students trained in eyeball-estimation techniques estimate with much greater accuracy than do students taught by traditional methods (Hurlburt, 1993).

Eyeball-estimation of the standard deviation is described on pages 74–80.

Eyeball-calibration

eyeball-calibration: quick, repeated practice in observing the characteristics of distributions

Eyeball-calibration exercises provide massed practice, not only in eyeball-estimation but also in direct recognition of the relationship between data and statistics. Eyeball-calibration examples are flagged with the eyeglass and ruler symbol shown here.

Eyeball-calibration is useful with or without eyeball-estimation. For example, Chapter 11 presents pairs of histograms and asks whether the independent-sample t test null hypothesis should be rejected. No quantitative estimation is required. Students learn to observe the degree of overlap of the distributions and to predict directly whether or not H_0 should be rejected. Thus, they gain awareness of the magnitude of difference required for rejecting null hypotheses. Students also have the option of using these figures for eyeball-estimation practice in estimating the sample statistic, the magnitude of t, and t's approximate value.

Eyeball-calibration exercises for independent sample t tests are provided on pages 228–231.

Progressive Cumulative Review

progressive cumulative review gradual, incremental, comparative recap of previously learned concepts

Students of statistics who do well on quizzes and midterm exams may nonetheless perform poorly on a cumulative final. Why? Because traditional statistics textbooks fail to incorporate practice in one of the most important skills, the ability to discriminate between procedures. The student who uses a typical text knows that all the problems in the t test chapter require t, all the problems in the ANOVA chapter require ANOVA, and so on. The student therefore gets no practice in deciding which test to use.

Comprehending Behavioral Statistics remedies this omission by including progressive cumulative review exercises. In each chapter, cumulative review exercises present, in random order, problems of the types found in that and previous chapters. Rather than compute, students are asked to state which null hypothesis is appropriate and to describe the characteristics of the appropriate statistical test.

A typical chapter cumulative review is on pages 314–315.

Cumulative review exercises are progressive in that the complexity of required discriminations increases gradually with each successive chapter.In Chapter 10, for example, the student discriminates among three easy options: finding the area under a normal distribution, creating a confidence interval, or testing a hypothesis. The task becomes slightly more complex in Chapter 11, where the student must also discriminate between testing a hypothesis about the mean of one group or the means of two groups. This step-by-step pattern of slightly increasing complexity continues throughout the text. By the end of the course, the student has become proficient in making complex discriminations.

I began developing cumulative review exercises for my graduate students. The exercises were so effective that I started using them with sophomores more than ten years ago. My sophomores' performance on cumulative exams now surpasses that of the graduate students I taught prior to using cumulative review exercises.

ESTAT Computer Simulation Package

ESTAT complements *Comprehending Behavioral Statistics*, but either can be used without the other.

ESTAT is the computer software designed to accompany *Comprehending Behavioral Statistics*. ESTAT (for ESTimating STATistics) is available in Windows and Apple Macintosh formats. *Comprehending Behavioral Statistics* can be used independently of computer software. Students who use ESTAT, however, will benefit from its innovations: eyeball-estimation exercises, eyeball-calibration exercises, and the most user-friendly computational package available.

ESTAT provides practice in eyeball-estimation and eyeball-calibration by generating and displaying data, inviting the student to eyeball-estimate a statistic, and then providing immediate feedback on the accuracy of the estimate. For example, in one of the standard deviation exercises, "sdest," ESTAT displays a histogram and asks the student to eyeball-estimate the standard deviation. When the student clicks a button, the actual standard deviation appears in both graphic and numeric form. Another click and ESTAT produces a new histogram from a randomly generated infinite series of data sets. Context-sensitive help is always available via a click, as is a step-by-step tutorial.

Compared with typical homework, ESTAT exercises are more efficient and cultivate better comprehension. Students who use a traditional textbook to do standard deviation homework spend one to three hours calculating the standard deviations of about three distributions, perhaps six if they also use a workbook. They compute their answers and check the results in the back of the book, spending almost no time developing comprehension of the relationship between those standard deviations and their distributions.

By contrast, students who use the ESTAT sdest standard deviation laboratory, with its infinite stream of Monte Carlo histograms, spend about a minute on each cycle of observation, estimation, and feedback. In less than half an hour, students can encounter more than 15 distributions, and *all* of that time is spent developing comprehension of the relationship between those standard deviations and their distributions. Plenty of homework time remains for computation, which is now based on a solid conceptual foundation.

Comprehending Behavioral Statistics invites students to use ESTAT where appropriate with a prompt:

You may wish to use the ESTAT exercise "sdest" at this time.

For those who do not have access to computers, ESTAT-like exercises are also included in the Study Guide.

ESTAT also includes a statistical computational package that is the most user-friendly available. ESTAT provides all relevant statistics automatically, freeing the student from the need to figure out how to ask the computer to display any statistic. For instance, if the data consist of three or more groups, ESTAT automatically displays an ANOVA. If the data consist of two groups, ESTAT automatically displays the independent-sample t; if the groups have equal numbers of observations, ESTAT also automatically displays the dependent-sample t, the correlation coefficient, and the regression equation.

Because statistics are displayed automatically, ESTAT elicits a decision process that is the reverse of the process required by other programs. Typical programs require students to decide which statistic to *request* from among *many* that might be available. With ESTAT, students decide which statistics to *use* from among *a few* that are automatically displayed.

As a result, ESTAT's computational package dispels anxiety for the beginning student whose grasp of statistical concepts is not yet secure. Typical computational packages escalate anxiety because any complexity of the computer interface substantially compounds the student's insecurity. Nothing is more discouraging than the failure to elicit an appropriate result from the computer, particularly when the student does not know whether to attribute that failure to unfamiliarity with the interface or to unfamiliarity with the statistics themselves. By contrast, students can immediately interact successfully with ESTAT. Its interface actively facilitates every task and elicits in students the desire to explore and master statistical concepts.

Instructor's Flexibility in Using the Text

Comprehending Behavioral Statistics is organized so that the instructor may choose which of its innovations to use. For example, the eyeball-estimation material in the hypothesis-testing chapters is presented in optional sections at the ends of those chapters. It can easily be assigned, made optional for selected students, or omitted altogether.

A complete set of ancillary materials is available and includes these items:

- ESTAT eyeball-estimation and computation computer program, available in Windows or Apple Macintosh formats. (*Comprehending Behavioral Statistics* can be used with or without computer software.)
- Student Study Guide (including ESTAT-like exercises for students without access to computers).
- Instructor's Manual (250 pages, including a test bank with 1000 items).
- Computerized test bank (1000 items for Windows or Macintosh)

Organization for the Convenience of Students

Comprehending Behavioral Statistics incorporates numerous features that enhance the student's learning and convenience:

- There are more than 400 accurately drawn figures (two to three times more than in most textbooks).
- End-of-chapter exercises are graduated from simple to complex and include examples from journals.
- Definitions of statistical terms and symbols are given both in the margins of pages where they initially appear and in the Glossary at the back of the book.
- Most frequently used formulas and tables are reproduced on the inside covers.
- "Info notes," flagged with the international information symbol ℹ, give useful comments and cross-reference information.
- Statistical tables have colored edges for easy accessibility (Appendix A).
- A reference review of basic arithmetic is included (Appendix B).
- All statistical formulas are listed together for ready reference (Appendix C).
- Results of exercise subcomputations, not just final results, are given in the answers to exercises (Appendix D).

What's New in the Second Edition?

New Material

- A new section presents a priori comparisons (see pages 326–333).
- A new section presents repeated measures ANOVA (see pages 333–340).
- Computer exploration exercises are now included in most chapters (for example, see pages 240–241). These exercises can be used with any computer statistics package (SPSS, SAS, BMDP, etc.).

- A Connections section added at the ends of most chapters typically has four parts: a cumulative review of the text so far; a discussion of how the material in that chapter is reported in journal articles; detailed instructions for operating the Windows and Macintosh versions of both ESTAT and SPSS 6.1 Student Version, including annotated examples of SPSS output; and homework tips (for example, see pages 232–235).

Substantially Altered Material

- The discussion of practical significance and effect sizes has been amplified substantially and distributed across all hypothesis-testing chapters. This material is illustrated more profusely here than in other textbooks (for example, see pages 191–194).
- The discussion of statistical power has been simplified, expanded, and integrated into Chapters 9 and 12 (see pages 172–173 and 257–270).
- Computations have been simplified wherever possible. For example, the ANOVA notation has been simplified (see pages 292–297), and the material on interpolation is now optional (see pages 122–123).
- Eyeball-estimation material for the hypothesis-testing chapters has been moved to optional sections at the end of each chapter (for example, see pages 241–244).

Acknowledgments

I am grateful to the many student and faculty users of the first edition who have provided extremely helpful suggestions. I am particularly grateful to those faculty who have contributed repeated or extended comments, among them Chris Heavey, University of Nevada, Las Vegas; Douglas W. Matheson, University of the Pacific; Peter Yarensky, University of New Hampshire; Nicholas DiFonzo, Rochester Institute of Technology; and Susan Campbell, Middlebury College. In addition, I thank the following reviewers for their insightful comments: Edem Avakame, Temple University; Claire Kibler, State University of New York, Binghamton; Lee A. Kirkpatrick, College of William and Mary; Louis Matzel, Rutgers University; John B. Murray, St. John's University; and Katherine Van Giffen, California State University, Long Beach. I also thank Lisa Coffey for many suggestions.

Finally, I thank the talented and dedicated staff at Brooks/Cole for their continuing commitment to this project. In particular, I'm grateful to Terry Thomas for attending to the myriad details of revising and producing a book; to Marlene Thom for shepherding the project so expertly through production; to Kelly Shoemaker and Lisa Thompson for creating such a beautiful interior design and cover; to Cat Morrison for taking care of all the permissions issues; and last but by no means least, to Jim Brace-Thompson for his encouragement and support. It is a pleasure to work with such a wonderful team of professionals.

Russell T. Hurlburt

CHAPTER 1

Introduction

LEARNING OBJECTIVES

1. What is an inductive statement and how can the truth of an inductive statement be analyzed?
2. What are examples of inductive statements that you deal with every day?
3. What is research design and how does it relate to statistical reasoning?
4. What is the "Pygmalion effect"? What three characteristics of the experimental/statistical method do experimental explorations of the Pygmalion effect illustrate?
5. What is the difference between a population and a sample? Between a parameter and a statistic?
6. How do we determine the probability that an event will occur?

This chapter points out that the word *true* has different meanings; one is inductive, which means that the truth or falsehood of a statement can be assessed by collecting and analyzing data. Statistics is the best tool available for analyzing data, and mastering statistical analysis is worth the effort involved.

This chapter presents a classic example of an experiment that uses statistical analysis and then uses that example to highlight the distinction between populations and samples. It also describes the basic notions of probability.

One of the principal aims of all education is to become more skilled in differentiating true statements from false ones. That may seem so obvious as to require no explanation, but learning about truth is not simple. Part of the difficulty is that *truth* has different meanings in different contexts. Consider these four statements, all of which use the word *true*:

1. The conclusion of the following syllogism is true:
   Major premise: All men are mortal.
   Minor premise: Socrates is a man.
   Conclusion: Therefore Socrates is mortal.
2. William Tell's arrow sped true to its mark and split the apple in two.
3. "For aught that I could ever read,
   Could ever hear by tale or history,
   The course of true love never did run smooth."[1]
4. It is true that one can get across town faster on Elm Street than on Oak Street.

All four statements use the word *true*, but the four meanings are not identical. Philosophers have labored at the task of determining whether there is a fundamental meaning of *truth* that underlies all four of our examples, but they have reached no uniform consensus.

1.1 Inductive Statements

inductive statements: statements whose truth can be assessed by collecting and analyzing data

Statement 4 is an example of what philosophers call *inductive statements*—that is, statements whose truth is assessed by observing a series of examples, by collecting and analyzing data.

We all frequently evaluate inductive statements, as in these examples:

- My golfing partner says, "Keep your left arm straight and you will hit straighter." I try it a few times and conclude that he is right.

[1] W. Shakespeare, *A Midsummer Night's Dream*, act 1, sc. 1, line 132.

- I look at the sky and listen to the radio in an attempt to decide whether it might rain tomorrow.
- I decide to use Crest toothpaste because it is said to be more effective than a nonfluoridated brand.
- I do not study for the Graduate Record Examination because I don't believe studying for it will raise my score.
- I choose to fly because I believe air travel is safer than driving.
- I change my spark plugs every 20,000 miles even though the car runs well because previous experience has shown that spark plugs last about 20,000 miles.
- I eat less red meat because I think eating red meat increases my chance of getting heart disease.
- The Surgeon General says that smoking cigarettes causes cancer.
- I think that the defense of the Pittsburgh Steelers is better this year than it was last year.
- I believe that in basketball, the home team has an advantage, but I wonder how big that advantage is.
- I believe that most beer drinkers can't distinguish between Miller and Bud; that is, most "preference" is really the result of advertising hype.
- I think the ozone layer is deteriorating.
- I believe that Anacin works faster than Bayer.

Some of these situations are rather trivial, some have life-or-death or worldwide implications, some of the opinions are true and some are false, but all are based on (implicit or explicit) evaluations of inductive statements. In fact, most of what we know about the world is the result of inductive processes. Like it or not, aware of it or not, skillful at it or not, we all engage in inductive reasoning almost constantly.

1.2 Statistical Reasoning

Statistics is the best set of tools available for deciding whether inductive statements should be considered "true." Statement 4 in our first four examples and all the opinions in the preceding list are best supported or discarded by using statistical procedures. Every time we exercise an inductive process (which is hundreds of times a day), we use statistical reasoning more or less skillfully. Statistical reasoning is not something foreign or separate from us, but rather something that we do all the time. Thus, this book is not intended to teach you something entirely new, but rather to increase your skill at doing what you already do, perhaps unknowingly and unskillfully.

Let us take the Oak Street/Elm Street trips as an example. I state that the Elm Street route is faster than Oak Street. You doubt my conclusion, so you propose a "race": We will leave at the same time, you will take Oak Street and I will take Elm, and we both agree to observe the speed limits.

Suppose we undertake this race and I "win." Should we conclude that the Elm Street route is in fact faster than Oak? "Perhaps," you say, "but not for certain." Maybe I was just lucky with the traffic lights; maybe you were unlucky because that truck blocked traffic while it backed up into the supermarket; maybe it depends on the time of day. You conclude, "We need more races to decide for sure."

When you think like that, you are reasoning like a statistician. You are recognizing that travel time is a "variable" that can take different values, some longer and some shorter. You are recognizing that many unpredictable things (such as traffic light timing and truck behavior) influence that variable, causing the value of the variable (the travel time itself) to become larger or smaller. In the language of statistics, you are recognizing that the travel time variable is "distributed," and that is one of the essential observations of statistics, to which we will return in Chapter 3.

You know already that we need more races; in this book you will learn something about *how many more* races we need to conclude rationally that the Elm Street route is faster than Oak Street. Furthermore, you will become more precise about what such a "rational" decision actually implies. I hope this example illustrates that the science of statistics is simply a refinement of techniques that we all use every day.

1.3 Rational Decision Making

We have seen that engaging in rational inductive processes entails using statistical reasoning. We should note in passing that modern behavioral science has developed *two* specialties that are part of the rational inductive process: research design and statistical reasoning. Research design is the science of making observations about the real world, considering such questions as how many observations to make, under what conditions to make them, and what should be held constant. Statistical reasoning begins with the data that are the results of those observations and prescribes the rules by which rational statements about those data can be made. Statistical reasoning and research design are intertwined, dependent skills; one cannot be a good research designer without being a good statistician, and vice versa. They are, however, usually taught as two separate skills, and this book follows that practice. We will focus on statistical reasoning and not discuss research design here.

Although inductive rationality (and therefore skill in statistical reasoning) is indeed valuable in many situations, it *cannot* claim to be the ultimate form of human truth seeking. Statements 1, 2, and 3 of the four statements near the beginning of the chapter do *not* involve inductive (statistical) reasoning. For example, statistics is of no use in determining whether your own or someone else's love is true, and yet the truth in such a situation may be of vital importance. That determination must be made on some grounds other than statistical. Thus, it seems to me, it is wise to be skillfully rational when the situation calls for it and to be artfully irrational when some other situation calls for that.

Although objective rationality is not necessarily the primary access to ultimate truth, it is our primary access to the truth about the real events in our world, and it is therefore one of the most valued skills we know. To be skillfully rational requires using statistical reasoning competently, and this book is dedicated toward that end.

1.4 A Classic Example: Pygmalion in the Classroom

> You see, really and truly, apart from the things anyone can pick up (the dressing and the proper way of speaking, and so on), the difference between a lady and a flower girl is not how she behaves, but how she's treated. I shall always be a flower girl to Professor Higgins, because he always treats me as a flower girl, and always will; but I know I can be a lady to you, because you always treat me as a lady, and always will.
>
> —Eliza Doolittle in George Bernard Shaw's *Pygmalion*

Pygmalion effect: people act in accordance with others' expectations

George Bernard Shaw was an acute observer of human nature, and Eliza Doolittle's important observation has come to be known as the "Pygmalion effect": People act in accordance with others' expectations.

However, the Pygmalion effect, though compelling and in agreement with our intuition, may or may not in fact be true. To test the truth of the Pygmalion effect requires the skillful rationality of the experimental method and its competent statistical analysis. Let's take a look at one of the more influential studies in the history of psychology, education, and medicine as an example of this method. We shall return to this example frequently throughout the textbook.

Robert Rosenthal, a Harvard psychologist, and his colleagues set out to determine whether the Pygmalion effect exists in a variety of settings. For example, they told one group of students who were training rats in an experimental psychology laboratory course that their rats had been specially bred to be extremely quick at learning to run a maze (that is, they were "maze-bright"), and they told another group of students in the same class that their rats were "maze-dull" (Rosenthal & Fode, 1963). Actually, the rats had been *randomly* assigned to the students and were neither particularly maze-bright nor maze-dull. The rats whose handlers thought they were maze-bright did in fact learn their mazes faster than those whose handlers thought they were maze-dull. The students were not intentionally trying to speed up or slow down their rats' learning; nonetheless, through some apparently unconscious mechanism, they did influence that learning. Thus, the Pygmalion effect applies in the rat lab: If rat handlers expect quick learning, they get quick learning.

Rosenthal is perhaps best known for a study of the Pygmalion effect at "Oak School," an elementary school in the old section of a city with a clear ethnic mixture. The question was: Did teachers' expectations of children affect the performance of those children? Rosenthal and Jacobson (1968) used the following scheme to manipulate teachers' expectations of their students.[2] In the spring of 1964, they administered the "Harvard Test of Inflected Acquisition" ("HTIA") to all the children of Oak School who might return the following fall. They explained to each teacher that the HTIA was a new test that could predict future "academic spurts" in students. They told the teachers that the HTIA was an experimental test that was not a perfect predictor of spurts; however, students who scored in the top 20% on the HTIA were more likely to "spurt" or "bloom" in the next year than were the other children. They further explained that their reason for administering this test at Oak School was simply as a final check on the validity of the HTIA, a project sponsored by the National Science Foundation.

That fall, when the students returned to school, Rosenthal and Jacobson gave teachers a list of "the top 20% scorers on the HTIA.... Teachers were told only that they might find it of interest to know which of their children were about to bloom. They were also cautioned not to discuss the test findings with their pupils or the children's parents" (p. 70).

This study involved some deception: The "top 20% scorers" were *not* really the highest scorers on the HTIA but were actually selected by using a table of random numbers; that is, students were chosen with no information whatever about their performance on the HTIA, their actual ability, or their previous performance in the classroom. (We'll discuss the use of tables of random numbers in Chapter 7.)

The result of this procedure was that teachers came to think of 20% of their students as "bloomers" and 80% of their students as what we will call the "others." Actually, there was no difference between the bloomers and the others; a totally arbitrary random procedure had assigned the label "Bloomer" to some children and the label "Other" to the remaining children.

You will recall that the HTIA was administered to all the students as part of the cover story for assigning the label "Bloomer." Actually, the "HTIA" was a standardized test of intelligence called the Tests of General Ability (TOGA) that yielded an IQ score for each child. The teachers were unaware that the TOGA (or "HTIA") was actually an IQ test, and they were not given the children's actual TOGA IQ scores.

A year later, Rosenthal and Jacobson administered the TOGA to the same children again. They defined "intellectual growth" as the difference between a child's current ("posttest") TOGA IQ and his or her "pretest" TOGA IQ from the original testing (positive scores indicate an increase in IQ). Pretest IQ, posttest IQ, and intellectual growth scores for the bloomers are listed in Table 1.1. (A similar table exists for the other children, but it is too long to show here.) The first student identified as a bloomer was Kathy. Her score on the original TOGA was 105, her score on the TOGA a year later was 125, so her intellectual growth (or IQ gain) was $125 - 105 = 20$ IQ points.

Rosenthal's question was whether there was more intellectual growth in the bloomer children than in the other children. Finding the answer requires inductive reasoning: the collection and analysis of data such as those in Table 1.2. That table shows the intellectual growth (IQ gain) scores for both the bloomers and the other children (you should recognize that the first column of Table 1.2 is the same as the last column of Table 1.1). The highest intellectual growth score (Mario's 69) is in the bloomer group, which might lead us to think

[2] We have simplified our explanation by presenting only a portion of Rosenthal and Jacobson's study.

TABLE 1.1 Intellectual growth (IQ gain) of Oak School second-grade bloomers*

Student	Pretest IQ	Posttest IQ	Intellectual growth (IQ gain)
Kathy	105	125	20
Tony	109	123	14
Mario	133	202	69
Louise	101	114	13
Juan	123	117	−6
Able	109	134	25
Patricia	89	90	1
Douglas	111	107	−4
Baker	108	132	24
Charlie	89	101	12
Delta	72	91	19
Echo	75	86	11

*These values can be inferred from statistics provided in Rosenthal and Jacobson (1968, pp. 75, 85–93, 187, 190, and 193). Rosenthal and Jacobson provided names for only seven of these students; I added Able, Baker, Charlie, Delta, and Echo for the missing names.

TABLE 1.2 Intellectual growth (IQ gain) of Oak School second-grade bloomers and other children*

Bloomers	Others			
20	3	−3	−10	−4
14	−2	4	15	−3
69	−15	20	15	6
13	1	8	8	11
−6	31	−6	6	13
25	10	−6	30	18
1	5	6	−5	4
−4	9	10	17	0
24	1	11	7	26
12	12	13	4	−1
19	14	14	19	14
11	−11	7	3	

*These values are from Rosenthal and Jacobson (1968). The Bloomers data are exact (see Table 1.1). The Others data are manufactured to match the classroom means and standard deviations that Rosenthal and Jacobson (p. 193) provided (they did not provide original data).

that being labeled a bloomer *does* improve intellectual growth. However, the next three highest intellectual growth scores (31, 30, and 26) are in the other children, which might lead us to think that *not* being labeled a bloomer improves intellectual growth. Furthermore, we are actually interested in *all* the children, not just the highest or lowest scorers, and there is quite a lot of overlap between the intellectual growth scores for bloomers and other children. For example, intellectual growth scores of 20, 19, 14, 13, 12, 11, 1, −4, and −6 occur in both the bloomers and the other children, which again would lead us to conclude that being labeled a bloomer does *not* have a particular advantage for improving intellectual growth.

Inspection of the data, then, does not lead us to an obvious answer to the question of whether positive expectancies lead to greater intellectual growth. You may feel that the bloomer scores are clearly higher as a group, but someone else might feel that the two sets of scores are about the same. Statistics is a set of tools designed to give us rational ways of deciding between such alternatives rather than relying on your own or someone else's feelings.

The decision about the Pygmalion effect, though perhaps not obvious, *is* important: If the Pygmalion effect does in fact exist, then students or teachers should not know the IQ scores of students because knowledge of a low score may decrease that student's performance. Do our data actually imply that the Pygmalion effect exists in the classroom? We shall see in Chapter 11 that the rational/statistical answer is yes.

In fact, many studies in many different situations have used statistical methods to demonstrate the existence of the Pygmalion effect, so the Pygmalion effect (also called "experimenter bias") is now an accepted fact, no longer in need of further empirical demonstration. Thus, for example, when researchers attempt to demonstrate the effectiveness of a new drug compared with a *placebo*, we demand that experimenter bias be eliminated by requiring that the researchers be "blind" to which subjects receive the actual drug and which subjects receive a placebo. Many school districts, as another example, have eliminated "tracking" (where students are divided into groups according to scores on IQ tests) because such groupings may affect teacher bias; some school districts have eliminated the use of IQ tests entirely. Thus, the experimental/statistical examination of the Pygmalion effect has led to important social consequences. Our lives today are in fact substantially altered by yesterday's application of statistical tests.

placebo: in a drug study, a substance that looks like the drug being tested but actually has no effect

blind: not knowing which subjects are assigned to which experimental condition

We will use this classic example to illustrate three characteristics of the experimental/statistical method. First, the inspiration for an experiment is a pre-experimental observation: George Bernard Shaw (and, of course, others) observed what he took to be a characteristic of human nature. Only later did Rosenthal and Jacobson seek to verify experimentally whether that observation was correct.

Second, Rosenthal and Jacobson's main interest was *not* in the intellectual growth patterns of the particular 59 students (12 bloomers plus 47 others) who participated in this study, but rather in the intellectual growth patterns of all students in general. Our statistical tools, then, must distinguish between samples (e.g., Rosenthal's 59 second-graders) and populations (e.g., all second-graders in the United States). We will begin to discuss the distinction between samples and populations in Section 1.5, and most of the text (Chapters 7–18) will discuss the general question of what can be inferred about populations when all we know is about samples.

The third characteristic of the statistical method is that it must allow us to derive meaningful results from data that are not perfectly consistent—that fluctuate more or less randomly from one person to another or from one occasion to another. The Pygmalion bloomers are a random sample from all the second-graders. We wonder whether our conclusions would be the same if a different random sample were used. Understanding this aspect of statistics requires us to understand some of the concepts of probability, which are discussed in Section 1.6, and the characteristics of random samples, which are described in Chapter 7. Furthermore, our statistical toolbox must include methods to measure the differences between individuals. We will discuss those tools starting in Chapter 3.

Nowadays we take the Pygmalion effect for granted because many experiments have collected data (not only in the rat lab and the classroom) that have been subjected to the kinds of analyses discussed in this textbook. Thus, our general acceptance of the Pygmalion effect is an example of the power of statistical analysis. We'll follow this example throughout the text.

1.5 Samples from Populations

We have seen that one of the main tasks of statistics is to use small samples to infer characteristics about larger populations. For example, Rosenthal and Jacobson wished to use the characteristics of their sample of 59 second-graders from Oak School to infer something about the characteristics of the population of second-graders in general. Now we must be clear about exactly what we mean by *populations* and *samples*.

population: all the members of the group under consideration

A *population* includes *all* the members of the group under consideration. Populations can be large (such as the residents of the United States, 270 million) or small (such as the students in a particular class, 15). Populations can be of people (such as people who are recovering alcoholics), of objects (such as the stars in our galaxy), or of events (such as baseball games played by the Yankees). What makes a particular group a population is *not* a characteristic of the group itself, but rather a characteristic of *our interest*. If we are interested in a group for its own sake, *not* as being representative of or selected from some other larger group, then the group about which we are interested is a population.

sample: some subset of the group under consideration

A *sample* is a subset of a population, a group that is interesting to us *not* on its own merits but because it somehow represents the larger population. For example, if we are interested in the voting patterns of the U.S. electorate and contact by telephone 500 voters and inquire how they plan to vote, then the 500 voters we contacted are not interesting to us *for themselves*, but only because they may be representative of the whole electorate. The 500 voters are a sample from the population of voters.

statistic: any measurement on a sample

parameter: any measured (or assumed) characteristic of a population

To be clear about the distinction between populations and samples, we use the term *statistic* to refer to any measurement on a sample and the term *parameter* to refer to any measurement on (or assumption about, see Chapter 8) a population.

Most often, it is too difficult, too costly, or impossible to measure all the elements of a population, so we select a sample of the population and measure just those elements. Almost always (but not necessarily), samples are much smaller than the parent populations.

Much of the science of statistics can be thought of as the procedures for using relatively small samples to infer the characteristics of large populations, and this text is designed to explore those procedures. Rosenthal and Jacobson, for example, used their small (47- and 12-person) samples to draw conclusions about the effect of teacher expectancies in the population in general. They would have preferred, of course, to have measured the intellectual growth of all the millions of students, but that would have been obviously impossible.

1.6 Probability

We have observed that in Rosenthal's data, the amount of intellectual growth differs from student to student. Some extraneous influences that might have affected intellectual growth during the year are help from parents with homework (or lack of it), good or bad parent models, and health or illness during testing. There are actually many such influences—far too many for us to be able to take directly into account. Because we can't measure them directly, we lump all such influences together into what we call "random" influences.

random: unpredictable given our current knowledge

Random means unpredictable at our current level of understanding. Perhaps if we knew how much homework help students got from their parents, and if we knew how stable their parents' interpersonal patterns were, and if we knew how healthy they were, and so on, we could reduce the size of this random effect. But there are almost always far too many things to measure, so our data almost always contain random effects.

The science that deals with the nature of randomness is called *probability theory*, and nearly all statistical tests rest squarely on probability concepts. Probability theory is a fascinating study in its own right, and its mastery can easily be the exclusive topic of a textbook. For our present purposes, we need only to understand its most basic elements.

probability of an event: the number of outcomes favorable to that event divided by the total number of possible outcomes

We begin with some basic terminology. In the language of probability theory, an *experiment* is a controlled operation that yields one of several possible *outcomes* or results.[3] An experiment might involve drawing a single card from a shuffled standard deck.[4] One possible outcome of this experiment would be "ace of spades," another would be "6 of diamonds," and so on. An *event* is a set of possible outcomes; thus, an event might be defined as "drawing a heart." The event of drawing a heart can be satisfied by any of the 13 outcomes ace of hearts, king of hearts, . . . , 2 of hearts.

---

[3]The term *experiment* has a somewhat different, more specific meaning in other contexts, such as in behavioral science research design.

[4]A standard deck has 52 cards divided into four suits (spades, hearts, diamonds, and clubs). Each suit has 13 cards (2, 3, 4, 5, 6, 7, 8, 9, 10, jack, queen, king, and ace).

If we assume that each outcome is equally likely, then the *probability of an event E* can be obtained from this formula:

probability of an event

$$P(E) = \frac{\text{number of outcomes favorable to } E}{\text{total number of possible outcomes}} \qquad (1.1)$$

Thus, for example,

$$P(\text{drawing a heart}) = \frac{13}{52} = \frac{1}{4}$$

$$P(\text{drawing the 7 of hearts}) = \frac{1}{52}$$

$$P(\text{drawing the 15 of hearts}) = \frac{0}{52} = 0$$

$$P(\text{drawing a heart or a spade or a diamond or a club}) = \frac{13 + 13 + 13 + 13}{52} = \frac{52}{52} = 1$$

These examples illustrate three important characteristics of probability:

1. The probability of an event that *cannot* occur is 0.
2. The probability of an event that *must* occur is 1.
3. $0 \leq P(E) \leq 1$; probabilities lie between 0 and 1 inclusive.

Probabilities are quite close to 0 for events that are unusual; for example, P(drawing the queen of spades) $= 1/52 = .02$. Probabilities are quite close to 1 for events that are very likely to happen; for example, P(drawing any card *except* the queen of spades) $=$ $51/52 = .98$. Probabilities are quite close to .5 for events that occur about half the time; for example, P(drawing a red card) $= (13 + 13)/52 = .5$.

When we determine probabilities, it is important to count *all* the possible outcomes of an experiment. That may seem obvious, but in practice such counting can be tricky. Consider the experiment of rolling dice. A die is called *fair* if $P(1) = P(2) = \cdots = P(6) = 1/6$. Now suppose you roll *two* fair dice. What is the probability that the two dice will sum to 4? Answering that question requires that we count the number of ways there are of rolling a sum of 4 on two dice. It may seem (incorrectly) that there are two such ways: rolling a 1 and a 3, and rolling a pair of 2s. However, the correct answer is three: rolling a 1 on the first die and a 3 on the second, rolling a 2 on the first die and a 2 on the second, and rolling a 3 on the first die and a 1 on the second. Because there are 36 total possible ways of rolling two dice (1–1, 2–1, 3–1, . . . , 6–1; 1–2, 2–2, . . . , 6–2; 1–3, etc.), P(rolling a 4 with two dice) $= 3/36 = 1/12$.

conditional probability: the probability of an event given that another event has occurred

Sometimes we put restrictions or *conditions* on the range of possible outcomes of an experiment. If we do so, we call it *conditional probability*. For example, we may ask: What is the probability of drawing a heart *on the condition that* the card we draw is known to be red? This condition reduces the number of possible outcomes (the denominator of the probability formula) from 52 to 26, so the conditional probability of drawing a heart on the condition that (or *given that*) the card is red is $P(\text{Heart} \mid \text{Red card}) = 13/26 = .5$. Note that we symbolize "given that" by a vertical line.

We can call the kinds of probabilities that we have been considering so far "discrete" in the sense that we could explicitly count all the possible outcomes. We can extend the notion of probability to situations where such counting is impossible or impractical. For example, suppose we are about to measure a person's intelligence quotient (IQ). We might ask, What is P(her IQ is higher than 130)? To answer that question using discrete probability, we would have to know the number of individuals whose IQs are higher than 130 and then divide that number by the total number of individuals. That would be impractical to do in real life; we will develop methods of approximating this probability in subsequent chapters. When we make a statement such as P(IQ is higher than 130) $= .02$, we mean that *if* we were to measure all people's IQs (which we won't because it is too impractical), we would find that 2% of all those individuals would have IQs over 130.

For reasons that will become clearer in Chapters 8 and 9, probabilities of .05 and .95 are the most widely mentioned values in statistics, so let's explore these values explicitly.

Suppose we have an urn that contains 20 identical balls except that 19 are red and 1 is white. We thoroughly mix the balls and then draw out one of them. What is the probability that the ball is white? Equation (1.1) indicates that $P(\text{White}) = 1/20 = .05$. Furthermore, $P(\text{Red}) = 19/20 = .95$.

Drawing a white ball out of such an urn is an "unusual" event. It is not impossible, but it won't happen very often—about 5 times out of every 100 attempts on the average. Drawing a red ball is not unusual; it would happen about 95 times out of every 100 over the long run.

Much of statistics depends on a definition of *unusual* such as this. We generally take *unusual* to mean "occurring with probability less than .05"; however, in some situations, we may prefer a stricter definition of *unusual*—perhaps a probability of less than .01.

1.7 A Note to the Student

I would like to impress upon you at the very beginning that statistics is fundamentally quite simple. In fact, there are basically only three major concepts to be mastered in this text. I'll state them here, even though the terms might not mean much to you yet. (1) What "the distribution of a variable" is and how to describe it, (2) what a "sampling distribution of means" is and how it is related to the distribution of the variable, and (3) what a "test statistic" is and how it is related to the sampling distribution of means. There are, to be sure, many important details to be learned, but once you grasp the three major concepts, the rest of statistics follows rather straightforwardly. It's worth memorizing those three concepts now (even if it is mere rote memory for the moment) to begin building the cognitive structure that we will elaborate throughout the remainder of the textbook.

Studying statistics—attempting to acquire statistical skills—is a valuable exercise. It is worthwhile to develop your ability to think rationally about empirical events. But skill acquisition requires work and practice, work and practice. I urge you to look forward positively to the prospect of engaging in this exercise. If I were your basketball coach, I would try to make running laps interesting (by using races, relays, music), but I would require that you run the laps regardless of whether you found them fun. I would try to impress upon you that good basketball players do not avoid calisthenics but instead look upon them as a discipline, a self-challenge.

The same is true about the effort required to learn statistics. As your statistics "coach," I have gone to great lengths to make the work of learning statistics as interesting, challenging, informative, and rewarding as possible, but you may still find that there are some "calisthenics" involved. I urge you not to avoid that work but to use it as a way to strengthen your mental discipline.

This mental discipline is itself worth striving for, as the world's greatest thinkers have maintained since the beginning of recorded history. For example, Buddha held 2500 years ago:

> The mind is wavering and restless, difficult to guard and restrain: let the wise man straighten his mind as a maker of arrows makes his arrows straight.
>
> Let a wise man remove impurities from himself even as a silversmith removes impurities from the silver: one after one, little by little, again and again.[5]

I urge you to recognize that the burn of annoyance when a computation does not work out is a sign of mental undiscipline, not the result of statistical ignorance, and to recognize that such undiscipline can be overcome with consistent practice.

---

[5] *The Dhammapada: The Path of Perfection*, vv. 33 and 239, translated from the Pali by Juan Mascaro (New York: Penguin Books, 1973).

Exercises for Chapter 1

Section A: Basic exercises (answers in Appendix D, page 505)

1. Define *inductive statement.*
2. Which of the following are inductive statements?
   (a) More Democrats favor socialized medicine than do Republicans.
   (b) That poem truly expresses how I feel.
   (c) Taking vitamin C reduces the frequency of colds.
   (d) I prefer new music to rap.
3. True or false: Statistical reasoning provides access to the ultimate truth.
4. Name the three major concepts to be mastered in this text.
5. What is the "Pygmalion effect"? What three characteristics of the experimental/statistical method do experimental explorations of the Pygmalion effect illustrate?
6. Define *sample* and *population.*
7. The University of Nevada, Las Vegas (UNLV) has 20,000 students, and I would like to know how much time UNLV students spend doing homework. My statistics class has 40 students in it, and I assume that they are representative of UNLV students in general. I ask students in my statistics class how many hours they spent doing homework during the last week. Is my statistics class a sample or a population? Why?
8. My statistics class has 40 students in it, and I would like to know how much time they spend doing homework. I ask them how many hours they spent doing homework during the last week. Is my statistics class a sample or a population? Why?
9. In an experiment involving drawing a card from a shuffled standard deck (13 cards in each of four suits), find these probabilities:
   (a) P(drawing the ace of spades)
   (b) P(drawing the 2 of clubs)
   (c) P(drawing a "face card" [10, jack, queen, king, or ace] in spades)
   (d) P(drawing a face card in any suit)
10. (a) In the dice game called craps, the shooter rolls two dice and wins immediately if he rolls either 7 or 11 as the sum of two dice. What is P(an immediate win)?
    (b) In craps, the shooter loses immediately if he "craps out"—that is, rolls 2, 3, or 12 as the sum of two dice. What is P(crapping out)?
11. Suppose we wish to select a birthday from the year 1993 "at random." We take 365 Ping-Pong balls (1993 is not a leap year), put them in a barrel, and mix them thoroughly. Find these probabilities:
    (a) P(selecting July 7)
    (b) P(selecting any July day)
    (c) P(selecting any February day)
    (d) P(selecting any Thursday) [*Hint:* The first day of 1993 is a Friday.]
    (e) P(selecting any Friday)
12. If we draw a card from a standard deck, find these probabilities:
    (a) P(drawing any ace)
    (b) P(drawing any ace | the draw is a face card)
    (c) P(drawing any ace | the draw is a spade)

Section B: Supplementary exercises

13. An American roulette wheel is a circular wheel with 38 indentations equally spaced around the circumference. The indentations are numbered 1 through 36, with the remaining two indentations numbered 0 and 00. Half of the numbers 1 through 36 are colored red, whereas the other half are colored black. The 0 and 00 indentations are green. The wheel is rotated in one direction, and a ball is rolled in the opposite direction until it comes to rest in one of the indentations, which is the winner. Assuming the wheel and ball are fair, find these values:
    (a) P(spinning black)
    (b) P(spinning red)
    (c) P(spinning green)
    (d) P(spinning "1")
    (e) P(spinning "23")
    (f) P("3" | the outcome is odd)
    (g) P("3" | the outcome is even)
    (h) P("0" | the outcome is green)
14. In roulette, what is the probability of spinning "3" on both of the next two spins?
15. Suppose you approach a roulette table where the winner was just "3." What is the probability that the next winner will also be "3"?
16. The players win when the spin is either red or black; the house wins when the spin is green. Out of every 1000 spins, about how often does the house win?

CHAPTER 2

Variables and Their Measurement

LEARNING OBJECTIVES

1. What is a variable?
2. What are the characteristics of the four measurement scales (nominal, ordinal, interval, and ratio) used to measure variables? Why do we lump interval and ratio variables together?
3. What is the difference between a continuous and a discrete variable?
4. What are the real limits of a measurement?
5. What are the three parts of the rule about rounding?
6. How is summation notation used to simplify the communication about sums?

This chapter describes some basic notions of mathematics that are necessary for understanding statistics. It defines variables and shows that there are four types (nominal, ordinal, interval, and ratio) based on their level of measurement. It also introduces summation ($\sum$) notation.

You may recall from the description of the study of Pygmalion in the classroom that Rosenthal and Jacobson (1968) led teachers to believe that the second-graders identified as "bloomers" would show IQ spurts but the other children would not spurt. In that study, we can identify three variables: the pretest IQ, the posttest IQ, and the intellectual growth (IQ gain). You will learn in this chapter that all three are interval/ratio variables.

Statistics can be thought of as the science of understanding data; data are the results of a series of measurements on one or more variables. Let us be clear about what these terms mean.

variable: a characteristic that can take several or many different values

A *variable* is any characteristic of the world that can be measured and that can take any of several or many different values. For example, height is a variable (defined as the number of inches a person is tall). It takes the value 70 inches when John is measured and 64 inches when Mary is measured.

Values of a variable may sometimes be assigned arbitrarily; for example, a person's gender is a variable that may take the values 1 for male and 2 for female (or 1 for female and 2 for male, or 27 for male and 136 for female, or any other values we choose).

measurement: the procedure for assigning a value to a variable

Measurement is the procedure that assigns values to the variable. For our purposes, it is enough to realize that a measurement rule must provide a unique and unambiguous result for every individual. Thus, in our gender example, although it makes no difference what values are assigned to male and female, the assignment must be *the same* for all males and all females. It is *not* satisfactory measurement to begin by assigning 1 to males and then, halfway through our data collection, change our minds and begin assigning 27.

2.1 Levels of Measurement

Statisticians distinguish four kinds of variables and measurement levels: nominal, ordinal, interval, and ratio. The main characteristics of these kinds of variables are given in Table 2.1.

nominal scale: classification of unordered variables

The *nominal scale of measurement* classifies objects into categories based on some characteristic of the object. Examples of nominal measurements of people are male/female, Republican/Democrat/Independent/decline to state, fraternity member/nonmember, and Californian/Ohioan/Nebraskan/Alaskan/and so on. In each case, the *measurement* is the placing of an individual into one of the categories in the measurement scale. We require that all measurement operations be *mutually exclusive*; that is, for example, a person cannot be both a Republican and a Democrat or both an Ohioan and a Nebraskan.

TABLE 2.1 Requirements for nominal, ordinal, interval, and ratio levels of measurement

Level of measurement	*Requirements*
Nominal	Categories are mutually exclusive.
Ordinal	Categories are mutually exclusive. Categories have logical order.
Interval	Categories are mutually exclusive. Categories have logical order. Equal differences in characteristic imply equal differences in value.
Ratio	Categories are mutually exclusive. Categories have logical order. Equal differences in characteristic imply equal differences in value. True zero point exists.

The order of categories in a nominal variable is *not* important. For example, it doesn't matter from a measurement standpoint whether we refer to the gender distinction as male/female or female/male. It may be useful to assign a numerical value to a nominal category; we might let male = 1 and female = 2. Doing that *does not change* the fact that the order is irrelevant; it would have made just as much measurement sense to let male = 2 and female = 1.

ordinal scale: measurement of variables that have an inherent natural order

The *ordinal scale of measurement* classifies objects into mutually exclusive categories based on some characteristic of the object (as does the nominal level of measurement), and furthermore it requires that this classification have some inherent, logical order. Examples of ordinal measurements of people are class standing (freshman/sophomore/junior/senior), class rank (first in class/second/third/etc.), grade in course (A/B/C/D/F), and level of depression (not depressed/slightly depressed/moderately depressed/severely depressed). As with nominal measurements, ordinal measurements must be mutually exclusive; we cannot assign both grades A and B or be both a freshman and a sophomore.

We may assign a numerical value to these categories, and if we do, the order of these categories *is* important (unlike in nominal variables where order is irrelevant). For example, it does make sense to assign freshman = 1, sophomore = 2, junior = 3, and senior = 4, but it does not make sense to assign sophomore = 1, senior = 2, freshman = 3, and junior = 4. Order is inherent in ordinal variables, and the values of the variables must reflect that order.

interval scale: a measurement scale for ordered variables that has equal units of measurement

The *interval level of measurement* classifies objects into mutually exclusive categories based on some characteristic of the object (as do the nominal and ordinal levels of measurement). It requires that this classification have some inherent, logical order (as does the ordinal level of measurement), and furthermore it requires that the width of all the categories be the same. An example of an interval variable is temperature as measured in degrees Celsius (°C). When we require that the intervals be equal, we mean, for example, that the temperature difference between 34°C and 35°C is the same as the difference between 77°C and 78°C. It follows that the distance between nonconsecutive measurements must also be equal if the measured differences are equal. For example, the temperature difference between 34°C and 37°C must be the same as that between 75°C and 78°C because both differences are 3 degrees.

ratio scale: an interval scale of measurement that has a true zero point

The *ratio level of measurement* has all the characteristics of the interval level, and furthermore it requires that the scale have a true zero point. Examples of ratio measurements are weight (expressed as number of pounds), height (number of inches), and time (number of minutes). A true zero point means that the thing being measured actually vanishes when the scale reads zero. Thus, for example, the variable weight measures the heaviness of an individual; when the weight is 0 pounds, there is in fact no weight.

Note that the existence of a true zero point on a scale does *not* imply the existence of an individual whose measurement is zero. There are no 0-pound humans, for example. The existence of a true zero simply means that *if* there were an individual with no weight, then the weight scale would read 0.

Temperature as measured in degrees Celsius is *not* a ratio scale because 0°C does not mean the absence of heat; 0°C is instead a relatively arbitrary point, the freezing point of pure water at sea level. The true zero of the Celsius scale (the complete absence of heat) is actually −273°C. By contrast, the Kelvin scale of temperature has absolute zero temperature as the 0K point of the scale (thus making the freezing point of water +273K). The Kelvin scale *is* therefore a ratio scale.

For most statistical procedures (including all those described in this book), there is no difference between the interval and the ratio levels of measurement. Therefore, we will lump interval and ratio scales together, referring to either as an *interval/ratio scale*.

interval/ratio variable: a variable measured at either the interval scale or the ratio scale of measurement

We must make the distinction among nominal, ordinal, and interval/ratio variables because statistics that are appropriate for variables measured at one level of measurement may not be appropriate for variables measured at a different level. For example, if Mary is 64 inches tall and John is 70 inches tall, it is quite reasonable to say that their average height is 67 inches. However, suppose we code political preference as Democrat = 1, Independent = 2, and Republican = 3 and that Mary is a Democrat and John is Republican. If we average Mary's and John's party preferences, we get $(1 + 3)/2 = 4/2 = 2$, which might lead us to the totally *unreasonable* conclusion that Mary and John's average political preference is "Independent." This absurdity arises because the average is an appropriate statistic for interval/ratio variables such as height but *not* for nominal variables such as political preference. We will return to this discussion in Chapter 4.

2.2 Continuous and Discrete Variables

continuous variable: one with an infinite number of values between adjacent scale values

Variables at the ordinal or interval/ratio level of measurement can be classified as either continuous or discrete (all nominal variables are discrete). A *continuous variable* is one that has in principle an infinite number of possible values between any two adjacent scale values. For example, height is a continuous variable because there is no limit to the refinement we can make in the measurement operation. If we measure more and more precisely, we can ascertain that Mary's height is 64.4 inches, or 64.37 inches, or 64.372 inches, or 64.3719 inches.

discrete variable: one with no possible intermediate values between two adjacent points

A *discrete variable* is one that has no possible intermediate values between two adjacent points. For example, the number of children a woman has is a discrete variable, with possible values of only whole numbers, not 1.5 or 2.993.

To restate: If, for any two values of a variable, you can imagine a meaningful intermediate value, then the variable is continuous. Otherwise, the variable is discrete.

Real Limits

When we say that Mary is 64 inches tall, we do not in general mean that she is *exactly* 64 inches tall; she could actually be 63.8 inches or 64.372 inches tall. All measurements made on continuous variables are only approximations because of the (theoretically) infinite number of possible measured values. When we say that Mary is 64 inches tall, we really mean (if we are to be precise) that Mary is taller than 63.5 inches and shorter than 64.5 inches. We sometimes say, then, that Mary's height is $64 \pm .5$ inches. The two values 63.5 and 64.5 are referred to as the *real limits* of the measured value of 64 inches. The real limits are the points that are half the smallest measuring unit above and below the measured value. For example, if Mary's time in the 100-yard dash is recorded as 10.6 seconds, the real limits of that measurement are 10.55 and 10.65 seconds, half the smallest measuring unit (.1 second) below and above the measured value of 10.6 seconds. Equally precisely, we can say that Mary's time is $10.6 \pm .05$ seconds.

real limits of a measurement: the points that are half the smallest measuring unit above and below the measured value

Significant Figures

Statistical computations on continuous variables often result in values that have some or many more digits than the numbers in the original data. For example, if Mary is 64 inches tall and both Natasha and Carla are 65 inches tall, then the average height of these three women is $(64 + 65 + 65)/3 = 64.66666666\ldots$ inches. Two questions should be answered about such a result: (1) How many of those digits should we report if that average height is to be the final result of our computations? (2) How many digits should we carry if that average is a subcomputation whose result will not be reported directly but will be used in some later computation?

In the behavioral sciences, there is no generally accepted answer to either of these questions. I (and many others) have these two suggestions:

1. Report in the final answer *two more* significant figures than were reported in the original data. Thus, because our height data were originally reported as whole numbers, our final answer should be reported as a value with two decimal places—that is, as 64.67 inches. If the original data had been reported in tenths, we would report a final answer with three decimal places, and so on.

2. Carry as intermediate subcomputations *at least three more* significant figures than were originally reported. Thus, the average height used as an intermediate value should be carried forward as 64.667 inches. It is usually simpler (and often better) to carry forward as many figures as your calculator will hold and round only when you report the final answer.

Rounding

Now that we have determined how many digits to report, we must decide how to determine the value of the final digit. Why, for example, did we write the final value of the average height as 64.67 and not as 64.66 inches?

The procedure for determining the final digit is called "rounding" and is generally accepted among all statisticians. The rule for rounding has three parts: (1) If the remainder beyond the last digit to be reported is less than 5 (or .5 or .05, etc., depending on where the decimal point happens to be), simply eliminate all remaining digits; thus, 4.347 rounds to 4.3 if we wish to report one decimal place. (2) If the remainder is greater than 5, increase the last reported digit by 1; thus, 6.867 rounds to 6.9. (3) If the remainder is *exactly* 5 (not 501 or 49, etc.), then round the last reported digit to the closest *even* number; thus, 3.55 rounds up to 3.6, but 3.65 rounds down to 3.6.

Some students find it easy to remember the following expression, which states part 3 of the rule in a different but equivalent way: "Even—leave it. Odd—up." For example, suppose we wish to round 7.3500 to one decimal place. The remainder is "500"—that is, exactly 5—so part 3 applies. Because the digit just before the remainder is "3," which is odd, we follow "Odd—up" and round up to 7.4. On the other hand, if we wish to round 81.65 to one decimal place, the remainder is "5" and the digit just before the remainder is "6," which is even, so we "Even—leave it" and round to 81.6.

Part 3 of the rule for rounding is necessary to prevent a systematic bias from entering our data as a result of rounding. See Box 2.1 for an explanation.

BOX 2.1 How part 3 of the rounding rule prevents bias

ⓘ This box may be omitted without loss of continuity.

Here is an example that shows how part 3 of the rounding rule prevents rounding from introducing a bias into computations. Consider a data set that consists of 100 two-decimal-place values: 2.79, 3.11, 1.90, 3.30, 2.35, ... We would expect, in these 100 values, about ten values that end in 0, about ten values that end in 1, ..., and about ten values that end in 9. Now consider the same data set rounded to one decimal place: 2.8, 3.1, 1.9, 3.3, 2.4, ... How does the *sum* of the rounded values compare to the sum of the original values? About ten values end in 9, such as 2.79. When we

(continued)

BOX 2.1 *(continued)*
round up, as we should (from 2.79 to 2.8), the rounded values will be larger than the originals by .01, so the sum of the rounded values will be too large by about ten times .01. That will be offset, however, by the (approximately) ten values that end in 1, such as 3.11. Those values will be rounded down (as from 3.11 to 3.1) by .01 each, and the sum of those rounded values will be too *small* by about ten times .01. Thus, the values that end in 1 will offset the values that end in 9. Similarly, the values that end in 2 will offset the values that end in 8, those that end in 3 will offset those that end in 7, and those that end in 4 will offset those that end in 6.

The values that end in 0 do not create any difference in the sum because the magnitudes of the original values (such as 1.90) are the same as the magnitudes of the rounded values (1.9).

That leaves the values that end in 5. If we round all of them upward (as from 2.35 to 2.4), then each of the (approximately) ten rounded values will be too large by .05, and the sum will be too large by about ten times .05. We therefore need the rule to ensure that about half of the values that end in 5 will be rounded upward, whereas the other half will be rounded downward.

Table 2.2 gives several examples of the rounding process. Note that the last column explains which of the three parts of the rounding rule was applied in obtaining the answer.

TABLE 2.2 Examples of rounding

Decimal places desired	*Original value*	*Rounded value*	*Rounding rule*
1	712.31	712.3	1
2	3.697	3.70	2
1	5.350	5.4	3
0	247.499	247.	1 (not 3!)
1	84.25	84.2	3
1	84.2501	84.3	2
2	.005	.00	3
0	800.501	801.	2 (not 3!)

2.3 Summation

Notation

It is often convenient to assign a symbol to take the place of a variable name. Most often we call the variable X; if there are two variables, we call them X and Y; and generally we call a third variable Z. When it is convenient, we use other symbols; for example, we might call height X or H.

A data set is a collection of measurements on one or more variables. It is often convenient to assign an *index variable*, usually i, that refers to a particular measurement's position in a data set. Table 2.3 shows the measurements on the variable height for five subjects. When we let $i = 1$, we refer to the first subject, John. When we let $i = 4$, we refer to the fourth subject, Carla. The order of data in a data set is usually quite arbitrary; we could arrange the subjects in alphabetical order, in which case $i = 1$ would refer to Carla. For sample data, we usually refer to the last value of i as n. Thus, n is the number of observations in the sample; $n = 5$ for the data in Table 2.3.

TABLE 2.3 Heights of five subjects

Subject	*Index*	*Height (in.)*
John	1	70
Mary	2	64
Natasha	3	65
Carla	4	65
Sam	5	68

We often put a *subscript* on the symbol for the variable when we wish to refer to a specific data point rather than the variable in general. Thus, X_1 refers to first value of X; in

Table 2.3 (assuming we are using X to denote height), $X_1 = 70$ inches, John's height, and $X_4 = 65$ inches, Carla's height. In general, X_i refers to the ith value of X.

Many of the computations in statistics involve summing all the entries in a data set. To make communication about such sums more compact, statisticians have developed summation notation. For purposes of this discussion, we will assume that we have made two measurements, X and Y, on each of six subjects, as shown in Table 2.4. We indicate the sum of all the values of the variable X as $\sum X_i$, which we read as "the sum of X sub i."[1] The character $\sum$ (pronounced "sigma") is the Greek capital S (for "sum"). For the data of Table 2.4, we can expand the indicated sum as follows:

$\sum$: symbol meaning "the sum of"

sum of values of a variable

$$\begin{aligned}\sum X_i &= X_1 + X_2 + X_3 + X_4 + X_5 + X_6 \\ &= 7 + 2 + 4 + 2 + 3 + 6 \\ &= 24\end{aligned} \tag{2.1}$$

TABLE 2.4 Measurements on variables X and Y

i	X_i	Y_i
1	7	5
2	2	12
3	4	4
4	2	10
5	3	9
6	6	9

In cases where it is unambiguous, we may drop the i subscript, understanding that $\sum X$ means $\sum X_i$. Occasionally, again when the situation is unambiguous, we use three dots (called "ellipses") to indicate a sum; for example, we may write Equation (2.1) as $\sum X_i = X_1 + X_2 + \cdots + X_6$.

Computations

The calculations required in statistics (and throughout this book) often require us to perform some calculations on the variable *before* we perform the summation. For example, for the data in Table 2.4, we may wish to calculate $\sum 3X$. Our procedure is to compute the product $3X$ for each value of X and *then* to sum those $3X$ values, as shown in Table 2.5. Note that the second column of Table 2.5 holds each of the values of $3X$, and the required sum is obtained simply by adding down that last column.

TABLE 2.5 Computing $\sum 3X$

X	$3X$
7	21
2	6
4	12
2	6
3	9
6	18
	$\sum 3X = 72$

Calculations may be even more complex. For example, $\sum 3X^2$ is obtained by first squaring the X values, as shown in the second column of Table 2.6, and then multiplying each squared X value by 3, as shown in the third column. Then $\sum 3X^2 = 354$ is obtained by simply adding down that last column.

See Box 2.2 for a note about the organization of computations.

TABLE 2.6 Computing $\sum 3X^2$

X	X^2	$3X^2$
7	49	147
2	4	12
4	16	48
2	4	12
3	9	27
6	36	108
		$\sum 3X^2 = 354$

BOX 2.2 A note regarding computations

The time you spend in solving statistics problems can be divided into two parts: time spent in performing the computations, and time spent in finding errors that you made in the computations. Many students tend to overlook the impact that error finding has on their total study time. There are two ways to minimize the error-finding time: work carefully and without distraction in the first place, and organize your work so that errors are easy to locate. Plan on making errors! To err is human! Failing to plan for errors is also human but not very smart!

Organize your computations so that you are performing only one operation at a time. For example, in computing $\sum 3X^2$ for the values of X in Table 2.6, it might be tempting to square the numbers and multiply them in the same step, going through a mental process such as: "7 squared is 49 times 3 is 147, 2 squared is 4 times 3 is 12, 4 squared is 16 times 3 is 48, ..." That seems to be more efficient because it eliminates writing down a column of values (the second column of Table 2.6); however, alternating back and forth between squaring and multiplying substantially

(continued)

[1] Our summation notation $\sum X_i$ is a simplification of the more complete notation sometimes used by statisticians $\sum_{i=1}^{n} X_i$, which is read "the sum of X_i for all values of i from $i = 1$ to n." Because in this book sums will always be taken over *all* the values of the variable (that is, in this book sums *always* run from $i = 1$ to n), we can use the simpler notation without ambiguity.

BOX 2.2 *(continued)*

increases the probability of making a mistake. It is much better to perform all the squarings first and then do all the multiplications.

You may be saying to yourself, "I can square and multiply in my head; I don't need that extra step!" and you may be right *most* of the time. But such a statement ignores the fact that if you do make a mistake, the mistake will be much harder to find when you have combined two computational steps.

I highly recommend that computations be performed in this manner: (1) List the data in a column, (2) create a new column for every subcomputation, and (3) perform a summation *only* by adding directly down one column.

Thus, to compute the sum $\sum 3X^2$, we (1) list the original data as shown in the first column of Table 2.6, (2) create a new column with the heading "X^2" and another new column with the heading "$3X^2$," and then (3) perform the summation by simply adding down that last column.

One operation per column leads to the lowest probability of making a mistake.

TABLE 2.7 Computing $\sum(X-4)^2$

X	$X-4$	$(X-4)^2$
7	3	9
2	−2	4
4	0	0
2	−2	4
3	−1	1
6	2	4
		$\sum(X-4)^2 = 22$

TABLE 2.8 Computing $\sum XY$

X	Y	XY
7	5	35
2	12	24
4	4	16
2	10	20
3	9	27
6	9	54
		$\sum XY = 176$

Let's consider a few more examples—first the sum $\sum(X-4)^2$. Following the suggestion of Box 2.2, we list the data, as shown in the first column of Table 2.7. Then we create a new column with the heading "$X - 4$" and perform all the subtractions; then we create another new column with the heading "$(X-4)^2$" to hold all the squarings. Finally, we perform the summation by adding down this last column, finding that $\sum(X-4)^2 = 22$.

Note that $\sum(X-4)^2$ is *not* equal to $\left[\sum(X-4)\right]^2$; that is, we do *not* obtain $\sum(X-4)^2$ simply by adding down the second column of Table 2.7 and then squaring that result. Adding down the second column gives $\sum(X-4) = 0$, and squaring that result gives $\left[\sum(X-4)\right]^2 = [0]^2 = 0$, which is *not* equal to $\sum(X-4)^2 = 22$.

Occasionally we perform computations with two variables at a time. For example, $\sum XY$ is obtained by first listing the values of X and Y as shown in Table 2.8. Then we create a new column headed "XY" and find $\sum XY = 176$ by simply adding down that column.

Note that $\sum XY$ is *not* equal to $\sum X \sum Y$. This can be verified by noting that $\sum X = 24$, as might be obtained by adding down the first column of Table 2.8, and $\sum Y = 49$, as might be obtained by adding down the second column. $\sum X \sum Y$ is thus $24(49) = 1176$, which is *not* equal to $\sum XY = 176$.

We will use summation notation in all chapters in this textbook, and it is crucial that you become confidently skilled in performing such computations.

2.4 Connections

Cumulative Review

Chapter 1 discussed statistics in general and described the Pygmalion in the classroom experiment that we will refer to frequently throughout this textbook. It also distinguished between populations (all the members of the group under consideration) and samples (some subset of that group) and between parameters (characteristics of populations) and statistics (characteristics of samples). It then defined the probability of an event as the number of outcomes favorable to that event divided by the total number of outcomes.

Computers

Many computer programs are available that perform statistical computations—far too many to be covered in any single book. This textbook provides examples from two such programs: ESTAT's datagen and SPSS 6.1 Student Version.

ESTAT's datagen, which I wrote, is designed especially for the student who is seeking to comprehend statistics. It allows easy access to all the computations typically encountered in introductory statistics so that the student can do as much or as little of a computation by hand as is desired. It is remarkably easy to use: Most students have no trouble getting it to do what they want within the first 5 minutes of operation. All statistics are presented automatically; there are no "statistics" menus to select or buttons to press. ESTAT's datagen is thus a specialized program: specialized for the acquisition of statistical comprehension by introductory statistics students. There is a price for this specialization, however; ESTAT's datagen is by no means a comprehensive program and thus does not perform many important (but more advanced) statistical analyses that a program like SPSS does.

SPSS is a family of statistical packages designed to analyze a wide variety of statistical data. There are many different versions of SPSS; the one we will use for examples is SPSS 6.1 Student Version. The instructions for SPSS presented here apply to both the Windows and the Macintosh versions of SPSS 6.1 Student Version. SPSS is extremely versatile, capable of performing many statistical tasks that go far beyond the needs of the introductory statistics student. The price for this versatility is that it is somewhat more difficult to use than ESTAT's datagen.

It is a dilemma whether to use datagen or a package such as SPSS in introductory statistics. Datagen makes the exploration and comprehension of the basics of statistics much more transparent. It is much easier for the student to see what is happening in a particular analysis. When the student moves beyond introductory statistics, however, datagen's limitations become more apparent—it does not perform factor analyses or two-way analysis of variance, for example—and then SPSS (or another complete package) is the better choice. We have found that students who start with datagen usually find that the later transition to SPSS (or other complete statistics package) is easy.

Both the exploration/comprehension and the tool-acquisition arguments are persuasive, so this textbook will provide examples from both ESTAT's datagen (version 2.0) and SPSS 6.1 Student Version.

Use datagen to compute $\sum X$ and $\sum Y$ for the data in Table 2.4:

1. Start ESTAT by clicking the `Estat` icon.
2. Start datagen by clicking the `datagen` button.
3. Enter the data from Table 2.4 into the datagen spreadsheet. To enter values going down a column, press `Enter` after each value. To enter values going across a row, press `Tab` after each value.
4. Note that the sum of each column appears automatically in the `datagen Descriptive Statistics` window on the line labeled "`Sum of observations`."

Now use datagen to compute $\sum XY$:

5. Click `Edit Variable`.
6. Click the `Create a new variable that is the product of two variables` button.
7. Click the `variable 1` button in the left box and the `variable 2` button in the right box and then click `OK`.
8. Note that a new column is created in the spreadsheet that holds the product of X and Y, and the sum of this column appears automatically in the `datagen Descriptive Statistics` window.

You can vary steps 6 and 7 to find any sum or product of variables.

SPSS

Use SPSS to compute $\sum X$ and $\sum Y$:

1. Start SPSS by clicking the `SPSS` icon.
2. Enter the data from Table 2.4 into the SPSS spreadsheet. To enter values going down a column, press `Enter` after each value. To enter values going across a row, press `Tab` after each value.
3. Ask SPSS to show sums by clicking `Statistics`, then `Summarize`, and then `Descriptives...`. Then click `Options...`, click the `Sum` box, and click `Continue`.

4. Request variables to be displayed by clicking ▶. Then click `var00002`, click ▶, and click `OK`.
5. Note that several statistics are displayed in the `!Output1` window; the sums are near the right-hand side of the window.

Now use SPSS to compute $\sum XY$:

6. Click `Window` and then click `Newdata`. Alternatively, click the spreadsheet button (looks like a grid) on the button bar near the top of the screen.
7. Click `var` in the third column.
8. Click `Transform` and then click `Compute....`
9. Create a new variable label: Type "product."
10. Request the multiplication: Click ▶, click `*`, click `var00002`, click ▶, and then click `OK`.
11. Note that a new column is created that holds the product of X and Y.
12. Click `Statistics`, then `Summarize`, and then `Descriptives....`
13. Click `product`, click ▶, and then click `OK`.
14. Note that several statistics are displayed in the `!Output1` window; the sums are at the right-hand side of the window.

Quite a few steps are involved in asking SPSS to perform this simple task. Fortunately, the effort expended here will make subsequent tasks easier. You can vary steps 7 and 8 to ask SPSS to create and evaluate any mathematical expression—an example of the versatility of SPSS.

Homework Tips

1. Check the list of learning objectives at the beginning of this chapter. Do you understand each one?

2. The hint given in Box 2.2 will in the long run save you a lot of computational time, and I strongly recommend it even if it seems at first like a waste of time. Almost all students are surprised and frustrated by how many computational errors they make. This procedure will help you avoid those costly errors.

3. Remember to include *at least three* extra significant figures in subcomputations.

4. This tip addresses what is perhaps the most frequent mistake in all introductory statistics computations. It is essential to be clear that $(\sum X)^2$ is *not* the same as $\sum X^2$. If there are three values of X—say, 2, 4, and 6—then $(\sum X)^2 = (2 + 4 + 6)^2 = (12)^2 = 144$. By contrast, $\sum X^2$ requires (if we follow the logic of Box 2.2) the creation of a new column headed "X^2" with the values 4, 16, and 36. Then $\sum X^2 = 4 + 16 + 36 = 56$. If this is difficult for you, work at it again and again until it becomes second nature. Otherwise, this mistake will "bite" you in nearly every chapter of the text.

Exercises for Chapter 2

Section A: Basic exercises (answers in Appendix D, page 505)

1. (a) Suppose I am interested in the effect of family size on life satisfaction. I code family size as 0 if four or fewer individuals live in the household and 1 if more than four individuals live in the household. Does such a coding constitute a measurement of family size? If not, why not?

   (b) Halfway through my study, I find that I am getting too many 0 family sizes, so I change my coding. I leave my original codings unchanged, but I now code family size as 0 if three or fewer individuals live in the household. Does such a procedure constitute measurement? If not, why not?

2. Identify the level of measurement (nominal, ordinal, or interval/ratio) for these variables:

(a) Weights of people (measured in pounds)
(b) The Likert scale (measured as 1 = strongly disagree; 2 = disagree; 3 = neutral; 4 = agree; and 5 = strongly agree)
(c) Basketball players' uniform numbers (0, 1, 2, . . . , 55)
(d) Clothing sizes (small, medium, large, extra large)
(e) Shades of nail polish (21 = "shameless rose," 25 = "pink in the afternoon," 33 = "orchid," 41 = "berry rich," 88 = "nouveau nude," etc.)
(f) Volumes of bottles (measured in fluid ounces)

3. Which of the following variables are discrete and which are continuous?
(a) Distance (in miles) from one city to another
(b) Number of bar presses a rat makes
(c) Number of students in your statistics class
(d) Fuel consumption (miles per gallon)

4. Consider the sample data given in the table.

i	X
1	6
2	1
3	9
4	−5
5	6
6	0
7	−3

(a) What value does X_4 have?
(b) What is n?
(c) Compute $\sum X$.
(d) Create a column headed "X^2" as described in Box 2.2 and compute $\sum X^2$.
(e) Compute $(\sum X)^2$.
(f) For these data, does $\sum X^2 = (\sum X)^2$?
(g) Create a column headed "$X + 10$" and compute $\sum(X + 10)$.
(h) For these data, does $\sum(X + 10) = \sum X + 10$?
(i) Create a column headed "$X^2 - 5$" and compute $\sum(X^2 - 5)$.

5. For the data of Exercise 4, consider the sum $\sum\left(\frac{3X_i^2 - 6}{4}\right)$.
(a) How many columns are required for this computation, following the suggestions of Box 2.2? What are their headings?
(b) Compute the sum.

6. Use the data in the accompanying table.
(a) Find $X_3 - Y_3$.
(b) Compute $\sum XY$.

X	Y
3	2
4	4
7	1
3	11

7. For the data of Exercise 6, consider the sum $\sum\left(\frac{2(X - Y)^2}{X + Y}\right)$.
(a) How many columns are required for this computation, following the suggestions of Box 2.2? What are their headings?
(b) Compute the sum.

8. For the data of Exercise 6, demonstrate each expression.
(a) $\sum X^2 \neq (\sum X)^2$
(b) $\sum(X + 5) \neq \sum X + 5$
(c) $\sum(X + 5)^2 \neq (\sum X + 5)^2$
(d) $\sum XY \neq \sum X \sum Y$

9. (a) For the data of Exercise 6, how many decimal places should you use in reporting the value of $(\sum X)/3$ as a final answer?
(b) What is that value? Be sure to round correctly.

10. Round each value to two decimal places.
(a) 14.637
(b) −14.637
(c) 2.152
(d) 3.40500
(e) 3.40501
(f) 3.315
(g) 3.325
(h) 3.335
(i) 27.43475

Section B: Supplementary exercises

11. Use the sample data in the table.

i	X_i
1	4
2	−5
3	2
4	−8
5	21
6	3
7	13
8	6

(a) What value does X_2 have?
(b) What is n?
(c) Compute $\sum X$.
(d) Create a column headed "X^2" as described in Box 2.2 and compute $\sum X^2$.
(e) Compute $(\sum X)^2$.
(f) For these data, does $\sum X^2 = (\sum X)^2$?
(g) Create a column headed "$X - 3$" and compute $\sum(X - 3)$.
(h) For these data, does $\sum(X - 3) = \sum X - 3$?
(i) Create a column headed "$X^2 - 9$" and compute $\sum(X^2 - 9)$.

12. For the data of Exercise 11, consider the sum $\sum\left(\frac{(X_i - 3)(2X_i + 4)}{2}\right)$.
(a) How many columns are required for this computation following the suggestions of Box 2.2? What are their headings?
(b) Compute the sum.

13. Consider the data in the accompanying table.
 (a) Find $X_4 - Y_4$.
 (b) Compute $\sum XY$.

X	Y
1.4	5
4.1	−2
−3.8	1
−6.0	−10
2.2	2

14. For the data of Exercise 13, consider the sum $\sum\left(\frac{(2X-Y)(X+2Y)}{X+Y}\right)$.
 (a) How many columns are required for this computation following the suggestions of Box 2.2? What are their headings?
 (b) Compute the sum.

15. For the data of Exercise 13, demonstrate each expression.
 (a) $\sum X^2 \neq (\sum X)^2$
 (b) $\sum(X+5) \neq \sum X + 5$
 (c) $\sum(X+5)^2 \neq (\sum X + 5)^2$
 (d) $\sum XY \neq \sum X \sum Y$

16. (a) For the data of Exercise 13, how many decimal places should you use in reporting the value of $(\sum X)/7$?
 (b) What is that value? Be sure to round correctly.

17. Round each value to one decimal place.
 (a) 13.55
 (b) 13.65
 (c) −4.1472
 (d) −3.5001
 (e) −3.6001
 (f) 3.14159
 (g) .02
 (h) 6.66
 (i) 12.5

Section C: Extending your comprehension (answers in Appendix D, page 506)

18. Create a data set that has five points and for which $\sum X = 2$ and $\sum X^2 = 20$. [*Hint:* You will need negative values of X.]

19. Let X be the number of days heart attack patients stay in the hospital on their first admission. Data for a sample of ten patients are shown in the table.

Patient	Duration
1	9
2	13
3	9
4	7
5	12
6	17
7	10
8	10
9	11
10	14

 (a) Evaluate $\sqrt{\frac{\sum(X_i - 11.2)^2}{9}}$.
 (b) Evaluate $\sum\sqrt{\frac{(X_i - 11.2)^2}{9}}$.
 (c) Explain why the values in parts (a) and (b) are different.

20. For the data in the table, calculate the following:

Subject	X	Y
1	23	72
2	43	92
3	26	71
4	29	81
5	37	78

 (a) $\sum X$
 (b) $\sum Y$
 (c) $\sum X^2$
 (d) $\sum Y^2$
 (e) $\sum XY$
 (f) $$\frac{\sum XY - \frac{\sum X \sum Y}{n}}{\sqrt{\left[\sum X^2 - \frac{(\sum X)^2}{n}\right]\left[\sum Y^2 - \frac{(\sum Y)^2}{n}\right]}}$$

 [*Note:* The computation required in part (f) is the most complicated computation ever required in this textbook.]

Section D: Cumulative review (answers in Appendix D, page 506)

21. Suppose you take the numbers 1 through 100 and write each one on a poker chip; then you put all the poker chips in a bag and mix them thoroughly. You draw one chip from the bag. What is the probability that you draw a chip with the property given?
 (a) A "34"
 (b) A "129"
 (c) An odd number
 (d) An even number
 (e) Greater than or equal to 75
 (f) Less than 10
 (g) Contains "7" as one (or more) of its digits
 (h) Has digits that add to 10

22. (*Exercise 21 continued*) Suppose I tell you that the first digit is a "3" and it is a two-digit number. What is the probability that the chip is "34"?

23. (*Exercise 21 continued*) Suppose I tell you that the chip is an even number. What is the probability that the chip is "34"?

CHAPTER

3 Frequency Distributions

LEARNING OBJECTIVES

1. How is a tabular frequency distribution constructed?
2. What are class intervals and what role do they play in the development of grouped frequency distributions?
3. What are two graphic methods of representing interval/ratio data from grouped frequency distributions?
4. What steps are involved in eyeball-estimating frequency distributions?
5. How are these terms used to describe the shape of distributions: unimodal, bimodal, symmetric, positively skewed, negatively skewed, asymptotic, normal?
6. How is a bar graph similar to and different from a histogram? For what kind of data is each used?

This chapter describes frequency distributions and how they are displayed: in tables (as ungrouped and grouped frequency distributions) and in graphs (as histograms and frequency polygons). It describes the eyeball-estimation of frequency distributions and discusses terminology for describing the shapes of distributions: unimodal, bimodal, symmetric, positively and negatively skewed, asymptotic, and normal.

Table 3.1 shows the grouped frequency distributions and Figure 3.1 shows the histograms for the intellectual growth scores described in Chapter 1. [You may recall that Rosenthal and Jacobson (1968) led teachers to believe that second-graders identified as "bloomers" would show IQ spurts, but the "other" children were not expected to spurt. Actually, the bloomers were a random sample of second-graders—no different, on average, from the other children.] The grouped frequency distributions and histograms make it easy to see that the most frequently occurring IQ gain among the bloomers was between 10 and 20 IQ points, whereas the most frequently occurring IQ gain among the other children was between 0 and 10 IQ points. It is also easy to see that there is considerable overlap between the bloomer and other children's intellectual growth.

We could have made the same observations from inspecting the data in Chapter 1, but it would have been much more difficult there. The frequency distribution displays that you will learn about in this chapter make it easier to see the characteristics of a data set.

Recall that in Chapter 1 we pointed out the three major concepts in statistics: (1) what a "distribution of a variable" is and how to describe it, (2) what a "sampling distribution of means" is and how it is related to the distribution of a variable, and (3) what a "test

TABLE 3.1 Grouped frequency distributions of intellectual growth (IQ gain) for Oak School bloomers and other children*

	Frequency	
IQ gain	*Bloomers*	*Others*
61 – 70	1	0
51 – 60	0	0
41 – 50	0	0
31 – 40	0	1
21 – 30	2	2
11 – 20	6	14
1 – 10	1	18
−9 – 0	2	9
−19 – −10	0	3
	12	47

*Based on a study by Rosenthal & Jacobson, 1968

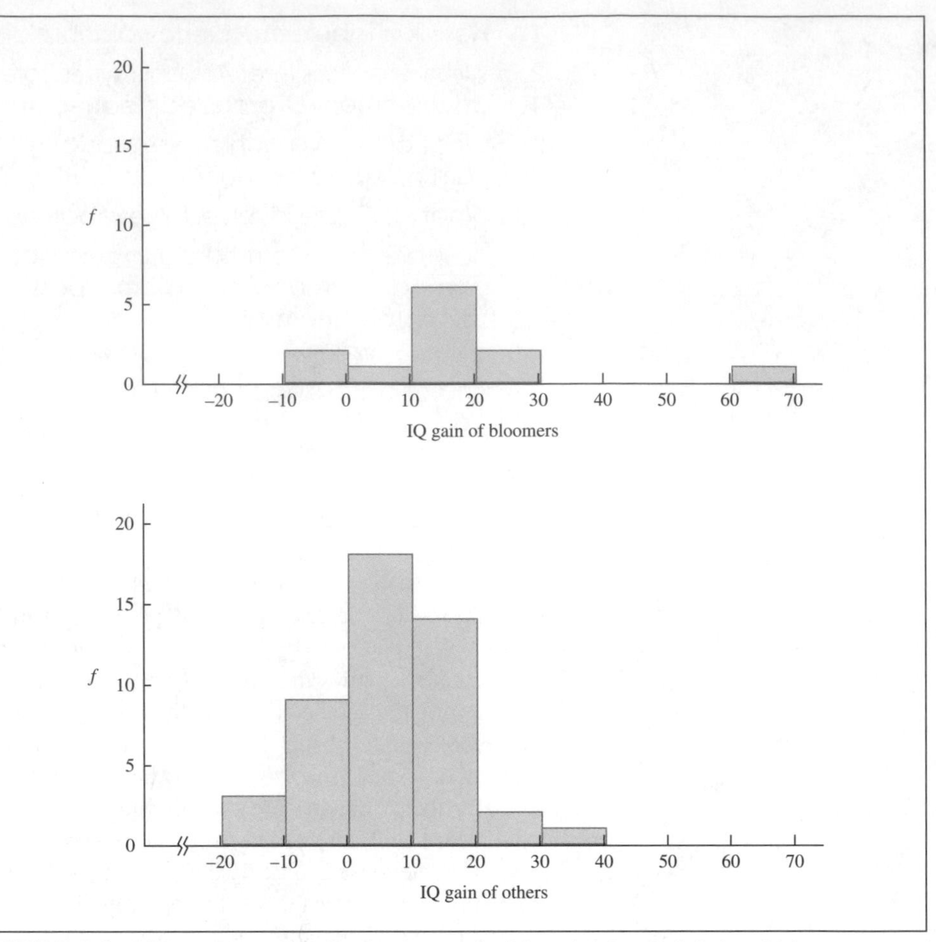

FIGURE 3.1 Histograms of intellectual growth (IQ gain) for Oak School bloomers and other children (Based on a study by Rosenthal & Jacobson, 1968)

statistic" is and how it is related to the sampling distribution of means. This chapter (and also Chapters 4 and 5) focuses on the first concept, describing the distributions of variables. Our first task is to convince ourselves that *understanding frequency distributions will make it simpler for us to think about and communicate about data*.

Suppose, for example, that we are interested in the weights of male students in our statistics class. There are 25 men in the class, and we measure each man's weight, with the results shown in Figure 3.2, where each man is represented by a square.

135	180	190	137	154
149	164	185	173	163
162	157	161	173	180
179	197	159	182	164
164	144	152	150	163

FIGURE 3.2 Weights (lb) of male students as they sit in the classroom

Now suppose our friend Jack asks us to tell him what the male students in our class weigh. We can give Jack a straightforward answer by listing the data: "The first student weighs 135 pounds, the next student weighs 180 pounds, the third student weighs . . . , and the last student weighs 163 pounds." This kind of listing is called an *enumeration*.

To respond to Jack's question in writing, we can provide him with a list of the men's weights, as shown in Table 3.2. Whether given in writing or orally, an enumeration is a perfectly accurate answer to Jack's question about student weights, but it is probably an undesirable answer both because it is too long (we give Jack more information than he wants) and because it does not highlight the important characteristics of the distribution (it does not, for example, make it easy to see which is the largest or the smallest weight, or which weight or range of weights occurs most frequently).

TABLE 3.2 List of male students' weights (lb)

135	180	190	137	154
149	164	185	173	163
162	157	161	173	180
179	197	159	182	164
164	144	152	150	163

Statisticians make the answers to questions such as Jack's more informative by using frequency distributions. As we will see, frequency distributions can be presented in the form of tables or graphs.

3.1 Distributions as Tables

tabular frequency distribution: an ordered listing of all values of a variable and their frequencies

frequency: the number of times the particular value of a variable occurs

A *tabular frequency distribution* is a presentation of the data as a table that lists the numerical values of a variable in a logical (usually descending) order and also lists the frequency of each value. A *variable*, as we saw in Chapter 2, is that characteristic of interest that can take on different values for each subject under consideration, and the *frequency* (usually abbreviated f) is the number of times the particular value of the variable occurs. Table 3.3 shows the tabular frequency distribution of male students' weights. The value 163 occurs twice ("has frequency 2"), as do 173 and 180, and the frequency of the value 164 is 3. All other frequencies for the weights listed in the table are 1. If a weight does not occur in the data of Table 3.2 (for example, 142 or 181), then its frequency is 0, and we omit it entirely in the Table 3.3 frequency distribution.

Note that the sum of the frequencies (25 in our example) shown at the bottom of the column is always equal to the number of entries in the original data set.

Note also that the frequency distribution shows the values of the variable under consideration (weight) *in order*, with the largest value first in the table. This presentation simplifies

TABLE 3.3 Frequency distribution of male students' weights

Weight (lb)	*f*	*Weight (lb)*	*f*
197	1	161	1
190	1	159	1
185	1	157	1
182	1	154	1
180	2	152	1
179	1	150	1
173	2	149	1
164	3	144	1
163	2	137	1
162	1	135	1
			25

to some degree our communication about the data. Now, if Jack wants to know about the weights of the men in our class, we can immediately say, "The lightest is 135 pounds, the second lightest is 137 pounds, the third lightest is 144 pounds, . . . , and the heaviest is 197 pounds." This is still a long and cumbersome answer, but the ordering makes it easier for Jack to gain some appreciation of where a particular weight stands in relation to the other weights in the list. However, although this frequency distribution is more informative than a simple enumeration of the individual weights, it has disadvantages: It still provides a long list of weights and their frequencies, and it is relatively difficult to identify weights with 0 frequencies.

grouped frequency distribution: a frequency distribution with adjacent values of the variable grouped together into class intervals

We alleviate those disadvantages by simplifying the frequency distribution into what is called a *grouped frequency distribution* or, equivalently, a "frequency distribution using class intervals," where successive weights are grouped together into class intervals and the frequencies are reported for the intervals, not for the individual weights, as shown in Table 3.4. Note that the sum of the frequencies is still 25, the number of original observations.

The use of class intervals (130–139 pounds, etc.) generally makes a frequency distribution easier to understand. For example, it is now clear that the most frequent weight is in the 160s.

There are some general rules to follow when choosing class intervals: Each interval must be the same width (in Table 3.4, 10 pounds); class intervals must be nonoverlapping and consecutive (for example, if the upper limit of one interval is 159 pounds, then the lower limit of the next must be 160 pounds); and it is generally desirable to have between 6 and 20 class intervals (exactly how many intervals is a matter of judgment, deciding which number of class intervals presents the data in the most informative manner).

Six steps are involved in creating a grouped frequency distribution:

steps for creating a grouped frequency distribution

Step 1. Find the range of the scores (highest score minus lowest score).

Step 2. Make a preliminary choice of the desired number of class intervals.

Step 3. Determine the interval width by dividing the range by the number of class intervals. Round the interval width in either direction to a convenient number, even if that means adjusting the number of class intervals you selected in Step 2.

Step 4. Determine the lower limit of the lowest interval. This should be chosen so that the lowest data point falls somewhere in the first interval and the intervals have convenient limits.

Step 5. Prepare a list of the limits of each class interval, beginning at the bottom of the table with the lowest score and proceeding upward. Be sure that the highest interval contains the highest score.

Step 6. Count the number of observations that occur in each interval, and enter that value as the frequency of the interval.

TABLE 3.4
Grouped frequency distribution of male students' weights using 10-pound class intervals

Class interval	f
190–199	2
180–189	4
170–179	3
160–169	7
150–159	5
140–149	2
130–139	2
	25

Note that the word *convenient* or *desired* appears in Steps 2, 3, and 4. This indicates that there is no simple rule governing the choices to be made; instead, some judgment must be made about the intended communication and the selections chosen to make the clearest communication.

We illustrate this procedure by preparing another grouped frequency distribution for our weight data. Suppose we inspect Table 3.4 and wonder whether we would have provided more information if we had used more class intervals than the 7 shown there. We decide to prepare a grouped frequency distribution using twice as many class intervals—14. We use the same steps:

Step 1. The highest weight is 197 pounds and the lowest is 135 pounds, so the range is $197 - 135 = 62$.

Step 2. Our preliminary choice of the number of class intervals this time is 14.

Step 3. The interval width is the range divided by the number of intervals, or $62/14 = 4.43$ pounds. We could round that to 4 or to 5 pounds, but we decide that 5 is more convenient because the lower limit of every second interval then ends in 0. That means we will have 13 instead of 14 intervals, but our statistical judgment is that using round numbers as interval limits more than offsets that disadvantage.

Step 4. The lowest interval must contain the lowest point (135), so we could choose any of the following values as the lower limit of that interval: 131, 132, 133, 134, or 135 (or any intermediate decimal value if we desired, but that might imply that we measured weights to a greater precision than whole pounds). We select 135 pounds because it is a round number. The lowest interval is thus 135–139 pounds. Note that that interval is 5 pounds wide (not 4 as it might appear) because it contains the five values 135, 136, 137, 138, and 139.

Step 5. The first interval is 135–139. The second interval must have the same width and be consecutive, so it is 140–144. We continue adding intervals as shown in Table 3.5, ending with the highest interval (195–199), which includes the highest point (197 pounds).

Step 6. We refer to our original frequency distribution (Table 3.3) and count the number of individuals whose weights fall into the listed classes. Two individuals (135 and 137) have weights in the first class (135–139), for example, so the frequency of that class is 2. We enter those frequencies in the right-hand column of Table 3.5, and the grouped frequency distribution using 5-pound intervals is completed. Double checking, we add down the list of frequencies; the sum must still be 25, the total number of data points.

TABLE 3.5
Grouped frequency distribution of male students' weights using 5-pound class interval

Class interval	*f*
195–199	1
190–194	1
185–189	1
180–184	3
175–179	1
170–174	2
165–169	0
160–164	7
155–159	2
150–154	3
145–149	1
140–144	1
135–139	2
	25

Note that in a grouped frequency distribution we include class intervals even if they have 0 frequency (for example, the interval 165–169 is included in Table 3.5). This is in contrast to an ungrouped frequency distribution (for example, Table 3.3), where values with 0 frequency are omitted.

Which of these two grouped frequency distributions is preferred, the one with 10-pound intervals or the one with 5-pound intervals? There is no definitive answer to that question. Choosing between them is a matter of judgment, depending on which distribution you think conveys more accurately the original data. If we wish to convey that the frequency of weights rises to a maximum in the 160-pound range and then decreases again, we would probably prefer the 10-pound intervals (Table 3.4). On the other hand, if we wish to demonstrate that no individuals weigh between 165 and 170 pounds, we would prefer the 5-pound intervals (Table 3.5). It is a matter of judgment and communication.

We must note, however, that some information is lost in a grouped frequency distribution such as Table 3.4 or Table 3.5. Although Table 3.4 tells us that five weights are between 150 and 159, we have no way of knowing whether those five individuals all weigh 150 or all weigh 159 or are spread over that range. Tables 3.4 and 3.5 have improved the clarity and ease of comprehending the data in comparison with Table 3.3, but they have sacrificed some detail in so doing.

3.2 Distributions as Graphs

Histogram

histogram: a graphical presentation of a grouped frequency distribution with frequencies represented as vertical bars; it is appropriate for interval/ratio data

The grouped frequency distributions we have shown as tables can also be represented graphically by histograms or frequency polygons. A *histogram* is a plot of the class intervals of the variable (weight, in our example) on the horizontal axis (sometimes called the *X*-axis or abscissa) and the frequency of each interval on the vertical axis (*Y*-axis or ordinate) of a graph.[1] Figure 3.3 shows the histogram obtained from Table 3.4, which used 10-pound class intervals. Note that each interval's frequency is displayed as a rectangle or "bar," and there is no space (only a single vertical line) separating the bars.

---

[1] There is no universally accepted way of labeling the *X*-axis of a histogram. Some statisticians support displaying only the midpoint of each class interval, so the *X*-axis of Figure 3.3 would have values 134.5, 144.5, 154.5, . . . centered under each bar; others support displaying the class intervals themselves, so the *X*-axis would have labels "130–139," "140–149," "150–159," and so on. The labelling procedure presented here is probably the most frequently used, but I recommend using whichever method makes the histogram the clearest.

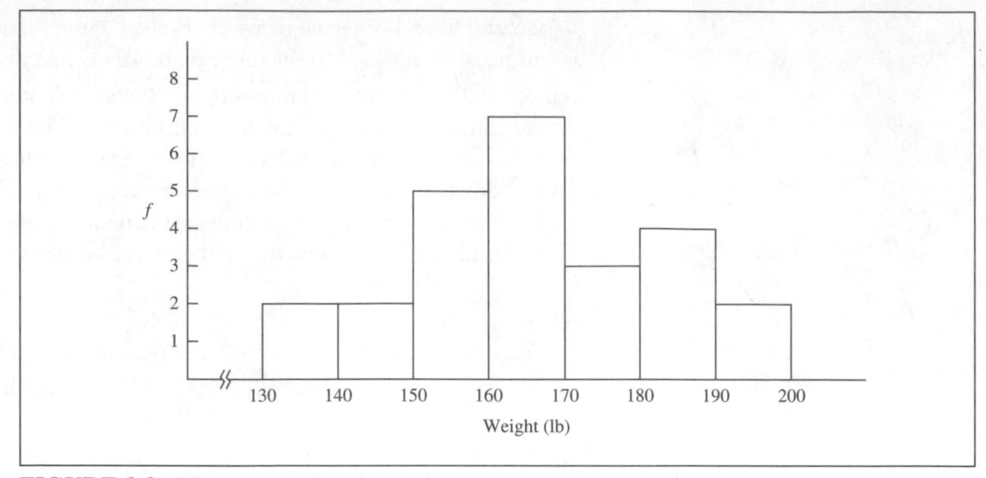

FIGURE 3.3 Histogram of male students' weights

Here are the steps used in creating a histogram:

steps for creating a histogram

Step 1. Begin with a grouped frequency distribution as described in the preceding section. For our weight example, that grouped frequency distribution, using 10-pound intervals (we could have used 5-pound intervals just as well), is shown in Table 3.4.

Step 2. Draw and label the axes of the histogram. It is conventional to make the vertical axis about two-thirds to three-quarters the length of the horizontal axis. The vertical axis should be labeled "f." The horizontal axis should be labeled with the name of the variable, "Weight" in our example, with the unit of measurement in parentheses, "(lb)." Enter values and tick marks on the horizontal axis that are equally spaced, convenient round numbers.

Step 3. It is conventional to have the vertical axis intersect the horizontal axis at the point where the value of the variable is 0. If this is not the case in your histogram, then indicate that with a break in the horizontal axis. The vertical axis in Figure 3.3 seems to intersect the horizontal axis at about 120 pounds, not 0, so a break is indicated in the axis.

Step 4. Draw vertical lines at the edges of the class intervals to form the edges of the histogram bars. Note that the edges of each bar are single lines, indicating that there is no space between the values that the bars represent. (See Step 6.)

Step 5. The height of each vertical bar should equal the frequency of the values in that interval.

Step 6. Provide an explanatory note, if necessary, about how values at the boundaries of the class intervals are handled in your histogram. For example, in the histogram of Figure 3.3, it is ambiguous whether the value 140 is included in the first vertical bar or in the second. To resolve this ambiguity, it may be desirable to include a statement such as: "The value 130 itself is included in the first bar, 140 is included in the second bar, and so on." If only a few values actually occur at the boundaries, then such a statement might easily be omitted. There is no universally accepted convention in this regard; clarity of presentation should be the guiding principle.

Now when Jack asks us about the weights of the male students in our class, we can answer by showing him this histogram. He can then immediately see in visual form the weights of our fellow students. This is a more concise summary for Jack because he can see at a glance that the lowest weight is about 130 pounds, the highest weight is about 200 pounds, and most of the weights cluster around 160 or 170 pounds.

The histogram based on class intervals presents a relatively complete, easily comprehensible picture of our data, and in fact histograms are among the most widely used ways of describing sets of data.

Frequency Polygon

frequency polygon: a graphical presentation of a grouped frequency distribution with frequencies represented as points; it is appropriate for interval/ratio data

Another equivalent graphical presentation of data is the *frequency polygon*. Recall that a histogram is a plot of a series of rectangular bars, where each bar's width is the width of the class interval and the bar's height is the frequency of that class interval (see Figure 3.3). A frequency polygon is simply a transformation of the histogram obtained by substituting a single point for each bar and then connecting the points with straight line segments. Figure 3.4 is the histogram of Figure 3.3 with the equivalent frequency polygon superimposed. Figure 3.5 then shows the histogram removed, leaving the frequency polygon by itself. The frequency polygon is thus a second form of graphical frequency distribution; it contains exactly the same information as does the histogram.

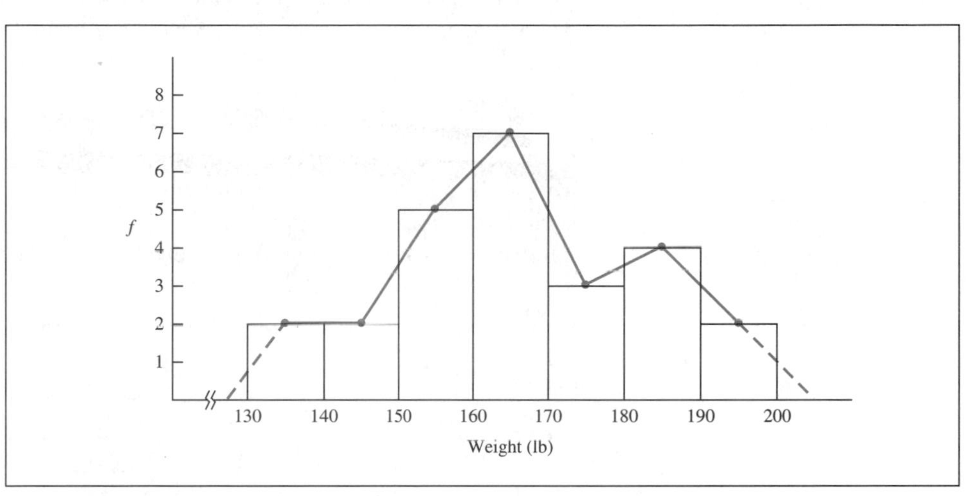

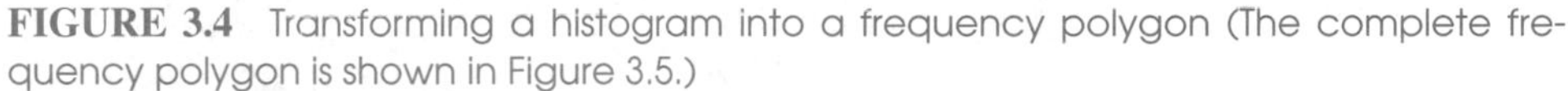

FIGURE 3.4 Transforming a histogram into a frequency polygon (The complete frequency polygon is shown in Figure 3.5.)

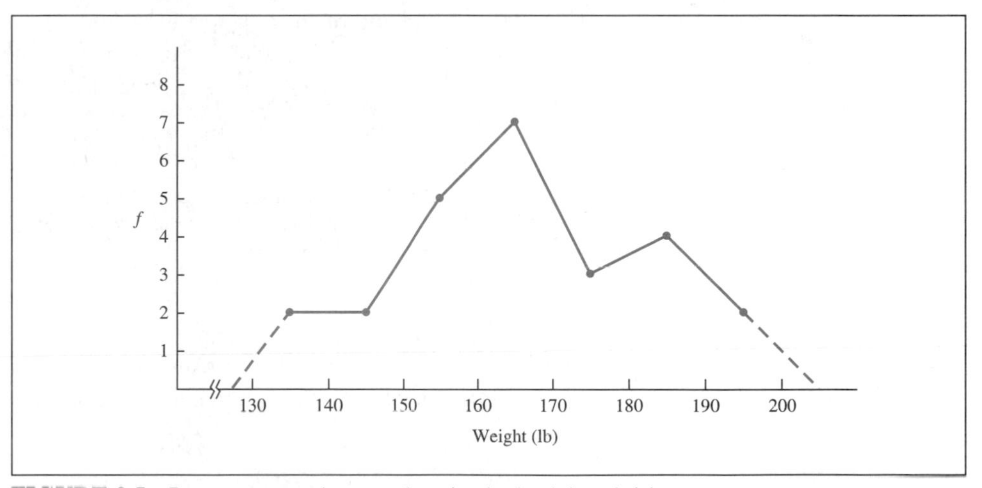

FIGURE 3.5 Frequency polygon of male students' weights

Note that the weight values on the *X*-axis of a frequency polygon are single points plotted at the *midpoints* of the class intervals rather than at the lower and upper limits of the class intervals as was the case in the histogram.[2] For example, the interval 160–169

[2]Many statisticians prefer to label the horizontal axis of a frequency polygon with the midpoints of the class intervals so that the dots of the frequency polygon have value labels directly beneath them. In that case, the horizontal axis of Figure 3.5 would have labels 134.5, 144.5, 154.5, and so on, with corresponding tick marks. Note that if the axis is labeled in this way, then there is some advantage in selecting class intervals that include an *odd* number of points because then the midpoint will be a whole number (unlike Figure 3.5, where each interval includes 10 points, so the midpoint ends in ".5").

pounds has the midpoint $(160 + 169)/2 = 164.5$. The dot of the frequency polygon is plotted directly above 164.5.

We can make three important observations about histograms and frequency polygons. First, they both contain the identical information found in a grouped frequency distribution table. Therefore, the choice of using a histogram, frequency polygon, or grouped frequency distribution table is based entirely on clarity of presentation. Use whichever presents your data most effectively.

relative frequency: frequency divided by the size of the group, expressed as a proportion or percentage

Second, we have plotted the frequencies on the vertical axis of the histograms and frequency polygons. In some cases, particularly when the number of observations is large, it is desirable to plot a measure of *relative frequency*, such as a proportion or percentage, on that axis. For example, because there are 25 men in our weight data, each man represents 1/25 or 4% of the total sample. Thus, if we wished, the label on the vertical axis in Figure 3.3, 3.4, or 3.5 could be altered by replacing "f" with "Percentage" and replacing the values "1, 2, 3, . . ." with "4%, 8%, 16%, ..." The new graph would convey the same information as the original.

Third, as sample sizes become larger, the width of each bar in a histogram (or the distance between each point in a frequency polygon) can be made narrower and narrower. Then the shape of the figure becomes, in general, smoother and smoother. Figure 3.6 shows such a histogram. If we imagine an infinitely large sample with infinitely narrow class intervals, then the histogram or frequency polygon becomes a smooth curve.

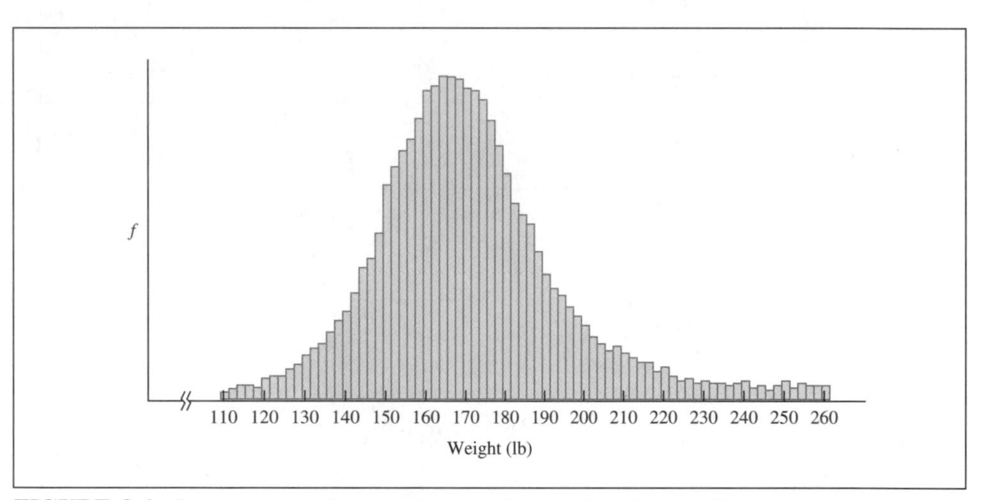

FIGURE 3.6 Frequency polygon of male students' weights with a very large sample size and very narrow class intervals

One other important manner of presenting frequency distributions of interval/ratio data, called a *stem and leaf display*, in some ways combines the best properties of tables and graphs. It is described in the optional section at the end of this chapter.

3.3 Eyeball-estimation

We have followed onc sct of data, the weights of a group of students, from the raw data stage (Table 3.2) to the histogram (Figure 3.3) and frequency polygon (Figure 3.5) graphical presentations of those data. We have seen that histograms and frequency polygons are simpler, more informative ways of communicating about a distribution of data than is the enumeration of the raw data themselves. It is our task for the rest of this chapter to become skilled in visualizing distributions even when we do not have the raw data to begin with. We want to be able to sketch a reasonably accurate frequency distribution "by eyeball," based on what we know about the world in general. (In this textbook, we will highlight descriptions of the process of "eyeball-estimation" by using the eyeglass logo shown at the beginning of this section.)

Now we are starting over with the weights of male students. Forget for the moment that we ever saw the data in Figure 3.2. Our new task is to attempt to sketch the distribution of weights of students in a statistics class *based on what we know about students in general*, not based on any data about any particular class that we might have seen. The steps in this sketching by eyeball procedure are listed here.

steps for eyeball-estimating a frequency distribution

Step 1. Draw and label the axes. A frequency distribution always has the variable that we are interested in plotted on the X-axis; in our case, the variable is "weight (lb)," which should be the label of the X-axis. Do not forget the unit of measurement ("lb") in the variable label because we will see in later chapters that keeping the units clearly in mind is essential to understanding statistics. The vertical axis of a frequency distribution is always the frequency and is usually labeled f. The vertical axis is customarily drawn to be about two-thirds or three-quarters the length of the horizontal axis. At the end of Step 1, our distribution should appear as shown in Figure 3.7. Note that our eyeballed distribution is an abstraction in the sense that it does *not* refer to any particular sample of values or to any particular sample size, so the f-axis does not have numerical values.

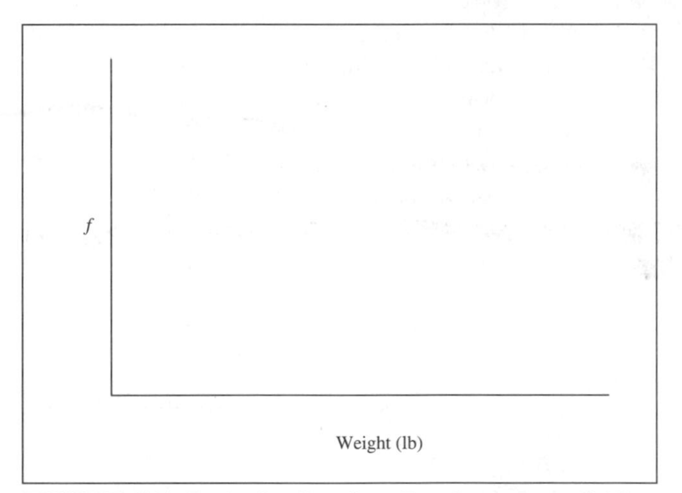

FIGURE 3.7 Eyeball-estimating the frequency distribution: Step 1, drawing and labeling the axes

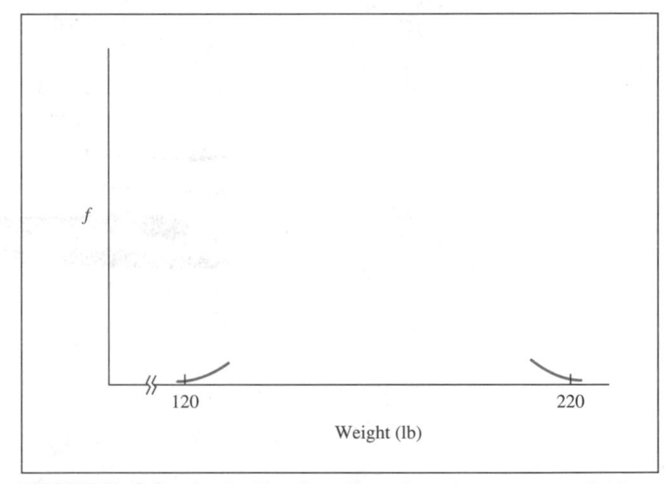

FIGURE 3.8 Eyeball-estimating the frequency distribution: Step 2, plotting minimum and maximum values

Step 2. Plot your estimates of the minimum and maximum values of the variable. What is the likely minimum weight of men in a college class? My estimate is about 120 pounds; your estimate might be a bit higher or lower depending on your experience. Our task here is to be approximate, not exact. If the lowest weight is thought to be 120 pounds, label that point near the left end of the X-axis. If we think that the lowest weight is 120 pounds, then the frequency of any value (weight) below 120 must be 0, and the plot of frequencies should begin to rise at about the value 120 pounds. Repeat the process for the largest weight likely to be found in a college class, putting that value near the right end of the X-axis. My estimate is 220 pounds. Figure 3.8 shows the plot with minimum and maximum values.

Step 3. Add intermediate values to the X-axis, using round numbers. Keep the intervals between the values equal, and make the distances between those values on the X-axis equal. In our example, we need round numbers between 120 and 220 with equal intervals between them; thus, 140, 160, 180, and 200 are appropriate and are equally spaced on the X-axis as shown in Figure 3.9.

Step 4. Plot estimate(s) of the mode(s) of the frequency distribution. The *mode* is the value of the variable (weight in our example) that occurs most frequently in a distribution; there can sometimes be more than one mode. What weight or small range of weights do you think is likely to occur most frequently? It seems to me that the most frequently occurring weight among college men is about 170 pounds (your estimate—and the

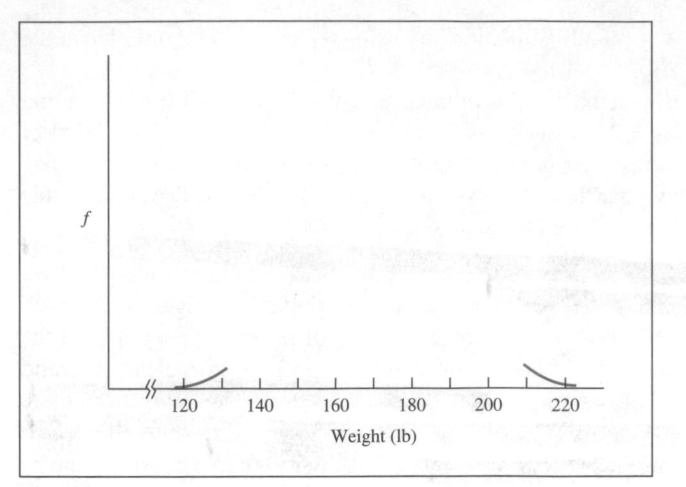

FIGURE 3.9 Eyeball-estimating the frequency distribution: Step 3, labeling intermediate values on the *X*-axis

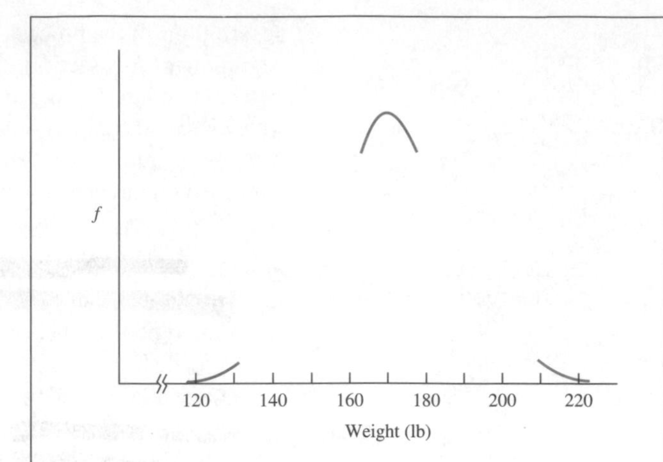

FIGURE 3.10 Eyeball-estimating the frequency distribution: Step 4, adding the mode

truth!—might be somewhat higher or lower). With this assumption, the frequency distribution must reach its maximum at 170 and decrease in both directions from there. Therefore, we locate a point above 170 pounds on our frequency distribution and sketch a little peak at that point to indicate that the frequency will decrease as weight either increases or decreases from 170 pounds (see Figure 3.10). The absolute height of this peak is *not* important in this sketch. If there are two or more modes, then the *relative* height *is* important: The more frequently occurring mode should be higher.

Step 5. Connect the parts of the sketch that you have just created, keeping the distribution smooth and continuous (unless there is good reason to have a discontinuity in the distribution). Nature as a rule does not change abruptly, and distributions of variables that occur in nature are therefore usually smooth. Figure 3.11 shows the result of our sketch.

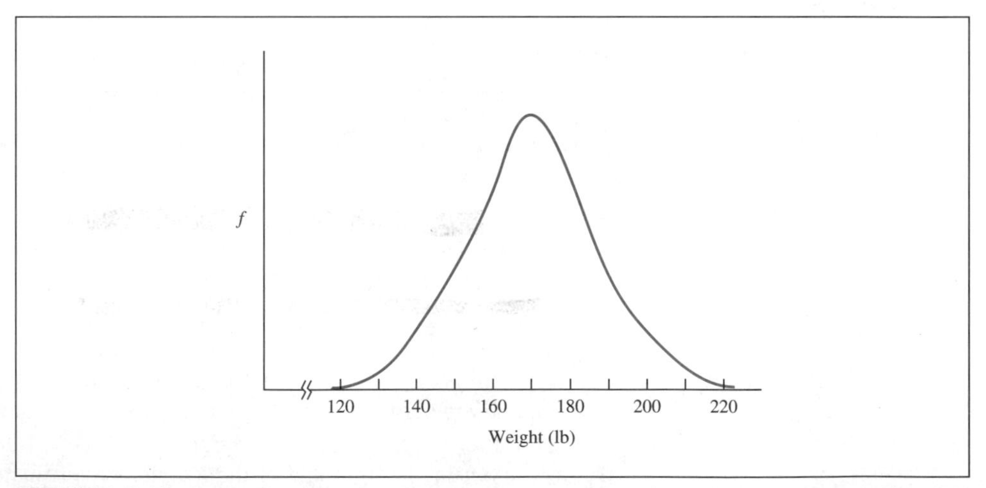

FIGURE 3.11 Eyeball-estimating the frequency distribution: Step 5, completing the sketch with a smooth curve

Our sketch of the distribution of men's weights in college classes is now complete. How does our sketch (which we created without considering any particular data) compare with the histogram and frequency polygon built from actual data and shown in Figures 3.3 and 3.5? First, our sketched curve is smoother than either the histogram or the frequency polygon. The rough edges or sudden changes in direction in the plots of Figures 3.3 and 3.5 are artifacts of the small sample size and wide class intervals. As we saw in Figure 3.6, the larger the sample size and the narrower the class intervals, the smoother the histogram and frequency polygon. Our eyeball-estimated frequency distribution is smooth, as if we had extremely narrow class intervals.

Second, the minimum and maximum values are not exactly the same (but close) in our estimate as those in the data. The difference could be because the data of Table 3.2 represented just one class, whereas our sketch represented classes in general. If Table 3.2 had been of a different single class, then the minimum and maximum values of the data would likely have been different. Alternatively, it is possible that we were somewhat mistaken in our estimates of the minimum and maximum values of this variable.

Third, both the sketch and the data have the mode at about 170 pounds. Our estimate of the peak for men's weights was corroborated by the data. We see that, in general, our sketch has roughly the same characteristics as the data.

3.4 The Shape of Distributions

Describing Distributions

As we sketch frequency distributions, it will be convenient to have some terms that describe the shapes of the distributions we are discussing.

The first two terms that we might use to describe the shape of a distribution refer to the number of modes the distribution has. You may recall that the mode is the most frequently occurring value in a distribution and lies directly below the highest point on a graphical frequency distribution. A distribution that has one mode is called *unimodal*. Our sketch of the distribution of men's weights in Figure 3.11 is unimodal.

unimodal distribution: a distribution that has one most frequently occurring value

bimodal distribution: a distribution that has two most frequently occurring values

A *bimodal* distribution has two modes. For example, we might expect the distribution of weights of college students in general to have two modes, one most frequently occurring weight for the men and one most frequently occurring weight for the women, as shown in Figure 3.12. Most distributions in nature are unimodal; bimodal distributions occur only when there is a distinct splitting of a population into two relatively discrete parts. Being bimodal does *not* imply *no* overlap between the distributions (for example, in our weight example some men are lighter than the female mode and some women are heavier than the male mode).

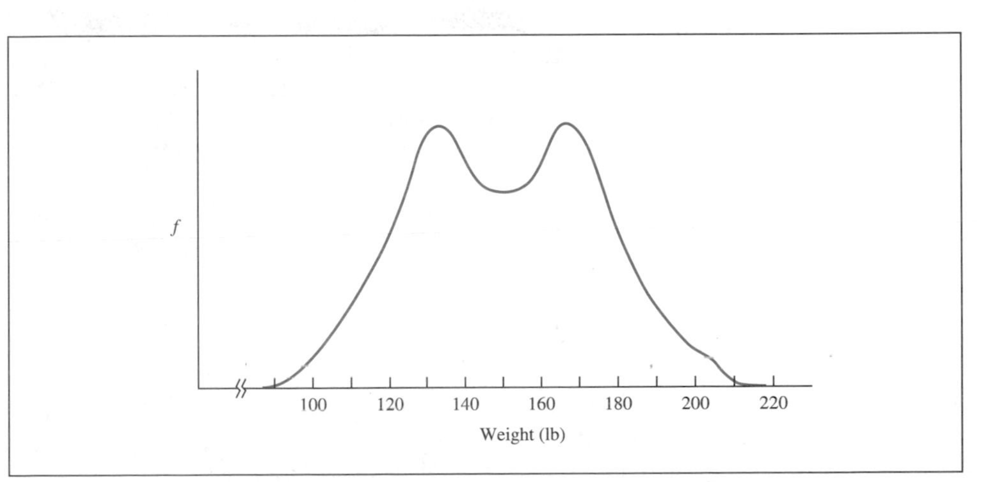

FIGURE 3.12 A bimodal distribution: weights of college students (both men and women)

symmetric distribution: a distribution whose left side is a mirror image of its right side

We call a distribution *symmetric* if the right side is a mirror image of the left side. The unimodal distribution of men's weights in Figure 3.11 is symmetric; the bimodal distribution of men's and women's weights in Figure 3.12 is also approximately symmetric.

We call a distribution *skewed* if one "tail" (the side of a distribution) is longer than the other; thus, skewed distributions must be asymmetric. We refer to distributions whose

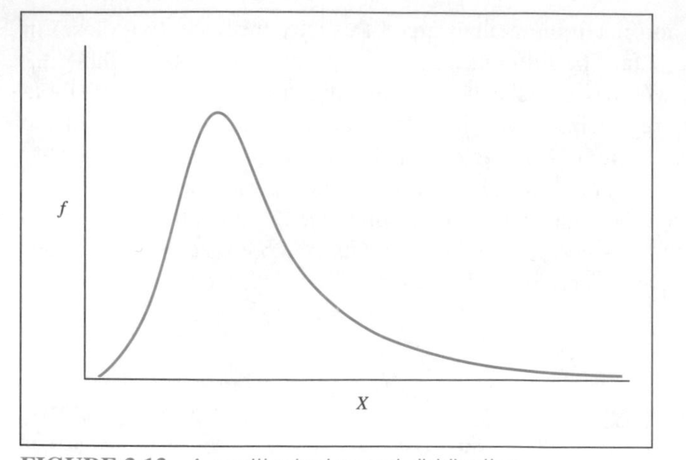

FIGURE 3.13 A positively skewed distribution

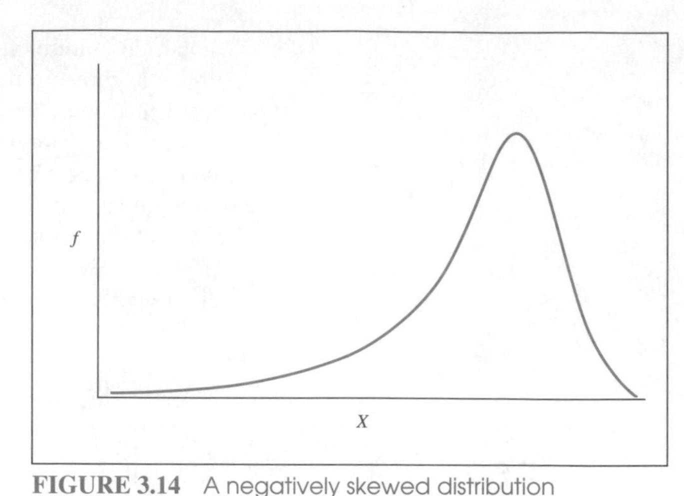

FIGURE 3.14 A negatively skewed distribution

positively skewed distribution: a distribution whose right tail is longer than its left tail
negatively skewed distribution: a distribution whose left tail is longer than its right tail
asymptotic: gradually approaching the *X*-axis

right-hand tail is longer than the left as *positively skewed* (see Figure 3.13), and to those whose left-hand tail is longer as *negatively skewed* (see Figure 3.14).

We call the tail of a distribution *asymptotic* if it gradually approaches the *X*-axis but never actually touches it. (To be truly asymptotic is therefore a mathematical abstraction; all actually occurring distributions have a lowest value below which the frequency is 0 and thus where the distribution actually meets the *X*-axis.) The distribution of Figure 3.14 is asymptotic in the left tail but *not* asymptotic in the right tail.

The Normal Distribution

normal distribution: a class of distributions that are unimodal, symmetric, and asymptotic

For reasons that will become clearer in Chapter 7, the distributions of many variables that occur in nature have a characteristic form called the *normal distribution*. The heights of people (or trees or dandelions), IQs of people, maze-running speeds of rats, and gains or losses on the stock exchange are all normally distributed variables. This section will acquaint you with the general shape of normal distributions, and later chapters will fill in the computational details.

Normal distributions are *unimodal*, *symmetric*, and *asymptotic*, which means that they have the general bell shape shown in Figures 3.15, 3.16, and 3.17. Thus, normal distributions can be tall or short, narrow or wide, but they must have the same general bell shape.

You may wonder why the normal distribution is called *asymptotic* when the curves drawn in Figures 3.15, 3.16, and 3.17 clearly touch the *X*-axis. Being asymptotic is a mathematical abstraction, impossible to represent accurately in those portions of the curve where the correct height is less than the width of the ink lines. Figures 3.15, 3.16, and 3.17 are

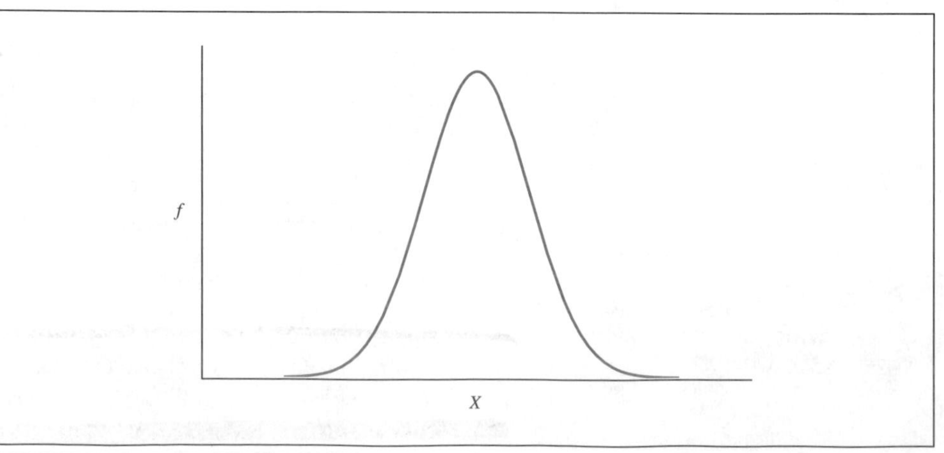

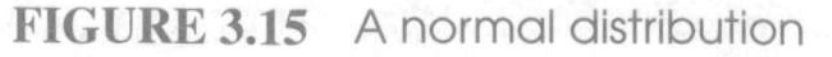

FIGURE 3.15 A normal distribution

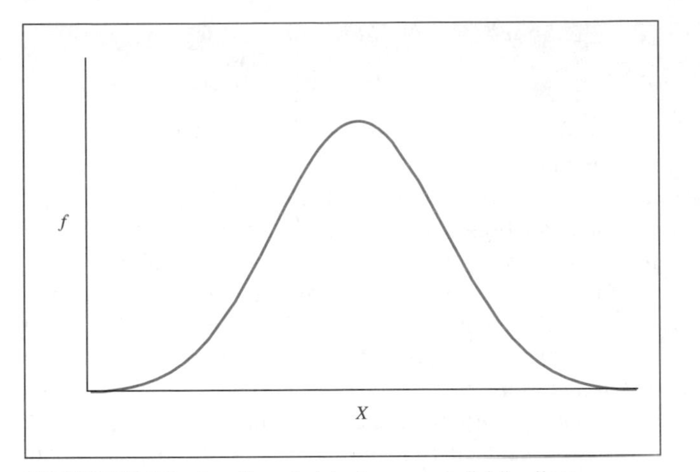

FIGURE 3.16 Another (wider) normal distribution

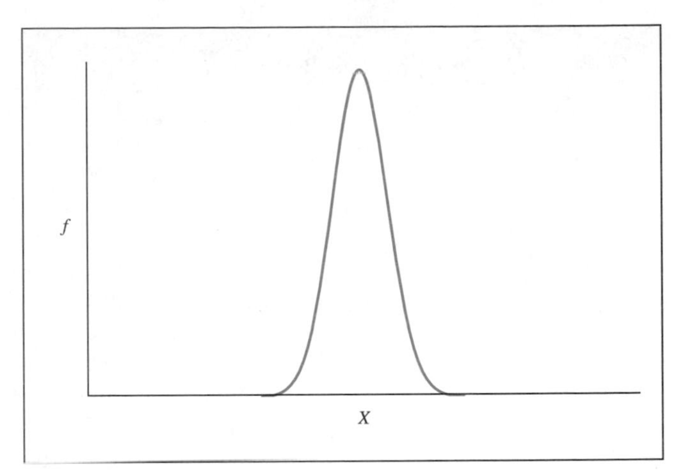

FIGURE 3.17 A third (narrower) normal distribution

in fact precisely drawn normal distributions. There is a widespread misconception, however, that normal distributions should seem to "float" above the *X*-axis. This misconception has been created by textbooks that incorrectly illustrate normal distributions by grossly exaggerating their asymptotic characteristic.

For the sake of contrast, we present several nonnormal distributions in Figures 3.18 through 3.21. Note that normal distributions must be unimodal, symmetric, and asymptotic, but not all distributions with those three characteristics are normal (for example, Figure 3.21).

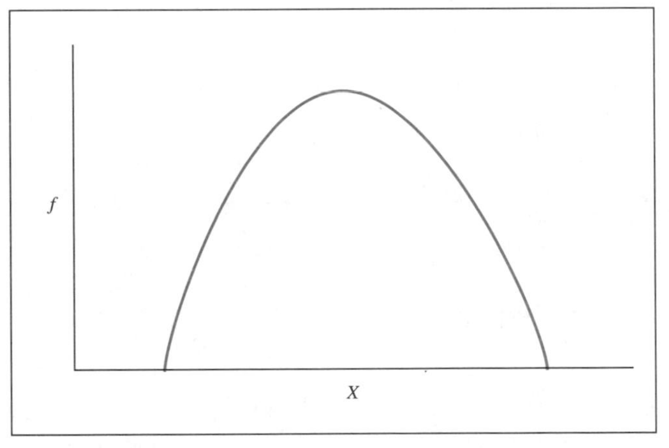

FIGURE 3.18 Nonnormal distribution (unimodal and symmetric but nonasymptotic)

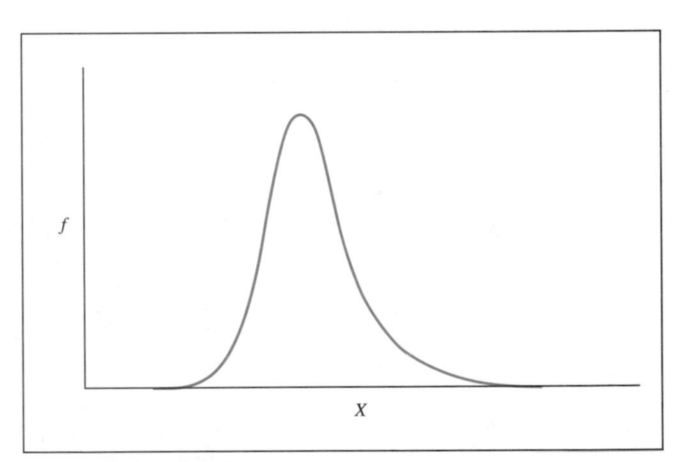

FIGURE 3.19 Nonnormal distribution (unimodal and asymptotic but asymmetric)

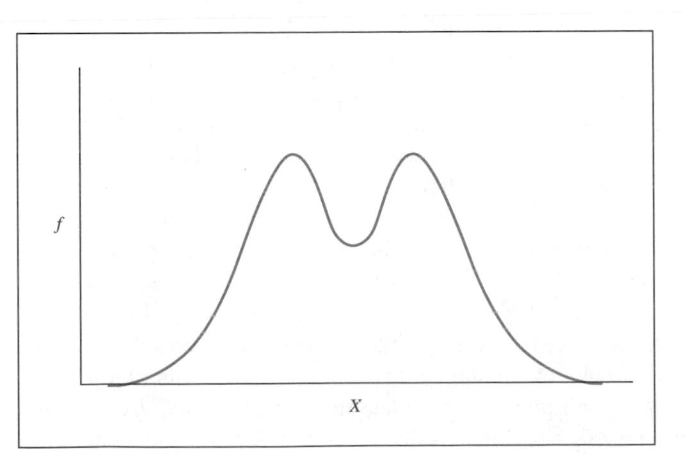

FIGURE 3.20 Nonnormal distribution (symmetric and asymptotic but bimodal)

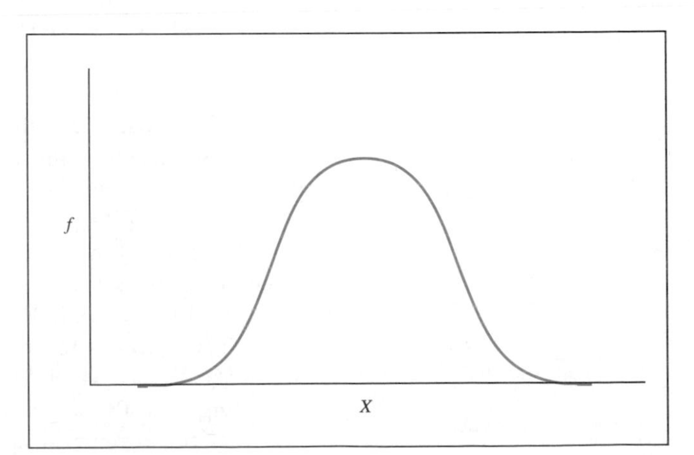

FIGURE 3.21 Nonnormal distribution (unimodal, symmetric, and asymptotic but too flat on top)

3.5 Eyeball-calibration

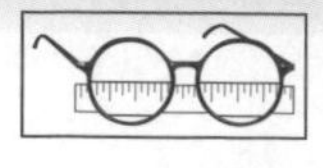

Let us have some practice in eyeball-estimating frequency distributions. (In this text, we call such practice "eyeball-calibration" and indicate it with the glasses-and-ruler logo shown at the beginning of the section.) Suppose we are interested in the distribution of the speeds of automobiles on a section of an interstate highway where the speed limit is 65 miles per hour (mph). In Step 1 of our procedure, we draw and label the axes. The horizontal axis is labeled "Speed (mph)" because that is the variable of interest, and the vertical axis is labeled "frequency," which we usually abbreviate as f. In Step 2, we ask ourselves what is the slowest likely speed on such an interstate, and we might guess about 50 mph. (Note that we are making some implicit assumptions here, such as the absence of hills or snowy road conditions. One of the tasks of science is to make such assumptions explicit, but that is a topic of research design, not of statistics itself.) The maximum speed is perhaps about 80 mph. We mark the minimum and maximum speeds on the X-axis. In Step 3, we add the intermediate values to the X-axis: 60 and 70 mph. In Step 4, we enter our estimate of the mode, which we might guess to be about 65 mph. Then in Step 5, we connect the points to form the complete sketch shown in Figure 3.22. Our sketched distribution is more or less normally shaped.

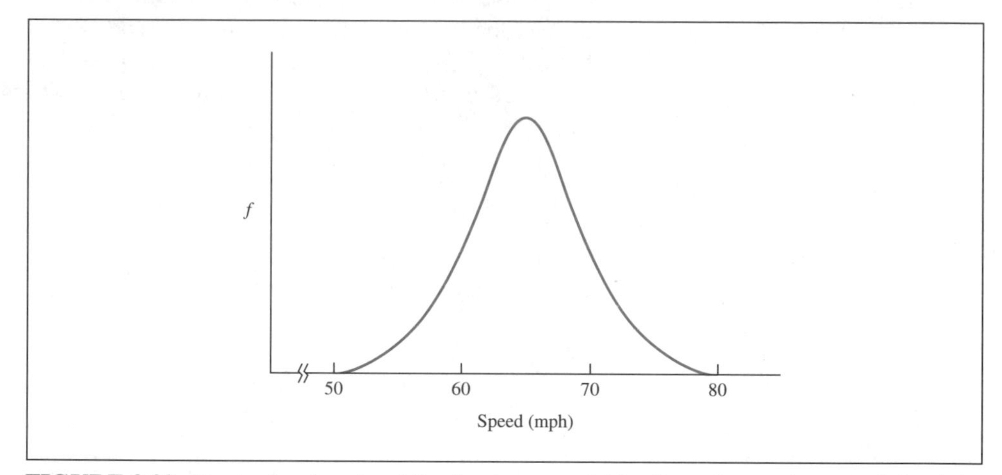

FIGURE 3.22 Eyeball-estimate of the frequency distribution of speeds on an interstate highway

As another example, let us eyeball-estimate the distribution of the lengths of blades of grass in a lawn that has just been mowed to a height of 3 inches. In Step 1, we draw and label the axes: The X-axis is "Length of blade (in.)" and the Y-axis is "frequency" or f. In Step 2, we plot the minimum length, which we guess to be about 2 inches, and the maximum length, which we take to be just slightly greater than 3 inches. The lawn has just been cut and any longer blades will have been snipped off, but we allow for some slightly longer blades that happened to be growing at an angle. In Step 3, we enter the intermediate values: 2.25, 2.50, and 2.75 inches. In Step 4, we eyeball-estimate the mode. The most frequently occurring length in this case seems to be 3 inches because any longer blades would have been cut to 3 inches. We enter the mode almost directly above the maximum value. In Step 5, we connect the points, and here we have reason to accept a rather sharp discontinuity in the shape of the distribution. There will be an abrupt decrease in the frequency of the lengths of blades at 3 inches because the mower cuts any blades that are longer than that. The result is shown in Figure 3.23. This distribution is extremely negatively skewed.

The exercises for this chapter will give you the chance to convince yourself that you can sketch by eyeball relatively accurate frequency distributions of all sorts of variables, ranging from the heights of basketball players to the lengths of McDonald's french fries. Being able to sketch an approximately accurate frequency distribution is evidence that you understand what a frequency distribution is and that you have some knowledge of the way the variable under consideration is distributed in the real world.

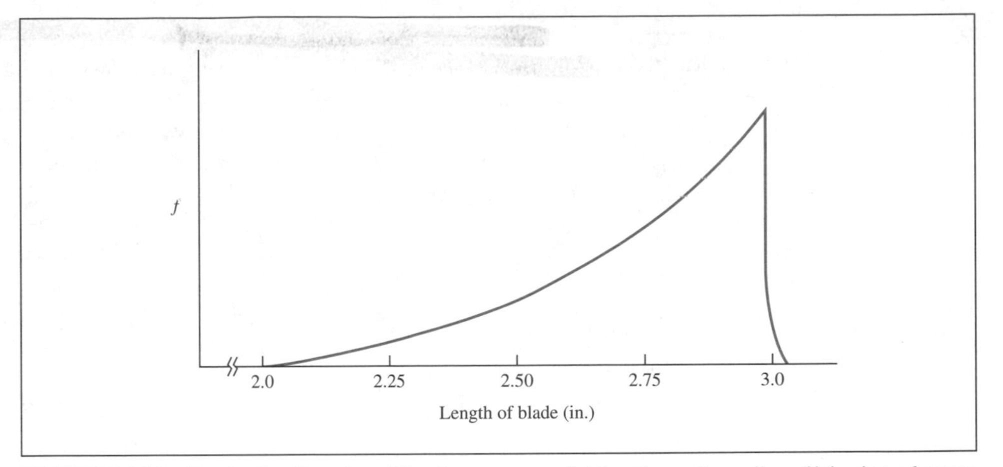

FIGURE 3.23 Eyeball-estimate of the frequency distribution of lengths of blades of grass in a recently mowed lawn

Statistics, to be sure, are frequently used to describe variables whose distributions we *do not* know; that's one way scientists learn about the real world. However, at this stage in learning to understand statistics, we are focusing on variables that are familiar. Your task is to recognize that there are variables with distributions everywhere you look, and that if you are at all familiar with the variable, you can describe its distribution with at least some accuracy. I want you, as you're walking across campus, to identify variables such as the height of pine trees, or the length of women's hair, or the number of books students carry, or the amount of time students take to walk from the library to the student union, or the number of miles on the odometers of cars in the parking lot, or the number of times the word *the* appears per page in your textbooks, or. . . . I want you to visualize the frequency distributions of those variables. I want you to say to yourself as you walk, "The shortest pine trees on this campus seem to be about 4 feet and the tallest about 40 feet, and the distribution seems to be symmetric with the mode at about 20 feet." "The women's shortest hair length is about 3 inches, the longest is about 30 inches, and the mode is about 8 inches, and I can see in my mind's eye the frequency distribution: It begins to rise at 3 inches, reaches a peak at 8 inches, and then tapers off, reaching 0 frequency at 30 inches." That is, I want you, for the moment, to practice seeing *distributions* of heights of trees, rather than the trees themselves, and *distributions* of the lengths of hair, rather than the hair itself.

We shall see throughout the remainder of this textbook that statistics is almost entirely the study of distributions and their characteristics. If we are to understand statistics, we must become confidently familiar with distributions, being able to "see" them accurately wherever we look.

3.6 Bar Graphs of Nominal and Ordinal Variables

Our discussions so far have focused on the graphical display of interval/ratio variables—that is, on histograms and frequency polygons. If a variable is nominal or ordinal, however, the use of a histogram or frequency polygon can be quite misleading.

TABLE 3.6
Frequency distribution of students' political preferences

Preference	*f*
Republican	375
Democrat	510
Independent	141
Decline to state	84

Suppose the campus newspaper polls students about their political preference, asking them whether they are Republican, Democrat, Independent, or decline to state. This is clearly a nominal variable, and the frequencies are shown in Table 3.6. If we (mistakenly) were to create a frequency polygon for these data, we would plot a point for Republicans, a point for Democrats, and so on, and then connect the points with continuous lines. The line between the Republican and the Democrat points would give the erroneous impression that measurements were made at intermediate points between those two parties. To avoid such misleading displays, we do *not* use histograms or frequency polygons for nominal or ordinal data, preferring instead the bar graph.

bar graph: a graphical presentation of a frequency distribution of nominal or ordinal data where frequencies are represented by separated bars

A *bar graph* is a graphical display of a frequency distribution where frequencies are represented as separated vertical bars, as shown in Figure 3.24. Note how the separation of the bars makes it clear that no intermediate values were measured. Note also that the order of the bars could be rearranged, with Democrats placed at the left, for example. This is because the political preference variable is measured at the nominal level, where order is not important.

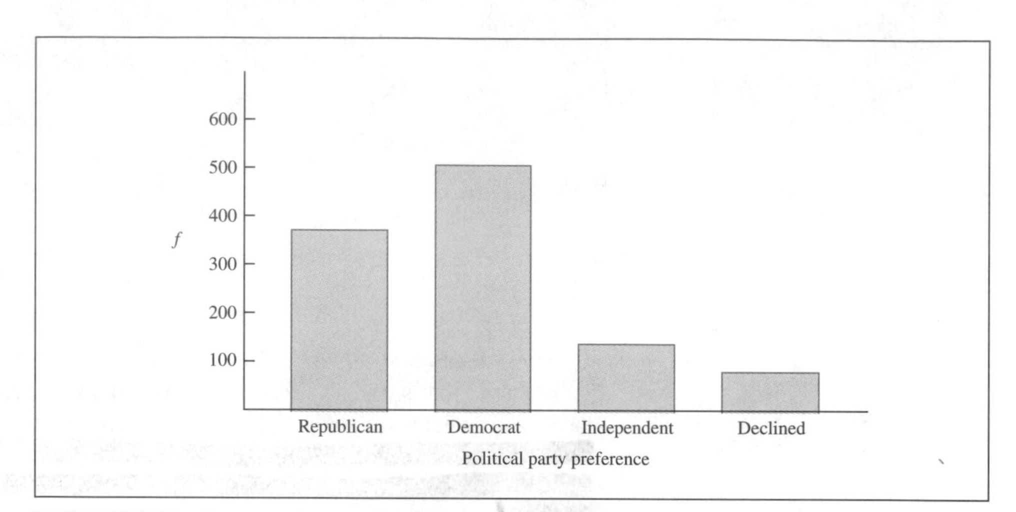

FIGURE 3.24 Bar graph of political party preferences

The frequency distributions of *ordinal* data are also customarily displayed as bar graphs, although histograms and frequency polygons are occasionally used. Using the bar graph for ordinal data has the advantage of making it clear that the intervals between points are not necessarily equal sized, but it has the disadvantage of implying that there are no intermediate values between the bars. The choice between the methods of display for ordinal data is a judgment call made for the individual frequency distribution, considering which kind of display gives the clearest communication about the data.

To review: Whereas histograms and frequency polygons are suitable for representing data measured only at the interval/ratio level, bar graphs are suitable for either nominal or ordinal data. And whereas the bars in histograms are directly adjacent, the bars in bar graphs are separated some distance from one another.

3.7 Connections

Cumulative Review

Chapters 1 and 2 distinguished between populations and samples and between parameters and statistics. Variables were defined as characteristics that could take any of several values, and we distinguished among nominal (categorization only), ordinal (ordered categorization) and interval/ratio (ordered categorization where the categories all have equal sizes) levels of measurement. Summation notation was introduced as one of the primary mathematical skills underlying statistics.

Computers

Use ESTAT's datagen to create a frequency distribution of the data in Table 3.2:

Note that ESTAT's datagen does not create frequency distributions or histograms, so this cannot be done directly. You can ask ESTAT (version 2.0 or later) to sort the data, so that creating a histogram is easier: Click any cell in the variable you wish to sort, click `Edit Variable`, click `Sort them in decreasing order`, and then click `OK`.

Use ESTAT's datagen to add 100 to all the values of a variable:

1. Enter any five values into the first column of the spreadsheet.
2. Click any cell in the first column, click `Edit Variable`, click `Add this constant to them`, enter 100 into the text entry cell, and then click `OK`.

Use SPSS to create a frequency distribution of the data in Table 3.2:

1. Enter the data from Table 3.2 by typing the values down a single column, using `Enter` between values. (Do *not* put the data in five different columns.)
2. Save these data so that you can use them later: Click `File`, then click `Save As....` Highlight the asterisk ("*") in the `File Name` cell, and then type "table3-2," so that the new file name is "table3-2.sav." Then click `OK`.
3. Recall this file to make sure it is saved correctly: Click `File`, then `Open`, and then `Data....` Click `table3-2.sav`. Click `OK`.
4. Click `Statistics`, then `Summarize`, then `Frequencies....`
5. Click ▶ and then click `OK`.
6. Note that the frequency distribution appears in the `!Output1` window. Check to make sure that the frequency column sums to 25 and that the table matches Table 3.3 in the textbook. Note that SPSS chooses to list the smallest values first, unlike the convention described in the text.

Use SPSS to create a histogram of these data:

7. Click `Graphs` and then `Histogram....`
8. Click ▶ and then `OK`.
9. Note that the histogram appears in the `Chart Carousel` window.

Use SPSS to add 100 to all the values of a variable:

1. Enter any five values into the first column of the spreadsheet.
2. Click `Transform` and then click `Compute....`
3. Enter "var00001" in the `Target Variable:` text entry cell.
4. Click ▶, click +, click `1`, click `0`, click `0`, and then click `OK`.
5. Click `OK` to `Change existing variable?`

Homework Tips

1. Check the list of learning objectives at the beginning of this chapter. Do you understand each one?

2. When you are preparing a grouped frequency distribution, make sure that all the class intervals have the same width. The most frequent mistake is to make the first or the last interval wider or narrower than the remaining intervals. If the intervals are equal, then the last digit in the interval limits (including the first and last interval) will form some uniform sequence, such as 0, 5, 0, 5,... or 0, 4, 8, 2, 6, 0, 4,....

3. Remember to plot the points in a frequency polygon at the midpoints of the class intervals.

Exercises for Chapter 3

Section A: Basic exercises (answers in Appendix D, page 506)

[*Hint:* Save your answers to these exercises. You will use them again for the exercises in Chapters 4 and 5. When sketching distributions, follow the steps of Figures 3.7–3.11. Do not forget to label your axes (including the unit of measurement).]

1. To measure the fitness of students, we select 20 students at random and ask them to ride a stationary bicycle as fast

as they can for 30 minutes. The bicycle has an odometer, and we record the number of miles each student rides. The data (in miles) are: 8, 15, 22, 17, 9, 16, 15, 15, 21, 12, 14, 16, 22, 17, 16, 11, 13, 15, 20, 15. Prepare these distributions:
(a) Frequency distribution
(b) Grouped frequency distribution with about eight groups
(c) Histogram
(d) Frequency polygon

2. (a) For the data of Exercise 1, prepare a histogram using about four groups.
(b) Which histogram is more informative, the one with eight groups or the one with four groups?

3. These data are the scores of all the students on the first exam in a statistics class: 81, 86, 91, 75, 96, 82, 88, 88, 71, 68, 72, 61, 84, 86, 95, 91, 84, 83, 91, 83, 89, 90. Prepare these distributions:
(a) Frequency distribution
(b) Grouped frequency distribution with about eight groups. The lower limit of the lowest group could be 60, 61, 62, or 63. Which of these choices makes the most sense? Why?
(c) Histogram
(d) Frequency polygon

4. Most professional basketball players are about 6 feet 8 inches (80 inches) tall. A few are as tall as 7 feet 2 inches, and a few are as short as 5 feet 8 inches. Sketch by eyeball the distribution of heights of professional basketball players. What terms can be used to describe this distribution?

5. Vibrato is the slight wavering in pitch that musicians add to musical notes to give them warmth. Most musicians' vibrato has a frequency of about 7 Hz (hertz, or cycles per second). A few are as fast as 9 Hz, and a few are as slow as 5 Hz. Sketch by eyeball the distribution of vibrato frequencies. What terms can be used to describe this distribution?

6. Coca-Cola bottling companies manufacture bottles of Coke that say "16 ounces" on the label. Bottlers know that consumer groups will complain if too many bottles contain less than 16 ounces, but they also don't want to waste money by putting too much Coke in the bottles. They know that their bottling machines are not perfectly accurate in dispensing exact amounts of liquid. They therefore set their machines to dispense Coke so that the modal amount of Coke is 16.5 ounces. Sketch by eyeball the distribution of amounts of Coke in "16 ounce" Coca-Cola bottles. What terms can be used to describe this distribution?

7. Sketch by eyeball the distribution of lengths of McDonald's french fries. What terms can be used to describe this distribution?

8. Royal Perfecto is a competitor of McDonald's, and they, like McDonald's, grow their own potatoes from which french fries are made. Royal Perfecto has discovered a way of growing potatoes that are perfectly cubical, 5 inches on each side. Furthermore, Royal Perfecto potatoes, when cut into strips and deep fried, retain their exact 5-inch lengths. Sketch the distribution of lengths of Royal Perfecto french fries. What terms can be used to describe this distribution?

9. Big Brothers is an organization that pairs young boys with adult men. The Los Angeles Big Brothers Club has a luncheon where boys aged 7 can come with their adult big brothers. Sketch the distribution of heights of all people (boys *and* their Big Brothers) who attend this luncheon. What terms can be used to describe this distribution?

10. (a) Eyeball-estimate the distribution of lengths of blades of grass on the lawn nearest the front door of the building where your statistics class meets. Use the procedure of Figures 3.7–3.11.
(b) Select 50 blades of grass at random and pull them from the lawn. Measure them and prepare a frequency distribution table.
(c) For the data of part (b), prepare a grouped frequency distribution table.
(d) For the data of part (b), prepare a histogram.
(e) For the data of part (b), prepare a frequency polygon.
(f) How do your graphs of parts (d) and (e) compare with your sketch in part (a)?

11. Suppose the college ground crew went on strike for two weeks and the lawn does not get mowed during that time. Sketch by eyeball the distribution of lengths of blades of grass at the end of the strike. How is this distribution similar to and different from the distribution you sketched in Exercise 10?

12. Suppose that the registrar at Hower University reports these enrollments in each of the colleges: Liberal Arts 3024, Science and Mathematics 1127, Performing Arts 752, Health Sciences 1452, Business and Economics 4320, and Education 2431.
(a) What kind of graphical presentation is appropriate for the frequency distribution at Hower University?
(b) Create that graphical distribution.

Section B: Supplementary exercises

13. For a study on human learning, a researcher created a list of 25 pairs of nonsense syllables (such as *dak, fom*). The first syllable of each pair is presented to the subject one at a time, and the subject is asked to say its paired associate. If the subject is incorrect, the correct syllable is then presented. One of the variables of interest in this study is the number of times such a list has to be presented until the subject can correctly give all 25 associates. The researcher randomly selects 50 students from an introductory psychology course and trains each student on the list until the 100% criterion is reached. The numbers of presentations of the list for the 50 students are: 8, 9, 7, 8, 16, 7, 10, 11, 16, 14, 12, 13, 12, 13, 12, 14, 8, 9, 17, 12, 5, 18, 14, 14, 12, 8, 11, 11, 9, 9, 18, 15, 11, 7, 9, 5, 6, 8, 10, 11, 11, 10, 14, 16, 6, 11, 15, 9, 19, 12. Prepare these distributions:

(a) Frequency distribution
(b) Grouped frequency distribution with six to eight groups
(c) Histogram
(d) Frequency polygon

14. A college is considering supplying personal computers for its students and wishes to know how fast the students can type. It takes a random sample of 55 students and administers a typing test. These are the results (in words per minute): 8, 24, 20, 20, 17, 18, 16, 19, 17, 29, 25, 27, 14, 5, 21, 36, 11, 16, 29, 20, 11, 17, 26, 10, 11, 5, 5, 19, 28, 7, 15, 8, 14, 32, 32, 12, 7, 12, 13, 30, 19, 16, 42, 26, 16, 30, 21, 8, 4, 23, 5, 15, 19, 9, 30. Prepare these distributions:
(a) Frequency distribution
(b) Grouped frequency distribution with six to eight groups
(c) Histogram
(d) Frequency polygon

15. A sociologist is interested in the economic impact of school events at Waytoo High School. She interviews couples who attend the senior prom and asks how much money each couple spent on this one evening, including tickets, clothing, and entertainment. The data (in dollars) are: 245, 190, 330, 225, 140, 120, 410, 395, 218, 264, 256, 302, 330, 310, 275, 272, 188, 380, 95, 160, 260, 265, 387, 342, 340. Prepare these distributions:
(a) Frequency distribution of money spent on the prom at Waytoo High
(b) Grouped frequency distribution with about eight groups
(c) Histogram
(d) Frequency polygon

16. The president of a sorority reports these grade point averages of sorority members: 3.2, 3.5, 3.2, 4.0, 2.7, 3.1, 3.1, 2.9, 3.7, 2.8, 3.6, 3.4, 3.8, 3.8, 2.8, 3.6, 3.9, 3.6, 3.3, 3.4, 3.3, 3.5, 3.4, 3.3, 3.7, 3.5, 3.9, 3.8, 3.1, 3.2. Prepare these distributions:
(a) Frequency distribution
(b) Grouped frequency distribution with about eight groups
(c) Histogram
(d) Frequency polygon

17. Sketch by eyeball the distribution of lengths of automobiles parked in your college parking lot. Describe (in words) this distribution.

18. Sketch by eyeball the distribution of lengths of songs played on a rock and roll radio station. Describe (in words) this distribution.

19. Sketch by eyeball the distribution of the prices of men's shirts in a local department store. Describe (in words) this distribution.

20. Sketch by eyeball the distribution of the prices of men's shirts in a local discount store. Describe (in words) this distribution.

Section C: Extending your comprehension

21. (a) Eyeball-estimate the distribution of times students spend getting a drink at the drinking fountain nearest your statistics classroom. Use the procedure of Figures 3.7–3.11.
(b) Watch 50 students as they approach the fountain and get a drink. Time each one and prepare a frequency distribution table.
(c) For the data of part (b), prepare a grouped frequency distribution table.
(d) For the data of part (b), prepare a histogram.
(e) For the data of part (b), prepare a frequency polygon.
(f) How do your graphs of parts (d) and (e) compare to your sketch in part (a)?

22. Choose any variable to measure. Make the variable be something that the other students in your class are familiar with and that will be convenient for you to measure. [*Hint:* My students have measured such variables as the number of unpopped kernels left in a movie theater's popcorn popper, number of cars passing through an intersection per green light, number of beers drunk by patrons at a local bar, amount of money spent by students in the cafeteria at lunch, amount of cash money carried by students in the student union, number of buttons on students' blouses or shirts, and number of pages of books in the library.]
(a) Make 50 observations of your variable and prepare a frequency distribution table.
(b) For the data of part (a), prepare a grouped frequency distribution table.
(c) For the data of part (a), prepare a histogram.
(d) For the data of part (a), prepare a frequency polygon.
(e) Without showing classmates your data, describe to them the variable you measured and ask them to sketch by eyeball the frequency distribution.
[*Hint:* I do this as a class exercise and use an overhead projector to display the graphs students prepare.]

Section D: Cumulative review (answers in Appendix D, page 507)

23. In the Australian gambling game of Two-Up, two coins are "spun" (thrown) simultaneously into the air. The person running the game calls out the outcome of the two coins using one of three terms: "heads" (both are heads), "tails" (both are tails), or "odds" (one head and one tail).
(a) What is the probability of "heads"?
(b) What is the probability of "tails"?
(c) What is the probability of "odds"?

24. Consider the data X_i : 14, 11, 12, 17, 13, 15. Find:
(a) X_4
(b) $\sum(4X_i^2 + 2X_i - 3)$

25. Sketch by eyeball the distribution of lengths of last names (number of letters in last name, including hyphenated surnames) for students in your college.

26. Identify the level of measurement (nominal, ordinal, or interval/ratio) for each of these variables:

(a) Number of pages in a textbook
(b) House address number as a measure of distance from the center of town
(c) Religious affiliation (1 = Protestant, 2 = Catholic, 3 = Jewish, etc.)

Section E: From the journals (answer in Appendix D, page 508)

27. Hearne (1989) contacted 50 individuals who considered themselves to be "psychics" or "mediums" and asked them to complete a questionnaire. Among other things, the questionnaire asked them to report the age at which they first recognized that they had psychic abilities. Hearne presented his results in a frequency distribution that is redrawn here. How would you characterize (in words) this distribution?

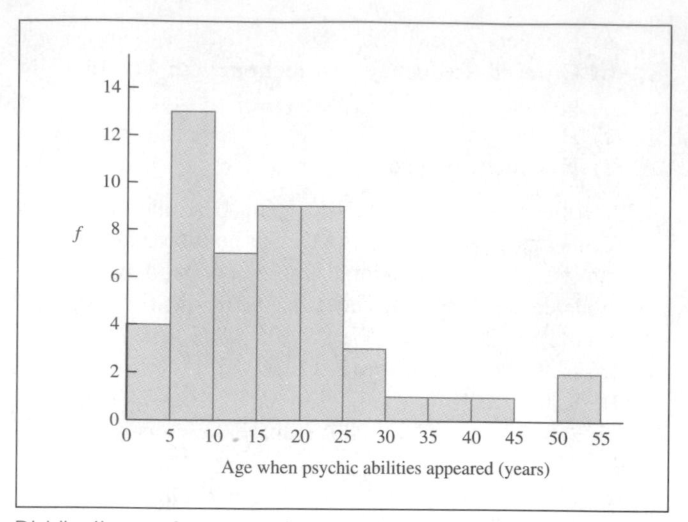

Distribution of age when psychic abilities appeared (Adapted from Hearne (1989))

Stem and Leaf Displays (Optional)

This section may be omitted without loss of continuity.

In this chapter, we described ways of presenting frequency distributions as tables and as graphs. John W. Tukey in 1977 proposed an alternative way of displaying a frequency distribution, called a *stem and leaf display*, which has three distinct advantages over either a table or a graph. First, tables and graphs are generally of grouped data, and some information is necessarily lost when data are grouped into class intervals. The stem and leaf display groups data but does not lose information. Second, the stem and leaf display has the visual appeal of a graph without the expense involved in preparing a figure. The stem and leaf display can be typeset just as easily as text. Third, the stem and leaf display makes the direct comparison of two data sets easier and more informative than either two tables or two graphical frequency distributions.

To create a stem and leaf display, divide each number in the data set into two parts: the *stem* and the *leaf.* The leaf is the last digit (or sometimes digits) in the data, and the stem is the first digit (or digits). Consider our men's weight data in Table 3.2. For the weight 135 pounds, we call the stem 13 and the leaf 5; for 180 pounds, the stem is 18 and the leaf is 0.

Next, put all the stems in an ordered column, as shown in Table 3.7. Following the convention we established in this chapter, we put the largest stem first. Include all stems even if there happen to be no leaves for a particular stem.

Then put all the leaves next to their stems, with one leaf for every data point. Thus, for 135 pounds, the leaf 5 is entered to the right of the stem 13. The complete stem and leaf display for the weight data is shown in Table 3.8.

TABLE 3.7 Stems for weight data

Stem
19
18
17
16
15
14
13

TABLE 3.8 Stem and leaf display for weight data

Stem	*Leaf*
19	07
18	0025
17	339
16	1233444
15	02479
14	49
13	57

Note that there are 25 leaves in the stem and leaf display, one for each of the 25 points in the original data. Note also that the stem and leaf display is quite similar to the histogram of Figure 3.3. If we turned Table 3.8 on its side and put bars around the leaves, we would almost reproduce Figure 3.3. Note also that we have not lost any information from the original data. For example, we can see from this display that there are no data points in the upper half of the 160–169 interval (because there are no leaves above 4 for the 16 stem). We could not possibly derive that fact from the histogram of Figure 3.3.

The stem and leaf display facilitates comparing two data sets when we list the leaves for one data set on the right and the other data set on the left of the same stems. For example, suppose we wished to compare the weights of students in our statistics class to those of another class. Such a display is shown in Table 3.9. Note that this side-by-side display (something relatively difficult to do with either a table or a graph) makes it easy to see that the weights of the students in class 2 (on the left in Table 3.9) are somewhat lower than the weights of the class 1 students.

TABLE 3.9 Comparing two groups: Stem and leaf displays for two classes' weights

Class 2 Leaf	*Stem*	*Class 1 Leaf*
1	19	07
94	18	0025
633	17	339
9510	16	1233444
9944422	15	02479
98731	14	49
532	13	57

It is possible to gain additional resolution in a stem and leaf display by dividing each stem into two, one for leaves 0 through 4 and the other for leaves 5 through 9, as shown in Table 3.10. This is entirely analogous to narrowing the class intervals from 10 pounds to 5 pounds in a table or graph. As with tables and graphs, the choice of the size of the stem depends on the judgment of the person who prepares the display. Here, the stem and leaf display in Table 3.8 seems more satisfactory than that in Table 3.10 because in 3.10 we have relatively few points spread over many stems.

TABLE 3.10 Stem and leaf display for weight data with 5-pound stems

Stem	*Leaf*
19	7
19	0
18	5
18	002
17	9
17	33
16	
16	1233444
15	79
15	024
14	9
14	4
13	57

You may ask: If the stem and leaf display has so many advantages over tables or graphs, why is it relegated to an optional section of this chapter? There are two reasons:

first, it is of relatively recent origin and as yet is not widely used in the behavioral science literature, and second, your understanding of other material in this book is not enhanced by understanding stem and leaf displays. Because of the philosophy of providing as little distraction from the main themes as possible, I moved this topic out of the mainstream.

Exercises for Optional Section

(Answers in Appendix D, page 508)

1. (a) How many weights from class 2 are shown in Table 3.9?
   (b) List those weights.
2. Create a stem and leaf display to show the following IQ scores in Mrs. Jones's classroom: 104, 100, 101, 131, 89, 95, 103, 85, 123, 117, 103, 106, 102, 121, 94, 93, 98, 100, 79, 100.
3. The IQ scores from Mr. Smith's classroom are: 106, 103, 98, 105, 86, 121, 118, 89, 115, 104, 89, 130, 117, 112, 93, 100, 111, 113, 126, 99.
   (a) Before you prepare any displays, inspect these data and try to determine whether Mr. Smith's class is brighter than, equal to, or not as bright as Mrs. Jones's.
   (b) Prepare a side-by-side stem and leaf display for the two classes.
   (c) Now which class do you think is brighter?
4. These data are the times (in seconds) for a group of runners in the 100-yard dash: 11.4, 11.3, 10.6, 10.3, 11.9, 10.9, 10.7, 11.4, 11.2, 10.6, 11.6, 11.2, 10.7, 10.5, 10.8. Prepare two stem and leaf displays, one with 1-second stems and one with .5-second stems. Which do you think is more informative? Why?

CHAPTER

4 Measures of Central Tendency

LEARNING OBJECTIVES

1. What is the purpose of obtaining a measure of central tendency?
2. What is the mode? With what types of data can it be used?
3. How is the mode eyeball-estimated?
4. What is the median? With what types of data can it be used?
5. How is the median eyeball-estimated?
6. What is the mean? With what types of data can it be used?
7. What are two techniques for eyeball-estimating the mean?
8. What criteria should be used in determining whether to report the mode, median, or mean?

Statisticians use the term *central tendency* to describe the center of a distribution. This chapter describes three measures of central tendency (the mode, the median, and the mean), shows how each can be eyeball-estimated and computed, and describes which of the measures is appropriate for the nominal, ordinal, or interval/ratio levels of measurement.

The central tendency of the intellectual growth (IQ gain) of the bloomers in Rosenthal and Jacobson's (1968) Pygmalion in the classroom data can be represented by the median of 13.5 and the mean of 16.5 IQ points, whereas the median and mean for the other children are both 7 IQ points. Thus, it is true to say that *on average*, the bloomer sample had greater intellectual growth than did the sample of other children.

We saw in Chapter 3 that grouped frequency distribution tables, histograms, and frequency polygons are easily comprehended characterizations of data. When Jack inquires about the weights of male students in our class, we can simply show him Table 3.4, Figure 3.3, or Figure 3.5 and the communication is relatively complete.

This communication is simple and direct, but the preparation required to reach it is rather cumbersome. We must carry the table or figure around with us at all times for the possibility that someone will inquire about the weights of our fellow students. Clearly we need a more "portable" method of communication. Because our goal is to simplify the task of communicating about data and because verbal communication is the simplest and most easily portable form of communication that we have at our disposal, we must develop a vocabulary that can convey most of the information that is available in Figure 3.3 or 3.5.

We shall see that *most of the information about most distributions is contained in three essential characteristics*: where the middle of the distribution is, how wide the distribution is, and what the shape of the distribution is. That these three characteristics are enough to describe distributions should come as no surprise to us when we reconsider our rules for sketching distributions in Chapter 3 (see pp. 33–34). Step 2 of those rules was to show on the *X*-axis how wide the distribution was: We plotted the minimum and maximum values. Then in Step 4 we gave some information about where the middle of the distribution was by plotting the mode (which, as we shall see, is one measure of the middle of a distribution). The shape of the distribution was to some degree implied by our choice of where to put the mode. If the mode was halfway between the minimum and maximum, the implication was that the distribution had a symmetric shape; if not, the shape was skewed. After we had identified the width, the center, and the shape, all that remained was to connect the dots and our sketch was complete.

Describing a distribution using only three concepts is extremely convenient because it is easily portable and compact. When Jack inquires about the weights of students, we can give a fairly complete answer in three short phrases. However, these three descriptors, though convenient, are only fairly complete; one does lose some information by using only three terms to describe a whole distribution.

Now that we recognize that referring to the "middle," the "width," and the "shape" of a distribution will simplify our communication about data, we must make these terms more explicit. We discussed the terminology for the shape of a distribution in Chapter 3 (symmetric, normal, positively skewed, negatively skewed, unimodal, bimodal). We will describe measures of the middle of a distribution in this chapter and measures of the width of a distribution in Chapter 5.

central tendency: the middle of a distribution

It is customary for statisticians to refer to attempts to specify the middle of a distribution as *measures of central tendency*. There are three important values we can use to measure central tendency: the mode, the median, and the mean. Which is most appropriate in a given situation depends on the level of measurement and the intent of the communication.

4.1 Mode

mode: a measure of central tendency appropriate for any variable; it is the most frequently occurring value in a distribution

We know from Chapter 3 that the *mode* is the most frequently occurring value in a distribution, and it can be used for nominal, ordinal, and interval/ratio variables. For example, we provided the frequency distribution (Table 3.6) and the bar graph (Figure 3.24) for the nominal variable of political preference, where the numbers of students in each political category were Republicans 375, Democrats 510, Independents 141, and those who declined to state 84. The mode is "Democrat" in this distribution because Democrat occurs most frequently in the sample. Note that the mode is the value of the variable ("Democrat"), *not* the number of times it occurs (510).

The mode can also be used to measure the central tendency of ordinal or interval/ratio variables. Table 3.3 (reproduced here as Table 4.1) presented the frequency distribution of male students' weights (an interval/ratio variable). We see that the value 164 pounds occurs three times and that no other value occurs three or more times. Thus, the value 164 is the mode of that distribution.

TABLE 4.1 Frequency distribution of male students' weights

Weight (lb)	*f*	*Weight (lb)*	*f*
197	1	161	1
190	1	159	1
185	1	157	1
182	1	154	1
180	2	152	1
179	1	150	1
173	2	149	1
164	3	144	1
163	2	137	1
162	1	135	1
			25

Eyeball-estimating the Mode

The mode is the value on the X-axis that lies directly below the highest point of a frequency distribution. Figure 4.1 shows a sketch of the distribution of female students' weights. It is clearly unimodal with $\text{mode}_{\text{by eyeball}} \approx 120$ pounds, the value on the X-axis directly beneath the highest point of the distribution. Figure 4.2 shows a common mistake: The mode is a value of the variable on the X-axis, *not* the highest point of the curve. (The highest point of the curve represents the *frequency* of the mode, not the mode itself.)

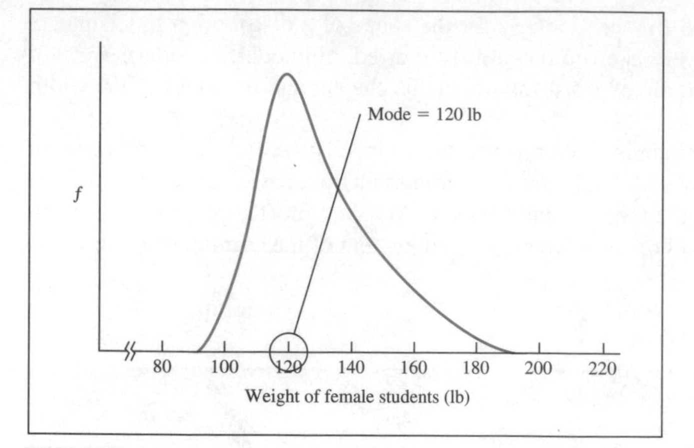

FIGURE 4.1 A unimodal distribution with mode shown correctly on the *X*-axis

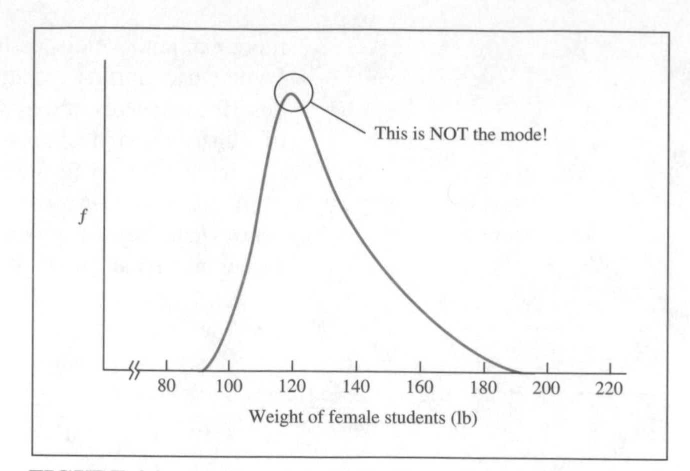

FIGURE 4.2 A unimodal distribution with mode shown incorrectly

Figure 4.3 shows a sketch of a distribution of *all* students, both men and women. This distribution is clearly bimodal, with a mode for women at 120 pounds and a mode for men at 166 pounds. Among this group of students, do you think there are more men or more women? That question is impossible to answer unequivocally, but the distribution does show that more students weigh between 150 and 180 pounds than between 110 and 130 pounds. Thus, it is a good guess that there are more men in the group that this distribution represents.[1]

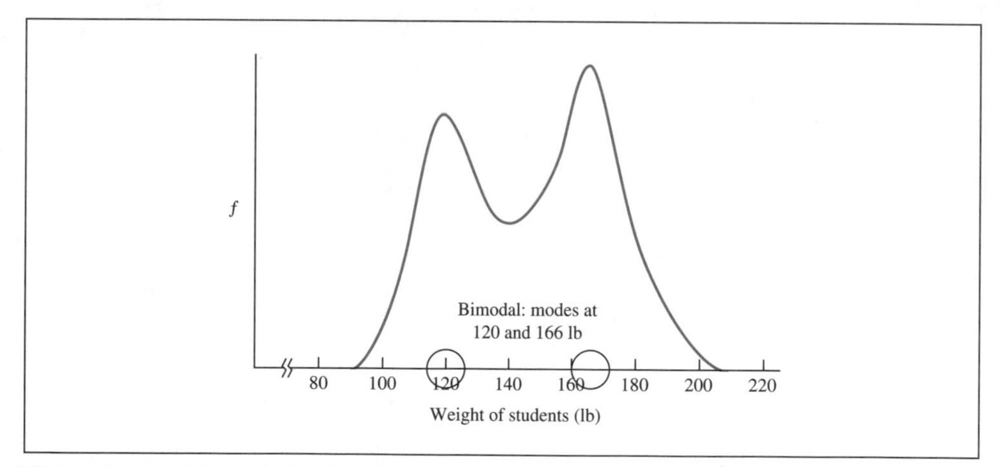

FIGURE 4.3 A bimodal distribution

Computing the Mode

There are several different ways of computing the mode of ordinal or interval/ratio data, and the different ways may lead to different results. We have given an example of one way: noting the highest frequency that occurs in the frequency distribution (see Table 4.1). Another method involves noting the highest frequency in a frequency distribution that uses

[1]Depending on the intent of the communication, it may be more desirable to call the distribution of Figure 4.3 unimodal (with mode at 166 pounds) rather than bimodal because 166 is the value with the highest frequency. Such ambiguity is one of the disadvantages of using the mode as a measure of central tendency.

TABLE 4.2 Frequency distribution of male students' weights (lb) using class intervals

Class interval	f
190–199	2
180–189	4
170–179	3
160–169	7
150–159	5
140–149	2
130–139	2
	25

class intervals; see Table 3.4 (reproduced here as Table 4.2). Inspection of that table shows that the frequency is 7 for the *modal interval* 160–169 pounds. It is customary to report the midpoint of the class interval as the value of the mode: $(160 + 169)/2 = 164.5$ pounds.

The modes as determined by these two methods are not identical: 164 versus 164.5 pounds. In some distributions, the discrepancy caused by using the two methods may be considerably larger. Furthermore, either method may produce more than one value that is a candidate for being called the mode. For example, if our class, whose distribution of weights is shown in Table 4.1, happened to have one more 180-pound student, we would have had *two* values with frequency 3. Should we say that this distribution is bimodal with modes at 164 and 180 pounds? We should note that adding this 180-pound person does not change the mode based on class intervals (the frequency of 180–189 becomes 5, which is still less than 7), so the distribution based on class intervals remains unimodal with mode 164.5 pounds. There is no generally accepted way to identify which mode is the *best* mode; one must make a judgment about how to communicate most effectively about the data.

4.2 Median

median: a measure of central tendency appropriate for ordinal or interval/ratio variables; it is the value that is the midpoint of a data set

The *median* is defined as that value of the variable that is the midpoint of the data set—that is, the value for which half the entries in the data set are larger and half are smaller. The median is appropriate for only ordinal and interval/ratio data because "larger" and "smaller" do not apply to nominal variables (for example, "Republican" is neither larger nor smaller than "Democrat").

The median can be derived from a frequency distribution by noting that it is that point on the distribution that divides the *area* under the distribution into halves. (We shall give a more complete consideration of the areas of regions under frequency distributions in Chapter 6, where we shall see that these areas have extremely useful properties. The median is our first example of that usefulness.)

Eyeball-estimating the Median

Reconsider Figure 3.3, the histogram of male students' weights. We show that figure again here as Figure 4.4, with a vertical line drawn through the point that divides the area of the distribution in half. Note that the areas both to the right and to the left of

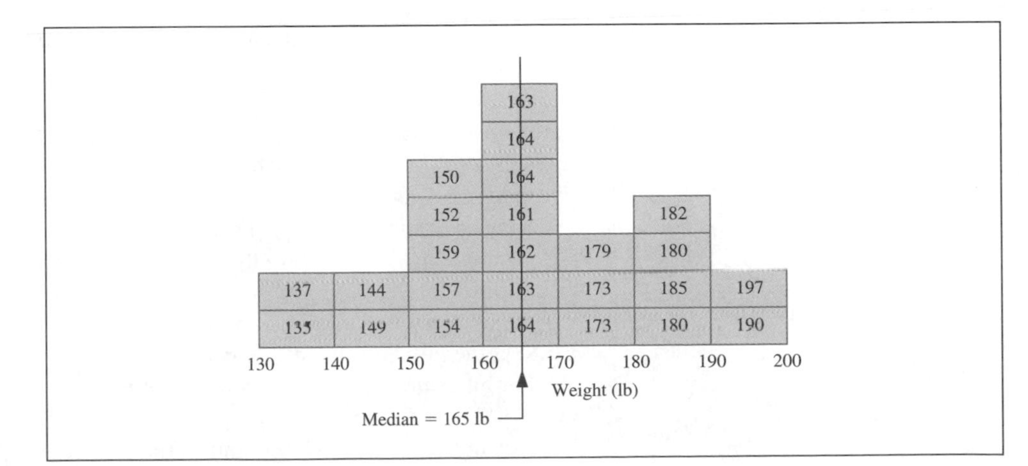

FIGURE 4.4 Weights of male students shown in order (redrawn from Figure 3.2) with median indicated

this vertical line are equal: 9 cells plus 7 half-cells, or $12\frac{1}{2}$ cells. The point on the X-axis through which that line passes is the median as inferred from this frequency distribution: $\text{median}_{\text{by eyeball}} \approx 165$ pounds.

Computing the Median

We now present the procedure for computing the median:

steps for computing the median

Step 1. Put all the entries in the data set in order, largest to smallest. If one value has frequency 3, for example, it should appear three times in the list. There should be the same number of items in the list as there are entries in the original data set.

Step 2. If the data set contains an odd number of entries, then the median is the middle number in the ordered list. For example, the data set shown in Table 4.1 has 25 values, so the middle, or 13th, value (163 pounds) is the median. If n is the number of values in the data set and n is odd, then the *location* of the median is given by the expression $(n + 1)/2$. In our example, $n = 25$ and the median is the $(25 + 1)/2 = 13$th score. Note that when counting scores in a frequency distribution, you must count each value as often as it actually occurs in the data set; for example, starting at the top of Table 4.1, you must count 180 and 173 each *twice* because they have frequency 2.

Note that the computed value of the median (163 pounds) is approximately equal to the value (165 pounds) we eyeball-estimated from the histogram shown in Figure 4.4.

Step 3. If the data set contains an even number of entries, then the median is halfway between the two values closest to the middle of the ordered list. For example, if a data set has six weights, 130, 142, 155, 162, 169, and 173 pounds, there is no middle weight. The median is therefore the value halfway between the two scores closest to the middle—that is, the sum of the $(n/2)$th and the $(n/2 + 1)$th values divided by 2. In this example, $n = 6$, so halfway between the 3rd and 4th values is $(155 + 162)/2$, or 158.5 pounds.

(Box 4.1 gives an optional, alternative way of determining the median.)

BOX 4.1 Another method for computing the median

 This box may be omitted without loss of continuity.

Some, though by no means all, statisticians prefer the following "linear" method for computing the median, which is more accurate if the distribution is approximately uniform. If the middle entry as described in Step 2, or either of the two values closest to the middle as in Step 3, is one of several identical values, then those identical values may be thought of as being uniformly spread out across the interval between their lower and upper real limits.

For an example where the number of points is odd, consider these values of X: 3, 4, 5, 6, 6, 6, 6, 7, 8. Step 2 dictates that the median should be the middle (5th) score, or 6. However, there are multiple (four) occurrences of the value "6," so the linear method states that we should think of those four values as uniformly spread out over the range of values from the lower real limit 5.5 to the upper real limit 6.5. Therefore, we divide the range from 5.5 to 6.5 into four equal parts (because there are four occurrences of the value "6"). These four parts are 5.5 to 5.75, 5.75 to 6.00, 6.00 to 6.25, and 6.25 to 6.5. Then we think of the four occurrences of "6" as being replaced by the midpoints of these parts—namely, 5.625, 5.875, 6.125, and 6.375—so our new "conceptual" list becomes 3, 4, 5, 5.625, 5.875, 6.125, 6.375, 7, 8. Then the median is 5.875, the middle value of the list.

For an example where the number of points is even, consider the data 3.1, 3.2, 3.2, 3.3, 3.4, 3.5. Step 3 dictates that the median is 3.25, halfway between 3.2 and 3.3. However, the linear method notes that the value 3.2 (which has real limits 3.15

(continued)

BOX 4.1 *(continued)*
and 3.25) occurs twice and so should be conceptually replaced by the midpoints of the regions 3.15 to 3.2 and 3.2 to 3.25 (3.175 and 3.225, respectively), making the conceptual list 3.1, 3.175, 3.225, 3.3, 3.4, 3.5. Then the median is 3.2625, halfway between 3.225 and 3.3.

4.3 Mean

mean: a measure of central tendency appropriate for interval/ratio variables; it is equal to the sum of all values of a variable divided by the number of values

When we use the term *average*, we are usually referring to what statisticians call the *mean* of a distribution. When, for example, you compute your average grade by adding all the individual grades and dividing by the number of grades, you are calculating the mean grade.

Eyeball-estimating the Mean

Balancing-point Method

The mean is the "balancing point" of a distribution. Imagine, for example, that the distribution of Figure 3.3 (which we show here as Figures 4.5, 4.6, and 4.7) were cut out of plywood. The mean of that distribution is the point where that piece of plywood would balance on a fulcrum. Thus, the mean (like the mode and median) is a point on the X-axis of the distribution (in our example, a particular weight).

We can eyeball-estimate the mean by imagining various points on the weight axis and observing whether the distribution would "tip" if we attempted to balance it at that point. For example, Figure 4.5 shows the histogram of male students' weights with a possible balance point at 150 pounds. If this distribution were made of plywood, the portion of the distribution to the right of the triangular fulcrum would outweigh the portion to the left, and the distribution would tilt downward on the right. Thus, 150 pounds must be too far to the left to be the mean.

So we try again, choosing a point farther to the right—say, 172 pounds—and sketching this possible fulcrum in Figure 4.6. Now the left side of the distribution seems to be too large, and the plywood distribution would tip down on the left side. Thus, our possible value of the mean, 172 pounds, is too far to the right.

Now we have a value, 150, too low (too far to the left) and another value, 172, too high (too far to the right), so we choose a point between the two values—say, 165 pounds—and sketch the fulcrum there (shown in Figure 4.7). This value seems to be about right—the distribution seems to balance—so we conclude that the $\text{mean}_{\text{by eyeball}} \approx 165$ pounds.

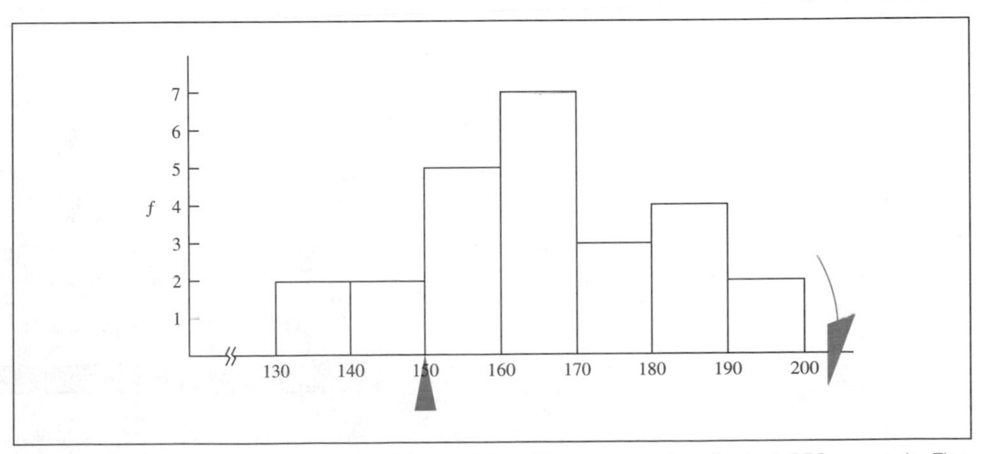

FIGURE 4.5 Histogram of students' weights with mean estimated at 150 pounds. The histogram would tilt down on the right, so the mean is too low.

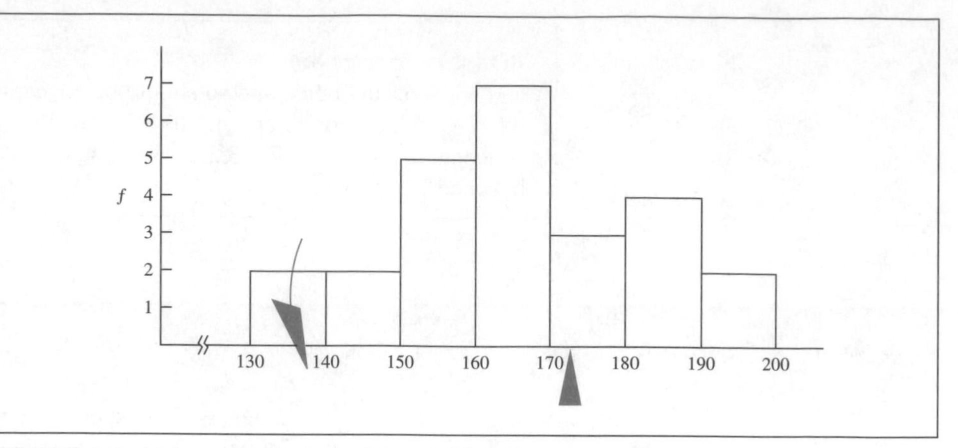

FIGURE 4.6 Histogram of students' weights with mean estimated at 172 pounds. The histogram would tilt down on the left, so the mean is too high.

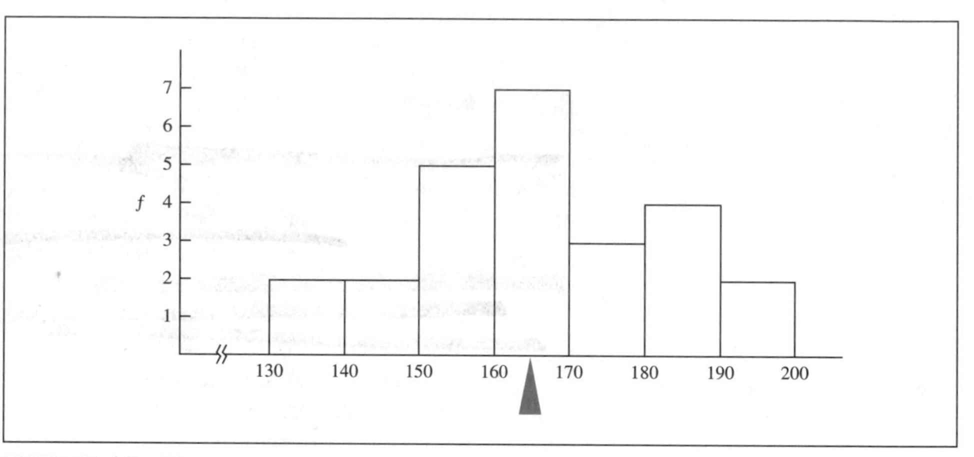

FIGURE 4.7 Histogram of students' weights with mean estimated at 165 pounds. The histogram would balance, so the mean is just right.

Eyeball-calibration

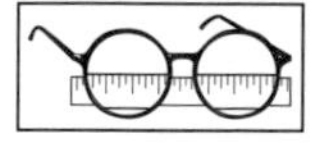

For practice, cover the bottom portion of Figure 4.8 and eyeball-estimate the balancing point; then convert this point to the numerical value of the mean according to the scale on the X-axis. The actual balancing point is shown at the bottom of the figure in the form of an inverted "T" display ($\perp$); the central vertical line is at the balancing point. The numerical value is given in the caption. Ignore until Chapter 5 the fact that the caption also gives the standard deviation.

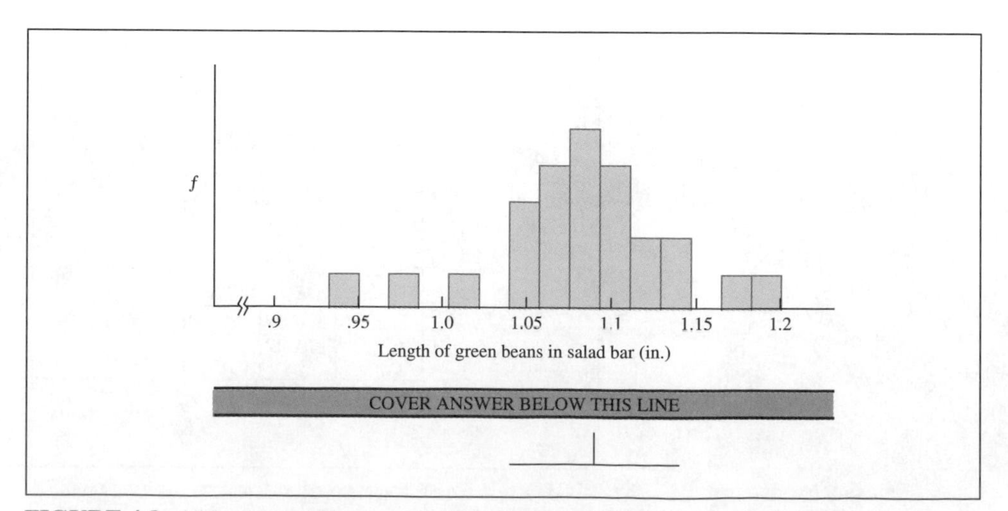

FIGURE 4.8 Histogram with mean 1.09 inches and standard deviation .05 inch

Figures 4.9–4.14 provide six more figures for eyeball-estimation practice. Cover the bottom portion of each and eyeball-estimate the mean. Note that the balancing-point procedure works as well with a frequency polygon as it does with a histogram.

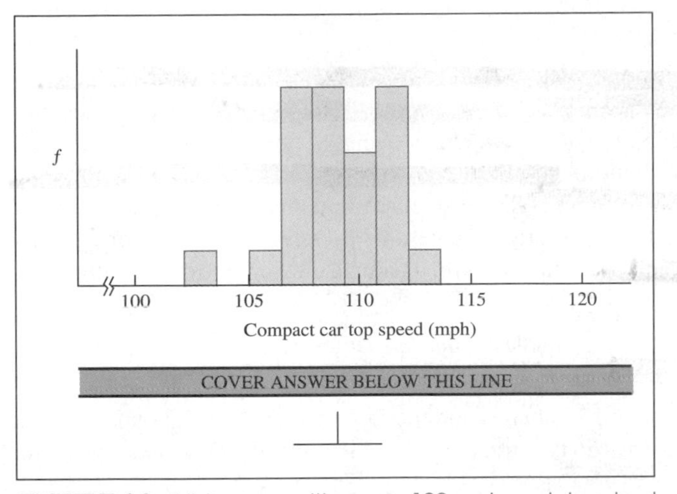

FIGURE 4.9 Histogram with mean 109 mph and standard deviation 2 mph

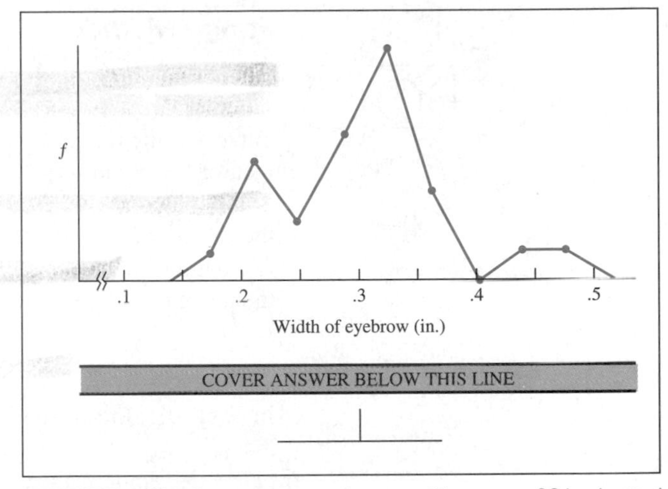

FIGURE 4.10 Frequency polygon with mean .30 inch and standard deviation .07 inch

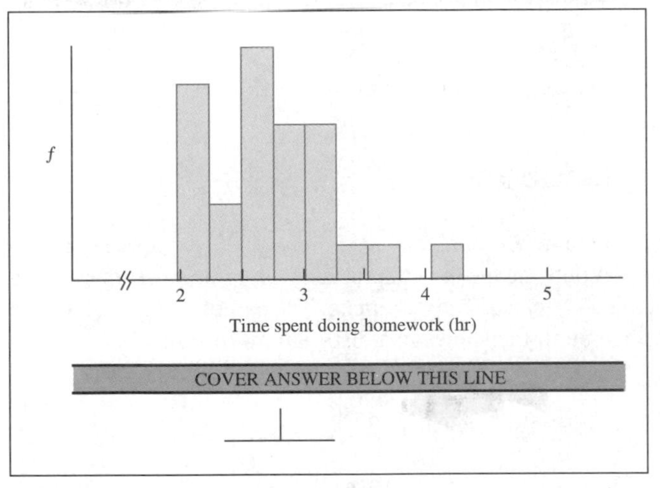

FIGURE 4.11 Histogram with mean 2.8 hours and standard deviation .45 hour

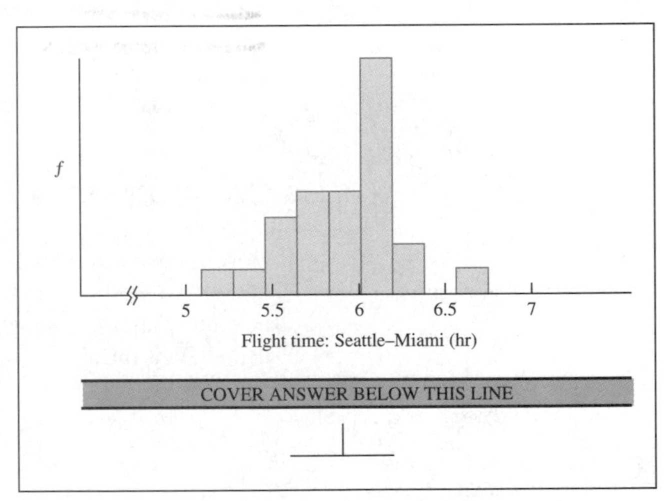

FIGURE 4.12 Histogram with mean 5.9 hours and standard deviation .3 hour

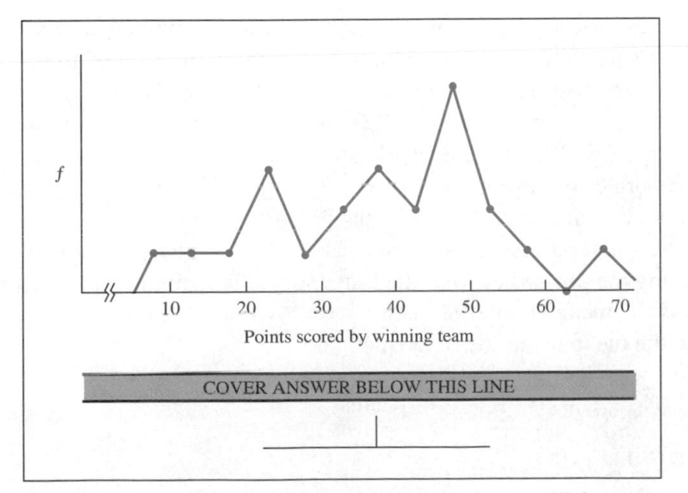

FIGURE 4.13 Frequency polygon with mean 37.3 points and standard deviation 15.2 points

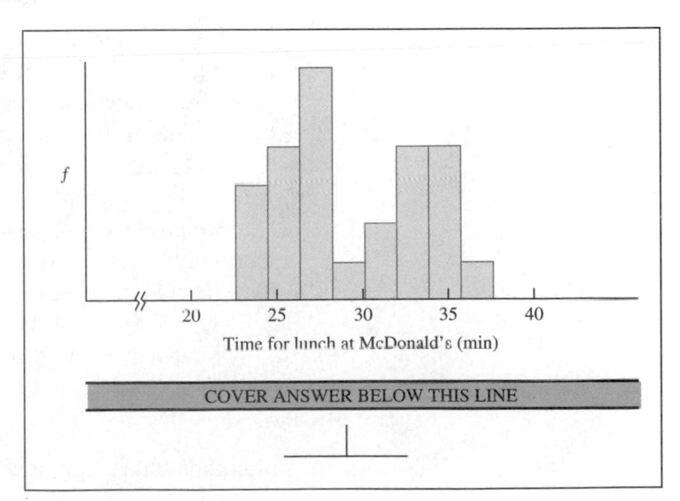

FIGURE 4.14 Histogram with mean 29 minutes and standard deviation 3.6 minutes

The ESTAT logo means that eyeball-calibration exercises appropriate to this section can be found in the ESTAT computer program or in the *Study Guide*. You may wish to use the ESTAT exercise "meanest" at this time.

Range Method

There is another "quick and dirty" way to eyeball-estimate the mean. It is most useful when a numerical, not a graphical, frequency distribution is available. Suppose we have access to Table 4.1, the frequency distribution of male students' weights, and we wish to get a rough idea of the mean of this distribution. The mean is roughly halfway between the smallest and the largest values in the distribution. As we shall see in the next chapter, the *range* is the distance between the smallest and largest values, so we refer to this way of eyeball-estimating as "the range method." In simplest form, the range method of eyeball-estimating the mean can be expressed as

$$\text{mean}_{\text{by eyeball}} \approx \frac{\text{smallest value} + \text{largest value}}{2}$$

In Table 4.1, the smallest value is 135 pounds and the largest value is 197 pounds, so the $\text{mean}_{\text{by eyeball}} \approx \frac{135+197}{2} = \frac{332}{2} = 166$ pounds, quite close to our eyeball-estimate from the balancing-point method.

We have, therefore, two methods (the balancing-point and the range) for eyeball-estimating the mean. Sometimes one will be more convenient to use, sometimes the other. The balancing-point method is generally more accurate and therefore preferred if both are equally convenient.

You may wish to use the ESTAT exercise "meannum" at this time.

Computing the Sample Mean

Now that we can eyeball-estimate the mean relatively accurately given a frequency distribution, we need to learn how to compute the mean with precision. Our computations should be merely an improvement on what we can already estimate. The sample mean, which we denote $\overline{X}$, is the arithmetic average of all the values in a distribution. In symbols,

mean of a sample

$$\overline{X} = \frac{\sum X_i}{n} \qquad (4.1)$$

The formula for the mean of a population is on page 59.

where

X_i = the ith value of the variable X

n = the number of observations in the sample

For the weights of male students, we can compute $\sum X_i$ by adding down the column of weights in Table 4.1, being sure that we add each weight the same number of times as its frequency (for example, we must add the weight 164 pounds *three times* because that weight occurs with frequency 3). For our data, $\sum X_i = 4117$ pounds and $n = 25$, so $\overline{X} = 4117/25 = 164.68$ pounds. Recall that we eyeball-estimated that mean to be about 165, so the computation corroborates our eyeball-estimate.

Suppose we wish to know how often a person visits the grocery store in a month. We randomly stop ten people as they enter a grocery store and ask them how many times they have been to the store (including the present visit) in the last 30 days. We get these data: 2, 5, 3, 4, 7, 3, 1, 1, 4, 4. What is the mean number of visits?

We can eyeball-estimate the mean using the range method:

$$\overline{X}_{\text{by eyeball}} \approx \frac{1+7}{2} = 4 \text{ visits}$$

or compute it using Equation (4.1):

$$\overline{X} = \frac{\sum X_i}{n} = \frac{2+5+\cdots+4+4}{10} = \frac{34}{10} = 3.40 \text{ visits}$$

Note that the discrepancy between the eyeball-estimate and the computed value can be larger when we use the range method than if we had constructed a histogram and used the balancing-point method. We would have noted on the histogram that the value "7" is a single large value and we would not have let that single value have such a large effect on our estimate.

4.4 Comparing the Mode, Median, and Mean

The mode, median, and mean are all measures of central tendency. How do we decide which to report? The first consideration is the level of measurement. Table 4.3 lists the appropriate measures of central tendency for each level of measurement. Applying an inappropriate measure can lead to absurd results. For example, the mode is the *only* appropriate measure of central tendency for nominal-level variables. Suppose that gender is our nominal variable and we measure it with female = 1 and male = 2. If there are 18 women and 25 men, then the frequency distribution is shown in Table 4.4. Because the mode is the most frequently occurring value, the mode for our data is 2, which indicates that there are more men.

TABLE 4.3 Appropriate measures of central tendency for each level of measurement

Level of measurement	*Permissible measure of central tendency*
Nominal	Mode
Ordinal	Mode or median
Interval/ratio	Mode, median, or mean

TABLE 4.4 Frequency distribution of student gender (female = 1, male = 2)

Gender	*f*
1	18
2	25
	43

The median and mean are *not* appropriate for nominal-level variables; they are computable (the numbers don't care what we do with them) but misleading. The mean for the data in Table 4.4 is $\overline{X} = \sum X_i/n = 68/43 = 1.58$, which would indicate that the average student is just a little more than halfway between female and male! To avoid such absurd statements we must recognize that the only appropriate measure of central tendency for nominal variables is the mode.

Consider an example that shows why it is misleading to use the mean as a measure of the center of ordinal data. A trucking company hauls ten packages that weigh 2, 2, 2, 7, 7, 9, 9, 400, 400, and 400 pounds. This company classifies packages into four weight classes: class 1: light (0–5 pounds), class 2: medium (6–8 pounds), class 3: moderate (9–10 pounds), and class 4: heavy (11 pounds or more). The frequency distribution of the weight class data is shown in Table 4.5. The weight class variable is thus ordinal (not interval/ratio because the categories are not of equal width: class 1 is 6 pounds wide, class 2 is 3 pounds wide, etc.). *If* we were to compute the mean of the weight classes (the numbers don't care if we compute the mean of ordinal data), we would find that the mean class is $(1 + 1 + 1 + 2 + 2 + 3 + 3 + 4 + 4 + 4)/10 = 25/10 = 2.5$, halfway between class 2 and class 3. We might then be led incorrectly to conclude that the average package weighed about 8.5 pounds. However, the actual mean weight is $(2 + 2 + 2 + 7 + 7 + 9 + 9 + 400 + 400 + 400)/10 = 1238/10 = 123.8$ pounds. The reason for the huge discrepancy between the mean derived (incorrectly) from the weight classes (an ordinal variable) and the mean computed (correctly) from the weights themselves (an interval/ratio variable) is that the actual widths of the ordinal weight classes here are markedly *unequal*: The "heavy" interval is 200 times as wide as the "moderate" interval. The mean is *not* an appropriate measure of central tendency if the data are ordinal.

TABLE 4.5 Frequency distribution of weight classes of packages

Weight class	*f*
1 light (0–5 lb)	3
2 medium (6–8 lb)	2
3 moderate (9–10 lb)	2
4 heavy (11+ lb)	3
	10

There is no such large discrepancy between the *median* of the weight class data and the median of the actual weights. The data as classes are 1, 1, 1, 2, 2, 3, 3, 4, 4, 4, and the median is 2.5 (halfway between the 5th value and the 6th), which, expressed as a weight, is

8.5 pounds. The median of the data expressed as weights is 8 pounds (halfway between 7 and 9 pounds). These two medians are quite close. The median *is* an appropriate measure of central tendency for ordinal data.

Although the mode, median, and mean are all appropriate measures of the center of the distribution of an interval/ratio variable, they are by no means identical measures. In our weight example of Table 4.1, the mode was 164 pounds, the median was 163 pounds, and the mean was 164.68 pounds. In some distributions, these three measures may be identical; in others, they may all be quite different.

To demonstrate this, let us consider distributions that are clearly unimodal. If such a distribution is symmetric, then the mode, median, and mean will all be identical, as shown in Figure 4.15. It follows (because normal distributions are unimodal and symmetric) that *normal* distributions have the property mode = median = mean.

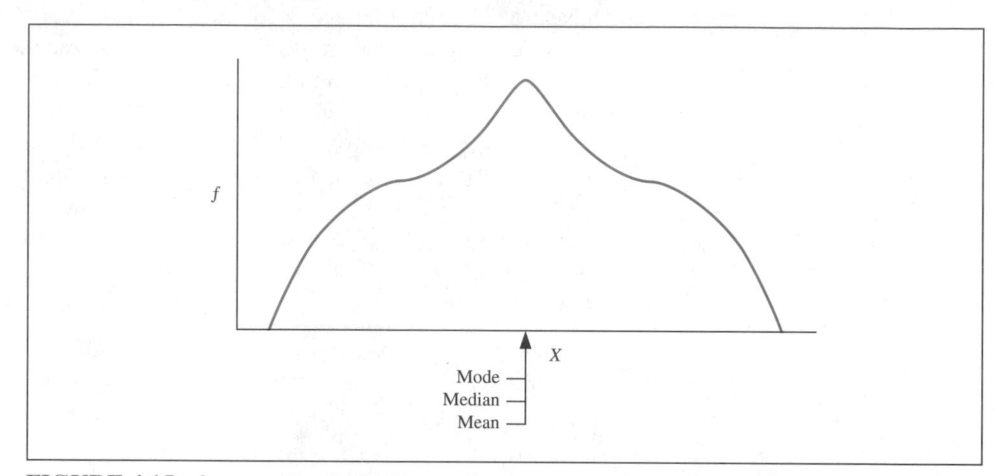

FIGURE 4.15 Symmetric distribution: mode, median, and mean are identical

If a unimodal distribution is not symmetric, however, the mean and median may be different from the mode, as shown in Figure 4.16, which is created from Figure 4.15 by stretching the right tail out. It is useful to think of the long tail of this positively skewed distribution as "pulling" the mean toward itself. In Figure 4.16, for example, the long right-hand tail causes the balancing point of the distribution, and thus the mean, to move away from the mode out to the right.

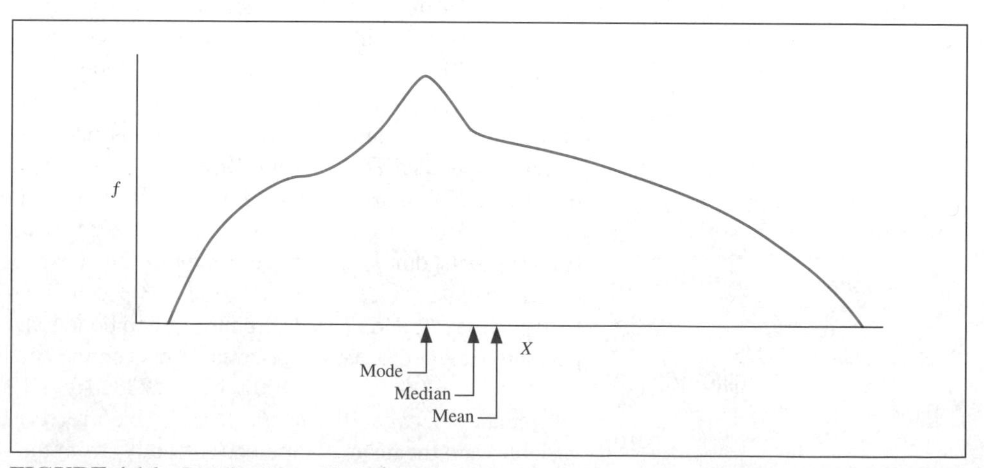

FIGURE 4.16 Distribution with long right-hand tail; mean and median shift to the right

Although positive skew pulls the mean in a positive direction (and negative skew pulls the mean in a negative direction), skew generally has relatively little effect on the median and even less effect on the mode. In skewed distributions, therefore, the median generally lies between the mode and the mean.

TABLE 4.6 Frequency distribution of employee incomes

Income	*f*
$200,000	1
20,000	1
19,000	1
14,000	3
	6

The three measures of central tendency can give very different results. Table 4.6 shows the incomes of workers in a small business that consists of the owner/president and five employees. Salary expressed in dollars is an interval/ratio variable because the difference between $1 and $2 is the same as the difference between $1001 and $1002, and so on. In Table 4.6, the mean salary is $\overline{X} = \sum X_i/n = \$281{,}000/6$ or $46,833; the median salary is $(X_3 + X_4)/2 = (\$19{,}000 + \$14{,}000)/2 = \$16{,}500$; and the modal salary is the most frequently occurring value, $14,000. If you were the owner of this company and wanted to convey that your company's employees were highly paid, you would say that your company's mean salary was $46,833 (and you would be technically correct but rather misleading). If, on the other hand, you were advising a friend who was applying for a job at that company, you would want to report the mode ($14,000) or the median ($16,500) as being more reflective of what most employees actually earn. The reason the mean and median of this distribution are so far apart is that the distribution is highly positively skewed, and, as usual, the skew pulls the mean toward the long tail while having relatively less effect on the median and even less effect on the mode. See Box 4.2.

BOX 4.2 Misleading with statistics

People sometimes criticize statistics as being a way of lying about data, but that is unfair. Statistics is a way of communicating about data, and as with any sophisticated way of communicating, there are choices to be made about what to say and how to say it. When the data are interval/ratio, one can use any (or all) of the three measures of central tendency, depending on what you wish to convey. Some ways of communicating may be deliberately misleading—for example, if the company president (see Table 4.6) intends to convey that his employees are all highly paid by saying the average salary is $46,833. But the fact that one can mislead with statistics is no more a criticism of statistics than the fact that one can mislead in English is a criticism of English. People mislead, not statistics.

We shall see in subsequent chapters that the mean is the most frequently used measure of central tendency for interval/ratio data. This is true for two reasons: (1) Every point in the data set contributes to the mean (unlike, for example, the median, which depends on the values of only one or two central points), and (2) many of the most powerful statistical techniques are based on the mean.

4.5 Computing the Population Mean

The formula for computing a population mean is

mean of a population

$$\mu = \frac{\sum X_i}{N} \tag{4.2}$$

The formula for the mean of a sample is on page 56.

where

X_i = the ith value of the variable X

N = the number of members in the population

There are two differences between this formula and Equation (4.1), the formula for the sample mean. First, we call the population mean μ instead of $\overline{X}$. Second, we denote the number of observations in a population with a capital N instead of a lowercase n.

We use the formula for the population mean, Equation (4.2), when the scores that we are considering are themselves the focus of our attention (that is, when the scores are a population; see the discussion of populations and samples in Chapter 1). We use the formula for the sample mean, Equation (4.1), when the scores are a subset (sample) of some larger group (the population) and our primary interest is in that larger population. Equations (4.1) and (4.2) yield identical numerical results for the same data.

These distinctions between sample and population may seem dry and uninformative at the present time. Beginning in Chapter 7, it will become quite clear why we need to be so careful in distinguishing between populations and samples; let's take a glance ahead.

Suppose that we are engineers working for Sylvania, a manufacturer of 100-watt light bulbs. We wish to know the average lifetime of all the light bulbs we produce. We could answer that question absolutely accurately by taking every single manufactured bulb, lighting it, and measuring the time until it burns out. If we then add up all the times and divide by the total number of bulbs, we would have a perfect measurement of μ, the mean of the population of light bulb lifetimes. Of course, we also would have no light bulbs left to sell, so we probably would not choose this technique. Instead, we take a sample of 1000 bulbs, burn them while measuring each bulb's lifetime, and then add up the times and divide by 1000. This gives us $\overline{X}$, the mean of the sample, which we believe will have a value close to μ. How close? That is one of the most basic questions in statistics, to which we will return in Chapter 7.

Statistics, simply stated, is primarily the science of saying something meaningful about populations when all we know about are samples. Almost all the remainder of this textbook will be variations on that theme.

4.6 Connections

Cumulative Review

In Chapter 1, we introduced the notion that there are basically three major statistical concepts to be mastered: (1) what the distribution of a variable is, (2) what a sampling distribution of means is, and (3) what a test statistic is. So far we have been focusing on the first of these concepts, the distribution of a variable: In Chapter 1 we discussed variables, in Chapter 2 we discussed their measurement, and in Chapter 3 we began to discuss their distributions. We noted there that if we know three characteristics (the shape, the center, and the width) of the distribution of a variable, we know quite a lot about that variable. This chapter has discussed ways of measuring the center of a distribution.

Journals

The sample mean, which we have called $\overline{X}$, is often called M (for "mean") in journal reports of experiments. The reason for this is historical: In older printing processes, it was quite difficult (and therefore expensive) to typeset the bar over the X. It no longer is difficult, but the tradition continues. Thus a report about the Pygmalion experiment might read something like "The bloomers (M = 16.5 IQ points) showed on average a higher intellectual growth than did the other children (M = 7.0 IQ points)."

Computers

Use ESTAT's datagen to determine the mean of the weight data that originally appeared in Table 3.2:

1. Enter the data from Table 3.2 by typing the values down a single column, using `Enter` between values. (Do *not* put the data in five different columns.)
2. Save these data because we will use them later: Click `Save`. Position the cursor after the asterisk ("*") in the File Name cell. Press `Backspace`. Type "`table3-2`." The file name should now read "`table3-2.txt`."

3. Open the file you just saved to make sure it was saved correctly: Click `Open`, click `table3-2.txt`, and then click `OK`.
4. Note that the mean appears automatically in the `datagen Descriptive Statistics` window in the row labeled "Sample mean."
5. Note that datagen does not compute the mode or median. Datagen (version 2.0 or later) can help you determine the mode and median by sorting the data for you: Click any point in the first column of the datagen spreadsheet to select the first variable, click `Edit Variable`, click the `Sort them in descending order` button, and then click `OK`.

Use SPSS to determine the mean of the weight data that originally appeared in Table 3.2:

1. In Chapter 3, you saved a file for the data of Table 3.2, calling that file "table3-2.sav." Now retrieve that file by clicking `File`, then `Open`, and then `Data...`. Click `table3-2.sav`, and then click `OK`.
2. Click `Statistics`, then `Summarize`, and then `Frequencies...`. Click ▶.
3. Request appropriate statistics: Click `Statistics...`, click the `Mean` checkbox, click the `Median` checkbox, click the `Mode` checkbox, and then click `Continue`.
4. Click `OK`.
5. Note that the mean, median, and mode appear in the `!Output1` window.

Note that if there are multiple modes, SPSS reports only the smallest value and does not inform you that there are multiple values with the same frequency.

Homework Tips

1. Check the list of learning objectives at the beginning of this chapter. Do you understand each one?

2. When you are asked to compute a mean, your first task is to determine whether you should use the population mean μ or the sample mean $\overline{X}$. If the data form a population (that is, contain all the values in which you are interested), then use the formula for μ. If the data form a subset of the population of interest, then use the formula for $\overline{X}$.

3. The formulas for μ and $\overline{X}$ are listed for your convenience on the inside front cover of the textbook.

Exercises for Chapter 4

Section A: Basic exercises (answers in Appendix D, page 508)

1. Use these values for X_i: 2, 4, 2, 6, 3, 1.
   (a) Eyeball-estimate the mean by the range method.
   (b) Compute the sample mean.
   (c) Compute the population mean.
   (d) Explain why the results of parts (b) and (c) are the same or different.
   (e) Determine the mode.
   (f) Determine the median.
2. Use the X_i values 3, 3, 4, 6, −2, 5, −2, 5, 4.
   (a) Eyeball-estimate the mean by the range method.
   (b) Compute the sample mean.
   (c) Compute the population mean.
   (d) Explain why the results of parts (b) and (c) are the same or different.
   (e) Determine the mode.
   (f) Determine the median.
3. Suppose you wish to tell a friend about the temperatures you experienced on your vacation in Florida in the first week of July. The daily high temperatures (in °F) were: 74, 82, 79, 82, 82, 83, 81. Do not forget to include units in your answers (here, °F).
   (a) Eyeball-estimate the mean by the range method.
   (b) Which formula is appropriate for computing the mean? Why?
   (c) Compute the mean.
   (d) Account for any discrepancy between your eyeball-estimate and computation.

(e) Which method of determining the median is appropriate? Why?
(f) Compute the median.
(g) Compute the mode.

4. Suppose you wish to vacation in Alaska in the first week of July, and you want to know what temperatures to expect. You go to the library and select July newspapers at random, one from each of the last six years, and look up the temperature in Alaska on each day. The daily high temperatures (in °F) were: 69, 58, 63, 66, 62, 66.
   (a) Eyeball-estimate the mean by the range method.
   (b) Which formula is appropriate for computing the mean? Why?
   (c) Compute the mean.
   (d) Account for any discrepancy between your eyeball-estimate and computation.
   (e) Which method of determining the median is appropriate? Why?
   (f) Compute the median.
   (g) Compute the mode.

5. In Exercise 1 in Chapter 3, we were interested in the fitness of students. We selected 20 students at random and asked them to ride a stationary bicycle as fast as they could for 30 minutes. The numbers of miles they rode were: 8, 15, 22, 17, 9, 16, 15, 15, 21, 12, 14, 16, 22, 17, 16, 11, 13, 15, 20, 15.
   (a) Eyeball-estimate the mean by the range method.
   (b) A histogram for these data (like the one you produced in Chapter 3) is shown here. Eyeball-estimate the mean using the balancing-point method.

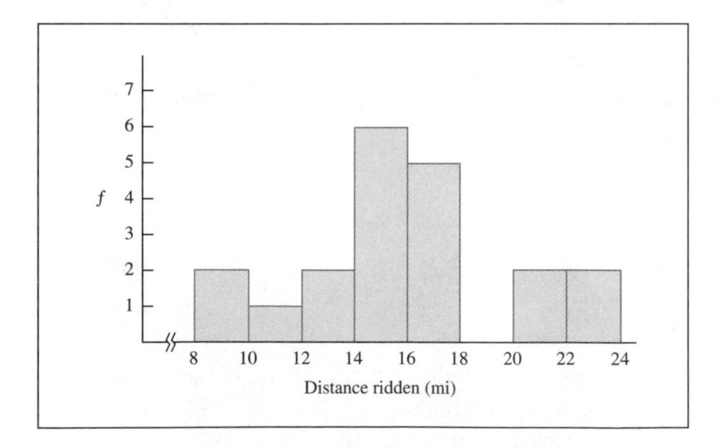

   (c) A frequency polygon for these data (like the one you produced in Chapter 3) is shown at the top of the next column. Eyeball-estimate the mean using the balancing-point method. Is your estimate approximately the same as in part (b)?
   (d) Which formula is appropriate for computing the mean? Why?
   (e) Compute the mean.
   (f) Eyeball-estimate the median from the histogram.
   (g) Which method of determining the median is appropriate? Why?
   (h) Compute the median.

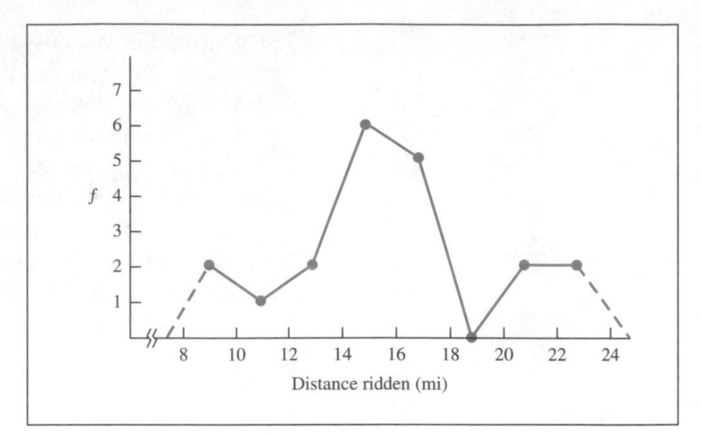

   (i) Eyeball-estimate the mode from the histogram.
   (j) Compute the mode.
   (k) Account for any discrepancies between eyeball-estimates and computations.

6. In Exercise 3 in Chapter 3, we presented the scores of all students on the first exam in a statistics class: 81, 86, 91, 75, 96, 82, 88, 88, 71, 68, 72, 61, 84, 86, 95, 91, 84, 83, 91, 83, 89, 90.
   (a) Eyeball-estimate the mean by the range method.
   (b) A histogram for these data (like the one you produced in Chapter 3) is shown here. Eyeball-estimate the mean using the balancing-point method.

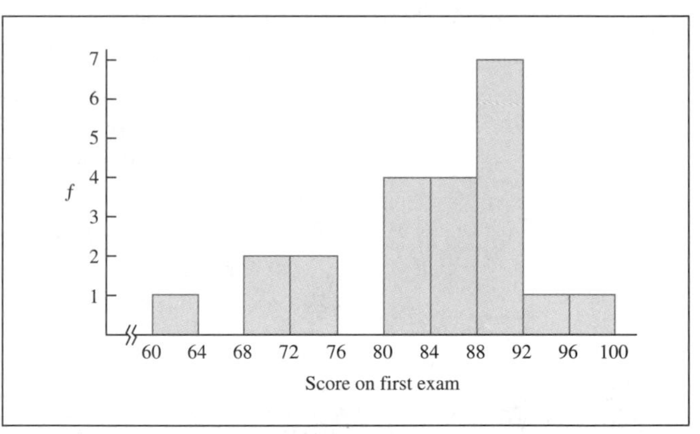

   (c) A frequency polygon for these data (like the one you produced in Chapter 3) is shown here. Eyeball-estimate the mean using the balancing-point method. Is your estimate approximately the same as in part (b)?

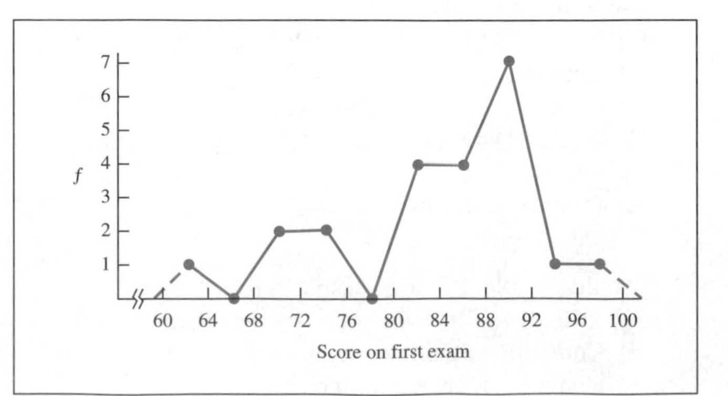

(d) Which formula is appropriate for computing the mean? Why?
(e) Compute the mean.
(f) Eyeball-estimate the median from the histogram.
(g) Which method of determining the median is appropriate? Why?
(h) Compute the median.
(i) Eyeball-estimate the mode from the histogram.
(j) Compute the mode from the data themselves.
(k) Compute the mode from the grouped frequency distribution.
(l) Account for any discrepancies between eyeball-estimates and computations.

7. Eyeball-estimate the mean using the balancing-point method for the histogram shown here.

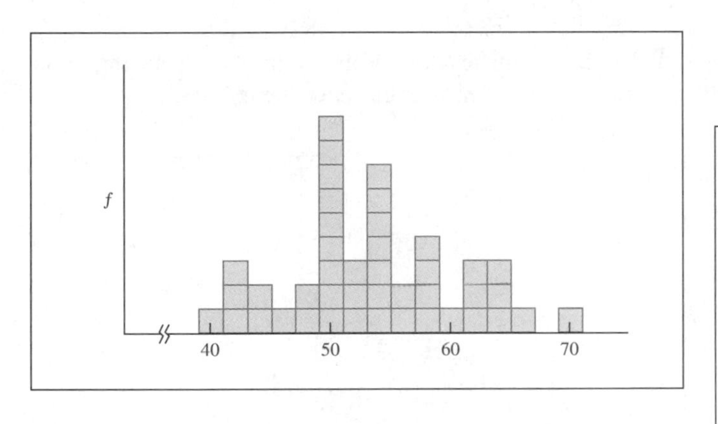

8. A sketch of the distribution of musicians' vibrato (Exercise 5 in Chapter 3) is shown here. Eyeball-estimate the (a) mean, (b) median, and (c) mode.

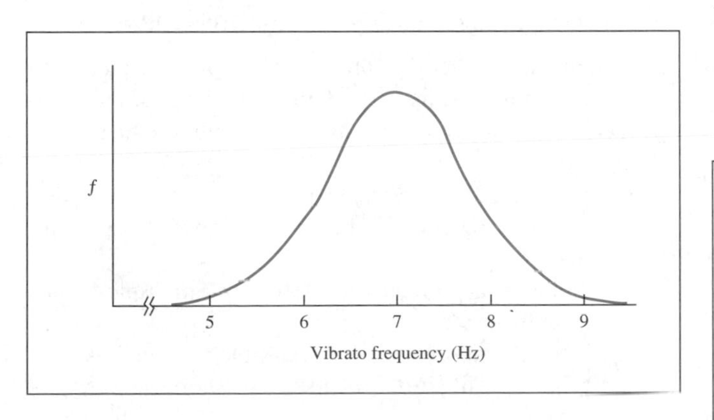

(d) Explain any differences in these three measures of central tendency.

9. A sketch of the distribution of the amounts of liquid in Coke bottles (Exercise 6 in Chapter 3) is shown here. Eyeball-estimate the (a) mean, (b) median, and (c) mode.

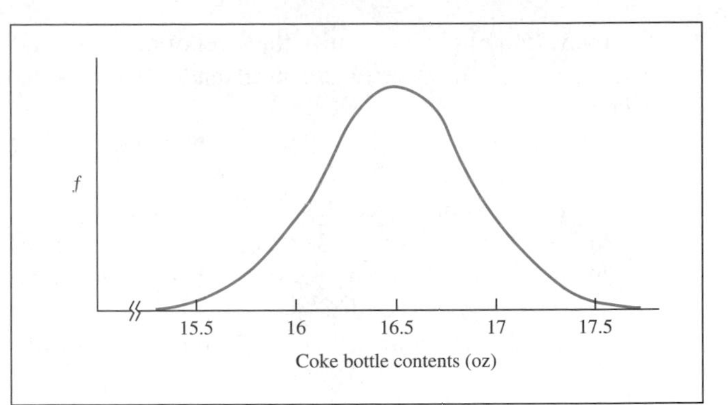

(d) Explain any differences in these three measures of central tendency.

10. A sketch of the distribution of the lengths of McDonald's french fries (Exercise 7 in Chapter 3) is shown here. Eyeball-estimate the (a) mean, (b) median, and (c) mode.

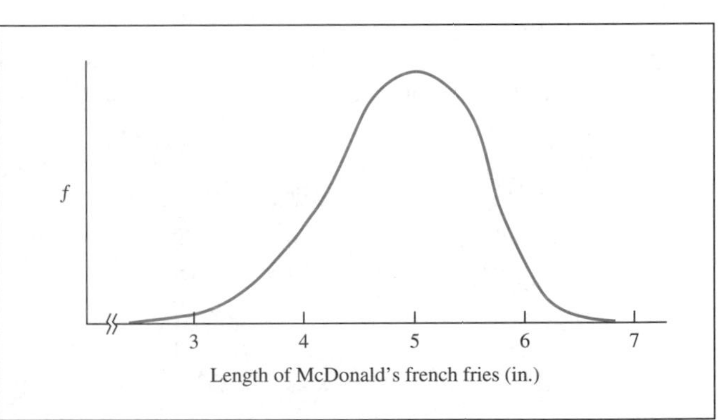

(d) Explain any differences in these three measures of central tendency.

11. A sketch of the distribution of the lengths of Royal Perfecto french fries (Exercise 8 in Chapter 3) is shown here. Eyeball-estimate the (a) mean, (b) median, and (c) mode.

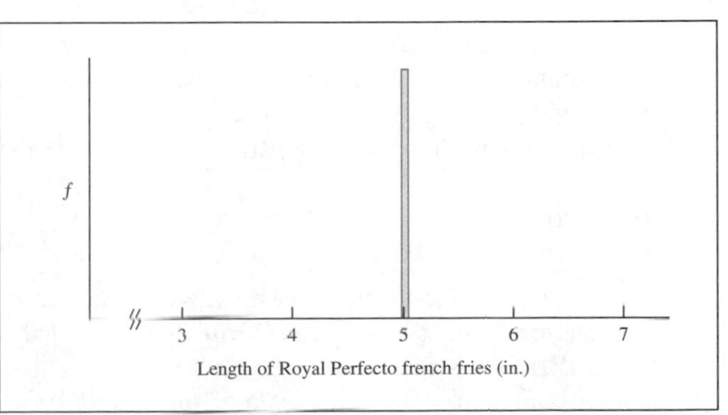

(d) Explain any differences in these three measures of central tendency.

12. For the variables given, identify the level of measurement and state which measure(s) of central tendency are appropriate.
    (a) Grit on sandpaper: extra fine, fine, medium, coarse, and extra coarse
    (b) Type of automobile: 1 = Chrysler product, 2 = General Motors product, 3 = Ford product, 4 = other
    (c) Number of dribbles a player makes before shooting a free throw in basketball
    (d) Flight number for an airline (e.g., Flight #112 is from Los Angeles to New York)
    (e) Teacher's rating of student's sociability: 1 = always plays alone, 2 = occasionally plays with others, 3 = usually plays with others, 4 = always plays with others
    (f) Number of pages in a book
13. Is the measurement described made on a sample or on a population?
    (a) You wish to know the temperature of all the stars in the sky. You select 15 at random and measure their temperatures.
    (b) You wish to know the heights of all the residents in Littletown, Kansas, and you measure each of the town's 100 residents.
    (c) You wish to know the heights of all residents in the United States. You have reason to believe that Littletown, Kansas, is representative of the United States, and you measure each of the 100 residents in Littletown.

Section B: Supplementary exercises

14. Consider the values of X_i: 2, 6, 8, 3, 3, 5, 7, 7, 7, 1.
    (a) Eyeball-estimate the mean by the range method.
    (b) Compute the sample mean.
    (c) Compute the population mean.
    (d) Explain why the results of parts (b) and (c) are the same or different.
    (e) Determine the mode.
    (f) Determine the median.
15. Use the data X_i: 12, 15, 16, 11, 16, 21, 14.
    (a) Eyeball-estimate the mean by the range method.
    (b) Compute the sample mean.
    (c) Compute the population mean.
    (d) Explain why the results of parts (b) and (c) are the same or different.
    (e) Determine the mode.
    (f) Determine the median.
16. Use your histogram (or frequency polygon) of the paired-associate experiment in Exercise 13 in Chapter 3. A researcher has presented lists of nonsense syllables to a random sample of 50 students from an introductory psychology course. The numbers of list presentations were: 8, 9, 7, 8, 16, 7, 10, 11, 16, 14, 12, 13, 12, 13, 12, 14, 8, 9, 17, 12, 5, 18, 14, 14, 12, 8, 11, 11, 9, 9, 18, 15, 11, 7, 9, 5, 6, 8, 10, 11, 11, 10, 14, 16, 6, 11, 15, 9, 19, 12. Eyeball-estimate and then compute (a) the mode, (b) the median, and (c) the mean.
17. Use your histogram (or frequency polygon) of the typing speeds in Exercise 14 in Chapter 3. A researcher had taken a random sample of 55 students and measured their typing speeds. The data (in words per minute) are: 8, 24, 20, 20, 17, 18, 16, 19, 17, 29, 25, 27, 14, 5, 21, 36, 11, 16, 29, 20, 11, 17, 26, 10, 11, 5, 5, 19, 28, 7, 15, 8, 14, 32, 32, 12, 7, 12, 13, 30, 19, 16, 42, 26, 16, 30, 21, 8, 4, 23, 5, 15, 19, 9, 30. Eyeball-estimate and then compute (a) the mode, (b) the median, and (c) the mean.
18. Eyeball-estimate the (a) mean, (b) median, and (c) mode of the distribution of lengths of automobiles you sketched in Exercise 17 in Chapter 3.
19. Eyeball-estimate the (a) mean, (b) median, and (c) mode of the distribution of lengths of rock and roll songs you sketched in Exercise 18 in Chapter 3.
20. Eyeball-estimate the mean using the balancing-point method for the histogram shown here.

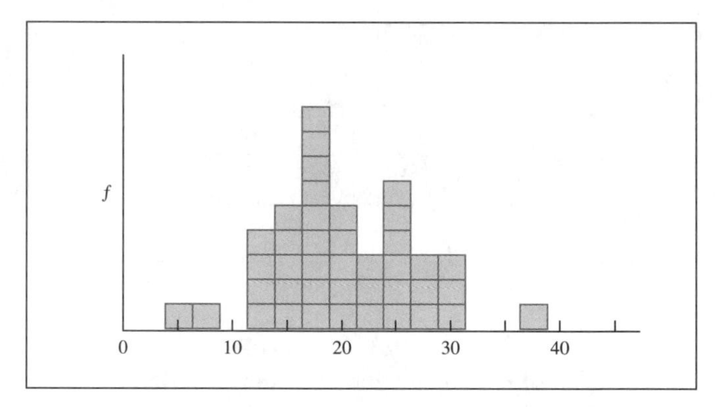

21. Eyeball-estimate the (a) mean, (b) median, and (c) mode of the distribution of the prices of shirts at the local discount store you sketched in Exercise 20 in Chapter 3.
22. Determine whether the following variables are measured at the nominal, ordinal, or interval/ratio level of measurement and state which measure(s) of central tendency are appropriate.
    (a) Letter grades (A, B, C, D, F) on an essay examination
    (b) Numerical grades (highest = 100) on an essay examination
    (c) Number of problems incorrect on a multiple-choice examination
    (d) Type of examination: 1 = multiple choice, 2 = true-false, 3 = fill in the blanks, 4 = short answer, 5 = essay

Section C: Extending your comprehension (answers in Appendix D, page 508)

23. Assume that the values in a data set are all whole numbers (integers) and that you compute the median as recommended in the text (*not* as in Box 4.1). Can you determine

whether there is an odd or even number of elements in the data if the median is known to be the following:

(a) 5.5
(b) 5

24. Assume ten members of a sample have the mean $\overline{X} = 5.7$; their data values are 3, 6, 4, 8, 6, 7, 5, 4, 6. On inspection of these data, you see that there are only nine data points. What is the value of the missing data point?

25. What happens to the sample mean in these cases:
    (a) All the values in a data set are increased by 2 (so the values in Exercise 24, for example, become 5, 8, 6, 10,...).
    (b) All the values in a data set are multiplied by 2 (so the values in Exercise 24, for example, become 6, 12, 8, 16,...).
    (c) All the values in a data set are squared (so the values in Exercise 24, for example, become 9, 36, 16, 64,...).

Section D: Cumulative review (answers in Appendix D, page 508)

26. Sketch by eyeball the distribution of the numbers of letters in the printed lines of this textbook (including only complete, margin-to-margin lines). Eyeball-estimate the mean, median, and mode of this distribution.

27. The student body at a particular university consists of 6400 men and 3600 women whose political party preferences are shown in the table. Suppose a student is selected at random. What is the probability that the student is:
    (a) Male?
    (b) Republican?
    (c) A female Republican?

    Suppose a *woman* is selected at random.
    (d) What is the probability that the student is Republican?
    (e) Explain why the answer in part (c) is different from the answer in part (d).

	Men	Women
Republican	2900	1700
Democrat	3500	1900
	6400	3600

28. Compute for X_i: 4, 2, 6, 3, 7, 3, 5:
    (a) $\sum X_i^2$
    (b) $(\sum X_i)^2$
    (c) Explain why your answers in parts (a) and (b) are different.
    (d) $\sum(X - 3)^2$

Section E: From the journals (answers in Appendix D, page 508)

29. Sobolski and coworkers (1988) were interested (among other things) in comparing the leisure-time physical activity patterns of middle-aged male factory workers in Belgium and Slovakia. They used 1995 Slovak and 2363 Belgian men in their study. Sobolski measured leisure-time physical activity with the Minnesota leisure-time physical activity questionnaire and determined the energy expended in each activity. Energy was measured in kilojoules (kJ) per minute. (The joule, like the calorie, is a measure of energy.) "Light" activities were defined as having energy expenditures of 8.4 to 16.8 kJ per minute, "moderate" activities expended 18.8 to 23.0 kJ per minute, and "heavy" activities required at least 25.0 kJ per minute. The "Total Activity Metabolic Index" (Total AMI) was determined for each man by ascertaining (from the Minnesota questionnaire) how much time per day (in minutes) was spent in each activity, multiplying by the energy expended per minute for that activity to get the energy expended each day in that activity, and then adding across all the activities the man reported. The resultant Total AMI is thus a measure of the leisure-time energy expended per day (expressed in kJ per day).

    The accompanying figure shows the relative frequencies of Total AMI for Belgian (light bars) and Slovak (dark bars) factory workers. Sobolski and colleagues reported that "the median of [Total AMI] is... higher in Slovakian subjects [than in Belgian subjects] (975 and 816 kJ per day [respectively])" (p. 8).

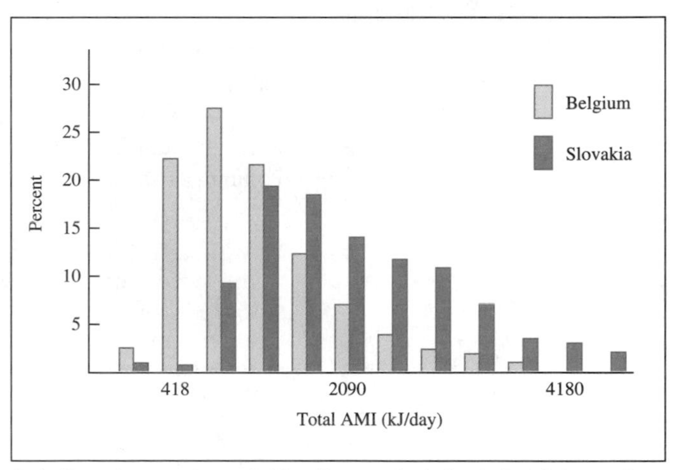

Relative frequency distributions of daily total leisure-time physical activity in two populations (Adapted from Sobolski et al. (1988))

    (a) Describe the distribution in workers in Belgium.
    (b) Describe the distribution in workers in Slovakia.
    (c) Is the modal Total AMI higher in Belgium or in Slovakia?
    (d) Is the median (by eyeball) higher in Belgium or in Slovakia? Does that agree with Sobolski's conclusion?
    (e) Eyeball-estimate the medians in Belgium and in Slovakia. Do your eyeballed medians correspond to the median values reported by Sobolski? Can you explain any discrepancies?
    (f) Is the mean (by eyeball) higher in Belgium or in Slovakia?

Section F: Computer explorations

30. Enter the following values into a statistics computation package (ESTAT's datagen, SPSS, etc.): 1.00, 2.00, 2.00, 3.00, 4.00, 5.00.
    (a) Have the computer compute the mean, median, and mode.
    (b) Change the value "5.00" to "500," as if you had forgotten to enter the decimal point. Describe what happens to the mean, median, and mode. Explain your results.
    (c) Experiment with values other than 500 and other original data sets until you get a feel for the effect of one "outlier" on the mean, median, and mode.
31. Use the original data of Exercise 30.
    (a) Add 1000 to each value in this data set. Describe what happens to the mean, median, and mode. Explain your results.
    (b) Use some constant other than 1000 and do step (a) again, first predicting what your results will be.

Computing Means from Frequency Distributions (Optional)

This section may be omitted without loss of continuity.

Equations (4.1) and (4.2) used to compute sample and population means can be awkward for large data sets because every element in the data set must be added individually, even if elements have the same values. For example, the value 165 in our weight data in Table 4.1 must be added three times because three men weighed 165 pounds. With large data sets (and therefore with high frequencies), it can be convenient to compute means from frequency distributions using these formulas:

mean of a sample computed from a frequency distribution

$$\overline{X} = \frac{\sum fX}{n} \tag{4.1a}$$

mean of a population computed from a frequency distribution

$$\mu = \frac{\sum fX}{N} \tag{4.2a}$$

where we sum over all the *different* values of X and where f is the frequency of those values.

For example, we can compute the sample mean for the data in Table 4.1 by creating a new column to hold the product of f and X (headed "fX") and then adding down that column, as shown in Table 4.7. Then

TABLE 4.7 Computation of the mean from the frequency distribution of male students' weights (Table 4.1 with new column headed "fX")

Weight (X)	f	fX
197	1	197
190	1	190
185	1	185
182	1	182
180	2	360
179	1	179
173	2	346
164	3	492
163	2	326
162	1	162
161	1	161
159	1	159
157	1	157
154	1	154
152	1	152
150	1	150
149	1	149
144	1	144
137	1	137
135	1	135
	$\sum f = 25$	$\sum fX = 4117$

$$\overline{X} = \frac{\sum fX}{n} = \frac{4117}{25} = 164.68 \text{ pounds}$$

exactly what we got from Equation (4.1) in Section 4.3.

For data sets where n is rather small, there is not much (if any) labor savings in using Equation (4.1a) rather than (4.1). When n is large, however, the labor savings can be dramatic.

Occasionally we wish to know the mean in situations where we do not have access to the original data but only to a grouped frequency distribution. Then we cannot compute the mean exactly, but we can obtain a close approximation of it using the following formulas:

approximate mean of a sample computed from a grouped frequency distribution

$$\overline{X} \approx \frac{\sum fX_{mid}}{n} \quad (4.3)$$

approximate mean of a population computed from a grouped frequency distribution

$$\mu \approx \frac{\sum fX_{mid}}{N} \quad (4.4)$$

where X_{mid} is the midpoint of the class interval, obtained exactly as in Chapter 3: midpoint = (lower limit + upper limit)/2. Thus, for example, we can compute the sample mean for the grouped data in Table 4.2 by (1) creating a new column for the midpoint of each interval ("X_{mid}"), (2) creating another new column for the frequency times the midpoint ("fX_{mid}"), and (3) adding down that "fX_{mid}" column, as shown in Table 4.8. Then

$$\overline{X} \approx \frac{\sum fX_{mid}}{n} = \frac{4132.5}{25} = 165.30 \text{ pounds}$$

TABLE 4.8 Computation of the approximate mean from the grouped frequency distribution of male students' weights (Table 4.2 with new columns for the midpoint ("X_{mid}") and for "fX_{mid}")

Class interval	X_{mid}	f	fX_{mid}
190–199	194.5	2	389.0
180–189	184.5	4	738.0
170–179	174.5	3	523.5
160–169	164.5	7	1151.5
150–159	154.5	5	772.5
140–149	144.5	2	289.0
130–139	134.5	2	269.0
		$\sum f = 25$	$\sum fX_{mid} = 4132.5$

Note that this mean computed from the class interval data is close, but not exactly equal, to the result we obtained with the original data in Equation (4.1) or (4.1a) because the method using class intervals does not take into account the exact positions of the values within each class interval.

Exercises for Optional Section

(Answers in Appendix D, page 509)

1. The data in Exercise 6 (page 62) can be displayed in a frequency distribution as shown on the following page.
   (a) Compute the population mean from Equation (4.2a).
   (b) Is the result of part (a) exactly the same as the result of Exercise 6(e)? Why or why not?

Score on first exam	f	Score on first exam	f
96	1	83	2
95	1	82	1
91	3	81	1
90	1	75	1
89	1	72	1
88	2	71	1
86	2	68	1
84	2	61	1
			22

2. The data of Exercise 6 (page 62) can be displayed as a grouped frequency distribution as shown here.
   (a) Compute the approximate population mean from Equation (4.4).
   (b) Is the result of part (a) exactly the same as the result of Exercise 1(a)? Why or why not?

Score on first exam	f
96–99	1
92–95	1
88–91	7
84–87	4
80–83	4
76–79	0
72–75	2
68–71	2
64–67	0
60–63	1
	22

5

CHAPTER

Measures of Variation

LEARNING OBJECTIVES

1. What are three measures of variation? What type of information do they provide?
2. Which measure(s) of variation can be used for the different levels of measurement (nominal, ordinal, or interval/ratio scales)?
3. What is the range? What are its advantages and disadvantages as a measure of variation?
4. What is the standard deviation? Why is it defined as a measure equal to the square root of the mean of the squared deviations?
5. What is an inflection point? How can it be used to eyeball-estimate the standard deviation of a normal distribution?
6. How can you eyeball-estimate the standard deviation from a table?
7. What is the computational formula for the standard deviation? Why is it often easier to use than the mean squared deviation formula?
8. How does the variance relate to the standard deviation?

This chapter describes measures of variation—that is, measures of the width of distributions: the range, the standard deviation, and the variance. It shows that the standard deviation is the square root of the mean of the squared deviations, and it demonstrates that the standard deviation can be eyeball-estimated from either a tabular frequency distribution or a histogram or frequency polygon. The chapter shows that the standard deviation can be computed directly from the mean squared deviation formula or (more easily) from the computational formula.

When we discussed the study on Pygmalion in the classroom at the beginning of Chapter 3, we created histograms to illustrate the distributions of the bloomer and other children's intellectual growth. Those histograms are given here in Figure 5.1. Now we would like to describe the widths of those distributions. The bloomer IQ gains are between −6 and 69 IQ points, so the range (as you will learn in this chapter) is 75 IQ points. The other children's IQ gains are between −15 and 31 IQ points, so their range is smaller: 46 IQ points. The standard deviation (another key concept in this chapter) of the other children is also smaller than that of the bloomers: 10.1 instead of 19.4 IQ points. Thus, both the range and the standard deviation indicate that the width of the other children's intellectual growth distribution is narrower than the bloomers' distribution. You will learn in this chapter the advantages and disadvantages of using the range and the standard deviation as measures of the width of distributions.

We have seen that specifying three characteristics of frequency distributions (the shape, the central tendency, and the width) conveys the essential features of distributions. Chapter 3 introduced the shapes of distributions, and Chapter 4 described in detail the measures (mode, median, and mean) of the central tendency. We now present the measures used to describe the "width" of a distribution, what statisticians call *measures of variation* or sometimes measures of dispersion: the range, the standard deviation, and the variance.[1] In all cases, the measures of variation convey something about how wide a distribution is: how far it is from the smallest point to the largest point, how far it is from the mean to a representative point, and so on.

variation: the width of a distribution; how much values of a variable differ from one another

As with the measures of central tendency discussed in Chapter 4, the level of measurement must be taken into consideration when you choose an appropriate measure of variation. Measures of variation are not appropriate *at all* when the data are nominal; only the range can be used for ordinal data. The range, standard deviation, and variance can all be used when the data are interval/ratio.

[1] For descriptions of other, less frequently used measures of variation, refer to other texts.

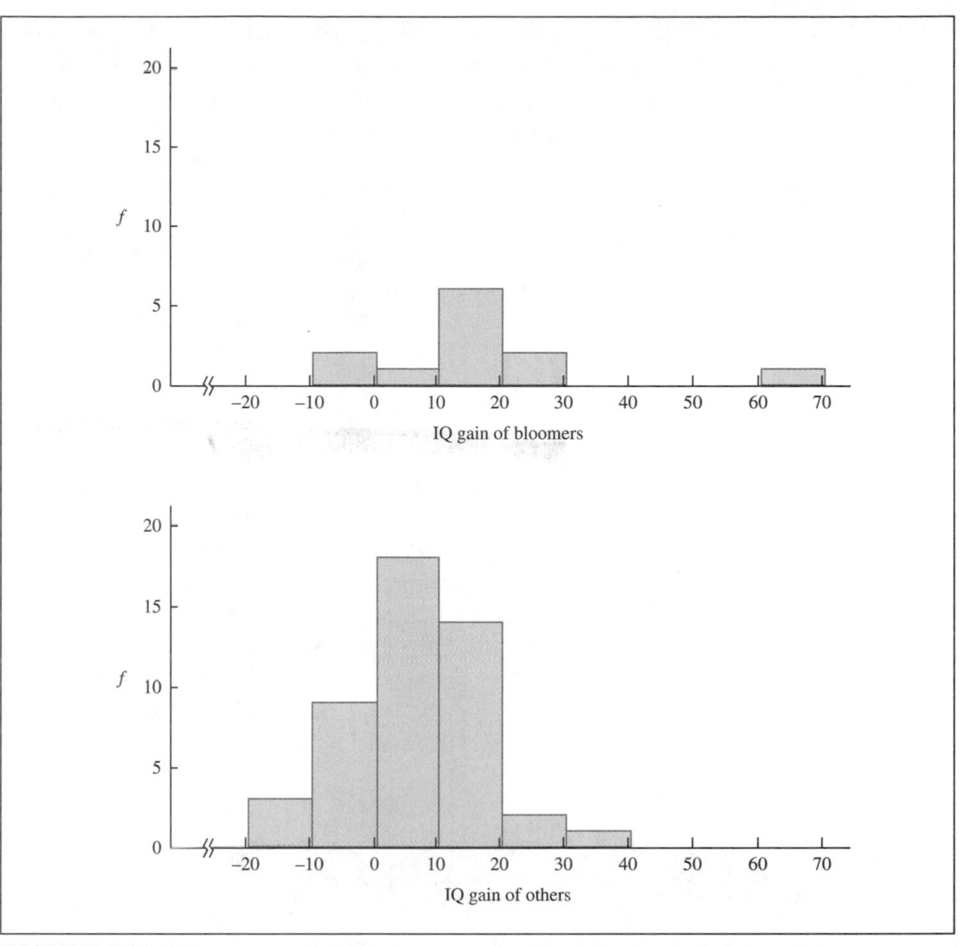

FIGURE 5.1 Histograms of intellectual growth (IQ gain) for Oak School bloomers and other children

5.1 Range

The *range* is the most straightforward measure of the width or variation of a distribution. It is given by the expression[2]

range

$$\text{range} = \text{highest value} - \text{lowest value} \tag{5.1}$$

For example, the weights of male students that were presented in Table 4.1 (and Table 3.3) have as the highest value 197 pounds and the lowest value 135 pounds. The range is therefore 197 − 135 = 62 pounds.

range: a measure of the width of a distribution equal to the highest value minus the lowest value

The range, like other measures of variation, is independent of measures of central tendency in the sense that two distributions can have the same mean but different ranges. Consider these data sets:

Data Set 1: 3, 3, 4, 4, 4, 5, 5

Data Set 2: 1, 2, 3, 4, 5, 6, 7

---

[2] Some texts define the range as (highest value − lowest value) + 1 to include both endpoints in a discrete distribution. However, the " + 1" is correct only when the data are all integers. It would be more correct to define the range as the upper real limit of the highest value minus the lower real limit of the lowest value, but that adds an unnecessary complication to a simple topic. Because statisticians are not unanimous, we will use the simpler definition given in Equation (5.1).

Both data sets have a mean (and also median) of 4, but the range of the first set is $(5 - 3) = 2$, whereas the range of the second is $(7 - 1) = 6$. The range thus conveys that the second data set is "wider" than the first in the sense that the points of Set 2 lie farther from the mean than do the points of Set 1, even though the means are the same.

The range is the simplest measure of variation, but it has the disadvantage of being crude because it depends on only two points in the entire distribution (the lowest and highest). For example, the following two data sets have the same range (6) and the same mean (4), even though in Set 3, most of the points lie very close to (in fact, exactly on) the mean, whereas in Set 4, most of the points lie quite far from the mean.

Data Set 3: 1, 4, 4, 4, 4, 4, 7

Data Set 4: 1, 1, 1, 4, 7, 7, 7

Thus, the range is affected only by the outermost points in a distribution; the positions of the inner points are immaterial.

Another disadvantage of the range as a measure of variation is that it is impossible to define for many important distributions, including the normal distribution that we introduced briefly in Chapter 3. Recall that the normal distribution is asymptotic, which means that it gradually approaches the X-axis without ever actually reaching it. Thus, in principle, the frequency of the normal distribution never becomes exactly 0 no matter how far you go out in the tail of the distribution; there is always one more point somewhere farther out in the tail. The lowest point is thus (in principle) infinitely far out in the left tail (and the highest point infinitely far out in the right tail), so the range is infinite in principle and therefore not an informative statistic for conveying the width of the normal distribution.

5.2 Standard Deviation

We need a measure of the width of a distribution that depends on the distance of *all* the points from the distribution's center, not just the endpoints of the distribution, and thus that is larger for Set 4 (above) than for Set 3 (because more points in Set 4 lie farther from the mean even though the ranges are equal). The *standard deviation* is such a measure, and we shall see that it uses the distance of every point from the distribution's mean.

BOX 5.1 The mean deviation and the average absolute deviation

This box may be omitted without loss of continuity.

This box explains why the average of the deviations is *not* a good measure of the variation of a distribution. (We will give the explanation for samples, but the logic is identical for populations.)

Equation (5.2) is the formula for the deviation of a point in a sample. The *mean deviation*, the average of all deviations, is then given by the expression

$$\text{mean deviation} = \frac{\sum \text{deviation}_i}{n}$$

However, it turns out that this mean deviation is *always* 0, as is shown by reconsidering Set 2, whose mean $\overline{X} = 4$:

Data Set 2: X_i: 1, 2, 3, 4, 5, 6, 7

Deviations in Set 2:

$\text{deviation}_i = X_i - \overline{X} = X_i - 4$: −3, −2, −1, 0, +1, +2, +3

Note that

$$\sum \text{deviation}_i = -3 - 2 - 1 + 0 + 1 + 2 + 3 = 0$$

(continued)

BOX 5.1 *(continued)*

and in fact $\sum \text{deviation}_i = 0$ for every sample or population. This is really just another way of defining the mean: The mean is by definition that unique point about which all deviations add to 0.

Because the positive and negative deviations always cancel one another, the next logical try in creating a useful measure of the width of a distribution might be to find the mean of the *absolute values* of the deviations; statisticians call this measure of a distribution's width the *average absolute deviation*. In symbols,

$$\text{average absolute deviation} = \frac{\sum |X_i - \bar{X}|}{n} = \frac{\sum |\text{deviation}_i|}{n}$$

The average absolute deviation makes logical sense, but absolute values are very difficult to deal with algebraically, so statisticians generally prefer the standard deviation. The average absolute deviation is presented here *only* so that you can see the intermediate steps in the logic behind the definition of the standard deviation.

deviation: the "distance" any point is from the mean

Let us call the distance of any point X_i from the distribution's mean the *deviation*. Then the deviation of the ith point is given by

deviation in a sample

$$\text{deviation}_i = X_i - \bar{X} \tag{5.2}$$

deviation in a population

$$\text{deviation}_i = X_i - \mu \tag{5.3}$$

It might seem that we could obtain a good measure of variation by computing the deviation_i for every point in the distribution and then calculating the mean of these deviations, thus finding the average distance of a point away from the mean. That is an attractive possibility, but Box 5.1 explains that because about half the deviations are positive and half negative, the average deviation is *always* 0 and so it is *not* a useful measure of a distribution's width. What we do instead is first to *square* each deviation (because all the squared deviations will then be positive and not cancel one another), then take the average of all the *squared* deviations, and then find the square root of that average. The standard deviation is the result of that process. Thus, *the standard deviation equals the square root of the mean of the squared deviations.*

standard deviation: a measure of the width of a distribution equal to the square root of the mean of the squared deviations

Mean Squared Deviation Formulas

We call formulas that show that the standard deviation is the square root of the mean of the squared deviations *mean squared deviation formulas*. They follow:

standard deviation of a sample (mean squared deviation formula)

$$s = \sqrt{\frac{\sum (X_i - \bar{X})^2}{n - 1}} \tag{5.4}$$

Equation (5.4a) on page 82 is identical but easier to compute.

where

s = sample standard deviation
$\bar{X}$ = sample mean
n = number of elements in the sample

and

standard deviation of a population (mean squared deviation formula)

$$\sigma = \sqrt{\frac{\sum (X_i - \mu)^2}{N}} \tag{5.5}$$

Equation (5.5a) on page 82 is identical but easier to compute.

where

σ = population standard deviation (Greek lowercase "s," pronounced "sigma")
μ = population mean
N = number of elements in the population

mean squared deviation formula for the standard deviation: a formula for computing the standard deviation that demonstrates that the standard deviation is in fact the square root of the mean of the squared deviations

Note that the mean squared deviation formula for the standard deviation of a population, Equation (5.5), has three differences from the mean squared deviation formula for a sample, Equation (5.4): The standard deviation itself is called σ instead of s (in keeping with our policy of labeling population parameters with Greek letters and sample statistics with Roman letters), the mean about which deviations are taken is the population mean μ instead of the sample mean $\overline{X}$, and the denominator is N instead of $n - 1$. The fact that the denominator of the formula for a sample has 1 subtracted from the sample size requires some explanation, but unfortunately that cannot be done until we discuss the concept of "unbiased estimation" in Chapter 8 (see Box 8.3). For now it must simply be memorized: The denominator of the formula for the standard deviation of a sample is $n - 1$.

Let us give a simple example of computing the standard deviation of a population using Equation (5.5). Assume that we have five values of X_i in our population, as shown in Table 5.1. The mean of that population is given by Equation (4.2):

$$\mu = \frac{\sum X_i}{N} = \frac{15}{5} = 3$$

The deviations $(X_i - \mu)$ are shown in the second column of Table 5.1, and the squared deviations $(X_i - \mu)^2$ are shown in the third column. The sum of the squared deviations is found by adding down the third column: $\sum(X_i - \mu)^2 = 10$. The mean of the squared deviations is thus

$$\frac{\sum(X_i - \mu)^2}{N} = \frac{10}{5} = 2$$

The standard deviation is the square root of the mean of the squared deviations, or

$$\sigma = \sqrt{\frac{\sum(X_i - \mu)^2}{N}} = \sqrt{2} = 1.41$$

TABLE 5.1 Computing the standard deviation of a population using the mean squared deviation formula

X_i	$X_i - \mu$	$(X_i - \mu)^2$
1	−2	4
2	−1	1
3	0	0
4	1	1
5	2	4
$\sum X_i = 15$	0	$\sum(X_i - \mu)^2 = 10$

sum of squares: the sum of the squared deviations

In passing, we note that the sum of the squared deviations, $\sum(X_i - \mu)^2$ [or $\sum(X_i - \overline{X})^2$ for a sample], is often called the *sum of squares*. It is one of the main building blocks in the analysis of variance and more advanced topics in statistics, as we shall see in Chapter 13.

Although the mean squared deviation formulas show directly how the standard deviation is defined, there are somewhat simpler methods of computing standard deviations, called the "computational formulas." We will discuss those later. In keeping with our philosophy that computation should be a refinement of our understanding, we will now explore the visual characteristics of the standard deviation and the techniques used to eyeball-estimate it. We begin by discussing the standard deviation of normal distributions, although the standard deviation can be computed not only for a normal distribution but also for any distribution whose variable is measured at the interval/ratio level of measurement.

Inflection Points of Normal Distributions

We have a sense of the general shape of normal distributions from Chapter 3 (unimodal, symmetric, and asymptotic). We noted in Chapter 4 that the mean, median, and mode of any normal distribution all lie at the same point: exactly in the center of the distribution (that is, the normal distribution's balancing point and median lie directly beneath the highest point of the distribution). We will need to notice here one additional characteristic of normal distributions: that they have two inflection points.

inflection point: the point on any curve where the curvature changes from upward to downward or from downward to upward

An *inflection point* is a point on a graph where the curvature changes. A normal distribution has two inflection points, one on each side of the mean. Consider Figure 5.2. If we follow the plot of the distribution beginning at the leftmost tail and moving toward the right, we see that the plot begins by curving upward; that is, the graph is concave up. As we approach the peak of the plot, the curve begins to level off, and the curvature shifts to being concave down. In every normal distribution, there are exactly two points (one on each side of the distribution) where the curvature changes from upward to downward.

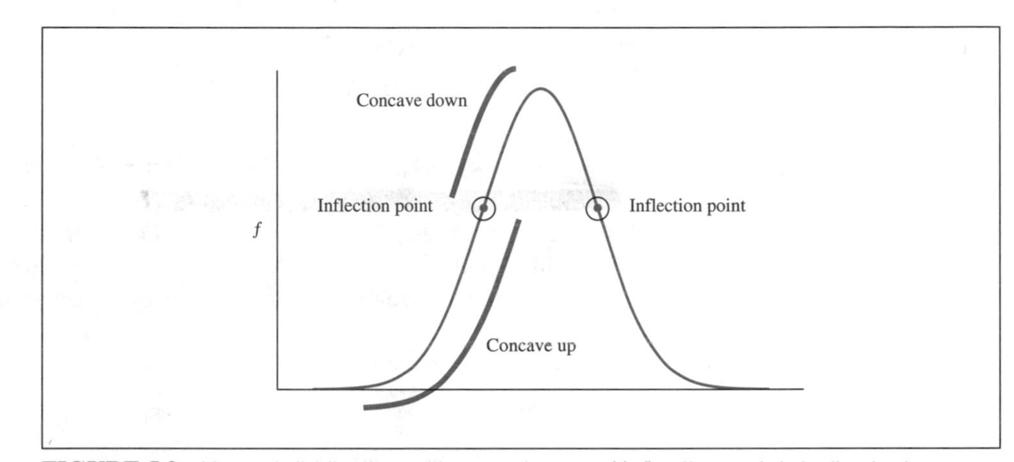

FIGURE 5.2 Normal distribution with curvature and inflection points indicated

There are exactly two inflection points in every normal distribution, regardless of whether it is wide or narrow, tall or short. Figure 5.3 shows another, wider normal distribution with the curvature and inflection points indicated. Notice that the distance from the mean (the center of the distribution) to the inflection point is greater in the wider distribution, and thus the distance between the mean and the inflection point can be taken as a measure of the width of a normal distribution. In fact, *in a normal distribution, the standard deviation is precisely the distance between the mean and either one of the inflection points*.

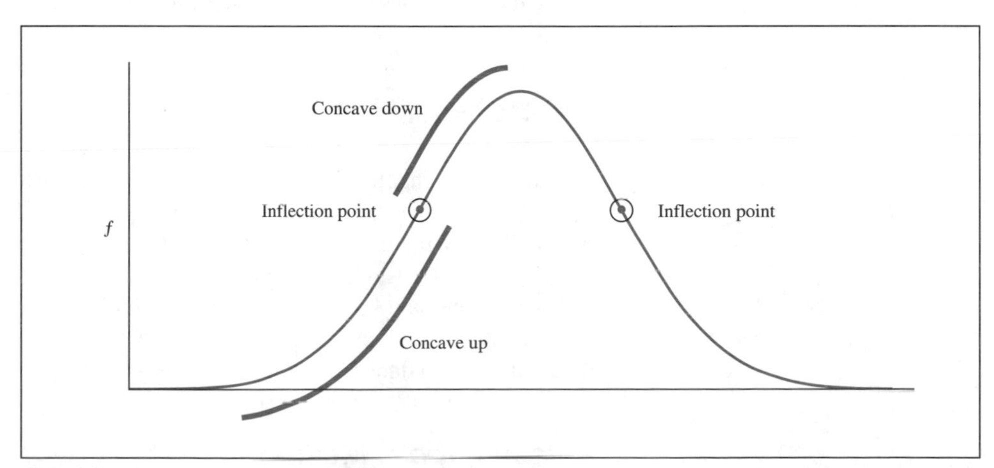

FIGURE 5.3 A wider normal distribution with curvature and inflection points indicated

We illustrate this in Figure 5.4, where we identify the inflection point, transpose the distance between the mean and that inflection point down to the X-axis, and measure the distance in the units of the X-axis. Thus, the standard deviation of this distribution is 1 cm.

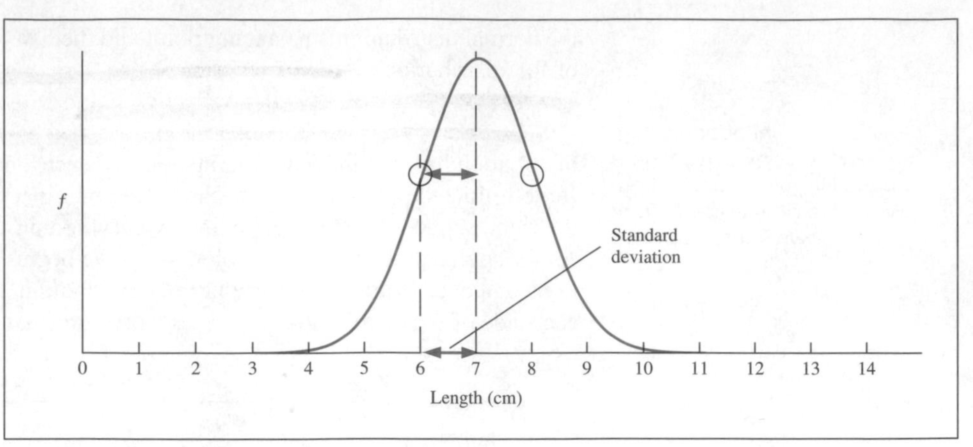

FIGURE 5.4 Normal distribution with standard deviation indicated

Note that the standard deviation is the length of a segment *of the X-axis* and therefore has the same units as the variable on the *X*-axis. Here the variable is measured in centimeters, so the standard deviation is also measured in centimeters. We repeat the process in Figure 5.5 with a wider distribution. Note that the standard deviation of a wider distribution is greater than the standard deviation of a narrow distribution.

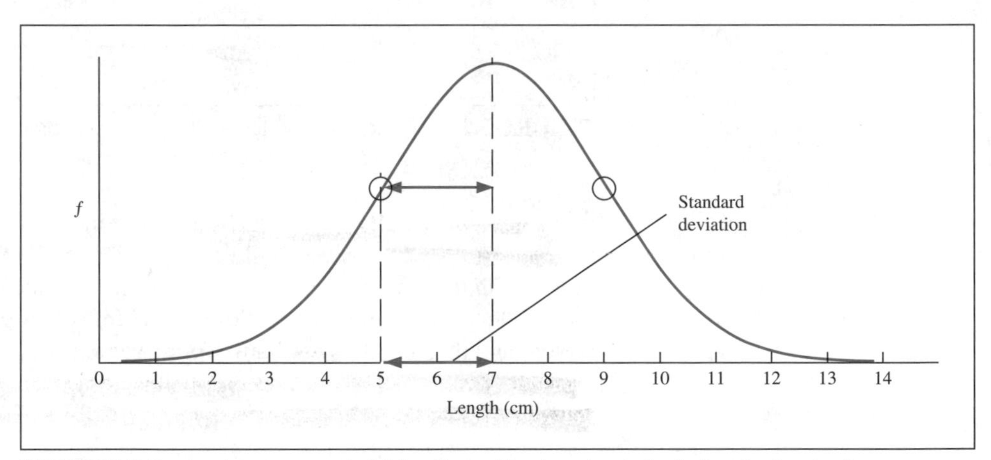

FIGURE 5.5 A wider normal distribution with standard deviation indicated

Eyeball-estimation

Recall that in Chapter 4 we presented two methods for eyeball-estimating the mean: one (called the balancing-point method) that began with a graphical frequency distribution and one (called the range method) that could be used with data presented in a table. The same is true for eyeball-estimating the standard deviation. We will discuss two methods of eyeball-estimating the standard deviation: one (which we call the inflection-point method) that begins with a graphical frequency distribution and one (which we call the range method) that is used with data presented in a table.

Inflection-point Method

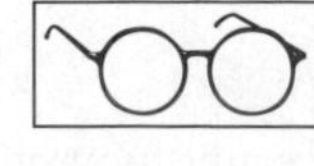

The inflection-point method of estimating the standard deviation begins with a histogram, a frequency polygon, or any graph of a frequency distribution. The procedure is to superimpose as best we can a normal distribution that reflects the shape of the histogram. Then we estimate the standard deviation of the superimposed normal distribution as described earlier.

This process is illustrated in Figure 5.6. Suppose we wish to estimate the standard deviation of the speeds of automobiles on a section of road where the speed limit is 55 mph.

steps for using the inflection-point method to eyeball-estimate the standard deviation

Step 1. Display the data as a histogram (or frequency polygon), as shown in the top portion of Figure 5.6.

Step 2. Superimpose a normal distribution on the histogram, as shown in the middle portion of Figure 5.6. The height of the normal distribution is immaterial; we have made its maximum height the same as the maximum height of the histogram, but we could have made it taller or shorter. Our task is to simulate, as best as our eyeball will allow us, the normal distribution from which the histogram was sampled. [*Hint:* The normal histogram should "slice through" the tops of many of the histogram's bars.]

Step 3. Locate the inflection points and transfer the distance between one of the inflection points and the mean to the *X*-axis, as shown in the bottom portion of Figure 5.6. We see that the right end of the standard deviation bar on the *X*-axis (which is at the mean of the distribution) is at approximately 57 mph, and the left end is at approximately 53 mph. The distance between them is our eyeball-estimate: standard deviation$_{\text{by eyeball}} \approx 57 - 53 = 4$ mph. Note that the standard deviation has the same units as the variable—in this case, miles per hour.

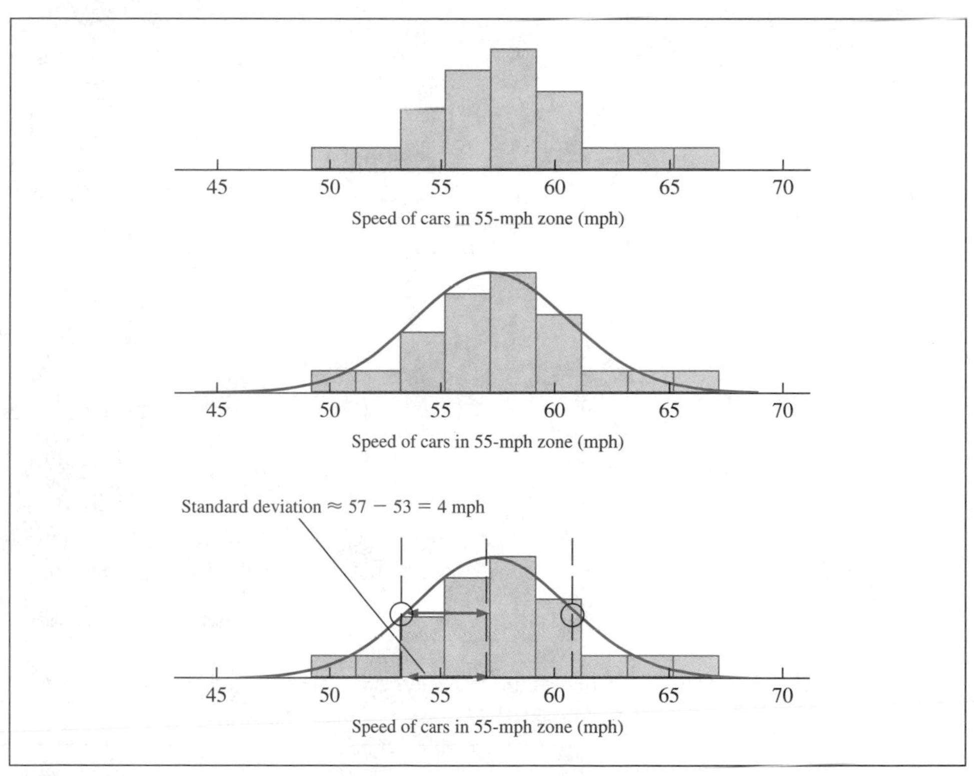

FIGURE 5.6 Three steps in eyeball-estimating the standard deviation by the inflection-point method. *Top:* original histogram. *Middle:* superimposed normal distribution. *Bottom:* inflection points identified and standard deviation indicated.

Figures 5.7, 5.8, and 5.9 illustrate this process for three additional distributions. Figure 5.7 shows the time spent by apprentice chefs preparing a certain recipe. Note that here we use the right-hand inflection point in eyeball-estimating the standard deviation, but we could have chosen the left just as effectively. The standard deviation$_{\text{by eyeball}} \approx 2.5$ minutes. Figures 5.8 and 5.9 provide two more illustrations.

Once you understand this process, return to Figures 4.8 through 4.14 (on pages 54 and 55), which you used in Chapter 4 as eyeball-calibration exercises for estimating the mean by the balancing-point method. Once again, cover the bottom portion of each figure; then use the inflection-point method to eyeball-estimate the standard deviation. That is, superimpose a normal distribution on each histogram, identify the inflection points, and

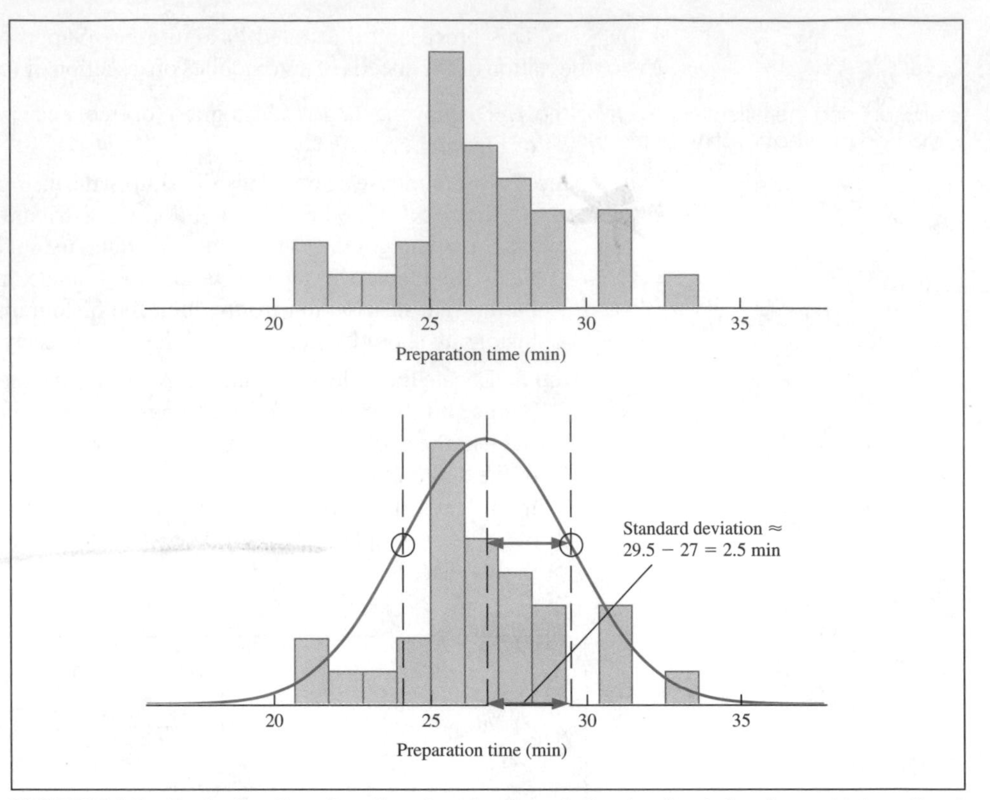

FIGURE 5.7 Eyeball-estimating the standard deviation by the inflection-point method

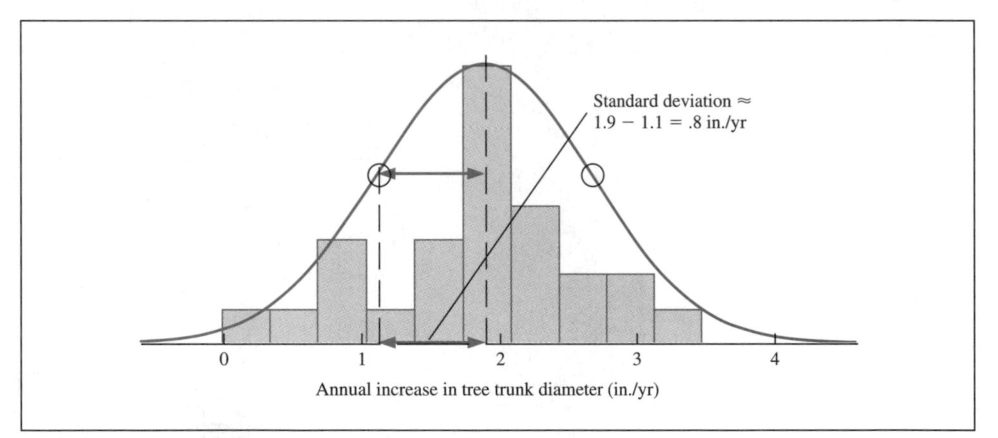

FIGURE 5.8 Eyeball-estimating the standard deviation

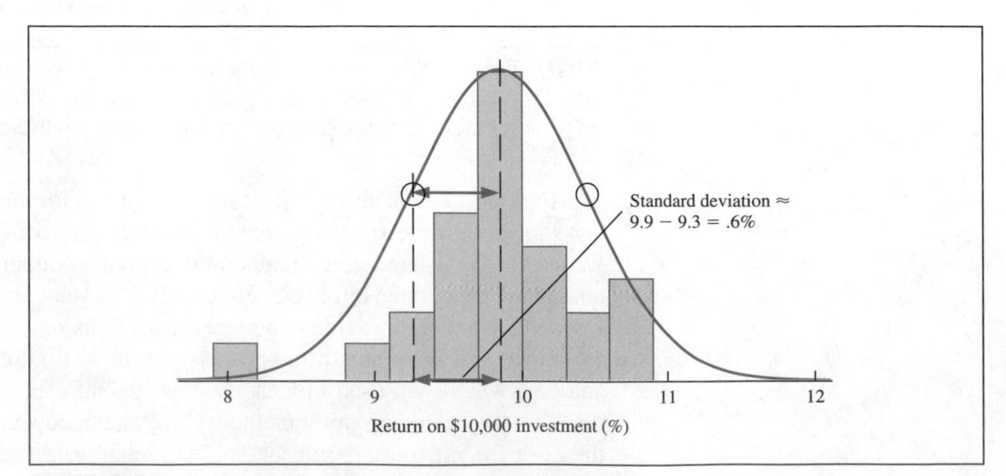

FIGURE 5.9 Eyeball-estimating the standard deviation

transpose the distance between one inflection point and the mean to the X-axis. When you uncover the bottom portion of each figure, you will see a figure shaped like an upside down "T." The vertical center indicates the mean (which we used in Chapter 4). The standard deviation is the distance from the mean to either one of the ends of the horizontal line (thus, the horizontal line is *two* standard deviations long, one in each direction from the mean). Each end of the horizontal line lies directly underneath the inflection point of a correctly superimposed normal distribution. Be sure to estimate the numerical value of the standard deviation (do not forget units). The actual numerical value is given in each figure caption. You should expect that your estimate will be close to but not exactly equal to the actual values provided.

ESTAT

You may wish to use the ESTAT exercise "sdest" at this time.

Range Method

We have seen that it is possible to estimate the standard deviation quite accurately from a histogram (or frequency polygon). As was the case when we eyeball-estimated the mean, it often happens that we wish to know the standard deviation of a data set when only the numerical data themselves, not a histogram, are available. We could, of course, construct a histogram from the data and then estimate the standard deviation. In fact, it is usually a good idea to construct a histogram of any data we are interested in; that's the best way to really understand what the data convey. However, it is also possible to get a rough estimate of the standard deviation from the numerical data themselves by a procedure we call the "range method."

The procedure is simple:

steps for using the range method to eyeball-estimate the standard deviation

Step 1. Scan the data for the lowest and highest values.

Step 2. Subtract the lowest value from the highest value to obtain the range.

Step 3. Divide the range by a number that we call the "range-to-standard-deviation conversion factor."

The range-to-standard-deviation conversion factor used in Step 3 is usually about 4, which tells us that the standard deviation is usually about one-fourth of the range.[3] That approximation holds for moderate sample sizes (between n of about 13 and n of about 30). If n is less than about 13, then the range-to-standard-deviation conversion factor is less than 4 (about 3, so that the standard deviation is about *one-third* of the range), and if n is greater than about 30, then the conversion factor is greater than 4 (about 5, so that the standard deviation is about *one-fifth* of the range).

Thus, the range-to-standard-deviation conversion factor depends on the sample size and is usually about 3, 4, or 5. Remember that the range method is a "quick and dirty" method of approximating the standard deviation, and keep in mind the spirit of eyeball-estimation: The object is to *approximate* the division of Step 3 "in your head," *not* to get an exact result using a calculator. The eyeballed standard deviation is "about a third of the range" or "about a quarter of the range" or "about a fifth of the range," *not* the exact result of the range divided by 3.0, by 4.0, or by 5.0. The accuracy required in eyeball-estimation is about the same as when four people go out to dinner and split the bill of $62. A quarter of $62 is $15 or $16—close enough.

TABLE 5.2 Range-to-standard-deviation conversion factor as a function of sample size

Sample size	*Conversion factor*
2	1.5
3–6	2.5
7–12	3
13–30	4
31–150	5
151–500	6
501+	6.5

My own Monte Carlo study, but see Pearson (1966, pp. 46–61, 189, 200).

Table 5.2 shows the range-to-standard-deviation conversion factors for all sample sizes. This table is useful as a reference, and we will refer back to it for small samples sizes in the optional material for Chapter 13, but for most purposes it is a bit of "overkill."

---

[3] Our discussion of normal distributions in Chapter 6 should give some idea of why the standard deviation is usually about one-fourth of the range. Here is a preview: We shall see in Chapter 6 that unless samples are very large, it is quite unusual to find points in any sample that are more than about 2 standard deviations above or below the mean. This implies that the range of most moderate samples extends from about 2 standard deviations below to 2 standard deviations above the mean; thus, the range of most samples is about 4 standard deviations. To find the approximate standard deviation of most samples, one simply divides the range by 4, as our discussion shows.

Our men's weights data (see Tables 4.1 and 3.3) have a highest value of 197 pounds and a lowest value of 135 pounds, so the range is 62 pounds. The number of data points is 25, so the appropriate conversion factor for that sample size is 4. The standard deviation is then about a quarter of 62—that is, a bit more than 15—so we have standard deviation$_{\text{by eyeball}} \approx 15$ or 16 pounds.

You may wish to use the ESTAT exercise "sdnum" at this time.

Using the Mean Squared Deviation Formula

Population

We have given a simple example of computing the standard deviation using the mean squared deviation formula, Equation (5.5). Now we look at a more realistic example: determining the standard deviation of the weights of our male students. The frequency distribution of these weights was first shown in Table 3.3 and is reproduced here as the first column of Table 5.3. Note that Table 5.3 lists all the points *singly*; that is, if a point has frequency 3 (e.g., 164 pounds), it is listed three times in the table. There are 25 points in the data set, so there are 25 lines in the table.

TABLE 5.3 Computing the standard deviation using the mean squared deviation formula

X_i	$X_i - \mu$	$(X_i - \mu)^2$
197	32.32	1044.5824
190	25.32	641.1024
185	20.32	412.9024
182	17.32	299.9824
180	15.32	234.7024
180	15.32	234.7024
179	14.32	205.0624
173	8.32	69.2224
173	8.32	69.2224
164	−.68	.4624
164	−.68	.4624
164	−.68	.4624
163	−1.68	2.8224
163	−1.68	2.8224
162	−2.68	7.1824
161	−3.68	13.5424
159	−5.68	32.2624
157	−7.68	58.9824
154	−10.68	114.0624
152	−12.68	160.7824
150	−14.68	215.5024
149	−15.68	245.8624
144	−20.68	427.6624
137	−27.68	766.1824
135	−29.68	880.9024
$\sum X_i = 4117$	$\sum(X_i - \mu) = .00$	$\sum(X_i - \mu)^2 = 6141.4400$

For the moment, let us assume that we are interested in describing the variation of weights *only* for our own class (*not* as representative of any other group), which means that we are considering our class to be a population.

Following our philosophy that computations should supply exact values for statistics we have already estimated, we first eyeball-estimate the standard deviation. Using the inflection-point method, we begin with the histogram we drew in Figure 3.3. We superimpose a normal distribution on the histogram, as shown in Figure 5.10. The distance from the

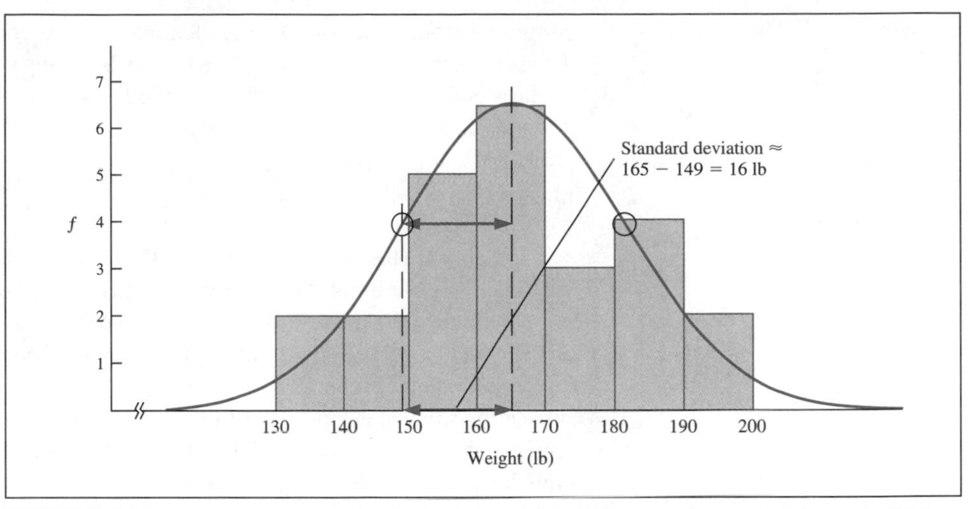

FIGURE 5.10 Histogram of male students' weights with superimposed normal distribution and estimated standard deviation

mean to the inflection point is the standard deviation$_{\text{by eyeball}}$ $\approx 165 - 149 = 16$ pounds. (If we had used the range method, we would have divided the range, 62, by 4, the conversion factor obtained from Table 5.2, to get about 15.5 pounds as the eyeball-estimate of the standard deviation.)

Now that we know the approximate value of the standard deviation, we can use the mean squared deviation equation (5.5) for the standard deviation of a population to compute it. Equation (5.5) is

$$\sigma = \sqrt{\frac{\sum(X_i - \mu)^2}{N}}$$

and to evaluate it, we need to create two new columns headed "$X_i - \mu$" and "$(X_i - \mu)^2$," as shown in Table 5.3. In this example, the X_i values are the students' weights themselves, as shown in the first column of Table 5.3. To create the two new columns, we must first determine the population mean μ from Equation (4.2): $\mu = (\sum X_i)/N$. $\sum X_i = 4117$ is found by adding down the first column. The population mean is thus $\mu = (\sum X_i)/N = 4117/25 = 164.68$ pounds.

The second column of Table 5.3 shows the deviation from the population mean $(X_i - \mu)$ for each point in the population. For the first point, $X_i - \mu = 197 - 164.68 = 32.32$. As a check, we compute the sum of the deviations $\sum(X_i - \mu)$ by adding down this column; that sum must equal 0 as was demonstrated in Box 5.1.

The third column of Table 5.3 lists the square of each deviation $(X_i - \mu)^2$. The sum of the squared deviations (or "sum of squares") $\sum(X_i - \mu)^2 = 6141.4400$ is obtained by adding down the third column. We substitute that value into Equation (5.5) to obtain the population standard deviation:

$$\sigma = \sqrt{\frac{\sum(X_i - \mu)^2}{N}} = \sqrt{\frac{6141.4400}{25}} = \sqrt{245.6576} = 15.67 \text{ pounds}$$

Recall that our eyeball-estimates were 16 pounds (by the inflection-point method) and 15.5 pounds (by the range method), which gives us confidence that we have performed the computations correctly.

Sample

Suppose that the men in our class whose weights we have been considering were selected randomly from the entire university, and it is this larger group of students in whom we are interested. We now consider this class to be a sample from a larger population, and we compute the standard deviation using the sample form of the mean squared deviation formula, Equation (5.4). The deviations are computed around $\overline{X}$ instead of μ, but numerically $\overline{X}$ as computed from Equation (4.1) is identical to μ as computed from Equation (4.2), so the deviations are identical for this group taken as a sample or as a population. The computation of the standard deviation for this sample is therefore identical to that shown above for a population, except the denominator is $n - 1 = 25 - 1 = 24$ instead of $N = 25$:

$$s = \sqrt{\frac{\sum(X_i - \overline{X})^2}{n-1}} = \sqrt{\frac{6141.4400}{25-1}} = \sqrt{255.8933} = 16.00 \text{ pounds}$$

We noted in the discussion of eyeball-estimation that the units of the standard deviation are the same as the units of the variable X. This is because, as in our example, if X is measured in pounds, then the mean $\overline{X}$ (or μ) is also measured in pounds, and the deviation $(X_i - \overline{X})$ or $(X_i - \mu)$ is measured in pounds. The squared deviations are then measured in pounds^2 ("pounds squared"; when we square a value, we also square the units of that value), and the sum of the squared deviations is therefore also in pounds^2. Because $n - 1$ (and N) are both unitless, the mean squared deviation has units of pounds^2. Taking the square root brings the units back from pounds^2 to pounds, so the standard deviation here is measured in pounds.

Using Computational Formulas

The mean squared deviation method of computing standard deviations, Equations (5.4) and (5.5), shows exactly how the standard deviation is derived: The standard deviation *is* the square root of the mean squared deviations. Somewhat simpler ways of computing standard deviations, called the "computational formulas," have the advantage of requiring fewer computations (and the disadvantage of making it somewhat less obvious what the standard deviation is measuring).

computational formula for the standard deviation: a formula for computing the standard deviation that is somewhat easier to use than the mean squared deviation formula

The mean squared deviation formulas are on page 73.

If we substitute $\sum X_i/n$ for $\overline{X}$ in Equation (5.4) [or $\sum X_i/N$ for μ in Equation (5.5)] and then algebraically simplify the resulting equations, we can show that the following formulas are identical to Equations (5.4) and (5.5). We call them Equations (5.4a) and (5.5a), respectively:

standard deviation of a sample (computational formula)

$$s = \sqrt{\frac{\sum X_i^2 - \frac{(\sum X_i)^2}{n}}{n-1}} \tag{5.4a}$$

standard deviation of a population (computational formula)

$$\sigma = \sqrt{\frac{\sum X_i^2 - \frac{(\sum X_i)^2}{N}}{N}} \tag{5.5a}$$

Note that there are three differences between Equations (5.4a) and (5.5a): The standard deviation is called either s or σ, the number of elements is either n or N, and the main denominator is either $n - 1$ or N. Those are the identical distinctions we made between Equations (5.4) and (5.5). Note also that the form of the two *numerators* is the same: The second term of the numerator is divided by the number of elements (*not* divided by the number of elements minus 1).

Population

Let us use the computational formula to compute the standard deviation of the simple example shown in Table 5.1; the data are redisplayed as the first column of Table 5.4. Recall that we used the mean squared deviation formula to find that $\sigma = 1.41$.

To use the computational formula, Equation (5.5a), we need to create another column in Table 5.4: the square of each X_i. We can then obtain $\sum X_i^2 = 55$ by adding down that column. The standard deviation is then

TABLE 5.4 Computing the standard deviation of a population using the computational formula

X_i	X_i^2
1	1
2	4
3	9
4	16
5	25
$\sum X_i = 15$	$\sum X_i^2 = 55$

$$\sigma = \sqrt{\frac{\sum X_i^2 - \frac{(\sum X_i)^2}{N}}{N}} = \sqrt{\frac{55 - \frac{(15)^2}{5}}{5}} = \sqrt{\frac{10}{5}} = 1.41$$

the same result we obtained from the mean squared deviation formula.

TABLE 5.5 Computing the standard deviation using the computational formula

X_i	X_i^2
197	38,809
190	36,100
185	34,225
182	33,124
180	32,400
180	32,400
179	32,041
173	29,929
173	29,929
164	26,896
164	26,896
164	26,896
163	26,569
163	26,569
162	26,244
161	25,921
159	25,281
157	24,649
154	23,716
152	23,104
150	22,500
149	22,201
144	20,736
137	18,769
135	18,225
$\sum X_i = 4117$	$\sum X_i^2 = 684{,}129$

Note that the computational formula is simpler to use than the mean squared deviation formula because it requires only one additional column (X_i^2) instead of the two required by the mean squared deviation formula [$(X_i - \mu)$ and $(X_i - \mu)^2$]. For simple data sets, this is not much of a savings, but for larger data sets, the savings can be considerable.

Sample

We demonstrate the computational formula for a sample using the weight data (reproduced here as the first column of Table 5.5). Recall that we used the mean squared deviation formula to find $s = 16.00$ pounds. To compute the standard deviation from the computational formula, Equation (5.4a), we need two sums: $\sum X_i$ and $\sum X_i^2$. These two sums are found by generating a table with columns headed X_i and X_i^2, as shown in Table 5.5.

We add down the first column to get $\sum X_i = 4117$ and down the second column to get $\sum X_i^2 = 684{,}129$. We substitute those numbers directly into Equation (5.4a):

$$s = \sqrt{\frac{\sum X_i^2 - \frac{(\sum X_i)^2}{n}}{n-1}} = \sqrt{\frac{684{,}129 - \frac{(4117)^2}{25}}{25-1}} = \frac{684{,}129 - 677{,}987.56}{24}$$

$$= \sqrt{255.8933} = 16.00 \text{ pounds}$$

which is the same value that we arrived at using Equation (5.4).

Note how many fewer computations are required for the computational formula [Equation (5.4a), two columns in Table 5.5] than for the mean squared deviation formula [Equation (5.4), three columns in Table 5.3].

5.3 Variance

variance: a measure of the width of a distribution equal to the mean of the squared deviations; it is the square of the standard deviation

The final measure of the variation of a distribution is the *variance*, which is simply the mean of the squared deviations. The variance is thus the square of the standard deviation; therefore, we call the variance of a sample s^2 and the variance of a population σ^2. The variance and the standard deviation contain exactly the same information about the variation of a distribution, but sometimes the standard deviation and sometimes the variance is more convenient to use. At first, the standard deviation is more convenient because we can easily visualize it as the distance from the mean to an inflection point of a normal distribution. Later, as in Chapter 13, the variance will become more convenient.

The formulas for the variance are the same as those for the standard deviation without the square root sign. As with the standard deviation, there are two pairs of formulas: for samples or populations, and mean squared deviation or computational formulas. The mean squared deviation formulas for the variance are:

variance of a sample (mean squared deviation formula)

$$s^2 = \frac{\sum(X_i - \bar{X})^2}{n - 1} \tag{5.6}$$

variance of a population (mean squared deviation formula)

$$\sigma^2 = \frac{\sum(X_i - \mu)^2}{N} \tag{5.7}$$

The computational formulas for the variance, which provide identical results but are easier to compute if data sets are large, are:

variance of a sample (computational formula)

$$s^2 = \frac{\sum X_i^2 - \frac{(\sum X_i)^2}{n}}{n - 1} \tag{5.6a}$$

variance of a population (computational formula)

$$\sigma^2 = \frac{\sum X_i^2 - \frac{(\sum X_i)^2}{N}}{N} \tag{5.7a}$$

Note that the units of the variance are the units of the standard deviation *squared.* For example, if the standard deviation is measured in seconds, then the variance is measured in seconds^2 (read "seconds squared").

For the male students' weights, if we assume the class is a population, then we can use either Equation (5.7) or (5.7a) to compute the variance. From Equation (5.7a), we obtain

$$\sigma^2 = \frac{\sum X_i^2 - \frac{(\sum X_i)^2}{N}}{N} = \frac{684{,}129 - \frac{(4117)^2}{25}}{25} = 245.66 \text{ pounds}^2$$

The variance can also be obtained by squaring the standard deviation (and the standard deviation is the square root of the variance). Thus, $\sigma = \sqrt{\sigma^2} = \sqrt{245.66} = 15.67$ pounds. We could have gotten the same result from Equation (5.5) or (5.5a).

5.4 Eyeball-calibration for Distributions

In this and the preceding two chapters, we have discussed the three major characteristics of distributions: their shape (in Chapter 3), their central tendency (in Chapter 4), and their variation (in this chapter). You should now be able to sketch distributions given just the three characteristics.

Before you look at the next page, sketch the following distributions for yourself, *being sure to label the axes.* Then check your results in the figures indicated.

1. A normal distribution with mean 15 inches and standard deviation 5 inches (Figure 5.11)
2. A bimodal distribution with modes at 30 mph and 50 mph and range 60 mph (Figure 5.12)
3. A unimodal distribution with mode \$120 and mean \$140; must this distribution be skewed? (Figure 5.13)
4. An extremely negatively skewed distribution with mean 60 seconds and range about 80 seconds (Figure 5.14)

5.5 Connections

Cumulative Review

In Chapter 1, we introduced the notion that the three basic concepts to master in introductory statistics are: (1) distributions of variables, (2) distributions of means, and (3) distributions of the test statistic. We are nearing the end of our examination of the first concept, the distributions of variables. We have seen that distributions of variables can be characterized by three features: their shape (Chapter 3), their center (Chapter 4), and their width (Chapter 5).

We have also seen that it is necessary to distinguish carefully between samples and populations because statistics is, in general, the science of saying something meaningful about populations when all we know about are samples.

Journals

The sample standard deviation, which we have called s, is often called *SD* in journal reports of experiments. Thus, a report about the Pygmalion experiment might read something like "The distribution of intellectual growth among the bloomers ($M = 16.5$, $SD = 19.4$ IQ points) showed considerable overlap with that of the other children ($M = 7.0$, $SD = 10.1$ IQ points)."

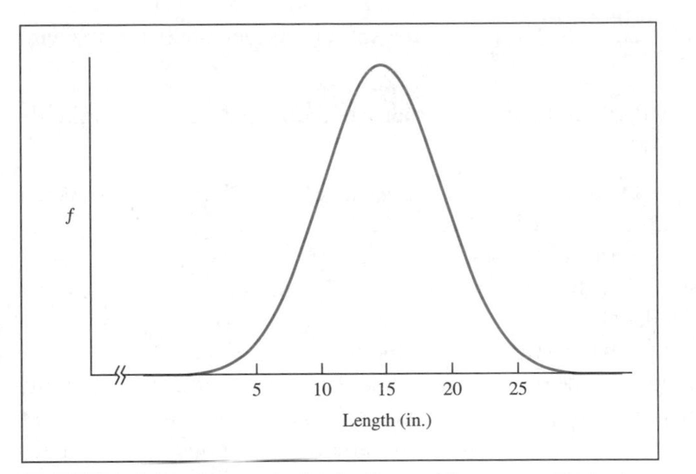

FIGURE 5.11 Normal distribution with mean 15 inches and standard deviation 5 inches

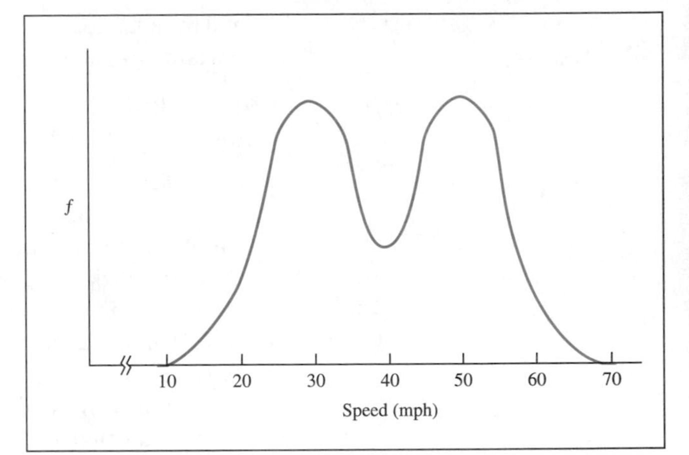

FIGURE 5.12 Bimodal distribution with modes at 30 mph and 50 mph and range 60 mph

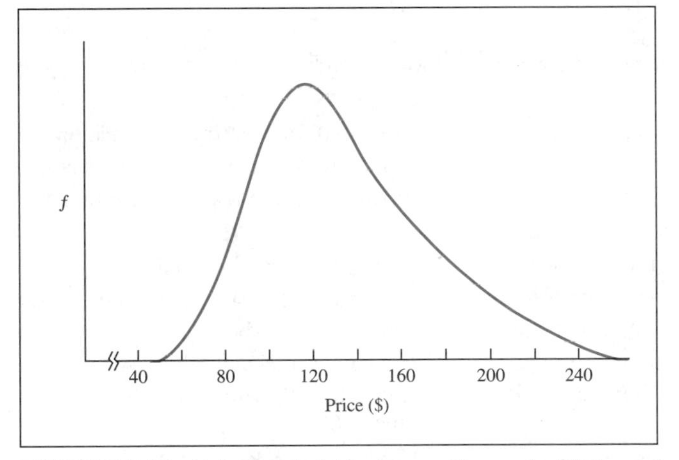

FIGURE 5.13 Unimodal distribution with mode $120 and mean $140

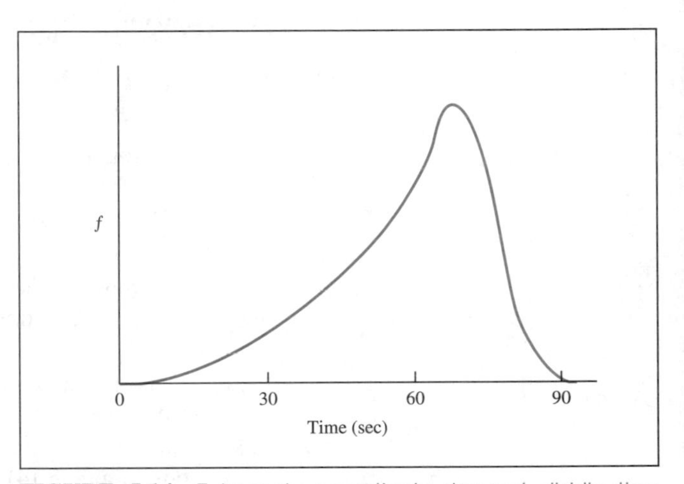

FIGURE 5.14 Extremely negatively skewed distribution with mean 60 seconds and range about 80 seconds

Occasionally "*SD*" refers to the population standard deviation σ, and it often requires some careful detective work to ascertain whether the author means s or σ.

We are calling s the "sample standard deviation"; some authors quite correctly refer to s as the "unbiased estimator of the population standard deviation."

Computers

Use ESTAT's datagen to determine the standard deviation, variance, and range of the weight data that originally appeared in Table 3.3:

1. Open the file you saved in Chapter 4: Click `Open`, click `table3-2.txt`, and then click `OK`.
2. Note that the sample standard deviation s appears automatically in the `datagen Descriptive Statistics` window in the row labeled "Standard deviation."
3. Note that datagen does not provide the variance. You must use your calculator to square the standard deviation.
4. Note that datagen does not compute the range. Datagen (version 2.0 or later) can help you determine the range by sorting the data for you: Click any point in the first column of the datagen spreadsheet to select the first variable, click `Edit Variable`, click the `Sort them in descending order` button, and then click `OK`. Use your calculator to compute the range: range = highest value − lowest value.
5. Note that datagen does not provide the population standard deviation. You can obtain the population standard deviation from the sample standard deviation using your calculator

and the equation $\sigma = s\sqrt{(n-1)/N}$. The population variance is the population standard deviation σ squared.

Use SPSS to determine the standard deviation and variance of the weight data that originally appeared in Table 3.3:

1. Open the weight data file that you saved in Chapter 2: Click `File`, then `Open`, and then `Data....` Click `table3-2.sav`. Click `OK`.
2. Click `Statistics`, then `Summarize`, and then `Descriptives....`
3. Click ▶, click `Options`, click the `Variance` checkbox and the `Range` checkbox, click `Continue`, and then click `OK`.
4. Note that the statistics appear in the `!Output1` window.
5. Note that SPSS does not provide the population standard deviation. You can obtain the population standard deviation from the sample standard deviation using your calculator and the equation $\sigma = s\sqrt{(n-1)/N}$. The population variance is the population standard deviation σ squared.

Homework Tips

1. Check the list of learning objectives at the beginning of this chapter. Do you understand each one?

2. The four equations for the standard deviation are listed for your convenience on the inside of the front cover. Simply remove the square root sign to convert a formula for s to a formula for s^2 (do the same for σ and σ^2). All the formulas in the textbook are listed in Appendix C.

3. The most frequent mistake students make in using the inflection-point method to eyeball-estimate the standard deviation is to make the superimposed normal distribution too wide. Here's a hint: The inflection point should never be more than about halfway between the mean and the most extreme point in the histogram. (This hint is actually a corollary of the range method, which says that the standard deviation is usually about one-fourth of the range.)

4. Remember that there are two ways to eyeball-estimate the standard deviation. Use the inflection-point method if you have a histogram or frequency polygon, but use the range method if you have a table.

5. Remember the spirit of eyeball-estimation: Determine the values only approximately and in your head, not using a calculator.

6. The main computational mistake that students make in this chapter is to fail to distinguish between $\sum(X_i - \mu)^2$ and $[\sum(X_i - \mu)]^2$. The mean squared deviation formulas for the standard deviation and variance use sums of the first kind, *not* of the second kind. To use the mean squared deviation formula, you need to create a column headed "$X_i - \mu$" and another column headed "$(X_i - \mu)^2$" and then sum down that last column. The same is true for the sample formulas, which substitute $\overline{X}$ for μ.

7. Remember that the units of the variance are the units of the standard deviation (and therefore the units of the variable) *squared*. Thus, if the variable is measured in pounds, then the standard deviation is in pounds and the variance is in pounds *squared*.

Exercises for Chapter 5

Section A: Basic exercises (answers in Appendix D, page 509)

1. Consider the data X_i: 3, 5, 3, 4, 5, 4, which represent a sample.
   (a) What is the range?
   (b) Eyeball-estimate the standard deviation using the range method.
   (c) Compute the standard deviation using the mean squared deviation formula.
   (d) Explain any discrepancy between your eyeball-estimation and computation.

(e) Compute the standard deviation using the computational formula.
(f) Is your result in part (e) the same as that in part (c)? Why?
(g) Compute the variance.

2. Consider the data X_i: 2, 5, 8, 6, 1, which represent a population.
(a) What is the range?
(b) Eyeball-estimate the standard deviation using the range method.
(c) Compute the standard deviation using the mean squared deviation formula.
(d) Explain any discrepancy between your eyeball-estimation and computation.
(e) Compute the standard deviation using the computational formula.
(f) Is your result in part (e) the same as that in part (c)? Why?
(g) Compute the variance.

3. Suppose you wish to tell a friend about the variability of temperatures you experienced on your seven-day vacation in Florida in the first week of July. You list these daily high temperatures (in °F): 74, 82, 79, 82, 82, 83, 81.
(a) What was the range of temperatures?
(b) Eyeball-estimate the standard deviation by the range method.
(c) Which formula is appropriate for computing the standard deviation? Why?
(d) Compute the standard deviation.
(e) Account for any discrepancy between your eyeball-estimate and computation.
(f) Compute the standard deviation using the mean squared deviation formula, and show that your result is identical to that obtained from the computational formula.
(g) Which formula is appropriate for computing the variance? Why?
(h) Compute the variance using the computational formula.
(i) Compute the variance using the mean squared deviation formula, and show that your result is identical to that obtained from the computational formula.

Do not forget to include units [here, °F for the standard deviation and $(°F)^2$ for the variance] in your answers.

4. Suppose you wish to vacation in Alaska during the first week of July, and you want to know what range of temperatures to expect. You go to the library and select July newspapers at random, one from each of the last six years, and find these daily high temperatures (in °F) on each day: 69, 58, 63, 66, 62, 66.
(a) What is the range of these temperatures?
(b) Eyeball-estimate the standard deviation by the range method.
(c) Which formula is appropriate for computing the standard deviation? Why?
(d) Compute the standard deviation.
(e) Compute the standard deviation using the mean squared deviation formula, and show that your result is identical to that obtained from the computational formula.
(f) Account for any discrepancy between your eyeball-estimates and computations.
(g) Which formula is appropriate for computing the variance? Why?
(h) Compute the variance using the computational formula.
(i) Compute the variance using the mean squared deviation formula, and show that your result is identical to that obtained from the computational formula.

5. Exercise 1 in Chapter 3 asked you to draw a histogram and frequency polygon for the following distances ridden (in miles) by a sample of 20 students who rode a stationary bicycle for 30 minutes: 8, 15, 22, 17, 9, 16, 15, 15, 21, 12, 14, 16, 22, 17, 16, 11, 13, 15, 20, 15.
(a) Eyeball-estimate the standard deviation by the range method.
(b) A histogram for these data (like the one you produced in Chapter 3) is shown here. Eyeball-estimate the standard deviation using the inflection-point method.

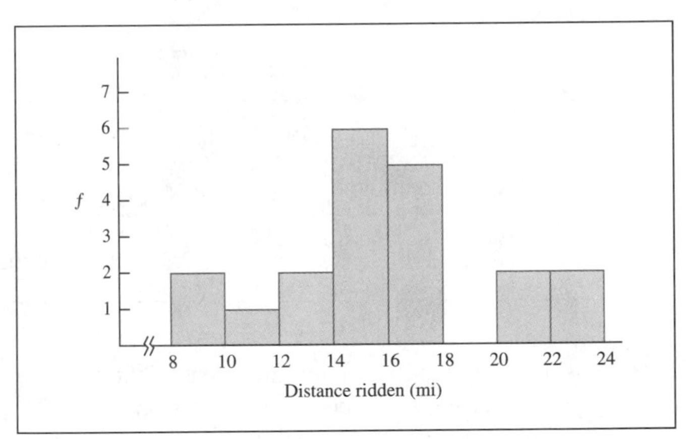

(c) A frequency polygon for these data (like the one you produced in Chapter 3) is shown here. Eyeball-estimate the standard deviation using the inflection-point method. Is your estimate approximately the same as in part (b)?

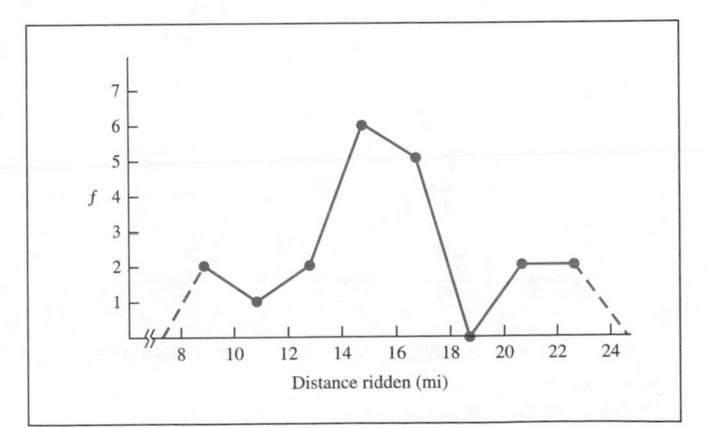

(d) Which formula is appropriate for computing the standard deviation? Why?
(e) Compute the standard deviation.
(f) Account for any discrepancies between eyeball-estimates and computations.
(g) Which formula is appropriate for computing the variance?
(h) Compute the variance.

6. In Exercise 3 in Chapter 3, we presented the scores of all students on the first exam in a statistics class: 81, 86, 91, 75, 96, 82, 88, 88, 71, 68, 72, 61, 84, 86, 95, 91, 84, 83, 91, 83, 89, 90. Assume that this class is your main interest (that is, you do not wish to assume that this class is representative of any larger group).
(a) Eyeball-estimate the standard deviation by the range method.
(b) A histogram for these data (like the one you produced in Chapter 3) is shown here. Eyeball-estimate the standard deviation using the inflection-point method.

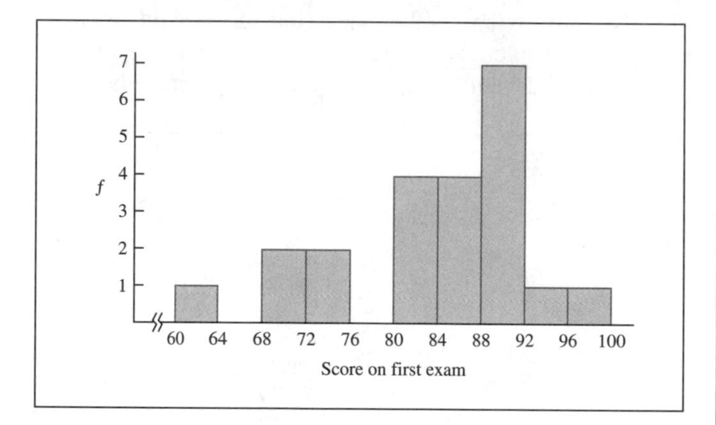

(c) A frequency polygon for these data (like the one you produced in Chapter 3) is shown here. Eyeball-estimate the standard deviation using the inflection-point method. Is your estimate approximately the same as in part (b)?

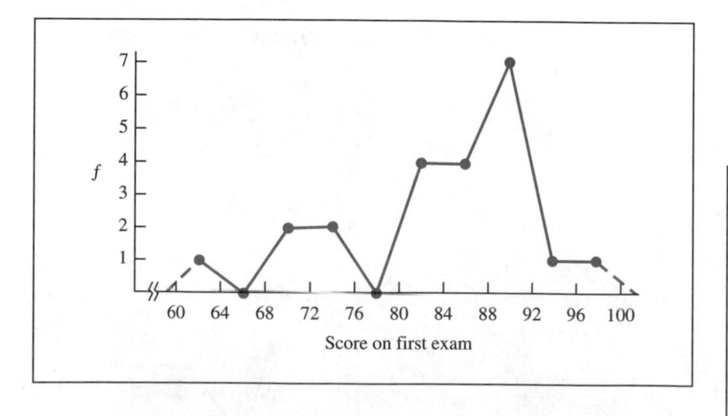

(d) Which formula is appropriate for computing the standard deviation? Why?
(e) Compute the standard deviation.
(f) Account for any discrepancies between eyeball-estimates and computations.
(g) Which formula is appropriate for computing the variance?
(h) Compute the variance.

7. Eyeball-estimate the standard deviation using the inflection-point method for the histogram shown here.

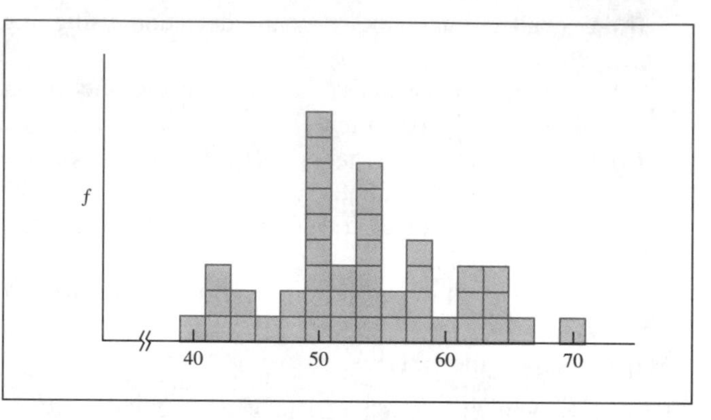

8. The distribution of musicians' vibrato (recall Exercise 5 in Chapter 3) is sketched here. Eyeball-estimate the standard deviation using the inflection-point method.

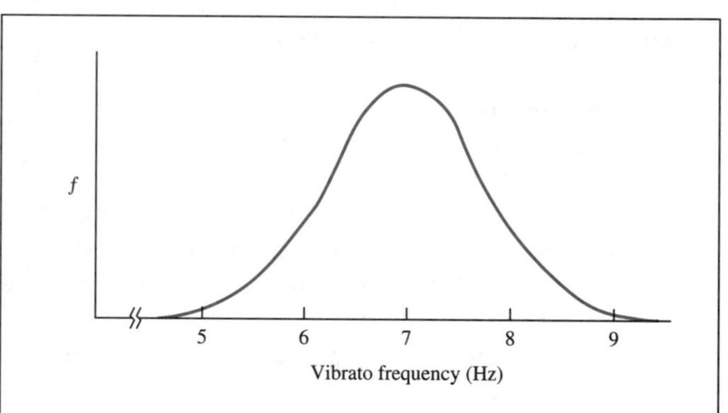

9. The distribution of the amount of liquid in Coke bottles (recall Exercise 6 in Chapter 3) is sketched here. Eyeball-estimate the standard deviation using the inflection-point method.

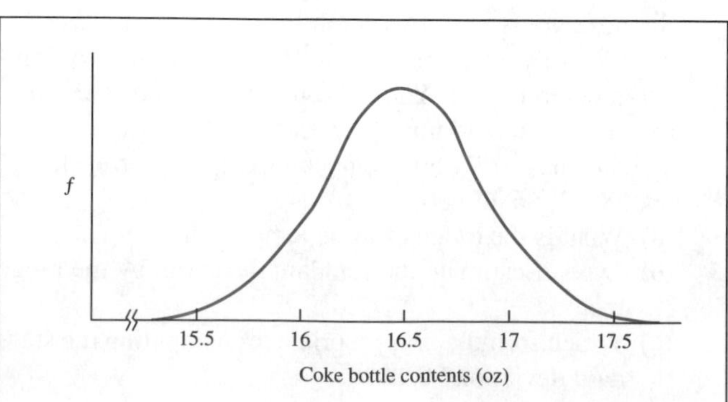

10. The distribution of the lengths of McDonald's french fries (recall Exercise 7 in Chapter 3) is shown here. Eyeball-estimate the standard deviation using the inflection-point method.

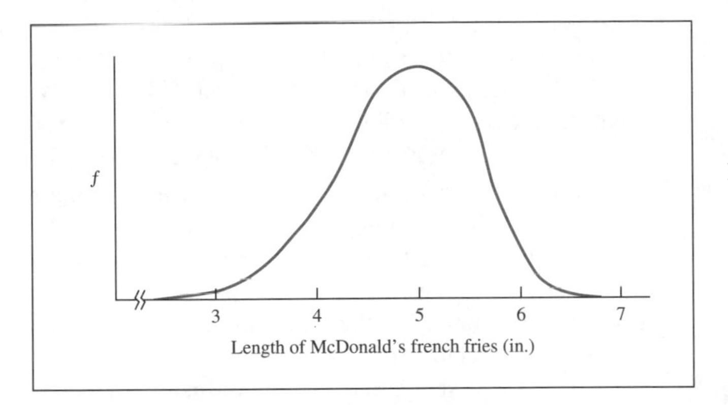

11. The distribution of the lengths of Royal Perfecto french fries (recall Exercise 8 in Chapter 3) is shown here. Eyeball-estimate the standard deviation using the inflection-point method.

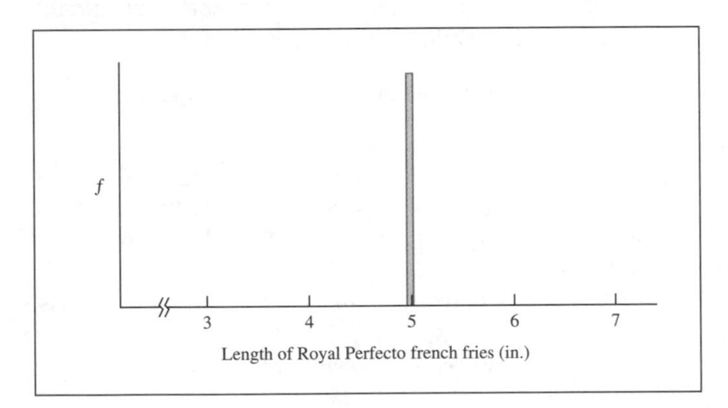

Section B: Supplementary exercises

12. Use the data X_i: 2, −4, 0, 1, 3, −2, which represent a sample.
    (a) What is the range?
    (b) Eyeball-estimate the standard deviation using the range method.
    (c) Compute the standard deviation using the mean squared deviation formula.
    (d) Explain any discrepancy between your eyeball-estimation and computation.
    (e) Compute the standard deviation using the computational formula.
    (f) Is your result in part (e) the same as that in part (c)? Why?
    (g) Compute the variance.
13. Use the data X_i: 3, 5, 7, 2, 9, 11, 4, 1, which represent a population.
    (a) What is the range?
    (b) Eyeball-estimate the standard deviation using the range method.
    (c) Compute the standard deviation using the mean squared deviation formula.
    (d) Explain any discrepancy between your eyeball-estimation and computation.
    (e) Compute the standard deviation using the computational formula.
    (f) Is your result in part (e) the same as that in part (c)? Why?
    (g) Compute the variance.
14. The accompanying table shows for a sample of ten heart attack patients the number of days they stayed in the hospital on their first admission.
    (a) Compute the mean.
    (b) Compute the range.
    (c) Eyeball-estimate and compute the standard deviation using the mean squared deviation formula.
    (d) Compute the standard deviation using the computational formula.
    (e) Compute the variance.

Patient	*Duration*
1	9
2	13
3	9
4	7
5	12
6	17
7	10
8	10
9	11
10	14

15. (*Exercise 14 continued*) Suppose that the hospital, for dishonest insurance purposes, adds one day to each patient's stay (so the data become 10, 14, 10, 8, . . .). How does that change each measure?
    (a) Mean
    (b) Range
    (c) Standard deviation
    (d) Variance
16. (*Exercise 14 continued*) Suppose that the hospital, again for dishonest purposes, *doubles* each patient's original stay (so the data become 18, 26, 18, 14, . . .). How does that change each measure?
    (a) Mean
    (b) Range
    (c) Standard deviation
    (d) Variance
17. For the hospital-stay data of Exercise 14, show that $\sum \text{deviation}_i = 0$ (where $\text{deviation}_i = X_i - \overline{X}$).
18. Assume that the data of Exercise 14 form a population instead of a sample. Compute these measures:
    (a) Mean
    (b) Range
    (c) Standard deviation
    (d) Variance
    (e) How do these values compare to the sample statistics you computed in Exercise 14?

19. The following data show the Minnesota Multiphasic Personality Inventory (MMPI) Depression Scale scores for all 15 patients on a particular hospital ward: 110, 119, 113, 111, 101, 135, 116, 130, 121, 105, 104, 69, 104, 105, 132. Assume that we are interested in describing the variation of the patients only on that particular ward.
    (a) Compute the range.
    (b) Eyeball-estimate and compute the standard deviation using the mean squared deviation formula.
    (c) Compute the standard deviation using the computational formula.
    (d) Compute the variance.

20. Eyeball-estimate the standard deviation using the inflection-point method for the histogram shown here.

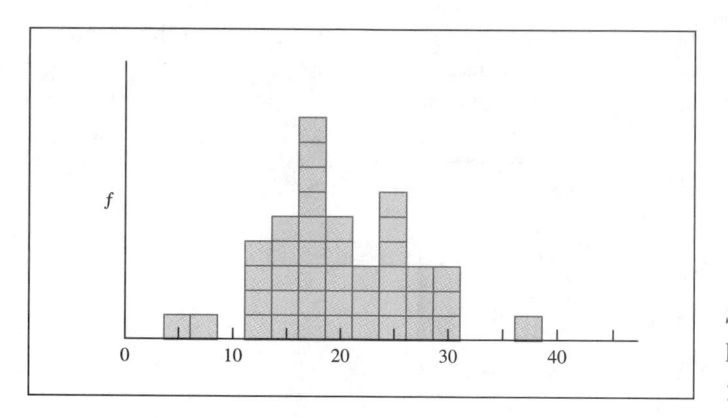

21. A researcher presented lists of nonsense syllables to a random sample of 50 students from an introductory psychology class. Use these numbers of list presentations from Exercise 13 in Chapter 3: 8, 9, 7, 8, 16, 7, 10, 11, 16, 14, 12, 13, 12, 13, 12, 14, 8, 9, 17, 12, 5, 18, 14, 14, 12, 8, 11, 11, 9, 9, 18, 15, 11, 7, 9, 5, 6, 8, 10, 11, 11, 10, 14, 16, 6, 11, 15, 9, 19, 12.
    (a) Compute the range.
    (b) Eyeball-estimate and compute the standard deviation using the mean squared deviation formula.
    (c) Compute the standard deviation using the computational formula.
    (d) Compute the variance.

22. A researcher has taken a random sample of 55 students and measured their typing speeds. Use these numbers of words per minute from Exercise 14 in Chapter 3: 8, 24, 20, 20, 17, 18, 16, 19, 17, 29, 25, 27, 14, 5, 21, 36, 11, 16, 29, 20, 11, 17, 26, 10, 11, 5, 5, 19, 28, 7, 15, 8, 14, 32, 32, 12, 7, 12, 13, 30, 19, 16, 42, 26, 16, 30, 21, 8, 4, 23, 5, 15, 19, 9, 30.
    (a) Compute the range.
    (b) Eyeball-estimate and compute the standard deviation using the mean squared deviation formula.
    (c) Compute the standard deviation using the computational formula.
    (d) Compute the variance.

Section C: Extending your comprehension

23. (a) Show for the hospital-stay data of Exercise 14 that $\sigma = s\sqrt{(10-1)/10}$.
    (b) Prove in general that for any data set, $\sigma = s\sqrt{(n-1)/N}$.

24. Prove that the mean squared deviation formula for the standard deviation of a sample, Equation (5.4), is equivalent to the computational formula, Equation (5.4a). [*Hint:* Substitute $\sum X_i/n$ for $\overline{X}$ in Equation (5.4).]

25. Use the distribution of the times students spend getting a drink of water collected in Exercise 21 in Chapter 3.
    (a) Eyeball-estimate by the range method the standard deviation of times spent.
    (b) Eyeball-estimate by the inflection-point method the standard deviation of times spent.
    (c) Compute the standard deviation of times spent.
    (d) Compute the variance of times spent.

26. Use the distribution of the variable you chose in Exercise 22 from Chapter 3.
    (a) Compute the standard deviation.
    (b) Present the histogram (or frequency polygon) to your fellow students so that they can eyeball-estimate the standard deviation. How well do they do?

Section D: Cumulative review (answers in Appendix D, page 509)

27. For each of the following variables, indicate whether the level of measurement is nominal, ordinal, or interval/ratio.
    (a) Pitch of musical notes, measured in hertz (cycles per second)
    (b) Class rank (valedictorian, salutatorian, third in class, etc.)
    (c) The Dewey Decimal System of organizing books in the library
    (d) Price in dollars
    (e) Political party (1 = Republican, 2 = Democrat, 3 = Independent)

28. The accompanying histogram shows the age at onset of a particular disease. Eyeball-estimate these measures:
    (a) Mode
    (b) Median
    (c) Mean
    (d) Standard deviation

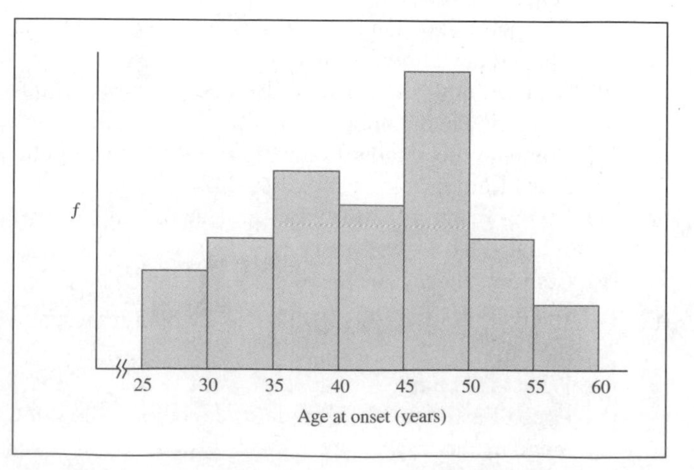

29. A computer company surveyed a random sample of 24 computer users to find out how many floppy disks they had. They found these numbers: 5, 58, 30, 9, 27, 5, 43, 11, 73, 32, 28, 12, 47, 5, 32, 19, 15, 9, 63, 60, 11, 44, 33, 8.
    (a) Find X_5.
    (b) Calculate $\sum X_i^2$.
    (c) Calculate $(\sum X_i)^2$.
    (d) Explain why the answers to parts (b) and (c) are different.

30. (*Exercise 29 continued*) For the floppy-disk data, eyeball-estimate and compute these measures:
    (a) Mode
    (b) Median
    (c) Mean
    (d) Standard deviation

Section E: From the journals (answers in Appendix D, page 509)

31. Columbino and Del Boca (1990) investigated the effects of taxes on the labor supply in Italy. As part of this investigation, they needed to know the numbers of hours per year that men and women in Italy worked. They selected (using rather complex selection criteria that need not concern us) a sample of 832 married couples; one selection criterion was that the husband was employed. They presented the two tables that we reproduce here. They also presented the descriptive statistics for these samples: For men, the mean was reported to be 2015 hours and the standard deviation 2052 hours; for women, the mean was reported to be 725 hours and the standard deviation 1188 hours.
    (a) Characterize (in words) the shapes of both distributions.
    (b) Are there problems with the way these frequency distributions are displayed (that is, violations of the rules for forming frequency distributions as we have discussed them)? If so, what are they?
    (c) Eyeball-estimate the means using the balancing-point method for both men and women. Do Columbino and Del Boca's values of 2015 hours and 725 hours seem reasonable? If not, why not?
    (d) Is the mean an informative statistic for the distribution of men? Of women? If not, why not?
    (e) Eyeball-estimate the standard deviation using the range method in both the men's and the women's samples. Do Columbino and Del Boca's values of 2052 hours and 1188 hours seem reasonable? If not, why not?
    (f) Prepare graphical frequency distributions for both men and women.
    (g) Eyeball-estimate the standard deviations using the inflection-point method. Do Columbino and Del Boca's values of 2052 hours and 1188 hours seem reasonable? If not, why not?

Sample frequency distribution of yearly hours worked by men

Hours	Freq.	Freq. (%)
320–504	2	.24
504–688	3	.36
688–872	1	.12
872–1056	6	.72
1056–1240	8	.96
1240–1424	7	.84
1424–1608	20	2.40
1608–1792	42	5.05
1792–1976	508	61.06
1976–2160	55	6.61
2160–2344	56	6.73
2344–2528	44	5.29
2528–2712	26	3.13
2712–2896	27	3.25
2896–3080	12	1.44
3080–3264	6	.72
3264–3448	4	.48
3448–3632	0	.00
3632–3816	3	.36
3816–4000	2	.24

Source: Columbino and Del Boca, 1990, p. 397

Sample frequency distribution of yearly hours worked by women

Hours	Freq.	Freq. (%)
0–0	494	59.38
0–187	1	.12
187–374	0	.00
374–562	7	.84
562–749	5	.60
749–936	13	1.56
936–1123	17	2.04
1123–1310	16	1.92
1310–1498	22	2.64
1498–1685	12	1.44
1685–1872	31	3.73
1872–2059	171	20.55
2059–2246	9	1.08
2246–2434	9	1.08
2434–2621	3	.36
2621–2808	6	.72
2808–2995	5	.60
2995–3180	4	.48
3182–3370	4	.48
3370–3557	1	.12
3557–3744	2	.24

Source: Columbino and Del Boca, 1990, p. 398

Section F: Computer explorations

32. Enter these values into a statistics computation package (ESTAT's datagen, SPSS, etc.): 1.00, 2.00, 2.00, 3.00, 4.00, 5.00.
    (a) Have the computer compute the range, standard deviation, and variance.
    (b) Change the value "5.00" to "500," as if you had forgotten to enter the decimal point. Describe what happens to the range, standard deviation, and variance. Explain your results.
    (c) Experiment with values other than 500 and other original data sets until you get a feel for the effect of one "outlier" on the range, standard deviation, and variance.

33. Use the original data of Exercise 32.
    (a) Add 1000 to each value in this data set. Describe what happens to the range, standard deviation, and variance. Explain your results.
    (b) Use some constant other than 1000 and redo part (a), first predicting what your results will be.

Computing the Standard Deviation from Frequency Distributions (Optional)

This section may be omitted without loss of continuity.

The computational formulas for the standard deviation and variance [Equations (5.4a), (5.5a), (5.6a), and (5.7a)] can be awkward when data sets are large and every element in the data set must be included individually, even if the elements have the same values. Therefore, for the same reasons we mentioned in Section 4.7, it can be convenient to compute standard deviations and variances from frequency distributions using these formulas:

standard deviation of a sample (computional formula for data in a frequency distribution)

$$s = \sqrt{\frac{\sum fX^2 - \frac{(\sum fX)^2}{n}}{n-1}} \tag{5.4b}$$

standard deviation of a population (computational formula for data in a frequency distribution)

$$\sigma = \sqrt{\frac{\sum fX^2 - \frac{(\sum fX)^2}{N}}{N}} \tag{5.5b}$$

variance of a sample (computional formula for data in a frequency distribution)

$$s^2 = \frac{\sum fX^2 - \frac{(\sum fX)^2}{n}}{n-1} \tag{5.6b}$$

variance of a population (computational formula for data in a frequency distribution)

$$\sigma^2 = \frac{\sum fX^2 - \frac{(\sum fX)^2}{N}}{N} \tag{5.7b}$$

where the sums are taken over all the *different* values of X and where f is the frequency of those values. These formulas yield identical results to Equations (5.4a), (5.5a), (5.6a), and (5.7a) [and also identical results to Equations (5.4), (5.5), (5.6), and (5.7)].

For example, we can compute the sample standard deviation for the weight data in Table 4.1 by creating a new column headed "fX" and adding down that column, and a new column headed "fX^2" and adding down that column, as shown in Table 5.6. Then

TABLE 5.6 Computation of the standard deviation from the fequency distribution of male students' weights (Table 4.1 with new columns headed "fX," "X^2," and "fX^2")

Weight (X)	*f*	*fX*	*X*^{2}	*fX*^{2}
197	1	197	38,809	38,809
190	1	190	36,100	36,100
185	1	185	34,225	34,225
182	1	182	33,124	33,124
180	2	360	32,400	64,800
179	1	179	32,041	32,041
173	2	346	29,929	59,858
164	3	492	26,896	80,688
163	2	326	26,569	53,138
162	1	162	26,244	26,244
161	1	161	25,921	25,921
159	1	159	25,281	25,281
157	1	157	24,649	24,649
154	1	154	23,716	23,716
152	1	152	23,104	23,104
150	1	150	22,500	22,500
149	1	149	22,201	22,201
144	1	144	20,736	20,736
137	1	137	18,769	18,769
135	1	135	18,225	18,225
	$\sum f = 25$	$\sum fX = 4117$		$\sum fX^2 = 684{,}129$

Equation (5.4b) yields

$$s = \sqrt{\frac{\sum fX^2 - \frac{(\sum fX)^2}{n}}{n-1}} = \sqrt{\frac{684{,}129 - \frac{(4117)^2}{25}}{25-1}} = 16.00 \text{ pounds}$$

exactly as we got from Equation (5.4a) in Section 5.2.

For data sets where n is rather small, as here, there is not much (if any) labor savings in using Equation (5.4b) rather than (5.4a). When n is large, however, the labor savings can be dramatic.

Occasionally we wish to know the standard deviation in situations where we do not have access to the original data but only to a grouped frequency distribution. In such instances, we cannot compute the standard deviation exactly, but we can obtain a close approximation of it using these formulas:

approximate standard deviation of a sample (computational formula for data in a grouped frequency distribution)

$$s \approx \sqrt{\frac{\sum fX_{\text{mid}}^2 - \frac{(\sum fX_{\text{mid}})^2}{n}}{n-1}} \tag{5.8}$$

approximate standard deviation of a population (computational formula for data in a grouped frequency distribution)

$$\sigma \approx \sqrt{\frac{\sum fX_{\text{mid}}^2 - \frac{(\sum fX_{\text{mid}})^2}{N}}{N}} \tag{5.9}$$

approximate variance of a sample (computational formula for data in a grouped frequency distribution)

$$s^2 \approx \frac{\sum fX_{\text{mid}}^2 - \frac{(\sum fX_{\text{mid}})^2}{n}}{n-1} \tag{5.10}$$

approximate variance of a population (computational formula for data in a grouped frequency distribution)

$$\sigma^2 \approx \frac{\sum fX_{\text{mid}}^2 - \frac{(\sum fX_{\text{mid}})^2}{N}}{N} \tag{5.11}$$

where X_{mid} is the midpoint of the class interval, obtained exactly as in Chapters 3 and 4: midpoint = (lower limit + upper limit)/2. Thus, for example, we can use Equation (5.8) to compute the sample standard deviation for the grouped data in Table 4.2 by (1) creating a new column for the midpoint (headed "X_{mid}") of each interval, (2) creating another new column for the frequency times the midpoint ("fX_{mid}"), (3) adding down that "fX_{mid}" column, (4) creating a new column headed "X_{mid}^2", (5) creating a new column headed "fX_{mid}^2", and (6) adding down that column as shown in Table 5.7.

TABLE 5.7 Computation of the standard deviation from the grouped frequency distribution of male students' weights (Table 4.2 with new columns for the midpoint ("X_{mid}"), "fX_{mid}", "X_{mid}^2", and "fX_{mid}^2")

Class interval	X_{mid}	f	fX_{mid}	X_{mid}^2	fX_{mid}^2
190–199	194.5	2	389.0	37,830.25	75,660.50
180–189	184.5	4	738.0	34,040.25	136,161.00
170–179	174.5	3	523.5	30,450.25	91,350.75
160–169	164.5	7	1151.5	27,060.25	189,421.75
150–159	154.5	5	772.5	23,870.25	119,351.25
140–149	144.5	2	289.0	20,880.25	41,760.50
130–139	134.5	2	269.0	18,090.25	36,180.50
		$\sum f = 25$	$\sum fX_{\text{mid}} = 4132.5$		$\sum fX_{\text{mid}}^2 = 689{,}886.25$

Then

$$s \approx \sqrt{\frac{\sum fX_{\text{mid}}^2 - \frac{(\sum fX_{\text{mid}})^2}{n}}{n-1}} = \sqrt{\frac{689{,}886.25 - \frac{(4132.5)^2}{25}}{25-1}}$$

$$= \sqrt{\frac{689{,}886.25 - 683{,}102.25}{24}}$$

$$= \sqrt{282.6667} = 16.81 \text{ pounds}$$

Note that this standard deviation computed from the class-interval data is close to, but not exactly equal to, the result we obtained from the original data with Equation (5.4), (5.4a), or (5.4b) because the method using class intervals does not take into account the exact position of the values within each class interval.

Exercises for Optional Section

(Answers in Appendix D, page 509)

1. The data in Exercise 6 (page 88) can be displayed in a frequency distribution as shown in the accompanying table.

Score on first exam	*f*	*Score on first exam*	*f*
96	1	83	2
95	1	82	1
91	3	81	1
90	1	75	1
89	1	72	1
88	2	71	1
86	2	68	1
84	2	61	1
			22

   (a) Compute the population standard deviation from Equation (5.5b).
   (b) Is the result of part (a) exactly the same as the result of Exercise 6(e)? Why or why not?
   (c) Compute the population variance from Equation (5.7b).
   (d) Is the result of part (c) exactly the same as the result of Exercise 6(h)? Why or why not?

2. The data of Exercise 6 (page 88) can be displayed as a grouped frequency distribution as shown in the accompanying table.
   (a) Compute the approximate population standard deviation from Equation (5.9).
   (b) Is the result of part (a) exactly the same as the result of Exercise 1(a)? Why or why not?
   (c) Compute the approximate population variance from Equation (5.11).
   (d) Is the result of part (c) exactly the same as the result of Exercise 1(c)? Why or why not?

Score on first exam	*f*
96–99	1
92–95	1
88–91	7
84–87	4
80–83	4
76–79	0
72–75	2
68–71	2
64–67	0
60–63	1
	22

CHAPTER 6

Using Frequency Distributions

LEARNING OBJECTIVES

1. What are percentiles and percentile ranks?
2. What is a standard score or z score?
3. What does "relative area" refer to and how is it useful in interpreting frequency distributions?
4. What is a normal distribution? What can be said about the areas of regions under the normal distribution?
5. What procedures are used for eyeball-estimating areas under the normal curve?
6. How is Table A.1 used for calculating areas under the normal curve?
7. What is interpolation?
8. What is a standardized variable and how does it relate to the unit normal distribution?

This chapter describes how to use the frequency distribution of a variable. The area of a region of a frequency distribution corresponds to the proportion of individuals whose value of the variable lies in that region. The chapter shows that all *normal* distributions have about 34% of their area between the mean and one standard deviation above (or below) the mean, and about 14% of their area between one and two standard deviations above (or below) the mean. The chapter shows how to use the table of areas under the normal distribution. It also describes standard scores (z scores) and explains how real-world variables can be transformed into standard scores.

We have observed since Chapter 1 that there are three major concepts to be mastered in statistics: (1) what "the distribution of a variable" is, (2) what a "sampling distribution of means" is, and (3) what a "test statistic" is. This is the last chapter in our focus on the distribution of a variable. Chapter 1 defined variables, Chapter 3 defined distributions and discussed their shape, Chapter 4 described their center, and Chapter 5 discussed their width. Our task in this chapter is to discover just how useful an understanding of distributions can be.

6.1 Points in Distributions

Suppose you have taken two exams in your statistics course; your scores were an 87 on the first exam and a 68 on the second. On which exam did you perform better? If your instructor grades on absolute criteria (for example, 90 or better is an "A," etc.), then clearly 87 is better than 68. However, if she grades "on a curve" and your 87 was the lowest score in the class on Exam 1 but your 68 was the highest score in the class on Exam 2, then 68 is clearly better than 87. It is thus often necessary to consider how you did relative to all those who took the tests—that is, to determine where a score fits in the distribution of all scores.

Percentiles and Percentile Rank

percentile: the score below which a specified percentage of scores in the distribution fall

percentile rank: the percentage of scores equal to or less than the given score

Table 6.1 lists all 25 scores on Exam 1 and all 20 scores on Exam 2, arranged in descending order for convenience. We can say that the score of 87 on the first exam is the "16th percentile" because 16% of all the scores are 87 or less. (Of the 25 total scores, four are 87 or lower, and $4/25 = .16$, or 16%.)

We can also say that the "percentile rank" of a score of 87 on the first exam is 16%. Thus, the terms *percentile* and *percentile rank* are inverses of each other: To obtain the percentile, you start with a percentage and end with a score, whereas to obtain the percentile rank, you

TABLE 6.1 Scores on Exams 1 and 2

Exam 1	*Exam 2*
100	73
96	70
95	69
95	69
95	69
95	68
95	66
95	63
95	63
94	63
93	63
93	63
93	62
92	62
92	62
91	60
91	60
91	59
89	58
89	57
88	
87	
86	
86	
81	
$\mu = 91.88$	$\mu = 63.95$
$\sigma = 4.023$	$\sigma = 4.307$

start with a score and end with a percentage. Some students find it convenient to remember that the percentile is a score, whereas the percentile rank (*of* a score) is a percentage.

The percentile rank of the score of 68 on the second exam is 75%. (Of the 20 total scores, 15 are 68 or lower, and $15/20 = .75$, or 75%.) Said the other way, the score of 68 is the 75th percentile.

Because 87 is the 16th percentile of Exam 1, whereas 68 is the 75th percentile of Exam 2, we conclude that the score of 68 is better than the score of 87.

The term *score* as used in this section should be understood to mean a value of any variable. Thus, for the male students' weights, *score* means "weight," and we could determine that the percentile rank of 161 pounds is 40%. (The data of Table 3.3 show 25 total individuals and 10 weigh 161 pounds or less, so $10/25 = .40$, or 40%.)

Standard Scores (z Scores)

Percentile and percentile rank are ways of specifying the position of a particular point within a distribution. These concepts are useful and widely used, but they have one major disadvantage: Equal differences in percentages do *not* reflect equal differences in values of the variable. Consider the scores on Exam 1, for example. When we move from a score of 81 to a score of 90 (an increase of 9 points), the percentile rank jumps from 4% to 28%—a 24-percentage-point increase. But when we move from a score of 91 to a score of 100 (also an increase of 9 points), the percentile rank jumps from 32% to 100%—a 68-percentage-point increase.

Thus, 90 is 24 percentage points greater than 81, but 100 is 68 percentage points greater than 91, even though both intervals are 9 points wide. That is a bit misleading. It would be desirable to create a measure that allows us to compare distributions but that retains the equal-interval characteristic of the underlying variable. Such a measure is the *standard score*—also called the z score.

standard score (or z score): a variable whose value counts the number of standard deviations a score is above or below its mean

raw score: a value of a variable in the original scale of measurement

A standard score (symbolized by z) is a variable whose value is the number of standard deviations a score is away from the mean. Thus, $z = 0$ denotes the mean itself, $z = 1$ denotes the point that is one standard deviation above the mean, $z = 2$ is two standard deviations above the mean, $z = -1$ is one standard deviation *below* the mean, $z = -1.5$ is one and a half standard deviations below the mean; and so on.

The following equations can be used to convert from the score of the original variable (sometimes called a "raw" score) to a standard score:

transformation from a raw score to a standard score in a population

$$z = \frac{X - \mu}{\sigma} \tag{6.1}$$

transformation from a raw score to a standard score in a sample

$$z = \frac{X - \overline{X}}{s} \tag{6.2}$$

For example, Exam 1 has mean $\mu = 91.88$ and standard deviation $\sigma = 4.023$ [those values were computed from Equations (4.2) and (5.5)]. The z score associated with your raw score of 87 can be computed from Equation (6.1):

$$z_{87} = \frac{X - \mu}{\sigma} = \frac{87 - 91.88}{4.023} = -1.21$$

which indicates that 87 is 1.21 standard deviations *below* (because z_{87} is negative) the mean of Exam 1.

Similarly, Exam 2 has mean $\mu = 63.95$ and standard deviation $\sigma = 4.307$. The z score associated with your raw score of 68 is computed from Equation (6.1):

$$z_{68} = \frac{X - \mu}{\sigma} = \frac{68 - 63.95}{4.307} = .94$$

which indicates that 68 is .94 standard deviation *above* (because z_{68} is positive) the mean of Exam 2. We conclude that your score of 68 on Exam 2 is better than your score of 87 on Exam 1 because $z_{68} = .94$ is greater than $z_{87} = -1.21$.

I leave it as an exercise for you to show that for Exam 1, the distance between z_{81} and z_{90} is the same as the distance between z_{91} and z_{100}, as we saw was desirable.

We say that z is "standardized" because it is always measured against its own "standard": the standard deviation of the variable. Thus, $z = 1$ is *always* one standard deviation above the mean; $z = -1.5$ is *always* one and a half standard deviations below the mean; and so on.

6.2 Areas Under Distributions

Percentiles, percentile ranks, and standard scores tell us the position of a point within its distribution. We shall see in the remainder of this chapter, however, that often the most useful information in frequency distributions is the *area* under portions of that curve. Let us reconsider Figure 3.3, our histogram for the weight data, which we have redrawn here as Figure 6.1. This figure is identical to Figure 3.3 except that we have stretched the vertical axis. Note that stretching the axis does not change the meaning of the histogram: Both histograms contain identical information. In Figure 6.2, we alter the new histogram to include explicitly a square for each student. Each of the 25 men in our statistics class is represented by a cell whose area is 1 square centimeter. The histogram of Figure 6.1 is obtained simply by removing the interior lines that formed the cells, leaving just the outlined vertical rectangles, which we called "bars" in Chapter 3. The bar whose base is from 160 to 169 pounds is seven cells high (it is composed of seven cells stacked) because seven men weigh between 160 and 169 pounds.

Useful as the height of a frequency distribution sometimes is, the *area* of regions under a frequency distribution will turn out to be even more useful. The rectangle we are considering has an area of 7 square centimeters (because each of the seven cells has area 1 square centimeter). The area is thus related to the frequency of occurrence: The area of a rectangle in square centimeters is the frequency of occurrence of a particular range of values.

Observe that the total area of this histogram is 25 square centimeters and that the cell for each individual is thus 1/25 of the total area.

The fact that we are measuring this histogram in square centimeters is not essentially important. If, for example, we rescaled this figure so that each cell was 1 inch on each side, the total area of the histogram would be 25 square inches. Each 1-square-inch cell in the rescaled histogram would still be 1/25 of the total area.

relative area: the proportional (or fractional) area under a frequency distribution

It is clear that exactly the same information is contained in both the large and the small histogram. What is generally most important in interpreting frequency distributions is the *relative* (or fractional) *area* contained in a part of the histogram. Regardless of whether the area of a histogram is measured in square inches or square centimeters, the area of one individual's cell is 1/25 of the total area, and the area of the rectangle at 160–169 pounds is still 7/25 of the total area. It is customary to call the total area of a frequency distribution 1.0 with no units (instead of 25 square centimeters as in our example). Then the area of the rectangle between 160 and 169 pounds is 7/25 of the total area (which is 1.0), or just 7/25 = .28.

Because they are "unitless," associated with no unit of measure, relative areas are convenient and allow us to draw histograms of any size. In our example, the area associated with one individual student is 1/25 of the histogram, or .04. The area associated with two students is 2(.04) = .08—that is, 2/25 of the entire histogram. We can generalize this observation: The area associated with 13 students is 13(.04) = .52, which is to say that just over half the area of this histogram is associated with just over half (13 of 25) the total number of students. The area associated with 5, or 20%, of the total number of students is 5(.04) = .20—that is, 20% of the total area.

The same kind of observation can be made about frequency polygons or frequency distributions in general, and it is one of the most important observations about distributions: *The relative area in any region of a frequency distribution is equal to the proportion of individuals whose variable value lies in that region.*

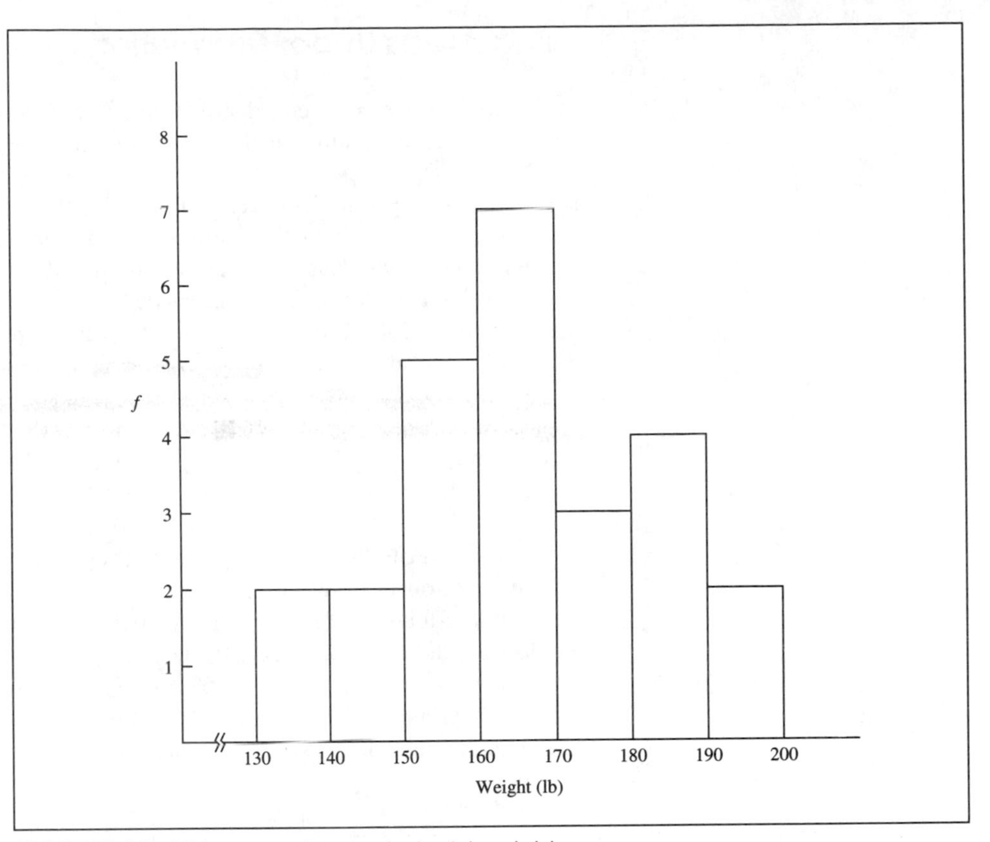

FIGURE 6.1 Histogram of male students' weights

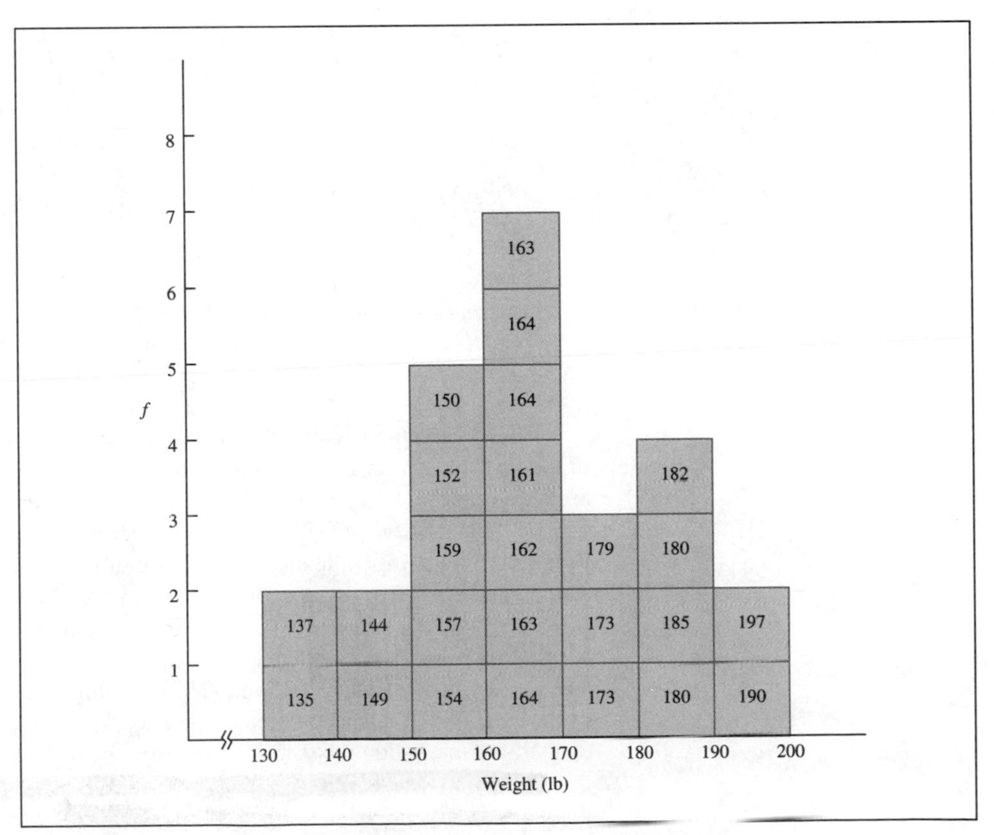

FIGURE 6.2 Figure 6.1 redrawn to show a cell for each student

Using Areas Under Distributions

In Figure 5.13, we sketched a distribution that was positively skewed with mode $120 and mean $140. Let us assume that this is a distribution of the prices of skis and ask the question: What percentage of skis cost $140 or less? The range of prices that we are interested in is less than or equal to $140 and so runs from the left end of the distribution up to (and including) $140. We show the distribution in Figure 6.3 and draw a vertical line at $140; we then "shade" with diagonal lines the region of the distribution that lies at $140 or below. We know that the relative area of the shaded region is equal to the proportion of all skis that cost less than $140, so the question is: What is the relative area (assuming the total area is 1.0) of the shaded region? To eyeball-estimate this area, we first ask: Is the shaded area about equal to, greater than, or less than *half* the total area? The answer seems to be greater, which means that the relative area must be greater than .5 (that is, greater than 50%). How much greater? It seems to be only slightly greater, so we might eyeball-estimate 60%. The actual answer is that about 57% of the distribution is shaded, which means that according to our distribution, 57% of all skis cost $140 or less.

You can verify that the shaded region is about 57% of the area under the distribution. The distribution in Figure 6.3 is "painted" with small dots. If you counted them, you would find a total of 133 dots. The number of dots in the shaded region is 76. Because the dots are uniformly arranged, the percentage of dots is approximately the same as the percentage of area, so $76/133 = .57$, or 57%. We shall see that there are more efficient ways of ascertaining areas than counting dots; the dots are presented here simply as a way of illustrating the relative sizes of the areas being discussed.

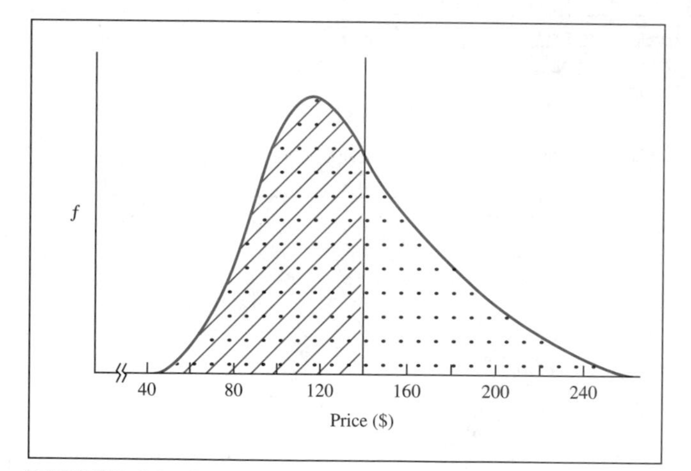

FIGURE 6.3 Frequency distribution of prices of skis with area less than $140 shaded

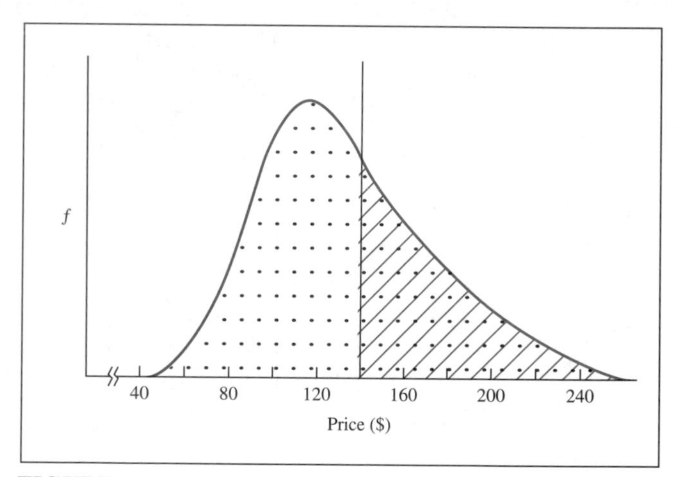

FIGURE 6.4 Frequency distribution of prices of skis with area above $140 shaded

What percentage of skis cost *more* than $140? Here we must shade the right-hand part of the distribution, as shown in Figure 6.4. Our eyeball tells us immediately that the shaded area is somewhat less than half the total area (and therefore less than .5), perhaps about .4. The actual answer is .43. We could have predicted this answer because the region shaded in Figure 6.4 is simply the unshaded region of Figure 6.3, and we know that the sum of the shaded and unshaded regions must be 1.0: $.57 + .43 = 1.0$.

Perhaps we are interested in knowing what percentage of skis cost between $100 and $140. Our procedure is to shade the relevant region—here between the two values $100 and $140—and estimate the area. This shading is shown in Figure 6.5. Our task is to eyeball-estimate the shaded area, which seems to be somewhat less than .5; perhaps our eyeball-estimate should be .45. As counting the dots shows, the actual area is 41%.

Relative areas under distributions correspond to relative frequencies of occurrence regardless of the shape of the distribution. The examples we have considered thus far come from a distribution that is moderately positively skewed.

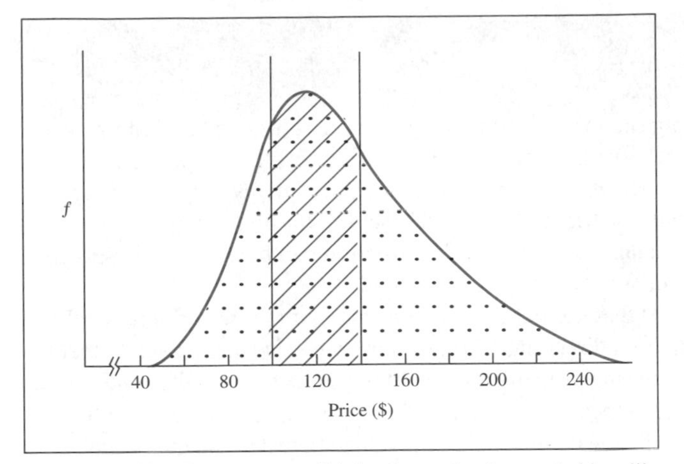

FIGURE 6.5 Frequency distribution of prices of skis with region between \$100 and \$140 shaded

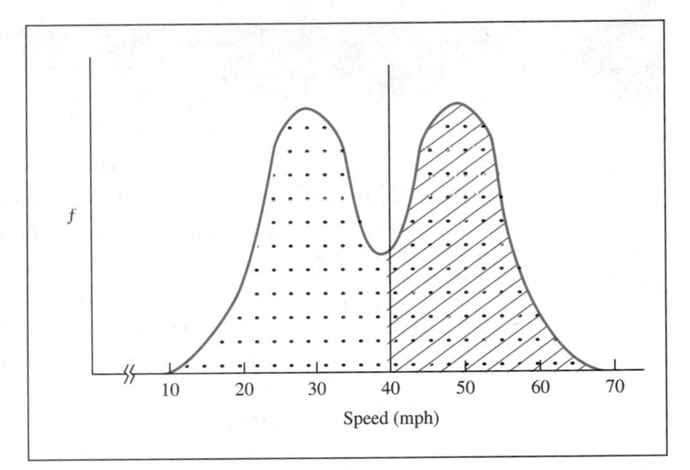

FIGURE 6.6 Frequency distribution of speeds of vehicles with region above 40 mph shaded

We had an example of a bimodal distribution in Figure 5.12, which we can take to represent the speeds of vehicles on a particular (uphill) stretch of road. This figure, with modes at 30 mph and 50 mph, indicates that basically two kinds of vehicles are on this stretch of road: trucks, which are progressing slowly (in the neighborhood of 30 mph), and cars, which are moving faster (about 50 mph).

We might be interested in knowing what percentage of vehicles in this distribution are moving fast, and we define "fast" as a speed greater than or equal to 40 mph. To answer this question, we must shade the region above 40 mph under this distribution and estimate its relative area. This is shown in Figure 6.6. The shaded area seems to be just slightly less than half of the total area; we might eyeball-estimate it to be .45. In fact (as counting the dots would show), the area is .49.

We have been asking questions of the form: Given a range of values of the X-axis, what is the relative frequency of occurrence of those values? That question can also be asked in reverse: Given a relative frequency, what is the range of values of X? For example, suppose we are interested in knowing how fast the fastest 10% of drivers travel on our particular stretch of road. We want to shade an area, and we know that it must be (1) at the right end of the distribution (to include the "fastest" drivers) and (2) 10% of the total area (because we want the fastest "10%"). Now, 10% is one-fifth of the upper half of this distribution, and what appears to be the upper 10% of the area is shaded in Figure 6.7. This region is bounded by the vertical line at 56 mph, so we can say that (by eyeball) the fastest 10% of the drivers travel at 56 mph or faster. Counting the dots would show that the 10% area actually begins at about 55.5 mph.

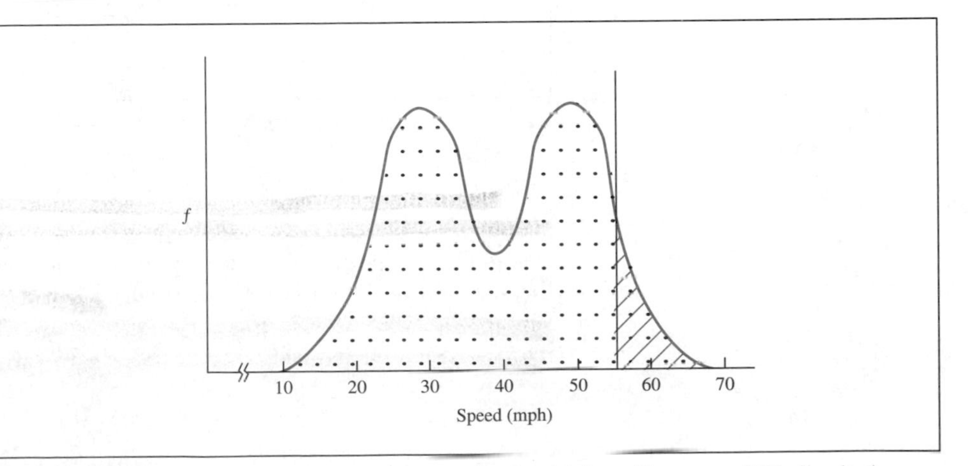

FIGURE 6.7 Frequency distribution of speeds of vehicles with upper 10% shaded

6.3 Areas Under Normal Distributions

normal distribution: any of a family of symmetric, unimodal, asymptotic distributions obtained by specifying values of μ and σ in the equation shown in Box 6.1

We have seen in the preceding section that areas under distributions can be used to answer all sorts of questions about relative frequencies. This is true for all distributions, but it is especially useful for *normal distributions* for two reasons: First, many things in nature are in fact approximately normally distributed (we will discuss why that is so in Chapter 7, Section 7.2), and second, the shape of the normal distribution can be completely specified by just two numbers, the mean and the standard deviation.

We know (from Chapter 3) that normal distributions are symmetric and unimodal and asymptotically approach 0 in both directions. Furthermore, we saw (in Chapter 4) that the mean, median, and mode of a normal distribution are all the same point in the center of the distribution, directly below the highest point, and (in Chapter 5) that there are two inflection points (one above and another below the mean). The distance between the mean and either of the inflection points is the standard deviation of the distribution.

unit normal distribution: a normal distribution with mean $\mu = 0$ and standard deviation $\sigma = 1$

There is thus a family of normal distributions, the members of which are identical to one another except for two numbers, the mean and standard deviation. This family of curves has the precise mathematical formula shown in Box 6.1. Figure 6.8 shows a *unit normal distribution*—that is, a normal distribution whose mean is 0 and whose standard deviation is 1.[1] Note that we use "z" to label the axis because (as we saw above) z is the variable that counts the number of standard deviations a point is above or below the mean. Thus, $z = 1$ is one standard deviation above the mean, $z = -2$ is two standard deviations below the mean, and so on.

ⓘ This formula is shown *only* to demonstrate that the normal curves have precise mathematical definitions. We will not refer to it again in this textbook.

BOX 6.1 Formula for the normal distribution with mean μ and standard deviation σ

$$f = \frac{1}{\sigma\sqrt{2\pi}}e^{-(X-\mu)^2/2\sigma^2}$$

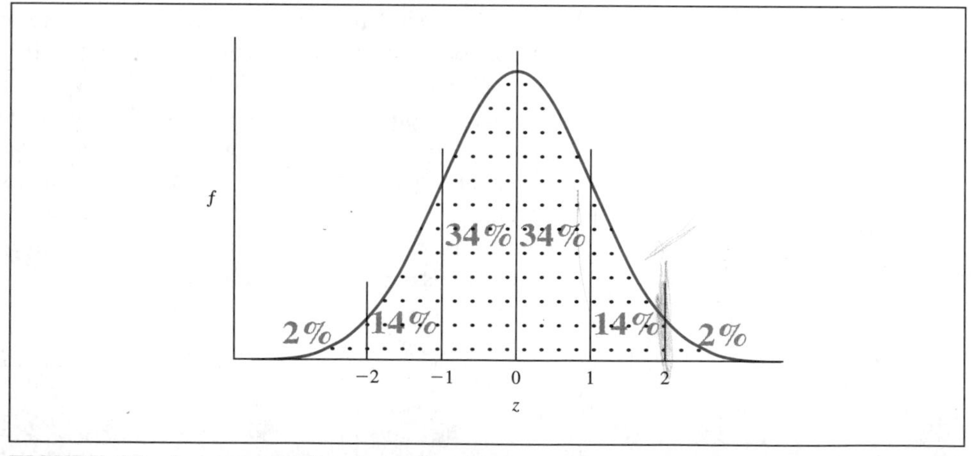

FIGURE 6.8 Areas under the normal curve

One of the major advantages of normal distributions is that the relative areas under regions of the normal curve are always the same, regardless of the values of μ and σ. As we saw in Chapter 5, there are wide and narrow normal distributions (depending on whether the standard deviation is large or small). Despite this fact, for all normal distributions, the area between the mean and the point one standard deviation above (or below) the mean is approximately 34% (more precisely 34.13%); the area between the points one and

[1] Because normal distributions are so frequently drawn inaccurately, it bears repeating (see Section 3.4) that all the normal distributions in this book are drawn with mathematical precision.

two standard deviations above (or below) the mean is approximately 14% (more precisely 13.59%), and the area beyond that is approximately 2% (more precisely 2.28%). In Figure 6.8 there are 132 total dots if you care to verify these percentages (at least approximately) for yourself.

This is a fact worth memorizing: For all *normal* distributions (tall or short, wide or narrow), the percentages of the area in standard-deviation bands are approximately 2%, 14%, 34%, 34%, 14%, and 2%.

Eyeball-estimation

The procedure for estimating areas under the normal curve is the same as that used for estimating areas under any distribution, except that we can be more accurate because we know the areas between integer values of z: about 2% when z is less than -2, 14% when z is between -2 and -1, 34% when z is between -1 and 0, 34% when z is between 0 and $+1$, 14% when z is between $+1$ and $+2$, and 2% when z is greater than $+2$, as shown in Figure 6.8.

For example, the area when $z \geq 0$ is shown shaded in Figure 6.9. This area is exactly .50 because we know that normal distributions are symmetric with the mean at the center and therefore have equal areas above and below the mean. We could also add up the areas known to occur in the shaded standard deviation regions: $.34 + .14 + .02 = .50$ or 50%.

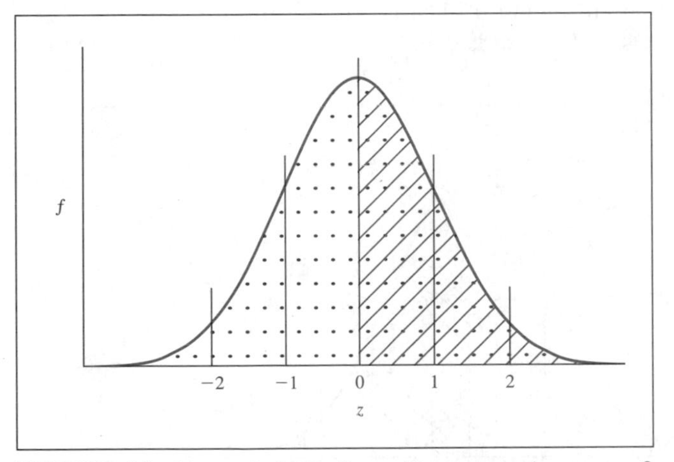

FIGURE 6.9 Normal distribution with region above $z = 0$ shaded

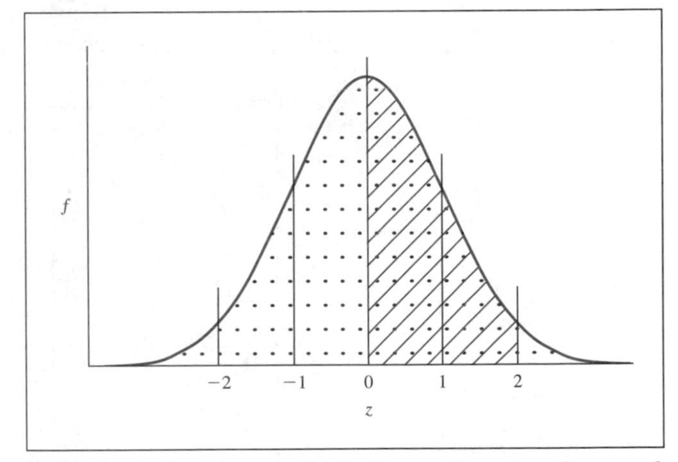

FIGURE 6.10 Normal distribution with region $0 \leq z \leq 2$ shaded

Figure 6.10 shows a normal distribution with the region $0 \leq z \leq 2$ shaded. We can accurately estimate this area to be $.34 + .14 = .48$. Figure 6.11 shows a normal distribution with the region $z \leq +1$ shaded. We can see that the area here is .50 (the area to the left of the mean) $+ .34 = .84$ or 84%.

Figure 6.12 shows a shading of a normal distribution that will be extremely important in later chapters of this book—namely, shading the extreme right and left ends of the distribution. The question to be answered is: If we wish to shade the most extreme 5% of a normal distribution, what values of z should we use? It is necessary to divide the extreme 5% into *two* regions, one in the right-hand and one in the left-hand tail, because both tails can lay equal claim to being extreme scores. Furthermore, because the normal distribution is symmetric, the area in the right-hand tail must equal the area in the left-hand tail, so each area must be 2.5%. We know that the region beyond $z = 2$ (or beyond -2) holds approximately 2% of the area, and because the desired area is slightly greater than that, the z value must be slightly closer to the mean, perhaps about ± 1.95. (Later we will see that the exact value is $z = \pm 1.96$.)

How do we estimate the areas associated with noninteger values of z? Figure 6.13 shows a normal distribution with the region $0 \leq z \leq .5$ shaded. It may (incorrectly) seem

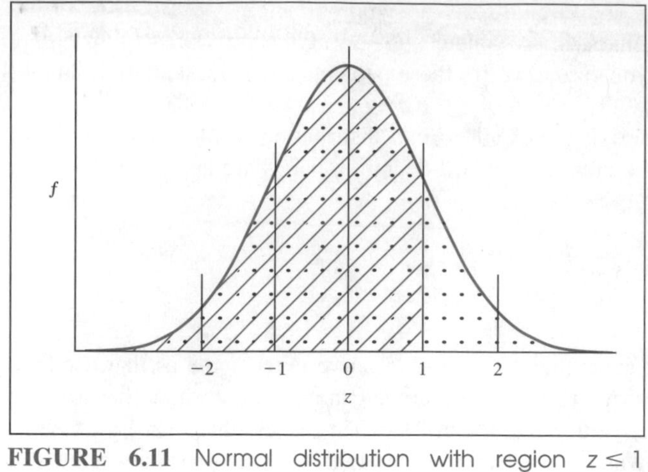

FIGURE 6.11 Normal distribution with region $z \leq 1$ shaded

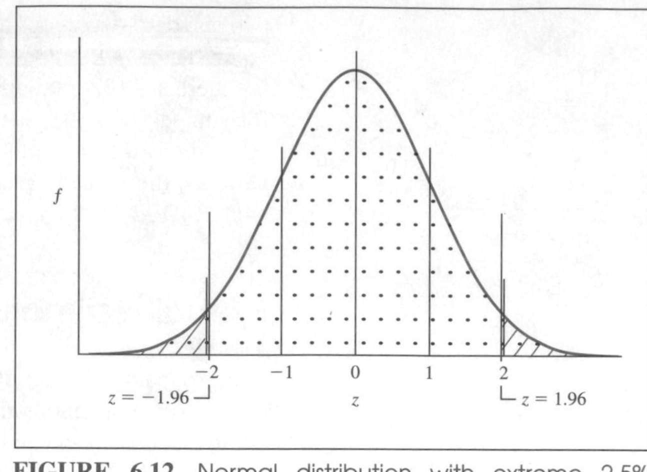

FIGURE 6.12 Normal distribution with extreme 2.5% shaded

that the area should be .17—that is, half of .34—because the range of z is half of the first standard deviation above the mean. However, inspection of Figure 6.13 shows that the shaded area is the "large half" of the first standard deviation area because the distribution is higher nearer the mean. Thus, the shaded region is larger than half of .34, so we take somewhat more than .17—say, .20—for our eyeball-estimate.

One final example is shown in Figure 6.14, a normal distribution with the region $.5 \leq z \leq 1.33$ shaded. We see that this area is composed of the "small half" of the first

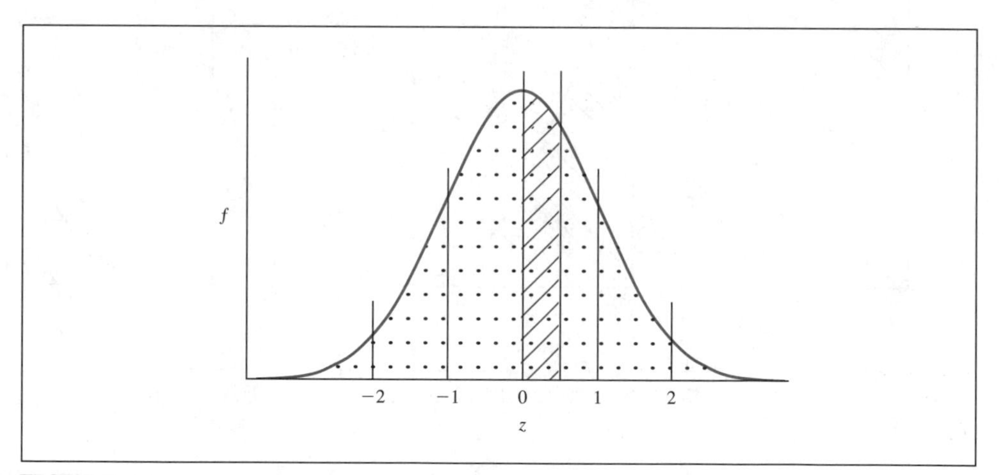

FIGURE 6.13 Normal distribution with region $0 \leq z \leq .5$ shaded

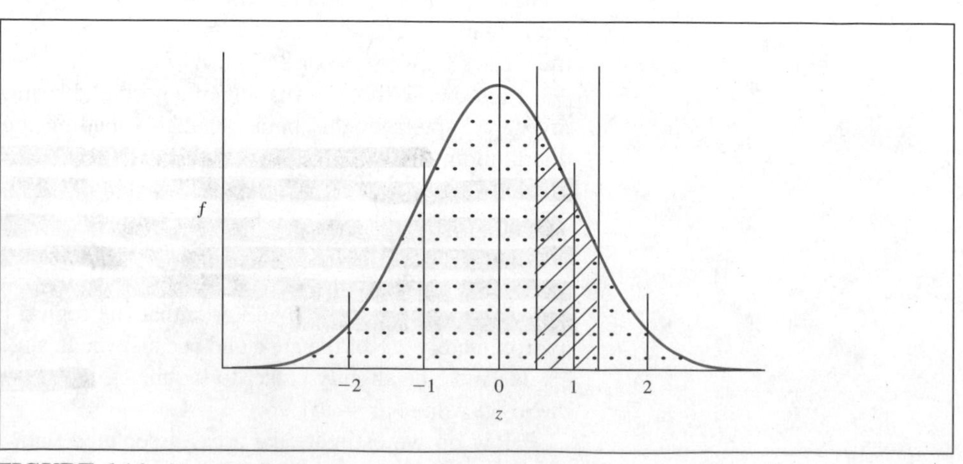

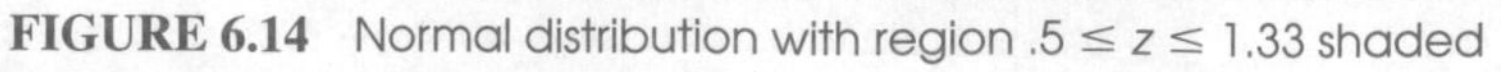
FIGURE 6.14 Normal distribution with region $.5 \leq z \leq 1.33$ shaded

standard deviation (somewhat less than half of .34—say, .14) and the "large third" of the second standard deviation (somewhat more than one third of .14—say, .06). Thus, our eyeball-estimate of the area of this region is .14 + .06 = .20.

Eyeball-calibration

Figure 6.15 contains six normal distributions, each with a shaded area. Your task is to eyeball-estimate the shaded areas. The answers are shown on the next page.

You may wish to use the ESTAT exercise "normal" at this time.

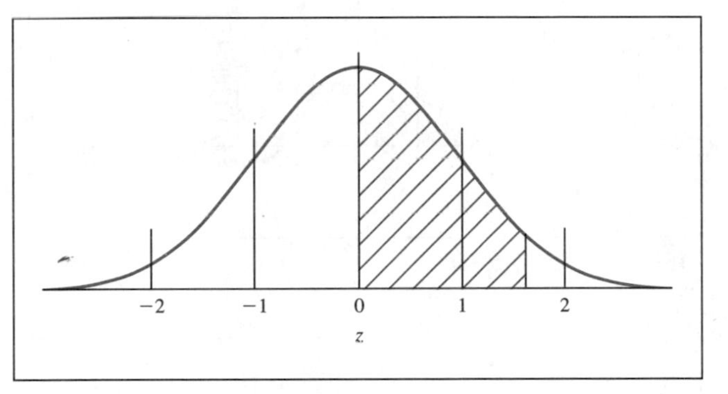

FIGURE 6.15a Eyeball-estimate the shaded areas (turn the page for answers)

FIGURE 6.15b

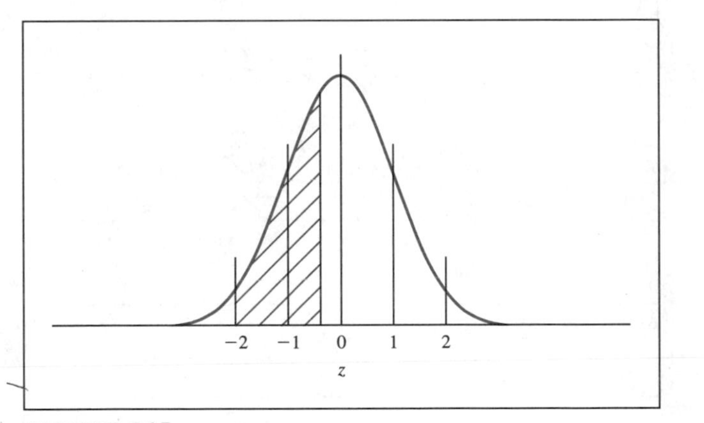

FIGURE 6.15c

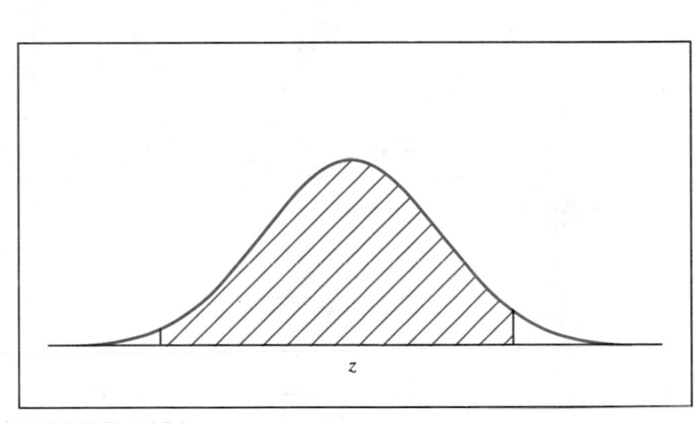

FIGURE 6.15d

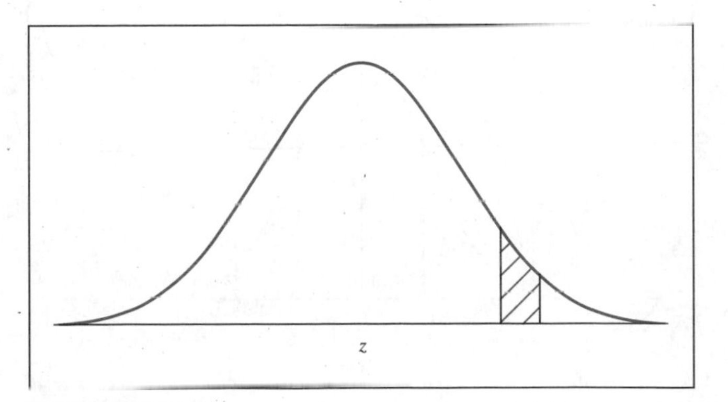

FIGURE 6.15e

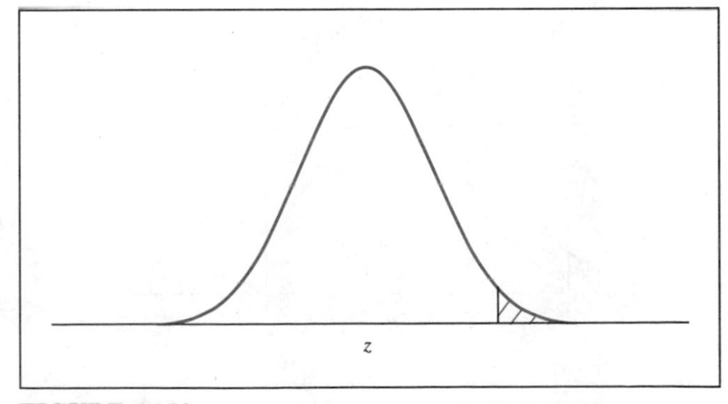

FIGURE 6.15f

Calculating Areas

We have learned to eyeball-estimate areas under normal distributions, and we should be able to achieve an accuracy of about ±2%. We can use Table A.1 in Appendix A, "Proportions of areas under the normal curve," to calculate the areas with more precision. That table gives areas *between* the mean and a given z score in the column headed A and areas *beyond* the given value of z in the column headed B.

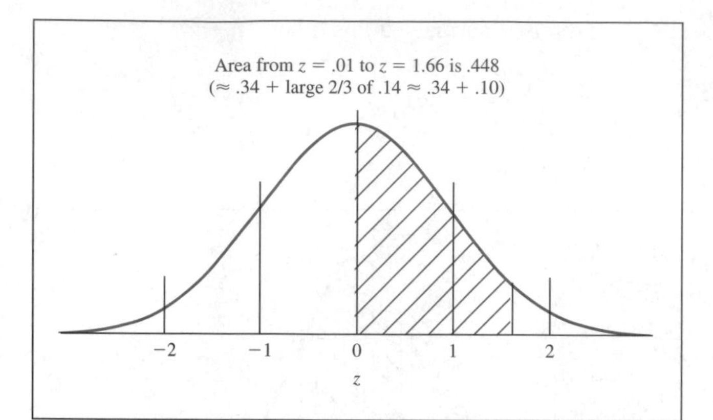

Answer to Figure 6.15a

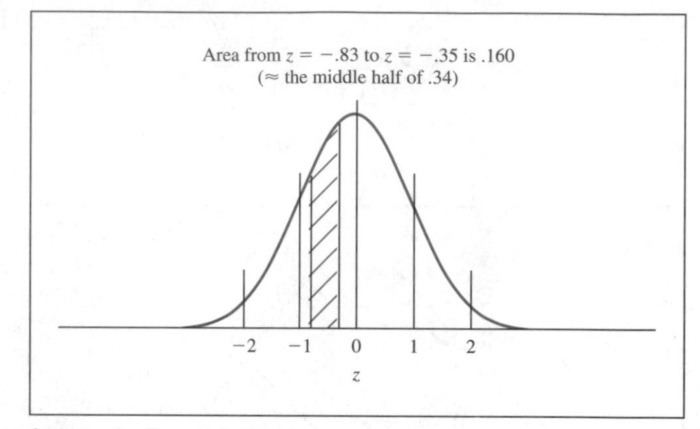

Answer to Figure 6.15b

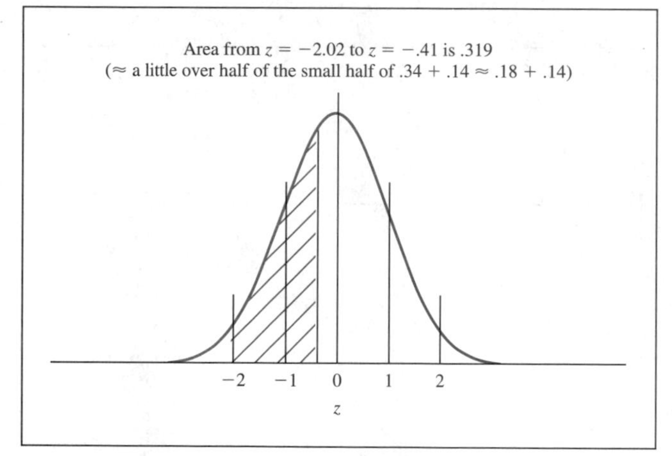

Answer to Figure 6.15c

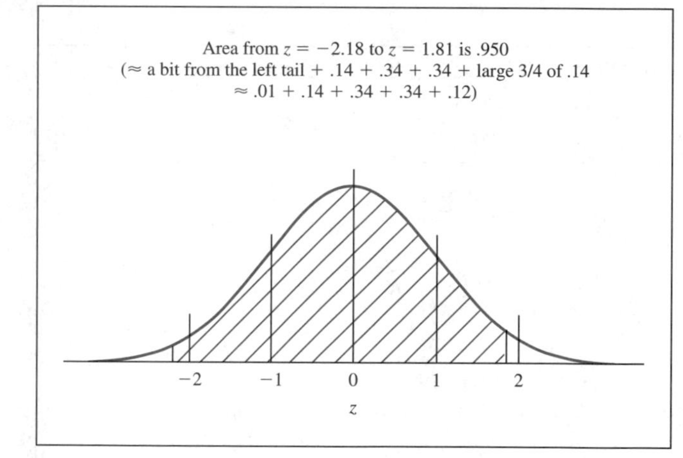

Answer to Figure 6.15d

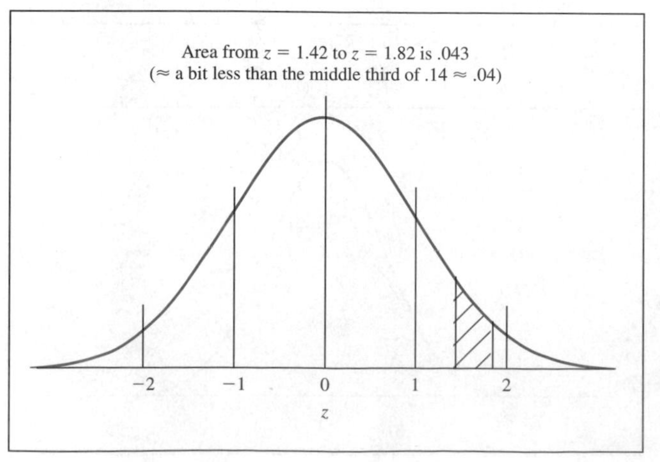

Answer to Figure 6.15e

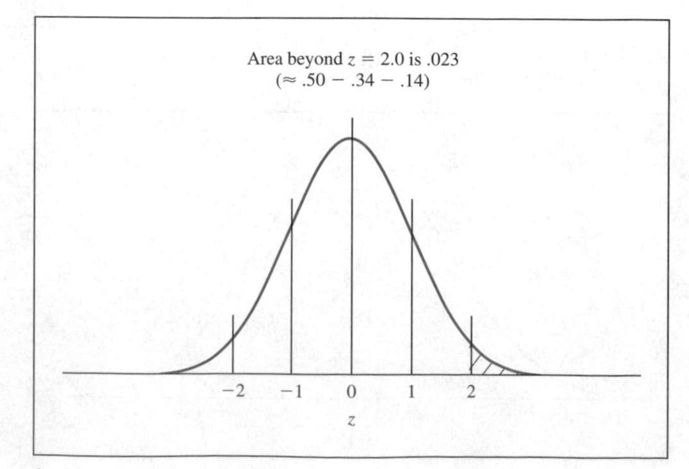

Answer to Figure 6.15f

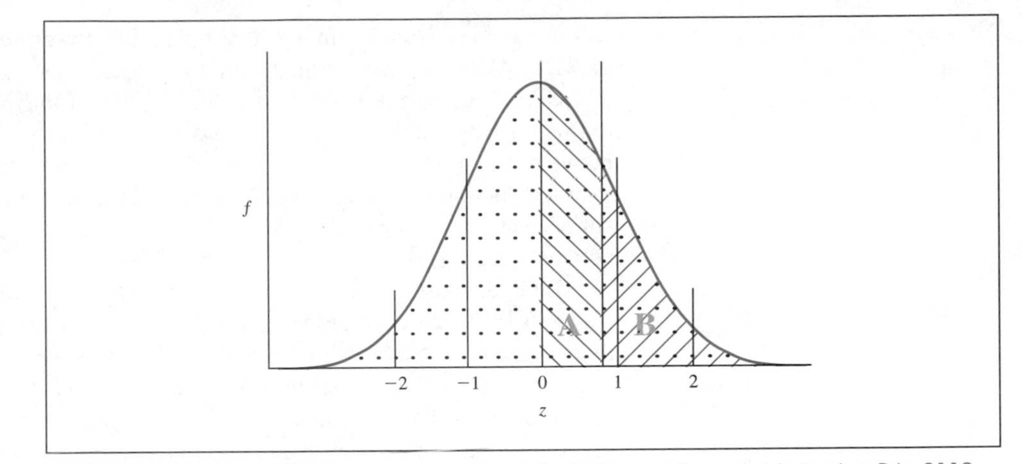

FIGURE 6.16 If $z = .80$, the area in region A is .2881 and the area in region B is .2119

Now consider Figure 6.16, which shows the region $0 \leq z \leq .8$ shaded and marked with the letter "A." Our eyeball should tell us that the area is the "large eight-tenths of .34," or about .29. Because $z = 0$ is the mean of a distribution, that region lies between the mean and $z = .8$. That is the kind of region we can look up directly in column A of Table A.1. We reproduce a portion of that table here for your convenience. We find $z = .80$ in the z

TABLE A.1 (Excerpts) Proportions of areas under the normal curve (Complete table in Appendix A)

z	Area between mean and z (A)	Area beyond z (B)	z	Area between mean and z (A)	Area beyond z (B)
.00	.0000	.5000	.97	.3340	.1660
.01	.0040	.4960	.98	.3365	.1635
.02	.0080	.4920	.99	.3389	.1611
.03	.0120	.4880	1.00	.3413	.1587
.04	.0160	.4840	1.01	.3438	.1562
.05	.0199	.4801	1.02	.3461	.1539
.06	.0239	.4761	1.03	.3485	.1515
.07	.0279	.4721	⋮	⋮	⋮
.08	.0319	.4681	1.14	.3729	.1271
.09	.0359	.4641	1.15	.3749	.1251
.10	.0398	.4602	1.16	.3770	.1230
⋮	⋮	⋮	⋮	⋮	⋮
.47	.1808	.3192	1.27	.3980	.1020
.48	.1844	.3156	1.28	.3997	.1003
.49	.1879	.3121	1.29	.4015	.0985
.50	.1915	.3085	1.30	.4032	.0968
.51	.1950	.3050	1.31	.4049	.0951
.52	.1985	.3015	1.32	.4066	.0934
.53	.2019	.2981	1.33	.4082	.0918
⋮	⋮	⋮	1.34	.4099	.0901
.67	.2486	.2514	1.35	.4115	.0885
⋮	⋮	⋮	⋮	⋮	⋮
.72	.2642	.2358	1.47	.4292	.0708
.73	.2673	.2327	1.48	.4306	.0694
.74	.2704	.2296	1.49	.4319	.0681
.75	.2734	.2266	1.50	.4332	.0668
.76	.2764	.2236	1.51	.4345	.0655
.77	.2794	.2206	1.52	.4357	.0643
.78	.2823	.2177	⋮	⋮	⋮
.79	.2852	.2148	2.00	.4772	.0228
.80	.2881	.2119	⋮	⋮	⋮
.81	.2910	.2090	2.50	.4938	.0062
.82	.2939	.2061			
.83	.2967	.2033			
⋮	⋮	⋮			

column and see from column A that the area between the mean and $z = .80$ is .2881 or 28.81%. We had eyeball-estimated that area to be about .29, quite close.

Note that Table A.1 gives areas for z scores of *only* 0 or greater. That is because the normal distribution is symmetric: Areas on the left side of the normal distribution (that is, between the mean and *negative* values of z) are identical to areas on the right side (positive z) of the distribution. Thus, if you need to find the area corresponding to a negative z score, simply enter the table at the z score with the negative sign eliminated. For example, the area between $z = 0$ and $z = -.80$ is also .2881.

Table A.1 also gives (in column B) areas beyond particular values of z. For example, Figure 6.16 shades the region beyond $z = .80$ and marks it with the letter "B." By eyeball, this area would be a bit more than .14 + .02, or about .20. To compute it, we enter Table A.1 at $z = .80$; column B shows that the area beyond $z = .80$ is .2119 or 21.19%.

Sometimes Table A.1 does not give the desired area directly, so we must look up two (or occasionally more than two) areas and add or subtract them. For example, the bottom part of Figure 6.17 shades the area $.5 \le z \le 1.33$ (we eyeball-estimated that area in Figure

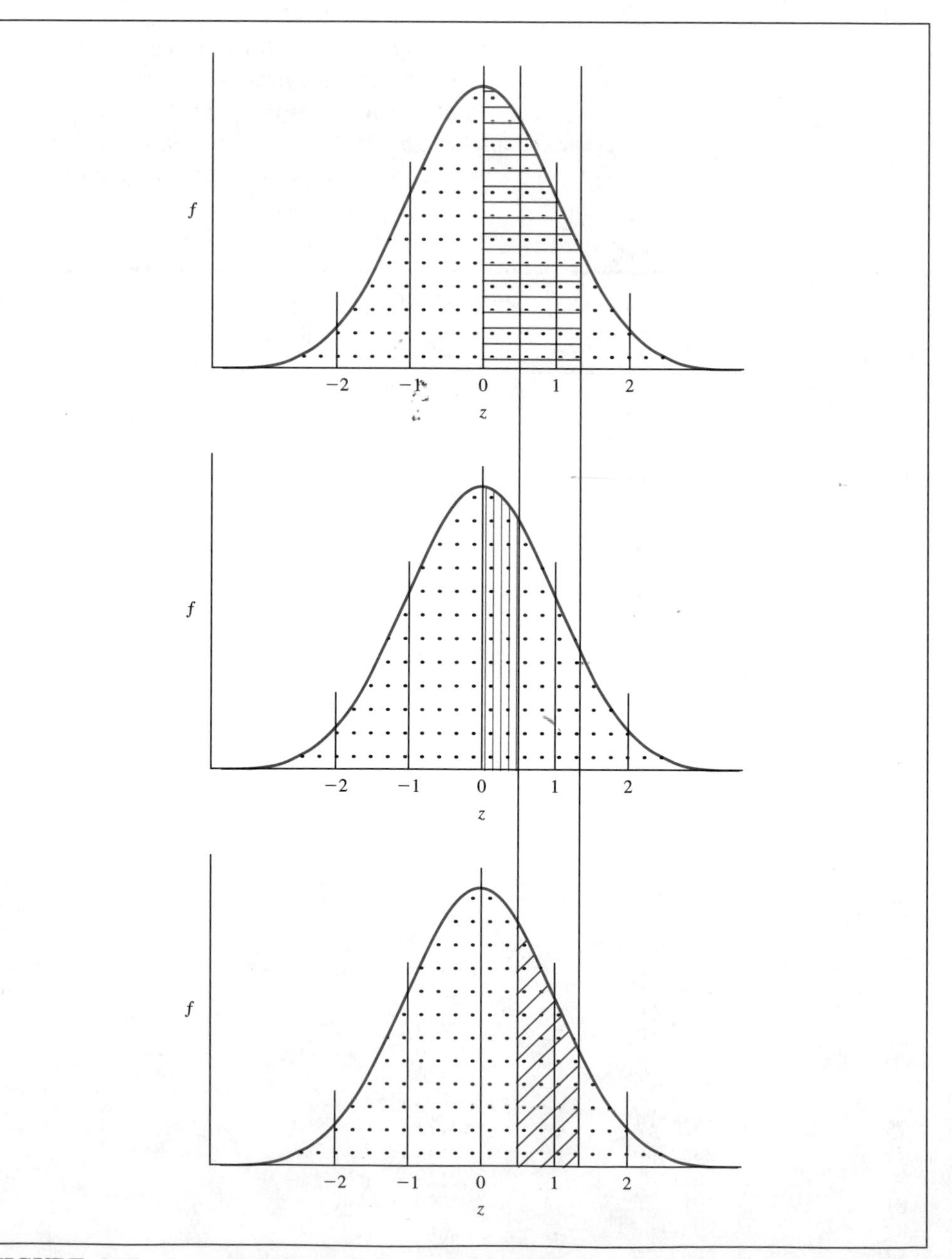

FIGURE 6.17 Areas looked up in Table A.1 to calculate area for $.5 \le z \le 1.33$. *Top:* $0 \le z \le 1.33$. *Middle:* $0 \le z \le .5$. *Bottom:* Result obtained by subtraction.

6.14). This region is neither between the mean and a z score (as would be found in Table A.1, column A) nor beyond a z score in the tail of the distribution (as would be found in column B), so Table A.1 will not give the area of this region directly.

To obtain the desired area, we must look up two areas in Table A.1 and then subtract them. First, we look up the area of the region $0 \leq z \leq 1.33$, as shown in the top part of Figure 6.17. That *is* an area we can look up in Table A.1; its value from column A is .4082. However, that value is larger than the desired area because it includes the region from 0 to .5. Therefore, second, we look up the region $0 \leq z \leq .5$ (shown in the middle part of Figure 6.17) in Table A.1 and find it in column A to be .1915. We then *subtract* these values to get the desired area: $.4082 - .1915 = .2167$. Our eyeball-estimate was .20, again quite close to the more precise result.

Sometimes areas that can be looked up in Table A.1 need to be added together, rather than subtracted, as illustrated in Figure 6.18, which shows the steps necessary to obtain the area of the region $-1.5 \leq z \leq .75$. The desired region is shown at the bottom of Figure

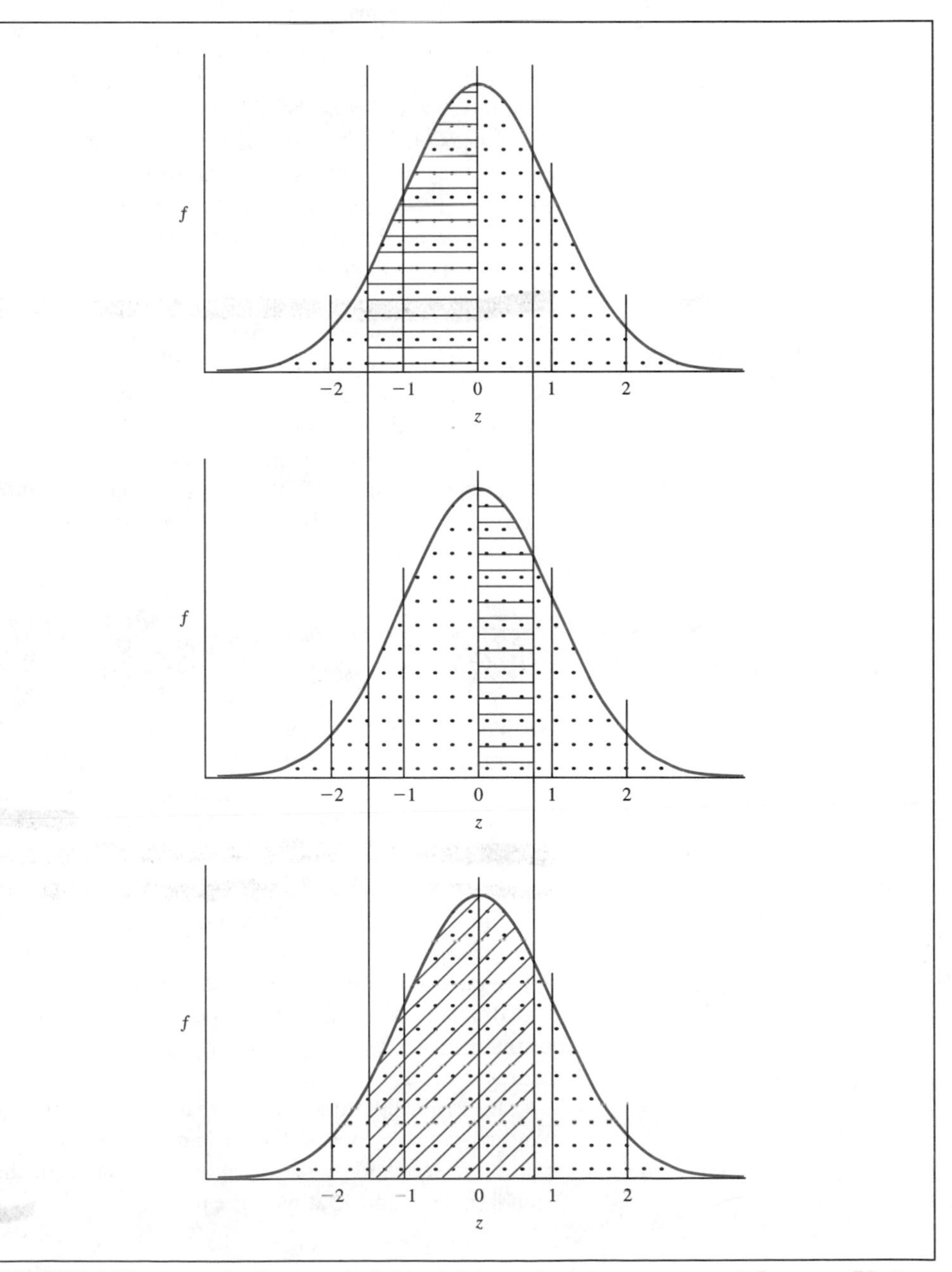

FIGURE 6.18 Areas looked up in Table A.1 to calculate area for $-1.5 \leq z \leq .75$. *Top:* $-1.5 \leq z \leq 0$. *Middle:* $0 \leq z \leq .75$. *Bottom:* Result obtained by addition.

6.18. Because the region extends in both directions from the mean, it is impossible to look up the area directly in Table A.1. We therefore look up separately the region below the mean, $-1.5 \leq z \leq 0$, which our eyeball tells us (see the upper portion of Figure 6.18) is .34 plus the large half of .14 (about .09), or .43. Table A.1, column A, shows it to be .4332. We then look up the region above the mean, $0 \leq z \leq .75$ (see the middle portion of Figure 6.18), which our eyeball tells us is the large three-fourths of .34, or about .28. Table A.1, column A, shows it to be .2734. The desired total area is obtained by *adding* these two areas: $.2734 + .4332 = .7066$. Our eyeball result is $.43 + .28 = .71$, again quite close to the more precise value.

Whether areas need to be added or subtracted will not be confusing to you if you sketch the normal distribution, shade the relevant areas, determine how you can construct the areas from those shown in columns A and B, and eyeball-estimate them *before* you turn to Table A.1.

There are often two or more ways to construct a given area. For example, the area of Figure 6.17 could just as easily have been found by first determining the area beyond $z = .5$: .3085 from column B. Then we could subtract the area beyond $z = 1.33$: .0918 from column B. The desired area is then $.3085 - .0918 = .2167$, just as we obtained earlier. The two methods give the same result because for any value of z, the sum of the entries in columns A and B must be .5000.

Occasionally the table of areas under the normal distribution (or any other table, for that matter) does not list the precise value we desire. For example, suppose we wish to know the area under the normal curve between the mean and $z = .503$. Table A.1 gives areas between the mean and $z = .50$ (.1915) and between the mean and $z = .51$ (.1950) but not for any intermediate values such as $z = .503$. We obtain the approximate areas for such intermediate values using a process called *interpolation*.

interpolation: finding a value located proportionately between two values in a table

For most purposes, eyeball interpolation will be adequate. To interpolate by eyeball, we note that the required z value (.503) is between .50 and .51 (the adjacent tabled values of z) but a bit closer to .50. The eyeballed area is therefore between .1915 and .1950 (the adjacent tabled areas) but a bit closer to .1915 (the value corresponding to .50), or about .1925. Thus, by eyeball interpolation we conclude that the area between the mean and $z = .503$ is about .1925. (Linear interpolation, a more accurate and repeatable method of interpolation, is described in the optional section at the end of this chapter.)

6.4 Other Standardized Distributions Based on z Scores

We have spent the last several sections on how to estimate and calculate areas under the *unit normal distribution*—that is, the distribution of z that has mean 0 and standard deviation 1. In many situations, however, users of normal distributions prefer to have scores that do not have negative values and decimal points. Then we transform the unit normal variable z into a *standardized variable*—that is, a normally distributed variable that has mean μ and standard deviation σ. We specify the values of μ and σ in advance.

standardized variable: a normally distributed variable that has mean μ and standard deviation σ

For example, modern intelligence tests are constructed so that their results are normally distributed, and we could report individual scores on such tests as z scores. If we did so, half the population would have negative intelligence scores (that is, $z < 0$) because the mean $z = 0$. The concept of negative intelligence seems contrary to reason, so most administrators of intelligence tests do not report z scores, preferring to report IQ (Intelligence Quotient) scores instead.

IQ is a standardized score that has mean $\mu = 100$ and standard deviation $\sigma = 15$.[2] We first consider how IQ scores are related to z scores. As we know, $z = 1$ refers to the point one standard deviation above the mean. In the IQ distribution with mean 100 and standard deviation 15, this is the point IQ = 115. In general, we can transform a z score to

[2] On a few modern IQ tests, $\sigma = 16$. The choice of the values 100 and 15 (or 16) is entirely arbitrary from a statistical point of view. Any other values would suffice, although there are historical reasons for choosing 100 and 15. Before the 1950s, IQ was defined differently and was not necessarily normally distributed.

transformation from a z score to a standardized score in a population

a standardized variable by using Equation (6.3):

$$X = z\sigma + \mu \tag{6.3}$$

transformation from a standardized score to a z score in a population

and transform a standardized variable back to a z score using Equation (6.1):

$$z = \frac{X - \mu}{\sigma} \tag{6.1}$$

where X is any standardized variable with mean μ and standard deviation σ.

Figure 6.19 illustrates the relationships between these equations and the table of the normal distribution (Table A.1). We call this figure the "know/want" diagram because you enter the diagram with the value that you currently *know* and follow the arrow(s) to obtain the value that you *want*.[3] Thus, if you know the raw score but want the z score, the figure shows (following the arrow at the lower left) that you must use Equation (6.1) to get it. Furthermore, the figure shows that it requires *two* steps to travel from a raw score to an area in Table A.1: First use Equation (6.1) to obtain z and then use z to look up the area in the table.

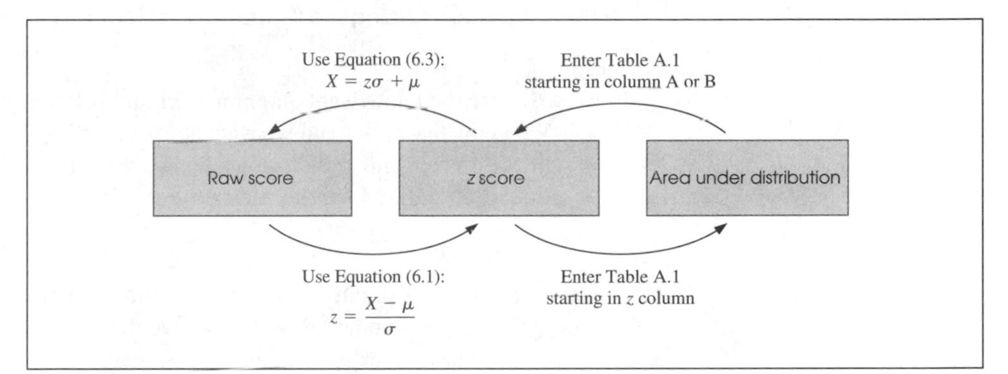

FIGURE 6.19 "Know/want" diagram of the relationships among raw scores, z scores, and tabled areas

Here is a general procedure for solving problems involving z scores, raw scores, and/or areas under normal distributions.

steps for solving problems involving z scores, raw scores, and/or areas under normal distributions

Step 1. Sketch the normal distribution with mean μ and standard deviation σ.

Step 2. Superimpose a unit normal z-axis, placing, of course, 0 under the mean of the variable and 1 under the point one standard deviation above the mean (which should also be under the inflection point).

Step 3. Shade the required area (if necessary).

Step 4. Eyeball-estimate the answer.

Step 5. Use the know/want diagram of Figure 6.19 and perform the required computations, look up the tabled values, or both.

We present two examples. First, suppose you wish to know the IQ score that is one standard deviation above the IQ mean.

Step 1. Sketch the normal distribution. For IQ, μ and σ are specified to be 100 and 15, respectively, as shown in Figure 6.20.

Step 2. Superimpose a unit normal z-axis. This is shown on the lower axis of Figure 6.20.

Step 3. Shade the required area (if necessary). This problem does not require an area.

Step 4. Eyeball-estimate the answer. The problem involves a score that is "one standard deviation above the IQ mean," so we know (because z counts the number of standard deviations a score is above the mean) that $z = 1$. We see from Figure 6.20 that the corresponding IQ is 115.

[3] I am indebted to Dr. Chris Heavey for this diagram.

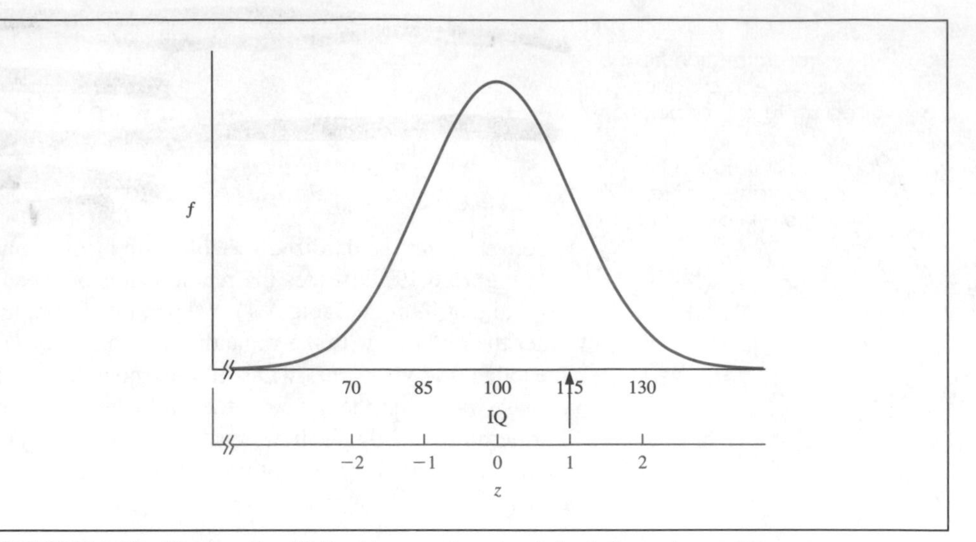

FIGURE 6.20 Finding the IQ that is one standard deviation above the mean

Step 5. Use the know/want diagram of Figure 6.19 to organize the required computations. We know that $z = 1$ and we want a raw score (IQ), so we enter the know/want diagram in the center and travel along the upper left-hand arrow. We will therefore apply Equation (6.3), where the variable X is IQ:

$$\text{IQ} = z\sigma + \mu = 1(15) + 100 = 115$$

Computation and eyeball-estimation give the same result (IQ = 115), so we conclude that the IQ that is one standard deviation above the mean is 115.

For our second example, suppose we believe (as some but not all observers do) that entry into a profession such as medicine requires an IQ of 110 or higher. We wish to know what percentage of the population is eligible to enter the medical profession (that is, has an IQ above 110). We follow the five steps:

Step 1. Sketch the normal distribution. For IQ, μ and σ are still 100 and 15, respectively, as shown in Figure 6.21.

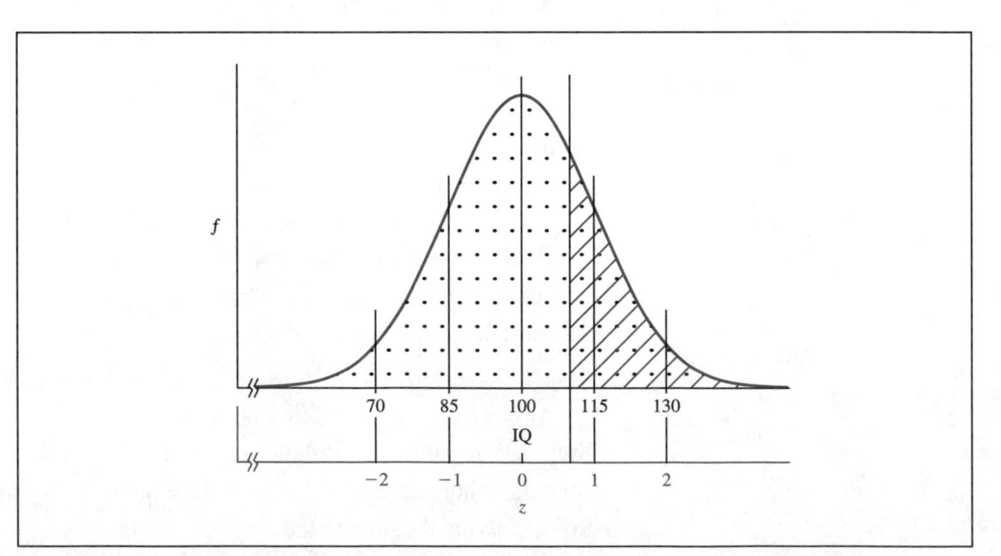

FIGURE 6.21 Calculating the percentage of individuals with IQ > 110.

Step 2. Superimpose a unit normal z-axis. This is done at the bottom of Figure 6.21.

Step 3. Shade the required area (if necessary). This problem does require an area—namely, the area above 110. We shade that region in Figure 6.21.

Step 4. Eyeball-estimate the answer. The shaded area is the small third of 34 (about .10) + .14 + .02, or about .26.

Step 5. Use the know/want diagram of Figure 6.19. Here we know the raw score (IQ = 110) and we want an area under a normal distribution, so the diagram shows that *two* steps will be required. First, following the arrow at the lower left of Figure 6.19, we transform the raw score IQ into z using Equation (6.1):

$$z = \frac{X - \mu}{\sigma} = \frac{110 - 100}{15} = .67$$

Then, following the arrow at the lower right, we enter Table A.1 (or its excerpt) with $z = .67$. Our answer will be in column B because we wish to know the area *beyond* .67. Table A.1 shows that the required area is .2514, or 25.14%. We check to see that this value is quite close to our eyeball-estimate in Step 4. Thus, we conclude that 25.14% of all individuals have an IQ higher than 110.

Many standardized scores other than IQ are also in common use—for example, college aptitude test scores such as SAT and GRE ($\mu = 500$, $\sigma = 100$), T scores ($\mu = 50$, $\sigma = 10$), stanines ($\mu = 5$, $\sigma =$ approximately 2), and intelligence test subscale scores ($\mu = 10$, $\sigma = 3$). Equations (6.1) and (6.3) apply to all these when the appropriate values for μ and σ are substituted.

It is often useful to transform from one standardized score to another—for example, transform IQ to GRE scores. This is done by using the know/want diagram *twice*: starting with the raw score IQ and transforming to z, and then starting with z and transforming to the raw GRE score. Thus, an IQ of 120 is transformed according to Equation (6.2) to $z = (120 - 100)/15 = 1.33$ because $\mu_{IQ} = 100$ and $\sigma_{IQ} = 15$. Then z of 1.33 is transformed according to Equation (6.3) to GRE = 1.33(100) + 500 = 633 because $\mu_{GRE} = 500$ and $\sigma_{GRE} = 100$.

Equations (6.1) and (6.3) assume that we know the population parameters μ and σ. We can perform entirely parallel operations when we must estimate μ and σ from their sample statistics $\overline{X}$ and s, as the following equations show:

transformation from a z score to a standardized score in a sample

$$X = zs + \overline{X} \tag{6.4}$$

transformation from a standardized score to a z score in a sample

$$z = \frac{X - \overline{X}}{s} \tag{6.5}$$

where X is any variable whose sample mean is $\overline{X}$ and whose standard deviation is s.

6.5 Relative Frequencies of Real-World Normal Variables

Many variables that occur in the real world are (at least approximately) normally distributed. The transformations described in Equations (6.1) and (6.3) allow us to make useful observations about the relative frequencies of particular values of such variables.

For example, suppose that the heights of female students at the University of South Hebrides (USH) are normally distributed with mean $\mu = 64$ inches and standard deviation $\sigma = 4$ inches. What percentage of the women are shorter than 62 inches? We can use our know/want procedure and the steps of the preceding section.

Step 1. Sketch the normal distribution. For USH women, $\mu = 64$ inches and $\sigma = 4$ inches, as shown in Figure 6.22.

Step 2. Superimpose a unit normal z-axis. This is the lower axis of Figure 6.22.

Step 3. Shade the required area (if necessary). We wish to know about the heights of women shorter than 62 inches, so we shade that region on Figure 6.22.

Step 4. Eyeball-estimate the answer. The shaded area is the small half of .34 (about .15) + .14 + .02 = .31, or 31%.

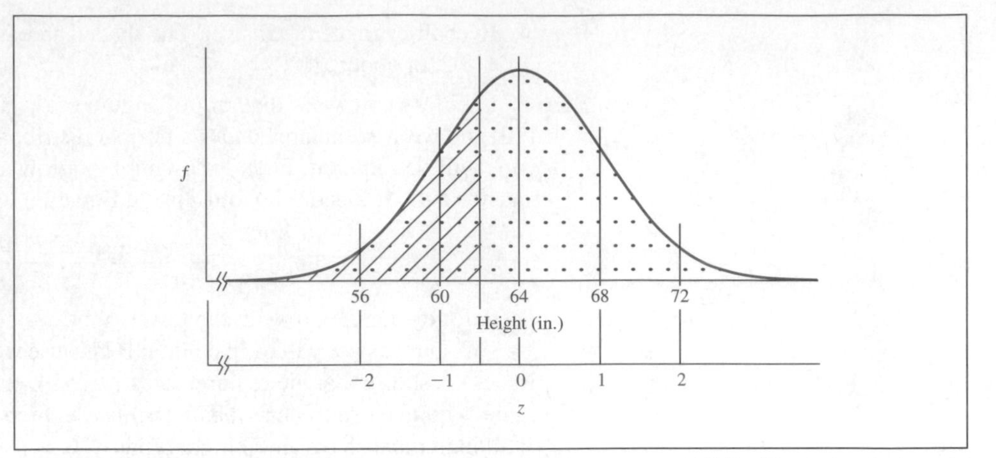

FIGURE 6.22 Frequency distribution of heights of women at USH with region below 62 inches shaded and with superimposed z-axis

Step 5. Use the know/want diagram of Figure 6.19. We know that the raw score (height) = 62 inches, and we want the area in the table. The know/want diagram shows that we must first obtain z according to Equation (6.1):

$$z = \frac{X - \mu}{\sigma} = \frac{62 - 64}{4} = -.5$$

Then we must find the area beyond that z value, so the know/want diagram shows that we will enter Table A.1 with the known z value. Note that we will ignore the sign of z because the normal distribution is symmetric. Column B (because we are interested in the area in the tail beyond z) of Table A.1 shows that the required area is .3085, or 30.85%.

Computation and eyeball-estimation give approximately the same result, so we conclude that 30.85% of USH women are shorter than 62 inches.

One final example: Suppose the Acme Manufacturing Company is a large business that employs thousands of people. Acme is proud that employees stay there for many years. They know that the distribution of terms of employment is normal with mean 14 years and standard deviation 6 years. The president of Acme asks you to find out the range of terms of employment for the middle 75% of employees. We see that this is the reverse of the kind of problem we have been considering. Here we know the area under the normal distribution (the middle 75%) and we want the raw score (term of employment). Therefore, we enter the know/want diagram (Figure 6.19) at the right-hand side, and we require two steps to the left. We follow the steps we have used before:

Step 1. Sketch the normal distribution. Term of employment is given to have $\mu = 14$ years and $\sigma = 6$ years, as we show in Figure 6.23.

Step 2. Superimpose a unit normal z-axis. This is the lower axis at the bottom of Figure 6.23.

Step 3. Shade the required area (if necessary). Because we are assuming that the distribution of terms of employment is normal, it is therefore symmetric, and the middle 75% can be divided into two areas: 37.5% above the mean and 37.5% below the mean, as illustrated in the top portions of Figure 6.23.

Step 4. Eyeball-estimate the answer. The shaded regions are each 37.5%, and we know that the area between the mean and $z = 1$ is about 34%. Therefore, each shaded region must have a boundary slightly farther from the mean than $z = \pm 1$—say, at about ± 1.1—so that slightly more than 34% of the area will be included. Our eyeball shows that $z \approx \pm 1.1$ corresponds to term of employment $X \approx 7$ and 21 years.

Step 5. Use the know/want diagram of Figure 6.19. We know that the area between the mean and z is to be 37.5%, so we follow the upper right arrow of the diagram. We enter

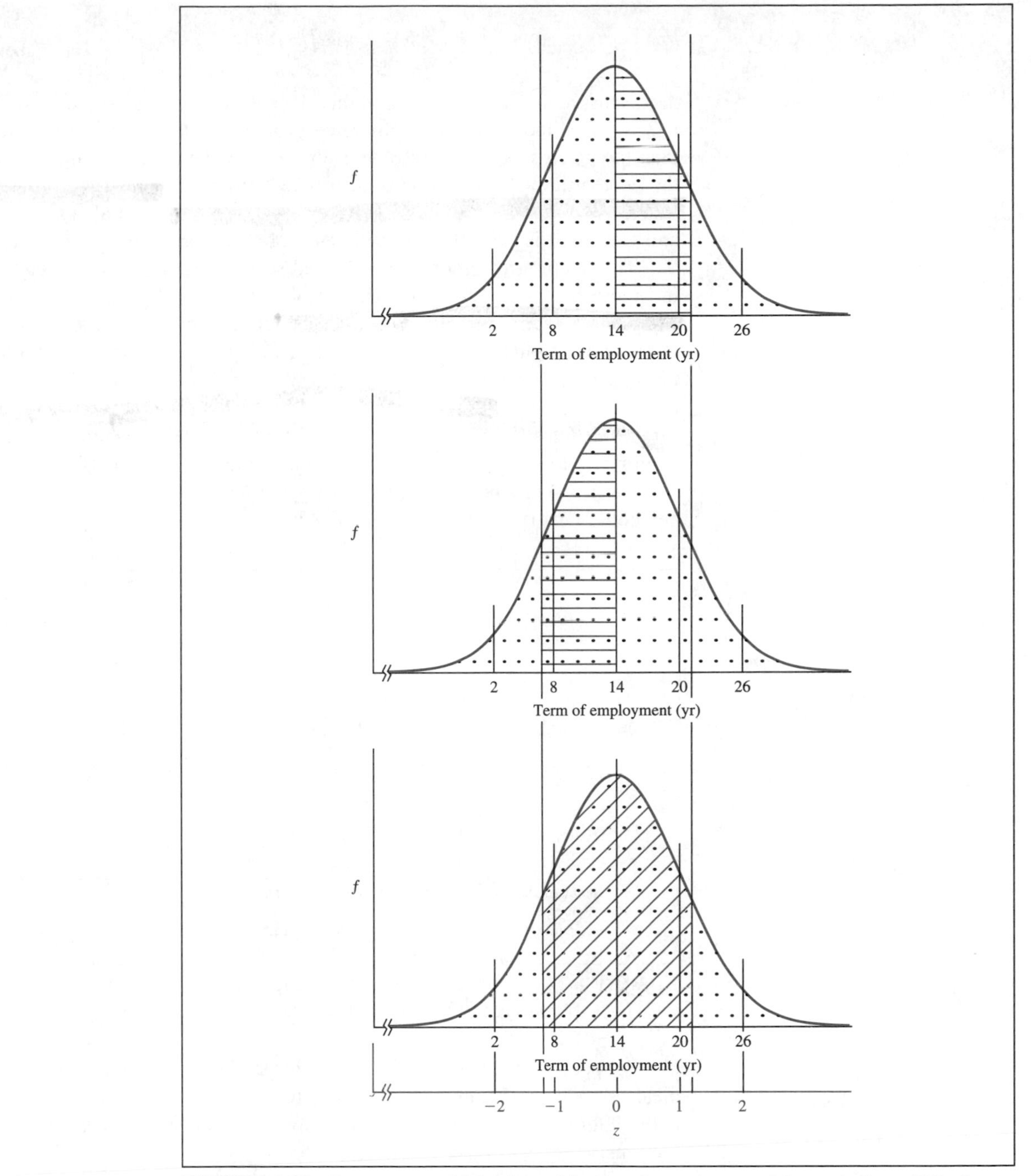

FIGURE 6.23 Frequency distribution of terms of employment (years). *Top:* .375 above the mean. *Middle:* .375 below the mean. *Bottom:* Middle 75% shaded.

Table A.1, column A (*not* the z column!) at .3750 and find that the corresponding z value is 1.15. Then we use the know/want diagram again, where the upper left arrow says we must use Equation (6.3) to obtain the raw score:

$$X = z\sigma + \mu = 1.15(6) + 14 = 20.9 \text{ years}$$

Following the same procedure for the lower area gives

$$X = z\sigma + \mu = -1.15(6) + 14 = 7.1 \text{ years}$$

Computation and eyeball-estimation give approximately the same result, so we conclude that the middle 75% of employees' terms of employment range from 7.1 to 20.9 years.

6.6 Percentiles and Percentile Rank in Normal Distributions

As discussed near the beginning of this chapter, we say that a score is "the 75th *percentile*" if 75% of all the scores in the reference group (population or sample) are equal to or less than the given score. Now we can use what we have learned in this chapter to determine the percentile ranks of normally distributed variables. Thus, we can say for unit normal distributions that $z = 0$ is the 50th percentile (that is, 50% of all z scores are 0 or smaller) and that $z = 1$ is the 84th percentile (84% of all z scores are 1 or smaller).

The relationship among standardized scores, z scores, and percentile ranks is shown in Figure 6.24 with IQ scores used as an example of the standardized score. Note that the relationship between standardized score and z score is linear; for example, a z-score increase of 1 from $z = 0$ to $z = 1$ results in a 15-point increase in IQ (from 100 to 115), and likewise a z-score increase of 1 from $z = 2$ to $z = 3$ also results in a 15-point increase in IQ (from 130 to 145). However, the relationship between percentile and z score (or between percentile and standardized score) is not at all linear; for example, a z-score increase of 1 from $z = 0$ to $z = 1$ results in a 34.13% increase in percentile (from 50.00% to 84.13%), whereas a z-score increase of 1 from $z = 2$ to $z = 3$ results in only a 2.15% increase in percentile rank (from 97.72% to 99.87%).

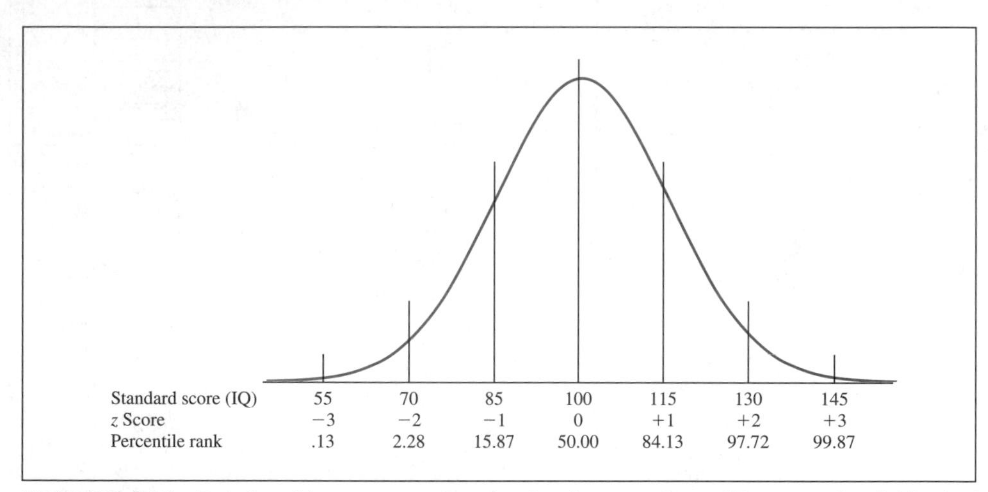

FIGURE 6.24 Relationships among standardized scores (here IQ, normally distributed with $\mu = 100$ and $\sigma = 15$), z scores, and percentile ranks

Suppose again that the heights of female students at USH are normally distributed with mean 64 inches and standard deviation 4 inches as was shown in Figure 6.22. What height is the 90th percentile? We follow the know/want diagram and the steps of the preceding sections:

Step 1. Sketch the normal distribution. USH height is normally distributed with $\mu =$ 64 inches and $\sigma = 4$ inches, as we show again in Figure 6.25.

Step 2. Superimpose a unit normal z-axis. This is the lower axis of Figure 6.25.

Step 3. Shade the required area (if necessary). Because the definition of *percentile* involves the percentage of scores equal to or lower than the given score and we wish to know the 90th percentile, we shade the lower 90% of Figure 6.25.

Step 4. Eyeball-estimate the answer. The area below the mean is (by symmetry) 50%, which leaves 40% above the mean. In the first standard deviation above the mean, 34% is accounted for, so we need to get 6% out of the next ($\approx$ 14%) region. Our eyeball tells us that we need z to be about 1.25 so that the "large quarter" of 14% is 6%. A z of 1.25 corresponds, as we can see by eyeball, to a height of 69 inches.

Step 5. Use the know/want diagram of Figure 6.19. We know that the area between the mean and z must be 40%, so we follow the upper right arrow of the diagram. We enter Table A.1, column A, at .40 and find that the corresponding z value is about 1.28. Then

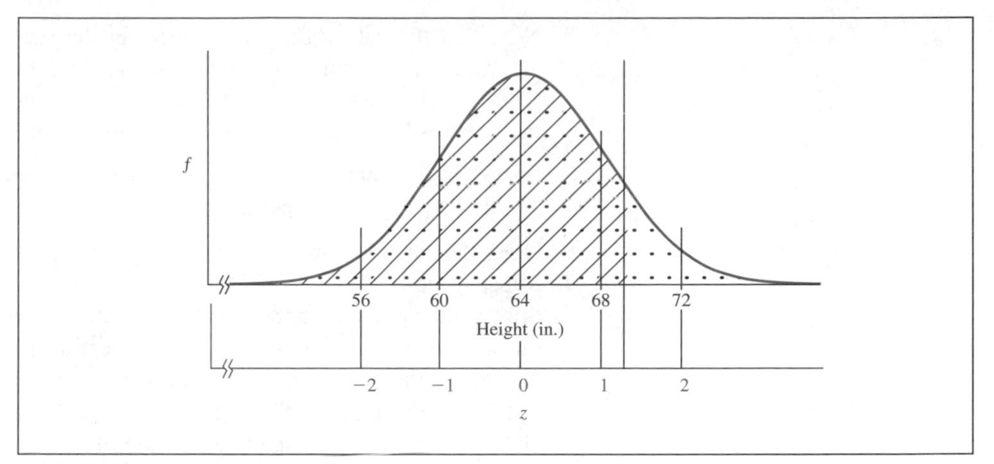

FIGURE 6.25 Distribution of heights at USH with lower 90% shaded, showing that the 90th percentile of height is 69.12 inches

we use the know/want diagram again, where the upper left arrow says we must use Equation (6.3) to obtain the raw score:

$$\text{Height} = z\sigma + \mu = 1.28(4) + 64 = 69.12 \text{ inches}$$

Thus, the height that is the 90th percentile is 69.12 inches.

6.7 Connections

Cumulative Review

In Chapter 1, we introduced the notion that the three basic concepts to master in introductory statistics are: (1) distributions of variables, (2) distributions of means, and (3) distributions of the test statistic. This chapter is the end of our examination of the first concept, the distributions of variables. We now know that distributions of variables can be characterized by three features: their shape (Chapter 3), their center (Chapter 4), and their width (Chapter 5).

Computers

ESTAT datagen

Use ESTAT's datagen to explore the characteristics of the z scores for the Exam 1 data from Table 6.1, first treated as a sample:

1. Enter the 25 values of the Exam 1 data from Table 6.1 into the first column of the datagen spreadsheet. Save these data for later.
2. Create a new column that is identical to the first column: Click anywhere in the second column; then click `Edit Variable`, click `Create a new variable that is the following combination of two variables`, and click `OK`. Alternatively, copy the first column to the clipboard and paste it into the second column.
3. Subtract the sample mean $\overline{X}$ (91.88 obtained from the `datagen Descriptive Statistics` window) from all the values in the second column: Click any cell in the second column, click `Edit Variable`, Enter −91.88 into the `Add this constant to them:` text entry cell, and click `OK`.
4. Divide all the values in the second column by the sample standard deviation s (4.106093 obtained from the `datagen Descriptive Statistics` window): Use your calculator to compute the inverse of 4.106093 (1/4.106093 = .2435405), click any cell in the second column, click `Edit Variable`, click the `Multiply them by this constant:` button, enter .2435405 into the `Multiply them by this constant:` text entry cell, and click `OK`.

5. Note that the values that now appear in the second column are the z scores of the values in the first column treated as a sample. Note in the `datagen Descriptive Statistics` window that the mean of the z scores is 0 and the standard deviation of the z scores is 1 (within the limits of roundoff error).

Now use ESTAT's datagen to explore the characteristics of the z scores for the Exam 1 data from Table 6.1, treated as a population:

6. Open the data file you saved in Step 1.
7. Repeat Step 2.
8. Subtract the population mean μ (91.88 obtained from Table 6.1) from all the values in the second column: Use the same procedure as Step 3.
9. Divide all the values in the second column by the population standard deviation σ (4.023 obtained from Table 6.1). Use your calculator to compute the inverse of 4.023 $(1/4.023 = .24857)$. Use the procedure of Step 4.
10. Note that the values that now appear in the second column are the z scores of the values in the first column treated as a population. Note in the `datagen Descriptive Statistics` window that the mean of the z scores is 0, but the standard deviation of the z scores is 1.02065 instead of 1. That is because datagen displays the *sample* standard deviation. To obtain the *population* standard deviation, use $\sigma = s\sqrt{(n-1)/N}$. Now $\sqrt{24/25} = .97978$, so $\sigma = 1.02065(.97978) = 1.00001$, as close to 1 as can be expected because of roundoff.

Use SPSS to explore the characteristics of the z scores for the Exam 1 data from Table 6.1, first treated as a sample:

1. Enter the 25 values of the Exam 1 data from Table 6.1 into the first column of the SPSS spreadsheet.
2. Obtain the sample mean and standard deviation by clicking `Statistics`, then `Summarize`, then `Explore...`, then ▶, and then `OK`. Note that the sample mean $\overline{X} = 91.88$ and the sample standard deviation $s = 4.1061$ appear in the `!Output1` window.
3. Create a new variable that is the original variable minus $\overline{X}$: Click `Transform` and then `Compute....` Then type "var00002" into the `Target Variable:` text entry cell. Click ▶. Then type "-91.88" into the `Numeric Expression:` window (so that the entire numeric expression reads "var00001-91.88"). Click `OK`.
4. Divide all the values in the second column by the sample standard deviation s (4.1061 obtained from Step 2): Click `Transform` and then `Compute....` Edit the `Numeric Expression:` window so that the entire numeric expression reads "var00002/4.1061." Click `OK`. When the query `Change existing variable?` appears, click `OK`.
5. Note that the values that now appear in the `var00002` column are the z scores of the values in the first column treated as a sample. Obtain the sample mean and standard deviation of the new variable var00002 by clicking `Statistics`, then `Summarize`, and then `Explore....` Click `var00002`, then ▶, and then `OK`. Note that the statistics for var00002 now appear in the `!Output1` window. The sample mean $\overline{X} = .0000$ and the sample standard deviation $s = 1.0000$.

Now use SPSS to explore the characteristics of the z scores for the Exam 1 data from Table 6.1, treated as a population:

6. Open the data file you saved in Step 1.
7. Repeat Step 3. The keystrokes are identical, but now we are treating 91.88 as the population mean μ.
8. Divide all the values in the second column by the population standard deviation σ (4.023 obtained from Table 6.1). Use the procedure of Step 4.
9. Note that the values that now appear in the var00002 column are the z scores of the values in the first column treated as a population. Obtain the mean and standard deviation of var00002 as you did in Step 5. Note that the statistics for var00002 now appear in the `!Output1` window. The mean of var00002 is .0000 but the standard deviation of var00002 is 1.0207 instead of 1. That is because SPSS displays the *sample* standard

deviation. To obtain the *population* standard deviation, use $\sigma = s\sqrt{(n-1)/N}$. Now $\sqrt{24/25} = .97978$, so $\sigma = 1.0207(.97978) = 1.00006$, as close to 1 as can be expected because of roundoff.

Homework Tips

1. Check the list of learning objectives at the beginning of this chapter. Do you understand each one?

2. There is often confusion between *percentile* and *percentile rank*. Remember that the percentile is a *score* and the percentile rank (*of* a score) is a *percentage*.

3. When you work with areas under normal distributions, the clearer your sketch of the area required, the less confusing (and less error-prone) will be your computations.

4. You will use Table A.1 frequently in this homework. It is found in Appendix A. Note that the tables have colored edges so that you can find them easily.

Exercises for Chapter 6

Section A: Basic exercises (answers in Appendix D, page 510)

1. What value of z corresponds to the following?
   (a) The mean
   (b) One standard deviation above the mean
   (c) Two standard deviations below the mean
   (d) One and a half standard deviations above the mean
2. Consider these data, which are the numbers of minutes students take to fuse a random dot stereogram: 3, 7, 2, 5, 6, 4, 9, 14, 4, 11. Remembering to consider the units of your answers, find the following:
   (a) 20th percentile
   (b) Percentile rank of 9 minutes
   (c) Percentile rank of 14 minutes
3. (*Exercise 2 continued*) We know that $\overline{X} = 6.5$ and $s = 3.808$.
   (a) What z score corresponds to 4 minutes?
   (b) What is the number of minutes that corresponds to $z = .657$?
4. Eyeball-estimate the area under the normal curve between $z = 0$ and $z = 1.5$.
5. (a) Sketch a normal distribution and shade the region between the mean and a point z such that z is positive and the shaded region is 47.5% of the total area of the distribution.
   (b) Eyeball-estimate the value of z where this region ends.
   (c) What is the value of z from Table A.1?
   (d) On the same sketch, enter the negative value of the z you obtained and shade the region between the two values of z.
   (e) What percentage of the total area is now shaded?
6. Sketch a normal distribution and shade the region that includes 5% in the right-hand tail. Eyeball-estimate the value of z where this region begins. What is its value from Table A.1?
7. Eyeball-estimate the areas under the normal distributions in the four figures shown here.

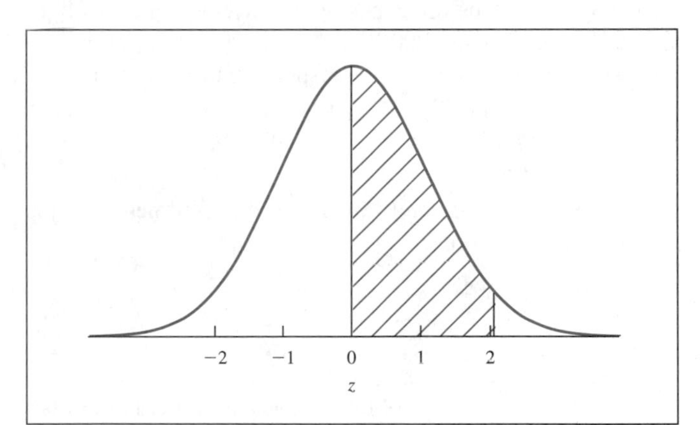

Exercise 7a

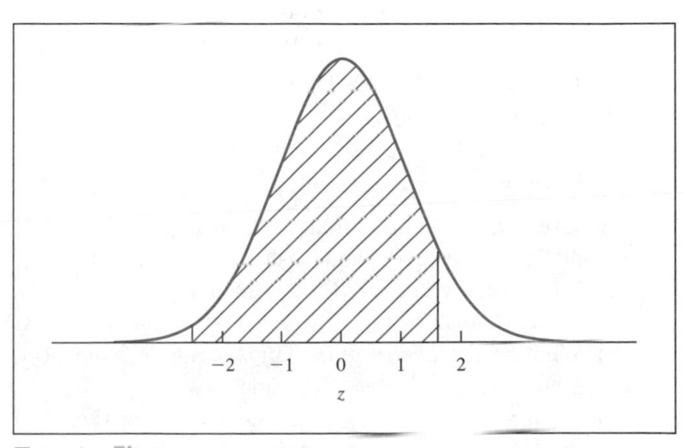

Exercise 7b

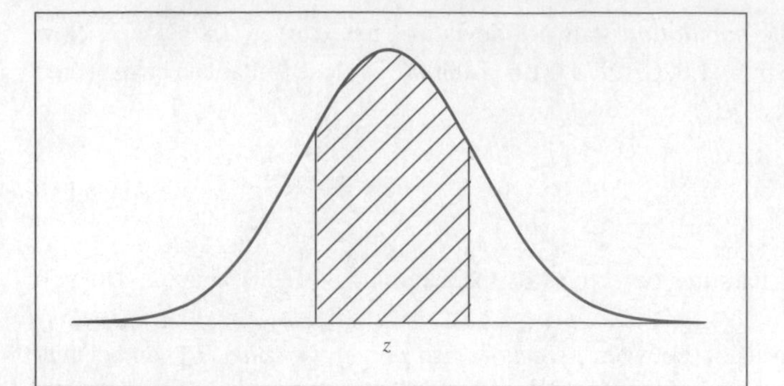

Exercise 7c

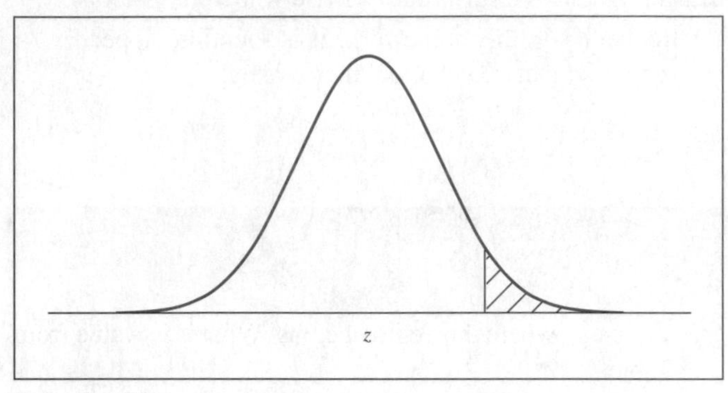

Exercise 7d

8. Suppose that the life expectancies in a population of light bulbs have mean 1000 hours and standard deviation 100 hours. What values of z correspond to these lifetimes?
   (a) 1200 hours
   (b) 900 hours
   (c) 1010 hours
   (d) 1000 hours
   (e) 895 hours
9. (*Exercise 8 continued*) What are the lifetimes of bulbs with these z scores?
   (a) 1
   (b) −2
   (c) 1.38
   (d) −.2
   (e) 0
10. (*Exercise 8 continued*)
    (a) What percentage of bulbs have a lifetime shorter than 700 hours? Use the steps of Section 6.4.
    (b) What percentage of bulbs have a lifetime longer than 1155 hours? Use the steps of Section 6.4.
    (c) What percentage of bulbs have a lifetime longer than 1155.3 hours? Use the steps of Section 6.4.
11. (*Exercise 8 continued*)
    (a) What bulb lifetime is the 99th percentile?
    (b) What is the percentile rank of 927 hours?
12. Suppose that the population of guinea pigs has mean weight 200 grams and standard deviation 40 grams.
    (a) What is the value of z for a guinea pig whose weight is 160 grams?
    (b) What percentage of guinea pigs weigh less than 160 grams? Use the steps of Section 6.4.
    (c) What percentage of guinea pigs weigh between 135 and 235 grams? Use the steps of Section 6.4.
13. Suppose that scores on a statistics exam are normally distributed with mean 80 points and standard deviation 9 points. What percentage of students score above 70 points? Use the steps of Section 6.4.
14. In a normal distribution where $\mu = 0$ and $\sigma = 1$, what are these percentiles?
    (a) 80th (b) 20th (c) 35th
15. In a normal distribution, what is the percentile rank of these z values?
    (a) −2 (b) 1.23 (c) .43
16. In Exercise 6 in Chapter 3, you sketched the distribution of the amounts of Coke in "16-ounce" Coke bottles, where the population mean $\mu = 16.5$ ounces. Then in Exercise 9 in Chapter 5 you eyeball-estimated the standard deviation of that distribution. Suppose that the actual standard deviation is $\sigma = .5$ ounce. What percentage of Coke bottles contain less than 16 ounces of Coke? Use the steps of Section 6.4.
17. A student's college achievement standardized score ($\mu = 500$, $\sigma = 100$) is 565. What would her score be if expressed as follows?
    (a) A z score
    (b) A T score ($\mu = 50$, $\sigma = 10$)
    (c) An IQ-type standardized score ($\mu = 100$, $\sigma = 15$)
    (d) A stanine ($\mu = 5$, $\sigma = 2$; round the result to an integer)

Section B: Supplementary exercises

18. By eyeball, what percentage of the curve is shaded in the accompanying figure? You can verify your answer, if you like, by counting the dots.

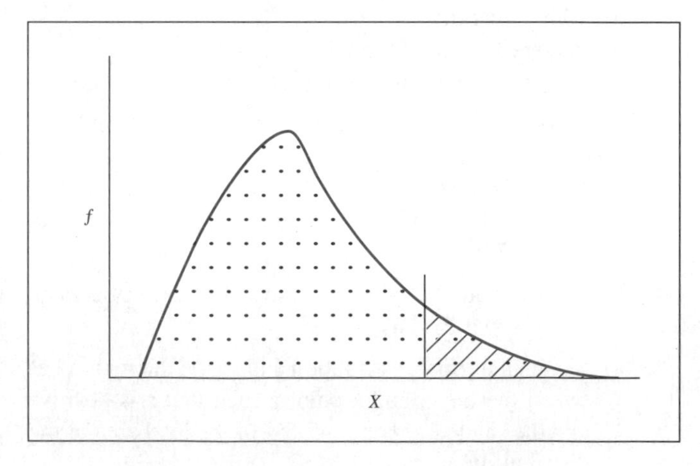

19. What area under the normal curve lies between $z = 0$ and these z values?
    (a) 1.13
    (b) 1.14
    (c) 1.135
    (d) 1.133
    (e) 1.138
20. Suppose that the sizes of houses in Oakdale are normally distributed with mean 1400 square feet and standard de-

viation 325 square feet. What are the z values for houses with these areas (in square feet)?
(a) 1600 (b) 1100 (c) 1425

21. (*Exercise 20 continued*) For the houses in Oakdale, what are the areas of houses that have these z values?
(a) 1.15 (b) −2.9 (c) .33

22. (*Exercise 20 continued*) Sketch the distribution of house sizes in Oakdale. Use the steps of Section 6.4 and determine the percentage of houses that are
(a) Larger than 1600 square feet
(b) Smaller than 1100 square feet
(c) Between 1200 and 1525 square feet
(d) Between 1460 and 1660 square feet

23. The ages of members in the Oakdale Senior Center are normally distributed with mean 70.4 years and standard deviation 4.6 years. What are the z values for individuals of these ages?
(a) 66 (c) 77
(b) 76 (d) 62

24. (*Exercise 23 continued*) How old is an Oakdale Senior Center member who has the given z score?
(a) −1.93 (b) .134 (c) 2

25. (*Exercise 23 continued*) Sketch the distribution of ages of Oakdale Senior Center members. Use the steps of Section 6.4 and determine the percentage of members who are
(a) Older than 65
(b) Younger than 75
(c) Between 68 and 72
(d) Between 72 and 80

26. A student's T score ($\mu = 50$, $\sigma = 10$) was 33. What would her score be if expressed as each of the following? [*Hint:* First compute z.]
(a) A college achievement standardized score ($\mu = 500$, $\sigma = 100$)
(b) An IQ-type standardized score ($\mu = 100$, $\sigma = 15$)
(c) A stanine ($\mu = 5$, $\sigma = 2$; round the result to an integer)

Section C: Extending your comprehension (answers in Appendix D, page 510)

27. In a normal distribution, what value of z has the same area between the mean and z as in the right-hand tail beyond z?

28. In a normal distribution, what value of z has the same area in the right-hand tail beyond z as the value of z itself?

29. In a normal distribution, what value of z has the same area between the mean and z as the value of z itself?

Section D: Cumulative review (answers in Appendix D, page 510)

30. These data are the numbers of days necessary for a sample of subjects to resynchronize their biological clocks after a long airplane trip in the eastward direction: 14, 12, 10, 13, 8, 12, 8, 13, 8, 13, 7, 3, 9, 8, 8, 15, 6, 16, 8, 8, 13. Eyeball-estimate and then compute these measures:
(a) Mean (d) Range
(b) Median (e) Standard deviation
(c) Mode (f) Variance

31. A bushel of tulip bulbs contains 225 bulbs, of which 100 will produce red flowers and 125 will produce yellow flowers. Tulips have either regular or feathery flowers, with the frequencies shown in the table. What is the probability that a bulb chosen at random from the bushel will have the given characteristic?
(a) Yellow
(b) Feathery
(c) Yellow and feathery
(d) Given that a bulb is yellow, what is the probability that it will also be feathery?

	Red	Yellow
Regular	72	91
Feathery	28	34
	100	125

32. Suppose that the amount of sleep per night in the population of 55-year-old individuals is normally distributed with mean 7.1 hours and standard deviation 1.2 hours. What are the z scores for these amounts of sleep?
(a) 6.1 hours (b) 9.1 hours (c) 8.5 hours

33. (*Exercise 32 continued*) How many hours of sleep are associated with these z scores?
(a) 1.43 (b) −2.01 (c) −.44

34. (*Exercise 32 continued*) What percentage of 55-year-olds sleep these amounts of time?
(a) Longer than 9 hours
(b) Less than 6 hours
(c) Between 8 and 9 hours
(d) Between 6.5 and 8.5 hours

Section E: From the journals (answers in Appendix D, page 510)

35. Schmidt (1992, p. 1174) wrote an article about "effect sizes" (exactly what effect sizes are need not concern us here; we will discuss them beginning in Chapter 9), which in Schmidt's situation are normally distributed with mean 0 and standard deviation (called "SE") .38. His main interest was in the effect sizes that lie in the upper 5% of this distribution. For this to be the case, Schmidt stated that "the effect size must be .62 or larger" (p. 1173). To illustrate this situation, he displayed the figure that is reproduced on the following page.
(a) Was Schmidt correct when he computed that 5% of effect sizes should be greater than .62?
(b) Is the figure drawn to scale? That is, does the area of the figure indicated as being "5%" contain approximately 5% of the total area?

(c) In the figure *as it is drawn*, approximately what by eyeball is the area of the region labeled "5%"?

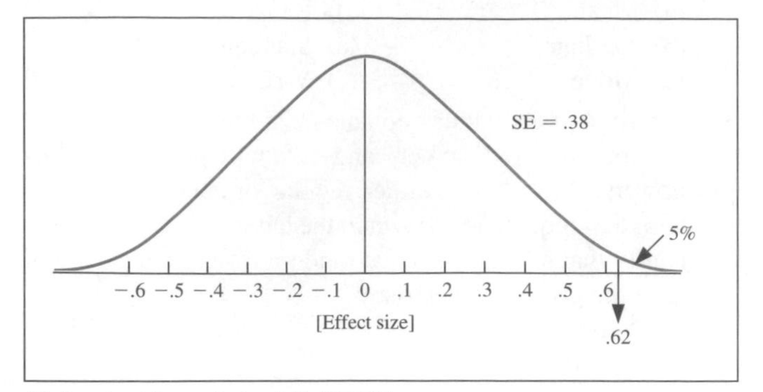

(Adapted from Schmidt (1992))

Section F: Computer explorations

36. Consider these values of X: 123, 109, 81, 97, 94.
    (a) Assuming that these are values from the population of IQ scores ($\mu = 100$, $\sigma = 15$), use the computer to compute z scores for each value.
    (b) Use the computer to compute T scores ($\mu = 50$, $\sigma = 10$) for each of the five z scores from part (a).
    (c) Verify that the highest IQ score corresponds to the highest T score, and the lowest IQ score corresponds to the lowest T score.

37. Consider these values of X: 1, 2, 3, 4, 5.
    (a) Assuming that these data are a sample, use the computer to compute z scores for each value.
    (b) Now add 10 to each original value of X and compute z scores for these new values. How do these z scores compare to those in part (a)?
    (c) Now multiply each original value of X by 10 and compute z scores for these new values. How do these z scores compare to those in part (a)?
    (d) What can you say in general about the effect of addition and multiplication on z scores?

Linear Interpolation (Optional)

This section may be omitted without loss of continuity.

When the table of the normal distribution (or any other table) does not list the precise value we desire, we usually use a process called linear interpolation to find a precise intermediate value. (Methods of interpolation other than linear are useful in particular circumstances, but they are beyond the scope of this book.)

For example, suppose we wish to know the area under the normal curve between the mean and $z = .555$. Table A.1 gives areas between the mean and $z = .55$ (.2088) and between the mean and $z = .56$ (.2123) but not for $z = .555$. We can obtain an approximate value for the area using interpolation.

Because the target value of z (.555) lies halfway between the two tabled values of z (.55 and .56), interpolation tells us that the desired area will also be halfway between the two corresponding areas (.2088 and .2123). Thus, the interpolated area is $(.2088 + .2123)/2 = .2106$.

Suppose we wish to know the area in the tail of the normal distribution beyond $z = 1.1075$. The table gives areas for $z = 1.10$ (.1357) and $z = 1.11$ (.1335), and we plot those two points (connected with a straight line) in Figure 6.26. Because $(1.1075 - 1.10)/(1.11 - 1.10) = .75$, the target value of z is three-quarters (.75) of the way from 1.10 to 1.11. The desired area will therefore be three-quarters of the way from .1357 to .1335. Because the distance from .1357 to .1335 is $.1335 - .1357 = -.0022$, three-quarters of that distance is $.75(-.0022) = -.00165$, so the interpolated area is .00165 *less* than .1357, or .1340.

Exercise for Optional Section

(Answers in Appendix D, page 510)

1. What area under the normal curve lies between $z = 0$ and these z values?
   (a) .70
   (b) .71
   (c) .705
   (d) .708

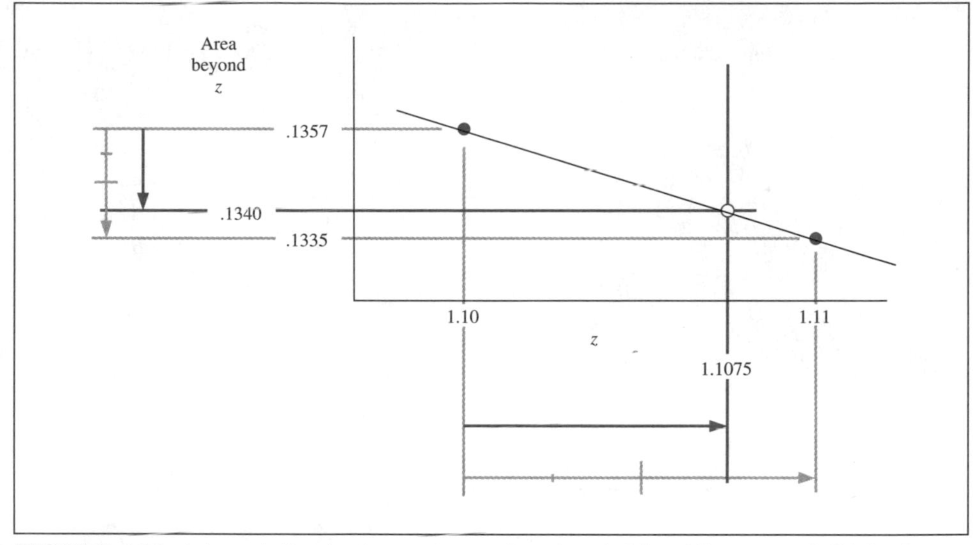

FIGURE 6.26 Interpolation between two values in a table of the normal distribution.

Samples and the Sampling Distribution of the Means

7

CHAPTER

LEARNING OBJECTIVES

1. What are representative samples? Give two ways that they can be chosen.
2. What are the dangers of choosing a sample that does not represent the population?
3. In what ways do samples usually differ from a parent population?
4. What is the sampling distribution of the means?
5. What are three important characteristics of the sampling distribution of the means as specified by the central limit theorem?
6. What two factors affect the magnitude of the standard error of the mean?
7. In what ways may the sampling distribution of the means be useful?

This chapter describes how to take random samples from populations and the characteristics of those samples. It describes one of the most important concepts in statistics: the sampling distribution of the means of samples. The central limit theorem states that this distribution of sample means: (1) is normally distributed (if n is sufficiently large) regardless of the shape of the distribution of the original variable, (2) has a mean equal to the mean of the variable, and (3) has a standard deviation (called the "standard error of the mean") equal to the standard deviation of the variable divided by the square root of the sample size. Because sample means are normally distributed, the table of areas under the normal distribution (Table A.1 in Appendix A) can be used to compute the relative frequency of particular ranges of sample means.

As we have seen since Chapter 4 in our examination of the data from the study on Pygmalion in the classroom, the mean of the Oak School "bloomers" intellectual growth was 16.5 IQ points. You may recall that this experiment selected the children to be called "bloomers" using a table of random numbers; you will learn the techniques of using such a table in this chapter. The sample thus obtained at Oak School is reproduced in Table 7.1.

TABLE 7.1 Intellectual growth (IQ gain) of Oak School second-grade bloomers*

Student	*Intellectual growth (IQ gain)*
Kathy	20
Tony	14
Mario	69
Louise	13
Juan	−6
Able	25
Patricia	1
Douglas	−4
Baker	24
Charlie	12
Delta	19
Echo	11

*Data reprinted from Table 1.1, based on Rosenthal & Jacobson (1968)

Suppose in a hypothetical situation that Maple School is identical to Oak School (identical building, identical teachers, identical students, etc.) and we run an identical study there. The particular Maple School children identified as "bloomers" might be George, Veronica, and so on. What would you expect the mean intellectual growth for the Maple School bloomers to be? The Maple School sample mean might turn out to be exactly 16.5 IQ points, just as at Oak School, but because the sample is chosen randomly in each school, you shouldn't be surprised if the Maple School sample mean turns out to be 14.7, or 19.3, or some other value instead of 16.5 IQ points. The Oak School bloomers and the Maple School bloomers are two samples from the same population, and you will learn to expect that two sample means from the same population will be similar but not identical. The central limit theorem will help us understand just how similar they can be expected to be.

Recall that the three major concepts involved in statistics are: the distribution of a variable, the sampling distribution of the means and how it is related to the distribution of the variable, and the test statistic and how it is related to the distribution of means. In Chapters 3–6, we explored the first of these concepts and became familiar with frequency distributions, their description, and use. In this chapter, we will turn to the second main skill in statistics: understanding the sampling distribution of the *means*. The concept of the "sampling distribution of the means" is actually quite simple. It is sometimes confusing to students encountering it for the first time, however, so let's outline the whole idea here and then spend the rest of the chapter elaborating that outline.

Suppose we draw four random samples from a population and then compute the sample mean of those four samples. We call that mean $\overline{X}_1$, shorthand for "the mean of the first sample of size four." Now we repeat the process: Draw another four samples and compute the sample mean of those four, which we call $\overline{X}_2$, shorthand for "the mean of the second sample of size four." Now we draw a third sample and compute another sample mean ($\overline{X}_3$), and another sample (yielding $\overline{X}_4$), and another and another until we have many sample means. If we created a histogram of all those $\overline{X}$ values, we would have a visual representation of a distribution, just as we did in Chapter 3. However, the distribution represented by that histogram would *not* be the distribution of values of a variable (as was the case in Chapter 3) but instead the distribution of values *of means* (of samples of size four). We call this distribution "the sampling distribution of the means" of samples of size four. We will explore the characteristics of this distribution in this chapter.

That's the whole concept of this chapter in a nutshell. Your comprehension of it at this point may be somewhat hazy; our task is to clarify all the aspects of that description. In order to do that, we first need to solidify our understanding of the distinction between samples and populations and then discuss how random samples are selected.

7.1 Random Samples

In Chapter 1 we distinguished between a population and a sample. A population is the group that is the focus of our interest, and a sample is some subset of that group, interesting not in its own right but because it represents in some way or is in some way similar to the larger population. It is implicit in this discussion that several or many different samples *could* be selected from any population if we so desired. For example, if the population of interest is the residents of the United States (approximately 270 million individuals), we could take a huge number (270,000) of different samples, each of size 1000, that would not have *any* of the same people in common.

representative sample: a sample of a population that reflects the characteristics of the parent population

random sample: a sample chosen so that each member of a population has an equal chance of being included in the sample

If we want a sample to reflect the properties of the parent population, then we must ensure that the sample is chosen so that the sample members are *representative* of the population members. For most statistical purposes, the best way to do that is to select a random sample from the population.

A *random sample* is a subset of a population chosen so that each member of the population *has an equal chance* of being selected to be in the sample, and the selection of each member is independent of whether or not any other particular member is selected. This means that investigators must set aside their own preferences or conveniences and use some purely chance method to select the sample from the target population.

The detailed consideration of the techniques one uses to assure representativeness is more properly a topic in research design than in statistics, but it is important here that we recognize how important choosing a representative sample is (see Box 7.1).

BOX 7.1 The risks of taking a nonrepresentative sample

The Literary Digest, in early November 1936, was one of the most prestigious, successful magazines in the United States and had been for decades. By 1938, it had disappeared completely, primarily because it took a nonrepresentative sample. Here's the story.

The magazine had built its reputation by taking and publishing opinion polls: "For nearly a quarter century, we have been taking Polls of the voters . . . in Presidential years . . . . So far we have been right in every poll . . . . Any sane person can not escape the implication of such a gigantic sampling of the popular opinion as is embraced in *The Literary Digest* straw vote . . . . *The Literary Digest* poll is an achievement of no little magnitude. It is a Poll fairly and correctly conducted."[1]

(continued)

[1] "Landon, 1,293,669; Roosevelt, 972,897," *The Literary Digest*, 122, October 31, 1936, p. 5.

BOX 7.1 *(continued)*

The magazine reported the results of "*The Digest*'s Poll of Ten Million Voters" that showed a landslide victory (55% to 41%) for Republican candidate Alf Landon over incumbent Democrat Franklin Roosevelt. A week later, the actual election outcome was a landslide *for Roosevelt* (62% to 38%). How could *The Literary Digest* have made such a large, embarrassing mistake?

Subsequent analysis demonstrated that the sample was *not representative* of the population of voters. The magazine had chosen its sample by obtaining lists of automobile owners and telephone owners, and it mailed poll ballots to those people. The population of automobile and telephone owners was generally upper or upper middle class; less wealthy individuals were less likely to have automobiles or telephones. Thus, all members of the population of interest (that is, the population of voters) did *not* have an equal chance of being selected (the wealthy had a better chance than did the poor).

Landon was indeed the candidate favored by the wealthy, but Roosevelt was the candidate favored by the people as a whole. *The Literary Digest*'s failure to understand that automobile and telephone owners were not necessarily representative of the whole population destroyed the magazine's credibility and led to its demise.

We must be extremely careful when choosing samples that we do not allow some unforeseen bias to creep into our selection procedure. There are many ways of choosing a representative sample from a population (see texts on experimental design). We will describe next the one that is the most straightforward: simple random sampling.

Simple Random Sampling

simple random sampling: a random sampling technique where all members of the population are treated equally regardless of their characteristics

Simple random sampling is the most straightforward and frequently used method of obtaining a representative sample. Conceptually, the names (or numbers) of all the individuals in the population are put into a hat and mixed thoroughly, and then a sample of the desired size is drawn out, with a thorough mixing between draws.

In actual practice, the use of a hat or other mechanical means of mixing may not be sufficiently random because names of adjacent individuals may "stick together"; an example of this difficulty is described in Box 7.2. The best way to avoid such risks is to use a table of random digits, such as the one in Table A.10, "4000 random digits," in Appendix A, or a computer programmed to generate sequences of random digits.

BOX 7.2 An example of the risks of taking a quasi-random sample

This example demonstrates the risks involved in using a quasi-random sample rather than an actual random sample. In the 1960s, the U.S. government reinstituted the military draft to supply troops for the Vietnam War. Previous drafts had had several undesirable consequences; for example, young men had no information about how likely it was that they would be drafted, thus making important life decisions difficult.

To help solve this problem, the government decided on a lottery system to be operated as follows: Each of the dates of the year would be marked on a ball and the 366 balls put into a barrel and mixed thoroughly. Then the balls would be drawn out, one at a time, and placed in order. Men whose birthdays fell on the date marked on the first-drawn ball would be drafted first. When that draft pool was exhausted, men whose birthdays fell on the second date would be called, and so on. If a man's birthday "draft number" was low, he could be assured of being drafted, but if his number was closer to 366, he could be confident of avoiding the draft.

(continued)

BOX 7.2 *(continued)*

The draw took place, and it was noticed that birthdates in January and February tended to cluster in the high draft numbers, whereas birthdates in November and December tended to cluster in the low draft numbers. Was this the result of chance or of inadequate mixing of the balls in the barrel? The ensuing debate was quite heated, with some favoring a redraw and some favoring letting the existing draw stand. Such a debate has no adequate statistical resolution, but it could have been avoided if the government had used the less dramatic but more defensibly fair method of creating the draft number list using a table of random numbers rather than the quasi-random shuffling of balls in a barrel.

A table of random digits is used as follows: Suppose we wish to have a sample of size 16 from the students who take statistics at XYZ University, and suppose further that there are a total of 80 statistics students. Conceptually speaking, we would write each student's name on a separate slip of paper, put the slips into a hat, mix them, and choose 16 slips. A more thoroughly random procedure is to assign a number from 1 to 80 to each student (it does not matter how these numbers are assigned; alphabetically is the most usual way). Then we take Table A.10, close our eyes, and point at any spot in the table. The random digit that we point at is called the *seed*, which becomes the beginning of our sequence of random numbers. Table 7.2 shows a section of the table of random digits, and our seed happens to be the "5" that is the eighth digit in the third row (shown in boldface type).

seed: the point of entry into a random number table, usually chosen in a quasi-random manner

TABLE 7.2 Excerpt from a table of random digits, including "seed" digit (shown in bold type)

46162	83554	94750	89923	37089	20048
70297	34135	53140	33340	42050	82341
39979	26**5**75	57600	40881	22222	06413
12860	74697	96644	89439	28707	25815
40219	52563	43651	77082	07207	31790

Because we want to select numbers between 1 and 80, we need to create a string of two-digit numbers, beginning with the seed digit: 57, 55, 76, 00, 40, 88, 12, 22, 22, 06, and so on. If there had been 800 students in our original population instead of 80, we would have needed to take *three* random digits at a time to form a string of three-digit random numbers: 575, 576, 004, 088, and so on. If there had been 1000 students, we would still need three-digit random numbers, assuming that we take the sequence "000" to indicate the 1000th student. If there had been 1001 students, then we would have needed four-digit random numbers.

The first member of our sample is selected by using the first of our string of two-digit numbers (57); thus, the 57th student is the first member of our sample. The 55th student becomes the second member of our sample, and the 76th student is the third. The next two-digit number is 00, but because we didn't use the number 0 (we numbered our students from 1 to 80), we simply discard the number 00 from our string. Next is 40, so we take the 40th student. The next is 88, but because there is no 88th student (the highest student number is 80), we discard both those digits. Then we take the 12th and the 22nd students. The next two-digit number is 22, but because we have already selected the 22nd student, we discard those digits also.

We continue in a similar manner, selecting sample members or discarding numbers from our string (if they are out of range or if they have already been used) until we have selected the 16 members we required. Notice that blanks in the table are irrelevant; they are there for ease in counting but are totally ignored in use. If we reach the end of the row in the table of random digits, we simply wrap around to the next row and continue.

This procedure guarantees that each student in our class has an equal chance of being selected for our sample. We selected the 22nd student not because she was rich or smart, or because she sat in the front row, or because she was eager to volunteer, but because a totally disinterested table of numbers told us to select the 22nd student. Box 7.3 describes a second frequently used way of choosing a sample: stratified random sampling.

BOX 7.3 Stratified random sampling

stratified random sampling: a sampling procedure where the population is divided into subgroups (strata) that have the same or similar characteristics, and then simple random samples are taken from each stratum

A second frequently used procedure for selecting random samples depends on our ability to identify "strata" or subpopulations within the total population that may be different from one another. We wish to ensure that our sample accurately represents this diversity. For example, we may know that statistics students include 20 freshmen, 30 sophomores, 20 juniors, and 10 seniors (a total of 80 students). Thus, the student population is 25% (20 of 80) freshmen, 37.5% (30 of 80) sophomores, 25% juniors, and 12.5% seniors. If we think that class standing is an important variable, we would probably want our sample also to contain 25% freshmen, 37.5% sophomores, 25% juniors, and 12.5% seniors.

In this example, the strata are the class standings. Again suppose that our total sample size is to be 16. We would then want our sample to contain $.25(16) = 4$ freshmen, $.375(16) = 6$ sophomores, $.25(16) = 4$ juniors, and $.125(16) = 2$ seniors.

Now that we have identified the strata and determined the sample size to be drawn from each stratum, we proceed to take a simple random sample *from each stratum*, exactly as we describe in the text. Thus, we number the freshmen from 1 to 20, identify a seed digit, and use two-digit numbers from a table of random digits until we have a freshman sample of size 4. We repeat the process for the sophomores, for the juniors, and then for the seniors.

Our stratified random sample has 16 members in it, just like our simple random sample. The difference is that the stratified sampling method guarantees that each stratum is proportionately represented in the sample. In simple random sampling, that is not always true.

Samples from Populations

Statistical inference, the topic of the remainder of this text, is basically the attempt to answer the question: What can we say about a population (whose parameters we do not know) when all we know are the statistics of a sample? To begin our understanding of that question, we will first ask it in reverse: What are the characteristics of samples when we know the parameters of the population from which those samples are drawn? Then we will return to the more interesting question.

We begin by considering Figures 7.1 through 7.6, which all show samples of size 25 drawn from the population of weights of sixth-grade boys in a particular school district. For purposes of this discussion, we assume that the population of weights is normally distributed with mean 99 pounds and standard deviation 4 pounds. Our task is to gain some sense of what might happen if we repeatedly drew random samples from this same population. Figures 7.1 through 7.6 show the population normal distribution ($\mu = 99$, $\sigma = 4$ pounds) superimposed on the histograms from six such samples.

Inspect the six samples. Are all identical to the parent population (normally distributed, $\mu = 99$ pounds, $\sigma = 4$ pounds)? No, they are not. The six samples are *not* identical to one another either (for example, the minimum weight in Figure 7.1 is about 92 pounds, whereas the minimum weight in Figure 7.2 is about 84 pounds), and they are *not* identical to the parent population (for example, the sample in Figure 7.1 is skewed negatively, whereas the sample in Figure 7.6 is bimodal). Our first conclusion is that the *shape* of the distribution of a sample does *not* necessarily reflect the shape of the distribution of the population from which it is drawn, at least when the sample size is relatively small. Stated another way, it

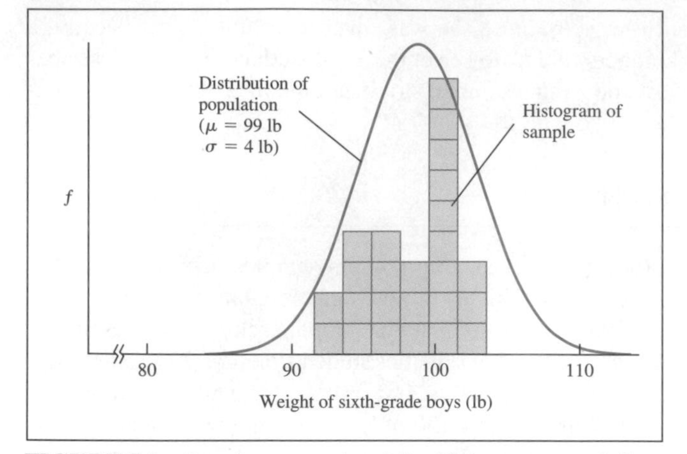

FIGURE 7.1 Random sample of size 25 from a population of weights of sixth-grade boys

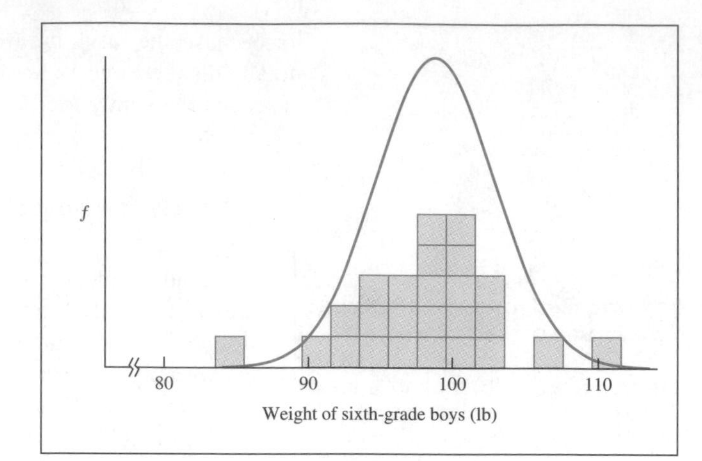

FIGURE 7.2 Another random sample of size 25 from a population of weights of sixth-grade boys

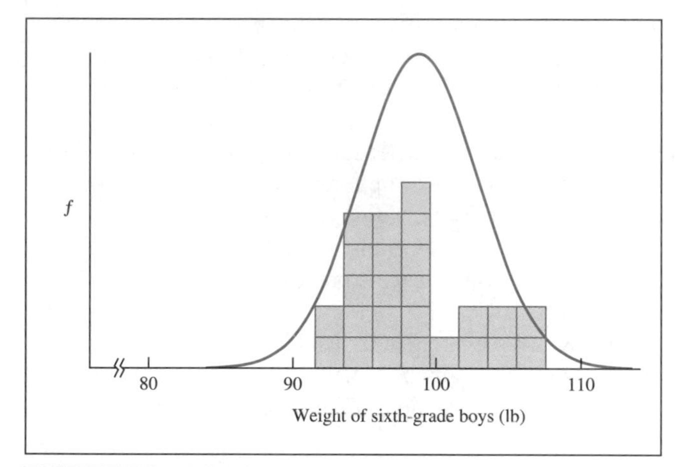

FIGURE 7.3 A third random sample of size 25 from a population of weights of sixth-grade boys

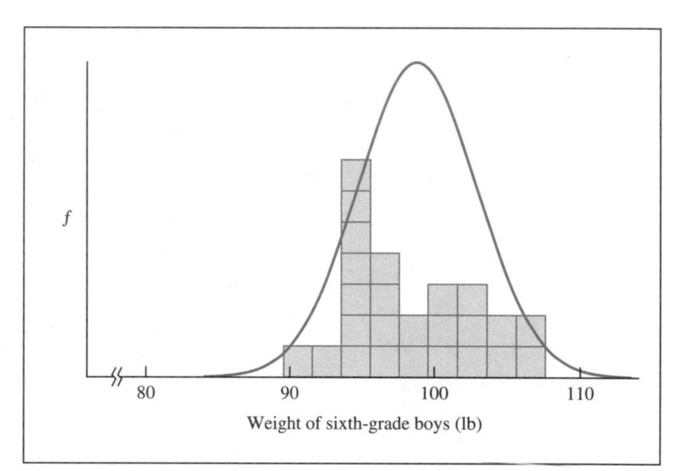

FIGURE 7.4 A fourth random sample of size 25 from a population of weights of sixth-grade boys

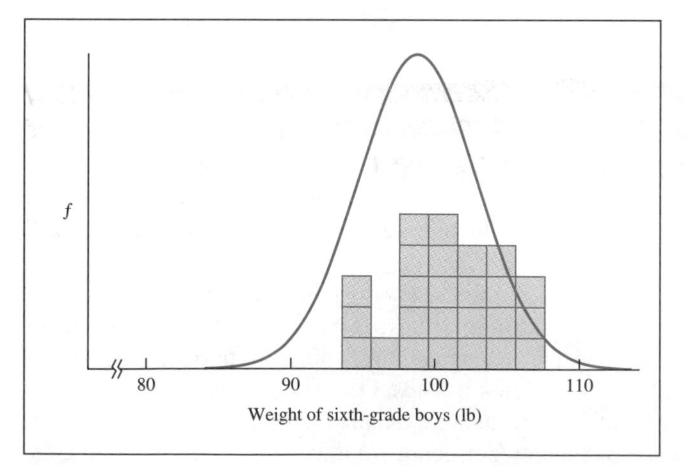

FIGURE 7.5 A fifth random sample of size 25 from a population of weights of sixth-grade boys

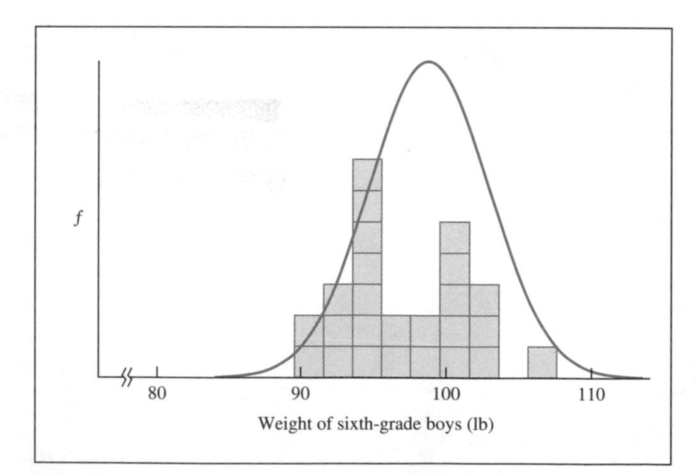

FIGURE 7.6 A sixth random sample of size 25 from a population of weights of sixth-grade boys

takes a relatively large sample size for the shape of the sample distribution to match reliably the shape of the parent population.

Now let us consider the *means* of the six samples shown in Figures 7.1 through 7.6. Eyeball-estimate the sample means in those figures by the balancing-point method; my estimates are 99, 98, 98, 99, 101, and 98 pounds, respectively. Those values *are* all rather similar and *do* reflect the *center* of the parent population normal distribution, where

$\mu = 99$ pounds. Thus, although the *shape* of the sample distribution *does not* approximate the shape of the parent population distribution, the *mean* of the sample *does* approximate the mean of the parent population. (We will leave to other texts the question of how well the variation of the sample reflects the variation of the parent population.)

The fact that the sample mean lies fairly close to the population mean is one of the most important characteristics of random samples, and most of the rest of this textbook will explore that fact. It cannot be said strongly enough: The concept of the sampling distribution of the means is one of the most important ideas in statistics.

7.2 The Sampling Distribution of the Means

In each of Figures 7.1 through 7.6, we drew a random sample of size 25 from a population that was normally distributed with mean 99 pounds and standard deviation 4 pounds. We saw that the mean of each sample was fairly close to 99 pounds. Now we must develop more precisely this concept of "fairly close."

In Figure 7.7, we present another six random samples of size 25 from the normal distribution with mean 99 pounds and standard deviation 4 pounds, but this time we explicitly show each sample mean by placing a triangular fulcrum at the balancing point of the sample. We can see that each of the sample means is near the middle of the normal population curve, but each sample mean is somewhat to the right or left of center.

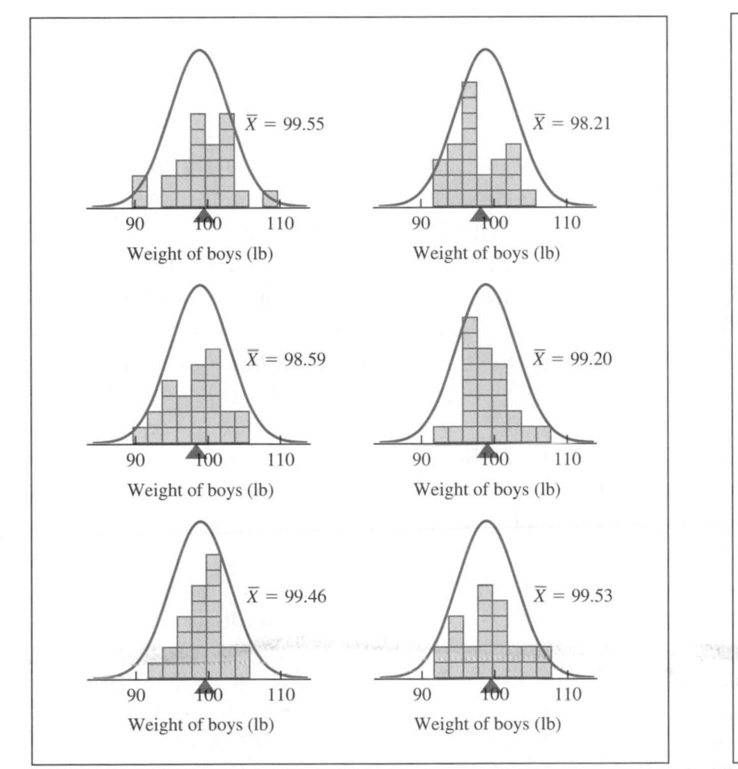

FIGURE 7.7 Six random samples of size 25 from the same population

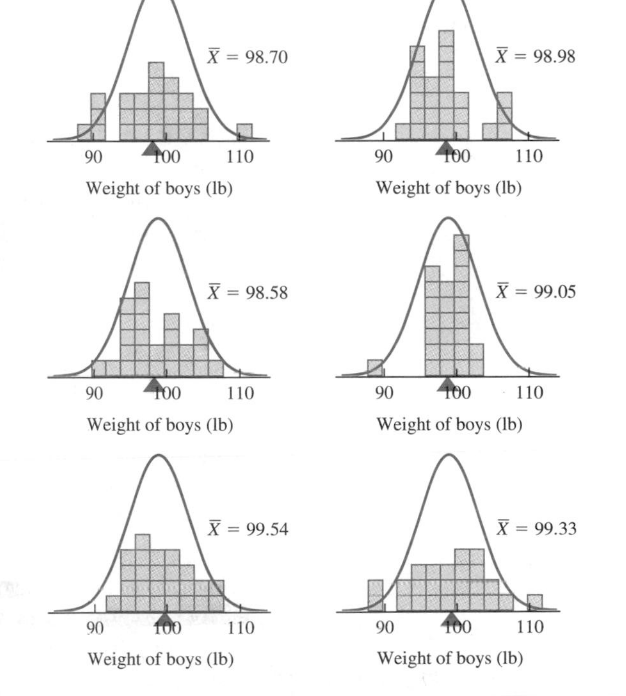

FIGURE 7.8 Six more random samples of size 25 from the same population

Figures 7.8 and 7.9 continue this same process, each showing six more 25-member samples from the same population, for a total of 18 samples where we have indicated the mean with a triangular fulcrum. Inspection of these three figures shows that the smallest sample mean is 96.71 pounds (in the bottom left corner of Figure 7.9) and the largest is 100.00 pounds (in the middle left of Figure 7.9).

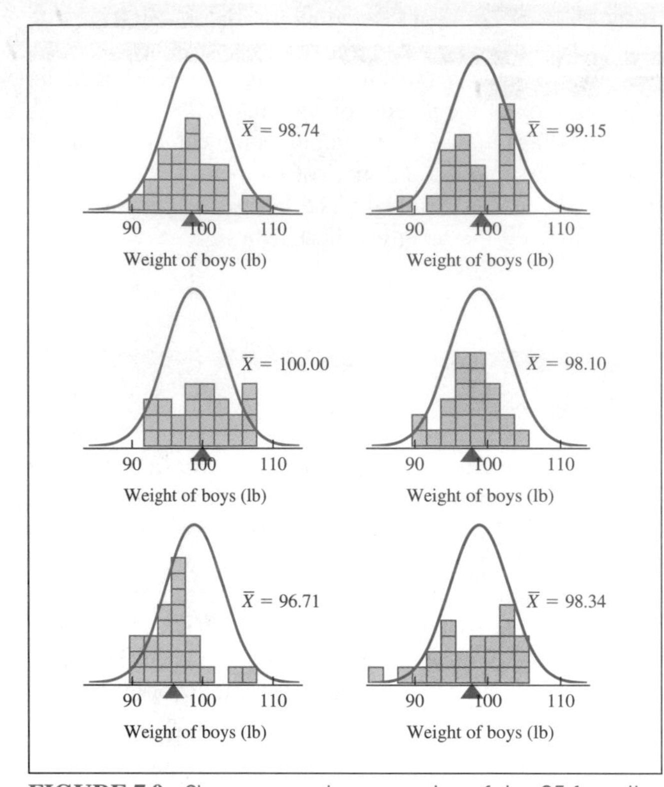

FIGURE 7.9 Six more random samples of size 25 from the same population

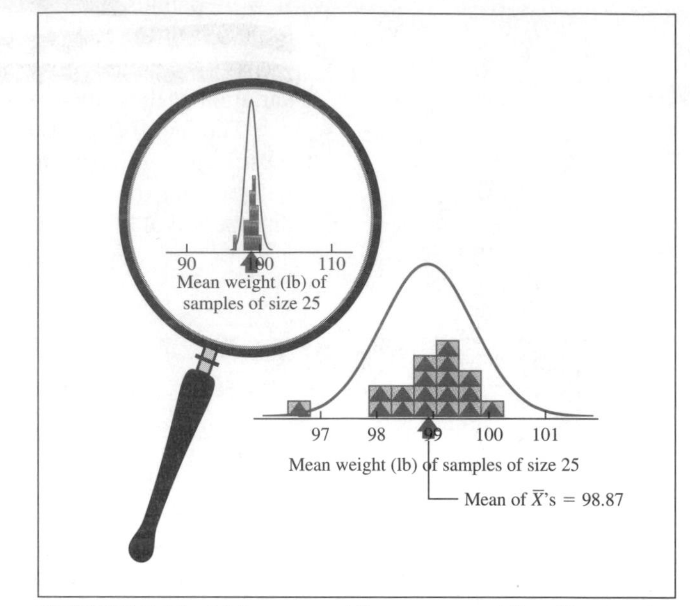

FIGURE 7.10 Histogram of the means of the 18 samples of size 25 shown in Figures 7.7, 7.8, and 7.9. The distribution of means as specified by the central limit theorem is superimposed. Note that this distribution of the means is five times as narrow as the original population, so it is magnified to show its details.

We now have 18 *means of samples of size 25*: 99.55, 98.21, 98.59, 99.20, and so on (and we could have more than 18 if we wished to continue to draw random samples of size 25 from the parent population). We take those 18 means and create a frequency distribution *of the means*, shown in Figure 7.10. Because the 18 means are all similar in magnitude, the distribution of those means is quite narrow. To see its details, we must magnify its width, as shown in Figure 7.10. In that figure, each of the sample means of Figures 7.7–7.9 is represented as a square, with the triangular fulcrum shown to emphasize that the squares in this distribution are *means* (each of 25 observations), not single observations. Thus, the square with the triangle at the extreme left of Figure 7.10 is the smallest of the 18 sample means, 96.71 pounds (from the bottom left of Figure 7.9), and there are 18 total squares, one for each of the 18 samples shown in Figures 7.7–7.9. This new distribution, because its elements are means rather than individual observations, is called a "distribution of means of samples of size 25."

Figure 7.10 has a normal distribution drawn over the sample means. As in all the other figures in this chapter, that normal distribution is the population from which each individual square is one sample. Thus, the normal distribution in Figure 7.10 is the distribution of all possible *means* of samples of size 25 that could be drawn from a population whose single elements have mean 99 pounds and standard deviation 4 pounds. We refer to this distribution as the *sampling distribution of the means* of samples of size 25. When we inspect Figure 7.10, we see that the mean of the sampling distribution of the means is quite close to the mean of the original population (99 pounds), but its standard deviation is *smaller*. The right-side inflection point is just under 100 pounds, so the standard deviation must be less than 1 pound.

sampling distribution of the means: The distribution formed by taking repeated samples from the same population, computing the mean of each sample, and forming the distribution of those sample means

We could, in principle, take all possible random samples of size 25 from the parent population, compute the sample means, and then compute the mean and standard deviation *of those means* to get the exact values of the mean and standard deviation of the sampling distribution of the means. Fortunately, there is an easier way, given by the central limit theorem, which we describe in the next section.

Take a look at the first sentence of the preceding paragraph; the word *mean* occurs five times and has three different meanings. Let us be clear about our notation:

$\overline{X}$ is the mean of a sample (here for samples of size 25).
μ is the mean of the population of original observations.
$\mu_{\overline{X}}$ (read "mew sub *X* bar") is the mean of the population of *means of samples* (of size 25).

We also need to be clear about terminology for the standard deviation:

s is the standard deviation of a sample.
σ is the standard deviation of the population of original observations.
$\sigma_{\overline{X}}$ (read "sigma sub *X* bar") is the standard deviation of the population of *means of samples* (of size 25). We give this standard deviation its own name: the *standard error of the mean*.

standard error of the mean: the standard deviation of the sampling distribution of the means

Thus, $\mu_{\overline{X}}$ is the mean of the distribution of all possible $\overline{X}$'s of samples of size 25, and $\sigma_{\overline{X}}$ (the standard error of the mean) is the standard deviation of that same distribution.

The Central Limit Theorem

central limit theorem: the theorem that describes the shape and parameters of the sampling distribution of the means

The *central limit theorem* is one of the fundamental theorems of statistics because it specifies the three important characteristics (shape, mean, and variation) of the sampling distribution of the means. It thus gives us a way of finding the values of $\mu_{\overline{X}}$ and $\sigma_{\overline{X}}$ without having to draw all possible samples. Thus the central limit theorem (which we will accept here without proof) states three characteristics of the distribution of sample means:

1. As the sample size n increases, the sampling distribution of the means of samples of size n approaches a normal distribution. This should be no surprise if the parent distribution is itself normal (as was the case in Figures 7.1–7.9). However, the theorem also states that even if the parent population is *not* normal (bimodal, skewed, etc.), the distribution of *means* of large samples drawn from the parent population *will* be normal. See Box 7.4.

BOX 7.4 Example of how the means from a nonnormal distribution approach normality, as predicted by the central limit theorem

The central limit theorem states that means of samples drawn from a nonnormal distribution will themselves be normally distributed if the sample size on which each mean is computed is sufficiently large. We illustrate that by considering a distribution that is clearly *nonnormal*: the outcomes of flipping a coin.

The two possible outcomes of a single flip are heads (which we will call $X = 1$) and tails (which we will call $X = 0$). The top portion of Figure 7.11 shows the results of an experiment that flipped 100 coins. This distribution is clearly not normal; in fact, it is completely bimodal.

Now consider the distribution of the *means* of *four* flips. The possible outcomes for the mean of four flips are all heads ($\overline{X} = 1$), three heads and a tail ($\overline{X} = .75$), two heads and two tails ($\overline{X} = .5$), one head and three tails ($\overline{X} = .25$) and all tails ($\overline{X} = 0$). The results of an experiment that flipped 100 sets of four coins, computing the mean for each set of four, is shown in the middle portion of Figure 7.11.

The bottom portion of Figure 7.11 shows the results of an experiment that flipped 100 sets of *ten* coins at a time and computed the mean of each set of ten, with possible outcomes $\overline{X} = 1.0$, $\overline{X} = .9$, $\overline{X} = .7$, $\overline{X} = .6$, and so on.

Observe that as the sample size increases (from one to four to ten), the distribution of means becomes more and more normal (unimodal and symmetric), as the central limit theorem predicts. The more coins we flip at a time, the more normal the distribution of means will be.

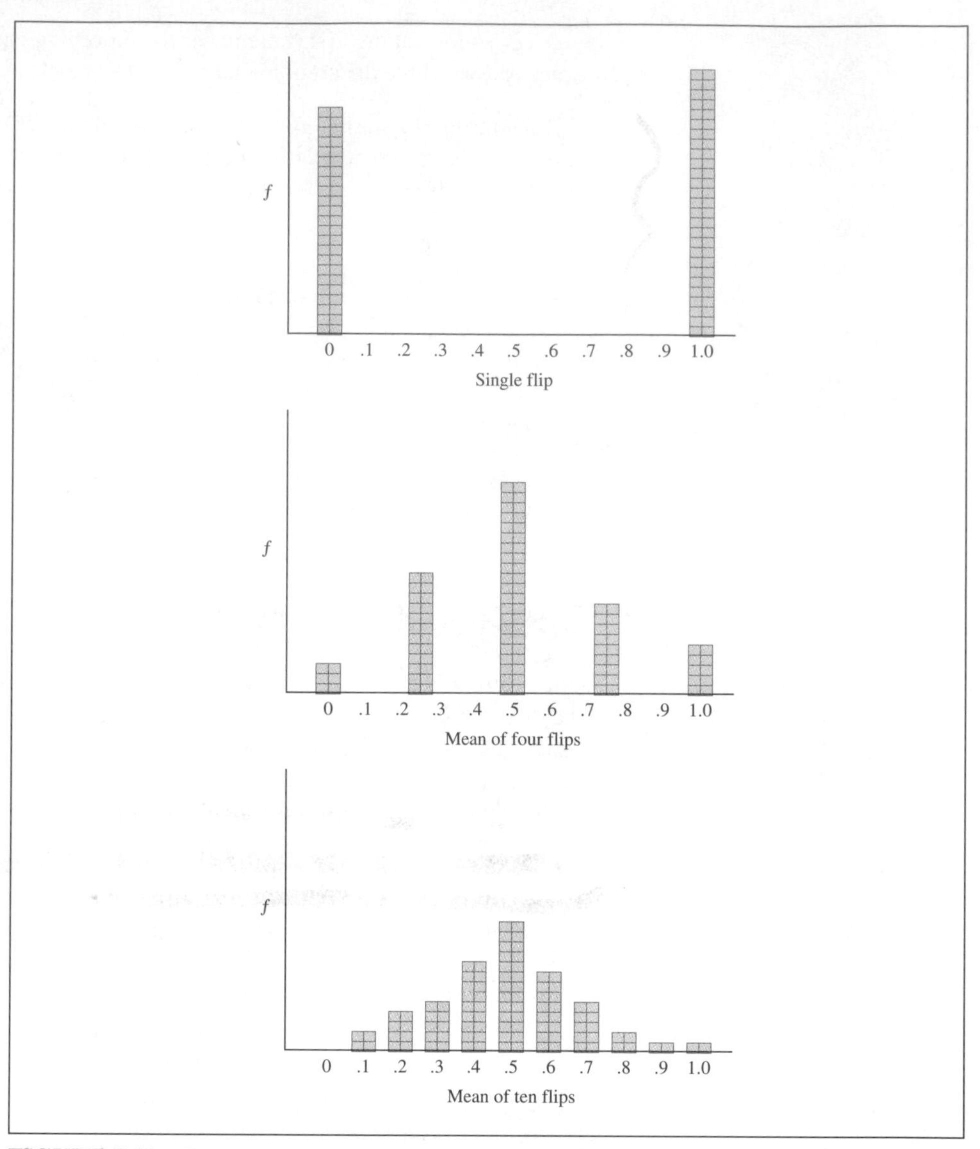

FIGURE 7.11 Outcomes of 100 flips. *Top:* Individual coins. *Middle:* Means of four coins. *Bottom:* Means of ten coins.

2. $\mu_{\bar{X}}$, the mean of the sampling distribution of the means, equals μ, the parent population mean. Therefore, we will drop the subscript and refer to the mean of the sampling distribution of means simply as μ.

3. $\sigma_{\bar{X}}$ (the standard error of the mean) is given by the formula

standard error of the mean in a population

$$\sigma_{\bar{X}} = \frac{\sigma}{\sqrt{n}} \tag{7.1}$$

Thus, the sampling distribution of means is *narrower* than the original population distribution by a factor of $\sqrt{n}$.

The fact that the central limit theorem holds that the means of samples of *any* distribution are normally distributed (if n is sufficiently large) is the reason the normal distribution is so important in statistics. An extension of this theorem shows why many occurrences in nature are at least approximately normally distributed. For example, the height of a tree is the result of the averaging of many factors: genetics, amount of sunlight, amount of rainfall, presence of particular nutrients in the soil, competition from other nearby plants, wind, temperature, humidity, and others. Even though genetic factors themselves may not be normally distributed, amount of sunlight may not be normally distributed, amount of rainfall may not be normally distributed, and so on, the *mean* of all of these factors taken together—that is, the total height of the tree—*is* (by an extension of the central limit theorem) normally distributed. Because many things in nature are the result of the averaging of many independent events, they have a distribution that is (at least approximately) normally distributed.

Factors That Affect the Magnitude of $\sigma_{\overline{X}}$: n and σ

We shall see in subsequent chapters that in many instances it will be desirable to have the standard error of the mean $\sigma_{\overline{X}}$ be as *small* as possible. Let us review how that can be accomplished. Recall that the central limit theorem, as expressed in Equation (7.1), shows that $\sigma_{\overline{X}} = \sigma/\sqrt{n}$. Thus, the magnitude of $\sigma_{\overline{X}}$ depends on only two factors: the standard deviation σ and the sample size n. $\sigma_{\overline{X}}$ will be small if σ is small or if n is large.

The standard deviation σ is in most cases quite difficult to manipulate; it is the standard deviation of the population, a state of nature outside of our direct control.[2] However, we can usually increase the sample size n by using more subjects. Because $\sqrt{n}$ appears in the denominator of Equation (7.1), increasing the sample size will have the desired effect of decreasing $\sigma_{\overline{X}}$. For example, multiplying the sample size by a factor of 4 (say, from 16 to 64) will reduce the standard error by a factor of 2, whereas multiplying the sample size by a factor of 100 will reduce the standard error by a factor of 10.

You may wish to use the ESTAT exercise "mdist" at this time.

Using the Sampling Distribution of the Means

Let us return to our example of sixth-grade boys' weights, where $\mu = 99$ pounds and $\sigma = 4$ pounds. We can now identify the shape, mean, and variation of the sampling distribution of the means of samples of size 25: It is normal with mean $\mu = 99$ pounds and standard error of the mean $\sigma_{\overline{X}} = \sigma/\sqrt{n} = 4/\sqrt{25} = .8$ pound by Equation (7.1). That information is the key that enables us to answer the original question of this section: When we said that $\overline{X}$ will be "fairly close" to μ, just how close is "fairly close"?

We must distinguish carefully between the distribution of the original *variable* and the distribution of the *means* of that variable. For example, when we take 25 samples from the population of the *variable* and then compute $\overline{X}$, that procedure is logically exactly the same as taking *one* sample (which itself is an $\overline{X}$) from the sampling distribution of the *means* of samples of size 25. Thus, the 25 samples of the variable (each displayed as a square) in the first distribution of Figure 7.7 have $\overline{X} = 99.55$; the value 99.55 is also *one* sample (displayed as a square with a triangular fulcrum painted on it) from the sampling distribution of *means* in Figure 7.10. We will see in subsequent chapters that this is the basis of nearly all experimental designs: We take a random sample (of size n) and note that that is exactly the same thing as taking a random sample of size *one* from the sampling distribution of means of samples of size n. Because it is so important, let us restate that for emphasis: A typical study involves taking *one* sample (one $\overline{X}$) from a population of means of samples of size n that have mean μ and standard error $\sigma_{\overline{X}} = \sigma/\sqrt{n}$.

How close to μ can we expect the $\overline{X}$ that we obtain from our experiment to be? We know that the $\overline{X}$'s are normally distributed with mean μ and standard error $\sigma_{\overline{X}}$. In our sixth-grade boys example, $\mu = 99$ pounds and $\sigma_{\overline{X}} = .8$ pound, as we saw above. Because the distribution of means is (from the central limit theorem) normally distributed, about 34% of all possible means will lie between μ and $\mu + \sigma_{\overline{X}}$—that is, between 99 and 99.8 pounds; and another 34% will lie between μ and $\mu - \sigma_{\overline{X}}$—that is, between 98.2 and 99 pounds. Furthermore, 14% will lie between $\mu + \sigma_{\overline{X}}$ and $\mu + 2\sigma_{\overline{X}}$—that is, between 99.8 and 100.6

[2] σ can be reduced—for example, by manipulating subject selection criteria—but such techniques are more properly the subject of research design.

pounds; and another 14% will lie between $\mu - \sigma_{\overline{X}}$ and $\mu - 2\sigma_{\overline{X}}$—that is, between 97.4 and 98.2 pounds.

All the rules that applied in Chapter 6 for computing areas under the normal curve also apply for this normal curve when we realize that a standard score can be computed for the distribution of means as follows:

standard score in the distribution of means in a population

$$z = \frac{\overline{X} - \mu}{\sigma_{\overline{X}}} \tag{7.2}$$

Let's review our findings from Chapter 6 and this chapter, still considering the population of sixth-grade boys' weights ($\mu = 99$ pounds and $\sigma = 4$ pounds). First, let us ask: What percentage of boys weigh less than 98 pounds? That is exactly the kind of question we asked in Chapter 6, and, as we did there, we first sketch the normal distribution and then shade the relevant area, as shown in Figure 7.12. By eyeball, the shaded area is the small three-quarters of 34% (say, 23%) + 14% + 2%, or approximately 39% by eyeball.

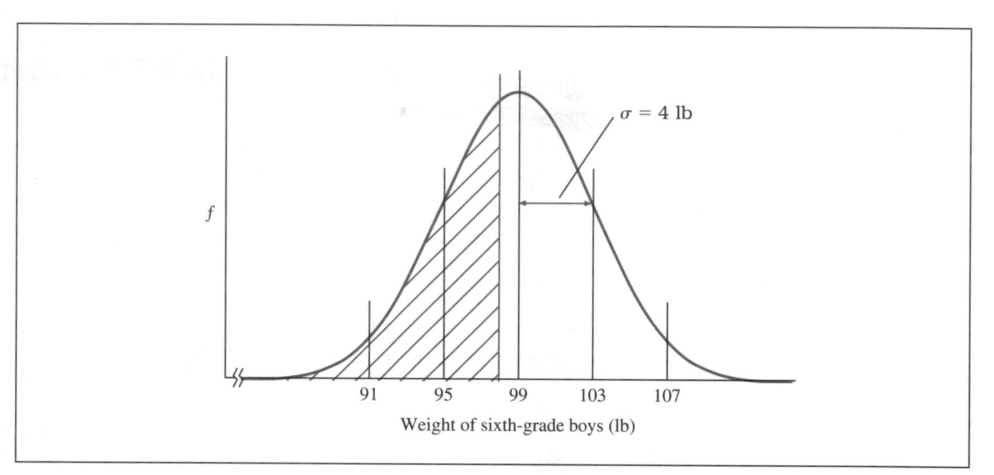

FIGURE 7.12 Distribution of sixth-grade boys' weights with region below 98 pounds shaded

By computation using Equation (6.1), we get

$$z_{98} = \frac{X - \mu}{\sigma} = \frac{98 - 99}{4} = -.25$$

Table A.1 entered at $z_{98} = .25$ gives the area beyond $z_{98} = .4013$, or 40.13% (we had eyeball-estimated 39%), so 40.13% of boys weigh less than 98 pounds.

As we have seen, Figure 7.12 and its discussion are a review of Chapter 6. Now let us ask a new question, one that is similar but distinctly different: What percentage of *samples of size 25* have *mean weight* less than 98 pounds? To answer this new question, we must sketch the distribution *of sample means*.

The central limit theorem tells us that the *mean* of this distribution of means is the same as the mean of the distribution of the original variable, so the center of the distribution of mean weights will have mean $\mu = 99$ pounds. However, the central limit theorem also tells us that the *standard error* of the distribution of means ($\sigma_{\overline{X}} = \sigma/\sqrt{n} = 4/\sqrt{25} = .8$ pound) is *smaller* than the standard deviation of the distribution of the variable ($\sigma = 4$ pounds)—smaller by a factor of $\sqrt{n}$. We show the distribution of means with these characteristics ($\mu = 99$ pounds and $\sigma_{\overline{X}} = .8$ pound) in Figure 7.13. Now we shade the relevant region (less than 98 pounds) in the figure. Our eyeball tells us that the shaded region is the small three-quarters of 14% (say, 10%) + 2%, or approximately 12%.

To compute this area, we would use Equation (7.2) to find z:

$$z_{98} = \frac{\overline{X} - \mu}{\sigma_{\overline{X}}} = \frac{98 - 99}{.8} = -1.25$$

Entering Table A.1 with $z_{98} = 1.25$ (taking the absolute value because the table shows only positive values of z), we find from column B that the area beyond 1.25 is .1056. Thus, 10.56% (we had eyeball-estimated 12%) of all possible samples have a mean less than 98

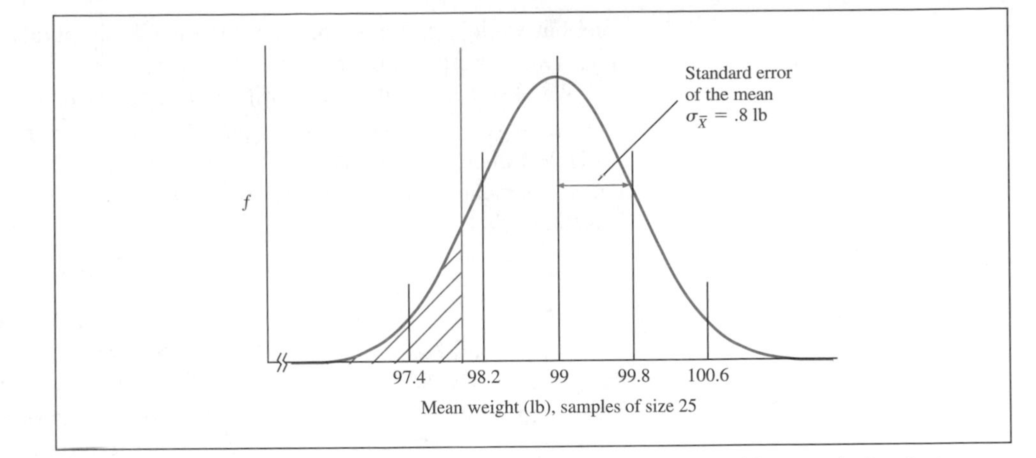

FIGURE 7.13 Distribution of mean weights with region below 98 pounds shaded

pounds. Note that this implies that whereas 40.13% of all *boys* weigh less than 98 pounds, only 10.56% of all possible *means of 25 boys* are less than 98 pounds.

Thus the shaded area under the distribution of *means* (10.56% in Figure 7.13) is much different from the shaded area under the distribution of the *variable* (40.13% in Figure 7.12), even though the regions have the *same* boundary (98 pounds). This is because the standard error of the distribution of means (.8 pound) is much smaller than the standard deviation of the variable (4 pounds).

The distinction between the distribution of the variable and the distribution of means is crucial to understanding nearly all of statistics, so we review it in Table 7.3, which summarizes the similarities and differences between Figures 7.12 and 7.13. Make sure you understand all the differences before proceeding further.

TABLE 7.3 Comparison of Figures 7.12 and 7.13

Characteristic	*Figure 7.12*	*Figure 7.13*
Question	What percentage of *boys* have *weight* less than 98 pounds?	What percentage of *samples of size 25* have *mean weight* less than 98 pounds?
Distribution	Of the variable (weight)	Of the sample means (*mean* weight)
X-axis label	Weight (lb)	Mean weight (lb), samples of size 25
Shape	Normal	Normal
Center	Mean $\mu = 99$ pounds	Mean $\mu = 99$ pounds
Width	Standard deviation $\sigma = 4$ pounds	Standard error $\sigma_{\bar{X}} = \sigma/\sqrt{n} = .8$ pound
Shaded region	Less than 98 pounds	Less than 98 pounds
Shaded area	40.13%	10.56%

We have one final example. Suppose that the manufacturers of Coca-Cola set their bottling machinery to put a mean of 16.5 ounces of Coke in each "16-ounce" bottle (see Exercise 6 in Chapter 3). If the standard deviation of the amount of Coke in each bottle is $\sigma = .5$ ounce, what percentage of bottles contain less than 16 ounces? We answered that question in Exercise 16 of Chapter 6, where we saw that the area required was to the left of one standard deviation below the mean—by eyeball, 14% + 2% = 16%, or more exactly, 15.87%. Now Coke is packaged in six-packs, and let us assume that each of the bottles in a six-pack is a random sample from the population of individual bottles (that is not precisely true, but we will assume it here for the sake of simplicity). What percentage of *six-packs* have a *mean* bottle amount of 16 ounces or less?

We sketch the sampling distribution of the *means of samples of size six*, which has mean $\mu = 16.5$ ounces and standard error $\sigma_{\overline{X}} = \sigma/\sqrt{n} = .5/\sqrt{6} = .204$ ounce (see Figure 7.14). We shade the region below 16 ounces, which we see begins about 2.5 standard errors below the mean, and eyeball the relevant area, which is the small half of 2%, or something less than 1%. By computation,

$$z_{16} = \frac{\overline{X} - \mu}{\sigma_{\overline{X}}} = \frac{16 - 16.5}{.204} = -2.45$$

(we had eyeball-estimated $z_{\text{by eyeball}} \approx -2.5$). We enter Table A.1 at $z_{16} = 2.45$ and find that the area beyond z_{16} is .0071, or .71% (we had eyeball-estimated somewhat less than 1%). Thus, only .71% of Coke six-packs have a mean bottle amount of 16 ounces or less even though 15.87% of individual bottles have 16 ounces or less.

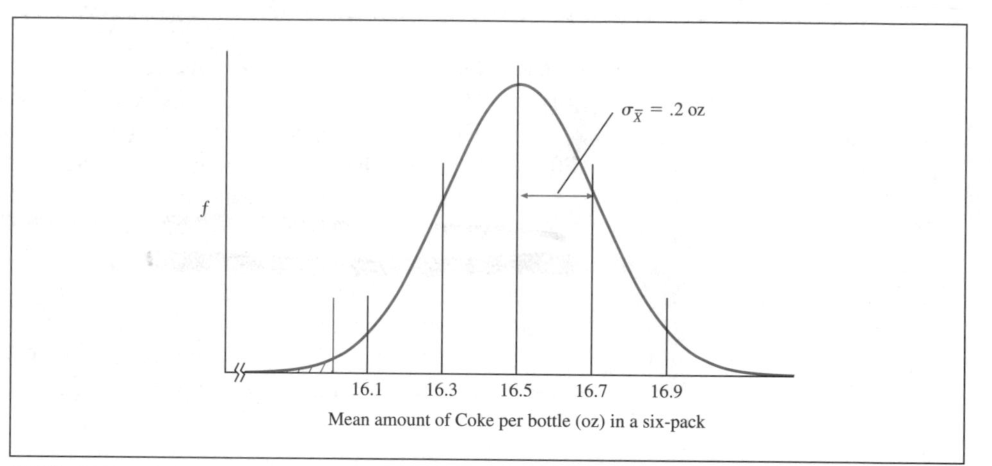

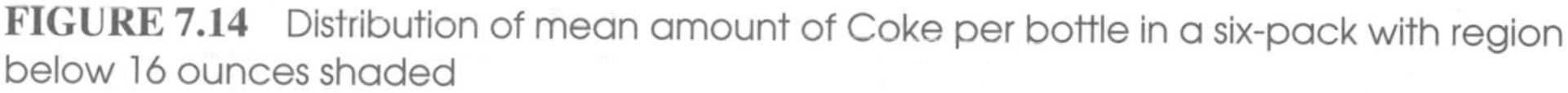
FIGURE 7.14 Distribution of mean amount of Coke per bottle in a six-pack with region below 16 ounces shaded

7.3 Connections

Cumulative Review

We have seen since Chapter 1 that the three basic concepts to master in introductory statistics are: (1) distributions of variables, (2) distributions of means, and (3) distributions of the test statistic. This chapter is the beginning of our examination of the second concept, the distributions of means. Previous chapters have discussed distributions of variables, which can be characterized by three features: their shape (Chapter 3), their center (Chapter 4), and their width (Chapter 5).

Computers

Use ESTAT's datagen to take a random sample of size $n = 5$ from the population of IQ scores ($\mu = 100$, $\sigma = 15$):

1. Click `Variable 1` at the top of the first column to highlight the entire column.
2. Click `Generate`, then click `OK`.
3. Note that datagen has taken a random sample of size five from the specified population. The statistics of this sample are automatically displayed in the `datagen Descriptive Statistics` window.

4. Is the mean of your sample 100? Is the standard deviation of your sample 15? Explain.
5. Repeat the process until you get a feel for the magnitude of the fluctuations of the sample mean and standard deviation.

Use SPSS to take a random sample of size $n = 5$ from the population of IQ scores ($\mu = 100$, $\sigma = 15$):

1. Enter "1" five times in column 1 (these are dummy values that will be replaced by the random sample).
2. Click `Transform` and then `Compute....`
3. Type "var00001" in the `Target variable` cell.
4. Type "rnd(rv.normal(100,15))" in the `Numeric expression` cell. Then click `OK`. The expression "rv.normal(100,15)" takes a random sample from the normal distribution with $\mu = 100$ and $\sigma = 15$; "rnd" rounds the sample value to an integer.
5. Click `OK` when asked "Change existing variable?"
6. Note that the sample appears in the spreadsheet.
7. To see the summary statistics for these data, click `Statistics`, then `Summarize`, and then `Descriptives...` as you did in Chapter 5.
8. Is the mean of your sample 100? Is the standard deviation of your sample 15? Explain.
9. Repeat the process until you get a feel for the magnitude of the fluctuations of the sample mean and standard deviation.

Homework Tips

1. Check the list of learning objectives at the beginning of this chapter. Do you understand each one?

2. The distribution of means is the basic concept upon which almost all of statistical inference (and thus most of the remainder of this textbook) rests. If you want to understand statistics, you must understand this chapter.

3. Many students find the concept of the distribution of means somewhat difficult at first. Once you firmly grasp the concept, however, you will see that it is quite simple (and will probably wonder why you fought it so hard).

4. Review Table 7.3.

5. Much confusion can be avoided if you clearly label the horizontal axes of your sketches. For most of this chapter, you have to choose one of two general forms: "Variable (units)" and "Mean variable (units), samples of size *n*." Make sure that every distribution is labeled with one of these forms, and remember that a variable has a standard deviation, whereas a mean has a standard error.

6. You will use Table A.1 frequently in this homework. It is found in Appendix A at the back of the book.

7. Review Table 7.3 one more time.

Exercises for Chapter 7

Section A: Basic exercises (answers in Appendix D, page 510)

1. Suppose you wish to take a random sample of size five from the alphabet. [*Hint:* E is the 5th letter, J is the 10th letter, O is the 15th letter, and T is the 20th letter.] Here is an excerpt from the table of random digits, and the seed turns out to be the underlined "2."

39256 38296 10015 29065 16845 65991 50876
12098 65095 61987 20607 76158 93647 16452

(a) How many digits must be used each time we enter the table of random digits?
(b) What is the result of the first entry into the table?
(c) What is the result of the second entry into the table?
(d) What is the result of the third entry?
(e) What is the result of the fourth entry?
(f) What is the result of the fifth entry?
(g) What is the result of the sixth entry?
(h) What is the result of the seventh entry?
(i) What is the result of the eighth entry?
(j) What is the result of the ninth entry?
(k) What is the result of the tenth entry?
(l) What is the result of the 11th entry?
(m) What is the result of the 12th entry?
(n) What is the result of the 13th entry?
(o) What is the result of the 14th entry?
(p) What is the result of the 15th entry?
(q) What is the result of the 16th entry?

2. Suppose you wish to take a random sample of size eight from the cards in a standard deck. Furthermore, you wish to ensure that the sample has two clubs, two diamonds, two hearts, and two spades. [*Hint:* Ace = 1, jack = 11, queen = 12, king = 13.] Here is an excerpt from the table of random digits, and the seed turns out to be the underlined "7."

23478 09094 12373 12990 38452 12031 11190
07734 09823 10487 08909 43276 10923 87023

(a) Should you take a simple random sample or a stratified random sample? Why?
(b) Which clubs are selected?
(c) Which diamonds are selected?
(d) Which hearts are selected?
(e) Which spades are selected?

3. Consider the central limit theorem. It describes three characteristics of the ________. Those three characteristics are the ________, the ________, and the ________.

4. Explain why the distribution under the magnifying glass in Figure 7.10 is narrower than the distributions in Figures 7.7, 7.8, and 7.9.

5. Suppose that the lifetimes of light bulbs are normally distributed with mean 1000 hours and standard deviation 100 hours.
(a) Sketch the distribution of light bulb lifetimes.
Light bulbs come in a four-pack, and we can assume that the four bulbs in the pack constitute a random sample of size four from the light bulb population. Now suppose we choose 100 four-packs at random, and we measure the lifetime of each of the four bulbs and compute the mean of the four.
(b) Sketch the distribution of mean lifetimes for these four-packs.
(c) What is the mean of this distribution of means?
(d) What is the standard deviation of this distribution of means called?
(e) What is the value of the standard deviation of this distribution?

6. (*Exercise 5 continued*)
(a) What percentage of all light bulbs can be expected to have a lifetime shorter than 850 hours?
(b) What percentage of all four-packs can be expected to have a mean bulb lifetime shorter than 850 hours?

7. Suppose that the weights of large eggs are normally distributed with mean 57 grams and standard deviation 3 grams.
(a) Sketch the distribution of egg weights.
Suppose we select 100 cartons (each containing 12 eggs) at random. The eggs in a carton can be assumed to be random samples from the population. For each carton, we measure the mean weight of the 12 eggs inside.
(b) Sketch what you would expect the distribution of carton mean egg weights of these 100 cartons to be.
(c) What would you expect the mean of these 100 carton means to be?
(d) What would you expect the standard deviation of these carton means to be?

8. (*Exercise 7 continued*)
(a) What percentage of all large eggs can be expected to be heavier than 60 grams?
(b) What percentage of cartons of eggs can be expected to have a mean egg weight heavier than 60 grams?

Section B: Supplementary exercises

9. The mathematical achievement stanine scores of all 180 sixth-graders at Danville School are shown in the table.
(a) Take a simple random sample of size $n = 10$ by using the table of random digits, Table A.10 in Appendix A.
(b) Compute the mean of this sample.
(c) Compute the standard deviation of this sample.

9	6	5	4	4	7	4	4	5	8
4	8	1	5	6	7	9	3	2	7
2	4	2	7	7	2	7	7	7	3
5	5	5	4	6	6	3	5	3	4
8	7	2	6	5	6	6	4	3	6
4	6	4	4	6	8	2	4	3	8
3	3	6	5	5	6	7	7	6	6
9	4	4	3	6	5	6	9	7	6
4	8	5	6	7	2	2	3	5	5
3	4	7	5	5	5	7	6	6	6
4	3	3	6	6	4	5	6	4	4
1	3	4	6	2	4	7	6	4	4
4	2	4	4	7	6	6	4	8	5
8	8	3	4	5	5	7	4	7	4
9	5	3	4	6	5	4	5	4	6

10. (*Exercise 9 continued*) Take a second random sample of size $n = 10$ from the stanine scores of the sixth-graders at Danville School.
(a) Compute the mean of this sample.
(b) Compute the standard deviation of this sample.

(c) Are the means and standard deviations you computed here and in Exercise 9 the same? Why?

11. Suppose that the oxygen consumption of individuals is normally distributed with mean 240 cubic centimeters per minute (cc/min) and standard deviation 21 cc/min.
    (a) Sketch the distribution of oxygen consumption in individuals.
    (b) What percentage of individuals have oxygen consumption greater than 250 cc/min? Shade the relevant area of the distribution, eyeball-estimate, and compute the percentage.

12. (*Exercise 11 continued*)
    (a) Suppose that groups of ten individuals are created by randomly selecting individuals from the population. Sketch the distribution of mean oxygen consumption in groups of size ten.
    (b) What percentage of such groups have mean oxygen consumption greater than 250 cc/min? Shade the relevant area of the distribution, eyeball-estimate, and compute the percentage.

13. Suppose that the ages at which children first walk alone are normally distributed with mean 11.5 months and standard deviation 1.1 months.
    (a) Sketch the distribution of first walking ages in children.
    (b) What percentage of children first walk at age 12.3 months or older? Shade the relevant area of the distribution, eyeball-estimate, and compute the percentage.

14. (*Exercise 13 continued*)
    (a) Suppose that kindergarten classrooms are created by randomly selecting individuals from the children population. Each kindergarten class has 20 children in it. Sketch the distribution of the mean first walking ages for these kindergarten classrooms.
    (b) What percentage of such classrooms have mean first walking age of 12.3 months or older? Shade the relevant area of the distribution, eyeball-estimate, and compute the percentage.

Section C: Extending your comprehension (answers in Appendix D, page 511)

15. For what sample size n do the standard deviation and the standard error of the mean have identical values?

16. For what sample size n is the standard error of the mean half as large as the standard deviation?

17. For what sample size n is the standard error of the mean one thousandth as large as the standard deviation?

18. For what sample size n is the standard error of the mean twice as large as the standard deviation?

Section D: Cumulative review (answers in Appendix D, page 511)

19. The table given here shows the numbers of mistakes made by all students who took a particular multiple-choice history examination.
    (a) Prepare a frequency distribution.
    (b) Prepare a grouped frequency distribution using six to eight groups.
    (c) Is this distribution skewed? If so, in which direction?

11	8	10	8	6	9	8	21	6	14
1	8	3	6	11	10	1	4	5	10
5	2	8	0	4	10	5	13	13	2
0	4	11	7	7	14	8	8	2	7
6	9	2	11	8	11	1	8	7	6
1	16	6	8	5	6	14	0	5	10
6	2	5	15	0	2	2	1	1	5
11	0	3	12	10	2	2	5	1	7
3	0	2	0	2	2	12	6	4	14
10	18	6	1	6	4	16	11	5	9

Labor savings for the student:

$\sum X_i = 653$ and $\sum X_i^2 = 6371$

20. (*Exercise 19 continued*)
    (a) Eyeball-estimate and compute the mean number of mistakes made.
    (b) Eyeball-estimate and compute the median number of mistakes made.
    (c) Eyeball-estimate and compute the modal number of mistakes made.
    (d) Do these three values have the relative sizes you would expect given the shape of the distribution?

21. (*Exercise 19 continued*) Eyeball-estimate and compute these measures:
    (a) Range
    (b) Standard deviation
    (c) Variance

22. (*Exercise 19 continued*) Using the table of random digits in Appendix A, take a random sample of size $n = 4$ from this group. Compute the sample mean. Then take another sample of size $n = 4$ and compute that sample mean. Repeat the process until you have taken nine separate random samples of size four and computed nine means.
    (a) What value would you expect for the mean of these nine? Why?
    (b) What value would you expect for the standard deviation of these nine means? Why?
    (c) The distribution of mistakes is quite skewed as you saw in Exercise 19. Would you expect the distribution of means of samples of size four to be just as skewed? Why or why not?

Section F: Computer explorations

23. (a) Have the computer take a random sample of size five from the population of IQ scores ($\mu = 100$, $\sigma = 15$; see the Computers portion of Section 7.3). Have the computer calculate the mean of this sample. Record this mean, either by writing it down or by entering it in the second column of the spreadsheet.

(b) Repeat part (a) eight more times so you have a total of nine means.

(c) Have the computer compute the standard deviation of these nine means. The central limit theorem says that the standard deviation of the *population* of sample means is $\sigma_{\overline{X}} = \sigma/\sqrt{n} = 15/\sqrt{5} = 6.71$. Is the standard deviation of your sample of nine means close to 6.71? (It is not expected to be exactly 6.71 because of sampling fluctuations.)

CHAPTER 8 Parameter Estimation

LEARNING OBJECTIVES

1. What is a point-estimate? Why is it referred to as an unbiased estimator?
2. What is the purpose of constructing a confidence interval?
3. What is the critical value of a statistic?
4. What advantage does a confidence interval have over a point-estimate?
5. What factors must be taken into account when constructing a confidence interval when σ is unknown compared to when σ is known?
6. What are the similarities and dissimilarities of the z distribution and the t distribution?
7. What are degrees of freedom?
8. What factors affect the width of a confidence interval?

We often wish to know the value of the population mean (the parameter μ) when the only information we have is a sample mean (the statistic $\overline{X}$). We can specify either a single value (called a "point-estimate") or a range of values (called a "confidence interval") that is likely to contain the population mean. This chapter introduces the family of t distributions, which are shaped approximately like the z distribution but depend on the "degrees of freedom" (which in turn depend on the sample size). The construction of the confidence interval uses the statistic z when σ is known, but the t statistic is used when σ is unknown. The level of confidence is the probability that we are correct when we assert that the population mean μ lies within the confidence interval. We can set any level of confidence we desire, but the higher the level of confidence, the wider the confidence interval.

As we have seen since Chapter 4 in our examination of the data from the Pygmalion study, the mean intellectual growth of the 12 Oak School bloomers (Kathy, Tony, Mario, etc.) was 16.5 IQ points. Our interest is not actually in this particular sample of students, however; our real interest is in the population of *all* second-graders whose teachers expect them to "bloom" in the next year. What is the mean intellectual growth of this entire population? We do not know, of course, and we can never know because it would be impossible to run this experiment with all possible second-graders. But you will learn in this chapter that it is possible to say with 95% confidence that the mean of this population lies between 4.2 and 28.8 IQ points. By contrast, we can say with 95% confidence that the mean of the population of second-graders whose teachers do *not* expect particular blooming lies between 4.0 and 10.0 IQ points.

In Chapter 7 we asked: What characteristics can we expect sample statistics to have if we draw repeated samples from a population with known parameters? We found that $\overline{X}$, the sample mean, fluctuates somewhat from sample to sample, but in general it lies relatively close to the population mean μ. How close? The $\overline{X}$ values are normally distributed with mean μ and standard error $\sigma_{\overline{X}} = \sigma/\sqrt{n}$. Also in Chapter 7 we observed that that process is the reverse of what we usually wish to do: What can we say about the (unknown) parameters of a population when all we know are the statistics of a sample? We turn to that question now.

8.1 Point-estimation

Suppose we wish to know the value of a (population) parameter; for example, we want to know the mean weight (the parameter μ) of all sixth-grade boys in the United States. We could identify all those boys (about 5 million), weigh each one, add the weights, and divide by N, and we would know μ precisely. That would be a prohibitively expensive task, however, so we take a random sample of size, say, $n = 25$, weigh those 25 boys, add the weights, and divide by 25 to obtain the sample statistic $\overline{X}$.

Now we know the statistic $\overline{X}$, but what we want to know is the parameter μ. Our first conclusion here (which we will accept without proof) is that $\overline{X}$ is the *best possible point-estimate* of μ that can be obtained from a sample of size n. To *point-estimate* is to compute a statistic and accept that computed value as an approximation of the unknown value of the parameter. Note that although $\overline{X}$ is the best point-estimate of μ, only rarely is it exactly equal to μ because of sampling fluctuations.

point-estimate: a computed statistic that approximates a parameter

When we say that $\overline{X}$ is the "best possible" point-estimate of μ, we do not mean that it is a perfect estimate. In fact, as we saw in Chapter 7, $\overline{X}$ is sometimes larger than μ and sometimes smaller than μ. "Best possible" indicates that $\overline{X}$ is an *unbiased estimator* of μ, which means that if we take a large number of samples, each of size n, half the $\overline{X}$ values will be larger than μ and the other half will be smaller. Best possible point-estimators, besides being unbiased, also have other characteristics that are beyond the scope of this text. For now, it is sufficient to note that although the sample median and sample mode are also *unbiased* estimators of μ, neither is the *best* estimator. The sample mean is a better point-estimator because in the long run, $\overline{X}$ will lie closer to μ more often than will either the median or the mode.

unbiased estimator: a point-estimate such that if we repeated the point-estimating process infinitely often, the same number of point-estimates would be too high as too low

We must be careful to distinguish clearly between the terms *point-estimate* and *eyeball-estimate*, which we have been using for the last several chapters. To *point*-estimate the mean is to compute $\overline{X}$ and accept that value as an approximation of μ; that is, point-estimation is a way of approximating a parameter from a statistic. To *eyeball*-estimate the mean, on the other hand, is to scan the data and make a quick guess of the mean's value; that is, eyeball-estimation is a way of approximating either a statistic or a parameter from visualizing the data.

8.2 Distribution of Sample Means

We now return to the sixth-grade boys example, where we would like to know μ but found it too expensive to measure directly. Suppose we take a random sample of size 25 and compute the sample mean, finding $\overline{X} = 99.55$ pounds. Now the best point-estimate we have of the population mean μ is 99.55 pounds. However, μ is probably not *exactly* 99.55 pounds (it could possibly be exactly 99.55, but that would be quite coincidental). How close to 99.55 pounds should we expect μ to be? To answer that question, we will review and extend the results of Chapter 7.

As we saw in Chapter 7, if a population has mean μ and standard deviation σ, the distribution of sample means is (at least approximately) normal with mean μ and standard error of the mean $\sigma_{\overline{X}} = \sigma/\sqrt{n}$. Those parameters of the distribution of means lead to one of the most important results in statistics: *The distance from the mean of a sample to the mean of the population from which that sample was drawn is probably less than about two standard errors of the mean.* There are two ideas in that statement that we must make more precise: *probably* and *about two*. We shall see that they are related.

Suppose we are about to take a random sample of size 25 from the population of weights of sixth-grade boys ($\mu = 99$ pounds, $\sigma = 4$ pounds). We have no way of knowing in advance whether the mean of this particular sample will be above or below 99 pounds. The randomness of the sample makes it impossible to predict that in advance. We do know that the sample mean will be a single sample from the sampling distribution of the mean that has mean $\mu = 99$ and standard error $\sigma_{\overline{X}} = \sigma/\sqrt{n} = 4/\sqrt{25} = .8$ pound. Thus, we can say something about *how close* the sample mean will lie to the population mean (the standard error is .8 pound), but we can't say in which direction it will lie.

We can define *probably* in any way we like, but to be precise, we must specify it mathematically: Perhaps we will specify that *probably* means "95 times out of 100." We could have chosen "99 times out of 100" or "999 times out of 1000" or any other precise definition, and we will see later what the ramifications of this choice are. For now, however, let us accept that *probably* means "95 times out of 100." Furthermore, "95 times out of 100" is a probability statement of the kind that we discussed in Chapter 1: If we have an urn that contains 100 thoroughly mixed balls, of which 95 are white and 5 are red, we can fairly say that we will *probably* obtain a white ball on the next draw.

We are in the process of clarifying the statement, "The distance from the mean of a sample to the mean of the population from which that sample was drawn is probably less than about two standard errors of the mean." If *probably* means "95 times out of 100," then how should *about two* be interpreted? This question asks us to determine the limits of a shaded region under the distribution of means, as shown in Figure 8.1. The shaded region must be the middle 95% of the distribution to agree with our definition of *probably* as meaning 95% of the time. We see that 95% of all possible sample means lie in the shaded region and, furthermore, we can see by eyeball that the limits of this shaded region are "about two" standard errors above and "about two" standard errors below the population mean. To be more precise, we should note that shading the center 95% involves shading 47.5% on each side of the mean (because the distribution is symmetric); equivalently, it means leaving 2.5% unshaded in each tail of the distribution. Just as we did in Chapter 6, we enter the table of the normal distribution (Table A.1 in Appendix A) at an area of .475 to find the precise value of z that divides the shaded from the unshaded region; we call this the *critical value of* z, or z_{cv} (sometimes $z_{cv.05}$ to emphasize that .05 remains unshaded in the two tails, .025 in each). Table A.1 shows that the required value is 1.96. By symmetry, and as is clearly evident from our sketch, $z_{cv.05} = \pm 1.96$; that is, the critical value lies 1.96 standard errors both above and below the population mean.

critical value of a statistic: the value of a statistic that marks the boundary of a specified area (such as .05 or .01) in the tail of a distribution

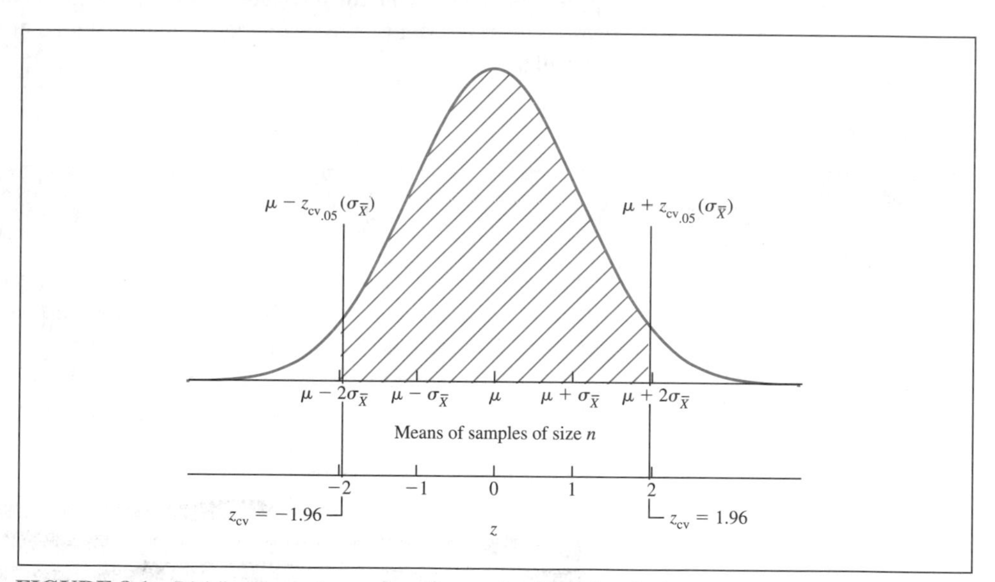

FIGURE 8.1 Distribution of means with center 95% shaded

We now have a precise formulation of the statement we set out to discuss: The distance from the mean of a sample to the mean of the population from which that sample was drawn is probably (95 times out of 100) about two (more precisely, 1.96) standard errors of the mean.

8.3 Confidence Intervals

We have been proceeding as if we knew the value of the population mean, and we have been considering what would happen if we drew repeated samples from that population. In actual practice, statistics are most often used when we *do not* know the population parameters (if we knew the parameters, we would not need to estimate them with statistics). One of the most frequent occurrences in science is the attempt to specify the value of the mean of a population (μ) when we do not have direct knowledge of it. We have seen that our best point-estimate of μ is the sample mean $\overline{X}$, but we observed that $\overline{X}$ is not likely to have *exactly* the precise value of the population mean μ. Because we cannot specify μ's value precisely, it would be desirable to specify the *range of values* in which μ can be expected to lie. This range of values is called a *confidence interval*, and the probability that μ lies in that interval is called the *level of confidence*.

95% confidence interval: a range of values that has a 95% probability of containing the actual value of the parameter

level of confidence: the probability that a given confidence interval contains the actual value of the parameter

To recapitulate: The population mean is some fixed number (that is, it is *not* a variable that can take on one of several values), but we do not know what that number is. The confidence interval is the range of values in which the population mean probably lies. As before, we must specify precisely *range of values* and *probably*.

Reconsider the conclusion of the preceding section. That conclusion is true even if we don't know the value of the population mean, so that conclusion can serve as the definitional statement of the confidence interval: The distance from the mean of a sample to the (unknown) mean of the population from which that sample was drawn is probably (95 times out of 100) about two (more precisely, 1.96) standard errors of the mean. In symbols, we can write that with the specified *level of confidence* (e.g., 95%),

confidence interval for the population mean when σ is known

When σ is unknown, see p. 155.

$$\underbrace{\overline{X} - z_{cv}(\sigma_{\overline{X}})}_{\text{lower limit}} < \mu < \underbrace{\overline{X} + z_{cv}(\sigma_{\overline{X}})}_{\text{upper limit}} \tag{8.1}$$

The confidence interval designated by Equation (8.1) is sometimes written (entirely equivalently) as "with the specified level of confidence, μ lies within $\overline{X} \pm z_{cv}(\sigma_{\overline{X}})$."

When σ Is Known

Let us take an example. Suppose we know that IQ scores are normally distributed with standard deviation $\sigma = 15$ points. We wish to know the mean IQ of all students at Upper Roanoke High School, a large private school noted for its academic achievement; that is, we wish to know the population mean μ. We could measure μ exactly by administering IQ tests to all 1200 students and taking the mean of those 1200 scores. That would involve more work and more expense than we can accept, so we take a random sample of 16 students and measure their IQs. The results are shown in Table 8.1. What can we say about the population mean IQ at Upper Roanoke High?

TABLE 8.1 IQ scores of students at Upper Roanoke High

110	115	108	130
105	125	125	120
120	122	116	106
108	136	124	118
$\sum X_i = 1888$		$\overline{X} = \sum X_i/n = 118.0$	

The procedure for eyeball-estimating the confidence interval is given in Box 8.1. Computing the confidence interval when σ is known involves direct substitution into Equation (8.1). For our Upper Roanoke High example, the standard deviation of IQ is assumed known to be $\sigma = 15$ IQ points. Therefore, the standard error of the means is obtained from Equation (7.1): $\sigma_{\overline{X}} = \sigma/\sqrt{n} = 15/\sqrt{16} = 3.75$ IQ points. The sample mean $\overline{X} = 118.0$ points, so the confidence interval is

$$\overline{X} - z_{cv_{.05}}(\sigma_{\overline{X}}) < \mu < \overline{X} + z_{cv_{.05}}(\sigma_{\overline{X}})$$
$$118.0 - 1.96(3.75) < \mu < 118.0 + 1.96(3.75)$$
$$110.65 < \mu < 125.35$$

BOX 8.1 Eyeball-estimating the confidence interval for μ when σ is known

The procedure for eyeball-estimating the confidence interval follows directly from our definitional statement: μ probably lies within about $2\sigma_{\overline{X}}$ of $\overline{X}$. The center of the confidence interval will thus be $\overline{X}$ (our best point-estimate of μ), and it will extend about two standard errors in each direction from that point. Therefore, this is our procedure for eyeball-estimating the confidence interval:

steps for eyeball-estimating a confidence interval

Step 1. Eyeball-estimate the sample mean $\overline{X}$ (as we have done in previous chapters).

Step 2. Divide the known standard deviation (σ) by $\sqrt{n}$ to obtain $\sigma_{\overline{X}}$.

Step 3. Multiply $\sigma_{\overline{X}}$ by about 2 (the critical value of z, actually 1.96) to obtain $z_{cv}(\sigma_{\overline{X}})$.

Step 4. Both subtract and add that result to $\overline{X}$. That will give the endpoints of the confidence interval—that is, points that are about two standard errors from the sample mean.

Keep in mind the spirit of eyeball-estimation: Perform these divisions, multiplications, and additions in your head, not on a calculator. The object is to get an *approximate* answer, to exercise your observational powers, *not* to get an exact answer.

Applying these steps to our IQ example: (1) We use the range method to find $\overline{X}_{\text{by eyeball}} \approx (136 + 105)/2 = 120$. (2) The standard error $\sigma_{\overline{X}} = \sigma/\sqrt{n} = 15/\sqrt{16} = 3.75$ IQ points. (3) Therefore, the true mean probably lies within about two standard errors [$\approx 2(3.75) = 7.5$ points] of 120. (4) Subtracting and adding 7.5 points to 120 give 112.5 and 127.5 IQ points as the endpoints of the confidence interval$_{\text{by eyeball}}$. The interval obtained by computation in the text is from 110.65 to 125.35 IQ points, which is quite close to our eyeball-estimated result.

Expressed in plain English: With 95% confidence we can say that the mean IQ of all 1200 students at Upper Roanoke High (that is, the population mean μ) is greater than 110.65 and less than 125.35 IQ points.

Notice that we are not saying *absolutely* that the population mean lies between 110.65 and 125.35; instead, we are specifying the degree of confidence we have in our assertion. When we say that we have 95% confidence, we mean that if we were to follow the same procedure of drawing a sample, computing the mean, and preparing a confidence interval 1000 times, we would be preparing 1000 different confidence intervals, and the population mean would lie within the confidence interval in about 950 cases (and outside the confidence interval about 50 times). In any given study, we do *not* know whether the population mean lies in our computed confidence interval; either it does or it doesn't (we will never know for sure). We *assert* that it does, and we will be correct in such assertions about 95% of the time.

We illustrate the process of obtaining the confidence interval with the 18 samples we took from the sixth-grade weight distribution in Chapter 7 (see Figures 7.7, 7.8, and 7.9). We reproduce Figure 7.7 (the first six samples) in Figure 8.2. When we developed those figures, we said that these were 18 separate studies, each of which drew a sample of size 25 from a distribution with $\mu = 99$ pounds and $\sigma = 4$ pounds. Now let us proceed as if we do not know the population mean μ of this distribution, but we wish to specify it with a confidence interval. (For the moment, we will assume that we *do know* that $\sigma = 4$ pounds. We will discuss later in this chapter how to proceed when σ is unknown.)

Our first study, with the data shown in the upper left corner of Figure 8.2, had a sample mean ($\overline{X}$) of 99.55 pounds. What can we say with 95% confidence about the population mean μ given $\overline{X}$? The best single point-estimate of μ is $\overline{X}$—that is, 99.55 pounds—and the true mean μ probably lies in the confidence interval centered on that value. Because $\sigma = 4$ pounds, the standard error of the distribution of weights of boys is $\sigma_{\overline{X}} = \sigma/\sqrt{n} = 4/\sqrt{25} = .8$ pound. We eyeball-estimate the resulting confidence interval

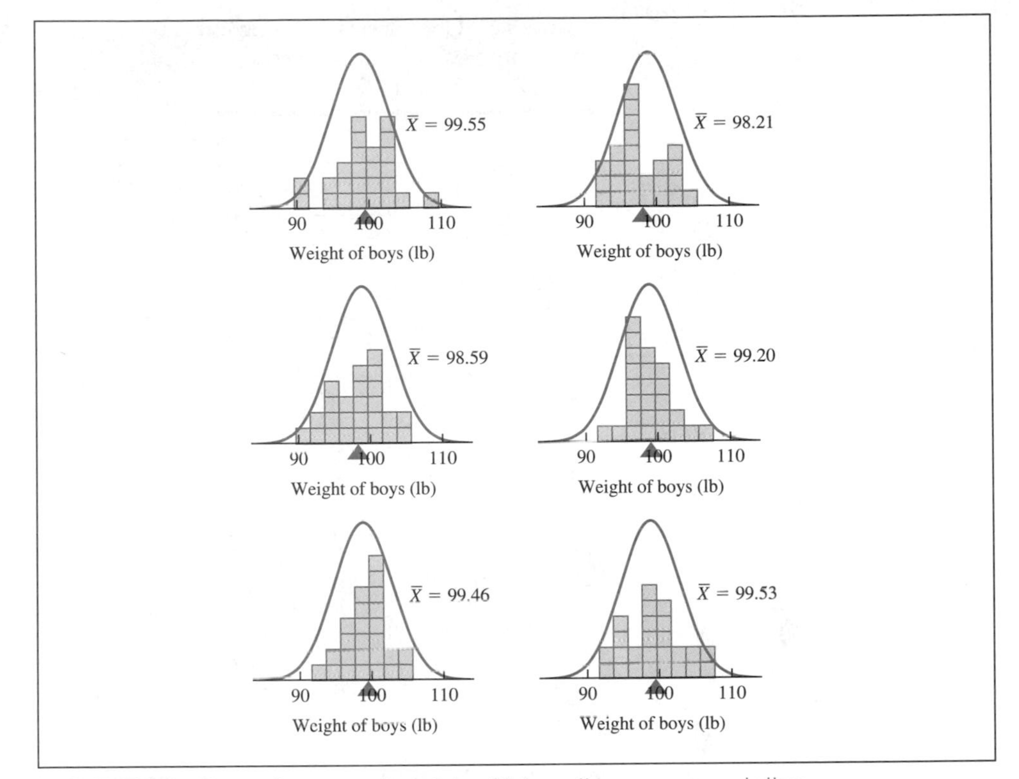

FIGURE 8.2 Six random samples of size 25 from the same population

in Box 8.2 and compute it from Equation (8.1) by multiplying the standard error by 1.96 and subtracting and adding the results to $\overline{X}$:

$$\overline{X} - z_{\text{cv}_{.05}}(\sigma_{\overline{X}}) < \mu < \overline{X} + z_{\text{cv}_{.05}}(\sigma_{\overline{X}})$$
$$99.55 - 1.96(.8) < \mu < 99.55 + 1.96(.8)$$
$$97.98 < \mu < 101.12$$

In plain English: We can say with 95% confidence that the population mean μ lies in the interval that extends from 97.98 to 101.12 pounds. (We happen to know that the population mean is actually 99 pounds and so can see that our assertion is true, but we are suspending that information for the moment.)

BOX 8.2 Eyeball-estimating the confidence interval for the first sample in Figure 8.2

The confidence interval$_{\text{by eyeball}}$ extends about two standard errors in both directions from the sample mean. The standard error is .8 and the sample mean is 99.55 pound, so the confidence interval$_{\text{by eyeball}}$ is 99.55 ± 2(.8), or from 97.95 to 101.15 pounds. The interval is computed in the text to be from 97.98 to 101.12 pounds, quite close to our eyeballed value.

We repeat the same procedure for the second sample in Figure 8.2. The sample mean $\overline{X}$ = 98.59 pounds, and when we subtract and add 1.96(.8), we find that the confidence interval based on that sample of size 25 is 97.02 to 100.16 pounds. Note that this confidence interval is *not* precisely the same as the first one; the width is identical, twice 1.96(.8), but the center is at the sample mean, which changes with each sample.

Table 8.2 shows the 18 confidence intervals computed from the 18 studies whose results are shown in Figures 7.7, 7.8, and 7.9. The two that we computed here are the first two in the table. Keep in mind that in a typical study, we would draw only *one* sample (of size 25 in the current example). Thus, for the first study, we would say with 95% confidence that μ lies

TABLE 8.2 Confidence intervals for the 18 samples of size 25 shown in Figures 7.7, 7.8, and 7.9

Study	$\overline{X}$	*95% confidence interval limits, lower–upper*
1	99.55	97.98–101.12
2	98.59	97.02–100.16
3	99.46	97.89–101.03
4	98.21	96.64– 99.78
5	99.20	97.63–100.77
6	99.53	97.96–101.10
7	98.70	97.13–100.27
8	98.58	97.01–100.15
9	99.54	97.97–101.11
10	98.98	97.41–100.55
11	99.04	97.47–100.61
12	99.33	97.76–100.90
13	98.74	97.17–100.31
14	100.00	98.43–101.57
15	96.71	95.14– 98.28
16	99.15	97.58–100.72
17	98.10	96.53– 99.67
18	98.34	96.77– 99.91

between 97.98 and 101.12 pounds; in the second study, we would say with 95% confidence that μ lies between 97.02 and 100.16 pounds; in the third study, we would say that μ lies between 97.89 and 101.03 pounds; and so on. In each study, we would *not* typically know whether we were correct or incorrect because the actual value of μ is typically not known. In our present *simulation*, we *do* know that $\mu = 99$ pounds, and so we know that our three confidence interval statements happen to be correct.

Now imagine that we performed just one study, study 15 in Table 8.2. Upon completion of that study (and totally ignorant, as we usually are, of the true population mean or any other estimates of it), we would say with 95% confidence that the population mean μ lies between 95.14 and 98.28 pounds. We would be *precisely as confident* of this result (that is to say, 95% confident) as we are of the results of any other study in this series, but here we know we are wrong; the confidence interval does *not* contain 99 pounds, the true value of μ. For every 100 studies we perform, we will reach the wrong conclusion (in the long run) in about five of them; that is what is meant by 95% confident. In a typical experiment, we *never* know whether our confidence interval *actually* contains the population mean; we can say only that it *probably* contains μ (that is, we are 95% confident that it contains μ).

Note that the experimenter here did nothing wrong. The results happened to be in error from the rare perspective of one who already knows that the true μ is 99 pounds, but the study was every bit as good as all the others. The experimenter just happened to be "unlucky" enough to draw a sample whose $\overline{X}$ was by random chance substantially lower than μ.

Changing the Level of Confidence

Perhaps you feel that being right 95 times out of 100 isn't good enough. If so, then we must change the definition of the word *probably* in our general statement of the confidence interval. You might like to define *probably* to mean a 99% level of confidence in your results. This requires that we resketch the distribution of means of samples of size 25 and shade the middle 99% (rather than 95% as before); this leaves .5% (.005) of the area unshaded in each tail. The distribution is shown in Figure 8.3.

If 99% is in the shaded center of this distribution, then 1% must be in the tails. We call the boundaries of this region $z_{\text{cv}_{.01}}$ to emphasize that .01 (that is, 1%) is in the tails of the distribution, .005 in each tail. These critical values must be somewhat larger than ± 2, by

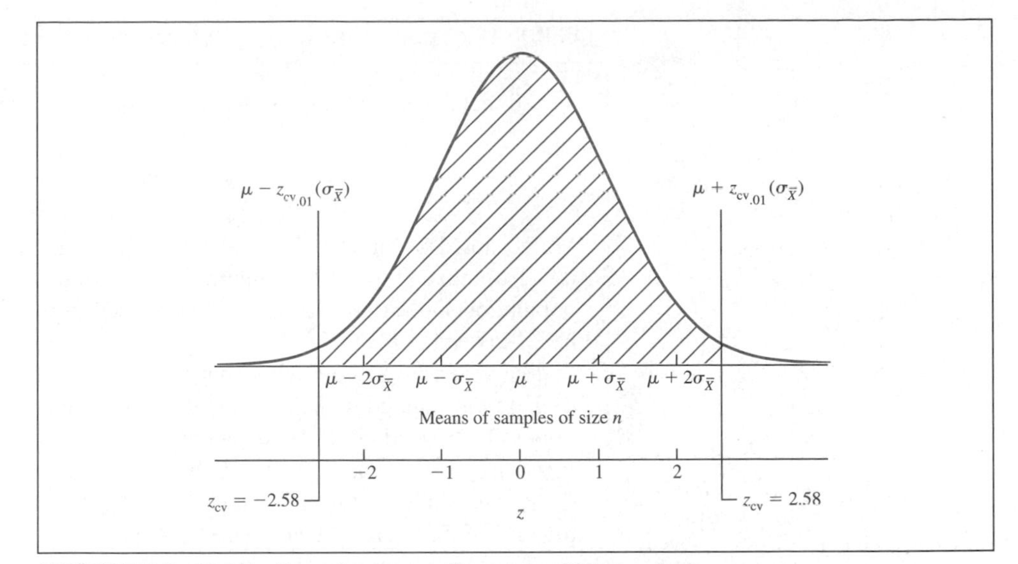

FIGURE 8.3 Distribution of means with center 99% shaded

eyeball about ±2.5. Entering Table A.1, column B, with an area of .005, just as we did in Chapter 6, shows that the more precise value is $z_{cv_{.01}} = \pm 2.58$.

The formula for the confidence interval remains the same except that $z_{cv_{.05}}$ is replaced by $z_{cv_{.01}}$: $\bar{X} - z_{cv_{.01}}(\sigma_{\bar{X}}) < \mu < \bar{X} + z_{cv_{.01}}(\sigma_{\bar{X}})$. For our first study of sixth-grade boys (upper left in Figure 8.2), the sample mean was $\bar{X} = 99.55$ pounds, and because $\sigma = 4$ pounds, $\sigma_{\bar{X}}$ remains $4/\sqrt{25} = .8$ pound. The 99% confidence interval is then $99.55 - 2.58(.8) < \mu < 99.55 + 2.58(.8)$, or $97.49 < \mu < 101.61$ pounds. In plain English, we can say that with 99% confidence, the population mean lies between 97.49 and 101.61 pounds.

When we computed the 95% confidence interval for this same sample (the first entry in Table 8.2), we found $97.98 < \mu < 101.12$ pounds. Note that the 99% confidence interval is *wider* than the 95% confidence interval. It is 2.58(2) = 5.16 standard errors wide, whereas the 95% confidence interval is 1.96(2) = 3.92 standard errors wide. This is surprising to many students, perhaps because it seems that a 99% confidence interval should be "better" than a 95% interval, and it seems (incorrectly) that "better" should mean more precise. The fact is that the *wider* we make an interval, the *more* confident we can be that a particular value lies somewhere within that interval. A 100% confidence interval would be infinitely wide; there would then be *no* chance that the parameter would lie outside the interval.

When σ Is Unknown

The confidence intervals we have been considering so far have all been based on the assumption that we knew in advance the value of the population standard deviation σ. This would be the case, for example, if we had done previous research with sixth-grade boys and that research had shown that the population standard deviation was 4 pounds. Or perhaps we were measuring IQ with a test whose standard deviation was known in advance to be 15 points. How do we construct a confidence interval if we do *not* know σ?

First, we should recognize that if we do not know σ, we cannot know $\sigma_{\bar{X}}$ either because $\sigma_{\bar{X}}$ is obtained from σ by dividing by $\sqrt{n}$. Therefore, we need a formula for the confidence interval that does *not* depend on $\sigma_{\bar{X}}$.

If we do not know σ, we must point-estimate it from the data we gathered in our sample, and the best point-estimator of σ is s, the sample standard deviation. Here again, *best* means, among other things, "unbiased," which implies that about half the time s is larger than σ and about half the time smaller. Note that s is *not* exactly equal to σ; it is merely the best (unbiased) point-estimate we have, given a sample of size n (see Boxes 8.3 and 8.4).

BOX 8.3 Unbiased estimators and the denominator of the standard deviation

Recall that Equation (5.4) for the sample standard deviation was

$$s = \sqrt{\frac{\sum(X_i - \overline{X})^2}{n - 1}} \tag{5.4}$$

We accepted on faith in Chapter 5 that the denominator in this equation must be $n - 1$ to make the sample standard deviation s an "unbiased point-estimator" of the population standard deviation σ. Now we can explain why the denominator must be $n - 1$ instead of n.

If we wish to know the standard deviation of a population, we would like to average the squared deviations about the population mean μ. That is, unfortunately, impossible if we don't know μ (which is usually why we are taking the sample in the first place), so we must take deviations about the sample mean $\overline{X}$ instead. The problem is that the sum of the squared deviations about $\overline{X}$ is *too small*—that is, smaller than the sum of the squared deviations about μ.

Table 8.3 illustrates the situation. The first column shows a random sample of size $n = 5$ from a population with mean $\mu = 10$. Usually we do not know μ, but here we are exposing it for illustrative purposes. The second column shows the deviations that we would like to take (but usually cannot) about the *population* mean μ; the third column shows those deviations squared. The fourth and fifth columns show the deviations about the *sample* mean and those deviations squared.

TABLE 8.3 A random sample from a population with mean $\mu = 10$, with deviations about the population and sample means

Original observations X_i	*Deviations from parameter* μ $X_i - \mu$	*Those deviations squared* $(X_i - \mu)^2$	*Deviations from statistic* $\overline{X}$ $X_i - \overline{X}$	*Those deviations squared* $(X_i - \overline{X})^2$
12	2	4	3	9
9	−1	1	0	0
11	1	1	2	4
9	−1	1	0	0
4	−6	36	−5	25
	$\sum(X_i - \mu) = -5$	$\sum(X_i - \mu)^2 = 43$	$\sum(X_i - \overline{X}) = 0$	$\sum(X_i - \overline{X})^2 = 38$

Note that the sum of the squared deviations about $\overline{X}$ (38) is *smaller* than the sum of the squared deviations about μ (43). $\sum(X_i - \overline{X})^2$ will always be smaller than $\sum(X_i - \mu)^2$ unless $\overline{X}$ happens to equal μ, in which case the two sums will of course be equal. This is because $\overline{X}$ is at the exact center of the sample (9.0 in our example), whereas μ may be some distance away from the center of the sample (10 in our example). The squared deviations about the exact center of a sample are in general smaller than the squared deviations about any other point (including the population mean μ).

To recapitulate: We would like to take deviations about μ, but we cannot because we don't know μ, so we take deviations about $\overline{X}$ instead. However, the sum of those squared deviations is too small.

If $\sum(X_i - \overline{X})^2$ is too small, then a standard deviation obtained from a simple average (dividing the sum of squared deviations by n) would also be too small; thus, this statistic would be a *biased* (that is, on average too small) estimator of σ. It turns out, as we shall see in Box 8.4, that decreasing the denominator from n to $n - 1$ makes the denominator smaller by precisely the same amount that the numerator was too small in the first place. When the numerator (which is too small) is divided by the $(n - 1)$ denominator (which is too small by the same amount), the resulting s of Equation (5.4) is just right; that is, s is an unbiased estimator of σ.

BOX 8.4 Degrees of freedom in the computation of a standard deviation or variance

degrees of freedom: the number of freely varying values in a given data set

In Box 8.3 we saw that decreasing the denominator of a standard deviation (or variance) from n to $n - 1$ was just the right correction to offset the fact that the numerator of s was too small. We call the denominator of a standard deviation the *degrees of freedom* and say that "the denominator must lose one degree of freedom for the computation of $\overline{X}$." Why is that?

Consider the second column of Table 8.3. In that column we are taking deviations about μ, and we say that these deviations have "five degrees of freedom" because each of the five deviations is independent of the others. For example, if we know that the first four deviations are 2, −1, 1, and −1, we still have no information whatever about the fifth deviation because the fifth value is independent of the previous four.

However, if we use $\overline{X}$ as the point about which deviations are taken, then as we saw in Box 5.1, $\sum(X_i - \overline{X})$ *must* be equal to 0. Once we know that $\sum(X_i - \overline{X}) = 0$, the deviations in the fourth column of Table 8.3 are no longer independent. For example, if we know that the first four deviations are 3, 0, 2, and 0, then the fifth deviation *must* be exactly −5 to make $\sum(X_i - \overline{X}) = 0$. Therefore, if we take deviations about $\overline{X}$, there are only *four* freely varying deviations; that is, there are now only four degrees of freedom.

Thus, there are *five* freely varying deviations about μ (the second column) but only *four* freely varying deviations about $\overline{X}$ (the fourth column).

For reasons that go beyond this text, when we compute the mean of the squared deviations (or any other mean, for that matter), we must divide by the number of degrees of freedom, not by the number of deviations. Thus, when we compute s from the deviations about $\overline{X}$, we must divide by $n - 1$, the degrees of freedom of those deviations.

We saw in Chapter 7 that if we know the population standard deviation σ, then the standard error of the mean $\sigma_{\overline{X}} = \sigma/\sqrt{n}$. The same kind of expression is true for the *estimate* of the standard error of the mean, $s_{\overline{X}}$, based on the sample standard deviation s. If s point-estimates σ, then $s_{\overline{X}}$ point-estimates $\sigma_{\overline{X}}$ and is computed from

standard error of the mean in a sample

$$s_{\overline{X}} = \frac{s}{\sqrt{n}} \qquad (8.2)$$

Thus, once we know s, we can obtain $s_{\overline{X}}$ by dividing by $\sqrt{n}$ and use that as the best point-estimator of $\sigma_{\overline{X}}$.

Given that $s_{\overline{X}}$ is the best point-estimator of $\sigma_{\overline{X}}$, it might seem (incorrectly, as we shall see) that the confidence interval for the mean might be $\overline{X} - z_{cv}(s_{\overline{X}}) < \mu < \overline{X} + z_{cv}(s_{\overline{X}})$, exactly parallel to Equation (8.1) except with s substituted for σ. However, that does *not* take into account the fact that $s_{\overline{X}}$ is an *imperfect* point-estimator of $\sigma_{\overline{X}}$, perhaps too large and perhaps too small. In Equation (8.1), only one value was being point-estimated: μ (point-estimated by $\overline{X}$). Now there are two: μ and $\sigma_{\overline{X}}$ (point-estimated by $s_{\overline{X}}$), so the confidence interval must be somewhat *wider* to account for the variabilities of both point-estimates.

Student's *t* Distribution

The observation that the confidence interval must be wider when σ is unknown was first made in the early 1900s by William S. Gossett, a chemist at an Irish brewery. Writing under the pen name "Student," he derived a family of distributions that were similar to the z distributions (unimodal, symmetric, asymptotic) but were sufficiently wider than z to account for the imperfection of using $s_{\overline{X}}$ as a point-estimate of $\sigma_{\overline{X}}$. We have come to call this family of distributions *Student's t*.

Student's *t* distribution: a family of distributions that, like z, are unimodal, symmetric, and asymptotic, but the exact shape (unlike z) depends on the degrees of freedom

How much wider must the distribution of t be than the distribution of z? That depends on how efficient s is as an estimator of σ. Clearly, if s can be said to be almost identical to σ, then t must be almost identical to z. Now the accuracy of s depends on the sample size: The larger the sample, the more precise the point-estimation. In fact, when the sample size is larger than about 120, s is almost indistinguishable from σ and therefore the t and z distributions are essentially identical. The smaller n is, the wider the t distribution must be because the efficiency of s as a point-estimator of σ decreases. Thus, the t distribution depends on n or, as we prefer to say, it depends on the number of *degrees of freedom*. As we saw in Box 8.4, the number of degrees of freedom is the size of the denominator in the computation of the variance. Because the relevant variance here is s^2, the degrees of freedom are $df = n - 1$.

Table A.2 in Appendix A (and inside the back cover) lists some frequently used critical values of t. We reproduce excerpts from it here as Table 8.4 to give an understanding of how much wider t is than z. First, we note that the heading of the second and third columns in Table 8.4 is "Level of significance for two-tailed test." That heading will be more understandable in later chapters. For now, we note that that heading is synonymous with "Total area beyond $\pm t$ including both tails." Thus, the middle column, headed ".05," represents the critical values of t that leave 5% of the area beyond $\pm t_{\text{cv}}$—that is, 2.5%—in each tail. As we did with z, we refer to this critical value of t as $t_{\text{cv}_{.05}}$ to call attention to the fact that 5% lies beyond the critical values (including both tails).

TABLE 8.4 Critical values of the t distribution for selected degrees of freedom and levels of significance .05 and .01 (excerpted from Table A.2)

	Level of significance for two-tailed test	
df	*.05*	*.01*
1	12.706	63.657
2	4.303	9.925
3	3.182	5.841
4	2.776	4.604
5	2.571	4.032
8	2.306	3.355
10	2.228	3.169
30	2.042	2.750
60	2.000	2.660
120	1.980	2.617
∞	1.960	2.576

The bottom row of Table 8.4 gives the critical values of t when the number of degrees of freedom (and therefore also n because $df = n - 1$) is infinite. In that case, s should be a perfect point-estimator of σ, and thus t should be equal to z. That is indeed true: The critical value of t when $df = \infty$ is 1.960, and that, as we have seen before, is precisely the same as the critical value of z; that is, $t_{\text{cv}_{df=\infty}} = z_{\text{cv}}$.

Beginning at the bottom of Table 8.4, note that as the number of degrees of freedom decreases, the critical value of t increases, very slowly at first. When df decreases from ∞ to 120, $t_{\text{cv}_{.05}}$ increases from 1.960 to only 1.980, and when df has decreased all the way to 5, $t_{\text{cv}_{.05}}$ has increased to only 2.571. Below $df = 4$, however, $t_{\text{cv}_{.05}}$ increases rapidly. Similar observations can be made about $t_{\text{cv}_{.01}}$ in the right-hand column of Table 8.4. Remember that t differs from z to account for the extent that s is likely to differ from σ. That means that s is a moderately good point-estimator of σ whenever df is greater than about 5, and s is an excellent point-estimator whenever df is greater than about 30.

To demonstrate that the t distribution widens as the degrees of freedom decrease, we show three different t distributions in Figure 8.4. When $df = \infty$, t is identical to z; when df is less than ∞, t becomes wider, thus putting a larger proportion of cases in the tails of the distribution. This figure also shows the 95% critical values that leave 2.5% in each tail. Note that as the t distribution widens (that is, as the degrees of freedom decrease), t_{cv} must increase so that .025 is left in each tail. Note also that there are critical values in the left-hand tails of these distributions (at values −1.96, −2.31, and −2.78), but to simplify Figure 8.4 these are not shown.

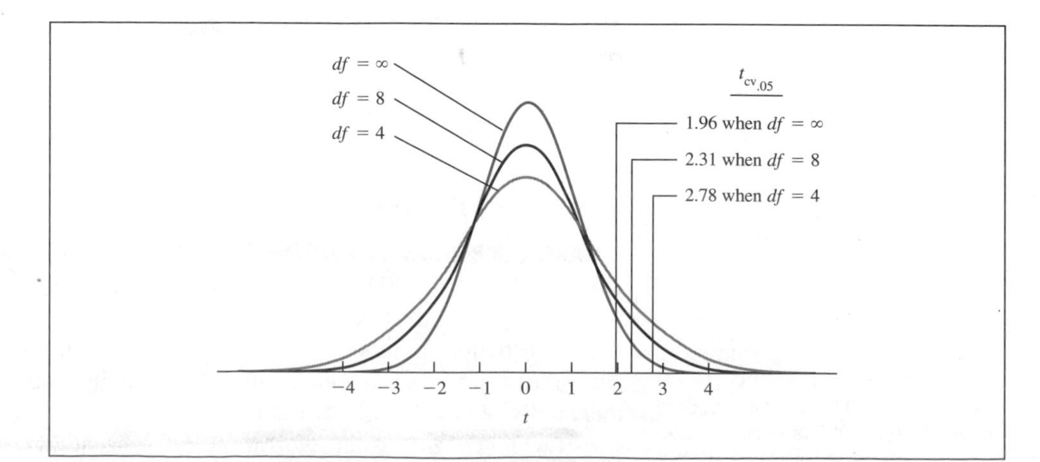

FIGURE 8.4 Student's t distribution when $df = 4, 8$, and ∞. The two-tailed values of $t_{\text{cv}_{.05}}$ are indicated only in the right-hand tail.

Computing the Confidence Interval When σ Is Unknown

We are now in a position to write the expression for the confidence interval for the mean under the assumption that we do not know the value of σ but instead must estimate it from the sample s:

confidence interval for the population mean when σ is unknown

$$\underbrace{\bar{X} - t_{cv}(s_{\bar{X}})}_{\text{lower limit}} < \mu < \underbrace{\bar{X} + t_{cv}(s_{\bar{X}})}_{\text{upper limit}} \qquad [df = n - 1] \tag{8.3}$$

When σ is known, see p. 147.

This expression is identical to the expression for the confidence interval when we *do* know σ, Equation (8.1), except that we substitute t for z and s for σ. Note that t_{cv} depends not only on the level of confidence (e.g., .05 or .01), as did z, but also on the degrees of freedom that we used to compute s; that is, $df = n - 1$.

As was the case with Equation (8.1), the confidence interval designated by Equation (8.3) is sometimes written (entirely equivalently) as "with the specified level of confidence, μ lies within $\bar{X} \pm t_{cv}(s_{\bar{X}})$."

To illustrate the procedure, let us assume that we wish to know the mean of the weights of the population of sixth-grade boys, and for now let us proceed as if we do *not* have any information about the population standard deviation. We determine that we will use a sample of size $n = 25$ and randomly select 25 sixth-graders and weigh them. The data are shown in Table 8.5.

TABLE 8.5 Weights (lb) of a sample of sixth-grade boys

X_i	X_i^2
93	8,649
99	9,801
101	10,201
103	10,609
99	9,801
104	10,816
99	9,801
91	8,281
95	9,025
101	10,201
103	10,609
103	10,609
99	9,801
103	10,609
92	8,464
101	10,201
93	8,649
99	9,801
99	9,801
102	10,404
109	11,881
103	10,609
103	10,609
98	9,604
97	9,409
$\sum X_i = 2489$	$\sum X_i^2 = 248{,}245$

We eyeball-estimate the confidence interval in Box 8.5 and compute it as follows: First, we find the sample mean $\bar{X} = \sum X_i / n = 2489/25 = 99.56$ pounds, so the computed center of our confidence interval (and the best point-estimate of μ) is 99.56 pounds. Then we compute the sample standard deviation:

$$s = \sqrt{\frac{\sum X_i^2 - \frac{(\sum X_i)^2}{n}}{n - 1}}$$

$$= \sqrt{\frac{248{,}245 - \frac{(2489)^2}{25}}{25 - 1}}$$

$$= \sqrt{\frac{440.1625}{24}} = 4.28 \text{ pounds}$$

and so $s_{\bar{X}} = s/\sqrt{n} = 4.28/\sqrt{25} = .856$ pound.

BOX 8.5 Eyeball-estimating the confidence interval for μ when σ is unknown

The procedure for eyeball-estimating the confidence interval when we do not know σ is identical to the method when σ is known (described in Box 8.1) with two exceptions: First, we must eyeball-estimate s (by methods used in the previous chapters) and use that value in place of σ; second, we must account for the fact that if df is very small, then t_{cv} can become quite a bit larger than 2.

We apply the steps of Box 8.1 to our example of Table 8.5.

Step 1. We eyeball-estimate $\bar{X}$ by the range method to be midway between 91 and 109; that is, $\bar{X}_{\text{by eyeball}} \approx 100$ pounds, which becomes the center of our confidence interval$_{\text{by eyeball}}$.

Step 2. Because we do not know σ, we eyeball-estimate s by the range method: Because the conversion factor from Table 5.2 is 4, the standard deviation is about a quarter of the range—that is, a quarter of 18—so $s_{\text{by eyeball}} \approx 4.5$ pounds. We next need the standard error: $s_{\bar{X}_{\text{by eyeball}}} = s_{\text{by eyeball}}/\sqrt{n} \approx 4.5/\sqrt{25} = .9$ pound.

(continued)

BOX 8.5 *(continued)*

Step 3. Now we must multiply our estimate of $s_{\overline{X}}$ (from Step 2) by our estimate of the critical value t. We know that t_{cv} is approximately 2 and somewhat greater when df is very small. Here, $df = n - 1 = 25 - 1 = 24$, so $t_{cv_{\text{by eyeball}}} \approx 2$ and $(t_{cv}s_{\overline{X}})_{\text{by eyeball}} \approx 1.8$.

Step 4. We add and subtract 1.8 to $\overline{X}_{\text{by eyeball}}$. Thus, the confidence interval$_{\text{by eyeball}}$ is approximately 100 ± 1.8, or $98.2 < \mu < 101.8$ pounds.

The text shows the computed confidence interval to be $97.79 < \mu < 101.33$, quite close to our estimate.

You may wish to use the ESTAT exercise "confide" at this time.

Now we determine t_{cv}. Assuming we wish to construct a 95% confidence interval, we enter Table A.2 in the two-tailed .05 column and the $df = n - 1 = 25 - 1 = 24$ row, and find that $t_{cv_{.05}} = 2.064$. Then we can substitute into Equation (8.3):

$$\overline{X} - t_{cv}(s_{\overline{X}}) < \mu < \overline{X} + t_{cv}(s_{\overline{X}})$$
$$99.56 - 2.064(.856) < \mu < 99.56 + 2.064(.856)$$
$$97.79 < \mu < 101.33$$

The interpretation of the confidence interval is the same as in the case where we knew σ: We can say with 95% confidence that the population mean μ lies between 97.79 and 101.33 pounds.

Note that the data we have been considering in Table 8.5 are the weights shown in the histogram in the upper left corner of Figure 8.2 and for which, near the beginning of the chapter, we constructed a 95% confidence interval assuming we knew that $\sigma = 4$ pounds. In that case, we said that (with 95% confidence) the population mean was between 97.98 and 101.12 pounds. As we see here, the confidence interval when σ is *unknown* is usually wider than the confidence interval when we know σ because t_{cv} is larger than z_{cv}. The only exception to this rule is when the sample we draw happens to have a standard deviation s that is unusually small. Then $s_{\overline{X}}$ is also small, and the confidence interval based on t and $s_{\overline{X}}$ can be narrower than the one based on z and $\sigma_{\overline{X}}$.

When the newspaper or television reports the results of an opinion poll (the proportion of voters who favor candidate Jones, for example), it generally also reports a "margin of error." Box 8.6 explains that the proportion plus or minus the margin of error is simply a special case of the confidence interval for a population mean.

BOX 8.6 Opinion polls: Using the confidence interval for a proportion

The confidence interval is used frequently in the press in political opinion polling. Newspapers frequently give reports such as: "A recent poll of 250 voters showed that candidate Jones has 64 percent of the votes. The margin of error is plus or minus 6 percent." "Margin of error" is the newspaper's way of referring to a 95% confidence interval for a population proportion, which is simply a special kind of population mean.

Let the variable X have only two possible values: $X_i = 1$ if the ith respondent says, "Yes, I favor candidate Jones," and $X_i = 0$ if the respondent says, "No, I do *not* favor candidate Jones." Then $\sum X_i$ = the *total number* of respondents who favor Jones. In our example, $\sum X_i = .64(250) = 160$ respondents who favored Jones.

We need some notation for proportions. We call p the *proportion* of respondents who say yes and q the proportion who say no. Thus, $p = \sum X_i/n = 160/250 = .64$, and $q = 1 - p = 1 - .64 = .36$. Note that by its definition $p = \overline{X}$. The population proportion (whose value we would like to know) is the parameter we call Π (Greek

(continued)

BOX 8.6 *(continued)*

capital "pi"), and Π is simply a special name for μ when we are discussing proportions. Thus, p (another name for $\overline{X}$) is the best point-estimator for Π (another name for μ).

Because in this case X_i can take values of only 0 or 1, we see that $X_i^2 = X_i$. That fact (and a little algebra) allows us to show that the numerator of the standard deviation is npq:

$$\sum(X_i - \overline{X})^2 = \sum(X_i^2 - 2X_i\overline{X} + \overline{X}^2) = \sum X_i^2 - 2\overline{X}\sum X_i + \sum \overline{X}^2$$
$$= np - 2p(np) + np^2 = np(1 - p) = npq$$

Then $s = \sqrt{\Sigma(X_i - \overline{X})^2/(n-1)} = \sqrt{npq/(n-1)}$, so

$$s_{\overline{X}} = \frac{s}{\sqrt{n}} = \sqrt{\frac{pq}{(n-1)}}$$

In the proportion situation, where $\overline{X} = p$, we see that $s_{\overline{X}} = s_p$, which we call the *standard error of a proportion*. Therefore

standard error of a proportion

$$s_p = \sqrt{\frac{pq}{n-1}} \tag{8.4}$$
$$= \sqrt{\frac{.64(.36)}{249}} = .03$$

The confidence interval is then defined exactly as in Equation (8.3):

confidence interval for a population proportion

$$p - t_{cv}(s_p) < \Pi < p + t_{cv}(s_p) \tag{8.5}$$

or, because $p = .64$ and $df = n - 1 = 249$,

$$.64 - 1.96(.03) < \Pi < .64 + 1.96(.03)$$

or approximately

$$.58 < \Pi < .70$$

Thus, we can say with 95% confidence that the proportion of the voter population who favor Jones is 64% ± 6%, which is exactly what the newspaper article reported.

Four Factors That Affect the Width of a Confidence Interval

In most cases, it is desirable that a confidence interval be narrow because the size of a confidence interval reflects our uncertainty about the actual value of the parameter μ. In general, the less uncertainty, the better. We review the four factors that affect the size of a confidence interval.

Increasing n Makes the Confidence Interval Narrower

As we have seen, a confidence interval is about $4\sigma_{\overline{X}}$ wide ($2\sigma_{\overline{X}}$ in each direction from $\overline{X}$), and $\sigma_{\overline{X}} = \sigma/\sqrt{n}$. That means that the width of the confidence interval is approximately inversely proportional to the square root of the sample size n. If we can quadruple the sample size, we can cut the width of a confidence interval in half. If we are constructing a confidence interval where σ is unknown, then increasing n also decreases t slightly, which also serves to make the confidence interval narrower.

Decreasing σ or s Makes the Confidence Interval Narrower

If we can reduce the population standard deviation σ (or the sample standard deviation s if σ is unknown), we can reduce $\sigma_{\overline{X}}$ (or $s_{\overline{X}}$) by the same factor, thus reducing the width of the confidence interval also by the same factor. In many situations, however, σ (or s) is a state of nature over which we have little or no control.

The Confidence Interval When σ Is Known Is Generally Narrower Than When σ Is Unknown

When we must use s to estimate σ, we must use t_{cv} in place of z_{cv}. Now t_{cv} is larger than z_{cv}, so the confidence interval using t is generally wider than the confidence interval using z. We say "generally" wider because it can happen, by sampling fluctuations, that s turns out to be smaller than σ, thus offsetting this effect.

Decreasing the Level of Confidence Makes the Confidence Interval Narrower

Decreasing the level of confidence from 99% to 95%, for example, changes z_{cv} from 2.58 to 1.96, thus making the confidence interval about 20% narrower. We do have total control over the level of confidence, and we must weigh the risks (the cost of saying that μ is in the interval when in fact with 5%, or 1%, probability it lies outside it) against the benefits of having the interval itself be narrow. That risk/benefit analysis is more a topic of research design than of statistics.

8.4 Connections

Cumulative Review

We have seen since Chapter 1 that the three basic concepts to master in introductory statistics are: (1) distributions of variables, (2) distributions of means, and (3) distributions of test statistics. Preceding chapters have explored the first two concepts; this chapter continues to explore the second idea because the confidence interval is a direct consequence of the distribution of means. This chapter also provides a first look at the third concept because, although we didn't use the term here, later chapters will refer to z and t as "test statistics."

Journals

A typical journal article report of a confidence interval might be: "We can say with 95 percent confidence that the true mean IQ gain for those children expected to bloom lies between 4.2 and 28.8 IQ points." Sometimes when a series of confidence intervals are reported, a journal article might report: "The group mean (with 95% confidence limits) for bloomers was 16.5 ($\pm$12.3) IQ points, whereas the mean for the other children was 7.0 ($\pm$3.0) IQ points." Sometimes such values in parentheses are standard errors ($\pm s_{\bar{X}}$) instead of the confidence interval width ($\pm t_{cv} s_{\bar{X}}$); careful attention may be necessary to determine which value the author intends.

Computers

Use ESTAT's datagen to determine the 95% confidence interval for the population mean of the weight data that originally appeared in Table 3.2:

1. Open the file you saved in Chapter 4 by clicking `Open`, `table3-2.txt`, and then `OK`.
2. Note that the limits of the 95% confidence interval appear automatically (version 2.0 or later) in the `datagen Descriptive Statistics` window in the row labeled "Confidence int. minimum" and "Confidence int. maximum." In versions prior to 2.0,

the confidence interval can be computed from the mean and standard error, both of which are provided.

3. Note that datagen does not compute a 99% confidence interval for you. However, it does automatically supply the standard error in the `datagen Descriptive Statistics`, in the row labeled "Standard error." You can multiply this value by the 99% critical value of t and add to and subtract from $\overline{X}$ to obtain the 99% confidence interval.

Use SPSS to determine the 95% confidence interval for the population mean of the weight data that originally appeared in Table 3.2:

1. Open the weight data file that you saved in Chapter 2 by clicking `File`, then `Open`, and then `Data....` Click `table3-2.sav` and click `OK`.
2. Click `Statistics`, then `Summarize`, and then `Explore....`
3. Click ►. Click `OK`.
4. Note that the lower and upper limits of the confidence interval appear in the `!Output1` window in a line that reads "95% CI for Mean (158.0769 − 171.2831)."
5. Note that you can request a 99% or any other confidence interval by clicking `Statistics...` in the `Explore` window and entering the desired confidence level.

Homework Tips

1. Check the list of learning objectives at the beginning of this chapter. Do you understand each one?

2. Repeat after me: "The confidence interval is the answer to the question, 'Where's μ?' "

3. Remember that the confidence interval has its center at the *sample* mean.

4. Remember that the confidence interval extends about two (that is, z_{cv} or t_{cv}) standard errors (that is, $\sigma_{\overline{X}}$ or $s_{\overline{X}}$) in each direction from $\overline{X}$.

5. Remember that we use z if we know σ and we use t if we do not know σ.

6. Students are often confused about what it means to "know σ." Knowing σ requires that you know the *population* standard deviation σ, which means that the problem must state something like "We know that the standard deviation of *all* students (not just the students in this sample) is...." Simply being able to compute s from the data does *not* mean that we know σ.

Exercises for Chapter 8

Section A: Basic exercises (answers in Appendix D, page 511)

1. Suppose we know that $\sigma = 3$ but we do not know μ. We would like to specify the interval that with 95% confidence contains μ. The data are X_i: 1, 4, 6, 9. Determine this confidence interval.

2. Suppose we know that $\sigma = 3$ but we do not know μ. We would like to specify the interval that with 95% confidence contains μ. The data are X_i: 1, 2, 3, 4, 5, 6, 7, 8, 9. Determine this confidence interval.

3. Is the confidence interval in Exercise 2 narrower or wider than the one in Exercise 1? Explain why.

4. Suppose we do *not* know σ and the data are the same as in Exercise 1. Determine the 95% confidence interval.

5. Is the confidence interval in Exercise 4 narrower or wider than the one in Exercise 1? Explain why.

6. Suppose we know that IQ scores are normally distributed with standard deviation $\sigma = 15$ points. Suppose we wish to know the mean IQ of those people who are born with low birth weight, which we define as 5.5 pounds or less.

We obtain a random sample of ten adults who had low birth weights and measure their IQs. The data are 114, 97, 87, 132, 67, 73, 95, 97, 105, 95 points.

(a) What is the best point-estimate of the mean IQ of adults who had low birth weights?
(b) Eyeball-estimate the 95% confidence interval.
(c) What can we say with 95% confidence about the IQs of adults who had low birth weights? (Compute this answer.)

7. (*Exercise 6 continued*) Suppose that the standard deviation of IQ in the general population is 15 points, but we suspect that it is not necessarily equal to 15 points in the population of low-birth-weight adults; that is, we cannot assume that we know σ.
   (a) How does that change the computation of the best point-estimate of the mean IQ of low-birth-weight adults?
   (b) How does that change the computation of the confidence interval?
   (c) Eyeball-estimate the 95% confidence interval.
   (d) What can we now say with 95% confidence about the IQs of adults who had low birth weights? (Compute this answer.)
   (e) How does your answer in part (d) compare with your answer in Exercise 6(c)? Why?

8. Suppose we wish to know the level of vocabulary of the 200 sixth-graders who have been in enriched classrooms ("enriched" classrooms in this school include computerized reading training). We elect to use the Vocabulary Subtest from the Wechsler Intelligence Scale for Children-Revised (WISC-R), which is standardized to have a standard deviation of three points. Previous research has shown that the standard deviation does not change for various groups that have been measured. We take a random sample of 24 children from the enriched classrooms, and when we measure those children on the WISC-R Vocabulary Test, we find that their mean vocabulary score is 11.6.
   (a) What is the best point-estimate of the mean vocabulary score among the 200 enriched sixth-graders?
   (b) Eyeball-estimate the confidence interval.
   (c) What can we say with 95% confidence about the mean vocabulary score of the 200 enriched sixth-graders? (Compute this answer.)

9. Suppose that a physicist wishes to know as precisely as possible the speed of light, which is a fixed constant. She performs an experiment to measure the speed of light, and the result is 30.2 billion meters per second.
   (a) Should she conclude that the speed of light is in fact 30.2 billion meters per second? Explain.
   (b) Is it possible to eyeball-estimate a confidence interval from this result? Why or why not?
   (c) Based on this study, how close to 30.2 billion meters per second should the actual speed of light be held to be?

10. (*Exercise 9 continued*) Suppose the physicist performs the same experiment six times. Assume that each time can be considered to be independent of the others; that is, assume that the six trials are six randomly chosen samples from the population of results if we performed the same experiment an infinite number of times. Her results are (in billions of meters per second): 30.1, 29.6, 30.3, 30.4, 30.0, 30.2.
    (a) Now what is her best point-estimate of the speed of light?
    (b) Eyeball-estimate the 95% confidence interval.
    (c) What can she say with 95% confidence about the actual speed of light? (Compute this answer.)
    (d) What can she say with 99% confidence about the actual speed of light? (Compute this answer.)
    (e) Is the confidence interval of part (c) wider or narrower than that of part (d)? Why?

11. Suppose we wish to know how much beer is served in each mug at Kelly's Tavern. We randomly select eight mugs and find that the contents of each (in ounces) are: 15.0, 16.5, 16.6, 15.3, 16.2, 16.2, 16.9, 16.3. (Labor savings for the student: $\overline{X}$ = 16.125 ounces, s = .650 ounce.)
    (a) What is the best point-estimate of the mean amount of beer in the population of Kelly's Tavern mugs?
    (b) Eyeball-estimate the 95% confidence interval.
    (c) What can we say with 95% confidence about the actual mean volume of beer in Kelly's Tavern mugs? (Compute this answer.)

12. (*Exercise 11 continued*) Suppose that we had sampled 16 mugs instead of 8 but that $\overline{X}$ was still 16.125 ounces and s was still .650 ounce.
    (a) What terms in the expression for the confidence interval are different from your answer in Exercise 11(c)?
    (b) Eyeball-estimate the 95% confidence interval.
    (c) What can we now say with 95% confidence about the actual mean volume of beer in Kelly's mugs? (Compute this answer.)
    (d) How does your confidence interval from part (c) compare with the one you computed in Exercise 11(c)?

13. See Box 8.6. Presidential candidate Harris wishes to know the percentage of the population of eligible voters who will vote for her in the upcoming election. She asks you to take a random sample of size 200 from that population and ask each sampled member whether he or she supports Harris (which you code as H), Albertson (which you code as A), or some other candidate (which you code as O). You gather these data:

    H A H H H A H A O H A H A H H A H H H A H A O H
    H O A H A H A H A H A H H H A H A O H H H O A H
    A H A H A H H H A H H H A H A O H O A H A H A H
    A H H O A H A H H H A H A O H H A H A H A H H O
    A H A H H H A H A O H H A H A H A H H O A H A H
    A H A H H O A H A H A H A H H O A H H A H H H A
    H A O H A H A H H A H H H A H A O H A H H O A H
    A H A H A H H H A H A O H H A H H A H H H A H A
    O H H O A H A H

(Labor savings for the student: There are 110 H's, 70 A's, and 20 O's.) What can you say to candidate Harris with 95% confidence about the percentage of voters in the population who support her?

Section B: Supplementary exercises

14. Suppose we know that $\sigma = 10$ but we do not know μ. We would like to specify the interval that with 95% confidence contains μ. The data are X_i: 10, 30, 20, 25. Determine this confidence interval.
15. Suppose we know that $\sigma = 10$ but we do not know μ. We would like to specify the interval that with 95% confidence contains μ. The data are X_i: 15, 20, 10, 30, 17, 24, 21, 15. Determine this confidence interval.
16. Is the confidence interval in Exercise 15 narrower or wider than the one in Exercise 14? Explain why.
17. Suppose we do *not* know σ and the data are the same as in Exercise 14. Determine the 95% confidence interval.
18. Is the confidence interval in Exercise 17 narrower or wider than the one in Exercise 14? Explain why.
19. Suppose we wish to know the mean speed of automobiles on a particular stretch of Interstate 40. Suppose further that we know that the standard deviation of automobile speed on interstates is $\sigma = 4.02$ mph. We randomly select 15 automobiles and measure their speeds: 62, 60, 68, 60, 63, 56, 70, 68, 60, 60, 67, 64, 59, 61, 65 mph.
    (a) What is the mean speed of this sample of automobiles?
    (b) What can we say with 95% confidence about the mean speed of all automobiles on this stretch of road? Eyeball-estimate and compute. State your result in symbols and in plain English.
    (c) What can we say with 99% confidence about the mean speed of all automobiles on this stretch of road? Eyeball-estimate and compute. State your result in symbols and in plain English.
    (d) Which confidence interval is wider, the 95% or the 99% interval? Why?
20. (*Exercise 19 continued*) Suppose we wish to know the mean speed of automobiles but we do *not* know σ. (Computational help for the student: $\sum X_i^2 = 59{,}509$.)
    (a) What can we say with 95% confidence about the mean speed of all automobiles on this stretch of road? State your result in symbols and in plain English.
    (b) What can we say with 99% confidence about the mean speed of all automobiles on this stretch of road? State your result in symbols and in plain English.
    (c) This example has been contrived so that the computed s is identical to σ in Exercise 19 (which of course would not be expected in real life). How do the confidence intervals computed here differ from those computed in Exercise 19?
21. The mathematical achievement stanine scores for all Danville sixth-graders were shown in a table in Exercise 9 in Chapter 7. You took a random sample of size $n = 10$ from this sample.
    (a) Eyeball-estimate and compute the 95% confidence interval for the mean.
    (b) Take another random sample of size $n = 10$ and again eyeball-estimate and compute the 95% confidence interval.
    (c) Are your two confidence intervals identical? Explain any discrepancy.
    (d) The true mean stanine score for these data is $\mu = 5.095$. Do your confidence intervals contain this value? If not, explain why not.
22. (*Exercise 21 continued*)
    (a) Suppose you took 100 separate samples, each with $n = 10$, and computed 100 confidence intervals. About how many would be expected to include the value 5.095?
    (b) Suppose you took 100 separate samples, each with $n = 25$, and computed 100 confidence intervals. About how many would be expected to include the value 5.095?
23. (*Exercise 21 continued*) What would the 100% confidence interval for these data be? [*Hint:* Stanines can take values only from 1 through 9.]
24. Suppose you wish to specify the mean heart rate in the population of long-distance runners. You take a random sample of 20 runners and measure their resting heart rates. The data (in beats per minute) are: 63, 59, 61, 67, 67, 59, 64, 67, 65, 69, 69, 60, 65, 65, 62, 68, 64, 62, 67, 70.
    (a) Eyeball-estimate and compute the 95% confidence interval. State your result in plain English.
    (b) Eyeball-estimate and compute the 99% confidence interval. State your result in symbols and in plain English.
25. Suppose your university is considering adopting a policy that would charge all students a flat "athletic activity fee" but would then allow students to attend any athletic event at no charge. There is some division of opinion about this proposed change, and you wish to know what percentage of students favor it. There are 20,000 students in your university, so contacting all of them would be prohibitively expensive. You contact 100 students randomly selected from a list of students supplied by the registrar and ask, "Do you favor the proposed athletic policy, yes or no?" The data are shown here. What can you say with 95% confidence about student opinion? State your answer in symbols and in plain English.

Yes Yes No Yes No No Yes No No Yes No Yes No Yes
Yes No Yes No No Yes No Yes No No No Yes No Yes
No No Yes No Yes No Yes No No No Yes Yes Yes Yes
Yes No No Yes No Yes Yes Yes No Yes No Yes Yes No
No No Yes Yes No No No No Yes Yes No No Yes No
Yes No Yes Yes No No Yes No Yes No Yes No Yes No
Yes Yes No No No No Yes Yes No No No Yes No Yes
No Yes

(Labor savings: There are 53 No's.)

Section C: Extending your comprehension (answers in Appendix D, page 512)

26. Suppose that the 95% confidence interval (*not* assuming that we know σ) for a certain μ is $10.5 < \mu < 13.5$ and that $n = 16$. Now suppose that n is 32 instead of 16, with all else remaining the same. What is the 95% confidence interval? [*Hint:* You have enough information; start by finding $\overline{X}$.]
27. Use the $n = 16$ confidence interval in Exercise 26.
    (a) By eyeball, approximately what is the 99% confidence interval?
    (b) Compute the 99% confidence interval.
28. The width of the 95% confidence interval in Exercise 26 is 3. What must n be (all else remaining the same) if we wish the width to be 1?
29. (a) Usually a confidence interval when σ is known is narrower than the same confidence interval when σ is unknown. Why?
    (b) Is it possible that the σ-unknown confidence interval is narrower? How?
30. (*Exercise 13 continued*) Suppose that when you give the report to candidate Harris, she says that the range of possible percentages is too wide to be acceptable. She wants you to give her a result that has the form "best point-estimate plus or minus 1%." Assuming that you retain the 95% level of confidence and assuming that the percentage favoring Harris remains about .55, how large a sample must you query to satisfy the candidate?

Section D: Cumulative review (answers in Appendix D, page 512)

31. The accompanying histogram represents the weights of infants in a particular nursery.

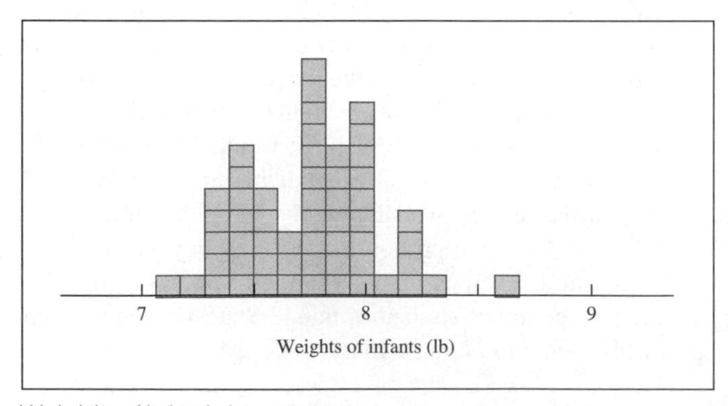

Weights of infants in a nursery

    (a) Eyeball-estimate the mean using the balancing-point method.
    (b) Superimpose a normal distribution.
    (c) Locate the inflection points.
    (d) Eyeball-estimate the standard deviation using the inflection-point method.
    (e) Approximately what is the variance?
    (f) Eyeball-estimate the mean using the range method.
    (g) Eyeball-estimate the standard deviation using the range method.
32. (*Exercise 31 continued*)
    (a) By eyeball, what is the standard error of the mean for samples of this size from the population of infant weights?
    (b) By eyeball, what is the critical value of t that leaves .025 in each tail for these data?
    (c) By eyeball, what is the 95% confidence interval for the mean of the population from which this sample was drawn?
33. The histogram in the figure represents the maximum speeds of entries in a motorboat race.
    (a) Eyeball-estimate the mean using the balancing-point method.
    (b) Superimpose a normal distribution.
    (c) Locate the inflection points.
    (d) Eyeball-estimate the standard deviation using the inflection-point method.
    (e) Approximately what is the variance?
    (f) Eyeball-estimate the mean using the range method.
    (g) Eyeball-estimate the standard deviation using the range method.

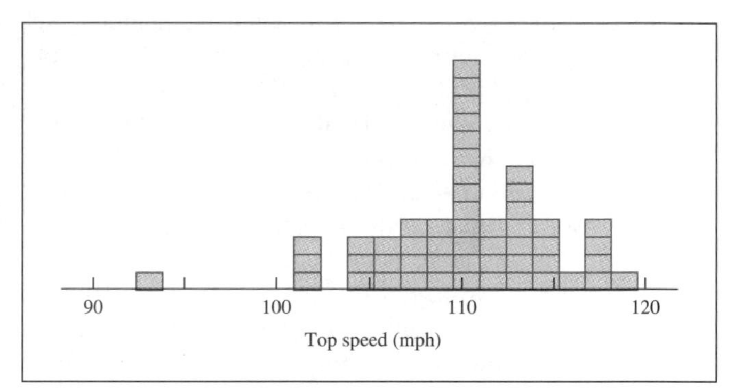

Top speed of motorboats

34. (*Exercise 33 continued*)
    (a) By eyeball, what is the standard error of the mean for samples of this size?
    (b) By eyeball, what is the critical value of t that leaves .025 in each tail for these data?
    (c) By eyeball, what is the 95% confidence interval for the mean of the population from which this sample was drawn?
35. Suppose that a sociologist surveys a random sample of 100 residents of Danville and asks how many years of formal education each resident has completed. The results are $\overline{X} = 11.9$ years and $s = 2.8$ years.
    (a) Sketch the distribution of years of education.
    (b) What percentage of Danville residents have more than 12 years of education? Shade the relevant area, eyeball-estimate, and compute.
    (c) What percentage have less than 10 years of education? Shade the relevant area, eyeball-estimate, and compute.

(d) What percentage have between 13 and 16 years of education? Shade the relevant area, eyeball-estimate, and compute.

36. (*Exercise 35 continued*) Suppose that Danville contains many townhouses in which four adults live, and further suppose that the four adults in each townhouse can be considered to be random samples from the Danville population.
   (a) Sketch the distribution of mean years of education in these townhouses.
   (b) What percentage of residents of townhouses have mean education more than 12 years? Shade the relevant area, eyeball-estimate, and compute.
   (c) What percentage have less than 10 years? Shade the relevant area, eyeball-estimate, and compute.
   (d) What percentage have between 13 and 16 years? Shade the relevant area, eyeball-estimate, and compute.

37. Account for the differences in your answers in Exercises 35 and 36.

Section E: From the journals (answer in Appendix D, page 512)

38. Folnegovic-Smalc, Folnegovic, and Kulcar (1990) sampled 360 patients from the population of 8069 schizophrenics hospitalized in Croatia, Yugoslavia, between 1962 and 1971. The mean age of onset was found to have the 95% confidence interval 22.9 to 26.7 years; the range was 9 to 48 years.
   (a) What was the mean age of onset for the 360 patients in this sample? (It can be computed from the confidence interval.)
   (b) What was the standard error of the mean?
   (c) Eyeball-estimate the standard deviation of age of onset for this sample using the range method.
   (d) What was the actual standard deviation of age of onset in this sample (computed from the confidence interval)?
   (e) Do Folnegovic-Smalc's results seem reasonable? Why or why not?

Section F: Computer explorations

39. (a) Use the computer to simulate taking a random sample of size $n = 5$ from the population of IQ scores, where $\mu = 100$ and $\sigma = 15$. Determine the 95% confidence interval. Does the confidence interval contain the point $\mu = 100$?
   (b) Repeat part (a) 20 times. How often is μ *outside* the confidence interval? How often would you expect it to be outside the confidence interval?

40. (a) Enter the values 1, 2, 3, 4, and 5 into a variable. Have the computer determine the mean, the standard deviation, and the 95% confidence interval limits.
   (b) Now duplicate those values so that you have ten values: 1, 2, 3, 4, 5, 1, 2, 3, 4, 5. How do the mean, the standard deviation, and the 95% confidence interval limits compare to those in part (a)?
   (c) Now duplicate all ten values so that you have 20 values (four replications of 1 through 5). How do the mean, the standard deviation, and the 95% confidence interval limits compare to those you found in parts (a) and (b)?
   (d) Now duplicate all 20 values so that you have 40 values (eight replications of 1 through 5). How do the mean, the standard deviation, and the 95% confidence interval limits compare to those you found in parts (a), (b), and (c)?
   (e) State a general rule about how the mean, the standard deviation, and the confidence interval limits change as the sample size gets larger but the data stay the same.

CHAPTER

Evaluating Hypotheses

LEARNING OBJECTIVES

1. What are descriptive and inferential statistics? How are they similar and dissimilar?
2. What is a hypothesis? Why is a hypothesis easier to prove false than it is to prove true?
3. What are null and alternative hypotheses? How are they similar and dissimilar?
4. How is the alternative hypothesis supported?
5. What are nondirectional ("two-tailed") hypotheses? What are directional ("one-tailed") hypotheses? What are the differences between nondirectional and directional hypotheses?
6. What is a Type I error? A Type II error?
7. What is the level of significance? How does it relate to a Type I error?
8. What is statistical power?
9. In what ways are Type I or Type II errors "expensive" to make?
10. What is a test statistic and how does it relate to hypothesis testing? What is the general formula for the test statistic?
11. How is practical significance different from statistical significance?

This chapter distinguishes between descriptive statistics, which are used to describe characteristics of distributions, and inferential statistics, which are used to make inferences or decisions. It describes the most frequently used inferential statistical procedure, the experiment, and shows that the experimental procedure requires the stating of two hypotheses (called the null and the alternative hypotheses) and an attempt to reject the null hypothesis. The chapter defines and describes other terms important to statistical inference: directional (one-tailed) and nondirectional (two-tailed) hypotheses, Type I and Type II errors, level of significance, test statistic, statistical significance, and practical significance.

We have seen since Chapter 4 that the mean intellectual growth of the Oak School bloomer sample was 16.5 IQ points. We also saw that the mean of the sample of other children was 7.0 IQ points. Clearly, the bloomer sample mean is greater than the other sample mean—16.5 is greater than 7.0—so we can conclude that Kathy, Tony, Mario, Louise, and the eight other bloomers had greater intellectual growth, on average, than did the 47 children in the other sample.

Our main interest is *not* in the particular 59 children (Kathy, Tony, Mario, etc.) in the Oak School samples, however. Our main interest is in the *population* of *all* the thousands of students whose teachers expect them to "bloom" and the *population* of *all* the thousands of students whose teachers have no particular expectations of them. Thus, the interesting question is whether one *population* mean is greater than another (values that we can never know), but all we know is about sample statistics.

Although we do not know the actual magnitudes of the two population means, we saw in Chapter 8 that we can use a sample mean and standard deviation (whose values we do know) to infer with confidence that the population mean lies within a particular interval. We will extend that logic in this chapter to frame questions about parameters that are unknown, such as: Is it reasonable to infer that the expected-to-bloom population mean is greater than the no-particular-expectation population mean? The answer to this question will be either "Yes, it is reasonable to conclude that the expected-to-bloom population has greater mean intellectual growth" or "No, it is not reasonable to conclude that"; we will answer the question in Chapter 11.

9.1 Descriptive Versus Inferential Statistics

Descriptive Statistics

descriptive statistics: the science of describing distributions of samples or populations

We have learned to describe distributions in terms of their shape (normal, skewed, bimodal, etc.), central tendency (mean, median, mode), and variation (standard deviation, variance, range). We can describe the distributions of samples by specifying their statistics and of populations by specifying their parameters. All these tasks are part of *descriptive* statistics in the sense that we are trying to *describe*—convey the characteristics of—some distribution. We will encounter other descriptive statistics in later chapters (most prominently the correlation coefficient, which describes the strength of the relationship between two variables and is discussed in Chapter 16).

Note that most of our descriptions (and most of the tasks of descriptive statistics) are quantitative: for example, when we wish to describe the width of a distribution, we give the numerical value of the standard deviation.

Inferential Statistics

inferential statistics: the science of using sample statistics to make inferences or decisions about population parameters

Inferential statistics is the science of drawing some conclusion or inference about a (population) parameter on the basis of information provided by a (sample) statistic. We have discussed two examples of inferential statistics: the point-estimates and confidence intervals of Chapter 8. When we use $\overline{X}$ to point-estimate μ, we are inferring something about μ—namely, that its magnitude is approximately $\overline{X}$. When we use a confidence interval, we are again inferring something about a parameter—namely, that it has a 95% probability of being in the given interval. Note that point-estimators and confidence intervals are, like descriptive statistics, quantitative (numerical) values.

hypothesis: an assumption or inference about a parameter or distribution that can be tested

The most widely used branch of inferential statistics (and the topic of most of the remainder of this textbook) is *hypothesis testing*, the science of deciding whether inferences (called *hypotheses*) about populations and their parameters should be accepted or not accepted. Whereas the final results of descriptive statistics are quantitative (how large, how wide, etc.), the final results of hypothesis tests are qualitative (yes, we should accept the hypothesis, or no, we should not accept it).

Let us begin with a simple example: Suppose we know that intelligence quotient (IQ) scores in the general population are normally distributed with $\mu = 100$ and $\sigma = 15$. Suppose we wish to know whether the students at Valley Junior High School can be considered to be a random sample from a population whose mean is equal to that of the general population. The answer to this question will be either, "Yes, it is reasonable to assume that these students can be considered to be a random sample from a population whose mean $\mu = 100$" or "No, it is not reasonable to assume that."

To determine whether these students can be considered to be a random sample from the general population, we take a random sample of size $n = 25$ and measure the IQs of these students. We plot the results on the histogram in the top portion of Figure 9.1. This figure superimposes the IQ population distribution (normal with $\mu = 100$ and $\sigma = 15$), and the answer to our question seems quite clear: Yes, the sample data in this histogram *are* consistent with the hypothesis that these students represent a sample from the general population.

On the other hand, if the data from our sample are the histogram at the *bottom* of Figure 9.1, we would conclude: No, these data are *not* consistent with the hypothesis. There are too many high-IQ students and not enough low-IQ students in these data for us to believe that these students are a random sample from a population with $\mu = 100$.

If the sample results turn out to be either of those in the middle two portions of Figure 9.1, however, our conclusion is not so obvious: We cannot immediately say yes or no to our hypothesis. There are more high-IQ students than low-IQ students to be sure, but perhaps that is just the result of sampling fluctuations (our sample, *by chance*, might contain a

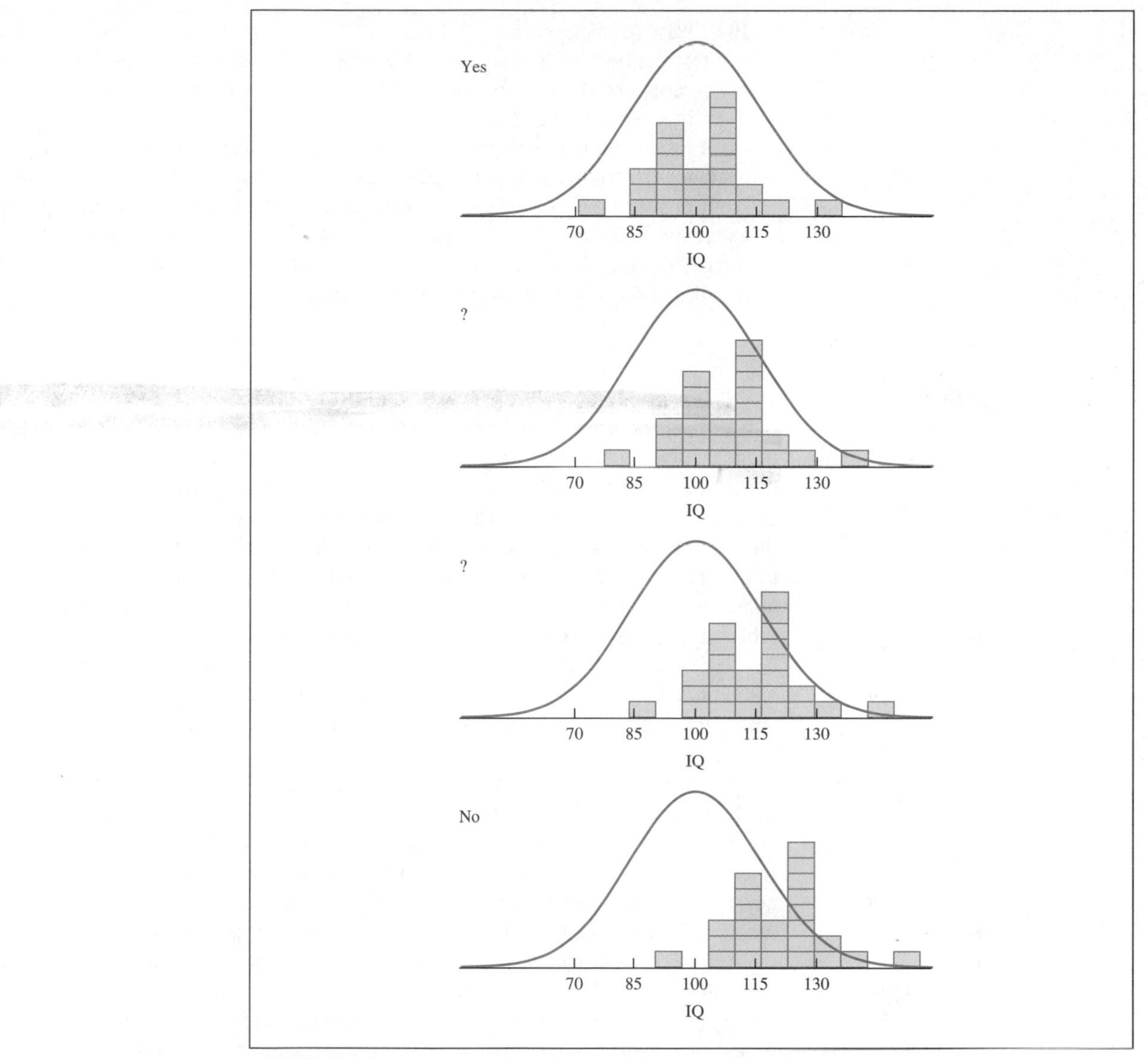

FIGURE 9.1 Does the sample in each histogram come from a population with $\mu = 100$, $\sigma = 15$?

few more high-IQ than low-IQ students). Clearly, we need some kind of decision rule that provides a rational dividing line between yes and no answers. The specification of that dividing line is the task of hypothesis testing.

9.2 Evaluating Hypotheses

A hypothesis is thus a statement about a population that can be tested using a sample statistic (see Box 9.1). To test a statement is to decide whether it is true or false, whether it can be supported or rejected.

BOX 9.1 Hypotheses about samples

We stated that hypotheses are statements about populations. Can hypotheses also be advanced about samples? Yes, they can, but statements about samples are trivially easy to evaluate. For example, if you assert that a *sample* mean $\overline{X}$ is 3.54 inches, we

(continued)

BOX 9.1 *(continued)*

can test the truth of that statement by simply adding all the values of the variable in the sample and dividing by the sample size. Then we can say with assurance either "yes, it is" or "no, it is not."

Statements about populations, on the other hand, are *not* typically trivial to evaluate. If you say, for example, that a *population* mean μ is 3.54 inches, we would have to measure the entire population (perhaps an impossible or prohibitively expensive task) to verify that statement absolutely. It is the task of inferential statistics to provide rules for evaluating these nontrivial hypotheses—that is, hypotheses about populations—using the properties of samples.

TABLE 9.1 IQs of a sample of 20 students at Gotham State University

Student's IQ
121
102
101
144
110
107
108
86
127
109
133
104
94
90
149
115
113
113
104
108
$\sum X = 2238$
$\overline{X} = 111.9$

It is easier to prove a statement false than it is to prove it true. The logic of science is based on the perhaps surprising fact that *disproving* hypotheses is easier than proving them. An example shows why this is true. Suppose someone asserts that the mean IQ at Gotham State University is $\mu = 112$. To test that hypothesis, we select a random sample of 20 students and measure each student's IQ as shown in Table 9.1. Then we compute the sample mean $\overline{X}$ because we know that $\overline{X}$ is the best point-estimate of μ. $\overline{X}$ turns out to be 111.9, so we might be tempted to accept the hypothesis $\mu = 112$ as true. However, someone else might have held the hypothesis that $\mu = 111.8$ (or 111.9, or 112.05, or any other value in the vicinity of 112). It would be very difficult (if not impossible) to determine which of these hypotheses is in fact true. Only one of these hypotheses *can* be true, but we cannot know which one. Thus, the attempt to prove a hypothesis true leads to some ambiguity: Perhaps the hypothesis is true, or perhaps it is false because some other similar hypothesis is itself true.

On the other hand, if someone asserts that $\mu = 70$, we can *confidently* conclude, on the basis that $\overline{X} = 111.9$, that the hypothesis $\mu = 70$ is *false*. Other hypotheses in the same vicinity (e.g., $\mu = 69$) are also concluded to be false. Note that we *can* be confident in the rejection of this hypothesis: The hypothesis is cleanly, unambiguously rejected, and other nearby hypotheses are also cleanly, unambiguously rejectable. Note also that to conclude with confidence that a hypothesis is false does not necessarily imply that the hypothesis is really false. There is always the possibility that in fact $\mu = 70$ but our sample "by luck" happened to include many unusually high IQs.

Therefore, the scientific method rests on the fact that it is much easier to reject a hypothesis than it is to prove it to be true. However, most often, in science, our main *interest* is in demonstrating that something is *true*; that is, we wish to do precisely that which is most difficult (or impossible) to do directly. How do we proceed?

Null and Alternative Hypotheses

Scientists deal with this dilemma by stating *two* hypotheses instead of one, and these two hypotheses are mutually contradictory. Then the scientist seeks to *reject* one of this pair of hypotheses. If one is rejected, the other must then be supported. Thus, support (which is difficult to do directly) for a hypothesis is acquired by rejection (which is relatively easy to do) of a contradictory hypothesis.

null hypothesis: the hypothesis that there is no effect or no difference

alternative hypothesis: the hypothesis that there is in fact an effect or difference

We call the hypothesis that we attempt to reject the *null hypothesis* and symbolize it by H_0; the other hypothesis of the pair, the one we wish to support, is called the *alternative hypothesis*, symbolized H_1. Almost always, the null hypothesis asserts that there is *no difference* or no effect or no change in the population parameter. It is this assertion of *nullity* of effect that gives the null hypothesis its name.

Let us take an example. Suppose we wish to know whether children who grow up in homes that are close to electric power stations have IQs different from those of children whose homes are not close to power stations. Some researchers claim that the magnetic fields associated with the power stations increase energy and might therefore increase IQ, whereas others claim that the magnetic fields interfere with normal cognitive processes and therefore might decrease IQ. To test whether living near a power station has an effect on a child's IQ, we create a pair of hypotheses:

$$H_0: \quad \mu = 100$$
$$H_1: \quad \mu \neq 100$$

Note that these two hypotheses are mutually exclusive: If the null hypothesis is true, then the alternative hypothesis is false, and vice versa. Furthermore, note that the null hypothesis is a statement of *no effect*: If H_0 is true, then living near power stations has not changed children's IQs. The aim of our study is to reject a hypothesis; in particular, we try to reject H_0. If we can reject the null hypothesis statement of no effect, then we can support the alternative hypothesis and conclude that $\mu \neq 100$—that is, that living near power stations *does* alter IQ.

The outcome of a study will be either that we reject the null hypothesis or that we fail to reject it. In actual practice, many researchers refer to the outcome of failing to reject H_0 as "accepting" H_0. As we saw in the preceding section, it is difficult (or impossible) to prove H_0 (or any other empirical statement) true. Therefore, strictly speaking, we *never* actually *accept* H_0; we simply sometimes do not have enough evidence to reject it. Because the fail-to-reject terminology is indeed rather awkward, however, we sometimes refer to this outcome as "accepting" H_0. When we do so, we will use the quotation marks to remind us that this term is a shorthand notation for "failing to reject."

The main topics of inferential statistics are how null and alternative hypotheses are generated and the rules for rejecting the null hypotheses in various situations. In all cases, the hypotheses are about *parameters*, and the decision whether to reject the null hypothesis is based on what is called the *test statistic*, which is a characteristic of the sample (as are all statistics) that is designed especially to test the particular null hypothesis under consideration. Thus, the remaining chapters are mainly discussions of how to create test statistics in differing research situations.

test statistic: a statistic specifically designed to facilitate the making of inferences

The procedure for rejecting a null hypothesis depends on the particular test statistic involved and on the "directionality" of the hypotheses and the "level of significance." We now turn to these two concepts.

Directional (One-tailed) and Nondirectional (Two-tailed) Hypotheses

Let us consider another example. Suppose Dr. Smith has reason to believe that large doses of vitamin B_{12} increase intelligence. We know from prior research that IQ scores are normally distributed with $\mu = 100$ and $\sigma = 15$ points. If we wish to test Dr. Smith's claim, we construct these two hypotheses:

$$H_0: \quad \mu \leq 100$$
$$H_1: \quad \mu > 100$$

This pair of hypotheses is the same as in the power station example except that the hypotheses here are *directional*, as indicated by the $>$ sign in the alternative hypothesis and the $\leq$ sign in the null hypothesis. We have reason to believe, before the experiment begins, that vitamin B_{12} will *increase* IQ, and the alternative hypothesis focuses our attention on that increase: H_1: $\mu > 100$. The null hypothesis in a directional experiment includes both those cases where vitamin B_{12} has *no* effect ($\mu = 100$) and those cases that are *contrary* to our expectations—in the vitamin B_{12} study, where it *decreases* IQ ($\mu < 100$). Therefore, we write the directional null hypothesis as H_0: $\mu \leq 100$.

directional (one-tailed) alternative hypothesis: where one direction of change (an increase or a decrease) is specified in advance

To recapitulate, as shown in Table 9.2, in a *nondirectional* test, the null and alternative hypotheses are of the form H_0: $\mu = a$ and H_1: $\mu \neq a$, whereas in a *directional* test, the null and alternative hypotheses are either of the form H_0: $\mu \leq a$ and H_1: $\mu > a$ (if the prior evidence pointed to an increase) or of the form H_0: $\mu \geq a$ and H_1: $\mu < a$ (if the prior evidence pointed to a decrease). A test with a directional alternative hypothesis is often

nondirectional (two-tailed) alternative hypothesis: where the hypothesized change can be either an increase or a decrease

TABLE 9.2 Hypotheses in nondirectional and directional tests

	Nondirectional test	*Directional test*		
Null hypothesis	H_0: $\mu = a$	H_0: $\mu \leq a$	or	H_0: $\mu \geq a$
Alternative hypothesis	H_1: $\mu \neq a$	H_1: $\mu > a$		H_1: $\mu < a$
	Use when prior evidence is inconclusive.	Use when prior evidence: Shows an increase		Shows a decrease

called a "one-tailed" test, and a test with a nondirectional alternative hypothesis is called a "two-tailed" test.[1]

There is considerable debate about whether, or under what circumstances, it is permissible or advisable to perform directional tests. Many observers hold that directional tests are never (or at best only rarely) justified in social science research. In this textbook, I will not take a stand in this debate because that is primarily an issue of research design. We consider both directional and nondirectional examples.

Type I and Type II Errors

We have seen that inferential statistics use a statistic (called the test statistic) to determine whether inferences about a parameter (hypotheses) are to be considered true or false. In particular, we attempt to ascertain whether or not null hypotheses should be rejected. The null hypothesis is an assertion about a state of nature. When we assert, for example, that $\mu = 100$, we are hypothesizing that nature is such that the IQ distribution has true mean $\mu = 100$. Now there are two possibilities regarding the truth value of our null hypothesis: Either H_0 is true (nature does indeed have the specified characteristic) or H_0 is false. However, the true state of nature is generally unknown; that's why we are stating hypotheses in the first place.

An experimenter collects a sample, computes a test statistic based on that sample, and then decides whether to reject or not to reject the null hypothesis. Thus, the only two possible alternatives for the outcome of an experiment are to "accept" H_0 or to reject H_0.[2] As we saw in the preceding paragraph, H_0 itself may be either true or false, entirely independent of whether we *say* it's true ("accept" it) or *say* it's false (reject it). There are, then, as shown in Table 9.3, four possible outcomes of an experiment: Two are correct (H_0 is true and we "accept" it, and H_0 is false and we reject it) and two are in error (the *Type I error*, where H_0 is true but we reject it, and the *Type II error*, where H_0 is false but we "accept" it).

Type I error: rejecting the null hypothesis when it is in fact true

Type II error: failing to reject the null hypothesis when it is in fact false

TABLE 9.3 Possible outcomes of an experiment as a function of the actual (but unknown) state of nature and the decision made

		Actual (but unknown) state of nature	
		H_0 true (No effect)	*H_0 false (Effect exists)*
Decision	"Accept" H_0	Correct	Type II error "bl**II**ndness"
	Reject H_0	Type I error "gull**I**bility"	Correct

[1] I recommend *against* calling a test "one-tailed" or "two-tailed" because that terminology, though descriptive for the next few chapters, is quite misleading in Chapters 13–15.

[2] Everything said about experiments in this chapter applies equally to quasi-experiments, a term we will define in Chapter 11.

Let's consider two examples that clarify the concepts of Type I and Type II errors. First, the HCAN is a home test to detect cancer cells. A null hypothesis in general is the assertion of no effect; in the HCAN example, the null hypothesis is the assertion that no cancer is present. Suppose that Sally takes the test and it says she has cancer, although actually she does not. Is this an error? Of course. What kind? The null hypothesis is in fact true (actually no cancer), but we reject it (the test says "cancer"), so this is a Type I error. What is the effect of this Type I error? Sally is distressed without cause. Second, Warren takes the HCAN test and it says he does *not* have cancer cells, when in fact those cells are present. Is this an error? Of course. The null hypothesis is false (cancer cells *are* actually present), but we fail to reject it (the test did not say "cancer"); therefore, this is a Type II error. What is the effect of this Type II error? Warren is falsely relieved when he should be seeking treatment. These examples make it clear that both Type I and Type II errors can be extremely important.

Some students are helped by the following device, which was inspired by a discussion in Rosenthal and Rosnow (1991, pp. 40–42): A Type I error is gull**I**bility (seeing something that isn't there); a Type II error is bl**II**ndness (missing something that is there; being "blind with two eyes").

We must recognize that in any given experiment, we *do not* know whether we are correct or whether we have made a Type I or Type II error because we do not know for certain the true state of nature (the true value of the parameter in question). Whenever we take a random sample, there is always the possibility that "by luck" a statistic based on that sample will be unusually large or unusually small, but we generally have no way of knowing whether that is the case in any particular experiment. The best we can do is limit the probability that Type I or Type II errors occur.

We must conclude, therefore, that we can never know for certain whether our experimental outcome is correct or in error. You may notice that that conclusion is quite similar to the one we reached in Chapter 8 when we considered confidence intervals. There, we said that the parameter either did or did not lie in the confidence interval. Because we were ignorant of the value of the parameter (that's why we were constructing the confidence interval in the first place), we could not know whether any particular confidence interval contained the parameter. The best we could do there was to state the level of confidence (e.g., 95%) we had in our statement that the parameter was within the confidence interval. We will see in the next section that the level of significance makes the same kind of probability statement about experimental results that the level of confidence did about confidence intervals results.

Level of Significance (α)

We have seen that there are two ways of being incorrect when testing hypotheses, Type I and Type II errors, and that these errors are unavoidable consequences of taking random samples. They are *not* the result of human mistakes but rather the characteristics of the experimental process.

level of significance: the probability (signified by α) of making a Type I error

The *level of significance* is the probability of making a Type I error; we generally call the level of significance α (Greek "alpha"). Thus, α is the probability that we will reject H_0 when it is in fact true; in other words, α is the probability that we will conclude that an effect exists when in fact there is none. α is the probability that we *say* that vitamin B_{12} increases IQ (that is, we reject H_0) when in fact it has no effect on IQ (that is, H_0 is true).

The level of significance is the complement of the level of confidence. We want the probability that we make a Type I error to be *small* (typically .05 or .01), whereas we want the probability that a confidence interval contains the parameter to be *large* (typically .95 or .99).

We want α to be small because Type I errors are expensive for the scientific (and the human) enterprise. Suppose, for example, we make a Type I error in our vitamin study. On the basis of our results, we report in a journal that doses of vitamin B_{12} increase IQ. That will be an error that we made in good faith because we had no way of knowing that this particular result was a Type I error; we obtained our results, believed we had supported the efficacy of vitamin B_{12}, and reported it. As a result, our readers will alter their behavior, perhaps focusing on a B_{12} diet while ignoring other avenues (such as reading enhancement programs) that might be effective in raising IQ. Sometime later, perhaps, someone will conduct many experiments and find that vitamin B_{12} has *no* effect on IQ. Those individuals who believed our earlier report (and therefore ignored other avenues) were done a possibly uncorrectable disservice. To prevent any further damage, we would want to find and contact the entire readership of our first report and inform them that our result was mistaken. That is clearly a highly expensive and difficult (if not impossible) thing to do.

α: the probability of making a Type I error (also called "the level of significance")

Because we know that Type I errors are unavoidable (due to sampling fluctuations) but undesirable, we wish to limit their occurrence. That means we wish to keep α (the probability that we make a Type I error) very small. In general, the scientific community rather arbitrarily chooses $\alpha = .05$ or $\alpha = .01$ as an acceptable probability of reporting false (Type I error) results.

If the level of significance is .05, then out of every 100 experiments we perform where the null hypothesis is in fact true, about five will incorrectly conclude that the null hypothesis should be rejected. We cannot know *which* five. That means that when you read a journal article that reports a result "at the .05 level of significance" (often written simply "$p < .05$"), there is a substantial (but small) probability that the result is due to a "lucky" sample (unusually large or unusually small due simply to sampling fluctuations) rather than to the presence of a real effect in nature.

It is sometimes *incorrectly* thought that the implication of the preceding paragraph is that about five out of every 100 journal articles actually make Type I errors. In fact, the number of Type I errors is smaller than that. Most scientific investigations take place in subject areas where the null hypothesis is *not* true. The reason a scientist conducts an experiment in the first place is because he or she has reason to believe that the alternative hypothesis is true. Holding the level of significance at .05 guarantees that in the long run *at most* five out of every 100 reports will be the result of Type I errors.

statistically significant: leading to the rejection of the null hypothesis

If the outcome of an experiment leads to the rejection of the null hypothesis, we say the outcome is "statistically significant" or that there was a "statistically significant difference"—that is, a difference greater than would be expected by chance alone. For example, if the bloomers' mean intellectual growth was $\overline{X}_{\text{bloomers}} = 16.5$ and the other mean was $\overline{X}_{\text{other}} = 16.4$, we would probably conclude that whereas the two sample means were numerically different, they were not *significantly* different. Chance alone, not the differential treatment caused by teachers' expectations, might easily account for a small difference between means.

Statistical Power

We want to be able to conclude at the end of most experiments that "We reject the null hypothesis" or "We believe an effect exists" or "We believe that vitamin B_{12} *does* increase IQ" or "We believe that high teacher expectations *do* lead to intellectual growth." However, we want that conclusion to conform to the facts; that is, we want the probability of *mistakenly* rejecting H_0 to be *small* and the probability of *correctly* rejecting H_0 to be *large*. The probability of mistakenly rejecting H_0 is the probability of a Type I error that we discussed in the preceding section; we keep it small by specifying that the level of significance α must be .05 or .01 or some other small number.

power: the probability of rejecting H_0 when H_0 is in fact false

The probability of *correctly* rejecting H_0 is called the *power* of a statistical test, and it is desirable that power be high. We use the term *power* in a way quite similar to its use in microscopes. We say a microscope is "powerful" to the extent that it allows the correct identification of objects, and we say that a statistical test is "powerful" to the extent that it correctly identifies situations where the alternative hypothesis is true (and where, of course, the null hypothesis is correspondingly false).

So statistical power is the probability of correctly rejecting the null hypothesis. Thus, power, like all probabilities, is a number between 0 and 1. We would like power to be close to 1, just as we would like the probability of *incorrectly* rejecting the null hypothesis (the level of significance α) to be close to 0.

Suppose that 100 different experimenters conducted 100 identical experiments to determine whether vitamin B_{12} increases IQ, and suppose that the statistical power of each of those experiments is .8. Furthermore, suppose that the null hypothesis (the actual state of nature) is in fact false: Vitamin B_{12} *does* increase IQ. We mortals do not know the actual state of nature, of course; that's why we conduct experiments. Based on these assumptions (power = .8, H_0 is false), however, we would expect that approximately 80% of those 100 experiments would correctly conclude that vitamin B_{12} increases IQ, whereas approximately 20% would incorrectly conclude that vitamin B_{12} has no effect. That is, 20% of the experiments would result in a Type II error—the failure to reject H_0 when it is in fact false. The probability of making a Type II error is sometimes called β (Greek "beta"; recall that the probability of making a Type I error is called α), and β is the complement of power: $\beta = 1 - \text{power}$ [the probability of mistakenly failing to reject H_0 (β) is one minus the probability of correctly rejecting H_0 (power)]. In our example, power = .8, so $\beta = 1 - .8 = .2$, which implies that 20% of such experiments end in a Type II error (as we have seen).

β: the probability of making a Type II error

We have explained that it is desirable to require that the level of significance α be some small but specific value such as .05 or .01 to protect us against making too many Type I errors. Would it make sense to require that power be some large but specific value such as .95 or .99 (or, entirely equivalently, to require that β have some small specific value such as .05 or .01)? No, it would not, because the scientific community does not in general need as much protection from Type II errors as it does from Type I errors. Here's why.

In most situations, we do not conduct 100 experiments to determine whether an effect exists; instead, we conduct one. Consider an investigator who conducts a single experiment to demonstrate that doses of vitamin B_{12} increase IQ. After the data are collected, she concludes that the null hypothesis should *not* be rejected; that is, she concludes that vitamin B_{12} has no demonstrable effect on IQ. The result is that the investigator will *not* report findings in a journal. Instead, the investigation may be considered an exploration of a blind alley. She had thought vitamin B_{12} was effective, but apparently it wasn't.

If subsequent research indicates that the conclusion of the original investigation was a Type II error, what is the expense of that error and who bears it? One major expense is the time lost in the original investigation, but the bearer of that expense is the original investigator, not the scientific community at large. There is *no* necessity of informing the community of a previous mistake because there was no report of findings in the first place. Thus, it is generally agreed that a Type II error is not as expensive to the scientific community as a Type I error, and for that reason, the scientific community does *not* place a limitation on its occurrence. There is thus *no* set limit on β, the probability of a Type II error, or (equivalently) on statistical power, the probability of correctly rejecting H_0. It is generally desirable to keep the power large (and thus β small), but how large (or small) is usually entirely at the discretion of the individual investigator. We will return to a discussion of power in Chapter 12.

The costs in human terms of Type I and Type II errors go well beyond the expense of correcting mistaken scientific reports. Consider a researcher who is testing a drug that may be effective in the treatment of AIDS. A Type I error would involve reporting that the drug is effective when in fact it is not. This can have cruel effects, such as raising false hopes or discontinuing the funding of some other research project that might actually be successful. On the other hand, a Type II error would involve failing to report a drug that is in fact effective. This would also have cruel results, depriving needy individuals of effective treatment. Which is more costly here, the Type I or the Type II error? There is no statistical answer to that question; it is a matter of complex human judgment.

The Courtroom Analogy

The logic of the null hypothesis is entirely analogous to the logic used in the U.S. legal system. John Q. Diddit, by all accounts a nasty guy, was found brutally stabbed to death on the morning after his 33rd birthday. His wife, May B. Diddit, was charged with murder. The parameter of interest (analogous to μ) is Mrs. Diddit's actual guilt. The actual value of that parameter (that is, whether Mrs. Diddit actually committed the murder) is unknown to us.

We conduct an experiment (that is, a jury trial) to determine what attitude we should hold about her guilt. We first state the null hypothesis "H_0: Mrs. Diddit is not guilty" (because the U.S. Constitution specifies that she is innocent until proven guilty). Our experiment (trial) will determine whether we will reject that hypothesis or not. We desire such proceedings to avoid making a Type I error (that is, to have a low probability of convicting an innocent person) but to be powerful (to have a high probability of convicting a guilty person).

The level of significance ("burden of proof") is specified before the experiment begins. The data (evidence and witnesses) are collected and analyzed (examined by the jury), and the test statistic (the "weight of the evidence") is compared to the critical value (the "beyond a reasonable doubt" standard). If the test statistic (weight of evidence) turns out to exceed the critical value (beyond a reasonable doubt), the jury will reject H_0 (find Mrs. Diddit guilty). If the data do not exceed the critical value, then the jury will fail to reject H_0 (find Mrs. Diddit not guilty).

The jury can make two possible decisions in the trial (ignoring the mistrial conclusion that is analogous to abandoning an experiment): a not guilty verdict (failure to reject H_0) or a guilty verdict (rejection of H_0). There are also "two actual states of nature," to use the Table 9.3 terminology: actually innocent (H_0 true) and actually guilty (H_0 false). Thus, there are four possible outcomes of a trial, entirely analogous to the four possible outcomes of an experiment shown in Table 9.3.

Suppose the jury decision is "guilty" (reject H_0). If H_0 is actually true (Mrs. Diddit in fact did not do it), then the jury made a Type I error (rejected H_0 when it was actually true). On the other hand, if H_0 is actually false, then the jury made a correct conviction.

Now suppose the jury decision is "not guilty" (failure to reject H_0). H_0 may be true (Mrs. Diddit in fact did not commit the crime), in which case the jury's decision was correct, or H_0 may be false (she is in fact the murderer), in which case the jury made a Type II error (failed to convict her even though she was actually guilty). No one other than Mrs. Diddit or the actual murderer knows for sure which alternative is correct. Our legal system instructs us to "accept" the null hypothesis—that is, to accord Mrs. Diddit all the legal benefits of innocence. Does "accepting" H_0 imply that Mrs. Diddit is actually innocent? No, it doesn't; the jury may have made a Type II error.

There is one important distinction between American jurisprudence and the experimental method. In jurisprudence, "accepting" a not guilty verdict is a permanent condition (no person may be tried for the same crime twice), whereas in experimentation, "accepting" the null hypothesis has no such permanence: Another experiment may be designed and carried out that may subsequently reject H_0. In part because of this incorrect connotation of permanence, many statisticians avoid the use of the phrase "accept H_0" in favor of the correct expression "fail to reject H_0."

Practical Significance

We have seen that when we conduct an experiment, we usually set up the situation so that the probability of a Type I error (that is, the level of significance) is $\alpha = .05$, $\alpha = .01$, or $\alpha =$ some other small number. This is important because it implies that if we obtain a statistically significant result (that is, we reject the null hypothesis), then we can be confident (95%, 99%, etc.) that the rejection of H_0 was not merely the result of chance alone.

However, statistical significance does not necessarily imply that the result is important in any practical sense. For example, it is almost always possible to obtain a statistically significant result by using a very large number of subjects in an experiment (for a more complete discussion of these issues, see the optional section at the end of this chapter).

practical significance: the degree to which a result is important

Clearly, if the result of an experiment is statistically significant, then the report of that experiment should include a discussion of its *practical significance* (the importance of the result) as well. (If the results are statistically *nonsignificant*, then there is usually no reason to report practical significance because the result can be ascribed "merely to chance alone.") There is as yet, however, no universally used method for reporting practical significance.

effect size: a measure of the magnitude of a result

This textbook will recommend, in succeeding chapters, that one or more measures of *effect size* be used in the discussion of practical significance. Effect size is measured in different ways in different contexts; in the example about Pygmalion in the classroom, it is based on the difference between the two sample means (effect size $= \overline{X}_{\text{bloomers}} - \overline{X}_{\text{other}} = 16.5 - 7.0 = 9.5$ IQ points). Thus, the Pygmalion effect size is the difference between the bloomers and the other children in mean intellectual growth. This effect size is, of course, the focus of our interest in the Pygmalion study because it measures in the most direct way possible the effect of teacher expectations. If the effect size were only one IQ point, for example, we might decide that even though such a result might be statistically significant (for example, if we had very large sample sizes), the result would have little or no practical significance because IQs that differ by only one point are basically indistinguishable.

The bottom line is that *both* statistical significance and practical significance are important, and our evaluation of experimental outcomes should consider both.

9.3 The Procedure for Evaluating Hypotheses

We have shown that an experiment is the attempt to reject the null hypothesis about a parameter that characterizes the distribution of a variable in a population. The experiment consists of drawing a random sample from that population, computing the sample statistic that is the best point-estimate of the parameter in question, and asking whether this sample statistic could reasonably be expected to come from the population if the null hypothesis is true. If the answer to that question is no, then we reject the null hypothesis; otherwise, we fail to reject it ("accept" it). To answer that question, we transform the statistic that point-estimates the parameter into the *test statistic*, which has a distribution whose probabilities (areas under the distribution) are known to us.

Three distributions are relevant to an experiment: the distribution of the *variable* (whose parameter is specified by the null hypothesis), the distribution of the *sample statistic* (which itself is the best point-estimate of the parameter), and the distribution of the *test statistic* (the transformation of the sample statistic into a distribution with known probabilities).

We can say that, in many situations, the test statistic has this general formula:[3]

general formula for the test statistic

$$\text{test statistic} = \frac{\text{sample statistic} - \text{population parameter}}{\text{standard error of the sample statistic}} \tag{9.1}$$

where "population parameter" is the particular parameter specified by the null hypothesis, "sample statistic" is the statistic that is the best point-estimate of that population parameter, and "standard error of the sample statistic" is the standard error of the particular sample statistic specified in the numerator (that is, the one that point-estimates the population parameter).[4] We will specify which distribution is appropriate for the test statistic in subsequent chapters as we explore the possible hypotheses that we may wish to test. We have encountered two of the possible test statistic distributions already (z and t), and we will see that there are others (e.g., F, χ^2, U), each appropriate in a particular experimental situation.

To understand why the test statistic has the form shown in Equation (9.1), let's observe that "the sample statistic minus the population parameter" [the numerator of Equation (9.1)] is a measure of the difference between the sample and the population that was *actually obtained* in the experiment. Furthermore, "the standard error of the sample statistic" [the denominator of Equation (9.1)] is a measure of the difference between the sample and the population *that one might expect by chance alone*. Thus, we can say in general that the test

---

[3] There are cases where test statistics do not have this form—for example, F in the analysis of variance (see Chapter 13).

[4] Note that the term *sample* in the phrase "sample statistic" and the term *population* in the phrase "population parameter" are (strictly speaking) redundant: A statistic is *always* a characteristic of a sample and a parameter is *always* a characteristic of a population. We include the redundancy for emphasis.

statistic is the difference actually obtained divided by the difference expected by chance alone.[5]

Let us now reconsider the experiment that tests whether living near a power station alters a child's IQ. This experiment is an example of a test for the mean of a distribution under the assumption that we know σ. How will the general formula for the test statistic, Equation (9.1), apply in this case? The null hypothesis in that experiment, as you may recall, was H_0: $\mu = 100$; thus, the population parameter to be used in the numerator of Equation (9.1) is μ. The sample statistic that point-estimates μ is $\overline{X}$, so that is the first term in the numerator. The denominator is then the standard error of that sample statistic—that is, the standard error of $\overline{X}$, $\sigma_{\overline{X}}$. Thus, the right-hand side of Equation (9.1) is $(\overline{X} - \mu)/\sigma_{\overline{X}}$. We will see in the next chapter that, if the null hypothesis is true, then the right-hand side of this equation is distributed as z, so our test statistic will be

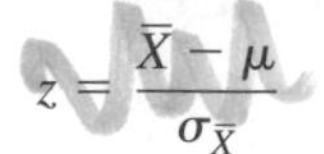

$$z = \frac{\overline{X} - \mu}{\sigma_{\overline{X}}}$$

Because we know the characteristics of z (that 34% of all instances fall between the mean and the first standard deviation, etc.), we can specify the probabilities of the possible outcomes of our experiment. In particular, we can specify a critical value of z, which we symbolize as z_{cv}, so that the probability of a Type I error is .05 or .01 or whatever value we choose.

Hypothesis evaluation follows a general procedure that is shown in the following outline. We provide the outline here and explore it completely in subsequent chapters.

outline of hypothesis evaluation

I. State the null and alternative hypotheses.
- A. Illustrate the null hypothesis by preparing three sketches.
  1. Sketch the distribution of the variable assuming H_0 is true.
  2. Sketch the distribution of the sample statistic (the one that point-estimates the population parameter specified by H_0), assuming H_0 is true.
  3. Sketch the distribution of the test statistic assuming H_0 is true.

II. Set the criterion for rejecting H_0.
- A. What is the level of significance?
- B. Is the test directional or nondirectional?
- C. Illustrate the criterion.
  1. Shade the rejection region(s) on the distribution of the sample statistic.
  2. Shade the rejection region(s) on the distribution of the test statistic.

III. Collect a sample and compute the observed values of the sample statistic and the test statistic.
- A. Indicate the observed value of the sample statistic on the sketch of its distribution.
- B. Indicate the observed value of the test statistic on the sketch of its distribution.

IV. Interpret the results.
- A. Decide whether the results are statistically significant; that is, decide whether to reject H_0.
- B. If the results are statistically significant, then consider whether they are practically significant.
  1. Determine the effect size.
  2. Illustrate the effect size by sketching the distribution of the variable assuming the point-estimate of the parameter is true.
  3. Consider the scientific, economic, political, and moral value of an effect of this size.
- C. Describe the results, including both statistical and practical significance, in plain English.

---

[5]This last formulation is actually somewhat more general than Equation (9.1); for example, F in the analysis of variance is a special case of this last formulation but not of Equation (9.1).

9.4 Connections

Cumulative Review

This chapter marks a dramatic turning point in the flow of this textbook. All the previous chapters were concerned with describing distributions or distributions of means, but now we begin to discuss hypothesis testing, the procedure that will occupy us for most of the remainder of the textbook (with the exception of parts of Chapters 17 and 18).

Computers

Because this chapter provides the theory of hypothesis testing, not computation, it has no computer comments.

Journals

Most journal articles report the level of significance with a statement such as: "This result was significant ($p < .05$)." In our terminology, this implies that the level of significance α was set at .05. Occasionally journals report so-called exact probabilities with a statement such as: "This result was significant ($p = .027$)." This can be interpreted as meaning that the null hypothesis would be rejected if the level of significance α had been set at any value greater than or equal to .027.

Homework Tips

1. Check the list of learning objectives at the beginning of this chapter. Do you understand each one?
2. This chapter covers a lot of material, and you shouldn't find it surprising if it seems difficult to master. The next five chapters are all variations on the themes spelled out here, so you can expect comprehension to deepen as you go along.
3. Some students find the distinction between Type I and Type II errors confusing. It is essential that you master that terminology; we will be using it throughout the remaining chapters. Remember the gull**I**bility (seeing something that isn't there) and bl**II**ndness (missing something that is there; "blind with two eyes") tips from Table 9.3.
4. Repeat after me: "The general formula for the test statistic is the sample statistic minus the population parameter divided by the standard error of the sample statistic."
5. Three things in this chapter that are worth memorizing are the general formula for the test statistic [Equation (9.1), see Homework Tip 4], the table of experimental outcomes (Table 9.3), and the four steps of the general hypothesis-testing procedure.

Exercises for Chapter 9

Section A: Basic exercises (answers in Appendix D, page 512)

1. List all the descriptive statistics we have considered so far in this text.
2. Suppose I give the same final exam every time I teach statistics and the mean score over all administrations is 81. This time when I teach statistics, I double the amount of homework I assign, and I attempt to ascertain whether that increases students' scores on the final exam.

(a) State the null and alternative hypotheses.
(b) Is this test directional or nondirectional? Why?

3. (*Exercise 2 continued*) Suppose I administer the final exam to the 25 students in my current (more homework) class and examine their final exam scores, finding that $\overline{X} = 83$. Can I conclude that more homework improves exam scores? Explain.

4. (*Exercise 2 continued*) Now suppose that I have enough information so that I do in fact reject the null hypothesis.
   (a) Is it possible that I have made a Type I error? Why or why not?
   (b) Is it possible that I have made a Type II error? Why or why not?

5. (a) What do we call the probability of a Type I error?
   (b) What do we call the probability of a Type II error?

6. Suppose we have stated a null hypothesis in some situation—for example, H_0: $\mu = 64$. There are two possibilities regarding the state of nature: Either H_0 is true or it is false. Now we run an experiment to test this hypothesis. The outcome of the experiment is either that we reject H_0 or that we fail to reject it. What are the four possibilities regarding the correctness or incorrectness of our decision? What are the probabilities of each outcome?

7. How is the level of significance related to the level of confidence?

8. State the general formula for the test statistic.

9. What are the four steps of the hypothesis-testing procedure?

Section B: Supplementary exercises

10. State the difference between descriptive and inferential statistics.

11. Describe why we attempt to reject a null hypothesis rather than directly try to support the experimental hypothesis.

12. A school has been gathering information about drug use in its student body by asking students to identify (by first initial only) their ten closest friends or associates at school and then indicate how many of these ten have used illegal drugs in the past 6 months. The school has been conducting such surveys for a number of years, and the results have always been about the same: $\mu = 3.4$ and $\sigma = 1.5$. This year the school has begun a "Just say NO" campaign and the administration would like to know whether it has been effective. Perhaps it has increased drug use, perhaps it has decreased it, or perhaps it has had no significant effect. The administration conducts the survey again.
    (a) State the null and alternative hypotheses.
    (b) Are these hypotheses directional or nondirectional? Why?
    (c) State in plain English what a Type I error would be in this situation.
    (d) State in plain English what a Type II error would be in this situation.

13. An insurance company is considering lowering its rates for drivers of automobiles that have air bags because it believes that the bags lower the medical costs it must pay for accident victims. It knows that in the past the average medical cost for a particular kind of accident has been $7400. It takes a sample of accidents involving air bags and determines the medical costs for each of those accidents. If air-bag accident costs are lower than the costs in the past, it will lower its insurance rates.
    (a) State the null and alternative hypotheses.
    (b) Are these hypotheses directional or nondirectional? Why?
    (c) State in plain English what a Type I error would be in this situation.
    (d) State in plain English what a Type II error would be in this situation.

14. An industrial psychologist believes that giving assembly-line workers some control over their situation (by asking them for suggestions about how assembly processes can be improved, etc.) will improve their performance. She decides to measure performance by counting the number of days per year a worker fails to report for work (calling in sick, etc.). She believes that giving the workers more control will decrease their absenteeism. She knows that in the past workers have been "sick" on average 15.7 days per year. She takes one assembly line, institutes the new procedures, and counts the days the workers on that line are absent.
    (a) State the null and alternative hypotheses.
    (b) Arc these hypotheses directional or nondirectional? Why?
    (c) State in plain English what a Type I error would be in this situation.
    (d) State in plain English what a Type II error would be in this situation.

15. Is it desirable to have α be large or small? Why?

16. Is it desirable to have β be large or small? Why?

Section D: Cumulative review (answers in Appendix D, page 512)

17. A certain kind of potato has mean weight 12 ounces with standard deviation 2.5 ounces.
    (a) Sketch the distribution of the weights of the potatoes.
    (b) What percentage of potatoes weigh more than 10 ounces? Shade the relevant area, eyeball-estimate, and compute.
    (c) What percentage weigh less than 10 ounces? Shade the relevant area, eyeball-estimate, and compute.
    (d) What percentage weigh between 13 and 16 ounces? Shade the relevant area, eyeball-estimate, and compute.

18. (*Exercise 17 continued*) Assume that potatoes are packaged 25 to a bag and that these 25 can be considered random samples from the potato population. Suppose that you inspect 100 bags of potatoes and for each bag, you compute the mean weight of the potatoes in it.

(a) What would you expect the mean of these 100 means to be?
(b) What would you expect the standard deviation of these 100 means to be?

19. (*Exercise 18 continued*) Suppose you inspect 1000 bags instead of 100.
(a) What would you expect the mean of these 1000 means to be?
(b) What would you expect the standard deviation of these 1000 means to be?

20. (*Exercise 18 continued*)
(a) Sketch the distribution of the mean weights of the potatoes in bags that contain 25 potatoes.
(b) What percentage of potato bags have potatoes with mean weight greater than 10 ounces? Shade the relevant area, eyeball-estimate, and compute.
(c) What percentage have mean weight less than 10 ounces? Shade the relevant area, eyeball-estimate, and compute.
(d) What percentage have mean weight between 13 and 16 ounces? Shade the relevant area, eyeball-estimate, and compute.
(e) Explain why these answers are different from those in Exercise 17.

21. Suppose you are a senator and you are considering voting for legislation that will increase the salaries of senators. You wish to know what percentage of voters in your state support such a measure. You have your office contact 400 voters selected at random and ask them whether they favor increasing senators' salaries. The results are shown here. What can you say with 95% confidence about the percentage of voters in your state who favor pay raises?

Responses (Y = Yes, N = No): NNYNYNNNYNNYN
YNNNYYNYNNNYNYNNNYNNYNYNNNYYNYN
YNNNNYNYNNNYNNYNYNNNYYNYNYNNNNY
NYNNNYNNYNYNNNYYNYNYNNNNYNYNNNY
NNYNYNNNYYNYNYNNNNYNYNNNYNNYNYN
NNYYNYNYNNNNYNYNNNYNNYNYNNNYYNY
NYNNNNYNYNNNYNNYNYNNNYYNYNYNNNN
YNYNNNYNNYNYNNNYYNYNYNNNNYNYNNN
YNNYNYNNNYYNYNYNNNNYNYNNNYNNYNY
NNNYYNYNYNNNNYNYNNNYNNYNYNNNYYN
YNYNNNNYNYNNNYNNYNYNNNYYNYNYNNN
NYNYNNNYNNYNYNNNYYNYNYNNNNYNYNN
NYNNYNYNNNYYNYNYNNYNNNNYNYNNNYN
NYNYNNNYYNYNYNN

(Labor savings for the student: There are 256 N's and 144 Y's.)

22. A nursing supervisor wishes to know how many requests patients make of their nurses per day. She takes a random sample of 40 patients and has nurses log each request that is made. The numbers of requests per day are: 17, 11, 13, 13, 12, 15, 10, 18, 20, 13, 11, 11, 6, 6, 13, 15, 18, 13, 10, 17, 9, 21, 16, 13, 10, 12, 15, 16, 7, 12, 17, 11, 10, 19, 13, 11, 13, 12, 17, 17. (Labor savings for the student: $\sum X = 533$, $\sum X^2 = 7617$, $\bar{X} = 13.325$, $s = 3.633$.) What can you say about the average number of requests per day in the hospital population? State in symbols and in plain English.

Why Statistical Significance Alone Is Not Enough (Optional)

This section may be omitted without loss of continuity.

Hypothesis testing has been the dominant feature of behavioral science research for a long time—so dominant, in fact, that many journal reports present descriptions of statistically significant results with no discussion at all of the practical significance of those results. That is, many journal articles omit Step IV.B of the hypothesis evaluation procedure presented in Section 9.3, as if the attainment of statistical significance (rejecting H_0) in itself is enough to indicate that the finding is important.

However, a science based on statistical significance (rejecting H_0) alone has four major flaws (and a host of minor ones—see Cohen, 1994; Gigerenzer, 1993; Gigerenzer et al., 1989; and Thompson, 1994):

1. The null hypothesis is never (or at least hardly ever) really true, so attempting to reject it is simply a proof of the obvious. In the electric power station study, for example, we stated the null hypothesis as H_0: $\mu = 100$. We would be extremely surprised if the population mean μ were actually exactly 100.000..., though; it is actually 99.372 or 101.594, or some other value close to 100.

2. If the null hypothesis is not exactly true, then it is *always* possible to *guarantee* a statistically significant result merely by increasing the sample size. Thus, you can nearly always "buy" statistical significance.

3. Flaw 2 implies that the statistical significance of a result has no direct relationship to the effect size. Small effects can be highly significant, whereas large effects can be nonsignificant.

4. The basic logic of significance testing is the reverse of what we would like it to be. To illustrate, consider the vitamin B_{12} experiment. The major hypothesis-testing consideration is the limitation of the probability of a Type I error. That is, the hypothesis-testing procedure keeps small ($< \alpha$) the probability that we would conclude that vitamin B_{12} increases a child's IQ (reject H_0) given that H_0 is true (vitamin B_{12} has no effect on IQ). In probability terms, we require that $P(\text{reject } H_0 \mid H_0 \text{ true}) < .05$. However, suppose we finish collecting our vitamin B_{12} data, analyze them, and reject H_0 at $\alpha = .05$. What is the probability that H_0 is actually true (vitamin B_{12} actually has no effect), given that we have just rejected it? In probability terms, what is $P(H_0 \text{ true} \mid \text{reject } H_0)$? That is arguably the most important question in experimentation. It may seem that the answer to that question is ".05 or less," but that is not the case, and the fact is we do not know, even at the end of the experiment, the probability that vitamin B_{12} has no effect. This fourth flaw is, therefore, that the two conditional probabilities $P(\text{reject } H_0 \mid H_0 \text{ true})$ and $P(H_0 \text{ true} \mid \text{reject } H_0)$ are *not* the same.

Consider this example as an illustration of Flaw 4, the fact that $P(\text{reject } H_0 \mid H_0 \text{ true}) \neq P(H_0 \text{ true} \mid \text{reject } H_0)$. Schizophrenia is a serious mental illness that occurs in about 1% of the population. Suppose we develop a test to determine whether a person has schizophrenia. As part of the validation study, we administer the test to 1000 randomly selected individuals, with the results shown in Table 9.4, which is set up in parallel to Table 9.3.

TABLE 9.4 Results of validation study of schizophrenia test

		Actual state of nature		
		H_0 true, actually normal	*H_0 false, actually have schizophrenia*	*Total*
Diagnosis	"Accept" H_0 (say "normal")	943 (correct)	1 (miss = Type II error)	944
	Reject H_0 (say "schizophrenia")	47 (false positive = Type I error)	9 (correct)	56
Total		990	10	1000

Of the 1000 total individuals, Table 9.4 shows that 990 are actually normal and 10 actually have schizophrenia, so we can see that the incidence of schizophrenia is $10/1000 = .01$—1% as expected. Table 9.4 also shows that of the 1000 individuals, 952 are diagnosed correctly (943 correctly "normal" and 9 correctly "schizophrenia") for a hit rate of $952/1000$, or 95.2%. This is indeed a good (in fact, better than almost all) psychological test.

For these data, what is $P(\text{reject } H_0 \mid H_0 \text{ true})$—that is, the probability of a Type I error? $P(\text{reject } H_0 \mid H_0 \text{ true})$ is the probability that we diagnose schizophrenia given that the person is actually normal, which is the number of Type I errors (47) divided by the number of normal individuals ($943 + 47 = 990$). Thus $P(\text{reject } H_0 \mid H_0 \text{ true}) = 47/990 = .047$, which is less than .05 as convention says is desirable.

By contrast, what is $P(H_0 \text{ true} \mid \text{reject } H_0)$? That is the probability that the person is actually normal given that we diagnose him as having schizophrenia, which is the number of Type I errors (47) divided by the number of individuals diagnosed as having schizophrenia ($47 + 9 = 56$). $P(H_0 \text{ true} \mid \text{reject } H_0) = 47/56 = .84$.

In this example, then, the probability of a Type I error [we diagnose "schizophrenia" given that the person is actually "normal"—that is, $P(\text{reject } H_0 \mid H_0 \text{ true})$] is small (.047), just as the hypothesis-testing model requires. However, the probability of the (arguably) more important error [the person is actually "normal" given that we diagnosed "schizophrenia"—that is, $P(H_0 \text{ true} \mid \text{reject } H_0)$]—is very high (.84). Clearly $P(\text{reject } H_0 \mid H_0 \text{ true}) \neq P(H_0 \text{ true} \mid \text{reject } H_0)$.

Now let us apply this idea to the vitamin B_{12} experiment, where we finished collecting data and rejected H_0 at $\alpha = .05$. What is the probability that H_0 is actually true, given that we have just rejected it? It may seem that the answer to that question is ".05 or less," but as our schizophrenia analogy shows, that is *not* the case and may in fact be much higher. It turns out that it is generally impossible to give a correct answer to that question, which is perhaps the most important issue in experimentation. The answer depends on the "base rate" of the phenomenon in question (e.g., schizophrenia occurs in 1% of the population), and without knowing the base rate, one cannot specify $P(H_0 \text{ true} \mid \text{reject } H_0)$.

The four flaws of hypothesis testing, taken together, have been articulated with increasing clarity since the late 1980s, but there is as yet no unanimous opinion about how to incorporate them into everyday scientific experimentation. My own opinion, as reflected in Step IV.B of the general hypothesis evaluation procedure, is that statistical significance and practical significance are both important and should both be reported. Furthermore, I believe that effect sizes should be reported in a way as clearly visualizable as possible, as is illustrated in subsequent chapters.

Exercises for Optional Section

(Answers in Appendix D, page 512)

1. State four reasons that a science based solely on statistical significance testing is inadequate.
2. (a) Give an example where $P(\text{reject } H_0 \mid H_0 \text{ true}) \neq P(H_0 \text{ true} \mid \text{reject } H_0)$.
   (b) What in plain English is $P(\text{reject } H_0 \mid H_0 \text{ true})$?
   (c) What in plain English is $P(H_0 \text{ true} \mid \text{reject } H_0)$?

CHAPTER 10

Inferences About Means of Single Samples

LEARNING OBJECTIVES

1. What is the distribution of the test statistic in a hypothesis test about μ when σ is known? When σ is unknown?
2. What three distributions illustrate the hypothesis-testing procedure?
3. How are the criteria for rejecting H_0 established?
4. What is the difference between the observed value of the sample statistic and the critical value(s) of the sample statistic? What is the rejection region?
5. How are the observed values of the sample statistic and the test statistic computed?
6. How is the decision to reject or fail to reject H_0 made?
7. How is a directional hypothesis test different from a nondirectional hypothesis test?
8. What is the relationship between hypothesis testing and confidence intervals?
9. Is statistical significance the same as practical significance?
10. What is the raw effect size? The effect size index? Cohen's nonoverlap statistic U?

This chapter applies the general procedure for evaluating hypotheses (described in Chapter 9) to the simplest research situation: evaluating a hypothesis about a population mean μ by taking a single sample. In such a situation, the two possible test statistics are z if the population standard deviation σ is known and t if σ is not known. The hypothesis-testing procedure is illustrated by sketches of three distributions: the distribution of the original variable (X), the distribution of the sample statistic ($\overline{X}$), and the distribution of the test statistic (z or t, depending on whether or not σ is known). If the observed value of $\overline{X}$ (called $\overline{X}_{obs}$) exceeds (in absolute value) the critical value of $\overline{X}$ (called $\overline{X}_{cv}$) or, equivalently, if the observed value of the test statistic (z_{obs} or t_{obs}) exceeds the critical value of the test statistic (z_{cv} or t_{cv}), then we reject the null hypothesis. Once the null hypothesis is rejected, we must consider the practical significance of the result.

We discussed the general principles of evaluating hypotheses in Chapter 9; it is now time to use these general principles with specific problems. We shall see in subsequent chapters that it is possible to test hypotheses about correlations, proportions, and other measures, but the most common tests are about the means of populations; we begin with these.

10.1 Evaluating Hypotheses About Means

Hypotheses about means can be classified according to whether or not the population standard deviation σ is known. As we saw in Chapter 8 when we considered confidence intervals, the test statistic is z if σ is known but is t if σ is unknown.

When σ Is Known

Let us take as an example of hypothesis evaluation the study of the effects of electric fields on IQ described in Chapter 9, where we wondered whether living near an electric field might increase or decrease a person's IQ. We know that in the general population, IQ is normally distributed with mean $\mu = 100$ and standard deviation $\sigma = 15$ points. Our investigation will consist of taking a random sample of 25 subjects who live near power stations and measuring each subject's IQ. Because we think it possible that the electric field might increase or decrease IQ, this test is *nondirectional*: We will reject the null hypothesis if our subjects' mean IQ is either substantially greater or substantially less than 100. We will follow the steps of the hypothesis-testing procedure given in Section 9.3 of Chapter 9.

I. State the Null and Alternative Hypotheses

Three sketches illustrate H_0:
Variable
Sample statistic
Test statistic

Eyeball-estimation of this procedure is shown on page 209.

The null hypothesis is the statement of no effect: If H_0 is true, then living near an electric power station neither increases nor decreases IQ, so IQ should be, on average, 100, just as it is in the general population. The alternative hypothesis says that living near a power station *does* affect IQ in one direction or the other, so the mean IQ of those residents should be either less than 100 or greater than 100. Thus,

$$H_0:\quad \mu = 100$$
$$H_1:\quad \mu \neq 100$$

To illustrate the null hypothesis, we sketch three distributions assuming H_0 is true. The first sketch is the distribution of the variable—in this case, IQ—assuming H_0 is true. The null hypothesis specifies that μ for this distribution must be 100; σ is, in this problem, assumed to be known to be 15. This distribution of IQ scores, assuming that H_0 is true, is thus normal with $\mu = 100$ and $\sigma = 15$ as shown in Figure 10.1.

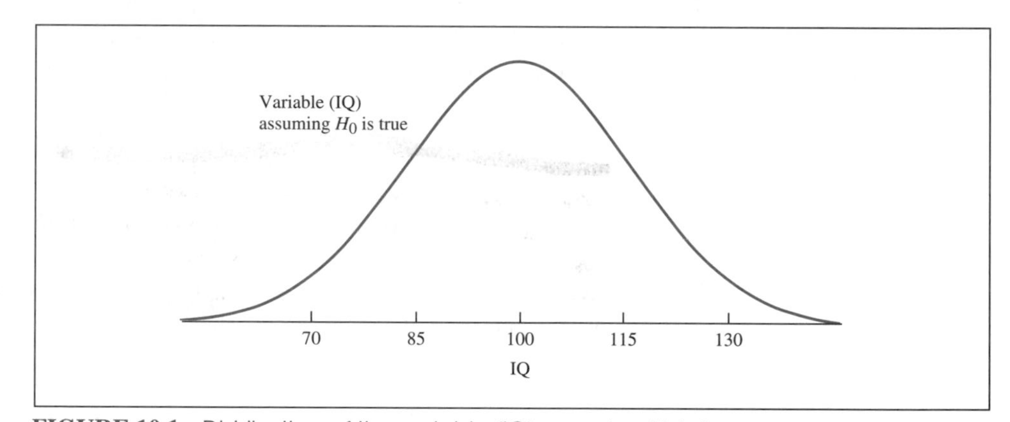

FIGURE 10.1 Distribution of the variable (IQ) assuming H_0 is true

The second sketch is the distribution of the sample statistic that point-estimates the population parameter specified by the null hypothesis. The population parameter is μ, and the best point-estimator we have of μ is (as we saw in Chapter 8) the sample mean ($\overline{X}$). The required sketch is therefore the sampling distribution of the means (the $\overline{X}$ values) assuming H_0 is true.

Recall from Chapter 7 that the central limit theorem states that the mean of this sampling distribution of means is the same as the mean IQ itself, which as specified by the null hypothesis is 100 points. Therefore, the mean of the sampling distribution of means must also be 100 points if the null hypothesis is true.

The central limit theorem also states that the standard deviation (which we call the standard error of the mean) of this sampling distribution of means is $\sigma_{\overline{X}}$. Because our sample size is 25, Equation (7.1) gives $\sigma_{\overline{X}} = \sigma/\sqrt{n} = 15/\sqrt{25} = 3$ points. Thus, the distribution of means of samples of size 25, assuming the null hypothesis is true, has mean $\mu = 100$ and standard error $\sigma_{\overline{X}} = 3$, as shown in the upper portion of Figure 10.2.

This sampling distribution of means shows what would happen if we repeatedly drew samples of size 25 from a population whose mean is 100 and whose standard deviation is 15 points. It shows, for example, that 68% of all the sample means would lie between 97 and 103 points (34% between the mean and one standard error below and above the mean), 96% of all the samples would lie between 94 and 106 points, and so on.

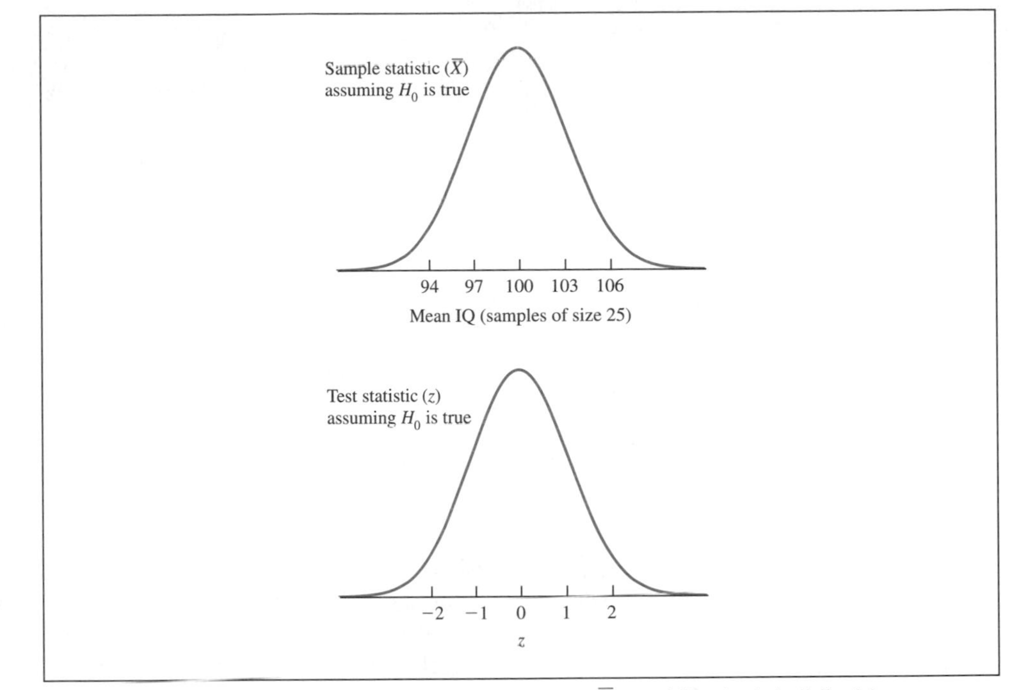

FIGURE 10.2 Distributions of the sample statistic ($\overline{X}$) and the test statistic (z)

The third sketch illustrating the null hypothesis is the distribution of the test statistic assuming the null hypothesis is true. Recall that the general form of the test statistic is

$$\text{test statistic} = \frac{\text{sample statistic} - \text{population parameter}}{\text{standard error of the sample statistic}} \tag{9.1}$$

The population parameter (the one specified by the null hypothesis) is μ, the sample statistic (which point-estimates that parameter) is $\overline{X}$, and the standard error of that statistic is $\sigma_{\overline{X}}$. The test statistic in this case is z as a result of the same logic that led to Equation (7.2) and the use of z in confidence intervals when σ is known. Thus, the test statistic for this one-sample hypothesis test for μ when σ is known is given by

test statistic (z), one-sample test for the mean when σ is known

When σ is unknown, see page 194.

$$z = \frac{\overline{X} - \mu}{\sigma_{\overline{X}}} \tag{10.1}$$

which in this example is

$$z = \frac{\overline{X} - 100}{3}$$

Because $\overline{X}$ is normally distributed with mean 100 and standard error 3 if the null hypothesis is true, this test statistic transforms $\overline{X}$ into z, which is normally distributed with mean 0 and standard deviation 1. That makes it possible to look up probabilities in Table A.1 in Appendix A, "Proportions of areas under the normal distribution." We sketch this third distribution in the lower portion of Figure 10.2. Note that this distribution is simply a change of scale from the distribution of the sample statistic in the upper portion of Figure 10.2. If the values of the sample statistic on the X-axis in the upper portion are transformed according to Equation (10.1), the result is the test statistic shown in the lower portion.

Because these two sampling distributions ($\overline{X}$ and z) are identical except for the scaling of the axes, it is convenient to illustrate them with one curve, as shown in Figure 10.3. Note that Figures 10.2 and 10.3 are identical, except that both axes are shown on the same curve in Figure 10.3.

Note further that these three distributions (the variable and the two sampling distributions) are sketched *before* we know the magnitude of $\overline{X}_{\text{obs}}$. Thus, the three distributions illustrate the null hypothesis in the sense that they show *all possible* values of the variable assuming the null hypothesis is true, all possible means of samples of size 25 assuming the

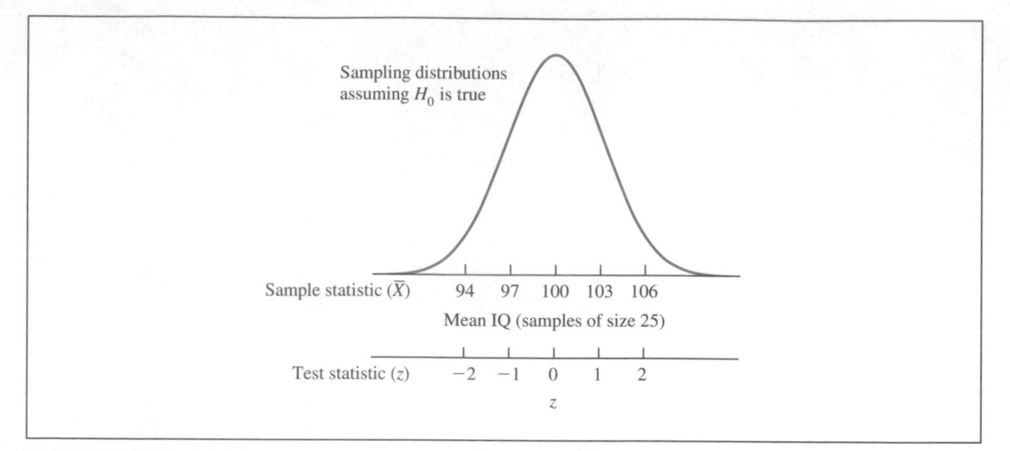

FIGURE 10.3 Distributions of the sample statistic ($\bar{X}$) and the test statistic (z) shown with two axes for the same curve. Identical to Figure 10.2.

null hypothesis is true, and all possible values of z assuming the null hypothesis is true. They do *not* show the characteristics of any particular sample.

The alert reader will have noticed that these three distributions are precisely the three major concepts of statistics that we have emphasized since Chapter 1: the distribution of the variable, the sampling distribution of the means, and the distribution of the test statistic.

II. Set the Criterion for Rejecting H_0

Level of significance?
Directional?
Shade rejection regions.

observed value of the statistic (here, $\bar{X}_{obs}$): a computed statistic of our sample (here, the sample mean)

Our investigation here involves taking a random sample of size $n = 25$ from the population of individuals who live near power stations and ascertaining $\bar{X}_{obs}$, the observed sample mean IQ. That is logically equivalent to taking *one* sample from the sampling distribution of *means of samples of size 25*. We know the standard deviation of this distribution—that is, the standard error of the mean, 3 points—but we do not, unfortunately, know the mean of this distribution (that's why we are performing the investigation). The null hypothesis says the mean is 100 points, and the alternative hypothesis says it is either greater than 100 or less than 100; it is an unknown state of nature. Our procedure is to *assume* that the null hypothesis is true (that is, to assume $\mu = 100$) and then *test* to see whether that assumption is reasonable.

If the null hypothesis is true, then our observed $\bar{X}_{obs}$, whatever that value turns out to be, is one sample from the sampling distribution of sample means that we sketched in the upper portion of Figure 10.2 (and the upper axis of Figure 10.3). Suppose $\bar{X}_{obs}$ turns out to be 102. Does it seem reasonable that the population mean μ of this distribution is 100? That is, does it seem reasonable that the null hypothesis is true? Yes, a sample mean of 102 is quite reasonable given a population mean of $\mu = 100$; it's only two-thirds of a standard error above 100. By contrast, suppose $\bar{X}_{obs}$ turns out to be 112. Does it now seem reasonable to assume that the null hypothesis is true? No, it does not; 112 is *four* standard errors away from 100 (the mean if the null hypothesis is true), too far out in the tail of the distribution to be expected. If $\bar{X}_{obs} = 112$, then we would *reject* the null hypothesis, whereas if $\bar{X}_{obs} = 102$, we would not reject it. Somewhere between 102 and 112 there must be a dividing line, which we call the *critical value of* $\bar{X}_{obs}$—that is, $\bar{X}_{cv}$. If an *observed* $\bar{X}_{obs}$ (the mean of our sample) exceeds this $\bar{X}_{cv}$, we will reject the null hypothesis. If $\bar{X}_{obs}$ does not exceed $\bar{X}_{cv}$, we will "accept" H_0. What is the magnitude of this critical value?

critical value of the statistic (here, $\bar{X}_{cv}$): a criterion set in advance; the beginning of the rejection region. If the observed value of the statistic lies in the rejection region, we reject the null hypothesis

This $\bar{X}_{cv}$ is a criterion by which we can either reject or not reject the null hypothesis—that is, whereby we can answer the question: Is it reasonable that the observed $\bar{X}_{obs}$ is a sample from a distribution of means with mean $\mu = 100$? If $\bar{X}_{obs}$ is "too far out in the tails" of the hypothesized distribution, we will reject the hypothesis that the distribution has mean $\mu = 100$. $\bar{X}_{cv}$ is the definition of "too far." We shall see that we can set the values of $\bar{X}_{cv}$ by adopting some level of significance—say, $\alpha = .05$.

When the level of significance is set at .05, we agree that we will say an $\bar{X}_{obs}$ is "too far" from μ as specified by the null hypothesis if $\bar{X}_{obs}$ falls in the outermost 5% of the tails of distribution. Because we will reject H_0 in favor of H_1 if $\overline{X}_{obs}$ is either significantly greater or significantly less than 100, we have what we called in Chapter 9 a nondirectional or two-tailed test, and the 5% must be equally divided between the left and right tails of the distribution—that is, 2.5% in each. Thus, there must be a critical value $\overline{X}_{cv}$ in both the left and the right tails. The logic that we used in constructing confidence intervals in Chapter 8 tells us that the critical value when $\alpha = .05$ (.025 in each tail) should be $z_{cv} = \pm 1.96$ standard errors above and below the mean; that is,

critical value of the sample mean when σ is known

$$\overline{X}_{cv} = \mu \pm z_{cv}(\sigma_{\overline{X}}) \tag{10.2}$$

When σ is unknown, see page 194.

In this example, $\mu = 100$ and $\sigma_{\overline{X}} = 3$, so $\overline{X}_{cv} = 100 \pm 1.96(3)$ points, or $\overline{X}_{cv} = 94.12$ IQ points in the left tail and $\overline{X}_{cv} = 105.88$ IQ points in the right tail.

rejection region: the portion of the distribution of the statistic (or test statistic) that lies beyond the critical value(s). If the observed value of the statistic (or test statistic) lies in the rejection region, we reject the null hypothesis

We can indicate the critical values on the distribution of the sample statistic, and it is customary to shade the *rejection region*, the portion of the sampling distribution that lies beyond the critical values. This is shown in Figure 10.4. If $\overline{X}_{obs}$ lies in the rejection region, we will reject the null hypothesis. If $\overline{X}_{obs}$ lies between the mean and the critical values (that is, *not* in the rejection region), we will *not* reject H_0.

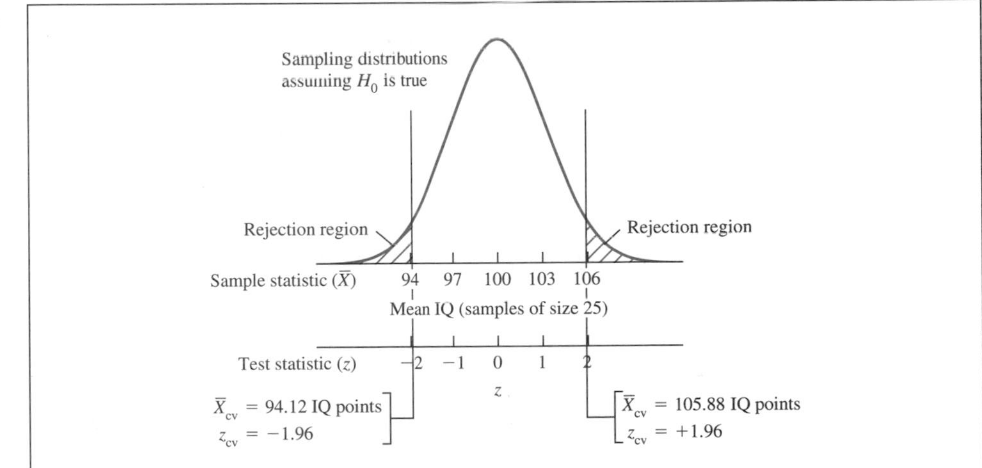

FIGURE 10.4 Distributions of the sample statistic and the test statistic with rejection regions shaded

observed value of the test statistic (here, z_{obs}): the value of the test statistic computed from our sample

critical value of the test statistic (here, z_{cv}): a criterion set in advance. If the observed value of the test statistic is greater (in absolute value) than the critical value, we reject the null hypothesis

We can also indicate the critical values on the distribution of the *test* statistic. Here, because the test statistic is z, the critical values are $z_{cv} = \pm 1.96$, as shown on the lower axis of Figure 10.4. The two axes in Figure 10.4 contain precisely the same information but highlight it somewhat differently. The distribution of means (upper axis) shows the values of the *sample statistic* that are critical: If the observed $\overline{X}_{obs}$ lies between 94.12 and 105.88 points, we will not reject. Displaying our procedure in this way provides some insight into how much larger (or smaller) than 100 the sample mean must be for H_0 to be rejected by our investigation. Clearly, if living near a power station does have an effect on IQ but that effect is relatively small (say, 2 IQ points, making the true state of nature $\mu = 102$ points), we would most likely not detect it with this experimental design.[1] On the other hand, the distribution of the *test statistic* (lower axis) is convenient because it shows directly the value of the test statistic that we look up in the table.

Note that we have specified the hypotheses and set the criterion for rejection *before* we collected the data—that is, before we know the value of $\overline{X}_{obs}$ in our sample. In order for the probability of a Type I error to be .05, the procedure for rejection *must* be specified in total ignorance of the research results.

[1] We could detect such a small effect if we used a larger sample size.

III. Collect a Sample and Compute the Observed Values of the Sample Statistic and the Test Statistic

Observed values: Sample statistic Test statistic

Now we are ready to collect the data, which are shown in Table 10.1. We first compute the observed value of the sample statistic from Equation (4.1):

$$\bar{X}_{\text{obs}} = \frac{\sum X}{n} = \frac{2587}{25} = 103.48$$

We indicate that value on the distribution of the sample statistic shown on the upper axis of Figure 10.5. We then compute the observed value of the test statistic according to Equation (10.1):

$$z_{\text{obs}} = \frac{\bar{X}_{\text{obs}} - \mu}{\sigma_{\bar{X}}} = \frac{103.48 - 100}{3} = 1.16$$

and we show that value on the test statistic (lower) axis of Figure 10.5.

TABLE 10.1 IQ scores of people who live near electric power stations

100	96	117	88	121
104	103	97	115	106
102	99	131	106	73
88	98	108	108	115
97	108	102	103	102

$\sum X_i = 2587$

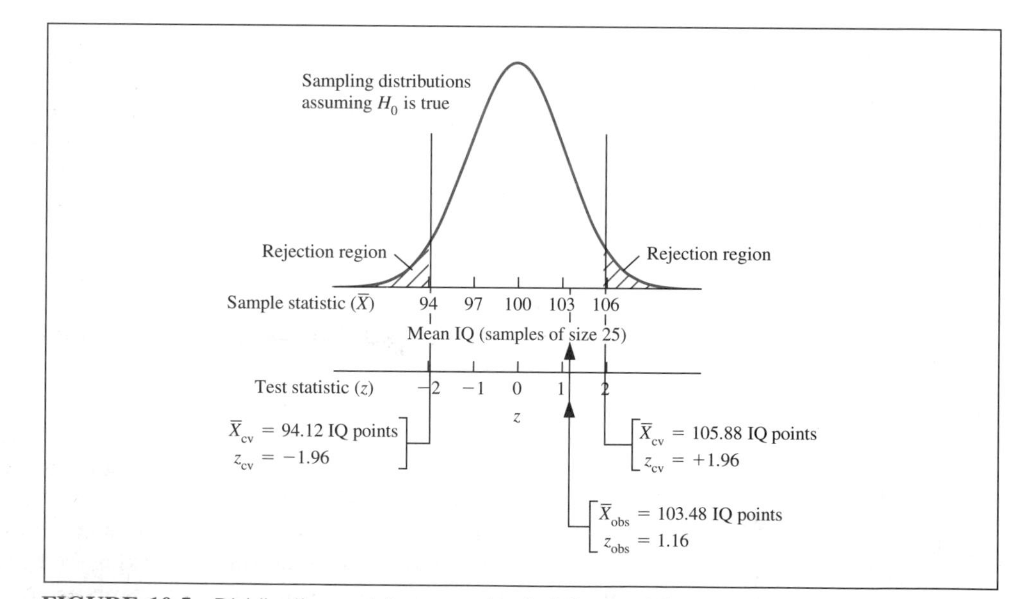

FIGURE 10.5 Distributions of the sample statistic and the test statistic with observed values indicated

IV. Interpret the Results

Statistically significant? Practically significant? State results clearly.

First we must decide whether the results are statistically significant—that is, whether to reject H_0. If the observed value of the sample statistic falls in its rejection region (or, equivalently, if the test statistic falls in its rejection region), we will reject the null hypothesis. In this example, the observed value of the sample statistic *does not* fall in the rejection region, so we *do not* reject H_0.

If the results had been statistically significant, we would have considered the practical significance of this result here. However, because we did not reject H_0, we can omit this step.

The concluding step is to state the results in plain English. Because we do not reject H_0, we do not support the alternative hypothesis. We have no evidence that people who live near power stations have IQs different from those of the general population whose mean $\mu = 100$.

We must make a few observations about this process. First, we notice that the magnitude of our observed $\overline{X}_{\text{obs}}$ (103.48 points) is *different* from 100 (the magnitude of μ as specified by H_0), and yet we concluded that we had *no* reason to decide that μ was different from 100. The reason is that even though $\overline{X}_{\text{obs}}$ is different from 100, it is *not* "significantly" different from 100 (that is, it does not lie in the rejection region). The extent to which $\overline{X}_{\text{obs}}$ is different from 100 could reasonably be explained by chance sampling fluctuations for a sample of this size ($n = 25$).

Second, we might ask: Does failing to reject the null hypothesis mean that μ is in fact actually equal to 100, as the null hypothesis specified? No, it does not. The actual value of μ might be 100, or 103.48, or 97.51, or any other value; we simply do not (actually, in most cases cannot) know its actual value. What we *do* know is that on the basis of these data, we are not justified in ruling out the hypothesis that $\mu = 100$, so the progress of science is best served if we "accept," at least for the moment, the notion that $\mu = 100$.

Third, we might ask: What would be the effect of changing the level of significance from $\alpha = .05$ to, say, $\alpha = .01$? The critical values of the test statistic z_{cv} would change from $z_{\text{cv}_{\alpha=.05}} = \pm 1.96$ when $\alpha = .05$ to $z_{\text{cv}_{\alpha=.01}} = \pm 2.58$ when $\alpha = .01$, because in the $\alpha = .01$ case, only .01 would be left in the two tails of z (.005 in each) and Table A.1 shows that the value of z that leaves .005 in each tail is 2.58. Furthermore (and equivalently), the critical value of $\overline{X}$ would now be ± 2.58 standard errors away from μ, so $\overline{X}_{\text{cv}_{\alpha=.01}} = 92.26$ and 107.44 points. (Recall that when $\alpha = .05$, the critical values of $\overline{X}$ are only ± 1.96 standard errors away from μ, so the $\overline{X}_{\text{cv}_{\alpha=.05}}$ values are 94.12 and 105.88 points.) Thus, when α is *smaller*, the critical values are *farther* from μ.

The critical value can be thought of as a hurdle: If $\overline{X}_{\text{obs}}$ is smaller than the hurdle, we do not reject H_0, but if it is larger, we do reject. Making the level of significance smaller makes the hurdle "higher" by increasing the required distance from the mean specified by H_0 (100).

A Directional (One-tailed) Example

Let us consider another example of the entire procedure. Suppose we know that scores on the Graduate Record Examination (GRE) have mean $\mu = 500$ and standard deviation $\sigma = 100$. I develop a training program similar to others that have been shown to increase GRE scores. I claim that my program will increase GRE scores also, and to evaluate this claim, I propose to take a random sample of ten prospective test takers, administer my training program, and then let them take the GRE and report their scores back to me. Should you believe the claim that my program increases GRE scores? We again follow the procedure presented in Section 9.3.

I. State the Null and Alternative Hypotheses

Three sketches illustrate H_0:
Variable
Sample statistic
Test statistic

H_0: $\mu \leq 500$ (the training program produces no increase)

H_1: $\mu > 500$ (based on previous research, I am claiming to *increase* the score)

Eyeball-estimation of this procedure is shown on page 209 .

Note that our interest as stated in this problem is whether the training program *increases* GRE scores, not merely whether it *changes* (increases or decreases) GRE scores. We call such a test directional. As we saw in Table 9.2 (p. 170), the alternative hypothesis should reflect our belief that the training program will *increase* GRE, so we write the alternative hypothesis as H_1: $\mu > 500$ (where the $>$ sign indicates the expected increase). The null hypothesis in this directional study includes the case ($\mu = 500$) where the training has no effect as well as the unlikely cases where the program's effect is in the direction opposite from what previous research has shown ($\mu < 500$). We combine those possibilities ($\mu = 500$ and $\mu < 500$) to form the single null hypothesis H_0: $\mu \leq 500$ because we have no desire to distinguish between $\mu = 500$ and $\mu < 500$; our interest is only whether there is a significant *increase* in GRE scores.

We can now sketch the three distributions that are relevant to hypothesis testing: (1) The distribution of the variable (in this case GRE score) assuming the null hypothesis is true (that is, assuming $\mu = 500$) is shown in Figure 10.6. This distribution of the variable illustrates the GRE scores of the population of all possible students (not just the ten students in the sample) who might receive the training program, assuming that the training program has no effect. (2) The distribution of the sample statistic (in this case, mean GRE score for samples of size ten, again assuming the null hypothesis is true) is shown in Figure 10.7 (upper axis). The central limit theorem specifies that this sampling distribution of the means has the same mean as the distribution of the variable ($\mu = 500$) but standard error smaller than the standard deviation of the variable by a factor of $\sqrt{n}$: $\sigma_{\overline{X}} = \sigma/\sqrt{n} = 100/\sqrt{10} = 31.623$. This sampling distribution of the means illustrates the population of the means of all possible samples of size ten that might be drawn from the population of GRE scores. (3) The distribution of the test statistic (in this case, z because we are assuming that we know σ), which has mean 0 and standard deviation 1 as always, is also shown in Figure 10.7 (lower axis). This distribution of the test statistic shows values that we can look up in standard tables of the normal distribution, such as Table A.1 in Appendix A.

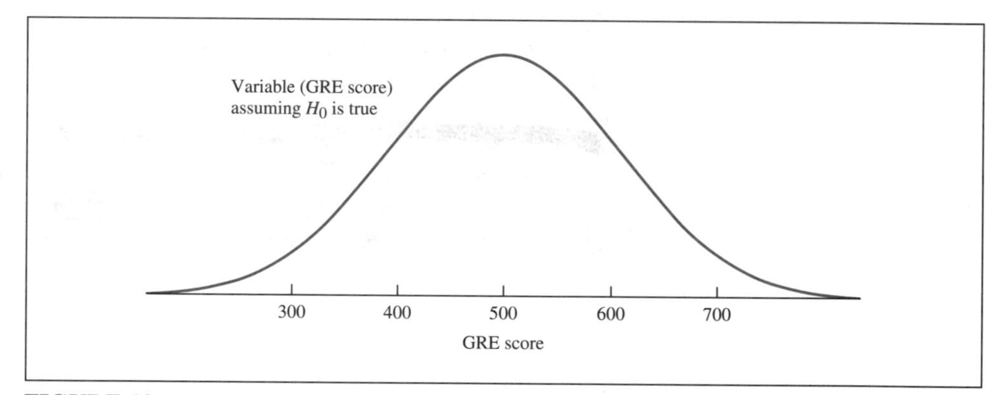

FIGURE 10.6 Distribution of the variable (GRE score) assuming H_0: $\mu = 500$

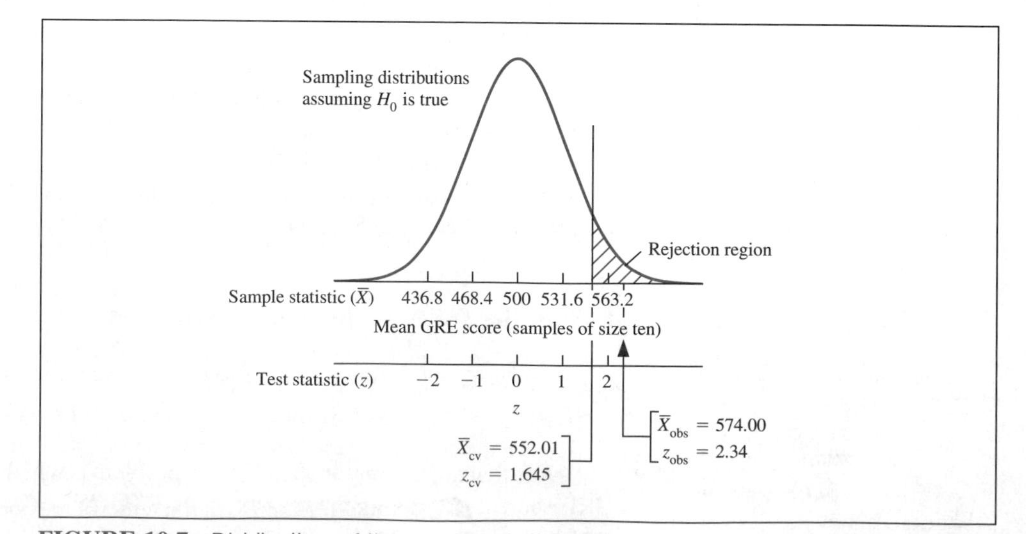

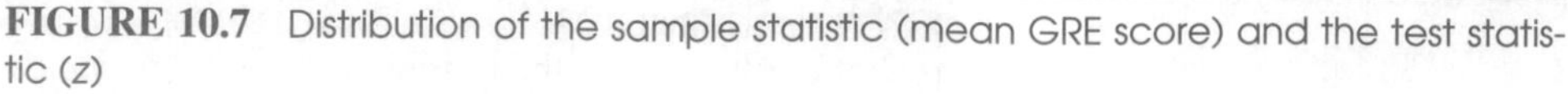

FIGURE 10.7 Distribution of the sample statistic (mean GRE score) and the test statistic (z)

II. Set the Criterion for Rejecting H_0

Level of significance?
Directional?
Shade rejection regions.

Because we did not specify a level of significance in the problem, we will assume that it is $\alpha = .05$. We also note that the test is directional: The alternative hypothesis has the direction specified as being an *increase* in GRE score, and therefore the rejection region is entirely in the right-hand tail [not divided between the two tails as in the previous nondirectional (two-tailed) example]. The critical value of the test statistic is $z_{\text{cv}_{\alpha=.05,\text{ one-tailed}}} = +1.645$ (which can be found from Table A.1 in Appendix A: 1.645 is the value of z that leaves .05 in the tail). We enter that value on the distribution of the test statistic z (the lower axis in Figure 10.7) and shade the rejection region, which lies beyond the critical value.

We also indicate a rejection region on the distribution of the sample statistic $\overline{X}$ on the upper axis in Figure 10.7. The rejection region begins at the critical value of the sample mean $\overline{X}_{\text{cv}}$, which can be found from Equation (10.2), except that here only the + sign is used because there is a rejection region only in the right-hand tail: $\overline{X}_{\text{cv}} = \mu + z_{\text{cv}}(\sigma_{\overline{X}}) = 500 + 1.645(31.62) = 552.01$.

III. Collect a Sample and Compute the Observed Values of the Sample Statistic and the Test Statistic

Observed values:
Sample statistic
Test statistic

We randomly sample ten individuals, administer the training procedure, and note their GRE scores, which are shown in Table 10.2. We compute the test statistic z as follows: The observed value of the sample statistic is $\overline{X}_{\text{obs}} = \sum X_i/n$, so we find that for the data in Table 10.2

TABLE 10.2
GRE scores

550
640
560
580
500
540
490
650
660
570
$\sum X_i = 5740$

$$\overline{X}_{\text{obs}} = \frac{5740}{10} = 574$$

We can enter that value on the upper axis of Figure 10.7. The test statistic is given by Equation (10.1), so

$$z_{\text{obs}} = \frac{\overline{X}_{\text{obs}} - \mu}{\sigma_{\overline{X}}} = \frac{574 - 500}{31.62} = 2.34$$

and we show that value on the lower axis in Figure 10.7.

IV. Interpret the Results

Statistically significant?
Practically significant?
State results clearly.

First we must decide whether the results are statistically significant—that is, whether to reject H_0. We check to see whether z_{obs} lies in its rejection region—whether z_{obs} is greater than z_{cv}—and the answer is yes: $z_{\text{obs}} = 2.34$ *is* greater than $z_{\text{cv}} = 1.645$. Entirely equivalently, we check to see whether $\overline{X}_{\text{obs}}$ lies in its rejection region—whether $\overline{X}_{\text{obs}}$ is greater than $\overline{X}_{\text{cv}}$—and again the answer is yes: $\overline{X}_{\text{obs}} = 574.00$ points *is* greater than $\overline{X}_{\text{cv}} = 552.01$ points. Both z_{obs} and $\overline{X}_{\text{obs}}$ contain the same information, so both must either be in or not be in the rejection region. Therefore, we reject the null hypothesis and conclude that our training program produces a statistically significant increase in GRE scores.

Because the results are statistically significant, we next consider whether they are practically significant.

Practical Significance: Effect Size and Nonoverlap

We have rejected the null hypothesis; that is, our results are statistically significant. Because we recall from Chapter 9 that statistical significance does not necessarily imply practical significance, we now wish to know the extent to which the effect being measured in this study (the effectiveness of our GRE training program) is actually important.

Because we have rejected H_0, we now support H_1; we now believe that the mean of our trained group is $\mu > 500$. But the question remains: By *how much* is μ greater than 500? We of course do not know μ exactly (that's why we are conducting the experiment), but our best point-estimate of μ is now the mean GRE score of the trained group: $\overline{X}_{\text{obs}} = 574.00$. Therefore, our best understanding of the distribution of all possible individuals who might

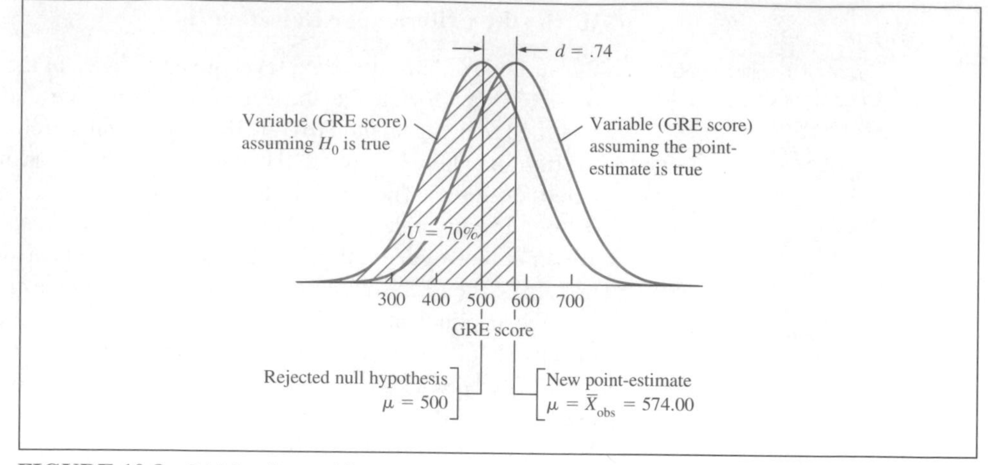

FIGURE 10.8 Distribution of the variable (GRE score) assuming the best point-estimate of μ is true—that is, $\mu = \overline{X}_{obs} = 574.00$

receive our training program is shown by the right-hand curve in Figure 10.8: Its mean is $\mu = \overline{X}_{obs} = 574.00$ and σ remains 100.

Note that the left-hand curve in Figure 10.8 represents the distribution of GRE scores specified by the null hypothesis *before* the experiment; it is identical to the curve in Figure 10.6. The right-hand curve shows the distribution of GRE scores as we believe them to be *after* the experiment—that is, after we have rejected the null hypothesis of $\mu = 500$ and take the best point-estimate of μ to be $\overline{X}_{obs} = 574.00$ GRE points.

The amount of overlap of these two curves is a measure of the practical significance of the GRE training. If the two curves lay directly on top of each other (total overlap), then the training program would be worthless. By contrast, if the *after* curve were so far to the right that there was almost no overlap, then almost all students who receive the training program would have higher GRE scores than any of the students who did not receive that training, so the practical significance would be very great.

Clearly, we need a measure of overlap, and several exist in the literature. We will discuss two of them: the effect size index d and Cohen's nonoverlap measure U.

raw effect size: the magnitude of an experimental result measured in the scale of its own experiment

The *raw effect size* is the magnitude of an experimental result measured in the scale of the original experiment. It measures how far the observed results depart from those specified by the null hypothesis. In our one-sample case, the raw effect size is the distance between the observed sample mean $\overline{X}_{obs}$ and the mean as specified by the null hypothesis; that is,

$$\text{raw effect size} = \overline{X}_{obs} - \mu = 574.00 - 500 = 74.00 \text{ GRE points}$$

The raw effect size is the most straightforward measure of practical significance: It states that the average person in the trained group has a GRE score that is 74.00 points higher than the average untrained person.

effect size index: a unitless measure of the magnitude of an experimental result

In many cases, it is useful to compare the effect size from one study with the effect size from another study, so the raw effect size is often divided by the standard deviation of the variable to convert it to a unitless number called the *effect size index*, d. For the experiments in the first portion of this chapter,

effect size index d, one-sample test for the mean when σ is known

When σ is unknown, see page 194.

$$d = \frac{|\overline{X}_{obs} - \mu|}{\sigma} \quad \text{or} \quad d = \frac{\overline{X}_{obs} - \mu}{\sigma} \qquad (10.3)$$

(nondirectional test) (directional test)

The effect size index for our GRE experiment is thus

$$d = \frac{\overline{X}_{obs} - \mu}{\sigma} = \frac{574.00 - 500}{100} = .74$$

We show d at the top of Figure 10.8. Note that d measures the number of standard deviations (.74) the observed $\overline{X}_{obs}$ is above the null hypothesized mean.

Let's consider an example that shows why d simplifies comparisons of effect sizes between experiments. Suppose that our training program may increase a person's IQ as well as GRE score, and another study, seeking to demonstrate that fact, begins with the null hypothesis H_0: $\mu = 100$ IQ points, collects some data, and finds that $\overline{X}_{obs} = 111.2$ IQ points. There the raw effect size is $\overline{X}_{obs} - \mu = 111.2 - 100 = 11.2$ IQ points. Does that imply that the GRE experiment (with raw effect size 74.00 GRE points) found a more powerful effect than the IQ experiment (with raw effect size 11.2 IQ points)?

No, it doesn't, as converting each raw effect size to the effect size *index d* will show. We have seen that d for the GRE experiment is $d_{GRE} = .74$, which indicates that $\overline{X}_{obs}$ is .74 standard deviation above its null hypothesis mean. For the comparison IQ experiment,

$$d_{IQ} = \frac{|\overline{X}_{obs} - \mu_{IQ}|}{\sigma_{IQ}} = \frac{111.2 - 100}{15} = .75$$

which indicates that $\overline{X}_{obs}$ is .75 IQ standard deviation above its IQ null hypothesis mean. The two d values make it obvious that the GRE experiment and the IQ experiment showed nearly identical effects (effect size index $d = .74$ versus .75) even though the raw effect sizes (74.00 GRE points versus 11.2 IQ points) were much different.

Now let's return to examining Figure 10.8. The effect size index d can be thought of as a measure of the amount of overlap between the distribution assuming that H_0 specifies μ and the distribution assuming that $\overline{X}_{obs}$ point-estimates μ. If $d = 0$, then the two curves totally overlap, which means that the average trained person has the same GRE score (500) as the average untrained person. By contrast, if $d = 2$, the average trained person has a GRE score of 700, two standard deviations above the null hypothesis mean, and the two curves would have substantially less overlap.

Cohen's nonoverlap *U*: the percentage of the null hypothesis population that is beyond the mean of the treated population

Clearly, the larger the effect size index, the less the overlap, and the less the overlap, the greater the practical significance. We can use a straightforward application of the skills we learned in Chapter 6 to measure the amount of *nonoverlap* of the two curves using *Cohen's nonoverlap* measure U.[2] U is the percentage of the population assuming H_0 is true that is less than the mean of the population assuming the new point-estimate of μ. This percentage is shaded and marked "U" in Figure 10.8. We can see by eyeball that the shaded area is about 75%; computation shows that the z score for 574.00 in the H_0 (left-hand) distribution is simply d: $z_{574.00} = (X - \mu)/\sigma = (574.00 - 500)/100 = .74 = d$. Just as in Chapter 6, we find the shaded area by entering Table A.1 at $z = .74$, and find from column A that the shaded area is $.5 + .2704 = .7704$. Therefore, Cohen's nonoverlap $U = 77.04\%$, which indicates that the average student who receives the GRE training program performs better than about 77% of individuals who do not receive it.

Let's consider how U is related to d. When $d = 0$ (treatment has no effect), $U = 50\%$ (the mean of the treated group is larger than the score of 50% of untreated individuals). We can see by eyeball that when $d = 1$, $U = 50\% + 34\% = 84\%$. When $d = 2$, $U = 50\% + 34\% + 14\% = 98\%$.

One more thing to note about U: We have said that "U is the percentage of the population assuming H_0 is true that is less than the mean of the population assuming the new point-estimate of μ." Because we are assuming that the distributions are normal, then because normal distributions are symmetric, U has the following entirely equivalent but opposite interpretation: U is the percentage of the population assuming the new point-estimate is true that is *greater* than the mean of the population assuming the null hypothesis is true. That interpretation could be illustrated in Figure 10.8 by removing the current shading (which is a region under the null hypothesis distribution) and shading instead a region under the point-estimate distribution that begins at 500 and extends out to the right tail. That region would be 77% of the area under the point-estimate distribution. The two interpretations of U are entirely equivalent, and the context must determine which one highlights the data more effectively.

---

[2] Cohen (1988, p. 21) defined three measures of overlap: U_1, U_2, and U_3. Our U is Cohen's U_3.

That's as far as the science of statistics can take us in the consideration of practical significance: In our GRE study, $d = .74$ and $U = 77\%$. Is it worthwhile to produce such an effect? That is not strictly a statistical question because its answer depends on the cost (in time required, dollars spent, stress endured, etc.) of the training program and the value (in opportunities created, salary increased, or self-esteem improved, etc.) of 74 GRE points. Those costs and values enter into complex judgments that are related to the size of the effect but go well beyond the effect size itself.

Before we leave our introduction to effect sizes, we should note that the literature often has some ambiguity in the use of the term *effect size*. Sometimes *effect size* means "raw effect size," whereas at other times, perhaps by the same author in the same paragraph, it means "effect size index." You must consider the context to determine whether the writer is referring to the raw or indexed effect size. Furthermore, there are measures of effect size other than the ones we will describe in this textbook. There is no general consensus on which are best. I chose the ones that are the most straightforward and clearest to visualize.

The last step in our hypothesis-evaluation procedure is to state the conclusions in plain English: The mean GRE score among students who have taken the training program (574.00) is statistically significantly greater than the mean GRE score in the student population (500). The effect size index $d = .74$ indicates that our training produces a .74 standard deviation increase in the GRE score, which implies that the average score of students who received our training exceeds 77% of untrained individuals.

When σ Is Unknown

We have been assuming that we know σ in advance of our investigation; we knew that IQ scores have $\sigma = 15$ and GRE scores have $\sigma = 100$. That is the simplest kind of hypothesis test to perform. It is much more common in practice, however, to test hypotheses when σ is *not* known in advance. How do we proceed?

If we do not know σ, we must point-estimate its value, and we saw in Chapter 8 that the sample standard deviation s is the best point-estimator of σ. We also saw in Chapter 8, when we were considering confidence intervals, that because s is not *exactly* equal to σ, we must replace the z distribution by a t distribution to account for the imprecision in point-estimating σ from s. That same logic applies here, and the formula for the test statistic *when we do not know* σ is obtained from Equation (10.1) by substituting s for σ and t for z:

test statistic (t), one-sample test for the mean when σ is unknown

When σ is known, see page 185.

$$t = \frac{\bar{X} - \mu}{s_{\bar{X}}} \qquad [df = n - 1] \tag{10.4}$$

The test statistic t is then compared to the critical value of t, t_{cv} (not to z_{cv} as before). Recall from Chapter 8 that identifying t_{cv} requires that we specify the number of degrees of freedom; for this test (just as for the confidence interval), $df = n - 1$. The critical value of the statistic (in this case $\bar{X}$) is also obtained by substituting s for σ and t for z:

critical value of the sample mean when σ is unknown

When σ is known, see page 187.

$$\bar{X}_{cv} = \mu \pm t_{cv}(s_{\bar{X}}) \tag{10.5}$$

The effect size index is then

effect size index d, one-sample test for the mean when σ is unknown

When σ is known, see page 192.

$$d = \frac{|\bar{X}_{obs} - \mu|}{s} \text{ (nondirectional test)} \quad \text{or} \quad d = \frac{\bar{X}_{obs} - \mu}{s} \text{ (directional test)} \tag{10.6}$$

With those exceptions, the procedure is the same as that when σ is known. Let us consider an example.

Some observers of the university scene report that students spend 1.5 hours studying for every hour in class. To see whether this observation is true of Georgia Southern University students, Richard Rogers asked a large number of upper-level psychology students to report honestly the number of hours they spent studying per week. A random sample of 16 of these students' study times are listed in Table 10.3.

Dr. Rogers assumed that students enroll on average in 14 hours of classes, so if the 1.5-study-hours-per-class-hour observation is true, then they should spend on average 21 hours per week studying. Is it reasonable to believe that Dr. Rogers's psychology students are samples from a population that studies on average 21 hours per week?

I. State the Null and Alternative Hypotheses

Three sketches illustrate H_0:
- Variable
- Sample statistic
- Test statistic

ⓘ Eyeball-estimation of this procedure is shown on page 210.

H_0: $\mu = 21$ hours (the population does in fact study 21 hours per week)

H_1: $\mu \neq 21$ hours (nondirectional because we do not specify whether Dr. Rogers's students study more or less than 21 hours on average)

When σ was known, we could sketch the three required distributions. Now this is as yet impossible because we do not know the standard deviation. We therefore must turn to the data in Table 10.3 to obtain s. By Equation (5.4) or (5.4a), $s = 7.78$ hours, so $s_{\bar{X}} = s/\sqrt{n} = 7.78/\sqrt{16} = 1.94$ hours.

TABLE 10.3 Time spent studying (hours)

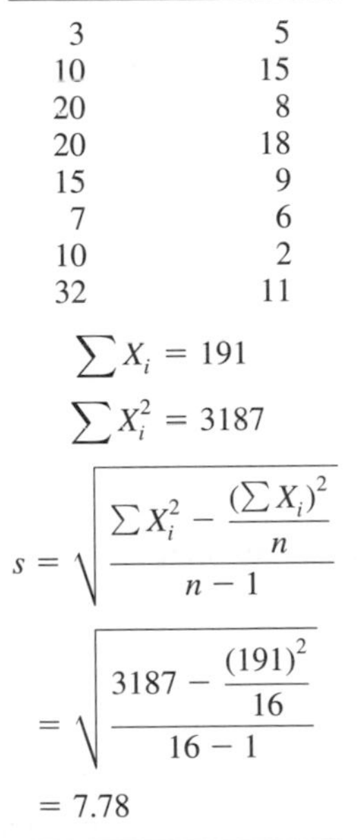

3	5
10	15
20	8
20	18
15	9
7	6
10	2
32	11

$$\sum X_i = 191$$

$$\sum X_i^2 = 3187$$

$$s = \sqrt{\frac{\sum X_i^2 - \frac{(\sum X_i)^2}{n}}{n-1}}$$

$$= \sqrt{\frac{3187 - \frac{(191)^2}{16}}{16-1}}$$

$$= 7.78$$

We can now prepare our three sketches. The variable (in this case, study time), assuming the null hypothesis is true, is shown in Figure 10.9. The sample statistic (the best point-estimate of the parameter specified by H_0—that is, the best point-estimate of μ) is $\bar{X}$. The sketch of this sample statistic (mean study time for samples of size 16) is shown on the upper axis of Figure 10.10. The test statistic is t as given by Equation (10.4); its distribution is also shown in Figure 10.10 (lower axis).

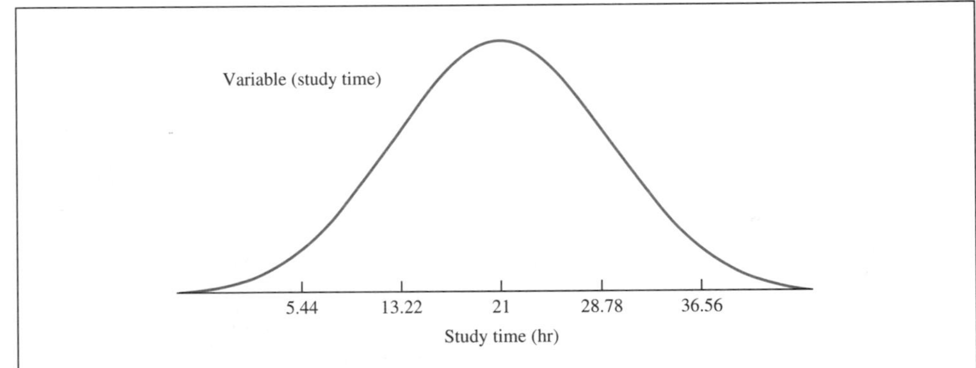

FIGURE 10.9 Distribution of subjects' study times assuming H_0: $\mu = 21$

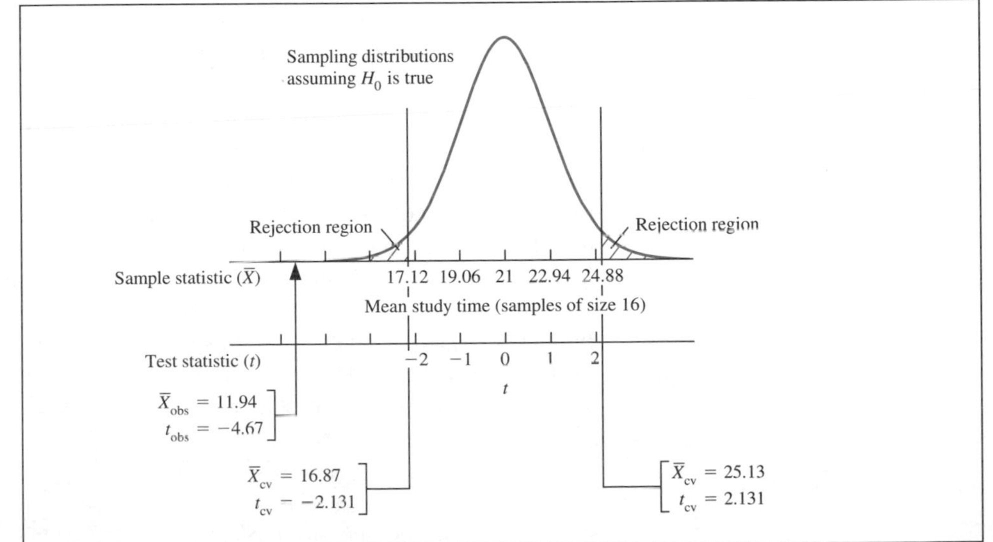

FIGURE 10.10 Distribution of the sample statistic and the test statistic with observed values indicated

II. Set the Criterion for Rejecting H_0

Level of significance?
Directional?
Shade rejection regions.

We did not specify the level of significance, so we choose $\alpha = .05$, and we noted above that this test is nondirectional because we did not specify whether study time might be more or less than 21 hours. In order to shade the rejection region on the distributions illustrated by the two axes of Figure 10.10, we must first know t_{cv}. When $n = 16$, $df = n - 1 = 15$. Table A.2 in Appendix A, entered for a two-tailed test with $\alpha = .05$ and $df = 15$, gives $t_{cv_{\alpha=.05, df=15}} = \pm 2.131$. [Note that if the desired df is not explicitly given in Table A.2, we must find the value of t_{cv} by interpolation. For example, if df were 34, then t_{cv} would be between 2.042 (the critical value when $df = 30$) and 2.021 (the critical value when $df = 40$) and slightly closer to the $df = 30$ value, so $t_{cv_{df=34}} = 2.034$.]

We can enter ± 2.131 as critical values on the t distribution on the lower axis of Figure 10.10 and shade the rejection regions. Then we can compute the critical values of $\overline{X}$ according to Equation (10.5): $\overline{X}_{cv} = \mu \pm t_{cv}(s_{\overline{X}}) = 21 \pm 2.131(1.94) = 16.87$ and 25.13 hours. We enter those values on the upper axis in Figure 10.10 and shade the rejection regions there.

III. Collect a Sample and Compute the Observed Values of the Sample Statistic and the Test Statistic

Observed values:
Sample statistic
Test statistic

The sample data are shown in Table 10.3. We compute the statistic and test statistic as follows: The observed statistic in this problem is the mean study time for our sample:

$$\overline{X}_{obs} = \frac{\sum X_i}{n} = \frac{191}{16} = 11.94 \text{ hours}$$

We can enter that value on the distribution of the sample statistic on the upper axis of Figure 10.10.

The observed value of the test statistic is given by Equation (10.4):

$$t_{obs} = \frac{\overline{X}_{obs} - \mu}{s_{\overline{X}}} = \frac{11.94 - 21}{1.94} = -4.67$$

which can be entered on the lower axis in Figure 10.10 directly below $\overline{X}_{obs}$, as it must if our computations are correct.

IV. Interpret the Results

Statistically significant?
Practically significant?
State results clearly.

Because the observed value of the statistic $\overline{X}_{obs}$ does in fact fall in its rejection region [that is, because $\overline{X}_{obs} = 11.94$ is beyond its lower critical value $\overline{X}_{cv} = 16.87$ (see Figure 10.10)], we do reject the null hypothesis. Entirely equivalently, because the observed value of the test statistic t_{obs} does in fact fall in its rejection region (that is, because $t_{obs} = -4.67$ is beyond its lower critical value $t_{cv} = -2.131$), we do reject the null hypothesis. Our conclusion is that the mean study time for Dr. Rogers's students is significantly different from (in fact less than) 21 hours.

Because the results are statistically significant, we must consider the extent to which they are practically significant. We have rejected H_0: $\mu = 21$, which implies that we now believe that μ has some value other than 21. Our best point-estimate of μ is now $\overline{X}_{obs} = 11.94$ hours. Assuming the standard deviation remains $s = 7.78$ hours, we can superimpose our new distribution on the one specified by the null hypothesis as shown in Figure 10.11. Here, the distribution specified by the null hypothesis is on the right, and the distribution using the point-estimate $\overline{X}_{obs}$ for μ is on the left.

The effect size index d is given by the nondirectional portion of Equation (10.6):

$$d = \frac{|\overline{X}_{obs} - \mu|}{s} = \frac{|11.94 - 21|}{7.78} = 1.16$$

which indicates that the mean as specified by $\overline{X}_{obs}$ is 1.16 standard deviations away from (in our case, below) μ as specified by H_0. The nonoverlap statistic U is found by entering Table A.1 with the value $z = d = 1.16$; then $U = .5 + .3770 = 87.7\%$, which indicates

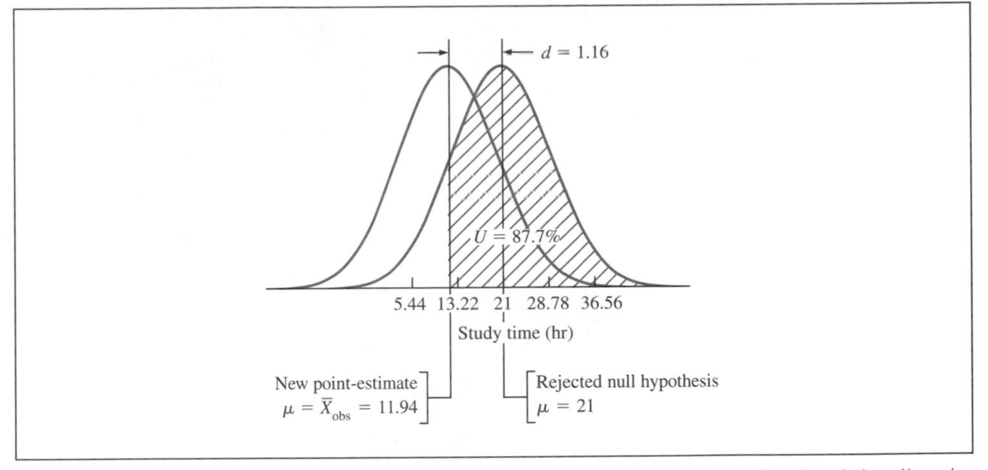

FIGURE 10.11 Distribution of the variable (study time) assuming the best point-estimate of μ is true—that is, $\mu = \overline{X}_{\text{obs}} = 11.94$ hours

that the mean study time of Dr. Rogers's students is less than 87.7% of the study times specified by H_0.

Stating our conclusion in plain English: The mean study time of the population of Dr. Rogers's students is significantly less than the mean assuming students spend 1.5 hours on homework for every hour they spend in class. In fact, the best estimate of Dr. Rogers's students' mean is 11.94 hours, or 1.16 standard deviations less than the 21 hours specified by the rule. This indicates that the average student of Dr. Rogers spends less time studying than would 87.7% of a population of students whose mean is actually 21 hours.

A few observations are in order: First, a good sketch of the results (such as Figure 10.11) is the best preparation for accurately stating the results.

Second, the fact that we take $\overline{X}_{\text{obs}} = 11.94$ hours to be the best point-estimate of μ does not mean that μ actually equals 11.94 hours. $\overline{X}_{\text{obs}}$ is a sample statistic and therefore will vary from sample to sample. We could, if we wished, create a confidence interval for μ exactly as we did in Chapter 8: With 95% confidence we can say that μ lies in the interval $\overline{X}_{\text{obs}} \pm t_{\text{cv}}(s_{\overline{X}}) = 11.94 \pm 2.131(1.94)$; that is, with 95% confidence we can say that μ lies between 7.81 and 16.07 hours.

Third, we might observe that the left-hand tail of the distribution, assuming the point-estimate is true, extends below the zero-point of the study time axis, which would indicate that a few students spend a negative amount of time studying. This absurd result illustrates one of the limitations of our method: We assume that distributions are normal but that is actually only an approximation. The present distribution in reality must be truncated at 0, which means that the actual distribution will be somewhat positively skewed. A corollary of this observation must be that U, which is based on the assumption of normal distributions, is not perfectly accurate but is only an approximation.

10.2 The Relationship Between Hypothesis Testing and Confidence Intervals

We have drawn two conclusions about the population mean μ of Dr. Rogers's students' study time: It is significantly different from 21 hours (that is, we rejected H_0), and we can say (as we just saw) with 95% confidence that it lies in the interval $\overline{X}_{\text{obs}} \pm t_{\text{cv}}(s_{\overline{X}}) = 11.94 \pm 2.131(1.94)$—that is, between 7.81 and 16.07 hours. We show in this section that the result of a nondirectional hypothesis test contains the same information as a confidence interval centered on $\overline{X}_{\text{obs}}$.

Figure 10.12 begins with the curve and upper axis of Figure 10.10 and overlays on it the confidence interval so that we can examine the hypothesis test and the confidence

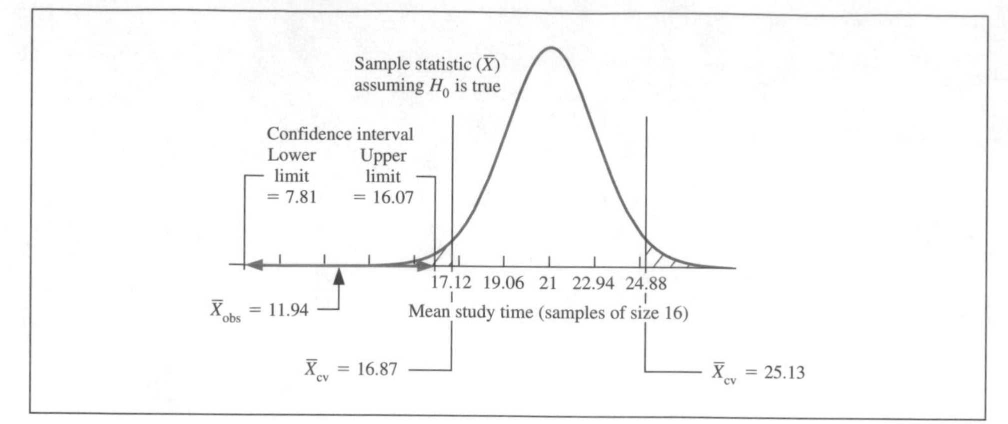

FIGURE 10.12 Distribution of the sample statistic (from Figure 10.10) with 95% confidence interval superimposed

interval at the same time. Consider the confidence interval: If the mean as specified by the null hypothesis ($\mu = 21$) *does not* lie in the confidence interval, then we can conclude with 95% confidence that $\overline{X}$ *did not* come from a population whose mean $\mu = 21$. That is identical to concluding that the null hypothesis is false.

Note that the width of the confidence interval is identical to the distance between the two values of $\overline{X}_{cv}$. The critical values of $\overline{X}$ are $t_{cv}(s_{\overline{X}})$ above and below μ as specified by H_0, while the ends of the confidence interval are $t_{cv}(s_{\overline{X}})$ above and below $\overline{X}_{obs}$. Thus, the following two statements are equivalent: "If μ as specified by H_0 lies outside the confidence interval, we reject H_0" and "If $\overline{X}_{obs}$ lies outside the middle 95% of the distribution of the statistic assuming H_0 (that is, lies in the rejection region), we reject H_0."

10.3 Statistical Significance Is Not Necessarily Practical Significance

Our hypothesis-testing procedure states that if the result of an experiment is statistically significant (that is, we reject the null hypothesis), then we should consider whether the result is practically significant. We provide here an extreme example to illustrate that a result can be statistically significant but have no practical significance.

Suppose that Dr. S. N. A. Koil develops a "subliminal hypnotic suggestion" device. It is a battery-operated object about the size of a pack of cigarettes that Dr. Koil sells on his infomercial for $29.95 by saying, "Just slip this device under your pillow and turn it on. It will emit the repetitive message 'I am becoming slimmer,' along with specially designed sound effects, at a volume you yourself can't hear but your brain can. Scientific study of thousands of individuals has shown that using the subliminal hypnotic suggestion device results in significant weight loss."

Pretty impressive! Worked for thousands of individuals!

Dr. Koil's "scientific study" was performed as follows: He knows from previous study that the mean weight of students at Acme Private School is $\mu = 120$ pounds, so he forms the null hypothesis H_0: $\mu = 120$ pounds. He gives 6400 randomly selected students at Acme a subliminal hypnotic suggestion device and a month later records all 6400 weights. He then computes the mean weight of this sample, finding $\overline{X}_{obs} = 119.8$ pounds and $s = 8.0$ pounds. He uses Equation (8.2) to compute $s_{\overline{X}} = s/\sqrt{n} = 8.0/\sqrt{6400} = .1$ pound and Equation (10.4) to compute $t_{obs} = (\overline{X}_{obs} - \mu)/s_{\overline{X}} = (119.8 - 120)/.1 = -2.0$. Because $df = 6399$, the critical value $t_{cv} = \pm 1.96$ when $\alpha = .05$, and because $t_{obs} = -2.0$ exceeds in absolute value $t_{cv} = \pm 1.96$, he rejects the null hypothesis. His conclusion: Users of the subliminal hypnotic suggestion device weigh significantly less than nonusers. Thus, Dr. Koil's claim that "Scientific study of thousands of individuals has shown that using the subliminal hypnotic suggestion device results in significant weight loss" is true.

We could criticize the design of this study on many grounds, but our focus here is on the size of the effect, which Dr. Koil has failed to tell us. We illustrate it in Figure 10.13, where our new best point-estimate μ is $\overline{X}_{obs} = 119.8$. From Equation (10.6), the effect size index $d = |\overline{X}_{obs} - \mu|/s = |119.8 - 120|/8.0 = .025$. This indicates that, on average, using this device will result in a weight loss of .025 standard deviation, or .2 pound. Cohen's nonoverlap statistic U is obtained by entering Table A.1 with $z = d = .025$, so $U = .5 + .010$, or 51%. Had Dr. Koil been more forthcoming, he would have stated "Scientific study of thousands of individuals has shown that using the subliminal hypnotic suggestion device results in an average weight loss of .2 pound, a very small but statistically significant effect. The average person using this device loses more weight than only 51% of nonusers."

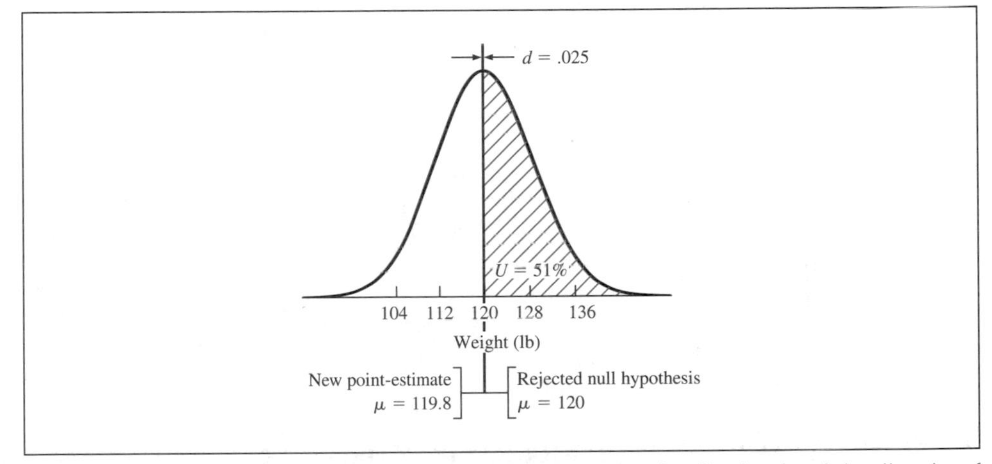

FIGURE 10.13 Distribution of the variable (weight) assuming the best point-estimate of μ is true—that is, $\mu = \overline{X}_{obs} = 119.8$ pounds

Note that when there is absolutely no effect, $U = 50\%$. The sentence, "The average person using this device loses more weight than 50% of nonusers" is a statement about an absolutely worthless device. An effect is important only to the extent that U is greater than 50%.

This example highlights the role of the number of subjects in an experiment: Using more subjects makes $s_{\overline{X}}$ smaller (because $\sqrt{n}$ is in the denominator); therefore, using more subjects makes rejecting the null hypothesis more likely. However, the number of subjects has no bearing on the effect size index.

10.4 One-sample *t* Test Eyeball-calibration

We now present a series of displays of one-sample t tests so that you can practice visualizing the results of those tests. Each of these seven displays shows the null hypothesis (indicated as a point on the X-axis) and a histogram of the sample data and asks whether we should reject the null hypothesis based on these data. The sample descriptive statistics and the results of the hypothesis test are also given, assuming a nondirectional test with $\alpha = .05$. The question to be answered in these displays is whether it is reasonable to suppose that the points shown in the histogram are random samples from a population whose mean μ is as specified by the null hypothesis.

The eyeball-calibration procedure is as follows: Eyeball-estimate $\overline{X}_{obs}$; it is the balancing point of the histogram. If your estimate of $\overline{X}_{obs}$ is *quite far away* from the mean as specified by the null hypothesis, then you should reject the null hypothesis.

How far is "quite far away"? The smaller the standard deviation (that is, the narrower the distribution), the smaller the required distance between μ and the sample mean. Furthermore, the larger the sample size (that is, the more blocks there are), the *smaller* the required distance between μ and the sample mean because the standard error is the standard deviation divided by $\sqrt{n}$. More details about how far is "quite far" may be found in the optional section at the end of this chapter, which describes the step-by-step procedure for eyeball-estimating one-sample t tests. Here our task is simply to get a sense of how far apart are μ (as specified by H_0) and $\overline{X}_{obs}$ (as eyeball-estimated from the histogram).

Figure 10.14 shows a histogram with the null hypothesis H_0: $\mu = 106$. Should we reject H_0? We eyeball estimate $\overline{X}_{obs}$ by the balancing-point method: The histogram seems to balance at about 109. Is 109 "quite far away" from 106 (as specified by H_0)? Yes, it seems to be, so we think we should reject H_0. If we performed the calculations for these data, we would find that we were correct: H_0 should in fact be rejected. (In case you wish to follow the calculations, here they are: $\overline{X}_{obs} = 109.74$, $s = 2.35$, $s_{\overline{X}} = .68$, $t_{obs} = 5.5$, $df = 11$, $t_{cv} = \pm 2.201$.)

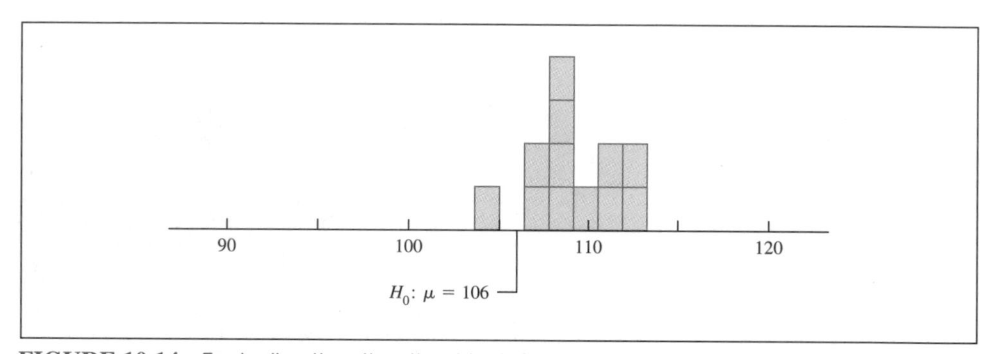

FIGURE 10.14 Eyeball-estimating the t test: Should we reject H_0?

This is simply an eyeball-calibration exercise: Make an educated guess about whether to reject H_0 and use the feedback from the actual answers to improve your understanding of how far "quite far away" actually is. If you wish to eyeball-estimate t step by step, see the optional section at the end of this chapter.

Figures 10.15 through 10.20 present a series of histograms for you to calibrate your eyeball. *Cover up the right-hand side of the next two pages* and determine by eyeball whether to reject H_0 for the examples shown. Then check to see whether the null hypothesis would be rejected by computation. If your eyeball disagrees with the computation, try to improve your understanding of how far "quite far away" actually is. (The actual values are provided in case you wish to follow along with the computations.)

You may wish to use the ESTAT exercise "ttest1" at this time. Follow the path labeled "Eyeball-calibration" or "Shortcut."

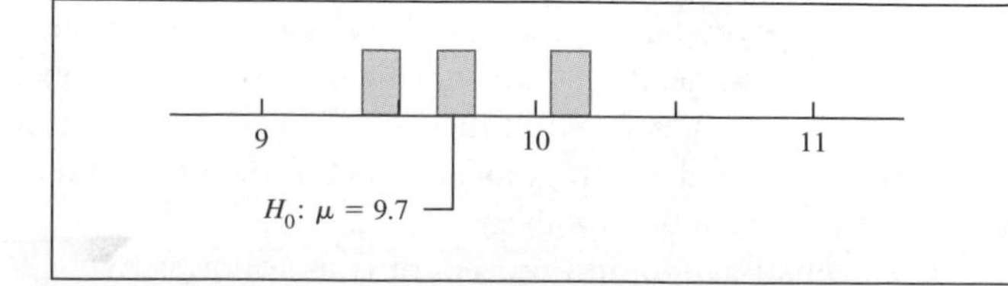

FIGURE 10.15 Should we reject H_0: $\mu = 9.7$?

Actual values	
Reject H_0?	No
$\overline{X}_{obs}$	9.87
μ	9.7
Numerator of t	.17
s	.323
$s_{\overline{X}}$ (denominator of t)	.186
t_{obs}	.912
df	2
t_{cv}	4.303

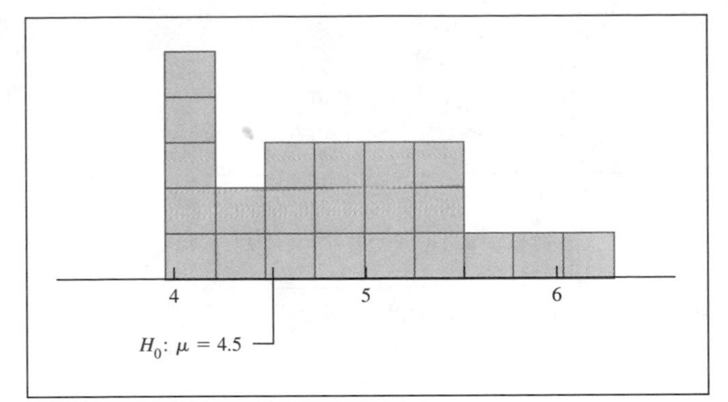

FIGURE 10.16 Should we reject H_0: $\mu = 4.5$?

Actual values	
Reject H_0?	Yes
$\overline{X}_{obs}$	4.89
μ	4.5
Numerator of t	.39
s	.540
$s_{\overline{X}}$ (denominator of t)	.115
t_{obs}	3.39
df	21
t_{cv}	2.080

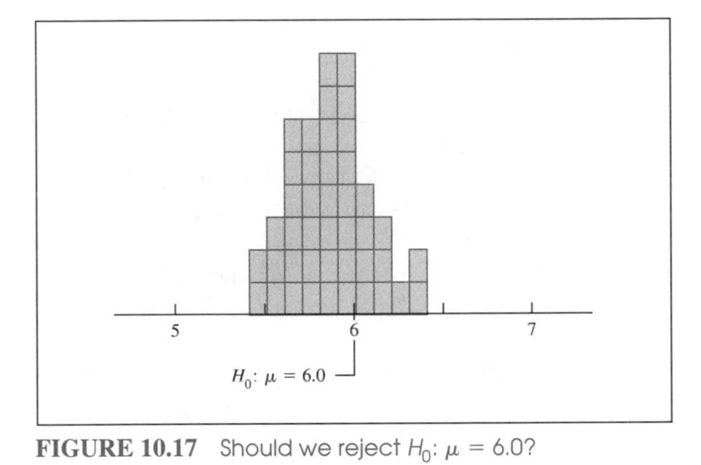

FIGURE 10.17 Should we reject H_0: $\mu = 6.0$?

Actual values	
Reject H_0?	Yes
$\overline{X}_{obs}$	5.91
μ	6.0
Numerator of t	−.09
s	.201
$s_{\overline{X}}$ (denominator of t)	.031
t_{obs}	−2.903
df	42
t_{cv}	2.020

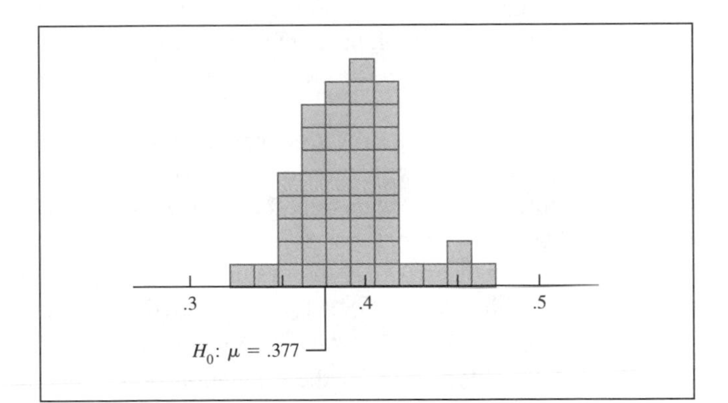

FIGURE 10.18 Should we reject H_0: $\mu = .377$?

Actual values	
Reject H_0?	Yes
$\overline{X}_{obs}$	.395
μ	.377
Numerator of t	.018
s	.0244
$s_{\overline{X}}$ (denominator of t)	.0035
t_{obs}	5.143
df	47
t_{cv}	2.020

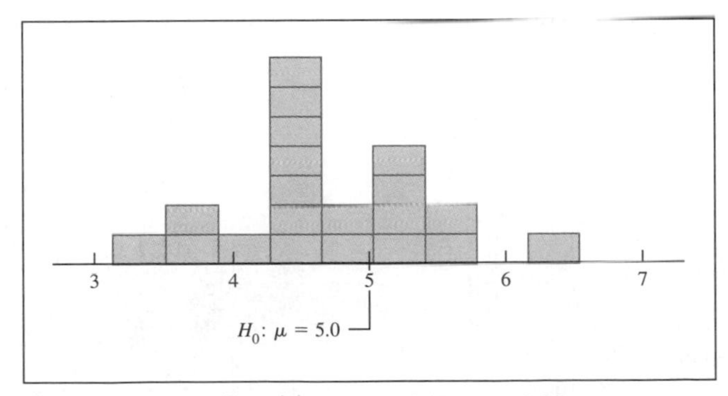

FIGURE 10.19 Should we reject H_0: $\mu = 5.0$?

Actual values	
Reject H_0?	No
$\overline{X}_{obs}$	4.805
μ	5.0
Numerator of t	−.195
s	.662
$s_{\overline{X}}$ (denominator of t)	.148
t_{obs}	−1.317
df	19
t_{cv}	2.093

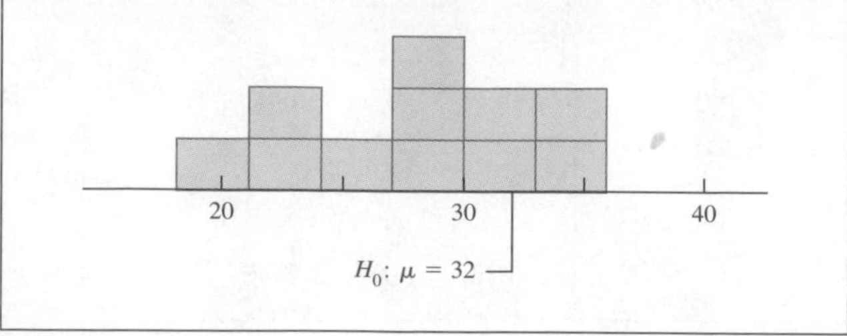

FIGURE 10.20 Should we reject H_0: $\mu = 32$?

Actual values	
Reject H_0?	Yes
$\overline{X}_{obs}$	27.96
μ	32
Numerator of t	−4.04
s	5.222
$s_{\overline{X}}$ (denominator of t)	1.574
t_{obs}	−2.567
df	10
t_{cv}	2.228

10.5 Connections

Cumulative Review

The book moved into its second phase with the beginning of Chapter 9. Until then we had focused on how to describe distributions, covering topics such as measures of central tendency and variation. With Chapter 9 we began the topic that will occupy us for most of the remainder of the textbook: hypothesis evaluation. Chapter 9 will serve as the template for hypothesis evaluation, discussing concepts such as null hypothesis (the statement of no effect), Type I error (rejecting H_0 when it is in fact true), level of significance (the probability of making a Type I error), and Type II error (failing to reject H_0 when it is in fact false).

One of the most valuable skills to be mastered in statistics is the ability to determine which statistical procedure is appropriate in a particular situation. For the remainder of the textbook, therefore, we will devote this cumulative review section to such discrimination practice. This is the basic procedure:

steps for discriminating among statistical procedures

Step 1: Determine whether the problem asks for (a) the area under a normal distribution, (b) a confidence interval, or (c) the evaluation of a hypothesis.

a. A problem that asks for the area under a normal curve usually contains wording such as "What percentage of . . . " Follow the procedures in Chapter 7.

b. A problem that asks for a confidence interval usually contains wording such as "What can we say with 95% confidence about . . . " or asks what can be inferred about a population mean when only the sample mean is known. Follow the procedures in Chapter 8.

c. A problem that asks for the evaluation of a hypothesis usually contains wording such as "Should we reject H_0?" or "Is there a difference between groups?" or "Did the treatment have an effect?" Proceed to Step 2.

Step 2: If the problem asks for the evaluation of a hypothesis, consult the summary of hypothesis testing in Table 10.4 to determine which kind of test is appropriate.

TABLE 10.4 Summary of hypothesis-testing procedures in Chapter 10

Design	*Chapter*	*Null hypothesis*	*Sample statistic*	*Test statistic*	*Effect size index*
One sample					
σ known	10	$\mu = a$	$\overline{X}$	z	$d = \frac{\overline{X}_{obs} - \mu}{\sigma}$
σ unknown	10	$\mu = a$	$\overline{X}$	t	$d = \frac{\overline{X}_{obs} - \mu}{s}$
Proportion	Exercise 7, Chapter 10	$\Pi = a$	p	t	—[a]

[a]Effect size index is not discussed in this textbook.

Table 10.4 shows that so far we have considered one basic kind of hypothesis test (one-sample test for the mean) that uses two different test statistics (z if σ is known or t if σ is unknown). We will extend this table throughout the remainder of the textbook.

Note that cumulative review exercises now have a new function.

Note that the exercises in Section D: Cumulative Review take on a new function beginning with this chapter: They are designed to give direct discrimination practice according to the steps presented here. See the description at the beginning of Section D of the exercises.

Journals

Reports of experiments in journals often do not explicitly identify the null hypothesis; it must be inferred from the description of the study. For example, the report might read something like "We tested the hypothesis that the mean study time for Dr. Rogers's students was different from 21 hours." We must work backward from such a statement to presume that the null hypothesis was H_0: $\mu = 21$ hours.

Reports of experiments often do not explicitly discuss practical significance or report effect size or overlap. That is, in the opinion of many (including me), a mistake that modern experimenters should remedy.

In the Journals section of Chapter 9 we discussed the so-called exact probabilities such as "$p = .027$." We said that this can be interpreted as meaning that the null hypothesis would have been rejected if α had been set at any value .027 or larger. Occasionally journal articles report "$p = .000$," which seems to imply that the null hypothesis would be rejected given *any* value of α, which is of course impossible. The culprit is computer packages such as SPSS, which provide exact probabilities and round them to three decimal places. For example, the SPSS output for Dr. Rogers's students gives the "exact" probability of obtaining $\overline{X}_{\text{obs}} = 11.94$ as ".000," which actually is some very small value such as .0002 rounded to three decimal places. You can avoid this difficulty in either of two ways: (1) Do not report exact probabilities and simply report "$p < .05$," or (2) report the largest possible exact probability that would round to .000—namely, "$p = .0005$."

Computers

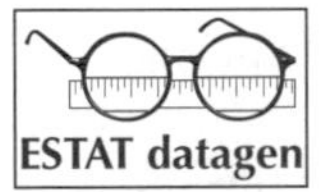

Use ESTAT's datagen to compute the t test for Dr. Rogers's study time data shown in Table 10.3, with H_0: $\mu = 21$ hours.

1. Enter the 16 values from Table 10.3 into *one* column of the datagen spreadsheet.
2. Note that datagen will not compute a one-sample t test. You can test a nondirectional hypothesis by using the 95% confidence interval that appears automatically in the `datagen Descriptive Statistics` window. Because $\mu = 21$ does not lie in the confidence interval (that is, does not lie between 7.81 and 16.07), we reject H_0.
3. See Homework Tip 6.

Use SPSS to compute the t test for Dr. Rogers's study time data shown in Table 10.3, with H_0: $\mu = 21$ hours.

1. Enter the 16 values from Table 10.3 into *one* column of the SPSS spreadsheet.
2. Click `Statistics`, then `Compare Means`, and then `One Sample T-Test....`
3. Click ▶, then enter `21` in the `Test Value:` window, and then click `OK`.
4. Note that the accompanying display, which I have annotated, appears in the `!Output1` window.
5. See Homework Tip 6.

Homework Tips

1. Check the list of learning objectives at the beginning of this chapter. Do you understand each one?

2. When you try to decide how to test a hypothesis, first determine whether you know σ: If you do, then use z as the test statistic and $\sigma_{\overline{X}}$ as its denominator. If not, use t and $s_{\overline{X}}$.

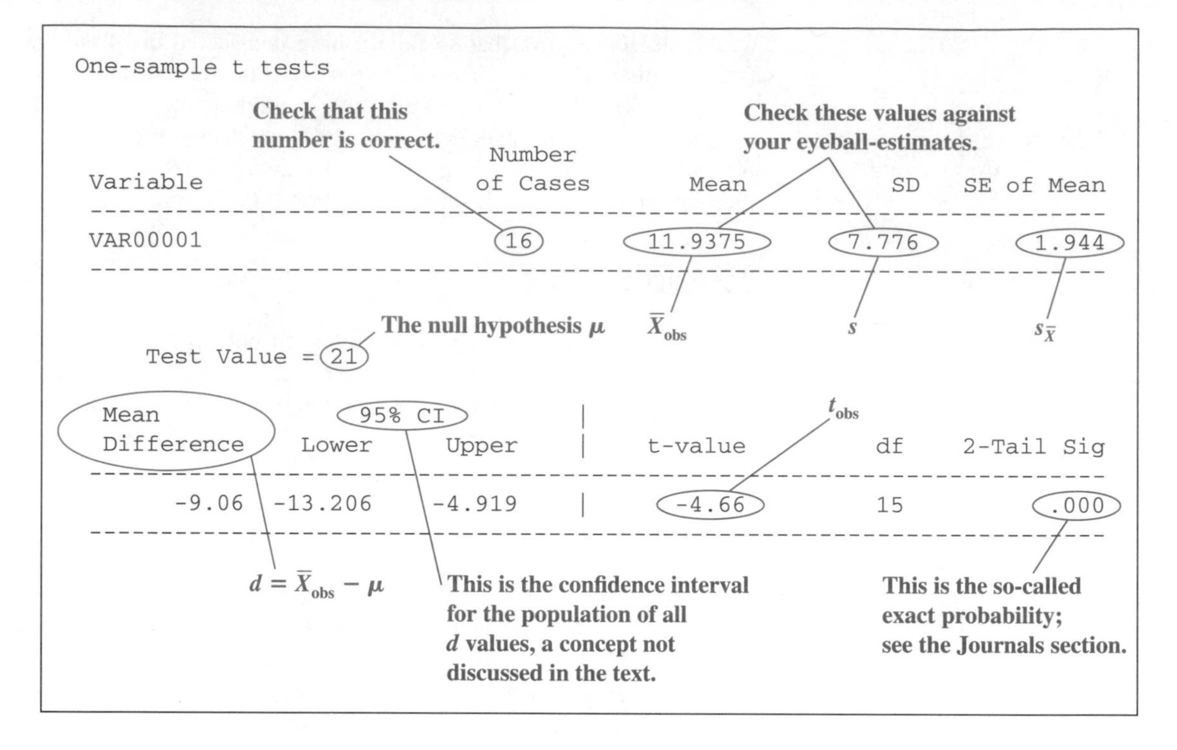

3. (This is a reprint of Homework Tip 6 from Chapter 8.) Students are often confused about what it means to "know σ." Knowing σ requires that you know the *population* standard deviation σ, which means that the problem must state something like "We know that the standard deviation of *all* students (not just the students in this sample) is . . . " Simply being able to compute s does *not* mean that we know σ. Being given s as a labor-saving device does *not* mean that we know σ.

4. Use the nondirectional (two-tailed) form of the hypothesis test unless the problem says something like "We have reason to believe that such a treatment will *increase* (or decrease). . . ," in which case use the directional (one-tailed) form.

5. Students sometimes have difficulty deciding which tail of a directional test should show the shaded rejection region. Shade the rejection region in the tail that makes the *alternative* hypothesis *true*.

6. Whenever you ask a computer to perform a task for you, you should (a) make sure the number of cases is correct and (b) eyeball-estimate enough statistics to convince yourself that the computer has interpreted the data in the way you expected.

7. Some students have difficulty determining the critical value of the sample statistic. Here are some tips to avoid confusion:

(a) The formula for the critical value of the sample statistic when we must use s to approximate σ (that is, when σ is unknown) is given in Equation (10.5): $\overline{X}_{\text{cv}} = \mu \pm t_{\text{cv}}(s_{\overline{X}})$. Note that this formula specifies that the critical value $\overline{X}_{\text{cv}}$ is a particular distance from μ as specified by H_0. That is, $\overline{X}_{\text{cv}}$ is $t_{\text{cv}}(s_{\overline{X}})$ or about two standard errors, from μ.

(b) Don't confuse the formula for the critical value of the sample statistic with the formula for the confidence interval, which was given in Chapter 8, Equation (8.3): $\overline{X} - t_{\text{cv}}(s_{\overline{X}}) < \mu < \overline{X} + t_{\text{cv}}(s_{\overline{X}})$. These formulas are similar, but the confidence interval specifies a particular distance from $\overline{X}$ whereas the critical value specifies a particular distance from μ.

(c) The key to understanding $\overline{X}_{\text{cv}}$ is to sketch the distributions of the sample statistic and the test statistic (for example, Figure 10.9). The critical value of the sample statistic $\overline{X}_{\text{cv}}$ is always directly above the critical value of the test statistic t_{cv}.

Exercises for Chapter 10

Section A: Basic exercises (answers in Appendix D, page 512)

1. Assume we know that $\sigma = 3$ and that the null hypothesis is H_0: $\mu = 10$. For the data X_i: 2, 4, 9, 7, what are the following?
   (a) The observed value of the statistic ($\overline{X}_{obs}$)
   (b) The observed value of the test statistic (z_{obs})
   (c) The critical values of the test statistic (z_{cv}) assuming the test is nondirectional and $\alpha = .05$
   (d) The critical values of the statistic ($\overline{X}_{cv}$)
   (e) Should we reject the null hypothesis? Why?

2. The Scholastic Assessment Test (SAT) is standardized to have scores that are normally distributed with $\mu = 500$ and $\sigma = 100$ in the population of high school students. Suppose I am interested in whether a period of meditation just before the SAT is taken alters students' scores on this test. I decide to explore this question by randomly selecting 16 students and training them to meditate just before the SAT test is administered; then I collect their scores, which are: 510, 500, 625, 490, 430, 700, 525, 610, 520, 480, 420, 420, 500, 505, 610, 520. (Labor savings: $\sum X_i = 8365$, $\sum X_i^2 = 4{,}464{,}575$.) Previous research has shown that meditation should not affect the standard deviation of SAT scores. Does meditation affect SAT performance? Use $\alpha = .05$. Omit the parts marked with an asterisk (*) unless you have read the optional section on eyeball-estimating one-sample tests.

   I. State the null and alternative hypotheses
   (a) What is the null hypothesis? The alternative hypothesis?
   (b) Is this test directional or nondirectional? Why?
   (c) What is the variable in this problem? Sketch the distribution of the variable assuming the null hypothesis is true.
   (d) What is the sample statistic in this problem? (Eyeball-estimate and then)* compute its standard error. Sketch the distribution of the sample statistic assuming the null hypothesis is true.
   (e) What is the test statistic in this problem? Sketch the distribution of the test statistic assuming the null hypothesis is true. You may simply append a new axis to a distribution you already sketched.

   II. Set the criterion for rejecting H_0
   (f) What is the level of significance?
   (g) How many degrees of freedom are there?
   (h) (Eyeball-estimate and then)* look up the critical value(s) of the test statistic. Enter it (them) on the appropriate distribution and shade the rejection region(s).
   (i) (Eyeball-estimate and then)* compute the critical value(s) of the sample statistic. Enter it (them) on the appropriate distribution and shade the rejection region(s).

   III. Collect a sample and compute statistics
   (j) (Eyeball-estimate and then)* compute the observed value of the sample statistic. Enter it on the appropriate distribution.
   (k) (Eyeball-estimate and then)* compute the observed value of the test statistic. Enter it on the appropriate distribution.
   (l) Do the critical values, rejection regions, and observed values have the same relative position on their respective distributions? Explain.

   IV. Interpret the results
   (m) Is this result statistically significant (that is, should we reject H_0?) Decide (by eyeball and)* by computation.
   (n) If the result was statistically significant, then determine the effect size index d and Cohen's nonoverlap statistic U. Illustrate these by superimposing the distribution of the variable assuming the best point-estimate of the parameter is true on your sketch in part (c). Indicate d and shade the U region as in Figure 10.8.
   (o) Describe the results, including both statistical and practical significance, in plain English.

3. (*Exercise 2 continued*) Suppose in our SAT study that we have no reason to believe that the standard deviation of meditators' SAT scores will be 100. Reconsider parts (a)–(o) of Exercise 2. Which of them have changed? How? Then answer the following:
   (p) What is the 95% confidence interval for the mean SAT score among meditators?
   (q) Does the confidence interval you computed in part (p) include the SAT score specified by the null hypothesis? Is that what you should have expected?

4. Suppose I am a race car driver and I know that my car's mean lap time at the Whitewater Raceway is 61.7 seconds. My mechanics install a new fuel injection system (which in other cars has increased performance and thus decreased lap times); I then run a lap in 61.0 seconds. Should I conclude that the new injection system has made a difference in my car?

5. (*Exercise 4 continued*) Now suppose that I run 21 laps, with the following times (in seconds): 61.0, 62.3, 62.1, 59.9, 60.5, 61.2, 61.3, 60.9, 61.4, 60.8, 62.4, 62.1, 60.5, 61.4, 61.0, 61.0, 62.1, 61.5, 62.0, 60.9, 62.1. (Labor savings for the student: $\sum X = 1288.4$, $\sum X^2 = 79{,}055.76$.) Assume that these 21 laps can be considered a random sample of size 21 from an infinite population of possible lap times with the new fuel injection system. Have lap times changed? Use $\alpha = .05$. Do parts (a)–(o) of Exercise 2, except substitute this part (c):

(c) What is the variable for this problem? (Eyeball estimate and then)* compute its standard deviation. Sketch the distribution of the variable assuming the null hypothesis is true.

6. (*Exercise 5 continued*) Suppose that we had decided to use $\alpha = .01$ instead of $\alpha = .05$. Reconsider parts (a)–(o) of Exercise 5. Which of them have changed? How? Then answer the following.
   (p) Explain why the decision in Exercise 6 is different from the decision in Exercise 5.

7. A politician has voted against abortion rights in the past but decides to change his voting pattern if significantly more than half the voters in his district are pro-choice. He takes a random sample of 100 voters and finds that 59 of these 100 are pro-choice. Should he change his voting pattern? [*Hint:* This requires a little thought. Use the general formula for the test statistic; the sample statistic is p, the sample proportion. The standard error of a proportion was derived in Box 8.6.]

Section B: Supplementary exercises

8. Assume we know that $\sigma = 2$, that the null hypothesis is H_0: $\mu = 7$, and that the alternative hypothesis is H_1: $\mu < 7$. Consider the data X_i: 6, 3, 8, 4, 4, 2.
   (a) What is the observed value of the statistic ($\overline{X}_{obs}$)?
   (b) What is the observed value of the test statistic (z_{obs})?
   (c) What is the critical value of the test statistic (z_{cv}) assuming thc test is directional and $\alpha = .05$?
   (d) What is the critical value of the statistic ($\overline{X}_{cv}$)?
   (e) Should we reject the null hypothesis? Why?

9. Recall from Exercise 12 in Chapter 9 that a school has been conducting surveys of drug use among peers for a number of years and the results have always been about the same: The number of friends who use drugs has $\mu = 3.4$ and $\sigma = 1.5$. This year the school has begun a "Just say NO" campaign, and the administration would like to know whether it has been effective. Perhaps it has increased drug use, perhaps it has decreased it, or perhaps it has had no significant effect. The administration conducts the survey again, randomly sampling 25 students, and obtains these results: 2, 3, 3, 0, 3, 3, 3, 1, 2, 3, 5, 0, 4, 4, 5, 2, 4, 2, 1, 4, 2, 4, 5, 2, 2. (Labor savings: $\sum X = 69$, $\sum X^2 = 239$.) Has the campaign had any effect? Use $\alpha = .05$. Do parts (a)–(o) of Exercise 2.

10. Recall from Exercise 14 in Chapter 9 that a particular assembly line has had an average absentee rate of 15.7 days per year. An industrial psychologist believes that giving assembly-line workers some control over their situation (by asking them for suggestions about how assembly processes can be improved, etc.) will decrease their absentee rate. She takes one assembly line, institutes the new procedures, and then counts the days per year the workers on that line are absent: 14, 17, 13, 13, 10, 18, 11, 18, 16, 16, 16, 23, 18, 18, 15, 15, 11, 14, 17, 10, 26, 10, 9, 21, 18, 10, 13, 10, 19, 6. (Labor savings: $\sum X = 445$, $\sum X^2 = 7181$.) Follow the same procedure as in Exercise 2, except substitute this part (c):
    (c) What is the variable in this problem? (Eyeball-estimate and then)* compute its standard deviation. Sketch the distribution of the variable assuming the null hypothesis is true.

11. Suppose that the mean grade point average (GPA) of applicants to XYZ University over the past ten years has been consistently $\mu = 2.74$. The university administration has a plan to encourage students with higher GPAs to apply to XYZU and decides to test this plan in ABC County, which in years past has had applicants quite typical of the university as a whole. The administration incorporates its plan and gets these GPAs of applicants from ABC County: 2.49, 2.85, 2.81, 3.10, 2.93, 2.59, 3.55, 2.85, 2.82, 3.18, 3.48, 3.82, 3.72, 2.53, 3.12, 2.05, 2.25, 3.42, 4.00, 2.89, 3.52, 2.53, 3.07, 2.58, 3.72, 2.50, 2.32, 3.56, 3.48, 3.10, 2.77, 2.75, 3.25, 3.11, 3.44. (Labor savings: $n = 35$, $\sum X = 106.15$, $\sum X^2 = 329.9425$.) Has the plan been effective? Use $\alpha = .05$. Follow the same procedure as in Exercise 2, except substitute this part (c):
    (c) What is the variable in this problem? (Eyeball-estimate and then)* compute its standard deviation. Sketch the distribution of the variable assuming the null hypothesis is true.

12. A company monitors the stress experienced by its employees by administering twice a year a short stress questionnaire. Past results have shown that experienced stress as measured by this questionnaire has mean 57.4 and standard deviation 14.4 points. Two of the company's directors have just been arrested for fraud, and the company wishes to know whether this event has increased stress in the company. It administers the questionnaire again to a random sample of 21 of its employees. It obtains these scores: 53, 51, 63, 55, 83, 51, 42, 73, 50, 63, 72, 55, 63, 62, 71, 78, 50, 67, 74, 64, 71. (Labor savings for the student: $n = 21$, $\sum X = 1311$, $\sum X^2 = 84,205$.) Have the arrests increased stress? Use $\alpha = .05$. Follow the same procedure as in Exercise 2, except substitute this part (c):
    (c) What is the variable in this problem? (Eyeball-estimate and then)* compute its standard deviation. Sketch the distribution of the variable assuming the null hypothesis is true.

13. The owner of a drugstore chain has 27 drugstores, each of which sells Feelgood, a drug that is supposed to reduce psychological and physical pain. He knows that the average drugstore in his chain sells 48 bottles of Feelgood a day. On a particular Wednesday, a news story about Feelgood is shown prominently on national television; the story is quite critical about the efficacy of the drug. The owner wonders whether the story will affect sales; perhaps the criticism of the drug will cause a decrease in sales, but perhaps the exposure itself will cause an increase in sales. He obtains these Thursday sales figures from each of his 27 stores: 63, 54, 50, 45, 58, 43, 54, 75, 38, 52, 45, 51, 45, 56, 52, 49, 30, 39, 55, 79, 64, 45, 35, 46, 46, 50, 29.

(Labor savings: $n = 27$, $\sum X = 1348$, $\sum X^2 = 70{,}810$.) Has the story affected sales? Use $\alpha = .05$. Follow the same procedure as in Exercise 2, except substitute this part (c):

(c) What is the variable in this problem? (Eyeball-estimate and then)* compute its standard deviation. Sketch the distribution of the variable assuming the null hypothesis is true.

14. A farmer grows wheat. He divides his farm into 18 equal parcels, and he knows from experience that he typically gets an average of 240 bushels of wheat from each parcel. The Growfast fertilizer seller claims that using the new fertilizer will improve his yield. He decides to test this by using Growfast for one year. If his land yields significantly more wheat in this year, he will switch permanently to Growfast. The farmer obtained these yields per parcel (in bushels): 288, 252, 211, 230, 296, 254, 256, 245, 264, 268, 253, 235, 255, 280, 274, 261, 228, 254. (Labor savings: $n = 18$, $\sum X = 4604$, $\sum X^2 = 1{,}185{,}438$.) Should he switch fertilizer? Use $\alpha = .05$. Follow the same procedure as in Exercise 2, except substitute this part (c):

(c) What is the variable in this problem? (Eyeball-estimate and then)* compute its standard deviation. Sketch the distribution of the variable assuming the null hypothesis is true.

Section C: Extending your comprehension (answers in Appendix D, page 514)

15. Assume that the null hypothesis is $\mu = 10$ and that σ is unknown. Create a data set that has five points such that we reject the null hypothesis if $\alpha = .05$ but fail to reject it if $\alpha = .01$.

16. Is it possible to create a data set where we reject H_0 using $\alpha = .01$ but fail to reject it when $\alpha = .05$? Explain.

17. Assume that the null hypothesis is $\mu = 10$ and that σ is unknown. Create a data set that has five points such that we reject the null hypothesis if the test is directional but fail to reject it if the test is nondirectional.

18. Is it possible to create a data set where we reject H_0 if the test is nondirectional but fail to reject it if the test is directional? Explain.

Section D: Cumulative review (answers in Appendix D, page 514)

ⓘ Note the new format of the cumulative review exercises.

Beginning now and continuing throughout the remainder of the textbook, the exercises in Section D: Cumulative Review have a new format designed to give you *discrimination training*—practice in determining which statistical procedure is appropriate. Being able to decide which statistical procedure to use in which situation is one of the most important skills to be acquired; these exercises will gradually build those discrimination skills. Note that long computations are *not* required; your task is to practice making the decision and then setting up the procedure.

For Exercises 19–24, choose the correct answer or fill in the blank. [***Note:*** **Computations are *not* required;** use $\alpha = .05$ unless otherwise instructed.]

(a) This problem requires:
 (1) Finding the area under a normal distribution [Skip parts (b)–(g) and proceed to (h).]
 (2) Creating a confidence interval
 (3) Testing a hypothesis

(b) The null hypothesis is of the form:
 (1) $\mu = a$
 (2) There is no null hypothesis

(c) The appropriate sample statistic is:
 (1) $\overline{X}$ (2) p

(d) The appropriate test statistic is:
 (1) z (2) t

(e) The number of degrees of freedom for this test statistic is ________. (State the number or *not applicable*.)

(f) The hypothesis test (or confidence interval) is:
 (1) Directional (2) Nondirectional

(g) The critical value of the test statistic is ________. (Give the value.)

(h) The appropriate formula for the test statistic (or confidence interval) is Equation (________).

ⓘ Note that computations are *not* required for any of the problems in Section D. See the instructions.

19. Politician Smith would like to know what percentage of the voters will vote for her in the upcoming election. She commissions a poll of 500 voters, each of whom is asked to identify the candidate he or she will vote for. Of these 500, 231 say they will vote for Smith. What can be said about the percentage of the entire population of voters who favor Smith?

20. The telephone company knows that its information operators require on average 9.4 seconds to respond to a request for a telephone number. It is considering installing a new computer system and will do so if it is convinced that the time per call with the new system is significantly shorter than 9.4 seconds. On a trial run, 100 calls are timed; the data (in seconds) are: 8.9, 11.0, 9.8, 7.8, 11.2, 8.5, 9.3, 8.2, 6.3, 11.0, 8.1, 7.3, 9.7, 8.5, 9.3, 7.4, 7.2, 9.0, 7.0, 5.3, 8.4, 7.5, 10.9, 9.3, 6.7, 8.9, 9.5, 7.9, 5.7, 6.9, 7.9, 10.7, 11.4, 9.0, 8.2, 9.6, 9.1, 7.2, 7.4, 11.3, 9.1, 9.3, 8.1, 9.0, 7.6, 9.3, 8.4, 7.2, 9.7, 9.9, 7.4, 7.5, 9.9, 9.8, 7.3, 8.6, 8.5, 10.3, 8.5, 7.8, 8.2, 7.5, 8.3, 10.0, 8.3, 8.2, 10.7, 10.7, 8.6, 8.1, 9.3, 7.3, 9.1, 10.0, 11.8, 10.6, 6.7, 10.5, 7.8, 9.1, 8.4, 7.7, 8.0, 8.8, 9.3, 8.1, 10.4, 9.4, 10.3, 9.8, 8.9, 8.2, 9.0, 7.7, 8.8, 10.7, 9.0, 6.8, 10.5, 7.2. Should the telephone company buy the new computers?

21. Patients with a particular kind of cancer have an average life expectancy of 2.3 years. A new medication is proposed that may alter life expectancy; perhaps it will increase it, perhaps it will decrease it. A random sample of 25 such

patients are chosen and given the medication. The numbers of years they survive are recorded: 3.2, 3.4, 1.4, 2.9, 2.6, 3.1, 3.9, 2.8, 3.0, 2.3, 3.5, 2.0, 3.2, 2.2, 3.6, 3.2, 3.2, 3.1, 2.7, 2.9, 3.5, 3.4, 3.9, 4.0, 1.8. Does the medication change life expectancy?

22. A sociologist knows that the income in a particular community has mean $24,000 and standard deviation $3000. What percentage of individuals earn less than $20,000?
23. Head Start is an enrichment program for young children. The program's administrators would like to know the mean intelligence (IQ) of its participants. They know that IQ has standard deviation 15 points. They take a random sample of 60 children and administer IQ tests. These are the scores: 82, 90, 105, 85, 108, 109, 68, 88, 76, 70, 76, 108, 119, 64, 73, 100, 78, 63, 95, 77, 71, 100, 76, 89, 83, 101, 86, 82, 103, 96, 90, 112, 101, 103, 96, 103, 100, 91, 109, 101, 88, 83, 72, 105, 98, 70, 87, 69, 80, 83, 73, 73, 97, 94, 68, 91, 55, 72, 91, 96. What can be said about the mean IQ of the population of Head Start children?
24. Head Start administrators would like to know whether this enrichment program is successful in increasing the IQs of children. They know that the IQs of children who enter the program have mean 88.0 and standard deviation 15.0. They take a random sample of 80 children who are completing the program and measure their IQs, finding that $\bar{X} = 88.8$. Has the program increased children's IQs?

Section E: Computer explorations

25. Have the computer take a random sample of size $n = 5$ from the population of IQ scores ($\mu = 100$, $\sigma = 15$). Then test the nondirectional null hypothesis H_0: $\mu = 96$. Did you reject H_0? Was it a "close call" or was the decision clear?
26. (*Exercise 25 continued*) Repeat the procedure of Exercise 25, except let $n = 200$. Now did you reject H_0: $\mu = 96$? Was it a "close call" or was the decision clear?
27. Explain the similarities and differences between your results in Exercises 25 and 26.
28. (a) Have the computer take a random sample of size $n = 10$ from the population of IQ scores ($\mu = 100$, $\sigma = 15$).
    (b) Add 20 points to each score in the sample (use `Edit Variable` in datagen or `Transform` in SPSS).
    (c) Test the nondirectional null hypothesis H_0: $\mu = 100$.
    (d) What constant must be added to or subtracted from all these values so that the null hypothesis is *just barely* rejected? [*Hint:* Think about the confidence interval.]

Eyeball-estimating One-sample *t* tests (Optional)

This section may be omitted without loss of continuity.

Beginning in this chapter and continuing throughout most of the textbook, I have placed all eyeball-estimation procedures in the optional section at the end of each chapter to allow maximum flexibility. I continue to recommend that you eyeball-estimate before you compute. Remember that the philosophy of eyeball-estimation is to perform all eyeball-computations in your head, *not* using a calculator. The object is to get a reasonable approximation, not an exact result. Use the procedure both as a quick check on the computation and, more important, as a way of quickly practicing the hypothesis-testing procedure.

General Eyeball Procedure

The eyeball-estimation of one-sample hypothesis tests follows the same general procedure that was described in Chapter 9 and used for computation earlier in this chapter.

steps for eyeball-estimating one-sample hypothesis tests

I. State the Null and Alternative Hypotheses

Use the same procedure we used when computing. Note that to sketch the distributions of the variable and the sample statistic by eyeball if σ is unknown, you must eyeball estimate $s_{\text{by eyeball}}$ and $s_{\bar{X}_{\text{by eyeball}}}$; see steps III.c and III.d.

II. Set the Criterion for Rejecting H_0

a. If σ is known, then $z_{\text{cv}} = \pm 1.96$ if the test is nondirectional or $z_{\text{cv}} = 1.645$ if the test is directional. You should memorize those values.
b. If σ is unknown, then $t_{\text{cv}_{\text{by eyeball}}}$ is somewhat greater than ± 1.96 (or 1.645 if directional); that is, the nondirectional $t_{\text{cv}_{\text{by eyeball}}}$ is about 2 unless *df* is quite small.

III. Collect a Sample and Eyeball-estimate the Observed Values of the Sample Statistic and the Test Statistic

a. The eyeballed sample statistic is $\overline{X}_{\text{by eyeball}}$, and we use either the range method or the balancing-point method (just as in Chapter 4; use whichever is more convenient) to obtain it.
b. The eyeballed test statistic follows the same formulas used for computing:

$$z_{\text{by eyeball}} = \frac{\overline{X}_{\text{by eyeball}} - \mu}{\sigma_{\overline{X}_{\text{by eyeball}}}} \quad \text{or} \quad t_{\text{by eyeball}} = \frac{\overline{X}_{\text{by eyeball}} - \mu}{s_{\overline{X}_{\text{by eyeball}}}}$$

(if σ is known) (if σ is unknown)

To obtain the *numerator* of the eyeballed test statistic, subtract the value of μ specified by H_0 from $\overline{X}_{\text{by eyeball}}$ to get $\overline{X}_{\text{by eyeball}} - \mu$.
c. If σ is given, use σ. Otherwise, eyeball-estimate $s_{\text{by eyeball}}$ by the range method or the inflection-point method (just as in Chapter 5; use whichever method is more convenient).
d. To obtain the *denominator* of the test statistic, eyeball-estimate $\sigma_{\overline{X}_{\text{by eyeball}}}$ (or $s_{\overline{X}_{\text{by eyeball}}}$) by dividing σ (or $s_{\text{by eyeball}}$) by $\sqrt{n}$. *Note:* Find the square root in your head by using the nearest perfect square (4, 9, 16, 25, 36, 49, 64, etc.) for the eyeballed value of n. Perform the division in your head also. You don't need an exact answer, just an approximation.
e. To obtain the test statistic $z_{\text{by eyeball}}$ (if we know σ) or $t_{\text{by eyeball}}$ (if we don't know σ), divide the numerator of the test statistic (from step b) by the denominator of the test statistic (from step d). Again, perform the division in your head.

IV. Interpret the Results

Exactly as by computation, if your eyeballed test statistic (from step e) is greater (in absolute value) than your eyeballed critical value (from part II), then reject$_{\text{by eyeball}}$ H_0.

Electric Fields and IQ Example

ℹ This eyeball-estimation procedure parallels the computation in the text on page 184.

We now apply this procedure to the eyeball estimation from the study of the effect of electric fields on IQ. You may recall that the question was whether living near a power line affected a person's IQ and the null hypothesis was H_0: $\mu = 100$.

I. State the Null and Alternative Hypotheses

Use the same procedure we used when computing. The statement of the problem provided σ, so we can use that to sketch the distribution of the variable. Just as when computing, we determine $\sigma_{\overline{X}_{\text{by eyeball}}}$ by dividing σ by $\sqrt{n}$, except that now the square root and the division are performed in your head: $\sigma_{\overline{X}_{\text{by eyeball}}} = \sigma/\sqrt{n} = 15/\sqrt{25} = 15/5 = 3$.

II. Set the Criterion for Rejecting H_0

Just as when computing, we let $\alpha = .05$. The test is nondirectional and σ is known, so $z_{\text{cv}} = \pm 1.96$.

III. Collect a Sample and Eyeball-estimate the Observed Values of the Sample Statistic and the Test Statistic

a. We can most conveniently use the range method to eyeball-estimate $\overline{X}_{\text{by eyeball}}$: $\overline{X}_{\text{by eyeball}} \approx (73 + 131)/2 = 102$.
b. Then the numerator of the test statistic is $\overline{X}_{\text{by eyeball}} - \mu = 102 - 100 = 2$.
c. We use σ because the problem provided it.
d. Then the denominator of the test statistic is $\sigma_{\overline{X}_{\text{by eyeball}}} = \sigma/\sqrt{n} = 15/5 = 3$.
e. $z_{\text{by eyeball}}$ is our eyeballed numerator (from step b) divided by our eyeballed denominator (from step d)—that is, $z_{\text{by eyeball}} \approx 2/3 = .67$.

IV. Interpret the Results

We compare $z_{\text{by eyeball}}$ to z_{cv}. Because .67 is not greater than 1.96, we do *not* reject$_{\text{by eyeball}}$ the null hypothesis.

As you may recall, the computed values were found in the text to be $\overline{X}_{obs} = 103.48$ and $z_{obs} = 1.166$, so we also did not reject H_0 by computation.

GRE Training Program Example

This eyeball-estimation procedure parallels the computation in the text on page 189.

We now show the eyeball-estimation of z for the GRE training program example. The question was whether the training program increased GRE scores, and the null hypothesis was H_0: $\mu = 500$.

I. State the Null and Alternative Hypotheses

Use the same procedure we used when computing. The statement of the problem provided σ, so we can use that to sketch the distribution of the variable. Just as when computing, we determine $\sigma_{\overline{X}_{\text{by eyeball}}}$ by dividing σ by $\sqrt{n}$, except that now the square root and the division are performed in your head: $\sigma_{\overline{X}_{\text{by eyeball}}} = \sigma/\sqrt{n} = 100/\sqrt{10} \approx 100/3 \approx 30$.

II. Set the Criterion for Rejecting H_0

Just as when computing, we let $\alpha = .05$. The test is directional and σ is known, so $z_{cv} = 1.645$.

III. Collect a Sample and Eyeball-estimate the Observed Values of the Sample Statistic and the Test Statistic

a. We can most conveniently use the range method to eyeball-estimate $\overline{X}_{\text{by eyeball}}$: $\overline{X}_{\text{by eyeball}} \approx (490 + 660)/2 = 575$.
b. Then the numerator of the test statistic is $\overline{X}_{\text{by eyeball}} - \mu = 575 - 500 = 75$.
c. We use σ because the problem provided it.
d. The denominator of the test statistic is $\sigma_{\overline{X}_{\text{by eyeball}}} = \sigma/\sqrt{n} = 100/\sqrt{10} \approx 30$.
e. $z_{\text{by eyeball}}$ is our eyeballed numerator (from step b) divided by our eyeballed denominator (from step d)—that is, $z_{\text{by eyeball}} \approx 75/30 = 2.5$.

IV. Interpret the Results

We compare $z_{\text{by eyeball}}$ to z_{cv}. Because 2.5 is greater than 1.645, we *do* reject$_{\text{by eyeball}}$ the null hypothesis.

As you may recall, the computed values were found in the text to be $\overline{X}_{obs} = 574.00$ and $z_{obs} = 2.34$, so we also did reject H_0 by computation.

Dr. Rogers's Study Time Example

This eyeball-estimation procedure parallels the computation in the text on page 195.

We now show the eyeball-estimation of t for the Dr. Rogers's study time example.

I. State the Null and Alternative Hypotheses

Use the same procedure we used when computing. The statement of the problem did not provide σ, so we must eyeball-estimate $s_{\text{by eyeball}}$ to sketch the distribution of the variable. The range method is most convenient: $s_{\text{by eyeball}} \approx (\text{highest} - \text{lowest})/4 = (32 - 2)/4 = 7.5$. To sketch the distribution of the sample statistic, we determine $s_{\overline{X}_{\text{by eyeball}}}$ by dividing $s_{\text{by eyeball}}$ by $\sqrt{n}$: $s_{\overline{X}_{\text{by eyeball}}} = s_{\text{by eyeball}}/\sqrt{n} = 7.5/\sqrt{16} = 7.5/4 \approx 2$.

II. Set the Criterion for Rejecting H_0

Just as when computing, we let $\alpha = .05$. The test is nondirectional and σ is unknown, so $t_{\text{cv}_{\text{by eyeball}}}$ will be somewhat greater than 1.96—say, ± 2.1.

III. Collect a Sample and Eyeball-estimate the Observed Values of the Sample Statistic and the Test Statistic

a. We can most conveniently use the range method to eyeball-estimate $\overline{X}_{\text{by eyeball}}$: $\overline{X}_{\text{by eyeball}} \approx (2 + 32)/2 = 17$.
b. Then the numerator of the test statistic is $\overline{X}_{\text{by eyeball}} - \mu = 17 - 21 = -4$.
c. Because the problem did not provide σ, we use the range method to eyeball-estimate $s_{\text{by eyeball}} \approx (\text{highest} - \text{lowest})/4 = (32 - 2)/4 = 7.5$.
d. Then the denominator of the test statistic is $s_{\overline{X}_{\text{by eyeball}}} = s_{\text{by eyeball}}/\sqrt{n} = 7.5/\sqrt{16} \approx 2$.
e. $t_{\text{by eyeball}}$ is our eyeballed numerator (from step b) divided by our eyeballed denominator (from step d)—that is, $t_{\text{by eyeball}} \approx -4/2 = -2$.

IV. Interpret the Results

We compare $t_{\text{by eyeball}}$ to $t_{\text{cv}_{\text{by eyeball}}}$. Because -2 is not greater than ± 2.1, we do *not* reject$_{\text{by eyeball}}$ the null hypothesis.

The text shows that by computation, $\overline{X}_{\text{obs}} = 11.94$, $t_{\text{obs}} = -4.67$, and $t_{\text{cv}} = \pm 2.131$, so we *should* have rejected H_0. Why the discrepancy between our eyeballed and computed results?

The data have one extremely high study time value (32), which makes $\overline{X}_{\text{by eyeball}}$ too large and therefore $t_{\text{by eyeball}}$ too small. This single large value has a greater effect on the eyeball-estimate than on the actual mean, so the eyeballed mean is too large (17 instead of 11.94).

Even though there is a discrepancy in our results, the eyeball-estimation procedure is still useful. By confronting the discrepancy, we gain a clearer understanding of the data.

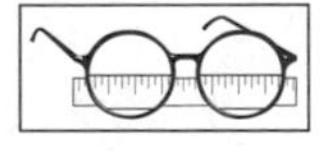

You should practice the step-by-step eyeball-estimation procedures described in this section on the histograms shown in Figures 10.15 through 10.20. Cover up the tables of actual values next to the histograms, and then eyeball-estimate t using the steps shown above.

ESTAT

You may wish to use the ESTAT exercise “ttest1” at this time. Follow the path labeled “Eyeball-estimation.”

Inferences About Means of Two Independent Samples

CHAPTER

LEARNING OBJECTIVES

1. What is a dependent variable?
2. What is an independent (or treatment) variable?
3. What is the null hypothesis for a two-independent-samples hypothesis test?
4. What is the nondirectional alternative hypothesis for a two-independent-samples hypothesis test?
5. What is the standard error of the difference between two means?
6. What is the formula for the test statistic for two independent samples?
7. What is the pooled variance? How is it eyeball-estimated?
8. How many degrees of freedom are there in a two-independent-samples hypothesis test?
9. How are the critical values determined for the sample statistic and the test statistic for two-independent-samples hypothesis tests?
10. (Optional) How are the sample statistic and the test statistic eyeball-estimated for two-independent-samples hypothesis tests?

This chapter describes the evaluation of hypotheses about the means of two independent groups—that is, testing null hypotheses of the form H_0: $\mu_1 = \mu_2$. The denominator of the resulting test statistic is called the "standard error of the difference between two means"; computing it requires that we first "pool" (that is, take a weighted average of) the variances of the two groups.

In Chapter 10, we discussed the evaluation of hypotheses about single samples—that is, attempts to reject null hypotheses of the form H_0: $\mu = a$, where μ is the mean of the population from which the single sample was drawn and a is the value of the mean that is specified by H_0. In this chapter, we describe how hypotheses are evaluated when we have *two* independent samples and we wish to know whether these two samples come from the same population.

11.1 Hypotheses with Two Independent Samples

We return to the conclusion of the study on Pygmalion in the classroom that we have been following since Chapter 1. As you recall, Rosenthal and Jacobson (1968) measured each Oak School second-grader's IQ at the beginning and at the end of the school year to obtain a measure of intellectual growth (IQ gain) for each student. Rosenthal and Jacobson randomly assigned the students to one of two conditions, the bloomers and others, and manipulated the teachers' expectations regarding these students: Teachers were led to expect that the bloomer children would show a spurt in intellectual growth, whereas they were given no such expectations for the other children.

Dependent and Independent Variables

dependent variable: the measured outcome of interest

independent (or treatment) variable: the variable whose value defines group membership

In the Pygmalion example (and all the remaining examples in this chapter), we measure each subject on *two* variables, which we call the "dependent" and "independent" variables. The *dependent variable* is the outcome variable that is of primary interest in the study, whereas the *independent* (or *treatment*) *variable* is the variable that defines group membership. The dependent variable in the Pygmalion example is intellectual growth; its value *depends* on the teachers' expectations and the individual characteristics of the subject. By contrast, the independent variable is the group—bloomers or others—to which the student is assigned. This assignment, as you recall, was made at random—that is, *independent* of any of the characteristics of either the student or the teacher (but see Box 11.1). We could assign a numerical value to the independent variable (e.g., 1 = bloomers, 2 = others), but that is not necessary.

BOX 11.1 Treatment variables and research design

The science of research design makes the important distinction between true experiments and other studies that are similar to true experiments but where the experimenter does not have total control over the treatment variable.

true experiment: a study where the experimenter assigns subjects to groups at random

In a *true experiment*, the experimenter has total control over the treatment variable and *randomly assigns* a level of that variable to each subject. This randomness guarantees that the treatment variable will be assigned to each subject entirely independent of any naturally occurring characteristics of that subject, which is why the treatment variable is called the "independent" variable in a true experiment. Our Pygmalion example is a true experiment; each subject is randomly assigned to either the bloomer group or the other group (that is, to one of the two values of the independent variable).

In some other studies (called quasi-experiments, *in situ* studies, or passive-observational studies—see texts on research design for distinctions among them) that have the same general form as true experiments, the experimenter does *not* have control over the treatment variable. Suppose we were interested in whether teacher expectations affect boys and girls differently. We select ten boys at random and ten girls at random and lead their teachers to expect that they will "bloom." Then we measure each child's intellectual growth (IQ gain). The treatment variable in this example (gender of subject) has two values (male and female). But here the assignment to groups is *not* random and *not* controlled by the experimenter; gender is a naturally occurring characteristic of the subject. Thus, this treatment variable is sometimes called a "status" or "classifying" variable.

From the standpoint of research design, the distinction between true experiments and these other studies is quite important because the ability to randomly manipulate the independent variable provides the strongest possible evidence for inferring that the different levels of the independent variable actually *cause* different values of the dependent variable. The evidence for causality in these other studies is much less clear (see, for example, Christensen, 1991, chaps. 4, 9, and 10). Some authors underscore this distinction by reserving the term *experiment* for only true experiments, calling the other designs "studies" instead. In this text, we use the term *experiment* to refer either to a true experiment or to these other studies.

Even though the inferences about causality that can be drawn from the two kinds of experiments are quite different, the statistical analyses of true experiments and the other studies are identical. That is, there is no *statistical* (or computational) difference between independent variables and status variables. We use the same statistical tests for both kinds of experiments. For this reason, status variables are sometimes (rather loosely) also referred to as "independent" variables, even though they actually depend very intimately on naturally occurring characteristics of the subjects.

Note that in the studies in Chapter 10, we had only one variable. For example, when we wished to know whether Dr. Rogers's students' study time was different from the 21 hours that might be expected, we measured students on only one variable: study time. The "variable" of Chapter 10 is what we are calling the "dependent variable" in this chapter.

The Null Hypothesis

The null hypothesis in the Pygmalion experiment is that teacher expectation has no impact on the intellectual growth of the student—that is, that μ_1, the mean intellectual growth of the population of students expected to bloom, is the same as μ_2, the mean intellectual growth of the population of students whose teachers have no such expectations. In symbols,

null hypothesis, two-independent-samples test for means (nondirectional test)

$$H_0: \quad \mu_1 = \mu_2 \tag{11.1}$$

Let us examine how this null hypothesis differs from the single-sample tests of Chapter 10. In those single-sample tests, we specified a *value*—that is, a number a such as 100 or 21—and attempted to ascertain whether the mean of the one population we were interested in had that particular value. In this chapter, we consider *two* populations instead of one, but we do *not* specify a value for *either* mean; we simply ask whether the two means are *equal*, regardless of their values. Thus, the null hypothesis for a two-sample test *is not* H_0: $\mu_1 = a$ and $\mu_2 = a$; that would be specifying the value for *each* of the two population means. Instead, the null hypothesis *is* H_0: $\mu_1 = \mu_2$—merely that the two means are equal *without* specifying their values.

Both the bloomers and the other students are considered to be random samples from their parent populations. Thus, the mean of the bloomers ($\overline{X}_1$) is the sample mean from a population whose mean is μ_1, and the mean of the other students ($\overline{X}_2$) is the sample mean from a population whose mean is μ_2. If the null hypothesis ($\mu_1 = \mu_2$) is true, we would expect $\overline{X}_1$ to be close to $\overline{X}_2$. However, we would *not* expect $\overline{X}_1$ to be *exactly* equal to $\overline{X}_2$ because some students, for a variety of reasons, will show more naturally occurring intellectual growth than other students, regardless of their teachers' expectations. The random assignment to groups makes it likely that about the same number of students with high naturally occurring intellectual growth are assigned to each group. Thus, the result of the random assignment is that if the null hypothesis is true, the sample mean intellectual growth in both groups will be about the same. If the null hypothesis is false, we expect one group's sample mean to be substantially larger than the other.

Our experiment is designed to answer the question: Can we confidently reject the null hypothesis? Let us explore some possible outcomes of this experiment and see whether we would reject the null hypothesis in each case, as shown in Figures 11.1 through 11.4.

Experimental Outcomes

Figure 11.1 shows one possible outcome of the Pygmalion study. You may recall that the actual Rosenthal and Jacobson (1968) study had a much smaller sample size for bloomers than for the other students (12 bloomers versus 47 others). We will return to that case in a moment, but for the present, let's simplify things and use a sample size of 12 for each group. If the data are as shown in Figure 11.1, we would immediately conclude that teacher expectation indeed does make a difference. All the bloomers show more intellectual growth than do any of the other children. Therefore, we would clearly reject the null hypothesis for these data.

If the data are as shown in Figure 11.2, however, we would immediately "accept" the null hypothesis. The overlap between the bloomers and the other children is nearly complete.

If the data are as shown in Figure 11.3, however, the conclusion is not so clear. There is *some* overlap between the intellectual growth scores, but the overlap is not complete. We must develop some criterion to determine whether or not to reject the null hypothesis.

These three sets of outcomes (Figures 11.1–11.3) are identical except for the difference between the sample means, $\overline{X}_1 - \overline{X}_2$, which is displayed as a horizontal arrow in each of the figures. In the first case, the difference between the means is large; in the second, it is almost zero; and in the third, it is of moderate size. Our first conclusion is that the larger the difference between the means, the more readily we will reject H_0.

Figure 11.4 shows another possible outcome of this experiment. Here, the sample mean of the bloomers ($\overline{X}_1$) is identical to their sample mean in Figure 11.3; the same is true for the other children (that is, $\overline{X}_2$ in Figure 11.4 is the same as it is in Figure 11.3). Therefore, $\overline{X}_1 - \overline{X}_2$ is exactly the same in Figure 11.4 as it is in Figure 11.3. There is no overlap between the histograms in Figure 11.4, however, so it appears that we are justified in rejecting H_0 for the Figure 11.4 data.

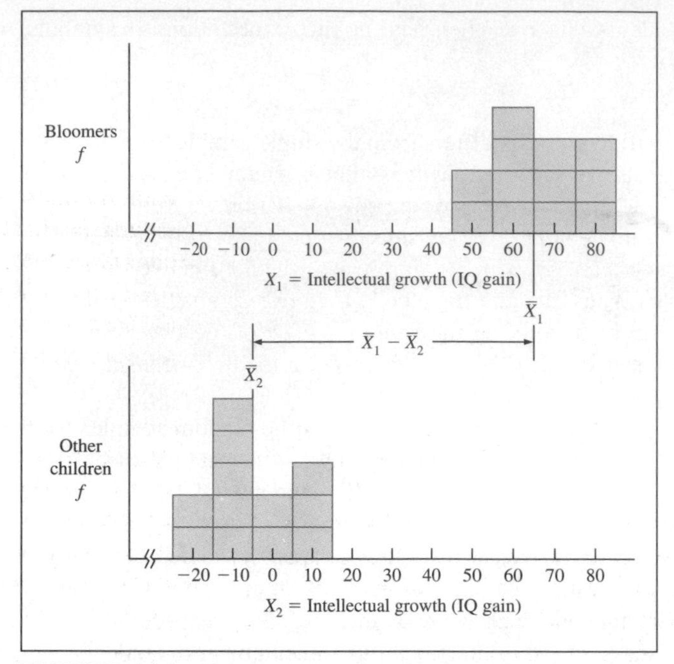

FIGURE 11.1 Histograms of intellectual growth in two groups. We should reject the null hypothesis.

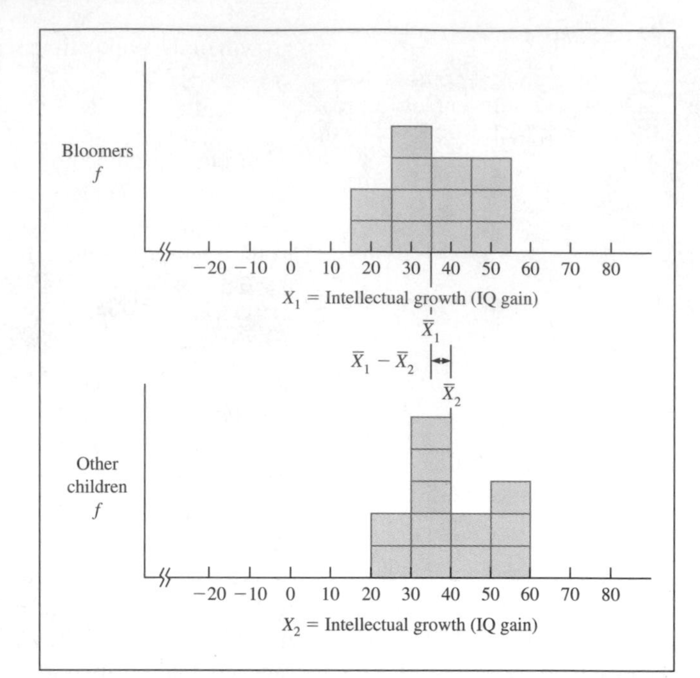

FIGURE 11.2 Histograms of intellectual growth in two groups. We should *not* reject the null hypothesis.

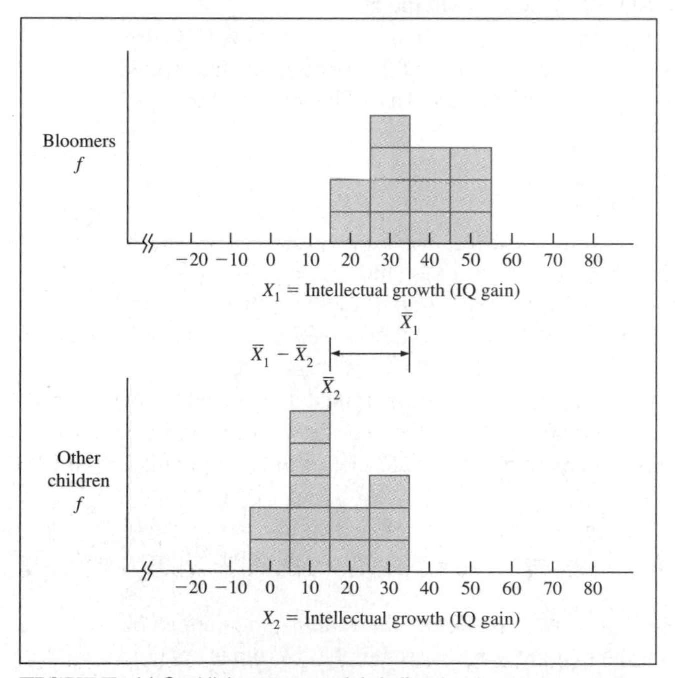

FIGURE 11.3 Histograms of intellectual growth in two groups. Should we reject the null hypothesis?

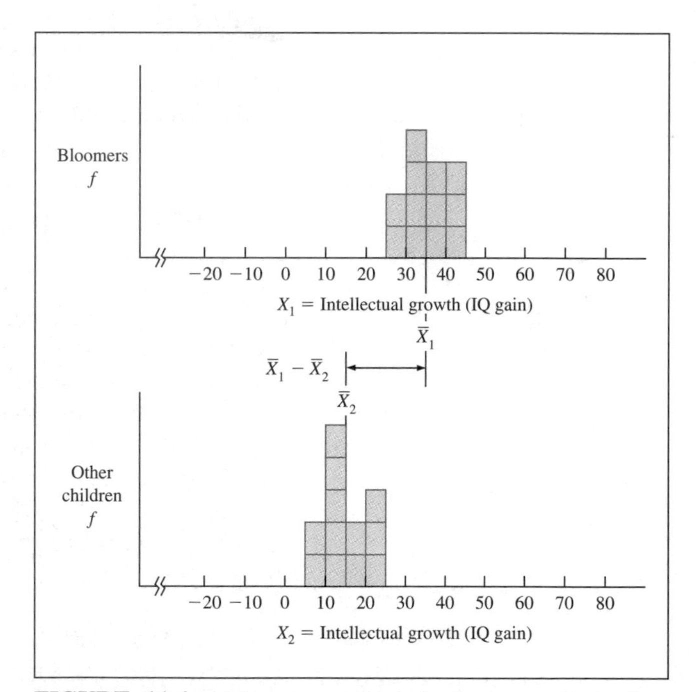

FIGURE 11.4 Histograms of intellectual growth in two groups. We *should* reject the null hypothesis.

How can the difference between the means be identical but the amount of overlap different? Because the standard deviation of the data for the bloomers in Figure 11.4 is *smaller* than the standard deviation of that data in Figure 11.3. In fact, our eyeball tells us that the Figure 11.4 standard deviation is *half* the Figure 11.3 standard deviation. The same is true for the two sets of data for the other children. The result is that there is almost *no* overlap between the two groups in Figure 11.4, even though the difference between the means is identical to that in Figure 11.3. Our second conclusion is then that the *smaller* the standard deviations of the two groups, the more readily we will reject H_0.

11.2 The Test Statistic

Based on these conclusions, and knowing that we wish to form a test statistic that becomes larger as we gain more and more confidence that the two samples come from different populations, we see that $\overline{X}_1 - \overline{X}_2$ should be a part of the *numerator* of the test statistic, so that as the difference between the two means gets larger, the test statistic becomes larger. Furthermore, a term related to the standard deviations must be in the *denominator* of the test statistic, so that as the standard deviations become smaller, the test statistic becomes larger.

To recapitulate: The Pygmalion experiment created two groups: the bloomers, whose teachers expect them to show high intellectual growth, and the other children, whose teachers have no such expectations. The null hypothesis for the test is H_0: $\mu_1 = \mu_2$, where μ_1 is the population mean of all students whose teachers have high expectations and μ_2 is the population mean of all other students. Recall that we are *not* specifying any numerical value (5.5, 5.75, 6.0, etc.) for either μ_1 or μ_2; the null hypothesis merely states that the two means are *equal*, whatever their value.

It is convenient, for reasons we will see in a moment, to write the null hypothesis for two-sample tests in this way:

null hypothesis, two-independent-samples test for means, alternative form

$$H_0: \quad \mu_1 - \mu_2 = 0 \tag{11.1a}$$

This statement of the null hypothesis is identical to Equation (11.1); it is obtained simply by subtracting μ_2 from both sides of the equation. This form is convenient because it makes somewhat more clear the fact that the population parameter in which we are interested is not μ_1 and not μ_2 but the difference between μ_1 and μ_2. Thus, we do not care whether μ_1 is small or large or whether μ_2 is small or large, but whether or not the parameter $(\mu_1 - \mu_2)$ equals 0.

Once we recognize that the (one) parameter of interest is the difference between two population means $(\mu_1 - \mu_2)$, we can return to the general form of the test statistic that we used in Chapters 9 and 10:

$$\text{test statistic} = \frac{\text{sample statistic} - \text{population parameter}}{\text{standard error of the sample statistic}} \tag{9.1}$$

The population parameter here, as we just saw, is $(\mu_1 - \mu_2)$. Recall that, in general, the sample statistic is the point-estimator of the parameter. The sample statistic that point-estimates $(\mu_1 - \mu_2)$ is the difference between the two sample means $(\overline{X}_1 - \overline{X}_2)$, where $\overline{X}_1$ is the mean of the sample drawn from the first population and $\overline{X}_2$ is the mean of the sample drawn from the second population.

the standard error of the difference between two means: the standard deviation of the distribution that would result from taking two random samples from the same population, computing the sample means, and subtracting one from the other to form the difference between two means; and then repeating the process indefinitely often

Recall that the denominator in the Equation (9.1) general expression for the test statistic is always the standard error of whatever particular statistic appears in the numerator, so in this case, the denominator is the *standard error of the difference between* $\overline{X}_1$ *and* $\overline{X}_2$, which we symbolize as $s_{\overline{X}_1 - \overline{X}_2}$. The standard error of the difference between two means is a key new concept to be mastered in this chapter, so we devote the next section to its understanding. For now, we can note that the standard error of the difference is directly related to the standard deviations of the two samples, which is also what we expected from our earlier discussion. As the standard deviations become smaller, the standard error of the difference between two means becomes smaller, and the test statistic becomes larger.

Thus, the test statistic, whose general form is "test statistic equals sample statistic minus population parameter divided by standard error of the sample statistic," becomes in this case

$$t = \frac{(\overline{X}_1 - \overline{X}_2) - (\mu_1 - \mu_2)}{s_{\overline{X}_1 - \overline{X}_2}}$$

However, the null hypothesis H_0: $\mu_1 - \mu_2 = 0$ implies that the population-parameter term $(\mu_1 - \mu_2)$ of the numerator disappears, which leaves this formula for the test statistic t in a two-independent-samples experiment:

test statistic (t), two-independent-samples test for means

$$t = \frac{\overline{X}_1 - \overline{X}_2}{s_{\overline{X}_1 - \overline{X}_2}} \qquad [df = n_1 + n_2 - 2] \qquad (11.2)$$

The alert reader might ask: Why are we using s and not σ and therefore using t and not z? Because we are not claiming here that we know both σ_1 and σ_2. It *is* possible to perform a two-sample test with z as the test statistic (and $\sigma_{\overline{X}_1 - \overline{X}_2}$ as the denominator) if we do in fact know both σ_1 and σ_2. However, that occurs only very rarely in practice, so we will not discuss the z test.

One further note: Our discussion of $s_{\overline{X}_1 - \overline{X}_2}$ in the next section will assume that the variance in the first population equals the variance in the second population (that is, $\sigma_1^2 = \sigma_2^2$). Because Equation (11.2) includes $s_{\overline{X}_1 - \overline{X}_2}$, that equation also assumes that $\sigma_1^2 = \sigma_2^2$. There are forms of the t test that do not depend on this equal-variance assumption, but we leave them for more advanced texts. We do, however, make two observations: (1) The assumption that the population variances are identical (that is, $\sigma_1^2 = \sigma_2^2$) does *not* imply that the *sample* variances need to be identical (that is, it is quite likely that $s_1^2 \neq s_2^2$ even though $\sigma_1^2 = \sigma_2^2$), and (2) the t test is relatively *robust*, which means that violations of the equal-variance assumption do not have a large effect on the probability of rejecting the null hypothesis.

robust: insensitive to violations of the assumptions

11.3 Standard Error of the Difference Between Two Means

We begin our consideration of the standard error of the difference between two means by recalling from Chapter 7 the definition of the standard error of *one* mean. If we (1) take a random sample of size n from some population and compute its mean $\overline{X}$; then (2) take another sample of size n and compute its $\overline{X}$; then (3) find yet another sample's $\overline{X}$; (4) find another and another, and so on; and finally (5) compute the standard deviation of all of these $\overline{X}$'s, that standard deviation is the standard error of the mean $s_{\overline{X}}$. We also saw (in Chapter 8) that we generally compute $s_{\overline{X}}$ from *one* sample, rather than from a long series of samples, using Equation (8.2): $s_{\overline{X}} = s/\sqrt{n}$.

The definition of the standard error of the *difference* between two means follows an analogous sequence: We (1) take *two* random samples from the same population (or from identical populations where $\mu_1 = \mu_2$), one of size n_1 and the other of size n_2, compute the *two* means $\overline{X}_1$ and $\overline{X}_2$, and then compute the difference between those two sample means ($\overline{X}_1 - \overline{X}_2$); then (2) take another two samples of sizes n_1 and n_2 and compute the difference between their means ($\overline{X}_1 - \overline{X}_2$); then (3) find yet another difference between a pair of sample means ($\overline{X}_1 - \overline{X}_2$); (4) find another difference and another, and so on; and finally (5) compute the standard deviation of all of these differences—that is, of all the ($\overline{X}_1 - \overline{X}_2$) values. Then that standard deviation is the standard error of the difference between two means $s_{\overline{X}_1 - \overline{X}_2}$. Note that each step in this paragraph is the same as in the preceding paragraph except that here we are taking *two* means and computing their *difference* in each step.

ESTAT

You may wish to use the ESTAT exercise "diffm" at this time.

In general, just as was the case for $s_{\overline{X}}$, we compute $s_{\overline{X}_1 - \overline{X}_2}$ from one pair of samples, rather than from a long series of samples. The computational formula for the standard error of the difference between two means is

standard error of the difference between two means

$$s_{\overline{X}_1 - \overline{X}_2} = \sqrt{s^2_{\text{pooled}}\left(\frac{1}{n_1} + \frac{1}{n_2}\right)} \qquad (11.3)$$

where s^2_{pooled} is the *pooled estimate* of σ^2, or the *pooled variance*, and is described next.

Pooled Variance

pooled variance: the weighted average of two (or more) variances; the weights depend on the respective sample sizes

The concept of the *pooled variance* is quite straightforward when we recognize that to *pool* two variances (one from each sample) is simply to take the *weighted average* of those sample variances, with the weights being proportional to their degrees of freedom. If we take two random samples (of sizes n_1 and n_2) from the *same* population, the variance of the first sample, s_1^2, is a point-estimate of the population variance σ^2; similarly, the variance of the second sample, s_2^2, is a point-estimate of the *same* population variance σ^2. If we have two independent point-estimates of the same variance, then the *average* of these two point-estimates (s_{pooled}^2) must be an even better point-estimate of σ^2 than either of the two sample variances (s_1^2 or s_2^2) individually because the averaged (pooled) variance is based on more samples (on $n_1 + n_2$ samples instead of on either n_1 or n_2).

If the sample sizes are equal (that is, if $n_1 = n_2$), then the degrees of freedom are equal, so the pooled variance is the simple average of the two sample variances:

pooled variance when sample sizes are equal

$$s_{\text{pooled}}^2 = \frac{s_1^2 + s_2^2}{2} \quad \text{[only when } n_1 = n_2\text{]} \tag{11.4}$$

When the sample sizes are *not* equal, the pooled variance is obtained by weighting each variance by its associated degrees of freedom:

pooled variance for equal or unequal sample sizes

$$s_{\text{pooled}}^2 = \frac{df_1 s_1^2 + df_2 s_2^2}{df_1 + df_2} \quad \text{[for any values of } n_1 \text{ and } n_2\text{]} \tag{11.5}$$

Let's use the Pygmalion data to see why we should weight the sample variances by the degrees of freedom. We show that data again in Table 11.1 and provide summary statistics for those same data in Table 11.2. Thus, we see that s_1^2 for the 12 bloomers point-estimates the population variance σ^2. Similarly, s_2^2 for the 47 other children also point-estimates the same population variance σ^2. Which of these sample variances is a *better* point-estimate of σ^2? Clearly s_2^2 because the sample size is larger. How much better? About four times better because there are about four times more degrees of freedom in the other sample ($47 - 1 = 46$) than in the bloomer sample ($12 - 1 = 11$). Equation (11.5) makes it clear that we give about four times more weight to s_2^2 (we will multiply it by 46) than to s_1^2 (we will multiply it by only 11) when we pool the variances:

$$\begin{aligned} s_{\text{pooled}}^2 &= \frac{df_1 s_1^2 + df_2 s_2^2}{df_1 + df_2} \\ &= \frac{11 s_1^2 + 46 s_2^2}{11 + 46} \\ &= \frac{11 s_1^2 + 46 s_2^2}{57} \end{aligned}$$

TABLE 11.1 Intellectual growth (IQ gain) of Oak School second-grade bloomers and other children*

Bloomers	*Others*			
20	3	−3	−10	−4
14	−2	4	15	−3
69	−15	20	15	6
13	1	8	8	11
−6	31	−6	6	13
25	10	−6	30	18
1	5	6	−5	4
−4	9	10	17	0
24	1	11	7	26
12	12	13	4	−1
19	14	14	19	14
11	−11	7	3	

*Based on Rosenthal & Jacobson (1968)

TABLE 11.2 Summary statistics for intellectual growth

Bloomers	*Others*
$n_1 = 12$	$n_2 = 47$
$\sum X_1 = 198$	$\sum X_2 = 329$
$\bar{X}_1 = 16.5$	$\bar{X}_2 = 7.0$
$\sum X_1^2 = 7406$	$\sum X_2^2 = 6963$
$s_1^2 = 376.273$	$s_2^2 = 101.304$
$s_1 = 19.398$	$s_2 = 10.065$

Thus, the pooled variance for the Pygmalion data is

$$s^2_{\text{pooled}} = \frac{11(376.273) + 46(101.304)}{57}$$

$$= 154.368 \text{ (IQ points)}^2$$

Because $df_1 = n_1 - 1$ and $df_2 = n_2 - 1$, Equation (11.5) can be rewritten in the entirely equivalent form:

pooled variance for equal or unequal sample sizes

$$s^2_{\text{pooled}} = \frac{(n_1 - 1)s_1^2 + (n_2 - 1)s_2^2}{n_1 + n_2 - 2} \tag{11.5a}$$

Despite the complexity of its computation, the main concept of the pooled variance is quite simple and worth restating: The variance of the first sample s_1^2 is one point-estimate of the population variance σ^2; the variance of the second sample s_2^2 is another point-estimate of the same population variance σ^2; and the weighted average of the two sample variances s^2_{pooled} is a better point-estimate of σ^2 than is either s_1^2 or s_2^2.

Interpreting the Standard Error of the Difference Between Two Means

We have had to detour slightly to describe the pooled variance; now we return to finish our discussion of the standard error of the difference between two means, which was given by Equation (11.3):

$$s_{\bar{X}_1 - \bar{X}_2} = \sqrt{s^2_{\text{pooled}}\left(\frac{1}{n_1} + \frac{1}{n_2}\right)}$$

For the Pygmalion data, we saw that $s^2_{\text{pooled}} = 154.368$ (IQ points)^2, so the standard error of the difference between two means is

$$s_{\bar{X}_1 - \bar{X}_2} = \sqrt{s^2_{\text{pooled}}\left(\frac{1}{n_1} + \frac{1}{n_2}\right)}$$

$$= \sqrt{154.368\left(\frac{1}{12} + \frac{1}{47}\right)}$$

$$= \sqrt{154.368(.10461)} = 4.019 \text{ IQ points}$$

To explore the meaning of the standard error of the difference between two means, we can simplify Equation (11.3) when $n_1 = n_2 = n$:

standard error of the difference between two means when sample sizes are equal

$$s_{\bar{X}_1 - \bar{X}_2} = \sqrt{s^2_{\text{pooled}}\frac{2}{n}} \quad [\text{only when } n_1 = n_2 = n] \tag{11.6}$$

When $n_1 \neq n_2$, use Equation (11.3).

This form of the expression for the standard error of the difference between two means can be transformed by some simple algebra into the expression

standard error of the difference between two means when sample sizes are equal, alternative form

$$s_{\bar{X}_1 - \bar{X}_2} = \left(\frac{s_{\text{pooled}}}{\sqrt{n}}\right)\sqrt{2} \quad [\text{only when } n_1 = n_2 = n] \tag{11.6a}$$

where $s_{\text{pooled}} = \sqrt{s^2_{\text{pooled}}}$. Equations (11.6) and (11.6a) are identical except for the order in which the square root is taken.

Equation (11.6a) can give us some insight into the meaning of the standard error of the difference between two means. The first term on the right side of this expression, $s_{\text{pooled}}/\sqrt{n}$, should look familiar to you; it is the expression for the standard error of the mean for samples of size n that we explored in Chapter 7, except that here we have the subscript "pooled" to indicate that the standard deviation is the result of pooling. Thus, *if the sample sizes are the same*, then the standard error of the difference between two means ($s_{\bar{X}_1-\bar{X}_2}$) is simply $\sqrt{2}$ times greater than the standard error of the mean ($s_{\bar{X}}$).

11.4 Evaluating Hypotheses About Means of Two Independent Samples

We now return to where this chapter began: evaluating the hypothesis that the means of two independent populations are the same. In the Pygmalion in the classroom example, we wish to know whether the intellectual growth of the bloomers has the same mean as that of the other children. We will follow the same procedure used in the preceding two chapters.

Procedure

Three sketches illustrate H_0:
Variable
Sample statistic
Test statistic

Eyeball-estimation of this procedure is shown on page 241.

I. State the Null and Alternative Hypotheses

We are inquiring whether the means of the two groups are equal. If subscript 1 refers to the bloomers and subscript 2 refers to the other children, then the hypotheses are:

$$H_0: \ \mu_1 = \mu_2 \quad \text{or} \quad \mu_1 - \mu_2 = 0$$
$$H_1: \ \mu_1 \neq \mu_2 \quad \text{or} \quad \mu_1 - \mu_2 \neq 0 \qquad \text{(11.1) and (11.1a)}$$

We treat this as a nondirectional example because it is possible that higher expectations could lead to *lower* intellectual growth. Recall that in our general hypothesis-testing procedure, we sketch three distributions (the dependent variable, the sample statistic, and the test statistic). In this case, there are actually two distributions of the dependent variable, the distribution of X in each of the two groups. This pair of distributions is shown in Figure 11.5. The null hypothesis specifies that the means of these two distributions are *the same*, but it does not specify a *value* for these means. The result is that we can sketch the distributions of the intellectual growth for the bloomers and the other children precisely aligned vertically

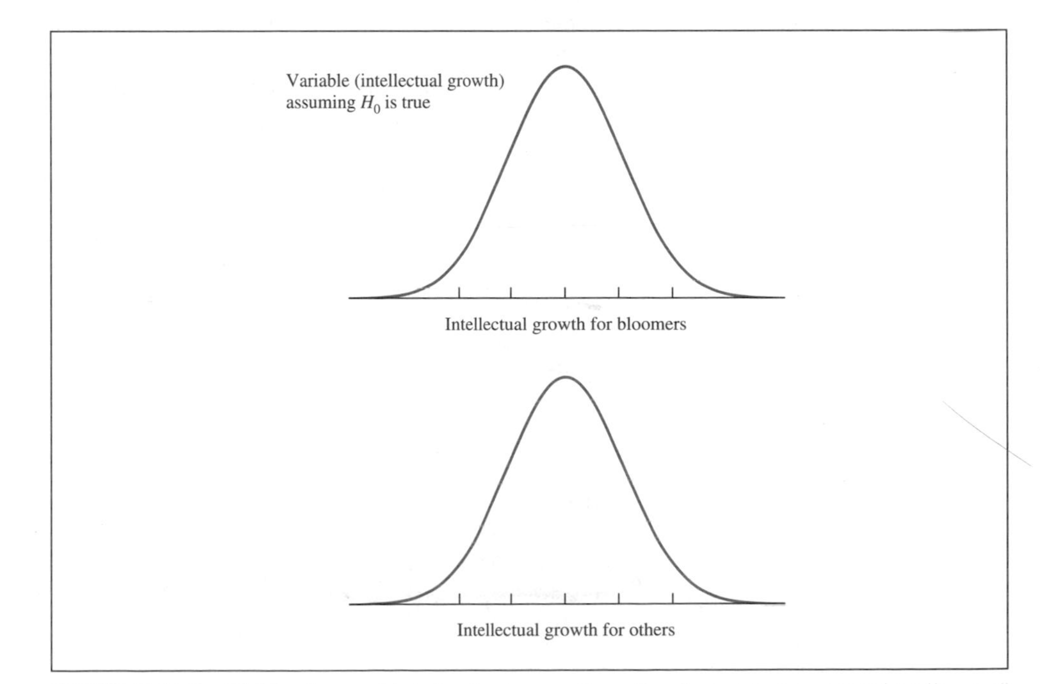

FIGURE 11.5 Distributions of intellectual growth in the two groups assuming the null hypothesis is true

(reflecting that the means are the same), but we cannot put values on the X-axis because the null hypothesis does not specify them.[1]

The second distribution we sketch is the distribution of the sample statistic, which in this case is the difference between the two sample means, $\overline{X}_1 - \overline{X}_2$. The null hypothesis *does* specify a value for the mean of this distribution—namely, $\mu_1 - \mu_2 = 0$. The standard deviation of this distribution is the standard error of the difference between two means, and because we do not know σ, we cannot know $\sigma_{\overline{X}_1-\overline{X}_2}$, so we must use the statistic that point-estimates it—that is, $s_{\overline{X}_1-\overline{X}_2} = 4.019$ IQ points, which we computed before. This distribution is shown on the upper axis of Figure 11.6.

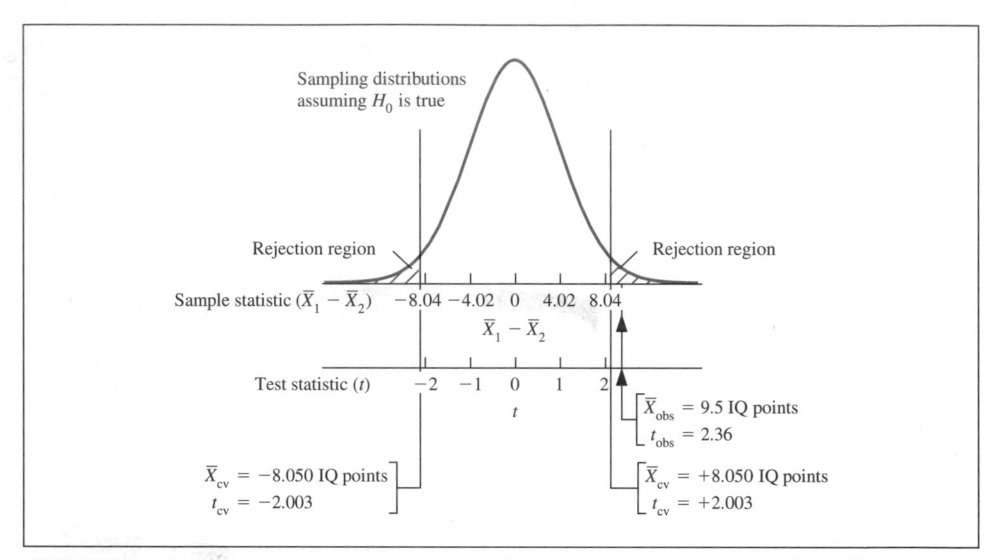

FIGURE 11.6 Distribution of the sample statistic and the test statistic with rejection regions and observed values indicated

Let us review what this distribution of the sample statistic tells us. If the null hypothesis is in fact true, then teacher expectations have no effect and we should expect the two groups to have sample means that are close to each other but not exactly equal because of sampling fluctuations. This distribution shows, for example, that approximately 68% of the time, the two means will lie within 4.019 IQ points of each other (in half of these, $\overline{X}_1$ will be larger than $\overline{X}_2$, and in the other half, smaller than $\overline{X}_2$).

The third distribution we sketch to illustrate our procedure is the distribution of the test statistic, which in this problem is t, shown on the lower axis of Figure 11.6. Note that a t value of 0 corresponds to a $(\overline{X}_1 - \overline{X}_2)$ value of 0, and a t value of 1 corresponds to a $(\overline{X}_1 - \overline{X}_2)$ value of 4.019 IQ points, one standard error of the difference between two means above 0.

II. Set the Criterion for Rejecting H_0

Level of significance?
Directional?
Shade rejection regions.

The appropriate test statistic to test this hypothesis, as we saw previously, is given by Equation (11.2):

$$t = \frac{\overline{X}_1 - \overline{X}_2}{s_{\overline{X}_1-\overline{X}_2}} \qquad [df = n_1 + n_2 - 2]$$

What are the critical values of t? We need to know the number of degrees of freedom, which we can see is $n_1 - 1 = 12 - 1 = 11$ from the first sample and $n_2 - 1 = 47 - 1 = 46$ from the second sample, for a total of $11 + 46 = 57$. We could also compute df from the formula $df = n_1 + n_2 - 2 = 12 + 47 - 2 = 57$.

[1]We have sketched these distributions with identical standard deviations because we have been assuming that $\sigma_1 = \sigma_2$.

We assume, as we usually do unless we specify otherwise, that $\alpha = .05$, and note that the test is nondirectional because we do not have any particular reason to believe that μ_1 is greater or less than μ_2. We look up the critical value of t. Table A.2 in Appendix A shows that with 57 degrees of freedom, $t_{\text{cv}_{\alpha=.05}} = \pm 2.003$. If the observed value of the test statistic t_{obs} lies outside that range, we will reject the null hypothesis. We show those critical values and the rejection regions of the test statistic on the lower axis in Figure 11.6.

We also wish to show the rejection regions on the distribution of the sample statistic $\overline{X}_1 - \overline{X}_2$. Those regions leave 2.5% in each tail, and the critical values are $\pm t_{\text{cv}}(s_{\overline{X}_1 - \overline{X}_2}) = \pm 2.003(4.019) = \pm 8.050$ IQ points, which are shown on the upper axis in Figure 11.6. If the null hypothesis in this problem is true, the two sample means will lie within 8.050 IQ points of each other 95% of the time; if the sample means are more than 8.050 IQ points apart, we will reject H_0.

The observed values and the critical values must always align vertically if the sketches are drawn accurately.

III. Collect a Sample and Compute the Observed Values of the Sample Statistic and the Test Statistic

Observed values: Sample statistic Test statistic

The sample data are shown in Table 11.1. The sample statistic is $\overline{X}_1 - \overline{X}_2$. We compute the sample statistic and the test statistic as follows: Using the values provided in Table 11.2, we find that the observed value of the sample statistic $(\overline{X}_1 - \overline{X}_2)_{\text{obs}} = 16.5 - 7.0 = 9.5$ IQ points. We show that value as $(\overline{X}_1 - \overline{X}_2)_{\text{obs}}$ on the upper axis of Figure 11.6.

By Equation (11.2), the statistic $t_{\text{obs}} = (\overline{X}_1 - \overline{X}_2)_{\text{obs}} / s_{\overline{X}_1 - \overline{X}_2} = 9.5/4.019 = 2.36$. We show that value on the lower axis in Figure 11.6 and note that it lies directly underneath the observed value in the distribution of the sample statistic.

IV. Interpret the Results

Statistically significant? Practically significant? State results clearly.

Because the observed value $t_{\text{obs}} = 2.36$ exceeds the critical value $t_{\text{cv}} = \pm 2.003$ [or, equivalently, because $(\overline{X}_1 - \overline{X}_2)_{\text{obs}} = 9.5$ IQ points exceeds $(\overline{X}_1 - \overline{X}_2)_{\text{cv}} = \pm 8.050$ IQ points], we do reject the null hypothesis. We conclude that there is a statistically significant difference between the children expected to bloom and the other children.

Because we found a statistically significant difference, we must consider whether this difference is practically significant. We have rejected H_0, so we no longer believe that $\mu_1 = \mu_2$, but we do not know the actual magnitude of either μ_1 or μ_2. Our best point-estimate of μ_1 is now $\overline{X}_1 = 16.5$ IQ points; similarly, our best point-estimate of μ_2 is $\overline{X}_2 = 7.0$ IQ points. We illustrate our new best estimates of μ_1 and μ_2 in Figure 11.7.

Figure 11.7 makes three assumptions: (1) The two point-estimates are true, (2) the distribution of intellectual growth is normal in both groups, and (3) the standard deviation of intellectual growth is the same in both groups (that is, $\sigma_1 = \sigma_2 = \sigma$), which implies that the best point-estimate that we have of σ is s_{pooled}. We computed the pooled variance $s^2_{\text{pooled}} = 154.368$ earlier. Therefore, $s_{\text{pooled}} = \sqrt{s^2_{\text{pooled}}} = \sqrt{154.368} = 12.425$. We should note that

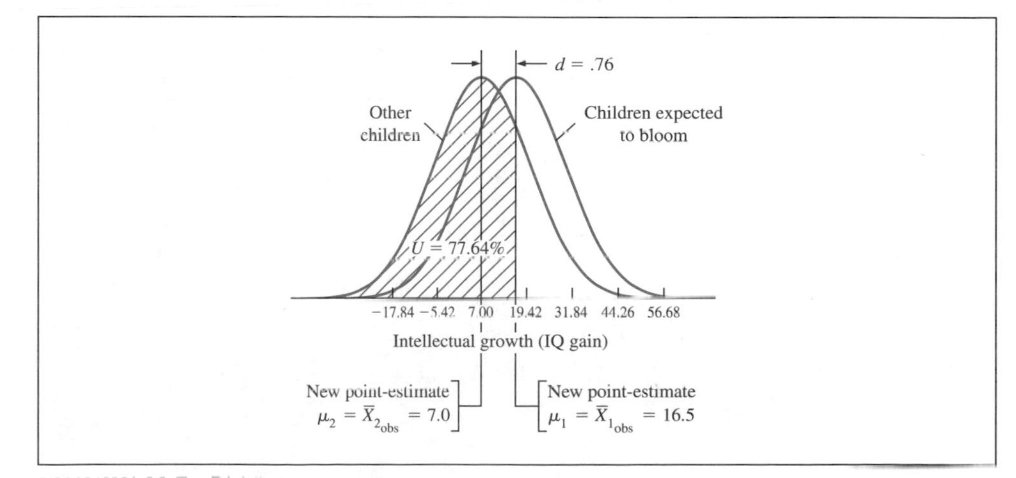

FIGURE 11.7 Distributions of the variable (intellectual growth) for two populations, assuming the best point-estimates of μ_1 and μ_2 are true

none of those assumptions is probably exactly true, but they are the best assumptions that we can make in most circumstances.

We compute the effect size index d for the two-independent-samples test of this chapter from

effect size index d, two-independent-samples test for means

$$d = \frac{|(\overline{X}_1 - \overline{X}_2)_{obs}|}{s_{pooled}} \quad \text{or} \quad d = \frac{(\overline{X}_1 - \overline{X}_2)_{obs}}{s_{pooled}} \tag{11.7}$$

(nondirectional test) (directional test)

where s_{pooled} is the pooled standard deviation. Our Pygmalion in the classroom example is nondirectional, so

$$\begin{aligned} d &= \frac{|(\overline{X}_1 - \overline{X}_2)_{obs}|}{s_{pooled}} \\ &= \frac{|16.5 - 7.0|}{12.425} \\ &= .76 \end{aligned}$$

which means that the mean intellectual growth of the bloomers is .76 standard deviation above the mean of the other children. We show that value on Figure 11.7.

We shade the region of nonoverlap on Figure 11.7 and compute Cohen's nonoverlap measure U by entering Table A.1 with $z = d = .76$ and finding that $U = .5 + .2764$, or 77.64%.

Now we can state our results in plain English: The mean intellectual growth of the bloomers ($\overline{X}_1 = 16.5$ IQ points) was statistically significantly greater than that of the other children ($\overline{X}_2 = 7.0$ IQ points), a difference of .76 standard deviation. In fact, the average intellectual growth of the child expected to bloom can be expected to be greater than 77.64% of the children whose teachers have no particular expectations.

A Directional (One-tailed) Example

Let us consider another example, this time a directional test. Suppose we know that High Brow University has a more selective admissions policy than does Simple University, and we wish to know whether that selectivity affects the quality of students who apply to the two schools. We decide to measure the quality of students by using their Graduate Record Examination (GRE) scores, so our research question is whether HBU applicants have higher GRE scores than do SU applicants. We take a random sample of 15 applicants from HBU and 10 applicants from SU. The data are shown in Table 11.3.

TABLE 11.3 GRE scores from samples of applicants

HBU	*SU*		*HBU*	*SU*
500	580	n	15	10
440	550	$\sum X$	8,540	4,970
390	510	$\sum X^2$	5,013,600	2,527,700
570	520	$\overline{X}$	569.33	497.00
520	570	s	104.02	80.01
800	430	s^2	10,820.96	6,041.11
540	440		$s^2_{pooled} = 9,091.45$	
640	500		$s_{\overline{X}_1-\overline{X}_2} = 38.93$	
630	550			
600	320			
680				
500				
490				
660				
580				

I. State the Null and Alternative Hypotheses

Three sketches illustrate H_0:
Dependent variable
Sample statistic
Test statistic

Eyeball estimation of this procedure is shown on page 242.

Because we have reason to predict that HBU scores will be higher than SU scores, we use a directional test, and the alternative hypothesis should reflect this prediction. Therefore, we use the "greater than" symbol and write the alternative hypothesis as H_1: $\mu_{HBU} > \mu_{SU}$ (or, equivalently, H_1: $\mu_{HBU} - \mu_{SU} > 0$). The null hypothesis in this directional study must include the case ($\mu_{HBU} = \mu_{SU}$) where the selectivity of admissions policies has no effect and also the unlikely cases where the selectivity effect is in the direction opposite from what we have predicted ($\mu_{HBU} < \mu_{SU}$). We combine those possibilities to form the single null hypothesis H_0: $\mu_{HBU} \leq \mu_{SU}$ (or, equivalently, H_0: $\mu_{HBU} - \mu_{SU} \leq 0$) because we have no desire to distinguish between $\mu_{HBU} = \mu_{SU}$ and $\mu_{HBU} < \mu_{SU}$; our interest here is only whether selectivity significantly *increases* GRE scores. Thus, these are our hypotheses:

$$H_0: \mu_{HBU} \leq \mu_{SU} \quad \text{or equivalently} \quad H_0: \mu_{HBU} - \mu_{SU} \leq 0$$
$$H_1: \mu_{HBU} > \mu_{SU} \quad \text{or equivalently} \quad H_1: \mu_{HBU} - \mu_{SU} > 0$$

We show the distributions of the two variables in Figure 11.8. Note that we cannot label the numeric values on the axes because we are not specifying values of μ_{HBU} or μ_{SU} in the null hypothesis, only that they are equal.

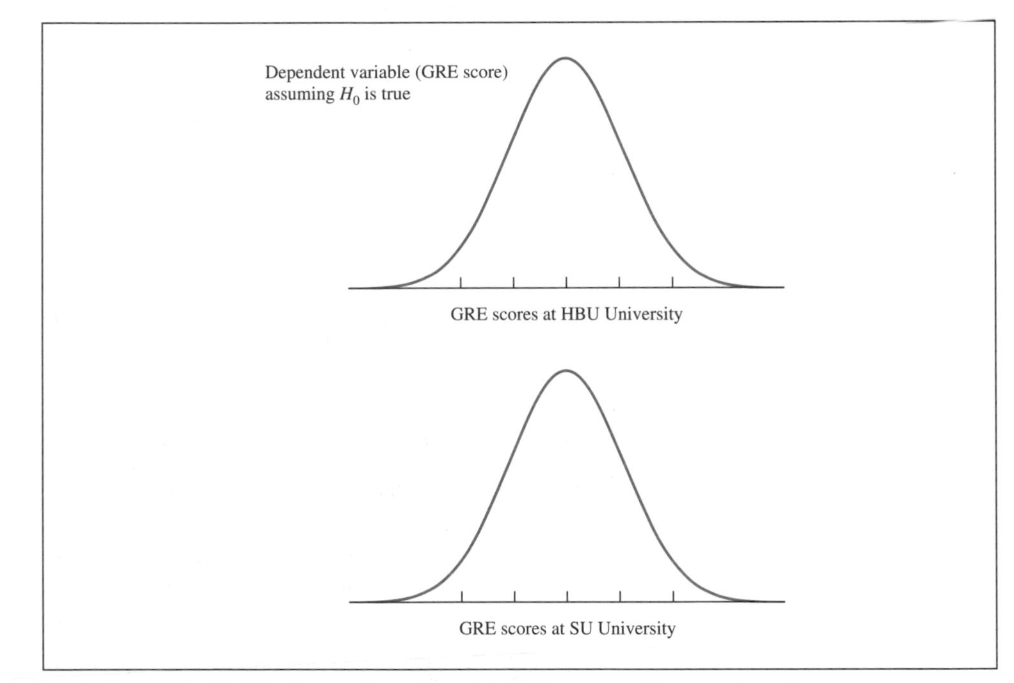

FIGURE 11.8 Distributions of applicants' GRE scores at two universities

Assuming that HBU is group 1 and SU is group 2, we show the distributions of the sample statistic $(\overline{X}_1 - \overline{X}_2)$ and the test statistic t in Figure 11.9. To put numeric values on the axis of $(\overline{X}_1 - \overline{X}_2)$, we must determine $s_{\overline{X}_1 - \overline{X}_2}$, which we compute as follows: From Table 11.3, $s_1 = 104.02$ points and $s_2 = 80.01$ points, so Equation (11.5a) gives

$$s^2_{\text{pooled}} = \frac{(n_1 - 1)s_1^2 + (n_2 - 1)s_2^2}{n_1 + n_2 - 2}$$

$$= \frac{(15 - 1)10{,}820.96 + (10 - 1)6401.11}{15 + 10 - 2}$$

$$= 9091.45 \text{ points}^2$$

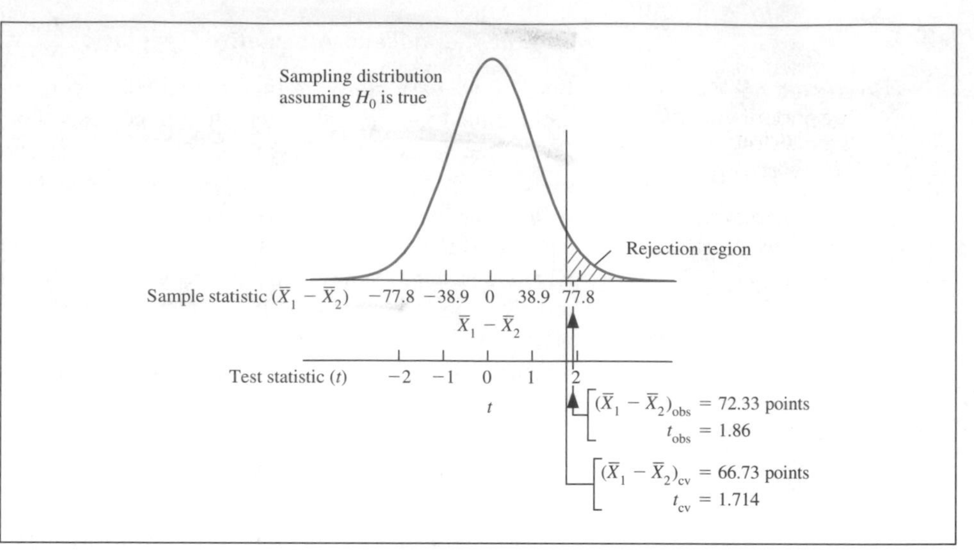

FIGURE 11.9 Distributions of the sample statistic and the test statistic with rejection regions and observed values indicated

Equation (11.3) then gives

$$s_{\bar{X}_1 - \bar{X}_2} = \sqrt{s^2_{\text{pooled}}\left(\frac{1}{n_1} + \frac{1}{n_2}\right)}$$

$$= \sqrt{9091.45\left(\frac{1}{15} + \frac{1}{10}\right)}$$

$$= 38.93 \text{ points}$$

Our sketch of the distribution of $(\bar{X}_1 - \bar{X}_2)$ thus has center 0 (as specified by the null hypothesis) and standard error 38.93 points.

II. Set the Criterion for Rejecting H_0

Level of significance?
Directional?
Shade rejection regions.

The test statistic is t as shown in Equation (11.2), with $df = n_1 + n_2 - 2 = 23$. If $\alpha = .05$, Table A.2 in Appendix A shows the directional critical value to be $t_{cv} = 1.714$. We show that value and its corresponding rejection region on the lower axis in Figure 11.9.

We also wish to show the rejection region on the distribution of the sample statistic $(\bar{X}_1 - \bar{X}_2)$. By computation,

$$(\bar{X}_1 - \bar{X}_2)_{cv} = t_{cv}(s_{\bar{X}_1 - \bar{X}_2}) = 1.714(38.93) = 66.73 \text{ points}$$

We show that value and its corresponding rejection region on the upper axis in Figure 11.9.

III. Collect a Sample and Compute the Observed Values of the Sample Statistic and the Test Statistic

Observed values:
Sample statistic
Test statistic

The sample data are shown in Table 11.3. The observed value of the difference between the two means is computed from the data in Table 11.3: $\bar{X}_1 = 569.33$, $\bar{X}_2 = 497.00$, and $(\bar{X}_1 - \bar{X}_2)_{obs} = 72.33$ points. We show this value on the upper axis in Figure 11.9.

We compute t_{obs} from Equation (11.2):

$$t_{obs} = \frac{(\bar{X}_1 - \bar{X}_2)_{obs}}{s_{\bar{X}_1 - \bar{X}_2}} = \frac{72.33}{38.9} = 1.86$$

We show this value on the lower axis in Figure 11.9.

IV. Interpret the Results

Statistically significant?
Practically significant?
State results clearly.

Because the observed value $t_{obs} = 1.86$ exceeds the critical value $t_{cv} = 1.714$ [or, entirely equivalently, because $(\overline{X}_1 - \overline{X}_2)_{obs} = 72.33$ points exceeds $(\overline{X}_1 - \overline{X}_2)_{cv} = 66.73$ points], we do reject the null hypothesis. We conclude that there is a statistically significant difference between the GRE scores of students at High Brow University and Simple University.

Because we found a statistically significant difference, we must consider whether this difference is practically significant. We have rejected H_0, so we no longer believe that $\mu_1 = \mu_2$, but we do not know the actual magnitude of either μ_1 or μ_2. Our best point-estimate of μ_1 is now $\overline{X}_1 = 569.33$ points; similarly, our best point-estimate of μ_2 is $\overline{X}_2 = 497.00$ points. We illustrate our new best estimates of μ_1 and μ_2 in Figure 11.10.

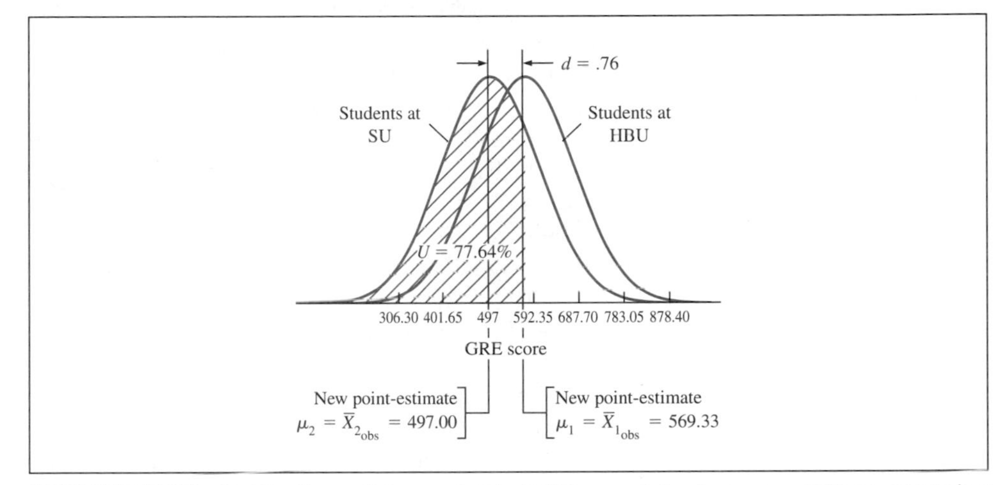

FIGURE 11.10 Distributions of the variable (GRE score) for two populations, assuming the best point-estimates of μ_1 and μ_2 are true

We compute the effect size index d for the two-independent-samples test of this chapter from Equation (11.7). Our example is directional, so

$$d = \frac{(\overline{X}_1 - \overline{X}_2)_{obs}}{s_{pooled}} = \frac{569.33 - 497.00}{95.35} = .76$$

which means that the mean GRE score of the HBU students is .76 standard deviation above the mean of the SU students. We show that value in Figure 11.10.

We shade the region of nonoverlap in Figure 11.10 and compute Cohen's nonoverlap measure U by entering Table A.1 with $z = d = .76$ and finding that $U = .5 + .2764$, or 77.64%.

Now we can state our results in plain English: The mean GRE score of the HBU students ($\overline{X}_1 = 569.33$) was statistically significantly greater than that of SU students ($\overline{X}_2 = 497.00$), a difference of .76 standard deviation. In fact, the average HBU student can be expected to have a GRE score greater than 77.64% of the students at SU.

11.5 Practical Significance Versus Statistical Significance Revisited

When we interpret the results in our hypothesis-evaluation procedure, we have been inquiring about practical significance as well as statistical significance. To measure practical significance, we have been computing d (and its related statistic U). To measure statistical significance, we have been computing t_{obs} (and asking whether t_{obs} is greater than t_{cv}).

The alert reader is perhaps wondering why we compute both d and t_{obs} when they are so similar. For example, when the sample sizes are equal ($n_1 = n_2 = n$) and the test is directional, Equation (11.7) gives d and Equation (11.2) gives t_{obs}:

$$d = \frac{(\overline{X}_1 - \overline{X}_2)_{obs}}{s_{pooled}} \quad \text{and} \quad t_{obs} = \frac{(\overline{X}_1 - \overline{X}_2)_{obs}}{s_{\overline{X}_1 - \overline{X}_2}}$$

Don't those similar equations measure approximately the same thing? No, they do not: d measures a "state of nature," whereas t_{obs} measures a "state of the experiment." Let's use the Pygmalion example to explore this distinction.

In that experiment, μ_1 is the mean of the population of all children whose teachers expect them to bloom, and μ_2 is the mean of all children whose teachers have no particular expectations. Therefore, children who are expected to bloom will show, on average, ($\mu_1 - \mu_2$) points more of an increase in IQ than will children whose teachers have no expectations. We do not have direct control over ($\mu_1 - \mu_2$): If teachers' expectations are important, ($\mu_1 - \mu_2$) will be large, whereas if teachers' expectations have little effect, ($\mu_1 - \mu_2$) will be small. Thus, we call ($\mu_1 - \mu_2$) a "state of nature"—a characteristic of the world over which our experiment has no control. When we find in the Pygmalion experiment that $(\overline{X}_1 - \overline{X}_2)_{obs} = 9.5$ IQ points, that implies that our best measure (that is, best point-estimate) of the state of nature ($\mu_1 - \mu_2$) is 9.5 IQ points. If we were to replicate the Pygmalion experiment with ten times more subjects, we would still expect the new $(\overline{X}_1 - \overline{X}_2)_{obs}$ to be about 9.5 (it could, of course, be more or less because of sampling fluctuations).

Furthermore, in the Pygmalion example, σ is the standard deviation of students' IQ gain, another state of nature over which our experiment has no control: Some students gain IQ points, other students lose IQ points, and σ is the standard deviation of all such gains or losses. Here, $s_{pooled} = 12.42$ IQ points is our best measure (that is, best point-estimate) of σ. If we replicate the Pygmalion experiment with ten times more subjects, we would still expect the new s_{pooled} to be about 12.42 (more or less because of sampling fluctuations).

Because both $(\overline{X}_1 - \overline{X}_2)_{obs}$ and s_{pooled} measure states of nature, their quotient d also measures a state of nature. d measures the effectiveness of teacher expectations ($\mu_1 - \mu_2$) in units of the standard deviation (σ), neither of which our experiment controls. If we replicate the Pygmalion experiment with ten times more subjects, we would still expect the new d to be about $9.5/12.42 = .76$ (more or less because of sampling fluctuations).

By contrast, the denominator of t_{obs}, $s_{\overline{X}_1 - \overline{X}_2}$, is *not* a measure of a state of nature over which our experiment has no control: $s_{\overline{X}_1 - \overline{X}_2}$ depends directly on the sample size of the experiment. Equation (11.6a) shows that when $n_1 = n_2 = n$, $s_{\overline{X}_1 - \overline{X}_2} = (s_{pooled}/\sqrt{n})\sqrt{2}$, and a similar (but less obvious) relationship holds when $n_1 \neq n_2$. Thus, if we replicate our Pygmalion experiment with ten times more subjects, we would expect the new $s_{\overline{X}_1 - \overline{X}_2}$ to be substantially *smaller*—smaller by about a factor of $\sqrt{10}$ (more or less because of sampling fluctuations). Because $s_{\overline{X}_1 - \overline{X}_2}$ is the denominator of t_{obs}, if we were to replicate the Pygmalion experiment with ten times more subjects, we would expect the new t_{obs} to be *larger*—larger by a factor of about $\sqrt{10}$ (more or less because of sampling fluctuations). Thus, t_{obs} does not measure a state of *nature* but instead measures a state of the *experiment*.

To recapitulate: d should be thought of as measuring a state of *nature*—how effective teacher expectations are (in units of the standard deviation), whereas t_{obs} should be thought of as measuring a characteristic of our *experiment*—how likely such a result is due to chance fluctuations if the null hypothesis is true. Both d and t_{obs} are important but for different reasons. If d is small, then teacher expectations are unimportant (regardless of whether the results are statistically significant). But if t_{obs} is small, then the results may be due simply to chance alone (regardless of the magnitude of our estimate of the effect size).

11.6 Two-sample *t* Test Eyeball-calibration

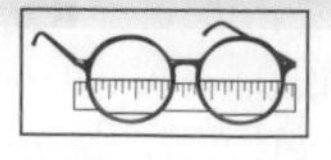

As in Chapter 10, we now present a series of pairs of histograms so that you can practice visualizing two-sample t tests. In each case, assume that the null hypothesis is H_0: $\mu_1 = \mu_2$, that the hypothesis is nondirectional (two-tailed), and that the level of significance $\alpha = .05$.

The optional section that follows the exercises at the end of this chapter describes the step-by-step procedure for eyeball-estimating two-sample t tests. Here our task is simply to get a sense of whether $\overline{X}_{1_{obs}}$ and $\overline{X}_{2_{obs}}$ are "far apart" as eyeball-estimated from the histograms.

The question to be answered in these exercises is whether it is reasonable to suppose that the points shown in the two histograms are random samples from two populations that have the same population mean.

The eyeball-calibration procedure is as follows: Eyeball-estimate the distance between $\overline{X}_{1_{obs}}$ (the balancing point of the first histogram) and $\overline{X}_{2_{obs}}$ (the balancing point of the second histogram). If your two estimates are *quite far apart*, then you should reject the null hypothesis.

How far is "quite far apart"? The larger the sample size and the smaller the standard deviation, the smaller the required distance between $\overline{X}_{1_{obs}}$ and $\overline{X}_{2_{obs}}$.

For example, Figure 11.11 shows two histograms. Should we reject H_0: $\mu_1 = \mu_2$? The two means seem rather far apart; the group 2 histogram seems to come from a population whose mean μ_2 is larger than μ_1. Therefore, it seems that we should reject H_0. If we performed the calculations for these data, we would find that we are correct: H_0 should in fact be rejected. (In case you wish to follow the calculations, they are summarized in Table 11.4.)

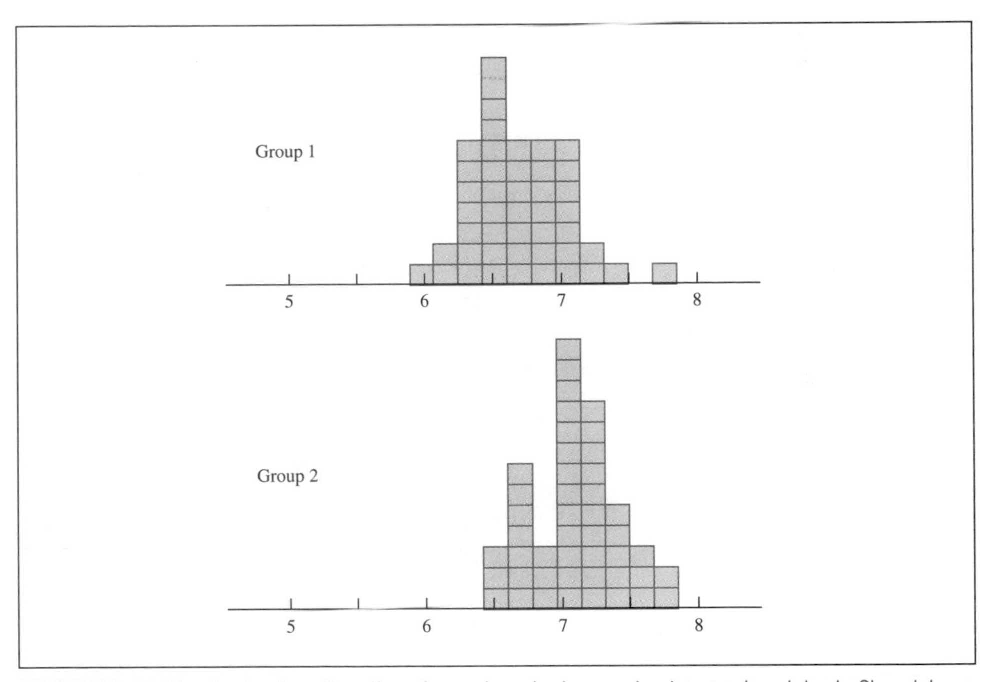

FIGURE 11.11 Eyeball-calibration for a two-independent-samples t test: Should we reject H_0?

TABLE 11.4
Statistics for the two histograms shown in Figure 11.11

Reject H_0?	Yes
$\overline{X}_{1_{obs}}$	6.75
$\overline{X}_{2_{obs}}$	7.10
$(\overline{X}_1 - \overline{X}_2)_{obs}$	−.35
s_{pooled}	.31
$s_{\overline{X}_1 - \overline{X}_2}$	.06
t_{obs}	−5.83
t_{cv}	1.99

This is simply an eyeball-calibration exercise: Make an educated guess about whether to reject H_0 and use the feedback from the actual answers to improve your understanding of how far "quite far apart" actually is. If you wish to eyeball-estimate t step by step, see the optional section.

Figures 11.12 through 11.17 present a series of pairs of histograms for you to calibrate your eyeball. *Cover up the actual values on the right* and determine by eyeball whether to reject H_0 for each example shown. Then check to see whether the null hypothesis would be rejected by computation. If your eyeball disagrees with the computation, try to improve your understanding of how far "quite far apart" actually is. (The actual values are also provided in case you wish to follow along with the computations.)

You may wish to use the ESTAT exercise "ttest2" at this time. Follow the path labeled "Eyeball-calibration" or "Shortcut."

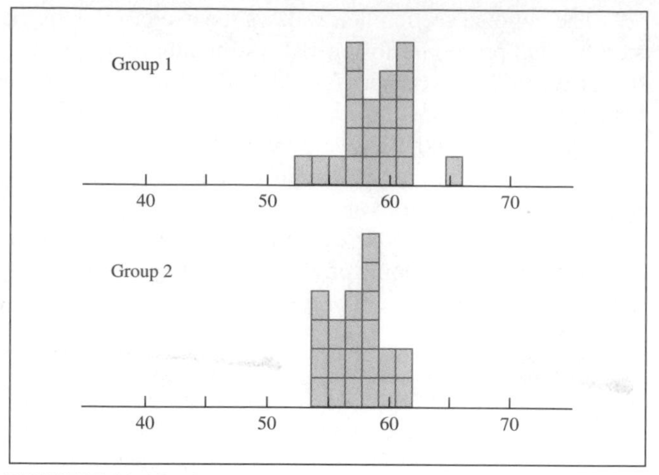

FIGURE 11.12 Should we reject H_0: $\mu_1 = \mu_2$?

Actual values	
Reject H_0?	No
$\bar{X}_{1_{obs}}$	59.10
$\bar{X}_{2_{obs}}$	57.99
$(\bar{X}_1 - \bar{X}_2)_{obs}$	1.11
s_{pooled}	2.26
$s_{\bar{X}_1 - \bar{X}_2}$	.70
t_{obs}	1.59
t_{cv}	2.03

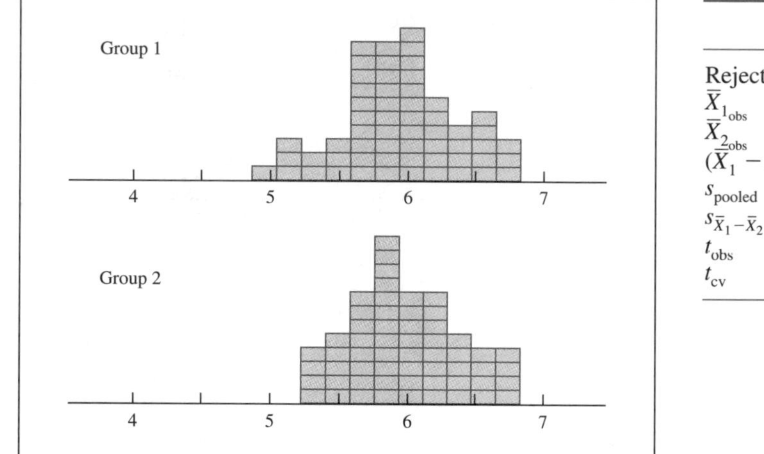

FIGURE 11.13 Should we reject H_0: $\mu_1 = \mu_2$?

Actual values	
Reject H_0?	No
$\bar{X}_{1_{obs}}$	5.99
$\bar{X}_{2_{obs}}$	6.03
$(\bar{X}_1 - \bar{X}_2)_{obs}$	−.04
s_{pooled}	.37
$s_{\bar{X}_1 - \bar{X}_2}$	.07
t_{obs}	−.57
t_{cv}	1.97

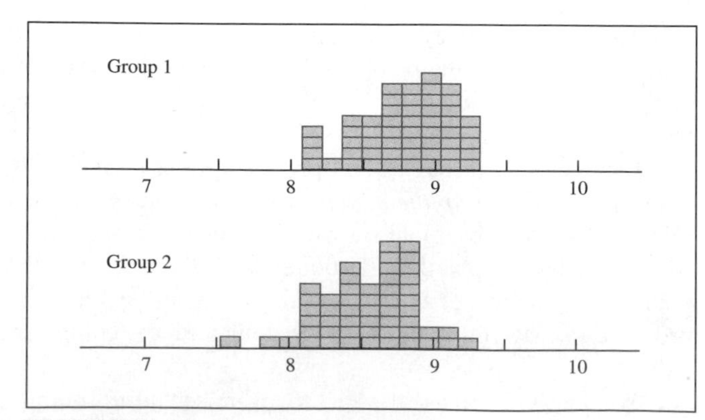

FIGURE 11.14 Should we reject H_0: $\mu_1 = \mu_2$?

Actual values	
Reject H_0?	Yes
$\bar{X}_{1_{obs}}$	8.85
$\bar{X}_{2_{obs}}$	8.64
$(\bar{X}_1 - \bar{X}_2)_{obs}$	.21
s_{pooled}	.30
$s_{\bar{X}_1 - \bar{X}_2}$	.06
t_{obs}	3.50
t_{cv}	1.98

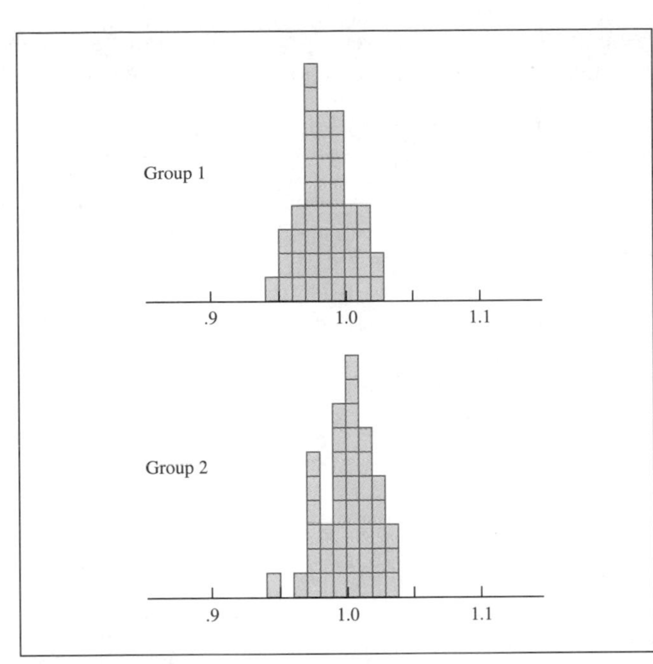

Actual values	
Reject H_0?	Yes
$\overline{X}_{1_{obs}}$	.99
$\overline{X}_{2_{obs}}$	1.01
$(\overline{X}_1 - \overline{X}_2)_{obs}$	−.02
s_{pooled}	.018
$s_{\overline{X}_1 - \overline{X}_2}$	.004
t_{obs}	−5.00
t_{cv}	1.991

FIGURE 11.15 Should we reject H_0: $\mu_1 = \mu_2$?

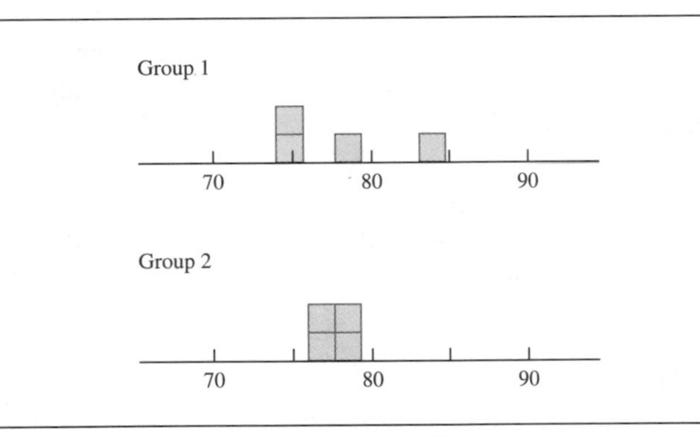

Actual values	
Reject H_0?	No
$\overline{X}_{1_{obs}}$	78.69
$\overline{X}_{2_{obs}}$	78.02
$(\overline{X}_1 - \overline{X}_2)_{obs}$	.67
s_{pooled}	2.96
$s_{\overline{X}_1 - \overline{X}_2}$	2.09
t_{obs}	.32
t_{cv}	2.447

FIGURE 11.16 Should we reject H_0: $\mu_1 = \mu_2$?

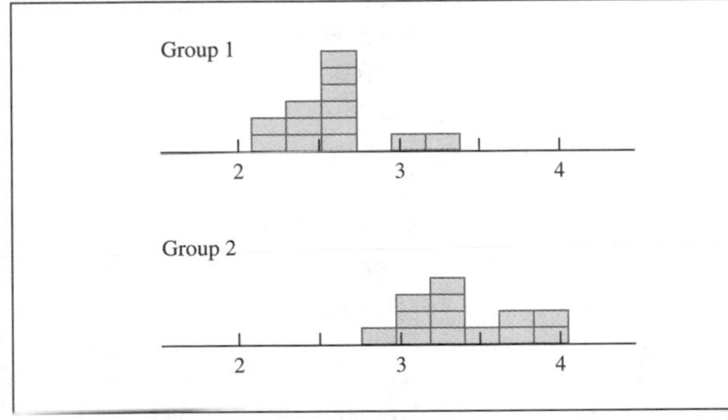

Actual values	
Reject H_0?	Yes
$\overline{X}_{1_{obs}}$	2.68
$\overline{X}_{2_{obs}}$	3.37
$(\overline{X}_1 - \overline{X}_2)_{obs}$	−.69
s_{pooled}	.30
$s_{\overline{X}_1 - \overline{X}_2}$	.12
t_{obs}	−5.75
t_{cv}	2.06

FIGURE 11.17 Should we reject H_0: $\mu_1 = \mu_2$?

11.7 Connections

Cumulative Review

We continue the discrimination practice begun in Chapter 10. You may recall the two steps:

Step 1: Determine whether the problem asks for (a) the area under a normal distribution, (b) a confidence interval, or (c) the evaluation of a hypothesis. See the Cumulative Review in Section 10.5 for hints about this step.

Step 2: If the problem asks for the evaluation of a hypothesis, consult the summary of hypothesis testing in Table 11.5 to determine which kind of test is appropriate.

TABLE 11.5 Summary of hypothesis-testing procedures in Chapters 10 and 11

Design	*Chapter*	*Null hypothesis*	*Sample statistic*	*Test statistic*	*Effect size index*
One sample					
σ known	10	$\mu = a$	$\overline{X}$	z	$d = \frac{\overline{X}_{obs} - \mu}{\sigma}$
σ unknown	10	$\mu = a$	$\overline{X}$	t	$d = \frac{\overline{X}_{obs} - \mu}{s}$
Proportion	Exercise 7, Chapter 10	$\Pi = a$	p	t	—[a]
Two independent samples					
For means	11	$\mu_1 = \mu_2$	$\overline{X}_1 - \overline{X}_2$	t	$d = \frac{\overline{X}_{1_{obs}} - \overline{X}_{2_{obs}}}{s_{pooled}}$

[a]Effect size index is not discussed in this textbook.

Table 11.5 shows that so far we have considered two basic kinds of hypothesis tests (the one-sample test for the mean and the two-independent-samples test for means). Note that this table is identical to Table 10.4, with the addition of the last line appropriate for this chapter. We will continue to expand this table progressively throughout the remainder of the textbook.

This would be a good place to review the general formula for the test statistic. Note that for each of the tests we have performed so far, the test statistic is the sample statistic minus the population parameter divided by the standard error of the sample statistic.

Journals

As you may recall, we found that the bloomers in our example about Pygmalion in the classroom had statistically significantly higher intellectual growth than did the other children. We reached that conclusion because at the $\alpha = .05$ level of significance, when $df = 57$, the critical value of t is $t_{cv} = \pm 2.003$. Because the observed test statistic $t_{obs} = 2.36$ exceeded the critical value, we rejected H_0.

When researchers report such results in professional journals, they frequently use M and SD in place of $\overline{X}$ and s, and put the degrees of freedom in parentheses. Thus, a journal report of the Pygmalion study might read, "Bloomers showed greater intellectual growth than did other children (bloomers $M = 16.5$ IQ points, $SD = 19.40$ IQ points; others $M = 7$ IQ points, $SD = 10.07$ IQ points; $t(57) = 2.36$, $p < .05$)."

Sometimes the analysis of variance is used to report the results of independent-sample tests. As we will see in Chapter 13, the analysis of variance of two-sample data is identical to the t test of two-sample data where the observed value of the analysis of variance (F_{obs}) is simply the *square* of t_{obs}. Then "$t(57) = 2.36$, $p < .05$" is (entirely equivalently) written as "$F(1, 57) = 5.57$, $p < .05$." Here "F with one and fifty-seven degrees of freedom" is the *square* of "t with 57 degrees of freedom" and $5.57 = (2.36)^2$. We will discuss this further in Chapter 13.

Sometimes researchers report the so-called exact probability, writing for the Pygmalion data "$p = .022$" instead of "$p < .05$." This can be interpreted as meaning that had the level of significance been set at any level .022 or greater (including .05), the null hypothesis would be rejected.

Occasionally if results are reported in a table, a single asterisk (*) is used to indicate that results are statistically significant when $\alpha = .05$. Two asterisks (**) are sometimes used if the results are significant at $\alpha = .01$, and three asterisks (***) are used if the results are significant at $\alpha = .001$. For example, imagine that the children in the Pygmalion study were measured on three different variables: intellectual growth, variable two, and variable three. It need not concern us what variables two and three actually are. The results could be reported as in Table 11.6. The table indicates that there is a significant difference between bloomers and other children in intellectual growth (at $\alpha = .05$) and in variable two (at $\alpha = .01$), but not in variable three at either level of significance.

More and more journal editors are also requiring that effect sizes be explicitly reported and practical significance be discussed for all statistically significant results. This in my view is a positive development in the field.

TABLE 11.6 Characteristics of bloomers and other children[a]

	Bloomers		*Other Children*		
	M	*SD*	*M*	*SD*	*t*
Intellectual growth	16.5	19.40	7.0	10.07	2.36*
Variable Two	10.3	4.51	5.3	3.97	3.79**
Variable Three	27.1	23.21	24.2	19.43	.44

$*p < .05$; $**p < .01$; $df = 59$
[a]Based on Rosenthal & Jacobson (1968)

Computers

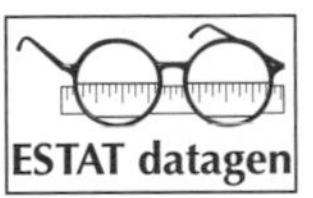

Use ESTAT's datagen to compute the t test for the GRE scores from High Brow University and Simple University. The data are shown in Table 11.3, with H_0: $\mu_1 = \mu_2$.

1. Enter the 15 values from the HBU column of Table 11.3 into the first column of the datagen spreadsheet.
2. Enter the 10 values from the SU column of Table 11.3 into the second column of the datagen spreadsheet.
3. Datagen automatically computes the two-sample t test, shown in the `datagen Statistics` window. Note that the subcomputations required to compute t by hand calculator are also automatically provided, either in the `datagen Descriptive Statistics` window or in the `datagen Statistics` window.
4. See Homework Tip 7.

Use SPSS to compute the t test for the GRE scores from High Brow University and Simple University. The data are shown in Table 11.3, with H_0: $\mu_1 = \mu_2$.

1. Enter the dependent variable (GRE score) into the first column of the SPSS spreadsheet. Thus, enter the first value ("500") from the HBU column of Table 11.3 into the upper left cell.
2. Enter the independent variable (group membership) into the second column of the SPSS spreadsheet. Thus, enter "1" into the first row of the second column of the SPSS spreadsheet to indicate that the value "500" comes from group 1 (that is, from the HBU group). Note that the Tab key will move you from the first column to the second.
3. Continue entering the dependent and independent variables by entering the remaining 14 HBU values into column 1 and a "1" into each row of column 2.
4. Now enter the values for the SU group. Enter the first SU value ("580") into the 16th row of the first column of the SPSS spreadsheet. Enter a "2" into the 16th row of the second column to indicate that this value comes from group 2 (that is, from the SU group).
5. Continue entering the dependent and independent variables by entering the remaining 9 SU values into column 1 and a "2" into each row of column 2.
6. Click `Statistics`, then `Compare Means`, and then `Independent-Samples T Test...`.
7. Click the upper ▶ to move var00001 into the `Test Variable(s)` window.

8. Click var00002 and then click the lower ▶ to move var00002 into the `Grouping Variable` window. (Note that "grouping variable" is SPSS's term for "independent variable.")
9. Click `Define Groups`. Enter "1" in the `Group 1` text entry cell; enter "2" in the `Group 2` text entry cell. Then click `Continue` and click `OK`.
10. The accompanying display, which I have annotated, appears in the `!Output1` window. Note that SPSS provides the exact probability (.076) for a nondirectional test. If we were performing a nondirectional test, we would conclude that we should *not* reject the null hypothesis because the exact probability is not less than .05. Note also that the exact probability for a directional test is always precisely half the exact probability for a nondirectional test. Thus, the exact probability for the directional test here would be $.076/2 = .038$, and we *would* reject the directional H_0 with $\alpha = .05$.
11. See Homework Tip 7.

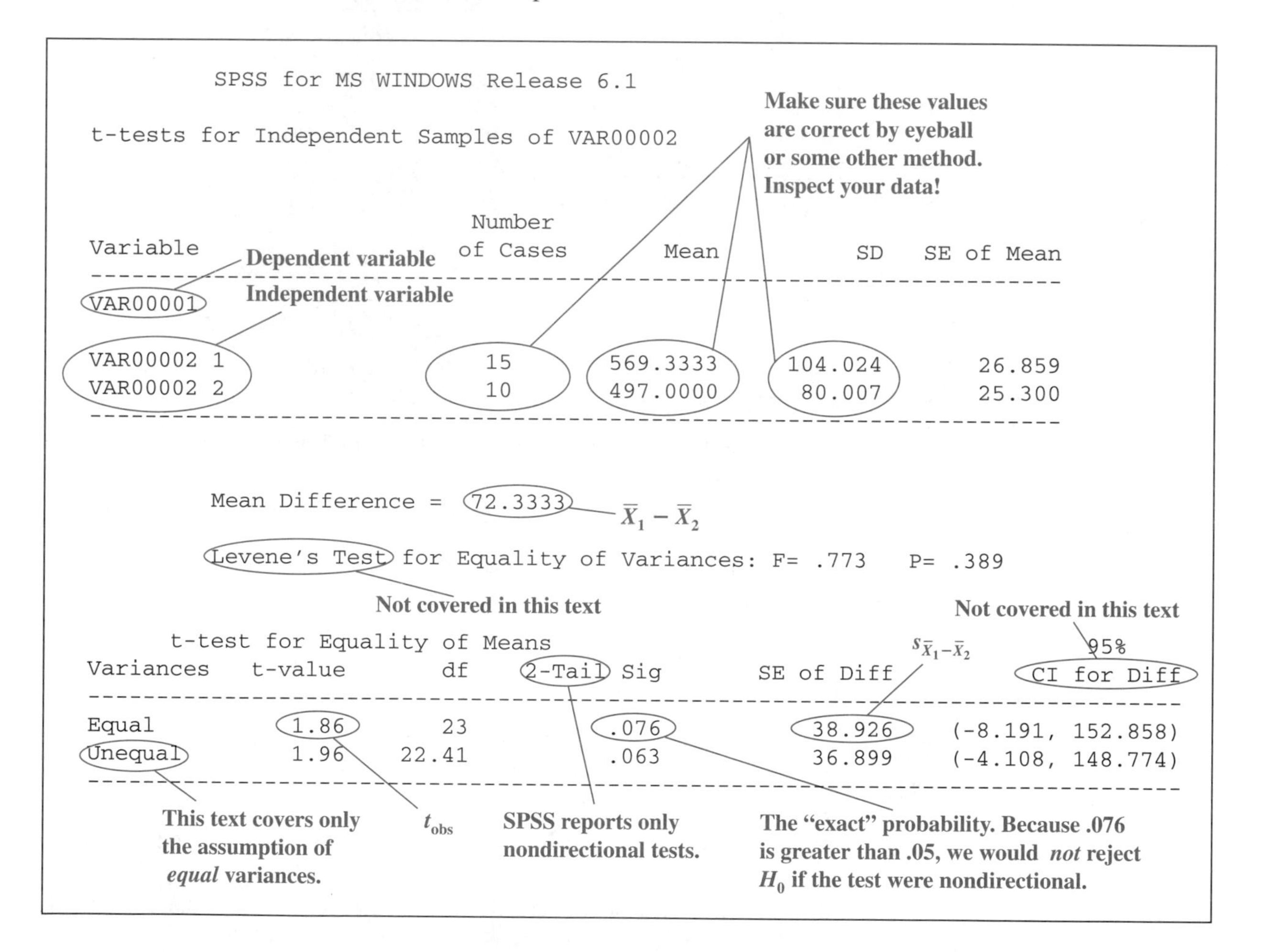

Homework Tips

1. Check the list of learning objectives at the beginning of this chapter. Do you understand each one?

2. You can use Equation (11.4) to compute the pooled variance whether or not $n_1 = n_2$. If the sample sizes are in fact equal, however, Equation (11.5) is easier to use.

3. You must compute the pooled variance [from Equation (11.4) or (11.5)] *before* you can compute the standard error of the difference between two means [Equation (11.3), (11.6), or (11.6a)].

4. Do not be scared off by the formula for the pooled variance—Equation (11.5) or (11.5a). Remember that to "pool" is simply to average variances (with appropriate weights if the sample sizes differ).

5. When you compute a pooled variance, it must be positive because it is the average of the sample variances, which themselves are all positive. Furthermore, the magnitude of your computed pooled variance must lie between the magnitudes of the two sample variances. Likewise, the magnitude of your computed pooled standard deviation must lie between the magnitudes of the two sample standard deviations. If any of these is not the case, then you have made a computational error.

6. The numerator of t is $\overline{X}_1 - \overline{X}_2$. Therefore, t will be negative if the first sample mean is smaller than the second sample mean.

7. Whenever you ask a computer to perform a task for you, you should (a) make sure the number of cases is correct and (b) eyeball-estimate enough statistics to convince yourself that the computer has interpreted the data in the way you expected.

Exercises for Chapter 11

Section A: Basic exercises (answers in Appendix D, page 514)

1. (a) Compute $s_{\overline{X}_1-\overline{X}_2}$ according to Equation (11.3).
   (b) Is it permissible to compute $s_{\overline{X}_1-\overline{X}_2}$ from Equation (11.6) or (11.6a) in this problem? Why or why not? If it is permissible, compute it.
   (c) How do your answers in parts (a) and (b) compare?

Table for Exercise 1

Sample 1	*Sample 2*
5	4
8	4
6	7
7	6
1	7

Table for Exercise 2

Sample 1	*Sample 2*
4	8
1	6
9	2
6	6
2	3
5	9
7	7
7	1

2. Assume that the data shown here represent WISC-R Vocabulary Subtest scores from random samples of size eight from two special education schools. Are the vocabulary levels of the two schools different? Use these steps.
   *Omit the portions marked with an asterisk (*) unless you have read the optional eyeball-estimation section at the end of this chapter.

I. State the null and alternative hypotheses

(a) What is the null hypothesis? The alternative hypothesis? Is this test directional or nondirectional? Why? Is this a one-sample test or a two-independent-samples test? Why?
(b) (Eyeball-estimate and then)* compute the pooled variance and the standard error of the difference between two means.
(c) What is the variable in this problem? Sketch the distributions of the variable assuming the null hypothesis is true.
(d) What is the sample statistic in this problem? Sketch the distribution of the sample statistic assuming the null hypothesis is true.
(e) What is the test statistic in this problem? Sketch the distribution of the test statistic assuming the null hypothesis is true. You may simply append a new axis to a distribution you already sketched.

II. Set the criterion for rejecting H_0

(f) What is the level of significance?
(g) How many degrees of freedom are there?
(h) (Eyeball-estimate and then)* look up the critical value(s) of the test statistic. Enter it (them) on the appropriate distribution and shade the rejection region(s).
(i) (Eyeball-estimate and then)* compute the critical value(s) of the sample statistic. Enter it (them) on the appropriate distribution and shade the rejection region(s).

III. Collect a sample and compute statistics

(j) (Eyeball-estimate and then)* compute the observed value of the sample statistic. Enter it on the appropriate distribution.
(k) (Eyeball-estimate and then)* compute the observed value of the test statistic. Enter it on the appropriate distribution.
(l) Do the critical values, rejection regions, and observed values have the same relative positions on their respective distributions? Explain.

IV. Interpret the results

(m) Is this result statistically significant (that is, should we reject H_0)?
(n) If the result was statistically significant, then determine the effect size index d and Cohen's nonoverlap statistic U. Illustrate these as shown in Figure 11.7.
(o) Describe the results, including both statistical and practical significance, in plain English.

3. A psychologist is interested in whether memory ability decreases with age. He takes a random sample of ten 30-year-old college graduates and gives them a list of 50 words to memorize. One week later, he tests those same subjects, asking each to recall as many words as possible

from the memorized list. He repeats the procedure with a random sample of 13 60-year-old college graduates. Is the memory of 60-year-olds diminished in comparison to that of 30-year-olds? The numbers of words recalled are listed in the table. Use parts (a)–(o) from Exercise 2.

	30-year-olds	*60-year-olds*
	28	28
	34	25
	25	30
	34	21
	33	30
	36	29
	33	33
	39	27
	37	27
	34	34
		31
		32
		26
Labor savings for the student:		
$\sum X_i$	333	373
$\sum X_i^2$	11,241	10,855

4. A psychologist is interested in whether viewing the movie *Schindler's List* changes individuals' anti-Semitism. She develops a test of anti-Semitism and selects two random samples of college students, both containing 15 individuals. She has one group watch *Schindler's List*, while at the same time the other group watches a different movie, equally long but with no ethnic content. She then administers the test of anti-Semitism to both groups. The anti-Semitism scores are shown in the table. Does viewing *Schindler's List* change anti-Semitism? Use parts (a)–(o) from Exercise 2.

	Schindler's List	*Other*
	65	67
	62	71
	53	59
	61	67
	48	53
	58	62
	72	75
	62	71
	49	49
	55	70
	69	61
	71	72
	56	70
	68	60
	68	62
Labor savings for the student:		
$\sum X_i$	917	969
$\sum X_i^2$	56,907	63,369

5. An experimenter performs two experiments to determine whether there is a statistically significant difference between two treatments, which we will call A and B. Summaries of the data are shown in the table.

	Experiment I		*Experiment II*	
	A	*B*	*A*	*B*
n	10	10	10	10
$\overline{X}$	45	40	45	40
s	5.0	5.2	10.0	10.4

(a) Is there a statistically significant difference between treatments A and B in experiment I? Use $\alpha = .05$.
(b) Is there a statistically significant difference between treatments A and B in experiment II? Use $\alpha = .05$.
(c) How does $(\overline{X}_1 - \overline{X}_2)$ in experiment I compare to $(\overline{X}_1 - \overline{X}_2)$ in experiment II? Given that, why are the results of parts (a) and (b) different?

6. A researcher performs two experiments to determine whether there is a statistically significant difference between two treatments, which we will call A and B. Summaries of the data are shown in the table.

	Experiment I		*Experiment II*	
	A	*B*	*A*	*B*
n	10	10	5	5
$\overline{X}$	45	40	45	40
s	5.0	5.2	5.0	5.2

(a) Is there a statistically significant difference between treatments A and B in experiment I? Use $\alpha = .05$.
(b) Is there a statistically significant difference between treatments A and B in experiment II? Use $\alpha = .05$.
(c) How does $(\overline{X}_1 - \overline{X}_2)$ in experiment I compare to $(\overline{X}_1 - \overline{X}_2)$ in experiment II? Given that, why are the results of parts (a) and (b) different?

7. Dr. Eva N. Flow wishes to market a magic pyramid that she claims makes babies sleep longer. She devises an experiment: She gets 3200 month-old babies and randomly assigns them to two groups, 1600 in each. She gives the magic pyramid to each mother in group 1 with the instruction to hang it from the crib when she puts the baby to sleep for the night. She gives the mothers in group 2 a placebo. Mothers are instructed to record the number of minutes the baby sleeps before waking up for the first time. The data are listed here.

	Group 1	*Group 2*
n	1600	1600
$\overline{X}$	184.9	183.6
s	18.3	16.5

(a) Can Dr. Flow correctly claim that there is a statistically significant difference in sleep times? (Compute t_{obs} and compare it with t_{cv}.)

(b) How much more sleep does the average group 1 baby get than the average group 2 baby?
(c) Compute the effect size d and the nonoverlap statistic U.
(d) The average group 1 baby gets more sleep than what percentage of the group 2 babies?
(e) Would you pay $29.95 for this "statistically significant" product?

Section B: Supplementary exercises

8. Suppose that two groups have these statistics:

$$\text{Group 1: } n_1 = 14, \bar{X}_1 = 54.2, s_1 = 13.8$$
$$\text{Group 2: } n_2 = 14, \bar{X}_2 = 61.3, s_2 = 15.1$$

(a) Show that the pooled variance as computed by Equations (11.5) and (11.5a) is identical to that computed by Equation (11.4).
(b) Show that the standard error of the difference between two means as computed by Equation (11.3) is identical to that computed by Equations (11.6) and (11.6a).
(c) Why is it permitted to use Equations (11.4), (11.6), and (11.6a) here?

9. An athletic trainer wishes to know whether isometric exercise (pressing against a fixed object) or dynamic exercise (moving a weight) is more effective in building strength. She assigns 20 individuals at random to two groups (ten per group) and gives one group a month of isometric exercises and the other group a month of dynamic exercises. At the end of the month, she ascertains each individual's strength by measuring the weight the individual can press. The data are given here. Is the effectiveness of isometric exercise more than, less than, or the same as that of dynamic exercise? Use parts (a)–(o) from Exercise 2.

Isometric group weight pressed (pounds): 95, 101, 116, 114, 109, 120, 108, 95, 110, 122.
(Labor savings: $\sum X_{i1} = 1090$, $\sum X_{i1}^2 = 119{,}632$.)
Dynamic group weight pressed (pounds): 90, 69, 86, 94, 95, 112, 100, 73, 97, 107.
(Labor savings: $\sum X_{i2} = 923$, $\sum X_{i2}^2 = 86{,}849$.)

10. (*Exercise 9 continued*)
(a) Show that the pooled variance as computed by Equations (11.5) and (11.5a) is identical to that computed by Equation (11.4).
(b) Show that the standard error of the difference between two means as computed by Equation (11.3) is identical to that computed by Equations (11.6) and (11.6a).

11. (*Exercise 9 continued*) Suppose 20 subjects were in *each* group instead of ten. We contrive the data so that each data point is repeated, as shown here:

Isometric group weight pressed (pounds): 95, 95, 101, 101, 116, 116, 114, 114, 109, 109, 120, 120, 108, 108, 95, 95, 110, 110, 120, 120.

Dynamic group weight pressed (pounds): 90, 90, 69, 69, 86, 86, 94, 94, 95, 95, 112, 112, 100, 100, 73, 73, 97, 97, 107, 107.

How does that change these values?
(a) Group means
(b) Group standard deviations
(c) Pooled variance
(d) Standard error of the difference between two means
(e) Numerator of t
(f) t_{obs}
(g) df_1, df_2, pooled df
(h) t_{cv}
(i) The conclusion

12. A researcher wishes to know whether taking larger doses of vitamin C reduces the occurrence of colds. He identifies 100 volunteers and randomly assigns them to two groups of 50 each. He gives vitamin C pills to the "vitamin" group for a year and gives pills that look identical to the vitamin C pills but that contain an inert substance to the "placebo" group, also for a year. He does not inform the participants whether they are taking the vitamin or the placebo (that is, the subjects are "blind"). He asks each subject to keep a record of the number of days he or she experiences colds during the next year. He obtains the results listed here. During the course of the year, several people move away or for other reasons discontinue participation in the study, which leaves 44 completers in the vitamin group and 47 completers in the placebo group. Does vitamin C reduce the frequency of colds? Use parts (a)–(o) from Exercise 2.

Vitamin C group: 11, 11, 14, 14, 3, 16, 11, 11, 10, 14, 6, 10, 13, 14, 15, 12, 17, 16, 10, 10, 16, 5, 6, 13, 24, 17, 11, 10, 17, 6, 12, 20, 20, 6, 10, 12, 12, 21, 11, 9, 14, 16, 20, 15.
(Labor savings: $n_1 = 44$, $\sum X_1 = 561$, $\sum X_1^2 = 8037$.)

Placebo group: 10, 18, 15, 25, 14, 8, 13, 17, 14, 22, 2, 10, 16, 15, 14, 18, 13, 8, 0, 18, 9, 13, 16, 17, 8, 14, 8, 18, 15, 16, 12, 15, 10, 8, 6, 18, 8, 19, 22, 8, 12, 18, 8, 25, 13, 5, 16.
(Labor savings: $n_2 = 47$, $\sum X_2 = 627$, $\sum X_2^2 = 9731$.)

13. There are two schools of thought in teaching bowling. Some say you should look at the "pocket" (the space between the first and third pins) as you release the ball, whereas others say you should look at a "spot," a point on the lane about 10 feet in front of the foul line. A bowling instructor decides to test these ideas with his new class of 24 students. He randomly assigns them to one of two groups. He teaches one group (the "pocket" group) to look at the pocket and the other (the "spot") group to look at the spot. At the end of the eight-week series of lessons, he has the members of each group bowl one game and records their scores. Is one method better than the other? Use parts (a)–(o) from Exercise 2.

Pocket group score summary:
$n_1 = 12, \overline{X}_1 = 127.2, s_1 = 14.6$
Spot group score summary:
$n_2 = 12, \overline{X}_2 = 134.4, s_2 = 11.8$

Section C: Extending your comprehension (answers in Appendix D, page 516)

14. (a) Generate two samples, each with $n = 5$, such that the two samples are significantly different from each other when $\alpha = .05$ but are *not* significantly different when $\alpha = .01$. [*Hint:* First determine $\overline{X}_1$, $\overline{X}_2$, s_1, and s_2; then generate the data that meet those specifications.]
    (b) Is it possible for two samples to be significantly different when $\alpha = .01$ but *not* significantly different when $\alpha = .05$?
15. (a) Generate two samples, each with $n = 5$, such that the mean of one sample is significantly larger than the mean of the other when a directional test is used but is *not* significantly different from the other when a nondirectional test is used. Use $\alpha = .05$. (See the hint in Exercise 14.)
    (b) Is it possible for the nondirectional test to be significant but the directional test to be not significant? Why?
16. (a) Flip a quarter, letting "heads" equal 1 and "tails" equal 0. Now flip the quarter four more times (for a total of five; that is, you are taking a random sample of size five). Record the sequence of 0's and 1's, and compute the mean $\overline{X}_{\text{quarter}}$ for those five flips. Now flip a nickel five times, again recording the data and computing the mean $\overline{X}_{\text{nickel}}$ for the nickel's five flips. Now compute the *difference* between those two means, $\overline{X}_{\text{quarter}} - \overline{X}_{\text{nickel}}$.
    (b) Compute the standard error of the difference between two means from Equation (11.3).
    (c) Repeat part (a) nine more times, thus obtaining ten differences between sample means—that is, ten values of $(\overline{X}_{\text{quarter}} - \overline{X}_{\text{nickel}})$.
    (d) Compute directly from the difference data the standard error of the difference between two means [which is simply the standard deviation of the ten $(\overline{X}_{\text{quarter}} - \overline{X}_{\text{nickel}})$ values obtained in part (c)].
    (e) How close was the standard error computed in part (b) to the standard error computed in part (d)? Explain any differences.

Section D: Cumulative review (answers in Appendix D, page 516)

For Exercises 17–22, choose the correct answer or fill in the blank. (**Note that computations are *not* required;** use $\alpha = .05$ unless otherwise instructed.)

(a) This problem requires:
  (1) Finding the area under a normal distribution [Skip parts (b)–(g) and proceed to (h).]
  (2) Creating a confidence interval
  (3) Testing a hypothesis about the mean of one group
  (4) Testing a hypothesis about the means of two independent groups

(b) The null hypothesis is of the form:
  (1) $\mu = a$ (2) $\mu_1 = \mu_2$
  (3) There is no null hypothesis.

(c) The appropriate sample statistic is
  (1) $\overline{X}$ (2) p (3) $\overline{X}_1 - \overline{X}_2$

(d) The appropriate test statistic is:
  (1) z (2) t

(e) The number of degrees of freedom for this test statistic is ________. (State the number or *not applicable*.)

(f) The hypothesis test (or confidence interval) is:
  (1) Directional (2) Nondirectional

(g) The critical value of the test statistic is ________. (Give the value.)

(h) The appropriate formula for the test statistic (or confidence interval) is given by Equation (________).

Note that computations are *not* required for any of the problems in Section D. See the instructions.

17. A dentist knows that children in a particular region have had an average of 2.84 cavities per year. She wishes to know whether giving children fluoridated vitamins will decrease the number of cavities. She randomly selects 37 children from her practice, gives them the fluoridated vitamins, and counts the number of cavities each has in the next year: 1, 1, 0, 0, 0, 0, 1, 2, 5, 0, 1, 0, 0, 3, 0, 2, 2, 0, 1, 0, 1, 0, 4, 1, 2, 1, 0, 0, 1, 0, 1, 0, 1, 0, 0, 0, 5. Does fluoride decrease the number of cavities?
18. A dentist knows that children in a particular region have had an average of 2.84 cavities per year. She wishes to know whether giving children fluoridated vitamins will decrease the number of cavities. She randomly selects 74 children from her practice and randomly assigns them to one of two groups. To the "fluoride" group she gives the fluoridated vitamins; to the "vitamin" group she gives vitamins that contain no fluoride. She counts the number of cavities each child has in the next year. Do fluoridated vitamins decrease the number of cavities compared with unfluoridated vitamins?

    Fluoride group: 1, 1, 0, 0, 0, 0, 1, 2, 5, 0, 1, 0, 0, 3, 0, 2, 2, 0, 1, 0, 1, 0, 4, 1, 2, 1, 0, 0, 1, 0, 1, 0, 1, 0, 0, 0, 5
    Vitamin group: 2, 1, 0, 2, 3, 0, 0, 2, 1, 0, 4, 0, 2, 5, 2, 2, 1, 0, 1, 0, 1, 2, 1, 5, 3, 2, 1, 0, 2, 0, 1, 0, 2, 0, 0, 0, 1
19. A traffic safety engineer wishes to know whether he can change the behavior of drivers in a particular neighborhood. He knows that the mean and standard deviation of automobile speeds on Elm Street have been constant for years: 39.0 mph and 4.2 mph, respectively. He in-

stalls brightly colored signs that say "Please! Drive more slowly" on Elm Street, and then takes a random sample of 59 automobiles and measures their speeds. He obtains these values (in mph): 38, 34, 30, 37, 42, 37, 40, 27, 39, 37, 41, 29, 34, 38, 34, 35, 34, 40, 35, 33, 39, 31, 28, 41, 32, 34, 38, 33, 37, 29, 37, 36, 35, 41, 32, 35, 36, 36, 35, 33, 39, 32, 38, 35, 32, 40, 39, 33, 38, 34, 32, 39, 29, 36, 43, 31, 33, 28, 32. Have the new signs decreased speed?

20. An ecologist wishes to know the percentage of carbon dioxide in the air of a particular city. She randomly selects 25 locations and measures the CO_2 percentage in each: 83.3, 80.1, 80.0, 83.4, 81.9, 81.0, 80.8, 75.8, 77.6, 85.7, 80.2, 77.5, 81.5, 81.9, 79.7, 82.1, 81.3, 83.2, 81.0, 75.1, 78.6, 82.9, 80.3, 81.6, 76.8. What can she say about the mean CO_2 percentage in the city?

21. A researcher wishes to know whether the way a person is dressed affects that person's ability to persuade people. He develops a message designed to persuade people that it is desirable to live in the Midwest. He videotapes an actor giving the persuasive speech. In one condition, the actor is casually dressed, whereas in the other, he is wearing a suit and tie. In all other respects, the two speeches are identical. He shows one or the other videotape to subjects and then gives each a rating scale that measures the desirability of the Midwest. Does dress make a difference?

    Casual: 39, 40, 49, 41, 27, 41, 47, 55, 49, 41, 39, 32, 32, 30, 49

    Formal: 61, 29, 45, 49, 40, 43, 62, 41, 42, 55, 41, 58, 53, 55, 61

22. Runners in a particular jogging club run the 440-yard dash with mean 59 seconds and standard deviation 4 seconds. Suppose runners are chosen at random from this club to form four-person relay teams. What percentage of these teams have mean 440-yard-dash times less than 55 seconds?

Section E: From the journals (answers in Appendix D, page 516)

23. Zeidner (1990) used a questionnaire to measure "trait anger" in a random sample of 923 Israeli adults. Among other things, he found that trait anger "scores for women are significantly above those of men" (p. 7). He reported the following results: $\bar{X}_{men} = 49.24$, $s_{men} = 10.12$, $n_{men} = 426$; $\bar{X}_{women} = 50.67$, $s_{women} = 9.85$, $n_{women} = 485$; $t_{df=909} = 2.17$, $p < .03$.
    (a) Are Zeidner's degrees of freedom computed correctly?
    (b) Do you have enough information to compute t? If so, compute it. Does your value correspond to Zeidner's?
    (c) Why does Zeidner report $p < .03$ instead of .05 or .01?

24. (*Exercise 23 continued*) Suppose you decide to replicate Zeidner's (1990) experiment, and for 15 men and 15 women you obtain the data shown in the table.
    (a) How do your means and standard deviations compare to Zeidner's?
    (b) Is $\bar{X}_1 - \bar{X}_2$ in your experiment larger or smaller than Zeidner's?
    (c) Compute t_{obs} for your data. Is your t_{obs} larger or smaller than Zeidner's? Why?
    (d) Is your t_{obs} significant?

Men	*Women*
44	58
41	50
53	50
51	52
47	45
65	45
50	28
72	64
32	44
41	54
36	58
52	35
63	64
37	48
48	78

25. Burger (1990) measured 60 male and female subjects' desire for control on the Desirability of Control (DC) Scale that he had developed. The top 15 scorers were the high-DC group, whereas the lowest 15 scorers made up the low-DC group. He brought subjects in pairs into the experimental session and had them engage in a structured conversation, discussing ten topics such as "the kinds of movies I like to see" and "my worst experience in school." Subjects then rated each other on a five-point scale that measured the amount of personal revealing done on each topic, from 1 = "not at all personally revealing" to 5 = "very personally revealing." Each subject's intimacy score was obtained by adding the ratings of the ten topics; thus, intimacy scores could range from 10 to 50. Burger reported that low-DC subjects had significantly higher intimacy scores ($\bar{X} = 34.87$) than did high-DC subjects ($\bar{X} = 28.20$); $F(1, 28) = 8.93$, $p < .006$. Now it can be shown that F with 1 and 28 degrees of freedom is identical to t^2 with 28 degrees of freedom. Thus, Burger's report is equivalent to reporting that $t_{df=28} = \sqrt{F} = 2.99$.
    (a) Are Burger's degrees of freedom computed correctly?
    (b) Is enough information presented to allow you to verify that t was computed correctly? If so, verify the computation of t.

26. (*Exercise 25 continued*) Suppose you decide to replicate Burger's (1990) study; your data are shown in the table.
    (a) How do your means compare to Burger's?
    (b) Is $\bar{X}_1 - \bar{X}_2$ in your experiment larger or smaller than Burger's?
    (c) Compute t_{obs} for your data. Is your t_{obs} larger or smaller than Burger's? Why?
    (d) Is your t_{obs} significant?

(e) How do you account for the fact that t_{obs} for these data is larger than Burger's, even though $\overline{X}_1 - \overline{X}_2$ here is smaller?

Low DC	High DC
35	29
35	31
33	33
36	26
35	25
28	26
33	29
35	33
37	29
34	29
33	25
31	28
43	28
36	29
30	39

27. Nonorganic failure to thrive (NOFT) is a frequent pediatric condition characterized by lack of weight gain at home without any major organic condition that could explain the lack of weight gain. Drotar and colleagues (1990) made home visits and rated the interaction patterns of 47 mothers of such infants and compared them with 47 mothers of normal infants. Ratings were made on six 9-point scales; the first was "Sensitivity–Insensitivity, which assessed the degree to which the mother perceived and responded appropriately to the infant's communication (a high rating indicated excellent maternal attunement to the child's signals, while a low score identified mothers who seemed more geared to their own needs than their child's)" (p. 45). The mean and standard deviation for the sensitivity–insensitivity scores for the NOFT group were 5.93 and 2.03, respectively, whereas for the comparison group, they were 7.36 and 1.23. These scores were reported to be significantly different ($t_{df=92} = -4.03$, $p < .01$).
   (a) Are Drotar's degrees of freedom computed correctly?
   (b) Is enough information presented to allow you to verify that t was computed correctly? If so, verify the computation of t.
   (c) Interpret Drotar's result in plain English.

28. (*Exercise 27 continued*) Suppose you replicate Drotar and colleagues' (1990) study by observing 15 mothers of NOFT infants and 15 mothers of normal infants. Your ratings are shown in the table.
   (a) How many degrees of freedom are there?
   (b) Compute t_{obs}.
   (c) Interpret your results.

Table for Exercise 28

NOFT	Normal
5	8
9	8
9	9
5	7
7	7
4	9
9	7
6	7
5	8
6	5
2	7
7	9
4	7
8	9
7	9

Section F: Computer explorations

29. (a) Have the computer create in the first column a random sample of size five from the population of IQ scores ($\mu = 100$, $\sigma = 15$). [*Hint:* We described how to do that in Section 7.3.]
   (b) Have the computer create in the second column another random sample of size five from the population of IQ scores.
   (c) Given that the null hypothesis is true for these two samples ($\mu = 100$ for each), what would you expect the value of t to be? Have the computer verify your answer. [*Hint:* Datagen will have automatically displayed t. If you are using SPSS, you will have to move the second sample to the bottom of the first column and create the independent variable (see Section 11.7).]
   (d) Repeat parts (a)–(c) several times until you get a feel for the values that t might take when the null hypothesis is true.

30. (*Exercise 29 continued*)
   (a) For the t test that you ended up with in Exercise 28, jot down the value of t.
   (b) Find the magnitude of the standard error of the difference between two means ($s_{\overline{X}_1 - \overline{X}_2}$) on the computer display and jot down this value also.
   (c) What would you expect to happen to t if you added the value of ($s_{\overline{X}_1 - \overline{X}_2}$) to the first five data points? Have the computer verify your prediction.

31. Ask the computer to perform a t test on these data:

   Sample 1: 1, 1, 1, 1, 1
   Sample 2: 2, 2, 2, 2, 2

   (a) Explain why the computer tells you that t is undefined for these data.
   (b) Change the last point in sample 2 from "2" to "2.01." Now what happens to t? Explain why.

Eyeball-estimating Two-independent-samples *t* Tests (Optional)

This section may be omitted without loss of continuity.

Remember that the philosophy of eyeball-estimation is to perform all eyeball-computations "in your head," *not* using a calculator. The object is to get a reasonable approximation, not an exact result. Use the procedure both as a quick check on the computation and, more important, as a way of quickly practicing the hypothesis-testing procedure.

General Eyeball Procedure

The eyeball-estimation of two-independent-samples hypothesis tests follows the same general procedure that we have been using since Chapter 9.

steps for eyeball-estimating two-independent-samples hypothesis tests

I. State the Null and Alternative Hypotheses

Use the same procedure we used when computing.

II. Set the Criterion for Rejecting H_0

$t_{cv_{\text{by eyeball}}}$ is somewhat greater than ± 1.96 (or 1.645 if directional); that is, the nondirectional $t_{cv_{\text{by eyeball}}}$ is about 2 unless *df* is quite small.

III. Collect a Sample and Eyeball-estimate the Observed Values of the Sample Statistic and the Test Statistic

The eyeballed test statistic follows the same formulas used for computing:

$$t_{\text{by eyeball}} = \frac{(\overline{X}_1 - \overline{X}_2)_{\text{by eyeball}}}{s_{\overline{X}_1 - \overline{X}_2{}_{\text{by eyeball}}}}$$

a. The eyeballed sample statistic is $(\overline{X}_1 - \overline{X}_2)_{\text{by eyeball}}$. We use either the range method or the balancing-point method to eyeball-estimate $\overline{X}_{1_{\text{by eyeball}}}$ and then $\overline{X}_{2_{\text{by eyeball}}}$. The sample statistic $(\overline{X}_1 - \overline{X}_2)_{\text{by eyeball}}$ equals the second eyeballed mean subtracted from the first. Note that the eyeballed sample statistic is also the numerator of $t_{\text{by eyeball}}$.

b. Eyeball-estimate $s_{1_{\text{by eyeball}}}$ by the range method or the inflection-point method; then eyeball-estimate $s_{2_{\text{by eyeball}}}$. The pooled standard deviation $s_{\text{pooled}_{\text{by eyeball}}}$ is the average of $s_{1_{\text{by eyeball}}}$ and $s_{2_{\text{by eyeball}}}$. (Note that to pool by computation is to average *variances*; for eyeball purposes it is sufficient to average standard deviations.)

c. To obtain the *denominator* of the test statistic, we proceed as if $n_1 = n_2 = n$ (use the approximate average of n_1 and n_2 if they are not actually equal). Then we eyeball-estimate following Equation (11.6a):

$$s_{\overline{X}_1 - \overline{X}_2} = \left(\frac{s_{\text{pooled}}}{\sqrt{n}}\right)\sqrt{2}$$

We estimated $s_{\text{pooled}_{\text{by eyeball}}}$ in step b. We first divide by $\sqrt{n}$. *Note:* Take the square root in your head by using the nearest perfect square (4, 9, 16, 25, 36, 49, 64, etc.) for the eyeballed value of *n*. Perform the division in your head also. We do not need an exact answer, just an approximation. Then we multiply by $\sqrt{2}$, which is 1.414, but 1.5 is close enough for eyeball purposes.

d. To obtain the test statistic $t_{\text{by eyeball}}$, divide the numerator of the test statistic (from step a) by the denominator of the test statistic (from step c). Again, perform the division in your head.

IV. Interpret the Results

Exactly as in computation, if your eyeballed test statistic (from step d) is greater (in absolute value) than your eyeballed critical value (from part II), then reject$_{\text{by eyeball}}$ H_0.

Pygmalion in the Classroom Example

We now apply that procedure to the eyeball-estimation of the study of Pygmalion in the classroom, which as you may recall, asked whether the mean intellectual growth (IQ gain) among the 12 bloomers was different from the mean intellectual growth among the 47 other children.

This eyeball-estimation procedure parallels the computation in the text on page 221.

I. State the Null and Alternative Hypotheses

Use the same procedure we used when computing.

II. Set the Criterion for Rejecting H_0

Just as when computing, we let $\alpha = .05$. The test is nondirectional and *df* is $11 + 47 = 58$, so $t_{\text{cv}_{\text{by eyeball}}}$ is just slightly greater than ± 1.96—say, about ± 2.

III. Collect a Sample and Eyeball-estimate the Observed Values of the Sample Statistic and the Test Statistic

a. We can most conveniently use the range method to eyeball-estimate the sample means:

$$\overline{X}_{1_{\text{by eyeball}}} \approx \frac{69 + (-6)}{2} = 31.5 \text{ IQ points}$$

$$\overline{X}_{2_{\text{by eyeball}}} \approx \frac{31 + (-15)}{2} = 8 \text{ IQ points}$$

Then $(\overline{X}_1 - \overline{X}_2)_{\text{by eyeball}} = 31.5 - 8 = 23.5$ IQ points. That will be the numerator of the test statistic.

b. The standard deviations are most conveniently eyeball-estimated using the range method:

$$s_{1_{\text{by eyeball}}} \approx \frac{69 - (-6)}{3} = \frac{75}{3} = 25 \text{ IQ points}$$

(we use 3 as the conversion factor because $n_1 = 12$)

$$s_{2_{\text{by eyeball}}} \approx \frac{31 - (-15)}{4} = \frac{46}{4} = 11.5 \text{ IQ points}$$

Then we pool by eyeball to obtain $s_{\text{pooled}_{\text{by eyeball}}} \approx (25 + 11.5)/2 \approx 18$ IQ points. If we were sophisticated here, we would notice that the second sample is much larger than the first sample, so the pooling should result in a value closer to 11.5 than to 25. Therefore, we adjust our estimate to be $s_{\text{pooled}_{\text{by eyeball}}} \approx 15$ IQ points.

c. To obtain the *denominator* of the test statistic, we proceed as if *n* were 25 (the closest perfect square to the average of the two sample sizes). Equation (11.6a) said $s_{\overline{X}_1-\overline{X}_2} = (s_{\text{pooled}}/\sqrt{n})$; therefore, $s_{\overline{X}_1-\overline{X}_{2_{\text{by eyeball}}}} \approx (15/\sqrt{25})1.5 = 4.5$ IQ points.

d. Now we divide the numerator (from step a) by the denominator (from step c) to obtain the test statistic: $t_{\text{by eyeball}} \approx 23.5/4.5 \approx 5$.

IV. Interpret the Results

We compare $t_{\text{by eyeball}}$ to $t_{\text{cv}_{\text{by eyeball}}}$. Because 5 is greater than 2, we *do* reject$_{\text{by eyeball}}$ the null hypothesis.

As you may recall, the computed values were found in the text to be $(\overline{X}_1 - \overline{X}_2)_{\text{obs}} = 9.5$ IQ points, $s_{\overline{X}_1-\overline{X}_2} = 4.02$ IQ points, $t_{\text{obs}} = 2.36$, and $t_{\text{cv}} = \pm 2.003$, so we also rejected H_0 by computation. The biggest discrepancy between the eyeballed and computed results was in $\overline{X}_1$, where the one unusually large value (69) had more effect on the eyeballed mean than it did on the computed mean.

High Brow University Example

We now apply the eyeball-estimation procedure to the High Brow University/Simple University study, which, as you may recall, asked whether the mean GRE score among 15 HBU students was significantly higher than the mean GRE score of 10 SU students.

I. State the Null and Alternative Hypotheses

This eyeball-estimation procedure parallels the computation in the text on page 224.

Use the same procedure we used when computing.

II. Set the Criterion for Rejecting H_0

Just as when computing, we let $\alpha = .05$. The test is directional and df is $14 + 9 = 23$, so $t_{cv_{\text{by eyeball}}}$ is slightly greater than 1.645—say, 1.7.

III. Collect a Sample and Eyeball-estimate the Observed Values of the Sample Statistic and the Test Statistic

a. We can most conveniently use the range method to eyeball-estimate the sample means:

$$\overline{X}_{1_{\text{by eyeball}}} \approx \frac{800 + 390}{2} = 595 \text{ points}$$

$$\overline{X}_{2_{\text{by eyeball}}} \approx \frac{580 + 320}{2} = 450 \text{ points}$$

Then $(\overline{X}_1 - \overline{X}_2)_{\text{by eyeball}} = 595 - 450 = 145$ points. That will be the numerator of the test statistic.

b. The standard deviations are most conveniently eyeball-estimated using the range method:

$$s_{1_{\text{by eyeball}}} \approx \frac{800 - 390}{4} \approx 100 \text{ points}$$

$$s_{2_{\text{by eyeball}}} \approx \frac{580 - 320}{3} \approx 90 \text{ points}$$

Then we pool by eyeball to obtain $s_{\text{pooled}_{\text{by eyeball}}} \approx (100 + 90)/2 = 95$ points. If we were sophisticated here, we would notice that the first sample is somewhat larger than the second sample, so the pooling should result in a value somewhat closer to 100 than to 90. Therefore, we adjust our estimate to be $s_{\text{pooled}_{\text{by eyeball}}} \approx 97$—actually, 100 would be close enough for eyeball purposes.

c. To obtain the denominator of the test statistic, we proceed as if n were 16 (the closest perfect square to the average of the two sample sizes). Because $s_{\overline{X}_1 - \overline{X}_2} = (s_{\text{pooled}}/\sqrt{n})\sqrt{2}$, we have $s_{\overline{X}_1 - \overline{X}_{2_{\text{by eyeball}}}} \approx (100/\sqrt{16})1.5 = 37.5$ points.

d. Now we divide the numerator (from step a) by the denominator (from step c) to obtain the test statistic: $t_{\text{by eyeball}} \approx 145/37.5$, or about 4.

IV. Interpret the Results

We compare $t_{\text{by eyeball}}$ to $t_{cv_{\text{by eyeball}}}$. Because 4 is greater than 1.7, we *do* reject$_{\text{by eyeball}}$ the null hypothesis.

As you may recall, the computed values were found in the text to be $(\overline{X}_1 - \overline{X}_2)_{\text{obs}} = 72.33$ points, $s_{\overline{X}_1 - \overline{X}_2} = 38.93$, $t_{\text{obs}} = 1.86$, and $t_{\text{cv}} = 1.714$, so we also rejected H_0 by computation. The biggest discrepancies between the eyeballed and computed results were in $\overline{X}_1$, where the one unusually high score (800) pulled the eyeballed mean too high, and in $\overline{X}_2$, where the one unusually low score (320) pulled the eyeballed mean too low.

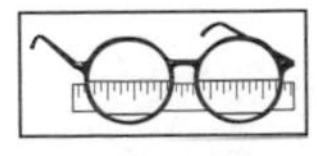

You should practice the step-by-step eyeball-estimation procedures described in this section on the histograms shown in Figures 11.12 through 11.17. Cover up the actual values on the right-hand side and then eyeball-estimate t using the steps shown above.

You may wish to use the ESTAT exercise "ttest2" at this time. Follow the path labeled "Eyeball-estimation."

12

CHAPTER

Inferences About Means of Two Dependent Samples and Statistical Power of t Tests

LEARNING OBJECTIVES

1. What are dependent samples? How are they different from independent samples?
2. What is a pretest–posttest design?
3. What is a difference score?
4. How is testing hypotheses about the means of two dependent samples similar to testing hypotheses about a single sample?
5. What is the standard error of the mean of the differences?
6. What is the power of a statistical test?
7. How is power related to the probability of a Type II error?
8. What five factors affect the magnitude of power?
9. How is power analysis used to determine the sample size of an experiment?

This chapter describes statistical inferences about two dependent samples—that is, about experiments in which the members of one group are statistically related to the members of the other group. For example, in the pretest–posttest design, the same subjects are measured twice; the group 1 data represent the measurements in condition 1, whereas the group 2 data represent the measurements on the same subjects in condition 2. Analysis of such data requires that for each subject (or each related pair) we compute a "difference score," which is the score in condition 1 minus the same (or related) subject's score in condition 2. These difference scores are then analyzed as if they were the original data in a one-sample test.

This chapter also describes the power of a statistical test—that is, the probability that an experiment will correctly reject the null hypothesis. To compute power requires assuming a specific value for the parameter set by the alternative hypothesis. Power is then the region under this alternative-hypothesis distribution that exceeds the critical value. Five factors can increase the power of a statistical test: increasing the sample size, increasing the effect size, decreasing σ, increasing α, and making the test directional instead of nondirectional. The experimenter in general has the most control over the sample size, and specifying the desired power can determine how many subjects should be used in a particular experiment.

dependent samples: two samples whose subjects are statistically related to each other

In Chapter 11, we considered testing hypotheses about the means of two samples when those two samples (which we frequently called "groups") were independent of each other. This chapter considers how hypotheses are tested when the two samples are *dependent*—that is, when the members of one group are statistically related to the members of the other group.

12.1 Dependent-sample Tests

pretest–posttest design: an experimental design where each subject is measured twice, once before and again after some treatment

Let us consider a few examples. The most straightforward dependent-sample test is the *pretest–posttest design*, where the *same* individuals are measured *twice*, once before and once after some treatment. Suppose we were interested in whether drinking caffeine increases a person's reaction time. Our design calls for us to measure a subject's reaction time, administer a dose of caffeine, and then measure that same subject's reaction time again. Note that in this design, each individual subject contributes *two* reaction time scores: one in the "pre" and one in the "post" condition. The scores in the two (pre and post) samples are related to each other. For example, if the first person happens to have a naturally fast reaction time, then the first reaction time score is likely to be fast in both pre and post samples, whereas if the fifth person happens to have a naturally slow reaction time, the fifth score is likely to be slow in both pre and post samples. We therefore refer to this as a "dependent-sample" test because the ith score in the second sample depends on or is related to the ith score in the first sample.

By contrast, we could design an *independent-sample* test (as described in Chapter 11) to determine the effect of caffeine on reaction time. Here we would randomly select two separate, independent groups of subjects, and give one group caffeine and the other group a placebo (which has no effect). Then we would measure the reaction times in all subjects (there is no pretest in this design). The scores in these two groups are *not* related to each other: The reaction time of the first subject in the first sample (group 1) in no way depends on the reaction time of the first subject in the second sample (group 2).

Dependent samples do not have to be scores from the same individuals. For example, if we were interested in determining whether married men's attitudes toward abortion differed from those of married women, we could randomly select married couples and form two samples, one of the husbands and one of the wives. These samples are *dependent* because the first man is married to the first woman and the ith man is married to the ith woman. In general, attitudes toward abortion are somewhat similar (not identical, to be sure) for husbands and wives, so if the ith man has a high score, then the ith woman will tend to have a high score also; this is therefore a dependent-sample test (sometimes called a "related samples" design). On the other hand, if we constructed two separate random samples, one of married men and another of married women, but in which the men and women were not married to each other, then the samples would be *independent*. If the ith man has a high score, that has *no* bearing on the ith woman's score.

Another frequently used design that employs the dependent-sample test is the "matched pairs" design, where subjects are measured on some important variable and then subjects with the most similar scores are paired off: the highest scorer with the second highest, the third highest with the fourth highest, and so on through the second lowest scorer with the lowest. Then, by a random decision, one of each pair is assigned to be a member of the first group and the other to the second group. In our reaction time study, for example, we could have given all our subjects a reaction time pretest and matched the two fastest subjects, the next two fastest subjects, and so on. Then for each pair, we would decide by some random method which subject would receive caffeine; the other subject of the pair would receive a placebo.

Note that for dependent samples, the *order* of the data in each sample *is* important. For the matched pairs example, if one member of a particular pair happens to be the seventh member of the first sample data set, then the other member of that same pair must also be the seventh member of the second sample data set. That is *not* true for the independent-sample tests of Chapter 11. Each data set in an independent-sample design may be shuffled at will and the results of the test will not change.

difference score: a subject's score in one condition minus that same subject's (or related subject pair's) score in the other condition

Understanding dependent-sample tests is straightforward when the concept of the *difference score* is understood. Remember that all data in dependent-sample designs are actually pairs of data, and the difference score for any pair is simply the difference between the scores of that pair. Thus, in a pretest–posttest design, the difference score is the subject's pretest score minus that subject's own posttest score. In our related samples design, the difference score is the husband's score minus his own wife's score. In a matched pairs design, the difference score is the score of a person in the first group minus that subject's own matched pair's score.

Therefore, we write the difference score for the ith pair, D_i, as

difference score for the ith individual

$$D_i = X_{i1} - X_{i2} \tag{12.1}$$

where X_{i1} is the ith score of the first sample, and X_{i2} is the ith score of the second sample.

Note that in a dependent-sample study, there must be the same number of data points in each group (because each point is half of a pair). For example, if there are 25 points in the first sample, there must be 25 points in the second sample (for a total of 50 data points arranged as 25 pairs).

Test your understanding of the distinction between independent samples and dependent samples in Box 12.1.

BOX 12.1 Distinguishing dependent and independent samples

For each of these examples, decide whether the samples are dependent or independent. The answers are in Box 12.2.

1. I wish to know whether the attitudes toward the U.S. justice system differ between blacks and whites. I obtain a random sample of 40 black individuals and another random sample of 40 white individuals and administer my Attitude Toward Justice scale.
2. I wish to know whether the attitudes of black individuals toward the U.S. justice system change as a result of the O. J. Simpson trial. I obtain a random sample of 40 black individuals before the trial begins and administer my Attitude Toward Justice scale. After the trial, I locate the same 40 individuals and administer my Attitude Toward Justice scale again.
3. The SPQ3R system is a way of studying that is intended to build active involvement with the subject matter. To see whether the SPQ3R system is effective, I decide to use an introductory psychology course that has 200 students. I will create two groups: Group I will get instruction in the SPQ3R system, whereas group II will get a study pep talk. I assign students to one of the two groups on the basis of their scores on the first exam. I consider the students with the top two scores on the first exam; I flip a coin and, if heads, put student 1 in group I and student 2 in group II but, if tails, I put student 2 in group I and student 1 in group II. Then I consider the two students with the next highest scores, and in the same way flip a coin to determine their assignment. I repeat the coin-flip procedure for each pair of students, assigning students to groups as a result of the flip. The outcome measure (dependent variable) will be the score on the final exam.
4. I wish to know whether the visual acuity of Republicans differs from that of Democrats. I obtain a random sample of 30 Republicans and measure each person's visual acuity. I obtain a random sample of 25 Democrats and measure each person's visual acuity.

12.2 Evaluating Hypotheses About Means of Two Dependent Samples

Analyzing a dependent-sample test involves forming the difference scores and then using those difference scores *as if they were the original data in a one-sample test.* Thus, the concept of the difference score is the only new idea to master in the first half of this chapter. Recall that Chapter 10 explored the evaluation of one-sample hypotheses. We will simply substitute "difference score" wherever Chapter 10 used the phrase "score on the original variable," and we will substitute "mean of the differences" wherever Chapter 10 used the term "mean." We then proceed as if we were performing a one-sample hypothesis test exactly as we did in Chapter 10.

An Example

Suppose that Nicoterm is a chewing gum designed to reduce cigarette smoking. Previous research has shown that similar preparations are effective. We decide to test Nicoterm's effectiveness using a pretest–posttest study with a random sample of 12 smokers.[1] The

[1]We should note that pretest–posttest studies do not provide enough experimental control to make them particularly desirable research designs. For example, undergoing the first measurement may alter a subject's response on the second measurement (making it possible, in our example, that it was the fact of counting cigarettes in the first week that caused the smoking reduction in the second week, not the use of Nicoterm itself). Other designs, such as the matched pairs design, are far superior to the pretest–posttest design. Because the pretest–posttest design is the most straightforward from a pedagogical point of view, however, we use it as an example.

BOX 12.2 Answers to examples in Box 12.1

1. Independent samples; the black individuals are different from the white individuals.
2. Dependent samples; this is a typical pretest–posttest design.
3. Dependent samples; this is an example of a matched pairs design.
4. Independent samples; the Republicans are different from the Democrats.

study will have two week-long periods. During the first week, we ask subjects to count the number of cigarettes they smoke; this number will be our "pre" score. During the second week, we ask them to chew the Nicoterm gum according to the manufacturer's directions and also to count the number of cigarettes they smoke during this week; that number will be the "post" score. The data are shown in Table 12.1. Note that the fourth column of that table shows the difference ($D_i = X_{i1} - X_{i2}$) in the cigarette smoking rates for each subject. For example, the first subject smoked 120 cigarettes in the week "before" Nicoterm use and 110 cigarettes "during" Nicoterm use. His difference score is therefore $D_1 = 120 - 110 = 10$ cigarettes. The analysis will then proceed as if the difference score column were the original data from a one-sample experiment as in Chapter 10. We will show the pre and post scores, compute the difference between them, and then, for the rest of the analysis, ignore the pre and post scores themselves, concentrating instead on the difference scores.

TABLE 12.1 Number of cigarettes smoked per week

Subject (i)	*Before* (X_{i1})	*During* (X_{i2})	*Difference* ($D_i = X_{i1} - X_{i2}$)
1	120	110	10
2	200	180	20
3	165	165	0
4	147	110	37
5	180	194	−14
6	106	106	0
7	235	241	−6
8	211	201	10
9	113	111	2
10	142	101	41
11	98	90	8
12	115	103	12
			$\sum D_i = 120$
			$\sum D_i^2 = 4094$
			$n = 12$
			$\bar{D} = 10.0$
			$s_D = 16.22$
			$s_{\bar{D}} = 4.68$

Null Hypothesis

Recall from Chapter 11 that the null hypothesis for a nondirectional, two-independent-samples test is H_0: $\mu_1 = \mu_2$ (if the hypothesis is nondirectional) or H_0: $\mu_1 \leq \mu_2$ or H_0: $\mu_1 \geq \mu_2$ (if the hypothesis is directional). Some statisticians use these same statements of the null hypothesis for a dependent-sample test. We prefer that the null hypothesis unequivocally indicate that a dependent-sample test is being performed, and so we define μ_D as the mean of the population of all difference scores. Then we can write a null hypothesis that is explicitly about difference scores and leaves no doubt that we are referring to a

null hypothesis, two-dependent-samples test for means

dependent-sample test:

$$H_0\colon \mu_D = 0 \quad \text{or} \quad H_0\colon \mu_D \leq 0 \quad \text{or} \quad H_0\colon \mu_D \geq 0 \tag{12.2}$$
(nondirectional) (directional)

In the Nicoterm example we are justified in choosing a directional test because the problem states that previous research has shown that preparations similar to Nicoterm have been effective in reducing smoking. The alternative hypothesis should reflect the direction of this effect. If Nicoterm is in fact effective, subjects should in general smoke more cigarettes "before" taking Nicoterm than "during" taking Nicoterm; that is, the X_{i1} values (in the "Before" column of Table 12.1) should in general be greater than the X_{i2} values (in the "During" column), and the difference scores $D_i = X_{i1} - X_{i2}$ should therefore in general be greater than 0. We use the "greater than" symbol in the alternative hypothesis, writing $H_1\colon \mu_D > 0$.

The null hypothesis in this directional study must include all those cases where the alternative hypothesis is not true—namely, $\mu_D = 0$ (Nicoterm has *no* effect) and also $\mu_D < 0$ (the unlikely cases where Nicoterm *increases* smoking—an effect in the direction *opposite* to what previous research has shown). We combine those possibilities to form the single null hypothesis $H_0\colon \mu_D \leq 0$ because we have no desire to distinguish between $\mu_D = 0$ and $\mu_D < 0$; our interest here is only whether Nicoterm significantly *decreases* the smoking rate.

The fourth column of Table 12.1 shows a sample of size 12 from the population of difference scores. We assume that the shape of the difference score population distribution is normal, and the null hypothesis tells us that its mean $\mu_D = 0$ (μ_D could also be less than zero, but that is considered unlikely). If indeed $\mu_D = 0$, then we would expect about half of the values of D_i to be positive and the other half to be negative. If H_0 is true, Nicoterm has no effect, and so about half the smokers would be expected to smoke more cigarettes during their second week of participation in the study, while the other half would be expected to smoke less.

Test Statistic

The test statistic for dependent-sample hypothesis tests follows the general form described in Chapter 9:

$$\text{test statistic} = \frac{\text{sample statistic} - \text{population parameter}}{\text{standard error of the sample statistic}} \tag{9.1}$$

The population parameter, specified by the null hypothesis, is μ_D. The sample statistic that point-estimates μ_D is $\overline{D}$, the mean of the sample difference scores, and its standard error is $s_{\overline{D}}$. Because we are assuming that we do not know σ_D, the test statistic is t. Inserting those symbols into the general formula for the test statistic gives

$$t = \frac{\overline{D} - \mu_D}{s_{\overline{D}}}$$

However, the null hypothesis specifies that $\mu_D = 0$ (ignoring the unlikely case the $\mu_D < 0$), so the last term in the numerator vanishes, leaving

test statistic (t), two-dependent-samples test for means

$$t = \frac{\overline{D}}{s_{\overline{D}}} \qquad [df = n - 1] \tag{12.3}$$

where n is the number of *pairs* of observations. $s_{\overline{D}}$ is given by the formula

standard error of the mean differences

$$s_{\overline{D}} = \frac{s_D}{\sqrt{n}} \qquad [\text{where } n \text{ is the number of } pairs] \tag{12.4}$$

where s_D is the sample standard deviation of the difference scores—that is,

standard deviation of the differences, mean squared deviation and computational forms

$$s_D = \sqrt{\frac{\sum(D_i - \overline{D})^2}{n - 1}} = \sqrt{\frac{\sum D_i^2 - \frac{(\sum D_i)^2}{n}}{n - 1}} \tag{12.5}$$

Note that Equation (12.4) is exactly parallel to Equation (8.2), which gave the sample standard error of the mean for the variable X_i as $s_{\overline{X}} = s/\sqrt{n}$. Furthermore, Equation (12.5) is exactly parallel to Equations (5.4) and (5.4a), which computed the standard deviation of a sample of variables X_i. These equations are identical except that the present equations substitute D for X everywhere it appears. Equations (8.2), (5.4), and (5.4a) began with raw data X_i (and computed their standard deviation s or standard error $s_{\overline{X}}$), whereas the present formulas begin with difference scores D_i (and compute their standard deviation s_D or standard error $s_{\overline{D}}$).

Again, as in Chapter 11, we are assuming we do *not* know σ_D; if we did (which in real life would be quite unusual), the test statistic would be z and the denominator of Equation (12.4) would be $\sigma_{\overline{D}}$, making the equation for the test statistic $z = \overline{D}/\sigma_{\overline{D}}$.

Evaluating the Hypothesis

We can now evaluate the hypothesis that the population mean of the difference scores is 0 (or less) by following the general hypothesis-evaluation procedure described in Chapter 9.

I. State the Null and Alternative Hypotheses

Three sketches illustrate H_0:
Variable
Sample statistic
Test statistic

Eyeball-estimation of this procedure is shown on page 280.

The null and alternative hypotheses for the dependent-sample test are

$$H_0: \quad \mu_D \leq 0$$
$$H_1: \quad \mu_D > 0$$

This test is shown as directional because previous research leads us to believe that the change will be in the direction of decreasing smoking. We show the alternative hypothesis with a "greater than" sign because we have defined the difference scores to be $X_{i1} - X_{i2}$. Thus, a decrease in smoking results in X_{i1} being greater than X_{i2}, with the corresponding difference being greater than 0. As with previous directional tests, we write the null hypothesis as H_0: $\mu_D \leq 0$ to include the null case ($\mu_D = 0$) and those other cases (considered unlikely on the basis of previous research) where $\mu_D < 0$.

As we have done before, we illustrate our hypothesis-evaluation procedure with sketches of three distributions. The first is the distribution of the variable assuming the null hypothesis is true. The variable in this problem is the individual difference score D; the mean of this distribution is specified by the null hypothesis to be $\mu_D = 0$ (ignoring the unlikely case that $\mu_D < 0$). The width of this distribution is σ_D, the standard deviation of the population of all D_i values. We do not know σ_D, so, as in the past, we have to point-estimate it using s_D, which is the *sample* standard deviation of the 12 difference scores that appear in the fourth column of Table 12.1. That table showed that $\sum D_i = 120$ and $\sum D_i^2 = 4094$, so Equation (12.5) gives

$$s_D = \sqrt{\frac{\sum D_i^2 - \frac{(\sum D_i)^2}{n}}{n-1}}$$

$$= \sqrt{\frac{4094 - \frac{(120)^2}{12}}{12 - 1}}$$

$$= 16.22 \text{ cigarettes}$$

We can now sketch the distribution of the variable D assuming H_0 is true in Figure 12.1. Inspection of that distribution tells us, for example, that, assuming the null hypothesis is true, about 34% of our smokers should be expected to decrease the amount they smoke in the second week from between 0 and 16 cigarettes (and 34% should increase by the same amount), about 14% should decrease by between 16 and 32 cigarettes, and so on. Remember that a positive difference score ($X_{i1} - X_{i2}$) indicates a decrease in smoking. Our question is whether the data we gathered support or reject the null hypothesis.

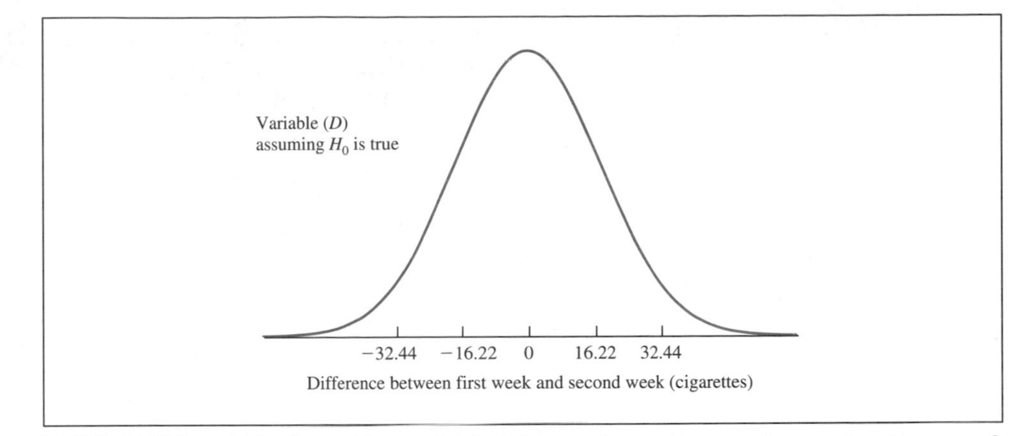

FIGURE 12.1 Distribution of the variable (change in smoking rate) assuming H_0: $\mu_D = 0$

The second distribution we sketch is the distribution of the sample statistic assuming H_0 is true—in this case, the distribution of $\overline{D}$. The mean of this distribution of sample means is $\mu_{\overline{D}}$. Recall the central limit theorem from Chapter 7, which says that the mean of the distribution of means is the same as the mean of the original variable—that is, $\mu_{\overline{X}} = \mu_X = \mu$. That theorem applies to the distribution of means of differences also, so $\mu_{\overline{D}} = \mu_D$. Now the null hypothesis states that $\mu_D \leq 0$, which implies that $\mu_{\overline{D}} \leq 0$. Therefore we draw the distribution of means of differences (the upper axis of Figure 12.2) with mean 0 (again ignoring the unlikely case that $\mu_{\overline{D}} < 0$).

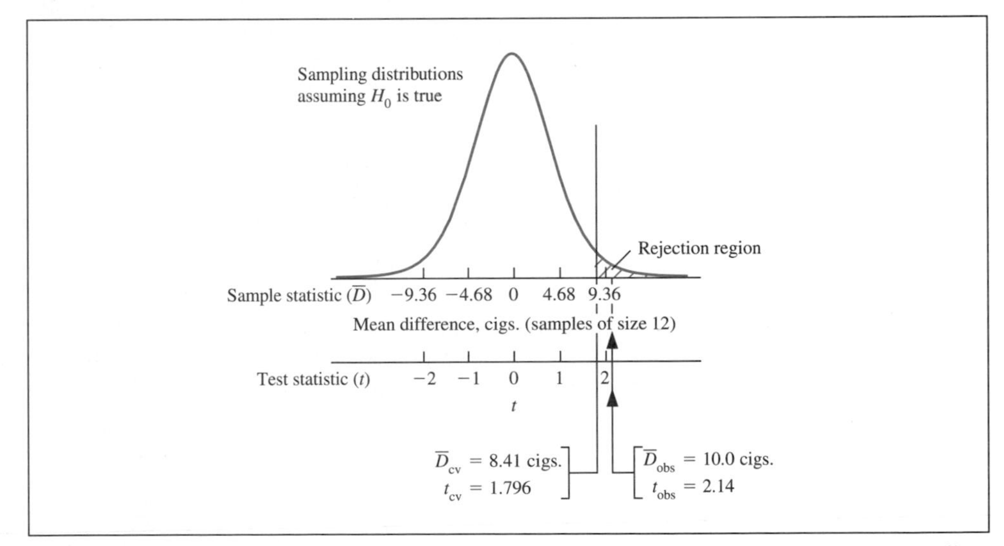

FIGURE 12.2 Distribution of the sample statistic and the test statistic with rejection regions and observed values indicated

standard error of the mean of the differences: the standard deviation of the distribution that would result from taking a random sample, measuring each member twice, computing difference scores for each member of the sample, and computing the mean of the difference scores for the sample; and then repeating the process indefinitely often

The standard deviation of this distribution is called the *standard error of the mean of the differences*, $s_{\overline{D}}$. Note that this terminology and notation are exactly parallel to those of Chapter 10. Wherever we wrote "mean" or "$\overline{X}$" in Chapter 10, we are now writing "mean of the differences" or "$\overline{D}$."

We compute $s_{\overline{D}}$, again according to the central limit theorem, from Equation (12.4) to be $s_{\overline{D}} = s_D/\sqrt{n} = 16.22/\sqrt{12} = 4.68$ cigarettes, and we show the distribution of the sample statistic $\overline{D}$ on the upper axis of Figure 12.2. We can interpret that sketch to demonstrate that, assuming the null hypothesis is true, approximately 34% of all *mean* differences for samples of size 12 will lie between 0 and an increase (or a decrease) of 4.68 cigarettes, and so on.

The third distribution of interest is the distribution of the test statistic t, which has mean 0 and standard deviation 1. It is shown on the lower axis of Figure 12.2.

Again note that the two (superimposed) curves in Figure 12.2 are constructed exactly parallel to those of Chapter 10. The terms "mean" and "$\overline{X}$" in Chapter 10 are replaced with "mean difference" and "$\overline{D}$," respectively.

II. Set the Criterion for Rejecting H_0

Level of significance?
Directional?
Shade rejection regions.

The appropriate test statistic to test this hypothesis was given by Equation (12.3):

$$t = \frac{\overline{D}}{s_{\overline{D}}} \qquad [df = n - 1]$$

The number of degrees of freedom is the number of *pairs* of data minus one; in this problem there are 12 pairs, so $df = n - 1 = 12 - 1 = 11$. Assuming that $\alpha = .05$, we find $t_{\text{cv}_{df=11}} = 1.796$ from Table A.2. Because we are interested in a *decrease* in smoking and a decrease is indicated by *positive* values of $D_i = X_{i1} - X_{i2}$, we show this critical value on the right-hand tail of the distribution of t on the lower axis of Figure 12.2.

We also wish to show the rejection region on the distribution of the statistic $\overline{D}$. The critical value of $\overline{D}$ is found just as in Chapter 10: $\overline{D}_{\text{cv}} = t_{\text{cv}}(s_{\overline{D}}) = 1.796(4.68) = 8.41$ cigarettes, shown on the upper axis of Figure 12.2.

III. Collect a Sample and Compute the Observed Values of the Sample Statistic and the Test Statistic

Observed values:
Sample statistic
Test statistic

The sample data are shown in Table 12.1. The observed value of the sample statistic is computed from $\overline{D}_{\text{obs}} = \sum D_i/n = 120/12 = 10.0$. We show that value on the upper axis in Figure 12.2. We then compute the value of the test statistic by Equation (12.3): $t_{\text{obs}} = \overline{D}_{\text{obs}}/s_{\overline{D}} = 10.0/4.68 = 2.14$, which we show on the lower axis of Figure 12.2.

IV. Interpret the Results

Statistically significant?
Practically significant?
State results clearly.

Because the observed value of the test statistic $t_{\text{obs}} = 2.14$ exceeds the critical value of the test statistic $t_{\text{cv}} = 1.796$ (or, equivalently, because $\overline{D}_{\text{obs}} = 10.0$ cigarettes exceeds $\overline{D}_{\text{cv}} = 8.41$ cigarettes), we do indeed reject the null hypothesis. Nicoterm has a statistically significant effect in reducing smoking rates.

Because we found a statistically significant difference, we must consider whether this difference was practically significant. We have rejected H_0, so we no longer believe that $\mu_D \leq 0$, but we do not know the actual magnitude of μ_D. Our best point-estimate of μ_D is now $\overline{D}_{\text{obs}} = 10.0$, and we illustrate that in Figure 12.3.

Figure 12.3 assumes that μ_D actually equals our best point-estimate of it—that is, $\mu_D = \overline{D}_{\text{obs}} = 10.0$ cigarettes—and that σ_D actually equals our best point-estimate of it—

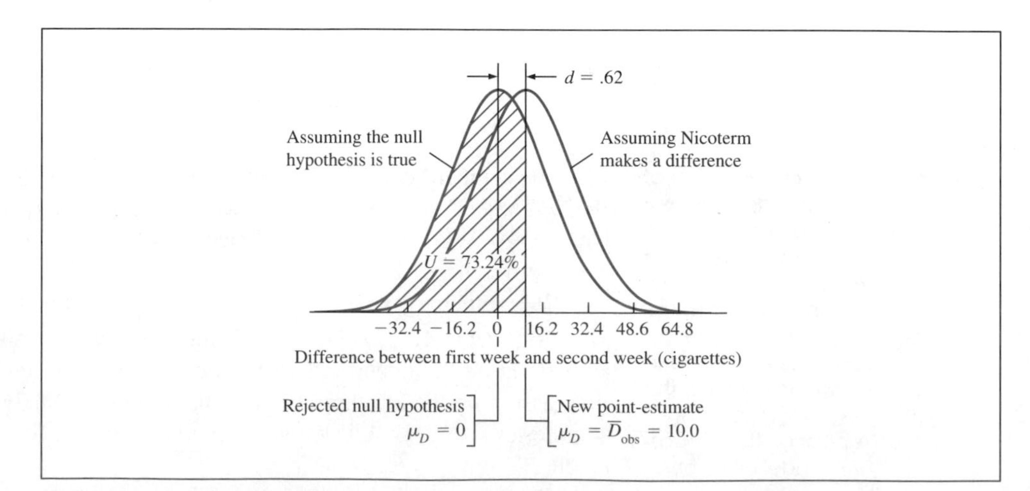

FIGURE 12.3 Distributions of the variable (difference in cigarette use), assuming the null hypothesis is true and assuming the new point-estimate is true

that is, $\sigma_D = s_D = 16.22$ cigarettes. These assumptions are probably not exactly true, but they are in most cases the best approximations we can obtain.

We compute the effect size index d for the two-dependent-samples tests of this chapter with these equations:

effect size index d, two-dependent-samples test for means

$$d = \frac{|\overline{D}_{\text{obs}}|}{s_D} \quad \text{(nondirectional test)} \qquad \text{or} \qquad d = \frac{\overline{D}_{\text{obs}}}{s_D} \quad \text{(directional test)} \tag{12.6}$$

where s_D is the standard deviation of the difference scores. Our Nicoterm example is directional, so

$$d = \frac{\overline{D}_{\text{obs}}}{s_D} = \frac{10.0}{16.22} = .62$$

which means that the average difference in smoking rates between Nicoterm users and nonusers is .62 standard deviation. We show that value on Figure 12.3.

We shade the region of nonoverlap in Figure 12.3 and compute Cohen's nonoverlap measure U by entering Table A.1 with $z = d = .62$ and finding that $U = .5 + .2324$, or 73.24%.

Now we can state our results in plain English: The mean reduction in weekly cigarette smoking (10.0 cigarettes) by using Nicoterm was statistically significantly greater than 0, a difference of .62 standard deviation. This implies that 73.24% of Nicoterm users can be expected to reduce the number of cigarettes they smoke.

12.3 Comparing Dependent- and Independent-sample t Tests

It is important to note that the same data that are used for a dependent-sample t test *could* also be used for an independent-sample t test. We illustrate this with the Nicoterm data we have been using (shown originally in Table 12.1). Instead of computing difference scores, we could compute means and standard deviations for each sample, pool the variances, and then compute the standard error of the difference between two means, exactly as we did in Chapter 11. The independent-sample summary statistics are shown in the first two columns of Table 12.2; the dependent-sample ("Difference") summary statistics are provided in the last column for comparison.

TABLE 12.2 Summary statistics for both an independent-sample t test and a dependent-sample t test using the Nicoterm data in Table 12.1

Without	*With*	*Difference*
$n_1 = 12$	$n_2 = 12$	$n = 12$
$\sum X_1 = 1832$	$\sum X_2 = 1712$	$\sum D = 120$
$\overline{X}_1 = 152.67$	$\overline{X}_2 = 142.67$	
$(\overline{X}_1 - \overline{X}_2)_{\text{obs}} = 10.0$		$\overline{D}_{\text{obs}} = 10.0$
$\sum X_1^2 = 302{,}378$	$\sum X_2^2 = 272{,}410$	$\sum D^2 = 4094$
$s_1^2 = 2062.97$	$s_2^2 = 2560.42$	
$s_1 = 45.42$	$s_2 = 50.60$	
$s^2_{\text{pooled}} = 2311.70$		
$s_{\text{pooled}} = 48.08$		$s_D = 16.22$
$s_{\overline{X}_1 - \overline{X}_2} = 19.63$		$s_{\overline{D}} = 4.68$
$t_{\text{obs}} = .51$		$t_{\text{obs}} = 2.14$
$df = 22$		$df = 11$
$t_{\text{cv}} = 1.717$, directional		$t_{\text{cv}} = 1.796$, directional
Fail to reject H_0		Reject H_0

It will be instructive to compare these computations and results with those from the dependent-sample analysis described in the previous section. First, in an independent-sample test, we begin with the means from each sample, rather than with the mean of the difference scores. But note that the observed difference between the sample means $(\overline{X}_1 - \overline{X}_2)_{\text{obs}}$ is always equal to the mean of the differences $\overline{D}_{\text{obs}}$ (10.0 in the Nicoterm example). Second, in an independent-sample test, we must compute a standard error of the difference between two means ($s_{\overline{X}_1-\overline{X}_2} = 19.63$) rather than the standard error of the mean of the difference scores ($s_{\overline{D}} = 4.68$). Third, the number of degrees of freedom in an independent-sample test increases from 11 to 22, which makes the independent-sample t_{cv} a bit smaller.

Note that the observed value of t is *much* smaller in our Nicoterm independent-sample analysis than it was in the dependent-sample analysis. Why is that?

Here are the formulas for the observed value of t side by side:

$$t_{\text{obs}} = \frac{(\overline{X}_1 - \overline{X}_2)_{\text{obs}}}{s_{\overline{X}_1-\overline{X}_2}} \qquad\qquad t_{\text{obs}} = \frac{\overline{D}_{\text{obs}}}{s_{\overline{D}}}$$

(independent samples) (dependent samples)

We saw that $(\overline{X}_1 - \overline{X}_2)_{\text{obs}}$ is always equal to $\overline{D}_{\text{obs}}$, so the two numerators must be identical. Therefore any difference between the independent-sample and the dependent-sample t_{obs} must be accounted for by differences in the denominators of the formulas for t_{obs}. In our Nicoterm example, $s_{\overline{X}_1-\overline{X}_2} = 19.63$ is much greater than $s_{\overline{D}} = 4.68$, so the independent-sample t_{obs} is much *smaller* than the dependent-sample t_{obs}.

Is $s_{\overline{X}_1-\overline{X}_2}$ always greater than $s_{\overline{D}}$? No, it is not. Therefore, t_{obs} for independent samples can be either larger or smaller than t_{obs} for dependent samples. The crucial factor is the degree to which the two samples are related. If, for most of the pairs, the second score is smaller than the first score by approximately the same amount (or if the second score of each pair is *larger* than the first score by approximately the same amount), then the difference scores (that is, the D_i values) will all be approximately the same size, and therefore s_D, the standard deviation of the D_i values, will be small. If s_D is small, then $s_{\overline{D}}$ will be smaller still, and because $s_{\overline{D}}$ is in the denominator of t, the dependent-sample t_{obs} will be large. On the other hand, if some of the difference scores are positive and some are negative, the variability of those scores, s_D, will be large, in which case $s_{\overline{D}}$ will also be large, so the dependent-sample t_{obs} will be small.

Another way of saying this is to note that if the X_{i1} data are strongly correlated with the X_{i2} data, then the dependent-sample t_{obs} is likely to be larger than the independent-sample t. We will discuss correlation in Chapter 16.

In many practical situations, we can choose between either independent- or dependent-sample designs. Making the decision between them is one of the major tasks of research design.

12.4 Dependent-sample t Test Eyeball-calibration

As in Chapters 10 and 11, we now present a series of histograms so that you can practice visualizing two-dependent-samples t tests. In Figures 12.4 through 12.10, the histograms represent difference scores. You are to assume that the null hypothesis is H_0: $\mu_D = 0$, that the hypothesis is nondirectional (two-tailed), and that the level of significance $\alpha = .05$.

The optional section following the exercises for this chapter describes the step-by-step procedure for eyeball-estimating two-dependent-samples t tests. Here our task is simply to get a sense of whether $\overline{D}_{\text{obs}}$ as eyeball-estimated from the histogram is "far away" from 0.

The question to be answered in these exercises is whether it is reasonable to suppose that the difference scores shown in a histogram are random samples from a population of differences that has population mean $\mu_D = 0$.

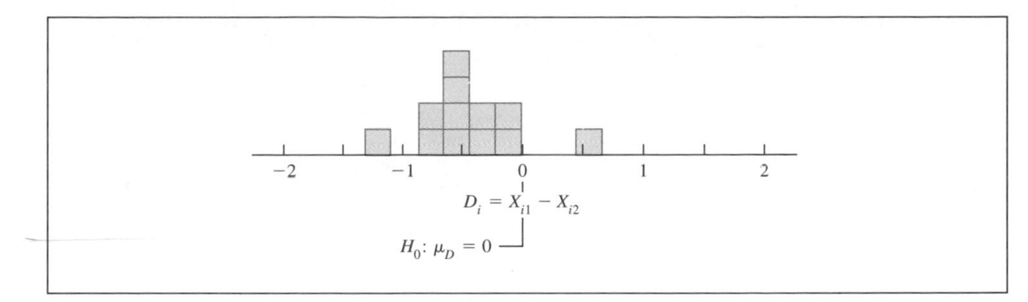

FIGURE 12.4 Should we reject H_0?

The eyeball-calibration procedure is as follows: By eyeball, determine whether $\overline{D}_{\text{obs}}$ is *quite far away* from 0. If it is, then you should reject the null hypothesis. How far is "quite far away"? The larger the sample size and the smaller the standard deviation of the differences, the smaller the required distance between $\overline{D}_{\text{obs}}$ and 0.

For example, Figure 12.4 shows a histogram of difference scores. Should we reject H_0: $\mu_D = 0$? The balancing point of the histogram seems to be substantially lower than 0, so it seems that we should reject H_0. If we performed the calculations for these data, we would find that we are correct: H_0 should in fact be rejected. (The computed values are $\overline{D}_{\text{obs}} = -.41$, $s_D = .40$, $s_{\overline{D}} = .116$, $t_{\text{obs}} = -3.53$, and $t_{\text{cv}} = \pm 2.201$.)

This is simply an eyeball-calibration exercise: Make an educated guess about whether to reject H_0 and use the feedback from the actual answers to improve your understanding of how far "quite far away" actually is. If you wish to eyeball-estimate t step by step, see the optional section at the end of the chapter.

The alert reader will recognize that this exercise is identical to the eyeball-calibration exercises in Chapter 10 with two exceptions: In Chapter 10, (1) the null hypothesis could have values other than 0, and (2) the points in the histograms represented scores rather than differences.

Figures 12.5 through 12.10 are a series of histograms for your eyeball-calibration practice. As before, *cover the actual values on the right* and decide whether to reject H_0 for the data shown in each histogram. Then check to see whether the null hypothesis would be rejected by computation. If your eyeball-estimate disagrees with the computation, try to improve your understanding of how far from 0 "quite far away" actually is. (The actual values are also provided in case you wish to follow along with the computations.)

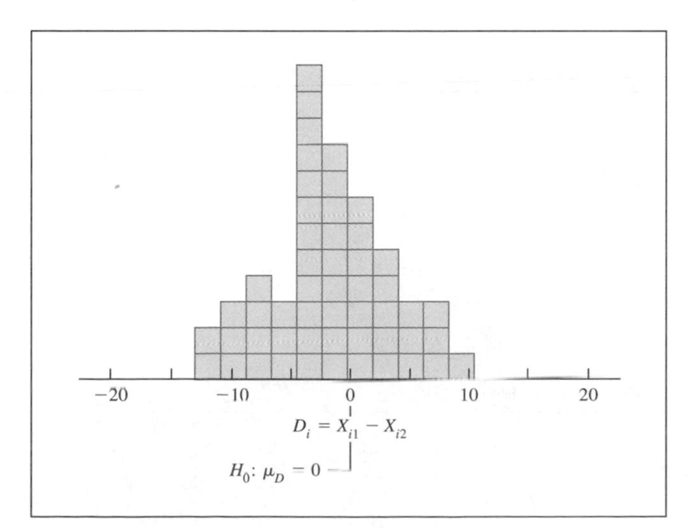

FIGURE 12.5 Should we reject H_0: $\mu_D = 0$?

Actual values	
Reject H_0?	Yes
$\overline{D}_{\text{obs}}$ (numerator of t)	−1.40
s_D	4.69
$s_{\overline{D}}$ (denominator of t)	.65
t_{obs}	−2.15
df	51
t_{cv}	±2.01

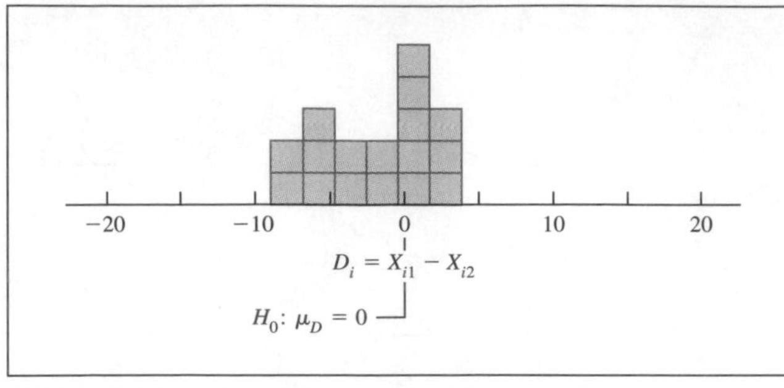

FIGURE 12.6 Should we reject H_0: $\mu_D = 0$?

Actual values	
Reject H_0?	No
$\bar{D}_{\text{obs}}$ (numerator of t)	−1.30
s_D	3.42
$s_{\bar{D}}$ (denominator of t)	.83
t_{obs}	−1.57
df	16
t_{cv}	±2.12

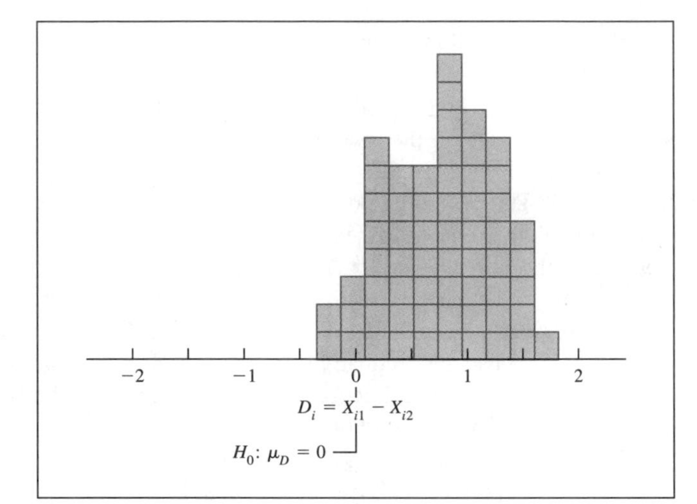

FIGURE 12.7 Should we reject H_0: $\mu_D = 0$?

Actual values	
Reject H_0?	Yes
$\bar{D}_{\text{obs}}$ (numerator of t)	.83
s_D	.45
$s_{\bar{D}}$ (denominator of t)	.06
t_{obs}	13.83
df	61
t_{cv}	±2.00

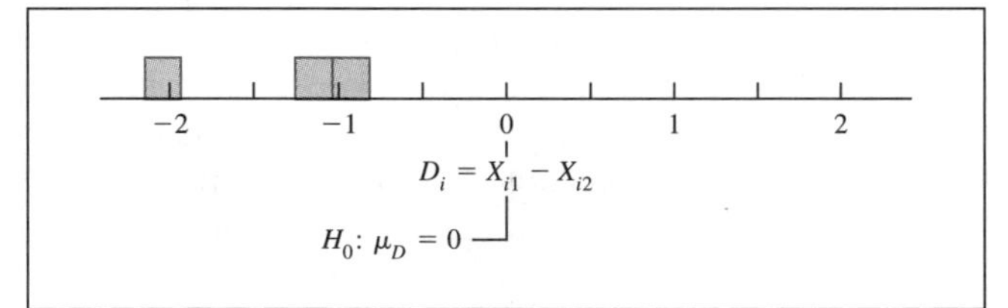

FIGURE 12.8 Should we reject H_0: $\mu_D = 0$?

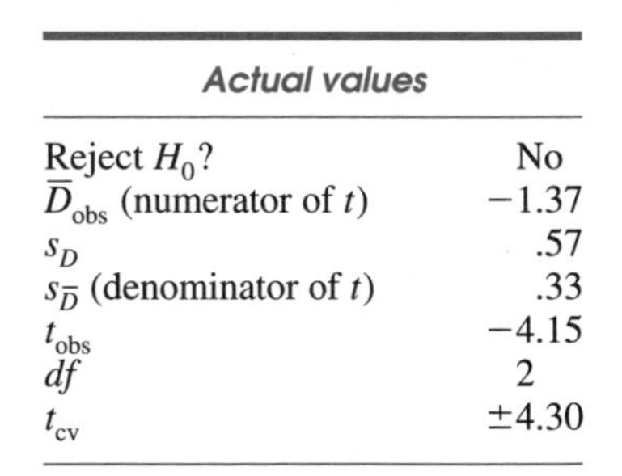

Actual values	
Reject H_0?	No
$\bar{D}_{\text{obs}}$ (numerator of t)	−1.37
s_D	.57
$s_{\bar{D}}$ (denominator of t)	.33
t_{obs}	−4.15
df	2
t_{cv}	±4.30

−2 −1 0 1 2

$D_i = X_{i1} - X_{i2}$

H_0: $\mu_D = 0$

FIGURE 12.9 Should we reject H_0: $\mu_D = 0$?

Actual values	
Reject H_0?	Yes
$\bar{D}_{\text{obs}}$ (numerator of t)	−.13
s_D	.36
$s_{\bar{D}}$ (denominator of t)	.05
t_{obs}	−2.60
df	47
t_{cv}	±2.01

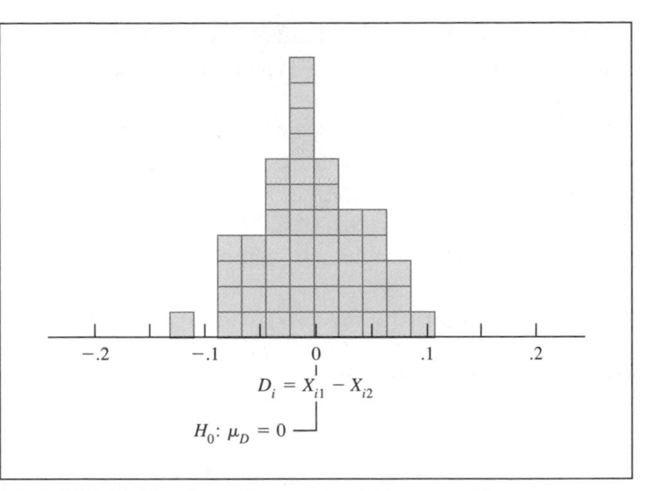

FIGURE 12.10 Should we reject H_0: $\mu_D = 0$?

Actual values	
Reject H_0?	No
$\overline{D}_{obs}$ (numerator of t)	−.002
s_D	.043
$s_{\overline{D}}$ (denominator of t)	.006
t_{obs}	−.32
df	47
t_{cv}	±2.01

12.5 Statistical Power

The present chapter has two parts. In the first part, which we have just concluded, we evaluated hypotheses in two-dependent-samples situations, thus completing the discussion of the evaluation of hypotheses about means using t and z statistics that we began in Chapter 9. In this second part of the chapter, we focus our attention on statistical power—a topic that applies to all statistical tests (those we have discussed since Chapter 9 as well as those we will discuss in the remainder of the textbook).

We saw in Chapter 9 that statistical power is the probability that an experiment will reject the null hypothesis when that null hypothesis is in fact false. For example, suppose the truth is that chewing Nicoterm reduces the number of cigarettes smoked by 16. Thus, the null hypothesis (H_0: $\mu_D \leq 0$) is in fact false and the alternative hypothesis (H_1: $\mu_D > 0$) is true—in fact, $\mu_D = 16$. We mortals are ignorant of that truth, however, so we decide to conduct an experiment to see whether Nicoterm reduces smoking. The statistical power of our Nicoterm experiment is the probability, ascertained before collecting any data, that the experiment we are about to conduct will reject the null hypothesis. (Note that the fact that the null hypothesis is actually false does not necessarily imply that our experiment will actually reject it; we may make a Type II error instead.) We noted in Chapter 9 that it was desirable for power to be high but that scientists do not conventionally specify how high.

power of a statistical test: the probability that a test will correctly reject the null hypothesis

Thus, *power* is the probability of correctly rejecting the null hypothesis. We will discuss this definition in two parts, first noting that power is a probability and second noting that it is the probability of correctly rejecting H_0.

The fact that power is a probability should not surprise us because nearly *every* statistical result is a statement of probabilities. We have seen that we are never perfectly confident of the outcome of a statistical test. For example, if we are speaking precisely, we do not say that a null hypothesis is *false*; instead, we say we *confidently reject* the null hypothesis and specify the probability that we are mistaken (e.g., at most .05 or .01). Similarly, at the outset of an experiment, we never know for sure whether we will reject H_0 or not. We call a test "powerful" if the *probability* of correctly rejecting H_0 is high.

The second thing to note about the definition of power is that it refers to a correct rejection of the null hypothesis. If H_0 is *correctly* rejected, then H_0 is false, and if H_0 is false, then the alternative hypothesis H_1 is true. Thus, another way of describing power is to say that it is the probability that we will *support* the *alternative* hypothesis when it is *true*.

You may recall that we identified in Chapter 9 the four possible outcomes of a statistical test: (1) H_0 is true and we "accept" that it is true (that is, we fail to reject it; this is a correct conclusion of no experimental effect); (2) H_0 is true but we reject it (a Type I error); (3) H_0 is false but we fail to reject it (a Type II error); and (4) H_0 is false and we reject it (a correct rejection of H_0). These four outcomes were shown in Table 9.3, which we reproduce and elaborate here as Table 12.3. Power is the probability of the *correct rejection of* H_0. In order for us *correctly* to reject H_0, it must be false; that state of nature is shown in the right-hand column of Table 12.3. Given that H_0 is false, we can either "accept" it or reject it; rejecting it is shown in the bottom row of Table 12.3. Thus, the outcome that corresponds to power is shown in the bottom right cell of Table 12.3.

TABLE 12.3 Possible outcomes of an experiment and their probabilities as a function of the true state of nature and the decision made

		Actual (but unknown) state of nature	
		H_0 true (no effect)	*H_0 false (effect exists)*
Decision	"Accept" H_0	Correct decision Probability $= 1 - \alpha$	Type II error Probability $= \beta = 1 -$ power
	Reject H_0	Type I error Probability $= \alpha$	Correct decision Probability $=$ power

In Chapter 9, we called the probability that we "accept" the null hypothesis when it is false β (that is, the probability of a Type II error). If we know that the outcome of a statistical test is in the right-hand column of Table 12.3 (that is, H_0 is false), then there are only two possible outcomes: correctly rejecting H_0 and committing a Type II error (failing to reject it). Now the probability of correctly rejecting H_0 (lower right cell in Table 12.3) is the statistical power. Because the sum of the probabilities in the upper and lower cells must be 1, the probability of the upper cell (incorrectly "accepting" H_0) must be $1 -$ power. Turning that equation around, we can see that power is one minus the probability of making a Type II error, or

power

$$\text{power} = 1 - \beta \tag{12.7}$$

We are noting here that power is a probability, which implies, like all probabilities, that we are ignorant of the outcome of our experiment. Thus, power is important to us primarily *before* we begin our experiment.[2] Power is a key element in the *planning* of experiments. We speculate (based on our previous experience with the field) about the true state of nature of our subject matter, and then we design an experiment that has a high probability of demonstrating that that supposedly true state exists.

An Example

Let us take as an example a one-sample, directional hypothesis test for means assuming that σ is known. Power can be defined in *any* hypothesis-testing situation (including the Nicoterm example we have been discussing), but the clearest illustration of power comes from the simplest examples. We therefore return to the GRE training program example from Chapter 10.

Recall that I claimed to have developed a training program that will raise students' scores on the Graduate Record Examination (GRE). I know that, in general, scores on the GRE are normally distributed with $\mu = 500$ and $\sigma = 100$. Suppose that before I collect any data, I believe (based on my previous experience with students who have taken my

[2] Power can be important after the data have been collected, too. For example, suppose we conduct an experiment that fails to reject H_0. If we perform a power analysis and determine that the experiment had high power, then we would be justified in concluding that the null hypothesis is quite likely to be at least approximately true. On the other hand, if the power analysis showed that the experiment had low power, then we would *not* be justified in drawing that conclusion because the likelihood of a Type II error is high. We leave further discussion of this important topic, usually called "proving the null hypothesis," to other texts (e.g., Cohen, 1988).

training program) that my GRE training program will increase GRE scores by 40 points. If this claim is true, then the distribution of GRE scores of students who have received my training program will have its mean shifted 40 points to the right; that is, the distribution will be normal with $\mu = 540$ and $\sigma = 100$ points. To test this claim, I will select a random sample of ten students, administer my training procedure, and have them take the GRE and report their scores back to me.

A power analysis asks this question: How likely is it that when we collect the data from this experiment, we will conclude that my training program actually increases GRE scores? Note that this question about power is asked *before* we conduct the experiment. It is a speculation into the future.

The null and alternative hypotheses for this experiment (as we saw in Chapter 10) are

$$H_0: \quad \mu \leq 500$$
$$H_1: \quad \mu > 500 \qquad \text{(directional)}$$

Note that, as we have seen, the hypothesis-testing procedure does not require us to specify a *particular* value for the parameter in the alternative hypothesis. To perform a test of a hypothesis, we allow μ as specified by H_1 to be any value greater than 500. Remember that in the typical hypothesis-testing situation, we do not know what the true value of μ is; if we did, we would not have to perform an experiment!

the "real" distribution: the distribution assuming the particular value of the parameter

In order to perform a *power analysis*, however, we *do* have to specify a particular value for the parameter, and we call that the *real value*. In the GRE example, we assume for the sake of the power analysis that my claim of a 40-point increase is true; that is, we assume that the real value of μ is 540 (and therefore *not* 500 as specified by H_0). Power, then, in this experiment, is the probability that we will correctly reject H_0 ($\mu = 500$) given that the real distribution has mean $\mu = 540$ points.

Illustrating Power

We illustrate these concepts in Figures 12.11 and 12.12. Figure 12.11 shows the distribution of the sample statistic ($\overline{X}$) assuming the null hypothesis is true. The mean of this distribution is set by H_0: $\mu = 500$ points. The standard deviation is the standard error of the mean of samples of size ten: $\sigma_{\overline{X}} = \sigma/\sqrt{n} = 100/\sqrt{10} = 31.62$ points. We have shaded the rejection region that begins at $\overline{X}_{cv} = \mu + z_{cv}(\sigma_{\overline{X}}) = 500 + 1.645(31.62) = 552.01$. Note that this distribution is drawn and the rejection region shaded *without* knowledge of the real value of μ.

This distribution illustrates that if we draw a sample whose mean is greater than 552.01, we will reject the null hypothesis; if the mean is less than 552.01, we will not reject H_0. All this is entirely a review of concepts we have been using since Chapter 10. Figure 12.11 is simply a redrawing of the sample statistic portion of Figure 10.7, with one change: We

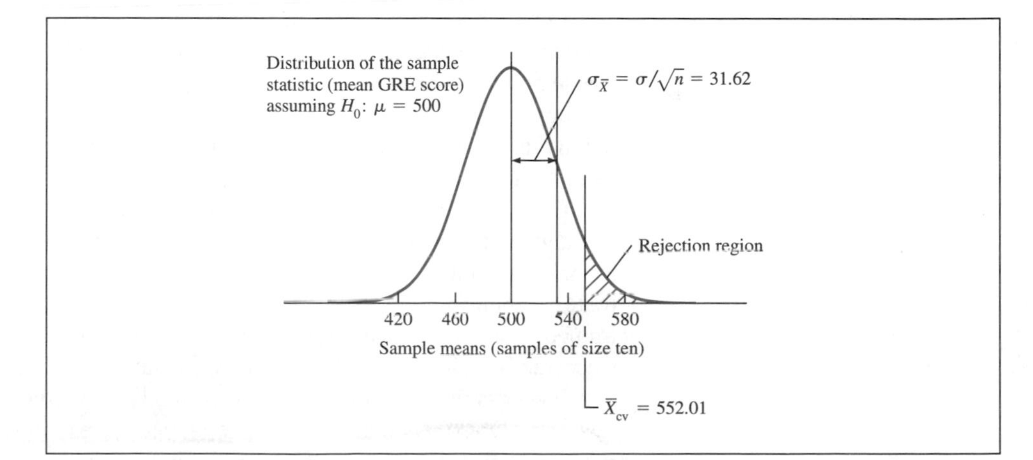

FIGURE 12.11 Distribution of the sample statistic (mean GRE score) assuming H_0, with rejection region shaded

have put round numbers on the X-axis to make the following discussion clearer. Note that the X-axis values in Figure 12.11 do *not* now correspond to standard errors: The standard error in Figure 12.11 is 31.62, and the value indicated on the X-axis is *not* 531.62 (as has been our more usual procedure), but 540.

Figure 12.12 takes the distribution of Figure 12.11 and overlays above it the real distribution, which has mean $\mu = 540$ points. [We assume it has the same standard deviation (100) and therefore the same standard error (31.62) as the distribution under H_0.] Remember that this upper distribution is the one that we are *assuming* is in fact the true distribution (which is why we call it the "real" distribution). The real distribution in Figure 12.12 illustrates that if we were to conduct this experiment repeatedly, half the sample means would lie above 540 points, about 34% would lie between 540 and 571.62 (because 540 + 31.62 = 571.62 is one standard error above the mean) points, and so on.

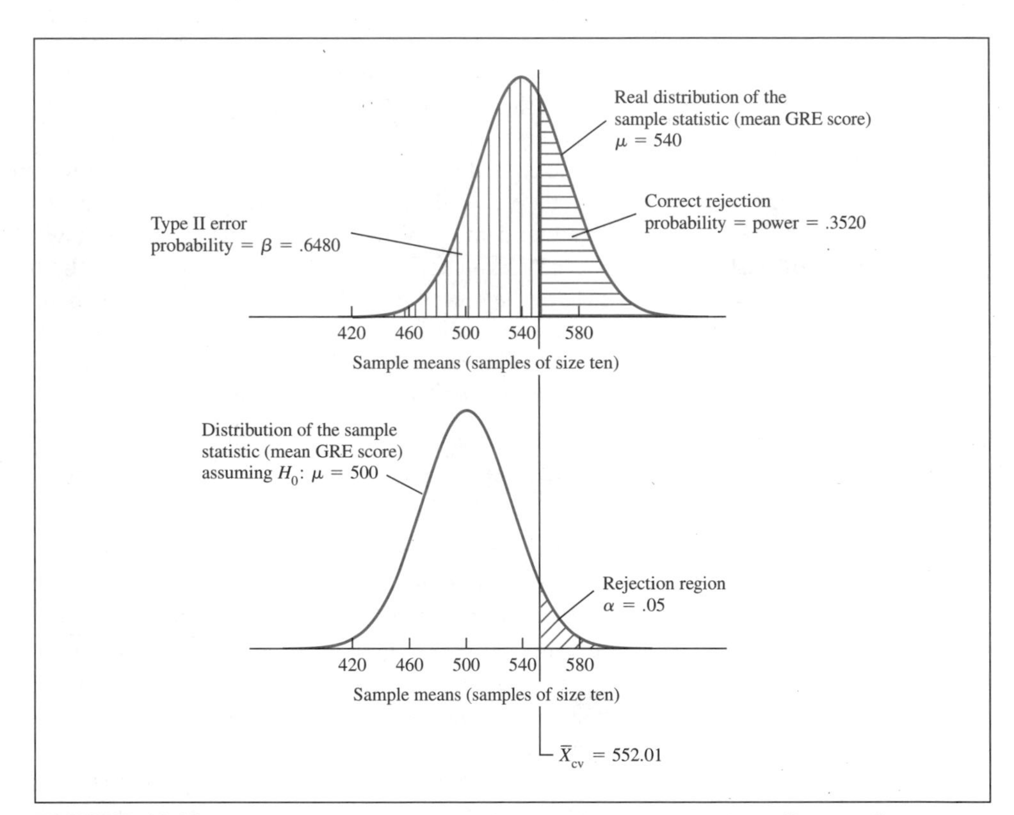

FIGURE 12.12 Distribution of the sample statistic assuming H_0 with rejection region shaded, and the real distribution with power region shaded

Remember that power is the probability of *correctly* rejecting the null hypothesis. If we are *correctly* rejecting, then the null hypothesis is false and the true state of nature is given by the real distribution. This means that power will be a region *under the real distribution* (the upper distribution in Figure 12.12). Furthermore, the only way we can *reject* H_0 is when $\overline{X}$ falls in the rejection region, which is defined as the region of the *distribution specified by the null hypothesis* (the lower distribution) when $\overline{X}$ is greater than $\overline{X}_{cv}$.

Even if the real distribution is true (that is, μ in fact does equal 540 points), we will not always reject the null hypothesis. We reject H_0 only when $\overline{X}_{obs}$ exceeds $\overline{X}_{cv}$—that is, only when $\overline{X}_{obs} > 552.01$ points. Some of the samples that we would draw from the real distribution of Figure 12.12 (more than half of them actually, as your eyeball should immediately tell you) have sample means smaller than 552.01, and in those cases, we would (incorrectly) fail to reject H_0 (that is, we would make a Type II error).

Power is the probability that we correctly reject the null hypothesis. That means that power is that region of the real distribution that corresponds to the rejection region of the null hypothesis distribution. The rejection region (of the null hypothesis distribution) is shaded with diagonal lines in Figure 12.12, just as it was in Chapter 10 (Figure 10.7); remember that the rejection region extends to infinity. The power region (of the real distribution) is shaded with horizontal lines.

To state it another way, the critical value of the sample statistic is a value determined by the distribution assuming H_0 (the lower distribution). That critical value divides the real (upper) distribution into two parts: the correct rejection region (shaded with horizontal lines in Figure 12.12 and whose probability is "power") and the Type II error (incorrect "acceptance") region (shaded with vertical lines and whose probability is β).

Power is the area of the tail of the real distribution that lies beyond the critical value of the statistic ($\overline{X}_{cv}$). We eyeball-estimate the area of power in Box 12.3 and look this area up in the table of the normal distribution exactly as we did in Chapter 6. The power area is a region under the real distribution, which is normal with $\mu = 540$ and $\sigma_{\overline{X}} = 31.62$ points. The power region is the area beyond $\overline{X}_{cv} = 552.01$ points. The z value in the real distribution associated with a sample mean score of 552.01 is obtained from Equation (7.2): $z_{552.01,\text{real}} = (\overline{X} - \mu_{\text{real}})/\sigma_{\overline{X}} = (552.01 - 540)/31.62 = .380$. Table A.1 shows that the area beyond $z_{552.01,\text{real}}$ is .3520. Thus, in this experiment, the probability that we will correctly reject H_0—that is, the power of the test—is .3520.

BOX 12.3 Eyeball-estimating power

These steps are necessary to eyeball-estimate power:

1. Sketch the distribution of the sample statistic assuming the null hypothesis is true (in our example, H_0: $\mu = 500$) and shade the rejection region. Note that in the one-sample case, this distribution is a distribution of means, and its standard error is the standard error of the mean $\sigma_{\overline{X}} = \sigma/\sqrt{n}$, as shown in Figure 12.12, lower curve.
2. Superimpose a sketch of the "real" distribution of the statistic that has the mean specified by the exact alternative hypothesis (in our example, $\mu = 540$) but the same standard error as the distribution assuming the null hypothesis (as shown in Figure 12.12, upper curve).
3. Shade the power region, which is the region under the real distribution that overlaps the rejection region (shaded with horizontal lines in Figure 12.12).
4. Eyeball-estimate the power (horizontally shaded) region using the method for eyeball-estimating areas under the normal distribution described in Chapter 6. In Figure 12.12, the power region appears to include slightly more than the small half of the first standard deviation (say, .17) plus the second standard deviation (.14) plus the right-hand tail (.02), so $\text{power}_{\text{by eyeball}} \approx .17 + .14 + .02 = .33$.

Power is computed in the text to be .3520, quite close to our eyeball estimate.

This means that even if my claim about the efficacy of my GRE training program were absolutely true (the true mean is in fact 540 points), the probability that the experiment I have designed will *conclude* that the claim is true (that is, will reject H_0) is only about .3520. If I were to perform the identical experiment 100 times, I would (correctly) reject H_0 only about 35 times, and about 65 times I would (incorrectly) fail to reject it. The statistical test in this experiment is not very powerful.

12.6 Factors That Increase Power

Experimenters rarely plan to perform 100 experiments to test the same hypothesis. Instead, they plan to perform one, and power is the probability that that single experiment will reject the null hypothesis if the variable does in fact have the real distribution. In our example, even though my training program does increase GRE scores by 40 points, the probability that my experiment will support that is only .3520. What can I do to increase that probability?

Five factors affect the power of a statistical test: the sample size, the effect size, the underlying standard deviation, the level of significance, and the directionality (number of tails) of the test. We will discuss each in turn.

Increasing the Sample Size Increases Power

The factor over which an experimenter generally has the most control is sample size. If I am willing to spend the time and money to process more subjects through my experiment, I can increase power.

For example, suppose in my GRE study that I use 16 subjects instead of 10. Figure 12.13 shows the new illustration of power. Note that for economy of space we have drawn the two distributions on top of each other rather than above and below, so that now the null hypothesis distribution and the real distribution share the same axis. Other than that, the format of Figure 12.13 is the same as Figure 12.12. Changing the sample size from 10 to 16 has made two changes from Figure 12.12. First, the standard error becomes smaller because $\sqrt{n}$ is the denominator: $\sigma_{\overline{X}_{n=16}} = \sigma/\sqrt{n} = 100/\sqrt{16} = 25$ points instead of $\sigma_{\overline{X}_{n=10}} = \sigma/\sqrt{n} = 100/\sqrt{10} = 31.62$ points. Second, the critical value of $\overline{X}$ becomes smaller because $\sigma_{\overline{X}}$ is smaller: $\overline{X}_{\text{cv}_{n=16}} = \mu_{H_0} + z_{\text{cv}}(\sigma_{\overline{X}}) = 500 + 1.645(25) = 541.125$ points as shown in Figure 12.13, instead of $\overline{X}_{\text{cv}_{n=10}} = 552.01$ points. The result is that power has increased. Computation shows that, under the real distribution, $z_{541.125,\text{real},n=16} = (541.125 - 540)/25 = .045$. Table A.1 shows the area beyond $z_{541.125,\text{real},n=16}$ to be .482, so power$_{n=16} = .482$. Recall that power$_{n=10} = .3520$.

Now suppose I use 100 subjects instead of 10 or 16. The standard error shrinks even more (to $\sigma_{\overline{X}_{n=100}} = \sigma/\sqrt{100} = 10$) and the critical value becomes even smaller $[\overline{X}_{\text{cv}_{n=100}} = 500 + 1.645(10) = 516.45]$ as shown in Figure 12.14. In this case, the power increases

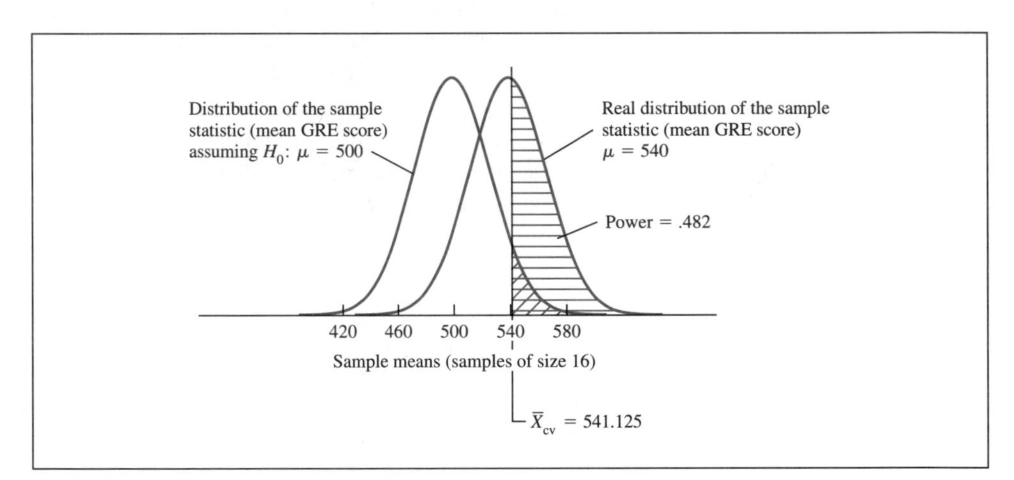

FIGURE 12.13 Distribution of the sample statistic assuming H_0 with rejection region shaded, and the real distribution with power region shaded. Like Figure 12.12 except the sample size is now 16.

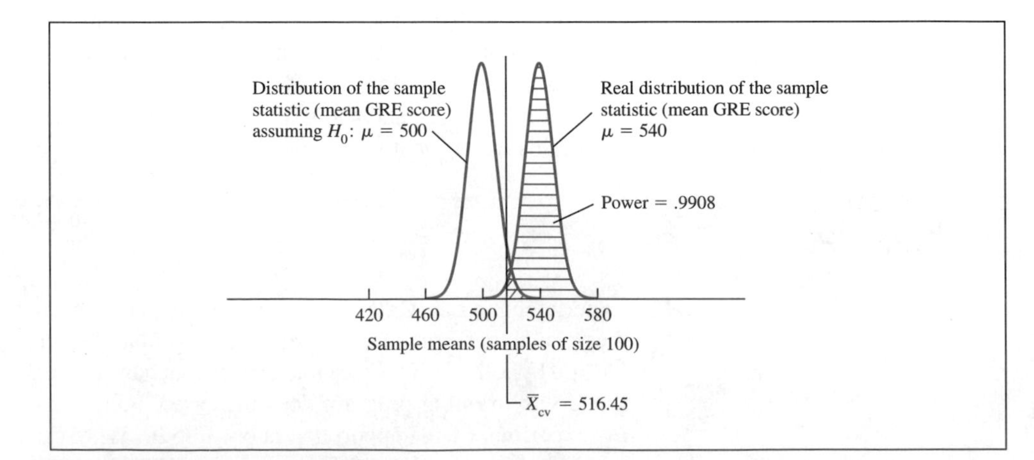

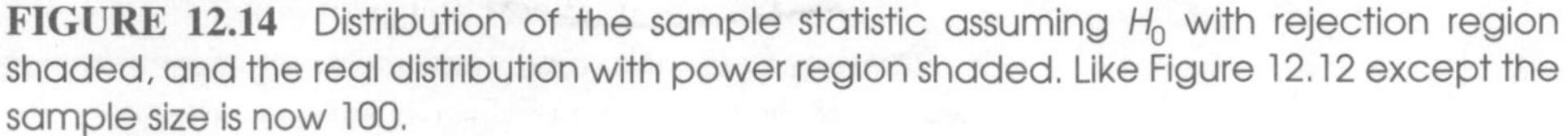

FIGURE 12.14 Distribution of the sample statistic assuming H_0 with rejection region shaded, and the real distribution with power region shaded. Like Figure 12.12 except the sample size is now 100.

dramatically. Computation shows $z_{516.45,\,\text{real},\,n=100} = (516.45 - 540)/10 = -2.355$. Table A.1 then shows power $= .4908 + .5000 = .9908$. This test is *very* powerful.

These examples illustrate a general rule: The more subjects, the higher the power.

Increasing the Raw Effect Size Increases Power

The population raw effect size in this case is $\mu_{\text{real}} - \mu_{H_0}$, the difference between the true mean of the real distribution (as specified by the exact alternative hypothesis) and the mean as specified by the null hypothesis. In our example, the raw effect size has been $540 - 500 = 40$ points. The raw effect size is the measure of the effectiveness of our treatment: how successful our training program is.

If we can increase the raw effect size, we can increase power. For example, if we can improve the GRE training program so that the average increase is 80 GRE points (that is, we double our raw effect size), the resulting illustration of power is as shown in Figure 12.15. The situation illustrated in Figure 12.15 is the same as in Figure 12.12 [same n (10), same standard error (31.62), same critical value (552.01), same rejection region (shaded with diagonal stripes as before)] except that now the raw effect size is 80 instead of 40 points, so the two distributions are somewhat more separated from each other. Note that the power area (shaded with horizontal stripes as before) is considerably larger. Calculation shows $z_{552.01,\,\text{real}} = (552.01 - 580)/31.62 = -.885$, so Table A.1 shows power to be $.3122 + .5000 = .8120$. Remember that with all else being equal, the power with a raw effect size of 40 points was .3520.

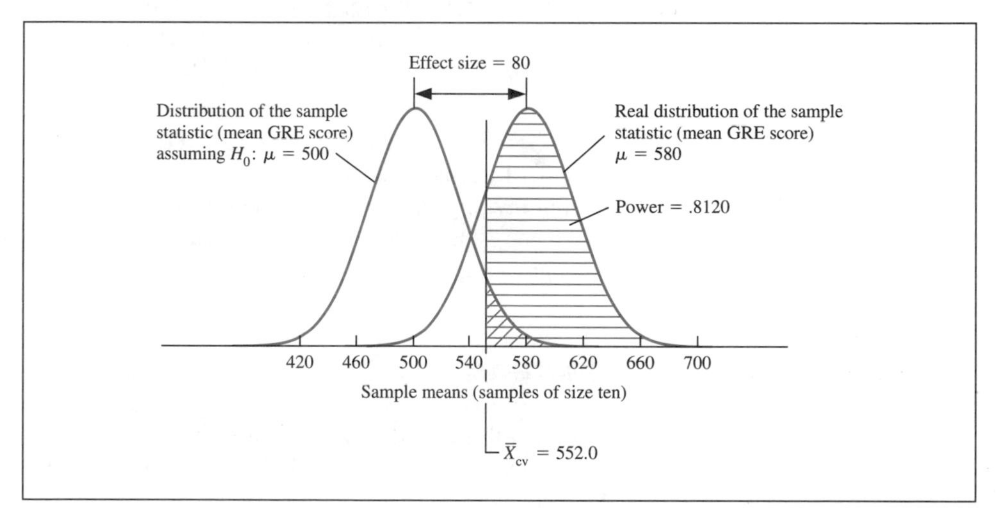

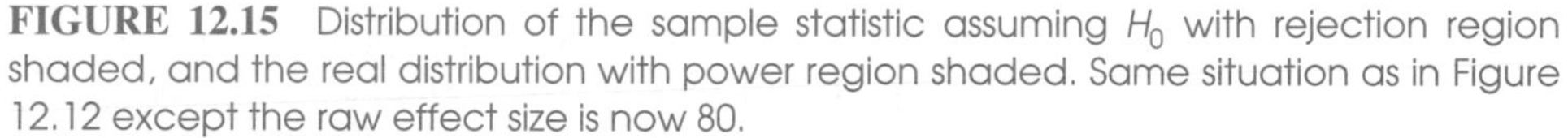

FIGURE 12.15 Distribution of the sample statistic assuming H_0 with rejection region shaded, and the real distribution with power region shaded. Same situation as in Figure 12.12 except the raw effect size is now 80.

In actual practice, it is often difficult arbitrarily to increase the raw effect size, which is an integral characteristic of the situation being investigated. In our example, it might be difficult to improve our training program so that the average increase is 80 points instead of 40. It is generally desirable, of course, to try to investigate problems where the raw effect size is large. Not only will the power of statistical tests be great in those situations, but the real-world impact will be large as well.

Decreasing σ Increases Power

The third factor that influences power is the standard deviation of the underlying distribution: If that can be reduced, power is increased. We might reduce the standard deviation in our experiment, for example, if it was conducted in a university that somehow excluded both very bright individuals (who would get high GRE scores) and the not so bright (who would get low GRE scores). Such a reduction in range would decrease the standard deviation of GRE scores to some number lower than 100 points.[3]

Let us assume for sake of illustration that σ decreases from 100 to 50. Then the standard error $\sigma_{\overline{X}}$ decreases from 31.62 to $\sigma_{\overline{X}_{\sigma=50}} = \sigma/\sqrt{n} = 50/\sqrt{10} = 15.81$ points, as shown in Figure 12.16. The critical value $\overline{X}_{\text{cv}}$ decreases to $500 + 1.645(15.81) = 526.01$ points. The power region is then much larger than its original value (.3520 in Figure 12.12). Calculation shows $z_{526.01,\text{real}} = (526.01 - 540)/15.81 = .885$, so Table A.1 shows power to be $.3120 + .5000 = .8120$.

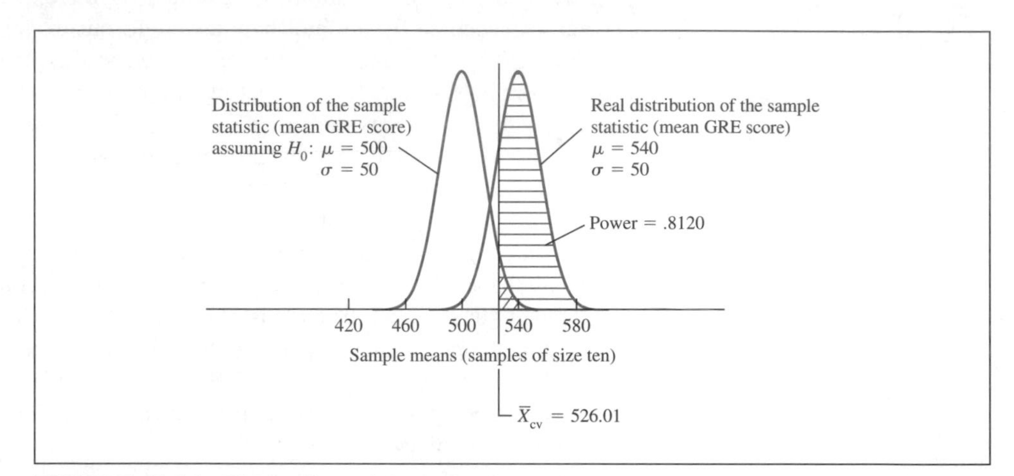

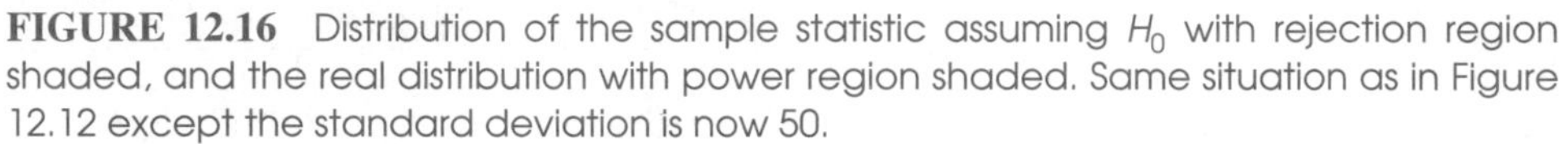

FIGURE 12.16 Distribution of the sample statistic assuming H_0 with rejection region shaded, and the real distribution with power region shaded. Same situation as in Figure 12.12 except the standard deviation is now 50.

Generally, the smaller the standard deviation, the larger the power. In actual practice, standard deviations, like raw effect sizes, are frequently characteristics of the situation being investigated and cannot be easily altered. There are exceptions, however; we have seen one—the restriction of range—and others would be described in a textbook on research design.

Increasing α Increases Power

The fourth factor that affects power is the level of significance α. The larger α becomes, the higher the power. We will illustrate this fact by showing that *decreasing* α from .05 to .01 *decreases* power.

Once again, we start with the distributions of Figure 12.12 [same raw effect size (40), same n (10), same σ (100), same standard error (31.62)], but this time we change α from .05 to .01 as shown in Figure 12.17. When α is .01, the critical value of z (directional) increases from $z_{\text{cv}_{\alpha=.05}} = 1.645$ to $z_{\text{cv}_{\alpha=.01}} = 2.326$. The critical value of the sample mean also shifts to the right, from $\overline{X}_{\text{cv}_{\alpha=.05}} = 552.01$ to $\overline{X}_{\text{cv}_{\alpha=.01}} = 500 + 2.326(31.62) = 573.55$ points, and the diagonally shaded rejection region under the null hypothesis is smaller. Moving the critical value to the right moves the left-hand boundary of the power region (horizontally shaded on the real distribution) to the right also, and power becomes smaller. Calculation shows $z_{573.55,\text{real}} = (573.55 - 540)/31.62 = 1.061$, so Table A.1 shows $\text{power}_{\alpha=.01} = .1443$ (recall that $\text{power}_{\alpha=.05} = .3520$).

It is always the case that increasing the stringency of the test (that is, making α, the probability of a Type I error, smaller) must be "paid for" by making the power of the test (the probability of correctly rejecting H_0) smaller.

---

[3]In practice, such a reduction might also shift the mean away from 500, so a one-sample test might not be appropriate. For the sake of illustration, we assume that such a shift does not take place.

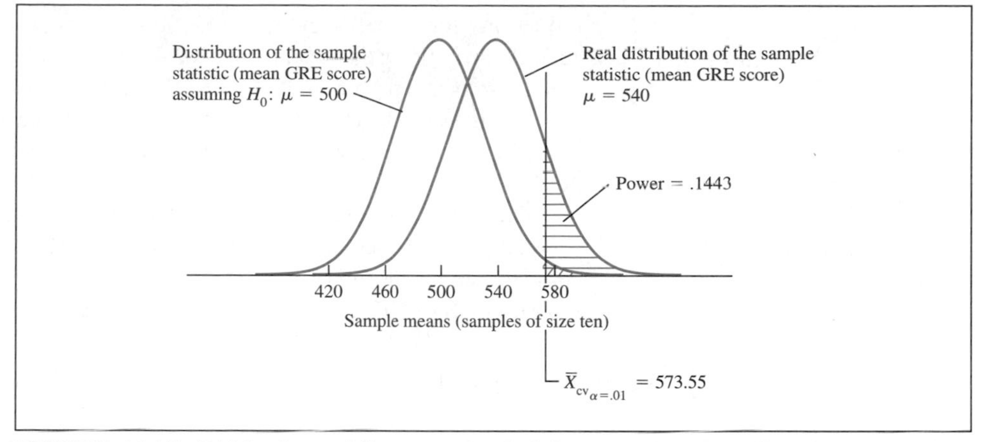

FIGURE 12.17 Distribution of the sample statistic assuming H_0 with rejection region shaded, and the real distribution with power region shaded. Same situation as in Figure 12.12 except the level of significance is now .01.

Changing from a Nondirectional to a Directional Test Increases Power

The fifth factor that affects the power of a test is the directionality of the test: directional (one-tailed) tests are more powerful than nondirectional (two-tailed) tests. Let us again start with the distributions of Figure 12.12 but change the test from directional to nondirectional as shown in Figure 12.18. The critical value of z changes from 1.645 to ± 1.96, so the critical values of $\overline{X}$ (assuming the null hypothesis is true) are now $500 \pm 1.96(31.62) = 438.02$ and 561.98. These values are farther out in the tails of the null-hypothesis distribution than the 552.01 value when the test was directional. That makes the power region in the right-hand tail smaller in the nondirectional (two-tailed) case than it was in the directional (one-tailed) case. Now $z_{561.98,\text{real}} = (561.98 - 540)/31.62 = .695$, so the nondirectional (two-tailed) power in the right-hand tail, from Table A.1, is .2435 compared to the directional (one-tailed) power of .3520.

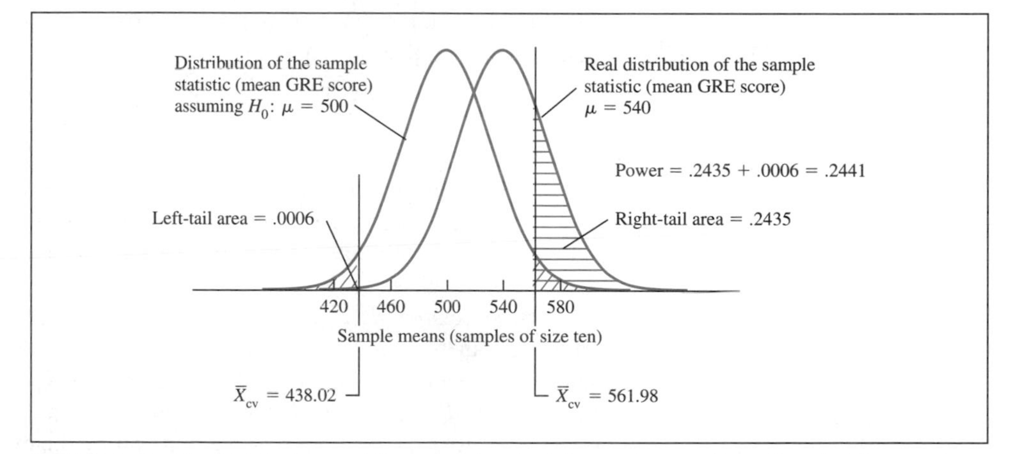

FIGURE 12.18 Distribution of the sample statistic assuming H_0 with rejection regions shaded, and the real distribution with power regions shaded. Same situation as in Figure 12.12 except the test is now nondirectional.

Note that Figure 12.18 shows a power region in the *left-hand* tail of the distributions also. Remember that power is the probability that a sample drawn from the real distribution will be extreme enough to cause rejection of the null hypothesis. It is possible (even though unlikely) that a sample from the real distribution (with $\mu = 540$) would be sufficiently extreme to lie beyond the lower critical value of 438.02, so we must calculate a power probability for the left-hand tail also. Now $z_{438.02,\text{real}} = (438.02 - 540)/31.62 = -3.225$, so Table A.1 shows that the power probability in the left-hand tail is .0006. The total power is the sum of the powers in both tails: $.2435 + .0006 = .2441$.

In this example, the total power comes overwhelmingly from one (the right-hand) tail. In cases where the raw effect size is small, the two tails may contribute more equally to power.

The GRE score example we have been using in our consideration of power has been a one-sample test of the kind described in Chapter 10. Power analyses can be performed in a parallel manner for the two-independent-samples tests of Chapter 11, the two-dependent-samples tests of the present chapter, and indeed for all other hypothesis tests.

You may wish to use the ESTAT exercise "power" at this time.

12.7 Using Power to Determine Sample Size

We have seen that we can increase power in five ways. Two of them narrow the width of the sampling distribution of means (increasing the sample size and decreasing the standard deviation), one widens the distance between the means (increasing the raw effect size), and two move the rejection region farther into the tail of the distribution (increasing α and making the test directional). High statistical power is desirable, but increasing it always has some kind of cost: We need more subjects, or a better training program, or to run a greater risk of a Type I error.

As we saw, the most direct control an experimenter has over statistical power is in the number of subjects who participate in the study. In actual practice, the most frequent application of power analysis is in determining how many subjects to use in an experiment.

To answer the question of how many subjects to use in a given experiment, it will be convenient to define the effect size index for the population in a one-sample test:

population effect size index, one-sample test

$$d_{\text{population}} = \frac{\mu_{\text{real}} - \mu_{H_0}}{\sigma} \qquad (12.8)$$

That should look familiar to you: It is very similar to the effect size index d that we defined in Chapter 10, Equation (10.3):

$$d = \frac{\overline{X}_{\text{obs}} - \mu}{\sigma}$$

When we recognize that $\overline{X}_{\text{obs}}$ is the sample point-estimate of the population μ_{real} and that μ is simply our Chapter 10 notation for μ_{H_0}, the mean specified by the null hypothesis, we can see that the effect size index d that we used in Chapter 10 is simply the sample statistic that point-estimates $d_{\text{population}}$.

The reason that $d_{\text{population}}$ is important in power analysis is that it combines two important determiners of power: the population raw effect size $(\mu_{\text{real}} - \mu_{H_0})$ and the population standard deviation σ. We have seen that increasing $(\mu_{\text{real}} - \mu_{H_0})$ *increases* power but that increasing σ *decreases* power. Furthermore, the alert reader will have noticed that, as far as power is concerned, the consequence of *doubling* the raw effect size (power equaled .8120 in Figure 12.15) was identical to the consequence of *halving* the standard deviation (power also equaled .8120 in Figure 12.16). We can use that observation to derive a general rule: Increasing $d_{\text{population}}$ increases the power of an experiment, and it does not matter whether the increase in $d_{\text{population}}$ comes from increasing its numerator or decreasing its denominator.

Thus, it generally makes most sense to perform a power analysis as a function of the effect size index rather than as a function of the raw effect size or the standard deviation separately.

Equation (12.8) gave the population effect size index for one-sample (Chapter 10) tests. Here are the population effect size indexes for two-independent-samples (Chapter 11) and two-dependent-samples (Chapter 12) tests:

population effect size index, two-independent-samples test

$$d_{\text{population}} = \frac{\mu_1 - \mu_2}{\sigma} \tag{12.9}$$

population effect size index, two-dependent-samples test

$$d_{\text{population}} = \frac{\mu_{D_{\text{real}}}}{\sigma_D} \tag{12.10}$$

Jacob Cohen, in his 25 years of writing about power analysis, has found it convenient to give the magnitude of the effect size index the labels "small," "medium," and "large." If $d_{\text{population}} = .2$, Cohen calls it a "small" effect, $d_{\text{population}} = .5$ is a "medium" effect, and $d_{\text{population}} = .8$ is a "large" effect. [See Cohen (1988) for a complete discussion or Cohen (1992) for a brief summary.] These labels are rather arbitrary but are useful as a way of comparing power across different studies. Table 12.4 gives the sample sizes required for t-test experiments to have power = .8 or power = .4 for small, medium, and large effect size indexes, given that the level of significance $\alpha = .05$ and the test is nondirectional. Note that you enter this table with the hypothesized population effect size index $d_{\text{population}}$ (small = .2, medium = .5, or large = .8) and the desired power (.8 or .4), and then read the sample size required.

For example, if we are about to perform a two-independent-samples t test (like those in Chapter 11), if this test is nondirectional with $\alpha = .05$, and if we believe that $d_{\text{population}} = .5$, then if we desire the power of the test to be .8, Table 12.4 indicates that we would have to have 64 subjects in each group. Inspection of Table 12.4 shows that increasing $d_{\text{population}}$ dramatically *decreases* the number of subjects required to obtain any given power.

TABLE 12.4 Number of subjects in each group necessary to obtain the required power for a nondirectional t test with level of significance $\alpha = .05$*

Test type	*Null hypothesis*	*Effect size index ($d_{\text{population}}$)*	*Power*	*Magnitude of $d_{\text{population}}$* Small	Medium	Large
One sample (Chapter 10)	H_0: $\mu = a$	$\frac{\mu_{\text{real}} - \mu_{H_0}}{\sigma}$	.8	196	33	14
			.4	73	13	8
Two independent samples (Chapter 11)	H_0: $\mu_1 = \mu_2$	$\frac{\mu_1 - \mu_2}{\sigma}$	.8	393	64	26
			.4	150	25	10
Two dependent samples (Chapter 12)	H_0: $\mu_D = 0$	$\frac{\mu_{D_{\text{real}}}}{\sigma_D}$	.8	196	33	14
			.4	73	13	8

*This table is derived from Cohen (1988), pp. 28–39.

Table 12.5 shows the impact that changing the level of significance or the directionality of the test has on the power of independent-sample (like Chapter 11) t tests. To continue the example of the preceding paragraph, if we change the level of significance from $\alpha = .05$ to $\alpha = .01$, we must increase the number of subjects in each group from 64 to 96. On the other hand, if we are justified in making the test directional, we can reduce the number of subjects from 64 to 50. The impact of such changes on one-sample and dependent-sample tests is quite similar.

TABLE 12.5 Number of subjects in each group necessary to obtain the required power for an independent-sample t test assuming a medium effect size, as a function of level of significance α and directionality of the test

	$\alpha = .05$		$\alpha = .01$	
Power	*Nondirectional*	*Directional*	*Nondirectional*	*Directional*
.8	64	50	96	83
.4	25	16	43	36

An Example

Suppose you are a surgeon who specializes in performing a particular kind of heart surgery. You have read medical journal reports that patients who undergo a month's exercise training prior to surgery may reduce the time it takes them to recover. Such reports, if true, would be important to your patients. You wish to know whether the exercise effect applies to your particular kind of heart surgery, so you decide to design a study where your patients are randomly assigned to one of two groups: Group 1 will undergo surgery with the normal hospital routine, whereas group 2 will receive exercise training during the month prior to hospital admission. The outcome measure will be the number of days until the patient has recovered sufficiently to return to work (or to other normal activity). The null hypothesis will be that the mean number of recovery days in group 1 is the same as the mean number of recovery days in group 2.

You decide that an independent-sample t test is appropriate to test this hypothesis, so now, in the design phase of the study, you must determine whether this test will be directional, its level of significance, and the number of subjects who will be included in each group. You survey the existing literature and decide that it is not conclusive regarding the effectiveness of such exercise, so it seems wisest to conduct a nondirectional test: Group 1 may have a larger or smaller mean recovery time than group 2.

Now you must decide the level of significance α. α is the probability of a Type I error, which in your case would be concluding that exercise training *does* make a difference when in fact it is inconsequential. Making a Type I error would be undesirable for your patients because you would advise them to delay surgery for a month (with the additional stress and inconvenience that that entails) so as to engage in exercise for no actual benefit. Thus, you would like α (the probability that you make that mistake) to be small. However, the smaller α is, the less powerful the test is, and it is desirable for the test to be as powerful as possible (see the next paragraph). You balance those two opposing forces and decide to let $\alpha = .05$. That is clearly a somewhat arbitrary choice: $\alpha = .06$ or .04 would have been just as good. It is not a totally arbitrary choice, however: $\alpha = .2$ would be too high and $\alpha = .001$ would be too low.

Now you need to decide on the power of the test. You would like to have the power be high because if exercise training does in fact have a surgical benefit, your failure to reach that conclusion would be undesirable for your patients: You would fail to advise them that they can reduce their recovery time by exercise before surgery. However, increasing power requires that you increase the sample size of your experiment, and that is also undesirable: If the null hypothesis turns out to be true, then you would have inconvenienced half your subjects (group 2—those who exercised) for no advantage to them at all. On the other hand, if the *alternative* hypothesis turns out to be true, then you would have subjected half your subjects (group 1—those who did not exercise) to a longer recovery time than was necessary. You balance those opposing forces and decide that the rational power is .8. If there is in fact a difference, you have a probability of .8 of finding it. That is clearly a somewhat arbitrary choice: power = .78 or .85 would have been just as good. It is *not* a *totally* arbitrary choice, however: power = .5 would be too low and power = .95 would be too high.

Now you need to determine the number of subjects required to attain that power, which requires that you know $d_{\text{population}}$. But if you knew $d_{\text{population}}$, there would be absolutely no reason to conduct the experiment! $d_{\text{population}}$ is the difference in mean recovery time divided by the standard deviation, and that is precisely what we *would like to*, but *do not*, know. So you must make an *educated guess* of $d_{\text{population}}$. That guess may be accurate or it may be too high or too low. But how to make it?

You recall that for independent-sample tests, $d_{\text{population}} = (\mu_1 - \mu_2)/\sigma$, so you will need to guess both numerator $(\mu_1 - \mu_2)$ and denominator (σ). $(\mu_1 - \mu_2)$ is the difference in recovery times in the two groups. You do not know that difference, so you reason as follows: If the difference $(\mu_1 - \mu_2)$ is only about one day, then your assessment is that the exercise training effect is not particularly important; it is probably not worth delaying surgery for a month to speed up recovery by one day. However, if the difference $(\mu_1 - \mu_2)$ is *two* days, then you think your patients would probably want to know about the effect and perhaps would rearrange their schedules accordingly. So you conclude that the *smallest* $(\mu_1 - \mu_2)$ that is important is about two days. If $(\mu_1 - \mu_2)$ is greater than two days, so much the better: The statistical test will be more powerful than you planned.

You also need to make an educated guess of the denominator σ, which is the standard deviation of the recovery time (assuming both groups have the same standard deviation). You ask your nursing staff to describe how long it takes patients to return to work, and their answer is sometimes as short as 15 days, sometimes about 30 days, but most often about three weeks. From that sketchy information you can eyeball-estimate the standard deviation by the range method: $\sigma_{\text{by eyeball}}$ is about a quarter of the range, or a quarter of 15 days, or about 4 days. (You also berate yourself for not keeping better records; if your charts all included the dates patients go back to work, you could compute, rather than eyeball-estimate, the standard deviation of the intervals between surgery and return to work.)

You then divide your estimate of the numerator by your estimate of the denominator to get your best sense of $d_{\text{population}} = (\mu_1 - \mu_2)/\sigma = (2 \text{ days})/(4 \text{ days}) = .5$, which is what Cohen would call a "medium" effect size. Now we can enter Table 12.4 with $\alpha = .05$, power $= .8$, a nondirectional test, and $d_{\text{population}} = .5$, and find that the required number of subjects in each group is 64. [If $d_{\text{population}}$ or the desired power had not been in Table 12.4, we would have referred to Cohen (1988), which gives the most complete power tables available.]

We illustrate the situation in Figure 12.19. The left-hand curve is the distribution of the differences between two means assuming the null hypothesis $\mu_1 - \mu_2 = 0$ is true, whereas the right-hand curve is the distribution of the differences between two means assuming the real hypothesis $\mu_1 - \mu_2 = 2$. We are assuming that $\sigma = 4$ days and there are 64 subjects in each group, so $\sigma_{\bar{X}_1 - \bar{X}_2} = (\sigma/\sqrt{n})\sqrt{2} = (4/\sqrt{64})\sqrt{2} = .707$ day by the population form of Equation (11.6a). With 64 subjects in each group, t has $63 + 63 = 126$ degrees of freedom, so Table A.2 shows that $t_{\text{cv}} = \pm 1.98$. The critical value of the difference between two means is therefore 1.98 standard errors of the difference between two means above and below 0—that is, $(\bar{X}_1 - \bar{X}_2)_{\text{cv}} = t_{\text{cv}}\sigma_{\bar{X}_1 - \bar{X}_2} = \pm 1.98(.707) = \pm 1.40$. The raw effect size is $(\mu_1 - \mu_2) = 2$ days, and because we are assuming that $\sigma = 4$ days, the standardized effect size $d_{\text{population}} = (\mu_1 - \mu_2)/\sigma = .5$. The power region is the region of the right-hand distribution that overlaps the rejection region (here almost entirely in the right tail). Your eyeball (and the table of the normal distribution) shows that power is indeed .8.

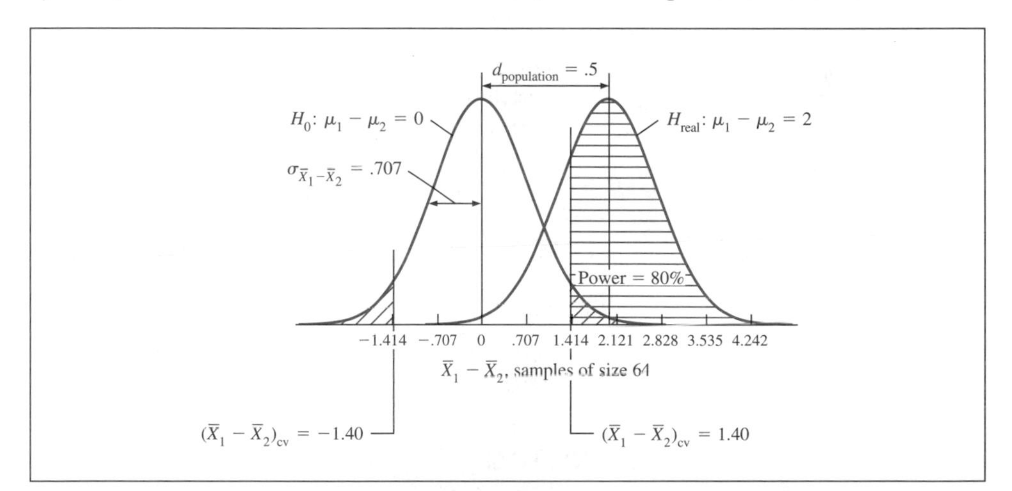

FIGURE 12.19 Power of the proposed exercise training surgery experiment

This has been a long example, so we recapitulate. We were planning a surgical experiment and we wanted to make a rational decision about how many subjects to use. We weighed the costs and benefits (primarily in terms of personal distress) of the possible experimental outcomes and decided that α should be .05 and power should be .8. Then we estimated the effect size index as best we could given the information we had available at the time. Our estimate of the effect size index, along with our required α and power, implied that we should use 64 subjects in each group.

Many observers (for example, Cohen, 1992) have criticized behavioral science research for the failure to perform power analysis and the tendency to perform experiments that are woefully underpowered. This example should make it clear that both the power and the level of significance are important considerations in research design.

12.8 Connections

Cumulative Review

We continue the discrimination practice begun in Chapter 10. You may recall the two steps:

Step 1: Determine whether the problem asks for (a) the area under a normal distribution, (b) a confidence interval, or (c) the evaluation of a hypothesis. See the Cumulative Review in Section 10.5 for hints about this step.

Step 2: If the problem asks for the evaluation of a hypothesis, consult the summary of hypothesis testing in Table 12.6 to determine which kind of test is appropriate.

Table 12.6 shows that so far we have considered three basic kinds of hypothesis tests (one-sample test for the mean, two-independent-samples test for means, and now two-dependent-samples tests for means). Note that this table is identical to Table 11.5, with the addition of the last line appropriate for this chapter. We will continue to expand this table progressively throughout the remainder of the book.

TABLE 12.6 Summary of hypothesis-testing procedures in Chapters 10–12

Design	*Chapter*	*Null hypothesis*	*Sample statistic*	*Test statistic*	*Effect size index*
One sample					
σ known	10	$\mu = a$	$\overline{X}$	z	$d = \dfrac{\overline{X}_{\text{obs}} - \mu}{\sigma}$
σ unknown	10	$\mu = a$	$\overline{X}$	t	$d = \dfrac{\overline{X}_{\text{obs}} - \mu}{s}$
Proportion	Exercise 7, Chapter 10	$\Pi = a$	p	t	—[a]
Two independent samples					
For means	11	$\mu_1 = \mu_2$	$\overline{X}_1 - \overline{X}_2$	t	$d = \dfrac{\overline{X}_{1_{\text{obs}}} - \overline{X}_{2_{\text{obs}}}}{s_{\text{pooled}}}$
Two dependent samples					
For means	12	$\mu_D = 0$	$\overline{D}$	t	$d = \dfrac{\overline{D}_{\text{obs}}}{s_D}$

[a] Effect size index is not discussed in this textbook.

Journals

The experiments reported in behavioral science journals often have very low power, which indicates (among other undesirable effects) that researchers are often willing to commit themselves to long research projects that have a low probability of success. The value of using a power analysis to determine the sample size as described in the last section of this chapter is not adequately appreciated by many researchers. (See Exercises 49 and 50.)

A typical journal report of a dependent-sample test might be: "Subjects on average smoked 10.0 fewer cigarettes per week while chewing Nicoterm, a significant reduction [$t(11) = 2.14$, $p < .05$]."

Computers

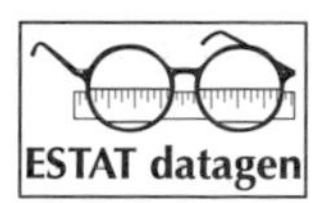

Use ESTAT's datagen to compute the dependent-sample t test for the Nicoterm data. The data are shown in Table 12.1, with H_0: $\mu_D = 0$.

1. Enter the 12 pairs of "Before" and "During" values from Table 12.1 into the first two columns of the datagen spreadsheet. Note that the Tab key will move you from the first column to the second column.
2. Datagen (version 2.0 or later) automatically computes the two-dependent-samples t test, shown in the `datagen Statistics` window in the section titled "Assuming the two variables are paired." The subcomputations required to compute t by hand calculator are also automatically provided, either in the `datagen Descriptive Statistics` window or in the `datagen Statistics` window.
3. See Homework Tip 4.

Use SPSS to compute the dependent-sample t test for the Nicoterm data. The data are shown in Table 12.1, with H_0: $\mu_D = 0$.

1. Enter the 12 pairs of "Before" and "During" values from Table 12.1 into the first two columns of the SPSS spreadsheet. Note that the Tab key will move you from the first column to the second column.
2. Click `Statistics`, then `Compare Means`, then `Paired-Samples T Test....`
3. Click `var00001` to make it the `Current Selections Variable 1`; then click `var00002` (on the Macintosh you must simultaneously press the Command key) to make it the `Current Selections Variable 2`.
4. Click ▶ to move "`var00001 - var00002`" into the `Paired Variables` window. Then click `OK`.
5. The accompanying display, which I have annotated, appears in the `!Output1` window. Note that SPSS does not provide the probability for directional tests, providing only the "2-tail Sig" exact probability. As we saw in Step 10 on page 233, the directional exact probability is precisely half the nondirectional exact probability.
6. See Homework Tip 4.

```
SPSS for MS WINDOWS Release 6.1

t-tests for Paired Samples

              Number of               2-tail
Variable        pairs        Corr      Sig         Mean          SD       SE of Mean
------------------------------------------------------------------------------------
VAR00001                                        152.6667       45.420        13.112
                 12          .949      .000
VAR00002                                        142.6667       50.601        14.607
------------------------------------------------------------------------------------

      Paired Differences
  Mean        SD     SE of Mean  |       t-value          df         2-tail Sig
---------------------------------|--------------------------------------------------
10.0000     16.220      4.682    |        2.14            11            .056
95% CI (-.306, 20.306)           |
```

Annotations: Make sure these values are correct by eyeball or some other method. Inspect your data! (Number of pairs, Mean, SD); VAR00001 = Before; VAR00002 = After; Corr: Pearson r (see Chapter 17); 2-tail Sig .000: Significance of Pearson r, *not* of t test; 10.0000 = $\bar{D}_{obs}$; 16.220 = s_D; 4.682 = $s_{\bar{D}}$; 2.14 = t_{obs}; 95% CI: Not covered in this text

Homework Tips

1. Check the list of learning objectives at the beginning of this chapter. Do you understand each one?

2. The only new concept in understanding dependent-sample tests is the difference score. Make sure you understand that.

3. Once you have computed difference scores, pretend you have never seen the original data and proceed exactly as you did in Chapter 10. That means that the number of degrees of freedom is the number of *pairs* minus one.

4. Whenever you ask a computer to perform a task for you, you should (a) make sure the number of cases is correct and (b) eyeball-estimate enough statistics to convince yourself that the computer has interpreted the data in the way you expected.

5. Make sure you are clear about the definition of statistical power: It is the probability of *correctly* rejecting H_0.

6. Note that the curves that we draw to illustrate power are distributions of the *sample statistic*, not distributions of the original variable.

7. When using the table of the normal distribution to compute power, you must compute z scores in the population assuming H_0 is true and z scores in the population assuming the real distribution is true. Be sure your notation explicitly differentiates those two different populations (e.g., z_{H_0}, z_{real}).

Exercises for Chapter 12

Section A: Basic exercises (answers in Appendix D, page 516)

1. The data in the table represent a pretest–posttest design.
   (a) Compute the difference scores (D).
   (b) Compute the sample statistic $\overline{D}_{\text{obs}}$.
   (c) Compute s_D.
   (d) Compute $s_{\overline{D}}$.
   (e) Compute the test statistic t_{obs}.
   (f) Should we reject H_0 if $\alpha = .05$ and the test is nondirectional?

Subject	*Before*	*After*
1	5	5
2	8	10
3	6	8
4	2	6
5	7	9

2. The data in this table are the identical data from Exercise 1 except the "After" values have been shuffled.

Subject	*Before*	*After*
1	5	10
2	8	5
3	6	9
4	2	8
5	7	6

   (a) Compute the difference scores (D).
   (b) Compute the sample statistic $\overline{D}_{\text{obs}}$.
   (c) Compute s_D.
   (d) Compute $s_{\overline{D}}$.
   (e) Compute the test statistic t_{obs}.
   (f) Should we reject H_0 if $\alpha = .05$ and the test is nondirectional?
   (g) Why is the result here different from that in Exercise 1?

3. TrimQuik is a pill advertised to facilitate weight loss. We collect a random sample of ten men and weigh them. We then administer TrimQuik and weigh the same individuals again. The weights (in pounds) are shown in the table. Use $\alpha = .05$. *Omit the portions marked with an asterisk (*) unless you have read the optional section on eyeball-estimation that follows the exercises.

Person	*Before*	*After*
1	190	184
2	191	166
3	164	172
4	181	192
5	187	174
6	178	153
7	175	173
8	164	158
9	198	176
10	153	152

I. State the null and alternative hypotheses

(a) What is the null hypothesis? The alternative hypothesis?
(b) Is this test directional or nondirectional? Why?
(c) What is the variable in this problem? (Eyeball-estimate and then)* compute its standard deviation. Sketch the distribution of the variable assuming the null hypothesis is true.
(d) What is the sample statistic in this problem? (Eyeball-estimate and then)* compute its standard error. Sketch the distribution of the sample statistic assuming the null hypothesis is true.
(e) What is the test statistic in this problem? Sketch the distribution of the test statistic assuming the null hypothesis is true. You may simply append a new axis to a distribution you already sketched.

II. Set the criterion for rejecting H_0

(f) What is the level of significance?
(g) How many degrees of freedom are there?
(h) (Eyeball-estimate and then)* look up the critical value(s) of the test statistic. Enter it (them) on the appropriate distribution and shade the rejection region(s).
(i) (Eyeball-estimate and then)* compute the critical value(s) of the statistic. Enter it (them) on the appropriate distribution and shade the rejection region(s).

III. Collect a sample and compute statistics

(j) (Eyeball-estimate and then)* compute the observed value of the statistic. Enter it on the appropriate distribution.
(k) (Eyeball-estimate and then)* compute the observed value of the test statistic. Enter it on the appropriate distribution.
(l) Do the critical values, rejection regions, and observed values have the same relative positions on their respective distributions? Explain.

IV. Interpret the results

(m) Is this result statistically significant (that is, should we reject H_0)? Decide (by eyeball and)* by computation.
(n) If the result was statistically significant, then determine the effect size index d and Cohen's nonoverlap statistic U. Illustrate these as shown in Figure 12.3.
(o) Describe the results, including both statistical and practical significance, in plain English.

4. The personnel director of a large corporation wishes to know whether changing from fluorescent lights to incandescent light bulbs will affect employee morale. He randomly selects 12 employees and administers to them the "Morale Scale," a paper-and-pencil test he has developed. He then has the work spaces of each employee equipped with incandescent light bulbs. One month later, he tests employees again on the Morale Scale. The scores are shown in the table. Has the type of lighting affected morale? Use $\alpha = .05$. Use parts (a)–(o) from Exercise 3.

Table for Exercise 4

Person	Fluores.	Incand.
1	84	86
2	66	73
3	71	66
4	80	74
5	78	69
6	77	77
7	75	70
8	84	78
9	71	70
10	80	74
11	83	76
12	66	63

Table for Exercise 5

	Fluores.	Incand.
	84	86
	66	73
	71	66
	80	74
	78	69
	77	77
	75	70
	84	78
	71	70
	80	74
	83	76
	66	63
Labor savings for the student:		
$\sum X_i$	915	876
$\sum X_i^2$	70,233	64,352

5. (*Exercise 4 continued*) The personnel director randomly selects 12 employees whose work spaces have incandescent lights and another 12 employees whose work spaces have fluorescent lights, and administers both groups the Morale Scale. The scores are shown in the table. Has the type of lighting affected morale? Use $\alpha = .05$. Use parts (a)–(o) from Exercise 3.

6. Note that the data used in Exercises 4 and 5 are identical. How do you account for the fact that the results are different?

7. An investigator wishes to test the hypothesis that "blondes have more fun." To investigate, he develops the Subjective Report of Fun Scale (high scores indicate more fun), selects 13 brunettes at random, and administers the SRFS to them. He then bleaches all 13 subjects' hair blonde, waits two weeks, and administers the SRFS again. The scores are shown in the table. Use $\alpha = .05$. Use parts (a)–(o) from Exercise 3.

Person	Brunettes	Blondes
1	70	50
2	47	56
3	57	63
4	61	73
5	65	51
6	57	65
7	58	60
8	58	63
9	62	64
10	56	66
11	52	61
12	60	83
13	51	71

8. (*Exercise 7 continued*) An investigator selects 13 brunettes at random and administers the SRFS to them. He then selects 13 blondes at random and administers the SRFS to

them also. The scores are shown in the table. Use $\alpha = .05$. Do blondes have more fun?

	Brunettes	Blondes
	70	50
	47	56
	57	63
	61	73
	65	51
	57	65
	58	60
	58	63
	62	64
	56	66
	52	61
	60	83
	51	71
Labor savings for the student:		
$\sum X_i$	754	826
$\sum X_i^2$	44,166	53,432

9. (*Exercises 7 and 8 continued*) Note that the data used in Exercises 7 and 8 are identical. How do you account for the fact that the results are different?

10. Suppose I have a pill that I believe increases a person's IQ by 5 points (IQ is normally distributed with $\mu = 100$ and $\sigma = 15$ points). To demonstrate my claim, I randomly sample 25 people from the population, administer my pill, and then test their IQs.
    (a) Sketch the distribution of the sample statistic (mean IQ, samples of size 25) assuming the null hypothesis is true, and shade a directional rejection region. (Use $\alpha = .05$.)
    (b) On the same axes, sketch the distribution of the sample statistic (mean IQ, samples of size 25) among those who have taken my pill, assuming my claim is true. Shade the power region.
    (c) By eyeball, what is the power of this test?
    (d) Compute the power.
    (e) What is the probability of making a Type II error in this experiment, assuming my claim is true?

11. (*Exercise 10 continued*) Suppose that instead of 25 subjects, I use 100 subjects.
    (a) Sketch the distribution of the sample statistic (mean IQ, samples of size 100) assuming the null hypothesis is true, and shade a directional rejection region. (Use $\alpha = .05$.)
    (b) On the same axes, sketch the distribution of the sample statistic (mean IQ, samples of size 100) among those who have taken my pill, assuming my claim is true. Shade the power region.
    (c) By eyeball, what is the power of this test?
    (d) Compute the power. How does that compare with the power in Exercise 10? Explain.
    (e) What is the probability of making a Type II error in this experiment, assuming my claim is true?
    (f) What can be said in general about the effect on power of increasing the sample size?

12. (*Exercise 10 continued*) Suppose that I use 25 subjects, but I decide to use $\alpha = .01$.
    (a) Sketch the distribution of the sample statistic (mean IQ, samples of size 25) assuming the null hypothesis is true, and shade a directional rejection region.
    (b) On the same axes, sketch the distribution of the sample statistic (mean IQ, samples of size 25) among those who have taken my pill, assuming my claim is true. Shade the power region.
    (c) By eyeball, what is the power of this test?
    (d) Compute the power. How does that compare with the power in Exercise 10? Explain.
    (e) What is the probability of making a Type II error in this experiment, assuming my claim is true?
    (f) What can be said in general about the effect on power of decreasing the level of significance?

13. (*Exercise 10 continued*) Suppose I use 25 subjects and $\alpha = .05$, but my pill actually increases IQ by 10 points instead of 5.
    (a) Sketch the distribution of the sample statistic (mean IQ, samples of size 25) assuming the null hypothesis is true, and shade a directional rejection region.
    (b) On the same axes, sketch the distribution of the sample statistic (mean IQ, samples of size 25) among those who have taken my pill, assuming my claim is true. Shade the power region.
    (c) By eyeball, what is the power of this test?
    (d) Compute the power. How does that compare with the power in Exercise 10? Explain.
    (e) What is the probability of making a Type II error in this experiment, assuming my claim is true?
    (f) What can be said in general about the effect on power of increasing the effect size?

14. (*Exercise 10 continued*) Suppose I use 25 subjects, the raw effect size is 5 points, and $\alpha = .05$, but I sample in a population where the standard deviation of IQs is $\sigma = 10$ instead of 15.
    (a) Sketch the distribution of the sample statistic (mean IQ, samples of size 25) assuming the null hypothesis is true, and shade a directional rejection region.
    (b) On the same axes, sketch the distribution of the sample statistic (mean IQ, samples of size 25) among those who have taken my pill, assuming my claim is true. Shade the power region.
    (c) By eyeball, what is the power of this test?
    (d) Compute the power. How does it compare with the power in Exercise 10? Explain.
    (e) What is the probability of making a Type II error in this experiment, assuming my claim is true?
    (f) What can be said in general about the effect on power of decreasing σ?

15. (*Exercise 10 continued*) Suppose I use a nondirectional test instead of a directional test; all else is the same as in

Exercise 10 (that is, raw effect size is 5 points, $\alpha = .05$, and $\sigma = 15$ points).

(a) Sketch the distribution of the sample statistic (mean IQ, samples of size 25) assuming the null hypothesis is true, and shade the nondirectional rejection region.
(b) On the same axes, sketch the distribution of the sample statistic (mean IQ, samples of size 25) among those who have taken my pill, assuming my claim is true. Shade the power regions.
(c) By eyeball, what is the power of this test?
(d) Compute the power. How does that compare with the power in Exercise 10? Explain.
(e) What is the probability of making a Type II error in this experiment, assuming my claim is true?
(f) What can be said in general about the effect on power of making a directional test nondirectional?

16. Assume that the population has a raw effect size of 20 points, the test is nondirectional, $\alpha = .05$, and $\sigma = 100$. Assume also that the experiment is a one-sample hypothesis test with test statistic t. How many subjects must I use if I desire power to be .8?

Section B: Supplementary exercises

17. Show for the data in the table that $\overline{D} = \overline{X}_1 - \overline{X}_2$. Does that mean that the numerators of the independent-sample and the dependent-sample t test will be identical for these data? In general, will the numerators of independent and dependent t tests be identical?

i	X_1	X_2
1	13	15
2	8	11
3	12	14
4	15	21
5	9	11
6	13	13

18. (*Exercise 17 continued*) Compute both the independent-sample t and the dependent-sample t for the data. Which of these t's is significant (nondirectional, $\alpha = .05$)?

19. (*Exercises 17 and 18 continued*) Suppose we shuffle the X_1 data. Does that change $\overline{X}_1$ or s_1? Suppose we then shuffle the X_2 data. Does that change $\overline{X}_2$ or s_2? Does either of these shufflings change the value of the pooled standard deviation s? Does either of these shufflings change the value of the independent-sample t?

20. (*Exercises 17–19 continued*) Suppose we shuffle the X_1 data and then shuffle the X_2 data. Does that change the value of the D_i's? Does that change the value of s_D? Does that change the value of $s_{\overline{D}}$? Does that change the value of the dependent-sample t?

21. (*Exercises 17–20 continued*) You should have concluded in Exercise 18 that the dependent t was significant, whereas the independent t was not. Can these data be shuffled so that *neither* t is significant?

22. (*Exercises 17–21 continued*) Can these data be shuffled so that the *independent*-sample t is significant but the *dependent*-sample t is not significant?

23. The school district would like to ascertain the effect of October's ending of daylight saving time on school attendance, reasoning that the extra hour of daylight in the morning would make it easier for some students to get to school. There are 14 schools in the district; it records the number of students absent on the last Thursday of daylight saving time and then again on the first Thursday of non–daylight saving time. The numbers of absences are shown in the table. Does the end of daylight saving time decrease school absence? Use parts (a)–(o) from Exercise 3.

School	Before	After
1	39	39
2	21	17
3	24	21
4	24	24
5	16	11
6	23	17
7	30	20
8	26	21
9	21	25
10	19	18
11	28	24
12	27	21
13	30	24
14	12	18

24. A politician admitted to his constituents that he had committed some indiscretions in his personal life and he is now trying to correct his past mistakes. He would like to know the effect this announcement has on voters. Perhaps his candor will increase voter respect, but perhaps his admissions will focus the voters' attention on his failings, thus decreasing voter respect. There are 12 counties in his state, and he conducts two polls in each county: one before his admission and one after. The table shows the proportions of voters who favor this candidate in each county. Has his admission affected the number of voters who favor him? Use parts (a)–(o) from Exercise 3.

County	Before	After
Elm	.61	.40
Oak	.34	.29
Aspen	.33	.35
Birch	.41	.48
Pine	.35	.35
Palm	.52	.53
Sycamore	.85	.61
Fir	.68	.51
Mahogany	.65	.44
Cedar	.54	.43
Juniper	.12	.22
Poplar	.80	.52

25. A company monitors the stress experienced by its employees by administering twice a year a short stress questionnaire. Past results have shown that experienced stress as measured by this questionnaire has mean 57.4 and standard deviation 14.4 points. Two of the company's directors have just been arrested for fraud, and the company wishes to know whether this event has increased stress among employees. It administers the questionnaire again to a random sample of 21 employees to ascertain whether the arrests increased stress. What is the power of this test if we assume that the true stress level has increased to 61.4 points?
    (a) Sketch the distribution of the statistic assuming the null hypothesis is true, and shade the rejection region(s).
    (b) On the same axes, sketch the distribution of the statistic, making the stated assumption about the true parameter. Shade the power region(s).
    (c) By eyeball, what is the power of this test?
    (d) Compute the power.
    (e) What is the probability of making a Type II error in this experiment, assuming the real claim is true (here, that mean stress equals 61.4)?
26. (*Exercise 25 continued*) About how many subjects would be necessary if we desired the power to be .8?
27. A particular assembly line has had an average absenteeism rate of 15.7 days per year. An industrial psychologist believes that giving assembly-line workers some control over their situation (by asking them for suggestions about how assembly processes can be improved, etc.) will decrease their absenteeism. She takes one assembly line, institutes the new procedures for those 30 employees, and counts the days the workers on that line are absent. How powerful is this test if we assume that the true mean absentee rate for workers under the new system is 14.0 days and that the standard deviation $\sigma = 4.47$ days? Use the steps of Exercise 25.
28. (*Exercise 27 continued*) How powerful would this test be if we used $\alpha = .01$ instead of $\alpha = .05$? Use the steps of Exercise 25.

Section C: Extending your comprehension (answers in Appendix D, page 520)

29. Is it possible to create your own data set with six pairs of points so that the *independent*-sample t is significant but the *dependent*-sample t is not significant? If so, do it.
30. Is it possible to create your own data set with six pairs of points so that the *dependent*-sample t is significant but the *independent*-sample t is not significant? If so, do it.
31. Is it possible to create your own data set with six pairs of points so that *both* the independent-sample and the dependent-sample tests are significant? If so, do it.
32. Is it possible to create your own data set with six pairs of points so that *neither* the independent-sample nor the dependent-sample test is significant? If so, do it.
33. Do you conclude from Exercises 29–32 that there exist data sets such that both, or neither, or the independent but not the dependent, or the dependent but not the independent tests are significant? What are the characteristics of those four kinds of data sets?
34. Suppose that $\alpha = .05$ and a test is directional. Suppose also that the power of this test is .05. What can be said about the raw effect size?
35. Suppose that $\alpha = .05$ and a test is directional. Suppose also that the power of this test is .50. What can be said about the raw effect size?
36. Can the power of a test ever be less than the level of significance α? If so, state the conditions.
37. Can the power of a nondirectional test ever be greater than the power of a directional test? If so, state the conditions.

Section D: Cumulative review (answers in Appendix D, page 520)

For Exercises 38–44, choose the correct answer or fill in the blank. (**Note that computations are *not* required;** use $\alpha = .05$ unless otherwise instructed.)

(a) This problem requires:
    (1) Finding the area under a normal distribution [Skip parts (b)–(g) and proceed to (h).]
    (2) Creating a confidence interval
    (3) Testing a hypothesis about the mean of one group
    (4) Testing a hypothesis about the means of two independent groups
    (5) Testing a hypothesis about the means of two dependent groups
    (6) Performing a power analysis (involving finding the area under a normal distribution). [Skip parts (b)–(h).]

(b) The null hypothesis is of the form
    (1) $\mu = a$
    (2) $\mu_1 = \mu_2$
    (3) $\mu_D = 0$
    (4) There is no null hypothesis.

(c) The appropriate sample statistic is:
    (1) $\overline{X}$ (2) p
    (3) $\overline{X}_1 - \overline{X}_2$ (4) $\overline{D}$

(d) The appropriate test statistic is:
    (1) z (2) t

(e) The number of degrees of freedom for this test statistic is ________. (State the number or *not applicable*.)

(f) The hypothesis test (or confidence interval) is:
    (1) Directional (2) Nondirectional

(g) The critical value of the test statistic is ________. (Give the value)

(h) The appropriate formula for the test statistic (or confidence interval) is given by Equation (________).

Note that computations are *not* required for any of the problems in Section D. See the instructions.

38. A golf ball manufacturer has available two kinds of synthetic rubber, compound A and compound B, with which to make the center of its golf balls. It wishes to know whether golf balls made of one compound travel farther than balls made of the other compound. It makes 36 of each kind of golf ball and hits them with a machine that simulates a driver. The distances each ball travels (in yards) are shown here. Is one type of ball superior to the other?

    Compound A: 233, 242, 237, 222, 251, 237, 237, 246, 227, 232, 230, 227, 225, 263, 263, 234, 248, 240, 237, 244, 234, 252, 227, 203, 249, 233, 257, 212, 237, 248, 252, 231, 240, 219, 215, 226

    Compound B: 212, 237, 209, 235, 255, 238, 220, 263, 221, 253, 217, 249, 222, 237, 220, 255, 212, 236, 250, 228, 210, 242, 221, 243, 231, 234, 237, 243, 220, 207, 269, 246, 216, 216, 221, 227

39. (*Exercise 38 continued*) The golf ball manufacturer also supplies balls made of the two compounds to 15 golfers, asking each to shoot 18 holes of golf with the compound A ball and 18 holes with the compound B ball. Both rounds of golf are to be shot on the same course. Half the players use the A ball first, and the other half use B first. All players report their scores for both balls. The scores are shown in the table. Is one ball superior to the other?

Player	Compound A	Compound B
1	97	96
2	96	104
3	96	94
4	103	111
5	79	88
6	85	88
7	100	102
8	96	99
9	98	98
10	87	99
11	93	99
12	87	103
13	94	99
14	86	89
15	102	102

40. (*Exercise 38 continued*) The golf ball manufacturer wishes to state how far its driving machine, on average, drives golf ball B. It takes a random sample of 36 balls and drives them with the machine. The driving distances (in yards) are: 212, 237, 209, 235, 255, 238, 220, 263, 221, 253, 217, 249, 222, 237, 220, 255, 212, 236, 250, 228, 210, 242, 221, 243, 231, 234, 237, 243, 220, 207, 269, 246, 216, 216, 221, 227. What can be said about the average distance of compound B balls in general?

41. Ping-Pong balls used in a Ping-Pong club have life expectancies that are normally distributed with mean 23 games and standard deviation 4 games. What percentage of Ping-Pong balls last for more than 30 games?

42. Suppose you run an all-you-can-eat buffet restaurant. You are considering adding pizza squares to your appetizer table. Your decision will be made on economic grounds: If the total food expense for your buffet is decreased by the addition of pizza squares (that is, if customers eat less of the expensive foods because they chow down on the cheaper pizza), then you will add pizza. For the next 20 days, you randomly assign each day to the pizza or no-pizza condition. The food expenses for each day (in dollars) are listed in the table. Has the introduction of pizza squares decreased food expense?

Pizza	No Pizza
595	807
568	837
667	835
581	618
683	846
697	647
703	679
446	698
746	767
698	664

43. A record store knows that its daily sales have mean \$905 and standard deviation \$16. It has just instituted a new advertising campaign and wishes to know whether this campaign has increased sales. Assuming that the actual daily sales have increased to \$913, for how many days should it sample in order to make the probability of concluding that sales have in fact increased greater than or equal to .8?

44. *(Exercise 43 continued)* Suppose that the record store actually records sales for the next 20 days. The sales (in dollars) are: 981, 1000, 1009, 985, 995, 983, 973, 1004, 1000, 971, 976, 974, 995, 985, 987, 981, 1000, 1004, 980, 991. Has the campaign increased sales?

Section E: From the journals (answers in Appendix D, page 520)

45. As part of a study on the influence of exercise on mood, Simons and Birkimer (1988) assigned 52 subjects to exercise classes that met twice a week for eight weeks. Before the researchers could conclude that exercise affected mood, they had to show that the exercise had some physiological effect. To show that exercise produced a change, they measured subjects' heart rates during a standard exercise in the first session, and then they measured the same subjects again during the same standard exercise in the last session. The mean heart rate dropped from 111.9 beats per minute at pretest to 98.7 at posttest ($t_{obs} = 9.12$, $df = 51$, $p < .001$). Simons and Birkimer provide a table that shows that the approximate standard deviation within groups is 18 beats per minute.
    (a) Are Simons and Birkimer's degrees of freedom computed correctly?

(b) Is enough information presented to allow you to verify that t_{obs} was computed correctly? If so, verify the computation of t_{obs}.
(c) Assuming that t was computed correctly, approximately what was the standard deviation of heart rate *difference scores*?
(d) Interpret this result in plain English.

46. (*Exercise 45 continued*) Suppose you replicate the results from Simons and Birkimer's (1988) study using 14 subjects. The heart rates (in beats per minute) are listed in the table.
(a) How many degrees of freedom are there?
(b) Compute t_{obs}.
(c) Interpret this result in plain English. (Use $\alpha = .05$.)

Subject	*Before*	*After*
1	88	72
2	133	139
3	127	90
4	111	96
5	86	101
6	109	139
7	112	108
8	108	106
9	128	105
10	122	80
11	124	47
12	116	109
13	101	116
14	113	105

47. Wu and colleagues (1988) investigated the behavioral patterns of heavy marijuana smokers by giving them marijuana cigarettes ("joints") that had either minimal tetrahydrocannabinol (THC) content (.004%, called here the placebo joints) or a 1.24% THC concentration (here called the potent joints). Did these smokers have different smoking behaviors while smoking the two kinds of joints? Wu et al. measured the number of puffs taken, the duration of each puff, the interpuff interval, the inhaled volume, and the breath-holding time. Does THC potency affect these variables? We will consider here only their breath-holding time data.

There were 15 subjects, and each was measured in both placebo and potent joint conditions. The breath-holding time statistics for the placebo joints were $\bar{X}_{placebo} = 14$ and $s_{placebo} = 9$ seconds, and for the potent joints $\bar{X}_{potent} = 15$ and $s_{potent} = 10$ seconds. "Data were analyzed using Student's t test for paired data; differences were considered statistically significant for $p < .05$" (p. 44).

Wu et al. reported that although there was a "marked difference in the intoxicating effects of the [placebo and potent] marijuana preparations [as self-reported by the subjects], no significant differences were noted between the two strengths of marijuana with respect to puff number, volume or duration, interpuff interval, inhaled volume, breath-holding time after inhalation of the smoke, or butt length" (pp. 44–45).
(a) Do Wu et al. provide enough information for you to ascertain the mean of the difference scores for breath-holding time? If so, what is it?
(b) Do Wu et al. provide enough information for you to ascertain what the value of t_{obs} was for breath-holding time? If so, what is it?
(c) How many degrees of freedom are there for these data?
(d) What is the maximum value of t_{obs} that this study could have produced?
(e) What is the smallest standard deviation of the difference scores that this study could have produced?

48. (*Exercise 47 continued*) Suppose you replicate Wu et al.'s (1988) study using 14 subjects. The breath-holding times (in seconds) are listed in the table.
(a) How many degrees of freedom are there?
(b) Compute t_{obs}.
(c) State your results in plain English.

Subject	*Placebo*	*Potent*
1	18.8	12.9
2	11.2	19.5
3	6.4	18.4
4	7.2	15.4
5	20.8	22.4
6	6.8	12.8
7	14.1	13.9
8	10.8	27.0
9	.9	12.7
10	15.8	16.6
11	17.6	3.6
12	37.4	36.4
13	35.6	25.8
14	9.1	24.0

49. In 1962, Jacob Cohen investigated the statistical power of all studies published in the 1960 volume of *Journal of Abnormal and Social Psychology*. He found that for studies that had a medium effect size, the median power was .46, a value that he thought was unconscionably low. His article (also published in the *Journal of Abnormal and Social Psychology*) encouraged experimenters in the future to design studies with higher power.

Sedlmeier and Gigerenzer (1989) wished to ascertain whether Cohen's exhortations had their desired effect of increasing power in subsequent studies. They replicated Cohen's study by examining all the studies in the 1984 volume of the same journal (by then called *Journal of Abnormal Psychology*) and found that instead of increasing, the median power had *decreased* to .37. They noted that only two of the 64 studies they reviewed even mentioned power, and none computed the power for its own hypothesis test.
(a) Approximately what is the median probability of a Type II error in the studies reported in the 1984 volume of *Journal of Abnormal Psychology*?

(b) Does it follow from Sedlmeier and Gigerenzer's findings that of all experiments designed to be included in the 1984 *Journal of Abnormal Psychology* approximately two-thirds (63%) would be expected to be a waste of the experimenter's time and money because the study would fail to reject the null hypothesis (and thus would not qualify for publication in that journal)?

50. Fagley (1985) discussed the fact that many authors attempt to "prove the null hypothesis" by conducting an experiment, failing to reject the null hypothesis, and then claiming that such a failure to reject supports the notion that the null hypothesis is true. In fact, failing to reject the null hypothesis is a likely event in cases where the power is low as well as in cases where the null hypothesis is true. He concluded that "the publication of research findings that are not statistically significant presents a novel problem in interpretation of research results. The contribution of nonsignificant results depends in part on whether the statistical test was powerful enough to detect an effect of 'meaningful' size. The primary responsibility rests with the authors of articles reporting nonsignificant results to demonstrate the worth of the results by discussion of the power of the tests" (p. 391).

    Suppose you are the manufacturer of VeryCheap Cola and you wish to "prove" that your cola is as good as Coke. Your method is to get a group of subjects to taste VeryCheap and rate it on a ten-point scale, and then get another group to taste Coke and rate it on the same scale. Your plan is to construct the experiment so that the null hypothesis will *not* be rejected even though, on average, you know that tasters prefer Coke to VeryCheap. Then you will advertise that VeryCheap is as good as Coke. Would it be to your (sleazy) advantage to have a large n or a small n in each group? Why?

Section F: Computer explorations

51. Ask the computer to perform a dependent-sample t test on these data:

    Sample 1: 1, 1, 1, 1, 1
    Sample 2: 2, 2, 2, 2, 2

    (a) Explain why the computer tells you that the dependent-sample t is undefined for these data.
    (b) Change the last point in sample 2 from "2" to "2.01." Now what happens to the dependent-sample t? Explain why.
    (c) Recall that in Exercise 31 in Chapter 11, you performed an independent-sample t test for these same data. Explain the similarities in your results.

52. (a) Have the computer create a random sample of size five from the population of IQ scores ($\mu = 100$, $\sigma = 15$). [*Hint:* We described how to do that in Section 7.3.]
    (b) Create a second set of five points that is identical to the first (use Copy and Paste or simply reenter the same data).
    (c) Have the computer determine the independent-sample t.
    (d) Have the computer determine the dependent-sample t.
    (e) Explain why t is 0 in part (c) but undefined in part (d).
    (f) Add 1 to each of the values in the second sample and recompute both the independent-sample and dependent-sample t. Explain how these results are similar to or different from those in parts (c) and (d).
    (g) Alter the last value in the second sample by adding .01 to that one value, and recompute both the independent-sample and dependent-sample t. Explain how these results are different from those in part (f).

Eyeball-estimating Dependent-sample *t* Tests (Optional)

This section may be omitted without loss of continuity.

Remember that the philosophy of eyeball-estimation is to perform all eyeball-computations "in your head," *not* using a calculator. The object is to get a reasonable approximation, not an exact result. Use the procedure both as a quick check on the computation and, more important, as a way of quickly practicing the hypothesis-testing procedure.

General Eyeball Procedure

The eyeball-estimation of two-dependent-samples hypothesis tests follows the same general procedure that we have been using since Chapter 9.

I. State the Null and Alternative Hypotheses

Use the same procedure we used when computing.

II. Set the Criterion for Rejecting H_0

$t_{\text{cv}_{\text{by eyeball}}}$ is somewhat greater than ± 1.96 (or 1.645 if directional); that is, the nondirectional $t_{\text{cv}_{\text{by eyeball}}}$ is about 2 unless df is quite small.

III. Collect a Sample and Eyeball-estimate the Observed Values of the Sample Statistic and the Test Statistic

The eyeballed test statistic follows the same formulas used for computing:

$$t_{\text{by eyeball}} = \frac{\overline{D}_{\text{by eyeball}}}{s_{\overline{D}_{\text{by eyeball}}}}$$

a. The eyeballed sample statistic is $\overline{D}_{\text{by eyeball}}$. We use either the range method or the balancing-point method to eyeball-estimate it. Note that the eyeballed sample statistic is also the numerator of $t_{\text{by eyeball}}$.
b. Eyeball-estimate the standard deviation of the differences $s_{D_{\text{by eyeball}}}$ by the range method or the inflection-point method.
c. To obtain the *denominator* of the test statistic, we follow the central limit theorem: $s_{\overline{D}} = s_D/\sqrt{n}$. We estimated $s_{D_{\text{by eyeball}}}$ in step (b). Now we eyeball-divide by $\sqrt{n}$. *Note:* Take the square root in your head by using the nearest perfect square (4, 9, 16, 25, 36, 49, 64, etc.) for the eyeballed value of n. Perform the division in your head also. We do not need an exact answer, just an approximation.
d. To obtain the test statistic $t_{\text{by eyeball}}$, divide the numerator of the test statistic (from step a) by the denominator of the test statistic (from step c). Again, perform the division in your head.

IV. Interpret the Results

Exactly as with computation, if your eyeballed test statistic (from step d) is greater (in absolute value) than your eyeballed critical value (from part II), then reject$_{\text{by eyeball}}$ H_0.

The Nicoterm Example

We now apply that procedure to the eyeball-estimation of the Nicoterm example that was calculated in the chapter. As you may recall, that experiment was a pretest–posttest study of 12 smokers, measuring the number of cigarettes smoked before and after using Nicoterm chewing gum.

I. State the Null and Alternative Hypotheses

Use the same procedure we used when computing.

This eyeball-estimation procedure parallels the computation in the text on page 250.

II. Set the Criterion for Rejecting H_0

Just as when computing, we let $\alpha = .05$. The test is directional and df is 11, so $t_{\text{cv}_{\text{by eyeball}}}$ is somewhat greater than 1.645—say, about 1.75.

III. Collect a Sample and Eyeball-estimate the Observed Values of the Sample Statistic and the Test Statistic

a. We can most conveniently use the range method to eyeball-estimate the mean of the difference scores:

$$\overline{D}_{\text{by eyeball}} \approx \frac{\text{highest } D_i + \text{lowest } D_i}{2} = \frac{41 + (-14)}{2} = 13.5 \text{ cigarettes}$$

That will be the numerator of the test statistic.

b. s_D, the standard deviation of the difference scores, is most conveniently eyeball-estimated using the range method. We use 3 as the conversion factor because $n = 12$, so

$$s_{D_{\text{by eyeball}}} \approx \frac{\text{highest } D_i - \text{lowest } D_i}{3} = \frac{41 - (-14)}{3} \approx 18 \text{ cigarettes}$$

c. To obtain the *denominator* of the test statistic, we divide $s_{D_{\text{by eyeball}}}$ by $\sqrt{n}$. Here $n = 12$ pairs, so we could take the square root of either 9 or 16 (the closest perfect squares). Using $\sqrt{9} = 3$ will make the subsequent division easier:

$$s_{\overline{D}_{\text{by eyeball}}} \approx \frac{s_D}{\sqrt{n}} \approx \frac{18}{3} = 6 \text{ cigarettes}$$

d. Now we divide the numerator (from step a) by the denominator (from step c) to obtain the test statistic: $t_{\text{by eyeball}} \approx 13.5/6 \approx 2.25$.

IV. Interpret the Results

We compare $t_{\text{by eyeball}}$ to $t_{\text{cv}_{\text{by eyeball}}}$. Because 2.25 is greater than 1.75, we *do* reject$_{\text{by eyeball}}$ the null hypothesis.

As you may recall, the computed values were found to be $\overline{D}_{\text{obs}} = 10.0$ cigarettes, $s_D = 16.22$ cigarettes, $s_{\overline{D}} = 4.68$ cigarettes, $t_{\text{obs}} = 2.14$, and $t_{\text{cv}} = 1.796$, so we also rejected H_0 by computation.

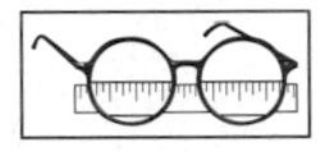

You should practice the step-by-step eyeball-estimation procedures described in this section on the histograms in Figures 12.4 through 12.10. Cover up the lists of actual values and then eyeball-estimate the dependent-sample *t* using the steps shown above.

Inferences About Two or More Means: Analysis of Variance

CHAPTER

LEARNING OBJECTIVES

1. What procedure is used for evaluating hypotheses about means of more than two samples?
2. What are pairwise and complex null hypotheses?
3. What are two reasons to prefer ANOVA instead of multiple *t* tests?
4. What is the mean square between groups?
5. What is the mean square within groups?
6. What is an *F* ratio?
7. How can a sum of squares be partitioned?
8. How does the mean square within groups relate to the pooled within-group variance?
9. What is a mean square?
10. What are the components of an analysis of variance summary table? How are they related to one another?

This chapter describes the analysis of variance: the statistical procedure for testing whether the means of two or more populations are equal. It is thus an extension of the two-independent-samples *t* test described in Chapter 11. The test statistic used in the analysis of variance is *F*, which is equal to a variance based on the sample means (called the mean square between groups, or MS_B) divided by the pooled within-group variance (also called the mean square within groups, or MS_W). Critical values of *F* are given in a table that requires the specification of two degrees of freedom: one for the numerator of *F* and the other for the denominator.

In Chapter 10, we evaluated hypotheses about the means of single samples (where the null hypothesis was H_0: $\mu = a$). Then in Chapter 11, we evaluated hypotheses about the means of two independent samples (where the null hypothesis was H_0: $\mu_1 = \mu_2$). This chapter extends that logic to the evaluation of hypotheses in cases where we have three independent samples and we wish to compare the means (where, as we shall see, the null hypothesis is H_0: $\mu_1 = \mu_2 = \mu_3$) or four samples (where the null hypothesis is H_0: $\mu_1 = \mu_2 = \mu_3 = \mu_4$), or even more than four samples. The testing procedure is called *analysis of variance* (*ANOVA*), and it can also be used when there are only two samples, in which case it is identical to the nondirectional independent-sample *t* test discussed in Chapter 11. (See Exercise 4.)

analysis of variance (ANOVA): the statistical procedure for testing hypotheses about two or more means

The analyses discussed in this chapter are independent-sample analyses, analogous to the independent-sample *t* tests of Chapter 11. Chapter 14 discusses dependent-sample ANOVAs, analogous to the dependent-sample *t* tests of Chapter 12.

13.1 Hypotheses with Three or More Samples

Suppose we have available to us *four* antidepressant drugs (Prozac, Elavil, Zoloft, and Tofranil), and we wish to know whether there is any difference in their effectiveness in reducing the symptoms of depression. We randomly select 24 depressed patients and then randomly assign each to one of four groups. We administer Prozac to the patients in group 1, Elavil to the patients in group 2, Zoloft to group 3, and Tofranil to group 4, and then we measure their depression symptoms on the Beck Depression Inventory (BDI), which is a psychological test on which higher scores indicate more severe depression. We will measure each subject on one dependent variable—namely, the subject's BDI score. The independent (or treatment) variable is group membership, and it can take the values 1, 2, 3, and 4. The data are shown in Table 13.1.

TABLE 13.1 Patients' Beck Depression Inventory scores

	Group 1 (Prozac)	*Group 2 (Elavil)*	*Group 3 (Zoloft)*	*Group 4 (Tofranil)*	
	29	35	23	33	
	32	34	25	36	
	26	28	26	30	
	28	32	26	32	
	25	37	25	32	
	28	35	19	32	
n_j	6	6	6	6	$n_G = 24$
$\overline{X}_j$	28.0	33.5	24.0	32.5	$\overline{X}_G = 29.5$
s_j^2	6.000	9.900	7.200	3.900	

Null Hypothesis

The null hypothesis is the assumption that drug type has no effect; that is, there is no difference in the antidepressant effects of Prozac, Elavil, Zoloft, and Tofranil. We write that null hypothesis in this simple form:

null hypothesis, analysis of variance, four groups

$$H_0: \ \mu_1 = \mu_2 = \mu_3 = \mu_4 \tag{13.1}$$

However, we must realize that the assumption of no drug effect (that is, the null hypothesis is true) implies that other possible null hypotheses are also true:

pairwise null hypothesis: a null hypothesis that compares one mean to another

$H_{0_1}: \ \mu_1 = \mu_2$ The effect of Prozac equals the effect of Elavil.
$H_{0_2}: \ \mu_1 = \mu_3$ The effect of Prozac equals the effect of Zoloft.
$H_{0_3}: \ \mu_1 = \mu_4$ The effect of Prozac equals the effect of Tofranil.
$H_{0_4}: \ \mu_2 = \mu_3$ The effect of Elavil equals the effect of Zoloft.
$H_{0_5}: \ \mu_2 = \mu_4$ The effect of Elavil equals the effect of Tofranil.
$H_{0_6}: \ \mu_3 = \mu_4$ The effect of Zoloft equals the effect of Tofranil.

complex null hypothesis: a null hypothesis that compares some combination of means to another combination of means

$H_{0_7}: \ \dfrac{\mu_1 + \mu_2}{2} = \dfrac{\mu_3 + \mu_4}{2}$ The average of the effects of Prozac and Elavil equals the average of the effects of Zoloft and Tofranil.

$H_{0_8}: \ \dfrac{\mu_1 + \mu_3}{2} = \dfrac{\mu_2 + \mu_4}{2}$ The average of the effects of Prozac and Zoloft equals the average of the effects of Elavil and Tofranil.

$H_{0_9}: \ \mu_1 = \dfrac{\mu_2 + \mu_3 + \mu_4}{3}$ The effect of Prozac equals the average of the effects of Elavil, Zoloft, and Tofranil.

This listing is clearly not exhaustive: Other possible null hypotheses—$H_{0_{10}}$, $H_{0_{11}}$, and so on—average the drug effects in still more ways.

In discussions later it will be useful to divide the possible null hypotheses into two groups. We will refer to the first six in our list as *pairwise null hypotheses* because they refer to the equality of one *pair* of means. We will refer to the remaining null hypotheses as *complex* because they refer to three or more means.

When we say that the null hypothesis $H_0: \mu_1 = \mu_2 = \mu_3 = \mu_4$ is true, we are implying that *all* the possible (pairwise and complex) null hypotheses—H_{0_1}, H_{0_2}, and so on—are also true.

omnibus null hypothesis: a null hypothesis that stands for all possible null hypotheses

Because $H_0: \mu_1 = \mu_2 = \mu_3 = \mu_4$ actually is many possible null hypotheses rolled into one, we call it an *omnibus null hypothesis*. *Omnibus* is the Latin word meaning "for all," and we call $H_0: \mu_1 = \mu_2 = \mu_3 = \mu_4$ an "omnibus" null hypothesis to emphasize that it stands "for all" the possible null hypotheses—H_{0_1}, H_{0_2}, H_{0_3}, and so on. We refer to the analysis of variance as an *omnibus test* because it tests all the possible null hypotheses at the same time.

Alternative Hypothesis

The best statement of the alternative hypothesis for the analysis of variance is: "The null hypothesis is false." An equivalent way of stating the alternative hypothesis is: "At least one of the possible (pairwise or complex) null hypotheses $H_{0_1}, H_{0_2}, \ldots$, is false."

Note that the following statement of the ANOVA alternative hypothesis is *not* correct: $H_1: \mu_1 \neq \mu_2 \neq \mu_3 \neq \mu_4$. That is *not* acceptable because it implies (incorrectly) that $\mu_1 \neq \mu_2$ *and* $\mu_2 \neq \mu_3$ *and* $\mu_3 \neq \mu_4$. In fact, the correct alternative hypothesis is true when *any one* of the pairwise alternative hypotheses is true *or* when any of the complex alternative hypotheses is true.

When we reject the ANOVA omnibus null hypothesis, we assert that at least one of the pairwise or complex null hypotheses is not true. The ANOVA procedure does *not* tell us which one(s). Chapter 14 provides procedures (called "post-hoc tests") for focusing on which one(s) of the possible pairwise or complex null hypotheses should be rejected.

13.2 Why Multiple *t* Tests Are Not Adequate

It may have occurred to you to ask why we need a new ANOVA procedure to analyze the antidepressant drug data. Wouldn't it be just as good to perform six *t* tests (just as in Chapter 11), one for each of the pairwise null hypotheses $H_{0_1}, H_{0_2}, H_{0_3}, H_{0_4}, H_{0_5}$, and H_{0_6}?

There are two main reasons that multiple *t* tests *cannot* substitute for ANOVA. First, the six pairwise null hypotheses represent only a portion of the null hypotheses being tested by the ANOVA procedure: the multiple *t* test procedure ignores the complex null hypotheses.

experimentwise probability of a Type I error: the probability of making one or more Type I errors in an experiment where more than one null hypotheses are being tested simultaneously

The second main reason that such a multiple *t* test technique is *not* acceptable is that multiple *t* tests have an inflated *experimentwise probability of a Type I error*. Recall that a Type I error rejects the null hypothesis when it is true, and that we limit that probability to some small value α, which we will set for this experiment to be .05. Now let us assume that *all* six of the pairwise null hypotheses listed earlier are in fact true. When we conduct our experiment, we will have a one-in-twenty chance (because $\alpha = .05$) of rejecting H_{0_1} when it is in fact true (that is, a one-in-twenty chance of making a Type I error). We will also have a one-in-twenty chance of rejecting H_{0_2} by Type I error, of rejecting H_{0_3}, and so on. In all, if we conduct six *t* tests, we will have *six* one-in-twenty chances of incorrectly rejecting a pairwise null hypothesis somewhere in this experiment.

If we have six one-in-twenty chances of incorrectly rejecting a null hypothesis somewhere in an experiment, what is the likelihood that we will *incorrectly* reject *at least one* of our null hypotheses? This likelihood is called the "probability of an experimentwise Type I error" (symbolized $\alpha_{\text{experimentwise}}$) because it measures the probability of making a Type I error somewhere, anywhere, within the whole experiment. How large is this experimentwise probability ($\alpha_{\text{experimentwise}}$)?

It is quite difficult to specify precisely the experimentwise probability in this (or any) study because it depends on the degree of interrelationship among the groups. Suffice it to say that $\alpha_{\text{experimentwise}}$ is *larger* than .05 and *smaller* than .265 (see Box 13.1). For the sake of argument, let us assume that $\alpha_{\text{experimentwise}} = .20$, and we can see why the multiple *t* test approach is a problem. Now the probability of making a Type I error *somewhere* in this experiment is not $\alpha = .05$ but rather $\alpha_{\text{experimentwise}} = .20$, four times as large as α. If we test these six null hypotheses in a situation where the null hypotheses are actually all true, then the probability of making a Type I error somewhere in the experiment is now *one in five*, not one in twenty. Given the expense of making Type I errors (described in Chapter 9), that probability is too high. Clearly, we need a method that keeps $\alpha_{\text{experimentwise}} = .05$ (or .01 or any other small value we care to choose). That method is the analysis of variance.

Note that in the ANOVA procedure there is *only one* (omnibus) null hypothesis, not six as in the multiple *t* test approach, so the α that we set for rejecting that (omnibus) null hypothesis is also $\alpha_{\text{experimentwise}}$, and we have avoided the inflation of the experimentwise Type I error probability that caused us to criticize multiple *t* tests.

BOX 13.1 The upper limit for $\alpha_{experimentwise}$

The formula for the probability of an experimentwise Type I error, assuming that all the tests are entirely independent of one another, is

$$\alpha_{experimentwise} = 1 - (1 - \alpha)^c$$

where c is the number of null hypotheses being tested simultaneously. In our example, $c = 6$ and $\alpha = .05$, so

$$\alpha_{experimentwise} = 1 - (1 - .05)^6$$
$$= .265$$

However, the six tests are *not* in general completely independent. For example, if the sample mean of a *single* group (say, $\overline{X}_1$) happens to be quite large, then *three* null hypotheses (H_{0_1}, H_{0_2}, and H_{0_3}) are all likely to be rejected. Thus, the actual value of $\alpha_{experimentwise}$ is somewhat smaller than .265. The exact probability is in general difficult to assess.

13.3 Logic of Analysis of Variance

Let us recall our experiment: We have four groups (that is, four levels of the treatment variable), and each group is administered a different drug. We wish to know whether there are differences in the effects of the four drugs on the subjects' levels of depression, and we measure level of depression using the Beck Depression Inventory (BDI), which is the dependent variable. The data are shown in the upper portion of Table 13.1. The lower portion of that table contains the summary statistics sample size, mean, and variance.

Given that this experiment is somewhat more complex than those we have encountered earlier (there are four groups instead of one or two), we must develop a notation that allows us to communicate unambiguously about these data. We use j to indicate the *column* number, so that $j = 2$ denotes the second column of Table 13.1. We use i to indicate the *row* number, so that $i = 3$ denotes the third row. We let X_{ij} (read "X eye jay") stand for an individual data point—that is, for an individual value of the dependent variable—and we say that X_{ij} belongs to the ith row and the jth column. Thus, X_{23} (read "X two three") is the data point in the second row and the third column; for the data in Table 13.1, $X_{23} = 25$. Similarly, X_{41} is the fourth point in the first column; in our data, $X_{41} = 28$.

When we present data for ANOVA, we generally use different *columns* to represent different *groups*. Thus, we speak of the jth group. If $j = 2$, for example, then the jth group is the second group, whose data are displayed in the second column. When j is used as a subscript, it always refers to the jth group. Thus, n_j is the number of data points in the jth group. For example, n_1 is the number of data points in the first group (in our data, $n_1 = 6$). The n_j values are displayed in the first row of the lower portion of Table 13.1. Similarly, $\overline{X}_j$ refers to the sample mean of the jth group. Thus, the mean of the fourth group is $\overline{X}_4 = 32.5$, computed just as we have since Chapter 4. s_j^2 refers to the variance of the jth group. Thus, the variance of the third group is $s_3^2 = 7.200$, computed just as we have since Chapter 5. The subscript i refers to the individual or subject number. Thus, X_{ij} refers to the value of X for the ith individual in the jth group.

grand mean: the mean of all the observations; the sum of all the observations in all the groups divided by the total number of observations

We use the subscript G (for "grand") to refer to all the data points in all the groups taken together. Thus, the "grand number of points" is the number of points in all groups combined: $n_G = n_1 + n_2 + n_3 + n_4 = 24$, and the *grand mean* is the mean of all the points in all the groups combined: $\overline{X}_G = (\sum X_{ij})/n_G = 29.5$. We refer to the total number of groups (which is also the total number of columns) as k; in our data, $k = 4$.

Let us reconsider the null hypothesis of our ANOVA example:

$$H_0: \quad \mu_1 = \mu_2 = \mu_3 = \mu_4 \tag{13.1}$$

which says that all four population means are the same. To test that hypothesis, we will draw four independent random samples and compute a sample mean for each: $\overline{X}_1$ will be the point-estimator of μ_1, $\overline{X}_2$ point-estimates μ_2, $\overline{X}_3$ point-estimates μ_3, and $\overline{X}_4$ point-estimates μ_4. If the null hypothesis is true, then these four sample means will all be "approximately" the same because all the values of μ_j are the same. On the other hand, if the null hypothesis is false (say, for example, that μ_3 is larger than the other population means), then the four sample means will *not* all be approximately equal ($\overline{X}_3$ would be expected to be "considerably" larger than the other $\overline{X}_j$ values).

Now even if H_0 is true, the sample $\overline{X}_j$ values will not all be *exactly* equal because sampling fluctuations will cause one sample $\overline{X}_j$ to be "slightly" larger and another "slightly" smaller, even if the population μ_j values are identical. The logic of the analysis of variance is simply a way of quantifying the terms *approximately* equal, *considerably* larger, and *slightly* smaller. In other words, ANOVA answers the question: How far apart must the sample means be before we are no longer willing to say that they are all "approximately" the same size?

The derivations that follow are exactly true only when all the n_j values are equal (in our example, $n_1 = n_2 = n_3 = n_4 = 6$), but the general logic is applicable to all ANOVA situations whether or not the n_j values are equal. Section 13.4 provides sum-of-square formulas that do *not* require equal n_j values. We discuss the equal-n_j situation first because it shows most clearly how the ANOVA procedure actually tests the null hypothesis.

ⓘ This paragraph is a "road map" for the next five pages.

Here is the logic in a nutshell; the following sections will expand on it. If we assume that H_0 is true, then all the groups are samples from populations that have the same population mean (μ) and the same population variance (σ^2). We will use our sample data to calculate two independent point-estimates of the population variance σ^2 from our sample data, which we will call MS_B and MS_W. (This is our first encounter with MS_B and MS_W; they will be explained shortly.) If the null hypothesis is true, then both MS_B and MS_W will be approximately equal to σ^2, the population variance they both point-estimate. However, as we shall see, if the null hypothesis is false, then MS_B will be substantially *larger* than σ^2, whereas MS_W will remain approximately *equal* to σ^2. To help us quantify the concept of "substantially larger than," we will take the *ratio* of these two variances, which we call $F = MS_B/MS_W$. If the two variances are approximately equal (as would be expected if the null hypothesis is true), then F will be approximately 1, but if MS_B is substantially *larger* than MS_W (as would be expected if the null hypothesis is false), then F will be substantially *larger* than 1. We test the null hypothesis by turning the preceding sentence around: If F (as obtained from the sample data) is approximately 1 (as would be expected if H_0 is true), then we will *not* reject H_0, but if F is substantially larger than 1 (as would be expected if H_0 is *false*), then we *will* reject H_0.

Many of these concepts are new and are explained clearly in subsequent sections. At this point, you should have only a general sense that we will use the sample data to determine two point-estimates of the population σ^2, compute F from the ratio of those two point-estimates, and then use that F to determine whether to reject H_0.

MS_B: Between-group Point-estimate of σ^2

The key to understanding ANOVA is to understand that it is a method of quantifying the notion of *how far apart sample means are from one another*. When the sample means are far apart, farther than we would expect by chance if the null hypothesis is true, we will reject H_0; when they are not far apart, we will not reject H_0. We simply need to quantify what we mean by "far apart." In the Table 13.1 data, for example, the sample $\overline{X}_j$ values are 28.0, 33.5, 24.0, and 32.5. Are they "far apart" or should such fluctuations be expected by chance if the null hypothesis is in fact true?

ⓘ The key to understanding this section is to keep clearly in mind that we are trying to measure how far apart the magnitudes of a set of means are.

This is precisely the same question we asked in Chapter 11 when we considered the logic of the t test, except that then we were considering only two groups. In the case of the t test, it was straightforward to measure how far apart the means were; we simply subtracted one from the other and used $\overline{X}_1 - \overline{X}_2$ as our measure of "far-apartness." Now, however, we have four means (and in other ANOVA problems we may have three, or five, or six, or more), and simple subtraction is not adequate. How can we measure the "far-apartness" or variation of the sample means?

We determined how to measure the far-apartness or variation of *simple variables* in Chapter 5, where we developed the concepts of range, standard deviation, and variance. We can use the same statistics to measure the variation of *means*, and of those three measures, it will turn out to be most convenient to use the variance. Then we can use the variance of the means to measure the far-apartness of the means. Therefore, we turn to a discussion of the variance of the means.

We can take Equation (5.6), the formula for the variance of variables:

$$s^2 = \frac{\sum(X_i - \overline{X})^2}{n-1} \tag{5.6}$$

and transform it into a formula for the variance *of means* by substituting $\overline{X}_j$ for X_i, and k (the number of means) for n. To emphasize that this is now a formula for the variance *of means*, we add an $\overline{X}$ subscript to s, giving:

variance of the sample means, definitional formula

$$s^2_{\overline{X}} = \frac{\sum(\overline{X}_j - \overline{X}_G)^2}{k-1} \tag{13.2}$$

where $\overline{X}_G$ is the grand mean; in our current example, $\overline{X}_G = (\sum X_{ij}/n_G)$.

Table 13.2 shows the computations. The grand mean $\overline{X}_G = \sum X_{ij}/n_G = 708/24 = 29.5$. Then $\sum(\overline{X}_j - \overline{X}_G)^2 = \sum(\overline{X}_j - 29.5)^2 = 57.50$, so Equation (13.2) gives

$$s^2_{\overline{X}} = \frac{\sum(\overline{X}_j - \overline{X}_G)^2}{k-1} = \frac{57.50}{4-1} = 19.1667$$

Note that this computation is identical to those we performed in Chapter 5, except here the data whose variance we compute are the four sample means.

TABLE 13.2 Computing the variance of the sample means

j	$\overline{X}_j$	$\overline{X}_j - \overline{X}_G$	$(\overline{X}_j - \overline{X}_G)^2$
1	28.0	−1.5	2.25
2	33.5	4.0	16.00
3	24.0	−5.5	30.25
4	32.5	3.0	9.00
$\sum \overline{X}_j = 118.0$		$\sum(\overline{X}_j - \overline{X}_G)^2 = 57.50$	

If all the $\overline{X}_j$ values are identical (and therefore all equal to the grand mean), $s^2_{\overline{X}} = 0$. On the other hand, if the $\overline{X}_j$ values are very different from one another, $s^2_{\overline{X}}$ will be large. The more different the $\overline{X}_j$ values are from one another, the larger the variance of the sample means $s^2_{\overline{X}}$ will be. Thus, the variance of the sample means gives us the measure of far-apartness of means that we have been looking for. All we need now is to know how large $s^2_{\overline{X}}$ must be in order for us to conclude that it is *too* large to be consistent with the null hypothesis. In our antidepressant study, is $s^2_{\overline{X}} = 19.1667$ *too* large for us to believe that the null hypothesis specified in Equation (13.1) is true?

To answer that question, we must know how large $s^2_{\overline{X}}$ might be if the null hypothesis is in fact true. If the null hypothesis is in fact true, then Prozac, Elavil, Zoloft, and Tofranil all have equal effects and so our four samples can all be considered random samples from the *same* population, which has mean μ and standard deviation σ (and therefore variance σ^2). As you may recall from Chapter 8, s^2 point-estimates σ^2, which means that s^2 is expected to be approximately equal to σ^2—perhaps somewhat larger, perhaps somewhat smaller because of sampling fluctuations. By analogy (which we will accept without proof),

$$s^2_{\overline{X}} \text{ point-estimates } \sigma^2_{\overline{X}}$$

Now the central limit theorem, Equation (7.1), tells us that $\sigma_{\bar{X}} = \sigma/\sqrt{n}$, where n is the size of each sample (remember that we are assuming that $n_1 = n_2 = n_3 = n_4 = n$). Squaring both sides gives $\sigma_{\bar{X}}^2 = \sigma^2/n$. Substituting that into our point-estimation tells us that

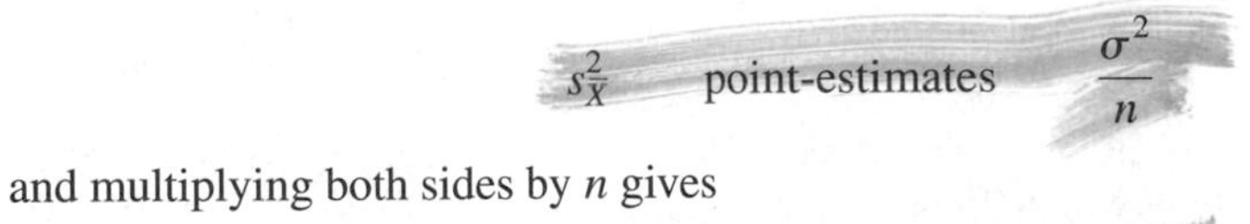

$$s_{\bar{X}}^2 \quad \text{point-estimates} \quad \frac{\sigma^2}{n}$$

and multiplying both sides by n gives

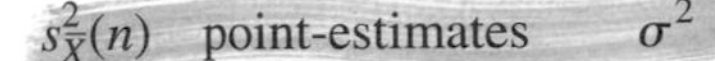

$$s_{\bar{X}}^2(n) \quad \text{point-estimates} \quad \sigma^2$$

Thus, if the null hypothesis is true, then $s_{\bar{X}}^2(n)$ can be expected to be about the same magnitude as σ^2, the variance of the population from which all four groups were drawn.

mean square between groups (MS_B): the between-group point-estimate of σ^2, based on the variation of the sample means

Because $s_{\bar{X}}^2$ depends on the differences between the group means, $s_{\bar{X}}^2(n)$ is sometimes called the "between-group point-estimate" of σ^2. For reasons that will become clearer later, the between-group point-estimate of σ^2 is also called the *mean square between groups* or MS_B; thus,

mean square between groups, definitional formula

$$MS_B = s_{\bar{X}}^2(n) \qquad [\text{only if } n_1 = n_2 = \cdots = n_k = n] \tag{13.3}$$

ⓘ Equation (13.3) assumes all $n_j = n$. For a computational formula without that assumption, see Equation (13.14) on page 296.

For our antidepressant drug data, $MS_B = s_{\bar{X}}^2(n) = 19.1667(6) = 115.00$.

Let's pause and review what we have learned so far. The variance of the sample means, $s_{\bar{X}}^2$, measures the distances between the sample means—the farther apart the means, the larger $s_{\bar{X}}^2$. If the null hypothesis is true, then the sample means should be close to one another, but because of sampling fluctuations, the sample means are *not* expected to be *exactly* equal. Therefore, the variance of these means, $s_{\bar{X}}^2$, should *not* be expected to be exactly 0, and in fact $s_{\bar{X}}^2(n)$, which we call MS_B, should be approximately equal to σ^2. For our data, $s_{\bar{X}}^2 = 19.1667$, so $s_{\bar{X}}^2(n) = MS_B = 115.00$, which is approximately equal to (that is, point-estimates) σ^2 if H_0 is true but is *larger* than σ^2 if H_0 is false. We would like to know whether MS_B is approximately equal to σ^2 or is larger than σ^2, but we do not know σ^2, so MS_B is only half the information we need. We need another independent point-estimate of σ^2.

MS_W: Within-group Point-estimate of σ^2

As we have seen, if the null hypothesis is true, then all four groups in this experiment are random samples from one population that has mean μ and variance σ^2. In that case, for the reasons that we discussed in Chapters 5 and 8, the variance within the first group, s_1^2, is a point-estimate of σ^2. Furthermore, s_2^2 also is a point-estimate of σ^2 (because we are assuming that all the samples come from the population with the same variance) and so also are s_3^2 and s_4^2. As we saw in Chapter 11, to "pool" variances is to take the weighted average of all the within-group variances, with the weights depending on the sample size. We saw there that the *pooled variance* is also a point-estimate of σ^2; in fact, it is the best point-estimate, better than any single sample variance by itself.

pooled within-group variance: the weighted average of the variances computed one group at a time; the weights depend on the sample size

The pooled variance when there are two equal-sized groups was given by Equation (11.5). We can generalize that equation to more than two groups to obtain the pooled variance for k equal-sized groups:

pooled variance, k equal-sized groups

$$s_{\text{pooled}}^2 = \frac{s_1^2 + s_2^2 + s_3^2 + \cdots + s_k^2}{k} \qquad [\text{only if equal } n_j] \tag{13.4}$$

ⓘ For equations that do not require equal-sized groups, see page 296.

For the antidepressant drug data, the individual sample variances s_j^2 are given at the bottom of Table 13.1. You should verify that $s_1^2 = 6.000$ is the variance of the six points in group 1, computed from Equation (5.6) just as in Chapter 5, and that the remaining s_j^2 values are obtained the same way. Then the pooled variance is (because the sample sizes are equal) the simple average of those sample variances:

$$s_{\text{pooled}}^2 = \frac{6.000 + 9.900 + 7.200 + 3.900}{4}$$

$$= 6.75$$

Just as in Chapter 11, this s^2_{pooled} is a point-estimate of σ^2. We call this the within-group point-estimate because each of the variances that are pooled together is calculated "within" a single group. This is also sometimes called the *mean square within groups*, or MS_W; thus,

mean square within groups, definitional formula

$$\text{MS}_\text{W} = s^2_{\text{pooled}} \tag{13.5}$$

mean square within groups (MS_W): the within-group estimator of σ^2, equal to the pooled within-group variance

For our depression inventory data, Equation (13.5) gives $\text{MS}_\text{W} = s^2_{\text{pooled}} = 6.75$. $\text{MS}_\text{W} = 6.75$ is thus our *second* point-estimate of σ^2 (MS_B was the first).

Note that if the null hypothesis is not true, then the *means* of each group become different from one another but the *variances* within the groups do not change. Invalidating the null hypothesis makes the means (the μ_j values) different from one another, but changing the mean of a distribution simply "slides" the whole distribution to the right or to the left and does not change the width (or variance) of the distribution. Thus, the pooled variance, MS_W, does *not* depend on whether or not the null hypothesis is true.

F Ratio

***F* ratio:** the variance based on the sample means (called the mean square between groups) divided by the pooled within-group variance (called the mean square within groups)

We have seen that if the null hypothesis is true, then we have two separate point-estimates of σ^2: MS_B, which is based on the group means, and MS_W, which is based on the variances within the groups. Furthermore, we know that MS_B is sensitive to whether or not the null hypothesis is true: MS_B is approximately equal to σ^2 if the null hypothesis is true, but it is expected to be substantially *larger* than σ^2 if the null hypothesis is *not* true (because the group means are expected to be relatively far apart). By contrast, MS_W, as we have seen, is *not* sensitive to whether or not the null hypothesis is true. It can be shown (but it is beyond the scope of this text) that these two point-estimates of σ^2 are *independent* of each other, in the sense that changing the group means (which *does* change MS_B) does *not* change the within-group variances (and therefore does *not* change MS_W).

We form the ratio of these two point-estimates and call it F:

test statistic (F), analysis of variance

$$F = \frac{\text{between-group point-estimate of } \sigma^2}{\text{within-group point-estimate of } \sigma^2} = \frac{\text{MS}_\text{B}}{\text{MS}_\text{W}} \tag{13.6}$$

If the numerator and the denominator are both point-estimates of σ^2 (that is, if the null hypothesis is true), then the numerator and the denominator will be approximately equal and F will be approximately 1—perhaps somewhat larger, perhaps somewhat smaller due to sampling fluctuations. On the other hand, if the numerator is *not* a point-estimate of σ^2 but is in fact substantially *larger* than σ^2, then F will be substantially larger than 1.

Thus, even though we do not know the magnitude of σ^2, we *do* know that F should be expected to be about 1 if the null hypothesis is true but substantially greater than 1 if the null hypothesis is not true. If we reverse the logic of the preceding sentence, we can use F to test the null hypothesis H_0: $\mu_1 = \mu_2 = \mu_3 = \mu_4$ as follows: If F is approximately equal to 1 (as would be expected if H_0 is true), we will *not* reject H_0, but if F is substantially larger than 1 (as would be expected if H_0 is false), then we *will* reject H_0. All that remains is to determine a logical meaning for "substantially larger than."

Determining a precise meaning for F being "substantially larger than" 1 is, of course, exactly analogous to the situations we have dealt with in Chapters 9–12, where we used the critical value t_{cv} (or z_{cv}) to determine whether the observed value of the test statistic t_{obs} (or z_{obs}) was "substantially larger than" 0. If it was, we rejected H_0. F is a test statistic (like z and t), and all we need to do now is determine the critical value F_{cv}. If the observed value of the test statistic F_{obs} exceeds F_{cv}, we will reject H_0.

For reasons that go beyond the scope of this text, this F test statistic (or, for that matter, any quotient of two independent point-estimates of the same variance) has a known and tabled distribution. F, like t, depends on the number of degrees of freedom it has, as shown in Figure 13.1. However, unlike t, one must specify the degrees of freedom *separately* for both the numerator and the denominator of F. As you may recall, the numerator of F is MS_B, which is n times the variance of the means $s^2_{\bar{X}}$. Because there were k means (4 in our

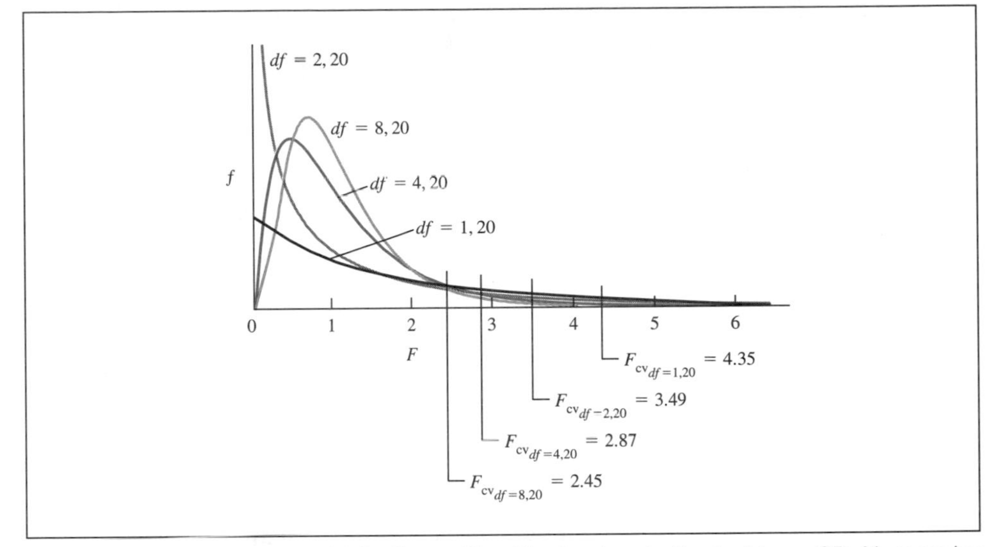

FIGURE 13.1 Selected F distributions with critical values indicated ($\alpha = .05$). Numerator degrees of freedom are 1, 2, 4, and 8, respectively; denominator degrees of freedom are 20 for all four distributions.

example), the degrees of freedom of this variance of the means is $k - 1$. Thus, the degrees of freedom for MS_B, which we call the "degrees of freedom between groups," is

degrees of freedom between groups (degrees of freedom for the numerator of F)

$$df_\text{B} = k - 1 \tag{13.7}$$

For our data, there are $k = 4$ groups, so $df_\text{B} = k - 1 = 3$.

The denominator of F, MS_W, is equal to the pooled variance s^2_pooled, so the "degrees of freedom within groups" (df_W) must have the same number of degrees of freedom as s^2_pooled. Remember (from Chapter 11) that when we pool variances, we *add* the degrees of freedom for each of the pooled groups, so

degrees of freedom within groups (degrees of freedom for the denominator of F)

$$df_\text{W} = df_1 + df_2 + \cdots + df_k \tag{13.8a}$$

Because df_1, the degrees of freedom for the first group, is $n_1 - 1$, and similarly for the remaining groups, that equation can be written in these equivalent forms:

degrees of freedom within groups, alternative form

$$df_\text{W} = (n_1 - 1) + (n_2 - 1) + \cdots + (n_k - 1) \tag{13.8b}$$

or, recalling that the grand number of points $n_\text{G} = n_1 + n_2 + \cdots + n_k$, we have

degrees of freedom within groups, alternative form

$$df_\text{W} = n_\text{G} - k \tag{13.8c}$$

For our data, each group has $n = 6$ points, and therefore each group has $n - 1$, or 5, degrees of freedom. When we pool all four groups together, we get $df_\text{W} = df_1 + df_2 + df_3 + df_4 = 5 + 5 + 5 + 5 = 20$.

When we specify the degrees of freedom for F, we customarily specify the numerator df first, so in our example we say "F with three and twenty degrees of freedom," or "$F_{3,20}$."

A few paragraphs ago we said that "F should be expected to be about 1 if the null hypothesis is true"—sometimes somewhat higher, sometimes somewhat lower. Figure 13.2 shows the F distribution with 3 and 20 degrees of freedom, assuming the null hypothesis is true. Your eyeball should verify the following statements: About half of the time, F (if H_0 is true) would be between 0 and 1 (that is, the area under the F curve below 1 is half the total area); about a third of the time, F (if H_0 is true) would be between 1 and 2; and about 95% of the time (if H_0 is true), F would be between 0 and about 3.10. Just as before, we call the value that cuts off 95% of the area of the test statistic the "critical value," and Table A.4 in Appendix A shows that $F_{\text{cv}_{3,20}} = 3.10$. Note that to use Table A.4, we need to specify *both* the numerator $df = 3$ (across the top of the table) *and* the denominator $df = 20$ (down the side of the table).

For our data, the observed value $F_\text{obs} = \text{MS}_\text{B}/\text{MS}_\text{W} = 115.00/6.75 = 17.04$. Just as in previous chapters, we ask whether $F_\text{obs} = 17.04$ is greater than $F_{\text{cv}_{3,20}} = 3.10$. In fact it

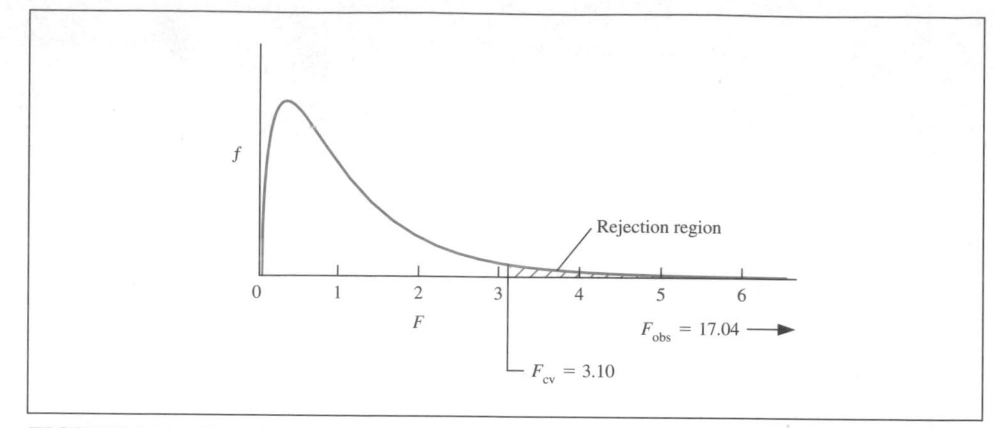

FIGURE 13.2 The ANOVA test statistic (F with 3 and 20 degrees of freedom) with rejection region shaded and observed value indicated

is, so we conclude that F is statistically significantly greater than 1—that is, that MS_B is *not* a point-estimate of σ^2 but instead is a point-estimate of some larger value. In short, we reject the null hypothesis.

ANOVA Is Nondirectional

Note that in Figure 13.2 we placed a critical value in only one tail of the distribution of the test statistic F, and thereby shaded only one rejection region. In the past several chapters, shading only one tail has implied that a hypothesis was directional or "one-tailed." That is *not* the case in ANOVA. The rejection region in ANOVA is *always* in the right-hand tail of the F distribution because a false null hypothesis always leads us to expect that MS_B (the numerator of F) will be *larger* than σ^2. For example, MS_B is large if $\overline{X}_1$ is substantially *larger* than $\overline{X}_2$ *or* if $\overline{X}_1$ is substantially *smaller* than $\overline{X}_2$.

There is some confusion here among practitioners, some calling F "one-tailed" (because the rejection region is in only one tail) and some calling F "two-tailed" (because it does not matter whether the means are smaller or larger than each other). This confusion can be avoided if the terms "directional" and "nondirectional" (which are *not* at all ambiguous) are always used, as in this textbook, instead of "one-tailed" and "two-tailed." F is always *nondirectional.*

A related observation about the F distribution is that it is positively skewed, *not* symmetric as are z and t. This is because F is always positive: It is the ratio of variances, both of which are positive, so F itself must be positive. There is no left-hand tail of F because the F distribution ends abruptly at 0.

13.4 Partitioning the Sum of Squares

Partitioning the sum of squares is of fundamental importance to advanced statistics, but it may be difficult at first. Keep in mind that the first goal of this section is to develop Table 13.3. Furthermore, the most important equations from this section are restated together on page 296.

We have found what we were looking for: a test statistic (F) that is sensitive to whether or not the null hypothesis is true and a critical value F_{cv} for that test statistic. Our exposition of the analysis of variance could end right here, except that we have not described the ANOVA summary table (the format in which ANOVA results are usually displayed), nor have we described why the point-estimates of σ^2 are called "mean squares." Furthermore, we have not presented equations that are applicable when the n_j are not equal.

The results of the analysis of variance are customarily displayed in a table called the "analysis of variance summary table"; the template for such a table is shown in Table 13.3. Our major task in this section is to describe the features of this table. The table may be unintelligible to you at this point, but by the end of this section we will have described all its characteristics and shown how we transform the template into Table 13.4, the ANOVA summary table for our antidepressant drug data.

TABLE 13.3 Summary table template for ANOVA

Source	SS	df	MS	F
Between	SS_B	df_B	MS_B	F_{obs}
Within	SS_W	df_W	MS_W	
Total	SS_T	df_T		

TABLE 13.4 Summary table for the ANOVA of drug data

Source	SS	df	MS	F
Between	345	3	115.00	17.04*
Within	135	20	6.75	
Total	480	23		

$*p < .05, F_{cv_{3,20}} = 3.10$

We begin our exploration of the ANOVA summary table by considering the column headed "SS." SS stands for "sum of squares," which itself is shorthand for "sum of squared deviations." We mentioned sums of squares in Chapter 5 when discussing the standard deviation and variance; furthermore, the sum of squared deviations (remember: that is the more formal way of saying "sum of squares") is an important topic in its own right because it is the foundation of much of advanced statistics. To understand the sum of squared deviations, we first discuss deviations themselves and then discuss squared deviations.

A deviation, as we learned in Chapters 4 and 5, is the distance a point is from the grand mean: $X_{ij} - \overline{X}_G$, where X_{ij} is the ith observation in the jth group and $\overline{X}_G$ is the grand mean (recall that the grand mean is the mean of *all* the observations in *all* the groups combined). We could (and will) also form deviations from any of the four group means $\overline{X}_j$. Unless specified otherwise, however, *deviation* will refer to a deviation from the grand mean.

To illustrate, we reprint the data from our drug study (originally shown as Table 13.1) as the first column of Table 13.5, which displays all four groups of X_{ij} values in one column. The deviations from the grand mean are shown in the second column of Table 13.5. Thus,

TABLE 13.5 Partitioning the sum of squares

(Note: grand mean $\overline{X}_G = 29.5$)	X_{ij}	Total: Deviation from grand mean $X_{ij} - \overline{X}_G$	$(X_{ij} - \overline{X}_G)^2$	Within groups: Deviation from group mean $X_{ij} - \overline{X}_j$	$(X_{ij} - \overline{X}_j)^2$	Between groups: Deviation of group mean from grand mean $\overline{X}_j - \overline{X}_G$	$(\overline{X}_j - \overline{X}_G)^2$
Group 1, Prozac ($\overline{X}_1 = 28.0$)	29	−.5	.25	1.0	1.00	−1.5	2.25
	32	2.5	6.25	4.0	16.00	−1.5	2.25
	26	−3.5	12.25	−2.0	4.00	−1.5	2.25
	28	−1.5	2.25	0.0	0.00	−1.5	2.25
	25	−4.5	20.25	−3.0	9.00	−1.5	2.25
	28	−1.5	2.25	0.0	0.00	−1.5	2.25
Group 2, Elavil ($\overline{X}_2 = 33.5$)	35	5.5	30.25	1.5	2.25	4.0	16.00
	34	4.5	20.25	.5	.25	4.0	16.00
	28	−1.5	2.25	−5.5	30.25	4.0	16.00
	32	2.5	6.25	−1.5	2.25	4.0	16.00
	37	7.5	56.25	3.5	12.25	4.0	16.00
	35	5.5	30.25	1.5	2.25	4.0	16.00
Group 3, Zoloft ($\overline{X}_3 = 24.0$)	23	−6.5	42.25	1.0	1.00	−5.5	30.25
	25	−4.5	20.25	1.0	1.00	−5.5	30.25
	26	−3.5	12.25	2.0	4.00	−5.5	30.25
	26	−3.5	12.25	2.0	4.00	−5.5	30.25
	25	−4.5	20.25	1.0	1.00	−5.5	30.25
	19	−10.5	110.25	−5.0	25.00	−5.5	30.25
Group 4, Tofranil ($\overline{X}_4 = 32.5$)	33	3.5	12.25	.5	.25	3.0	9.00
	36	6.5	42.25	3.5	12.25	3.0	9.00
	30	.5	.25	−2.5	6.25	3.0	9.00
	32	2.5	6.25	−.5	.25	3.0	9.00
	32	2.5	6.25	−.5	.25	3.0	9.00
	32	2.5	6.25	−.5	.25	3.0	9.00
			$SS_T = 480.00$		$SS_W = 135.00$		$SS_B = 345.00$

the first point in the first (Prozac) group is $X_{11} = 29$, as shown at the top of the first column of Table 13.5. As we have seen, the grand mean $\overline{X}_G = 29.5$, so the deviation of that point from the grand mean is $X_{11} - \overline{X}_G = 29 - 29.5 = -.5$, as shown at the top of the second column.

A *squared* deviation from the grand mean is then $(X_{ij} - \overline{X}_G)^2$, as shown in the third column of Table 13.5. Because the deviation of the first point is $-.5$, the squared deviation is $(-.5)^2 = .25$, as shown at the top of the third column.

total sum of squares (SS_T): the sum of the squared deviations from the grand mean

The sum of the squared deviations from the grand mean is found by adding down this third column. The "sum of the squared deviations from the grand mean" is customarily called "the total sum of squares" or "SS_T," so we can see that

total sum of squares, deviation formula

$$SS_T = \sum(X_{ij} - \overline{X}_G)^2 \tag{13.9}$$

For our data in Table 13.5, $SS_T = 480.00$ as shown at the bottom of the third column.

We locate SS_T in our ANOVA summary table template (Table 13.3), finding it at the intersection of the "SS" column and the "Total" row. We replace SS_T in the template with 480.00 in the ANOVA summary for our antidepressant drug data (Table 13.4).

Note that the first three columns in Table 13.5 are precisely the computations we would have used in Chapter 5 for computing the variance (or standard deviation). See Box 13.2.

BOX 13.2 Total sum of squares, total degrees of freedom, and total mean square

This box may be omitted without loss of continuity.

The total sum of squares is a concept that we mentioned in passing in Chapter 5. Equation (5.6), the formula for the variance of a sample, was

$$s^2 = \frac{\sum(X_i - \overline{X})^2}{n - 1} \tag{5.6}$$

If we substitute our current notation, we see that the variance of all the points taken together is simply the total sum of squares divided by the total degrees of freedom:

$$s^2 = \frac{\sum(X_{ij} - \overline{X}_G)^2}{n_G - 1} = \frac{SS_T}{df_T}$$

where df_T is the total degrees of freedom in the data set: the total number of observations in all groups combined minus one for the grand mean—that is, $df_T = n_G - 1$. (Recall that we discussed the concept of the degrees of freedom in Box 8.4, where we saw that you lose a degree of freedom whenever you compute deviations about a sample mean.)

We shall see that by analogy to other mean squares, we might refer to s^2 as MS_T. That would be entirely correct, but because MS_T is never used in ANOVA computation, the symbol MS_T is rarely (if ever) used.

Now we turn our attention to the derivation of SS_W and SS_B, the remaining sums of squares. If we transform SS_T by both adding and subtracting the sample mean $\overline{X}_j$, we obtain

$$SS_T = \sum(X_{ij} - \overline{X}_j + \overline{X}_j - \overline{X}_G)^2$$

and a little algebraic manipulation gives this fundamental but rather surprising result:

partitioning the total sum of squares

$$\underset{\text{(total)}}{SS_T} = \underset{\text{(within groups)}}{\sum(X_{ij} - \overline{X}_j)^2} + \underset{\text{(between groups)}}{\sum(\overline{X}_j - \overline{X}_G)^2} \tag{13.10a}$$

The SS equations all apply regardless of whether the n_j are equal. Note also that the (optional) computational formula (Equation (13.18) on p. 320) is easier to compute.

Now $X_{ij} - \overline{X}_j$ is the deviation of a point from its own *group* mean, which we refer to as a deviation *within* its group. These within-group deviations are shown in the fourth column of Table 13.5. For example, the first point in the first (Prozac) group is $X_{11} = 29$, and the Prozac group mean is $\overline{X}_1 = 28.0$. Therefore, the deviation of the first point from its group mean is $X_{11} - \overline{X}_1 = 29 - 28.0 = 1.0$, as shown at the top of the fourth column of Table 13.5.

The *squared* within-group deviations are shown in the fifth column of Table 13.5. Thus, for the first point, the squared within-group deviation is $(1.0)^2 = 1.0$, as shown at the top of the fifth column.

sum of squares within groups (SS_W): the sum of the squared deviations of an observation from its own group mean

The sum of the squared within-group deviations, obtained by adding down the fifth column, is called the "sum of squares *within* groups" or "SS_W." Thus,

sum of squares within groups, deviation formula

$$SS_W = \sum (X_{ij} - \bar{X}_j)^2 \tag{13.11}$$

so for our data in Table 13.5, $SS_W = 135.00$, adding down the fifth column. We locate SS_W in our ANOVA summary table template (Table 13.3), finding it at the intersection of the "SS" column and the "Within" row. Therefore, we replace SS_W in the template with 135.00 in the ANOVA summary for our antidepressant drug data (Table 13.4).

Now $\bar{X}_j - \bar{X}_G$ is the deviation of a point's *group mean* from the *grand mean*, which we refer to as the deviation *between* groups. These between-group deviations show how far apart the groups are from one another and are shown in the sixth column of Table 13.5. For example, the first point comes from the first (Prozac) group, whose mean is $\bar{X}_1 = 28.0$. Because the grand mean $\bar{X}_G = 29.5$, the deviation of the group mean from the grand mean for the first point (and for all the other points in the Prozac group) is $\bar{X}_1 - \bar{X}_G = 28.0 - 29.5 = -1.5$, as shown at the top of the sixth column.

sum of squares between groups (SS_B): the sum of the squared deviations of an observation's group mean from the grand mean

The *squared* between-group deviations are shown in the seventh (right-hand) column of Table 13.5. For the first point, $(-1.5)^2 = 2.25$, as shown in the seventh column of Table 13.5.

The sum of the squared between-group deviations, obtained by adding down the seventh column, is called the "sum of squares *between* groups" or "SS_B." Thus,

sum of squares between groups, deviation formula

$$SS_B = \sum (\bar{X}_j - \bar{X}_G)^2 \tag{13.12}$$

ℹ The SS equations all apply regardless of whether the n_j are equal. Note also that the (optional) computational formula (Equation (13.19) on p. 320) is easier to compute.

For our data in Table 13.5, $SS_B = 345.00$, adding down the right-hand column. We locate SS_B in our ANOVA summary table template (Table 13.3), finding it at the intersection of the "SS" column and the "Between" row. We replace SS_B in the template with 345.00 in the ANOVA summary for our antidepressant drug data (Table 13.4).

Now that we have defined SS_W and SS_B, we can rewrite Equation (13.10a) as the "partition" of the total sum of squares:

partitioning the total sum of squares

$$SS_T = SS_W + SS_B \tag{13.10b}$$

partition: divide into pieces with no remainder

where to *partition* is to divide into pieces with nothing left over. Thus, the analysis of variance procedure partitions the total sum of squares into two pieces: the sum of squares between groups and the sum of squares within groups. For our data, $SS_T = 480.00$ does in fact equal $SS_B + SS_W$: $345.00 + 135.00 = 480.00$, with no remainder.

The analysis of variance procedure partitions the total degrees of freedom in the same way. The total degrees of freedom is the total number of observations (in all groups combined) minus one (because we lose one degree of freedom for the grand mean; see Box 13.2). Thus,

total degrees of freedom

$$df_T = n_G - 1 \tag{13.13}$$

where $n_G = n_1 + n_2 + \cdots + n_k$ is the grand number of all the observations in all the groups. For our data, $n_G = 6 + 6 + 6 + 6 = 24$, so $df_T = n_G - 1 = 24 - 1 = 23$. We locate df_T in our ANOVA summary table template (Table 13.3), finding it at the intersection of the "df" column and the "Total" row. We replace df_T in the template with 23 in the ANOVA summary for our antidepressant drug data (Table 13.4). This total degrees of freedom can be partitioned into the within-group and between-group degrees of freedom, just as was the total sum of squares; thus

partitioning the total degrees of freedom

$$df_T = df_W + df_B \tag{13.13a}$$

Recall that MS_B, the between-group point-estimate of σ^2 (the numerator of F), was given by Equation (13.3) as $MS_B = s^2_{\bar{X}}(n)$—that is, the variance of k means times the size of each group. Because a sample variance always has one fewer degrees of freedom than the number of observations, the degrees of freedom between groups is, as we saw in Equation

(13.7), $df_B = k - 1$. For our data, with $k = 4$ groups, $df_B = 4 - 1 = 3$. We locate df_B in our ANOVA summary table template at the intersection of the "*df*" column and the "Between" row. We replace df_B in the template with 3 in the ANOVA summary for our antidepressant drug data.

Also recall that MS_W, the within-group point-estimate of σ_2 (the denominator of F), was given by Equation (13.5) as the pooled variance: $MS_W = s^2_{pooled}$. Because pooling adds degrees of freedom, the degrees of freedom within groups is, for our data, $df_W = (6 - 1) + (6 - 1) + (6 - 1) + (6 - 1) = 20$. We locate df_W in our ANOVA summary table template at the intersection of the "*df*" column and the "Within" row, and replace it with 20 in the ANOVA summary for our antidepressant drug data. That completes the SS and *df* columns, so we turn to the MS column.

mean square: a sum of squares divided by its corresponding degrees of freedom; the ANOVA procedure uses two mean squares: the mean square between groups and the mean square within groups

A *mean square* in general is a sum of squares divided by its associated degrees of freedom. Thus, the "mean square between groups" is

mean square between groups

This equation applies whether or not the sample sizes are equal.

$$MS_B = \frac{SS_B}{df_B} \tag{13.14}$$

For our data, $MS_B = 345.00/3 = 115.00$. We locate MS_B in our ANOVA summary table template at the intersection of the "MS" column and the "Between" row, and replace it with 115.00 in the ANOVA summary for our antidepressant drug data. [Recall from Equation (13.3) that if all the n_j are equal, then $MS_B = s^2_{\bar{X}}(n) = 115.00$. Equation (13.14) is generally easier to compute.]

Similarly, the "mean square within groups" is

mean square within groups

This equation applies whether or not the sample sizes are equal.

$$MS_W = \frac{SS_W}{df_W} \tag{13.15}$$

For our data, $MS_W = 135/20 = 6.75$. We locate MS_W in our ANOVA summary table template at the intersection of the "MS" column and the "Within" row, and replace it with 6.75 in the ANOVA summary for our antidepressant drug data. [Recall from Equation (13.5) that MS_W is equal to the pooled variance, but we must be clear that Equation (13.4) shows how to obtain s^2_{pooled} only if all the n_j are equal (if the n_j are not equal, obtain s^2_{pooled} by finding SS_W from Equation (13.11) and dividing by df_W to obtain $MS_W = s^2_{pooled}$).]

If the null hypothesis is true, then MS_B is the between-group point-estimate of σ^2 (that is, the numerator of F). MS_W is the within-group point-estimate of σ^2 (that is, the denominator of F). (See Box 13.2 for a note about MS_T.)

The test statistic F is the ratio of the two mean squares, as was given in Equation (13.6): $F_{obs} = MS_B/MS_W = 17.04$, just as in Section 13.3. F is shown at the upper right of the ANOVA summary table template, and we replace F with 17.04 in the antidepressant ANOVA summary table. That completes our transformation of the ANOVA summary table template (Table 13.3) into the ANOVA summary for our antidepressant data (Table 13.4).

We restate the important equations here for your convenience. The three sums of squares in ANOVA are computed from the following formulas, which are applicable whether or not the n_j are equal:

Summary of equations for the analysis of variance: These equations apply whether or not the sample sizes are equal. The computational formulas in the optional section at the end of the chapter are somewhat easier to compute.

$$SS_T = \sum (X_{ij} - \bar{X}_G)^2 \tag{13.9}$$

$$SS_B = \sum (\bar{X}_j - \bar{X}_G)^2 \tag{13.12}$$

$$SS_W = \sum (X_{ij} - \bar{X}_j)^2 \tag{13.11}$$

Furthermore, the following relationships apply within ANOVA summary tables:

$$SS_T = SS_W + SS_B \tag{13.10b}$$

$$df_T = df_W + df_B \tag{13.13a}$$

$$df_B = k - 1 \tag{13.7}$$

$$df_W = df_1 + df_2 + \cdots + df_K = n_G - k \qquad \text{(13.8a) and (13.8c)}$$

$$MS_B = \frac{SS_B}{df_B} \tag{13.14}$$

$$\text{MS}_\text{W} = \frac{\text{SS}_\text{W}}{df_\text{W}} \tag{13.15}$$

$$F = \frac{\text{MS}_\text{B}}{\text{MS}_\text{W}} \tag{13.6}$$

If you are given an ANOVA summary table with only about half the entries completed, you can usually fill in the table using these last seven equations.

13.5 Review of the Procedure

Our discussion of the ANOVA has been rather long, so we recapitulate here and at the same time note that the ANOVA hypothesis-testing procedure is the same as the procedure we used in earlier chapters. Our question is whether there is any difference in the effectiveness of four different antidepressant drugs (Prozac, Elavil, Zoloft, and Tofranil).

I. State the Null and Alternative Hypotheses

Three sketches illustrate H_0:
Variable
Sample statistic
Test statistic

There is no sample statistic for ANOVA.

$$H_0: \mu_1 = \mu_2 = \mu_3 = \mu_4 \tag{13.1}$$

H_1: The null hypothesis is false.

Eyeball-estimation of this procedure is shown on page 318.

Our usual procedure has been to sketch three distributions (of the variable, the sample statistic, and the test statistic) to illustrate the hypothesis-testing situation. You may recall from Chapter 11 that there were actually *two* variables whose distributions we could sketch—namely, X_1 and X_2 for the two independent groups (sketched in Figure 11.5). In our present analysis of variance example, *four* variables underlie the ANOVA (X_1, X_2, X_3, and X_4) and we show them in Figure 13.3. Just as in Chapter 11, the null hypothesis specifies that these variables have the same mean but does not specify what the value of that mean is, so we cannot label numeric values on the axes.[1]

The second sketch in our usual procedure is of the sample statistic, but the analysis of variance procedure does not involve a sample statistic that is different from the test statistic F, so we omit that sketch.

The third sketch is the test statistic F, as was shown in Figure 13.2. Note that F is not a symmetric distribution; in fact, it is positively skewed. Also note that the F test statistic does *not* follow the general format for the test statistic that we stated in Equation (9.1).

II. Set the Criterion for Rejecting H_0

Level of significance?
Directional?
Shade rejection regions.

Note that ANOVA is always nondirectional.

We choose $\alpha = .05$ and that will specify the critical value of F. Because our design has $k = 4$ groups (Prozac, Elavil, Zoloft, and Tofranil), $df_\text{B} = k - 1 = 4 - 1 = 3$. And because each group has $n_j = 6$ members, each group has $n_j - 1 = 5$ degrees of freedom. Pooling adds the degrees of freedom, so s^2_pooled has $5 + 5 + 5 + 5 = 20$ degrees of freedom. And because $\text{MS}_\text{W} = s^2_\text{pooled}$, MS_W also has 20 degrees of freedom. We therefore look up the critical value F_cv in Table A.4 with 3 degrees of freedom for the numerator (across the top of Table A.4) and 20 degrees of freedom for the denominator (down the left side of the table) and find that $F_{\text{cv}_{3,20}} = 3.10$. We showed this critical value and the rejection region in Figure 13.2. Note that in the ANOVA we reject *only* in the right-hand tail of the F distribution. This is because if the group means are different from one another, MS_B (the

[1] We draw each of the variables as having the same population standard deviation (that is, $\sigma_1 = \sigma_2 = \sigma_3 = \sigma_4$). That is an assumption of the ANOVA procedure (although it can be shown that ANOVA tests are "robust," meaning that violations of the equal-standard-deviation assumption do not have a large effect on the probability of rejecting the null hypothesis).

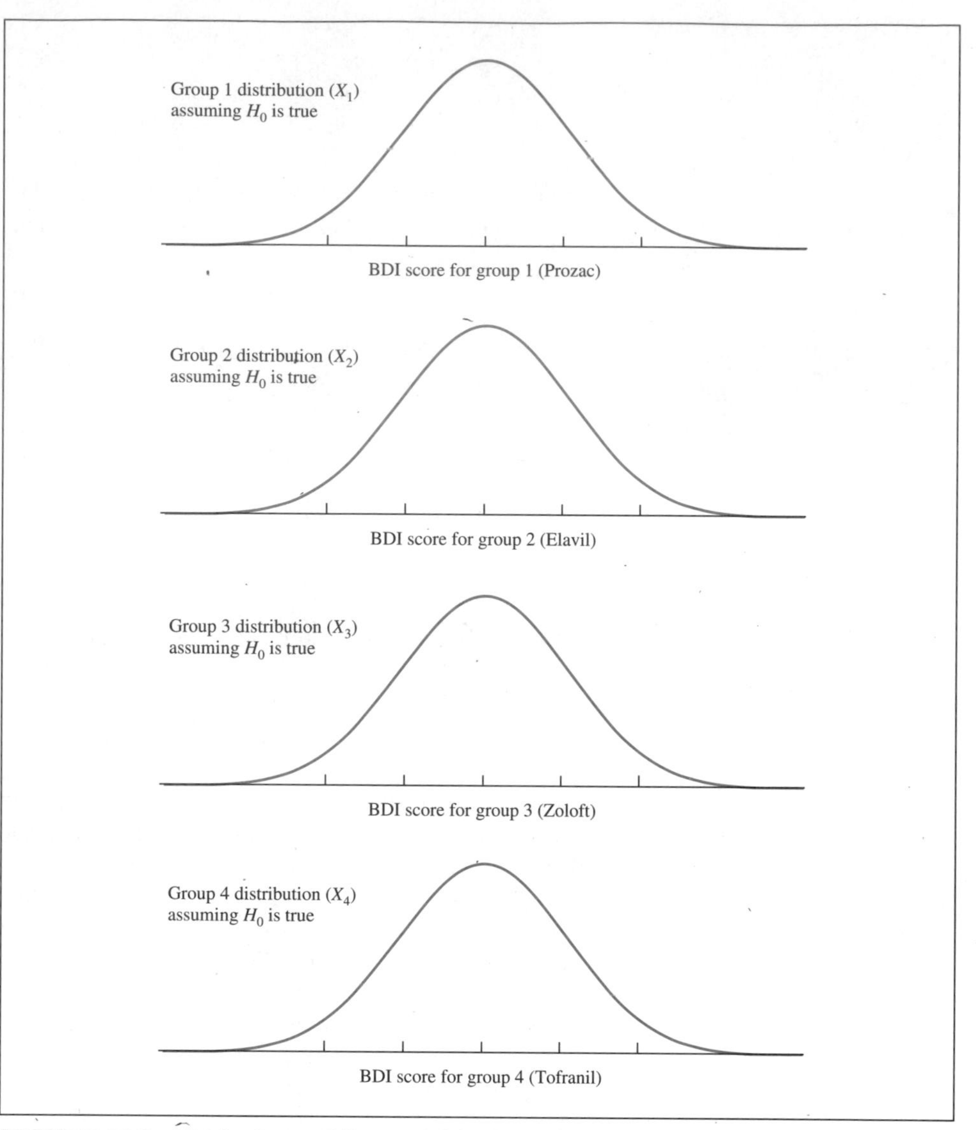

FIGURE 13.3 Distributions of the variable, Beck Depression Inventory (BDI) scores for the four groups, assuming the null hypothesis is true

numerator of F) will be large, regardless of whether any particular $(\overline{X}_i - \overline{X}_j)$ is positive or negative. Even though the rejection region is only in the right-hand tail of the F distribution, it makes sense to think of all ANOVA tests as being nondirectional because F will be large and positive if, for example, $\overline{X}_1$ is much larger than $\overline{X}_2$ *or* $\overline{X}_1$ is much smaller than $\overline{X}_2$.

III. Collect a Sample and Compute the Observed Value(s) of (the Sample Statistic and) the Test Statistic

Observed values: Sample statistic Test statistic

There is no sample statistic for ANOVA.

Here we depart slightly from our usual procedure and compute only the test statistic F (rather than a test statistic and a sample statistic) because in the ANOVA there is no sample statistic that is more informative than F. We compute the ANOVA results using Equations (13.3)–(13.6). The observed F ratio, $F_{\text{obs}} = 17.04$, is then shown on the distribution of the test statistic in Figure 13.2.

IV. Interpret the Results

Statistically significant? Practically significant? State results clearly.

Because the observed value $F_{\text{obs}} = 17.04$ exceeds the critical value $F_{\text{cv}} = 3.10$, we reject the null hypothesis and conclude that there is a statistically significant difference in the effectiveness of the four drugs.

Because we found a statistically significant difference, we must consider whether this difference is practically significant. In previous chapters, this has involved computing the standardized effect size d, which was defined in Chapter 10 as $d = (\overline{X} - \mu)/s$, in Chapter 11 as $d = (\overline{X}_1 - \overline{X}_2)/s_{\text{pooled}}$, and in Chapter 12 as $d = (\overline{D} - 0)/s_D$. Note that the numerator in all those definitions of d is the difference between the sample statistic and the parameter (in Chapters 10 and 12) or the difference between the two sample statistics (in Chapter 11).

Here in Chapter 13 such a straightforward difference is impossible to take because we have three or more means. Therefore, the most frequently reported measure of effect size for the analysis of variance is

ANOVA effect size as the ratio of standard deviations

$$f = \frac{s_{\overline{X}}}{s_{\text{pooled}}} \tag{13.16}$$

where $s_{\overline{X}}$ is the standard deviation of the k sample means and s_{pooled} is the square root of the pooled variance (that is, $s_{\text{pooled}} = \sqrt{s^2_{\text{pooled}}} = \sqrt{\text{MS}_\text{W}}$). For our data, $s_{\overline{X}} = \sqrt{s^2_{\overline{X}}} = \sqrt{19.1667} = 4.378$ and $s_{\text{pooled}} = \sqrt{6.75} = 2.60$, so $f = s_{\overline{X}}/s_{\text{pooled}} = 4.378/2.60 = 1.68$, which indicates that the distribution of the sample means is about 1.68 times as wide as the distribution of the original observations.

It can be shown that if the group sizes are equal, then Equation (13.16) is equivalent to

ANOVA effect size (the ratio of standard deviations) when sample sizes are equal

$$f = \sqrt{\frac{F}{n}} \quad \text{[only if all } n_j \text{ equal } n\text{]} \tag{13.16a}$$

which is easier to compute from an existing ANOVA summary table.

Unfortunately, such a measure of effect size is quite difficult (if not impossible) to visualize adequately. Cohen (1988) has suggested the rough convention of calling $f = .10$ a "small" effect size, $f = .25$ a "medium" effect size, and $f = .4$ a "large" effect size, but these conventions are at best only general guides. Nevertheless, our depression study $f = 1.68$ is clearly very large.

A measure of effect size that *is* possible to visualize (but not so widely used) is

ANOVA effect size by analogy to previous chapters

$$d_\text{M} = \frac{\overline{X}_{\text{max}} - \overline{X}_{\text{min}}}{s_{\text{pooled}}} \tag{13.17}$$

where $\overline{X}_{\text{max}}$ is the largest of the $\overline{X}_j$ and $\overline{X}_{\text{min}}$ is the smallest of the $\overline{X}_j$. This formula is identical to the effect size d from Chapter 11, except that d_M uses the maximum and minimum group means $\overline{X}_{\text{max}}$ and $\overline{X}_{\text{min}}$, whereas Chapter 11's d uses the (only) two group means $\overline{X}_1$ and $\overline{X}_2$. For our data, $\overline{X}_{\text{max}} = \overline{X}_2 = 33.5$, $\overline{X}_{\text{min}} = \overline{X}_3 = 24.0$, and $s_{\text{pooled}} = 2.60$, so $d_\text{M} = (33.5 - 24.0)/2.60 = 3.65$.

We can illustrate this effect size exactly as we did in Chapter 11, as shown in Figure 13.4. There we see that when we take s_{pooled} as the point-estimate of σ, the standard deviation within each group, and the sample means $\overline{X}_2$ and $\overline{X}_3$ as the point-estimates of μ_2 and μ_3, respectively, there is almost no overlap between the two distributions: The second group has much higher depression scores than does the third group; in fact, almost 100% of the group 3 distribution lies below the mean of group 2.

Using d_M as the measure of effect size has the advantage of visualizability but it has the disadvantage of depending on only two sample means (the largest and the smallest) and ignoring the relative positions of the other means.[2]

[2]There is yet another measure of effect size commonly reported for ANOVA, called the "proportion of variance accounted for," or R^2. $R^2 = \text{SS}_\text{B}/\text{SS}_\text{T}$, the sum of squares between groups divided by the total sum of squares. R^2 is thus the proportion of the total variance that can be accounted for ("explained") by the independent (treatment) variable and is a number between 0 and 1. For the antidepressant data, $R^2 = \text{SS}_\text{B}/\text{SS}_\text{T} = 345/480 = .72$. Cohen's (1988) suggestions for describing R^2 are: $R^2 = .01$ is "small," $R^2 = .06$ is "medium," and $R^2 = .14$ is "large," so the drug data have a very large effect size. For additional discussion of the proportion of the variance that can be accounted for, see Section 17.5 in Chapter 17.

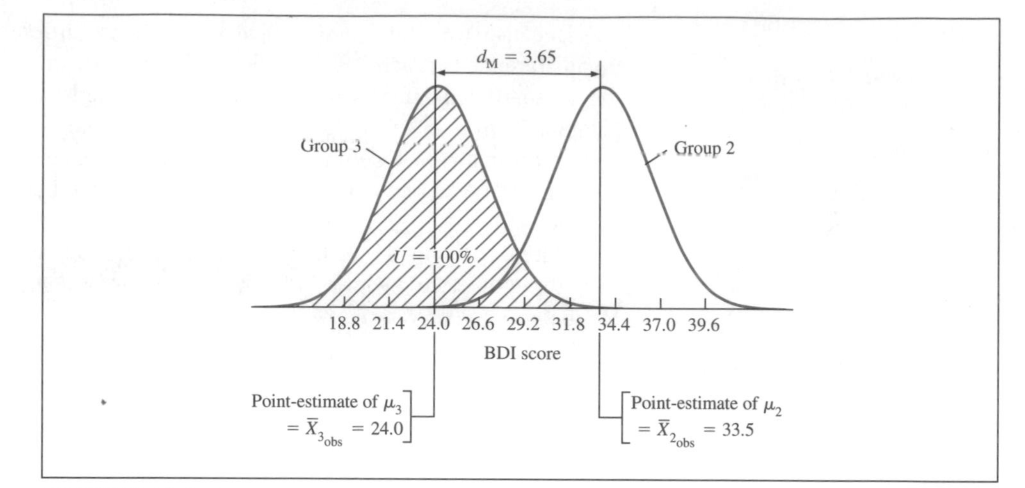

FIGURE 13.4 The largest and smallest population distributions as point-estimated from the depression data

Now that we have considered the practical significance, we can state our results in plain English: There is a statistically significant difference in the effectiveness of the four antidepressant drugs used in our study. The sample means on the Beck Depression Inventory ranged from 24.0 to 33.5, which (given the pooled standard deviation of 2.60) implies that there is almost no overlap between the smallest and the largest groups.

Note that the analysis of variance results tell us that there is a significant difference *somewhere* among these means, but it does *not* tell us where that difference is. We might look at the data and be tempted to conclude that group 3 is significantly lower than group 2, but we do *not* have a right to reach such a conclusion on the basis of the ANOVA alone. We must use the post hoc tests described in Chapter 14 to explore such conclusions.

13.6 Another Example

In our drug effectiveness study, subjects were assigned to groups randomly, which makes the study a true experiment. Analysis of variance can also be used for studies that are based on naturally occurring groups (see Box 11.1 in Chapter 11), as this example shows.

Suppose an ethologist is studying three groups of chimpanzees, which we call groups 1, 2, and 3. He is interested in a particular kind of grooming behavior and wishes to know whether the groups differ in the number of times they engage in that behavior each day. He observes five randomly chosen chimps of each group and counts the number of occurrences of this kind of grooming. The data are shown in Table 13.6, along with summary information. Is there a difference among the groups?

TABLE 13.6 Number of grooming behaviors in three groups of chimpanzees, and summary statistics

	Group 1	*Group 2*	*Group 3*
	4	6	4
	3	4	6
	5	3	7
	4	4	7
	4	4	6
n_j	5	5	5
$\bar{X}_j$	4.00	4.20	6.00
s_j^2	.500	1.200	1.500

I. State the Null and Alternative Hypotheses

Three sketches illustrate H_0:
Variable
Sample statistic
Test statistic

ⓘ As always in ANOVA, we omit the sketch of the sample statistic because there is no informative sample statistic distinct from F. Eyeball-estimation of this procedure is shown on page 319.

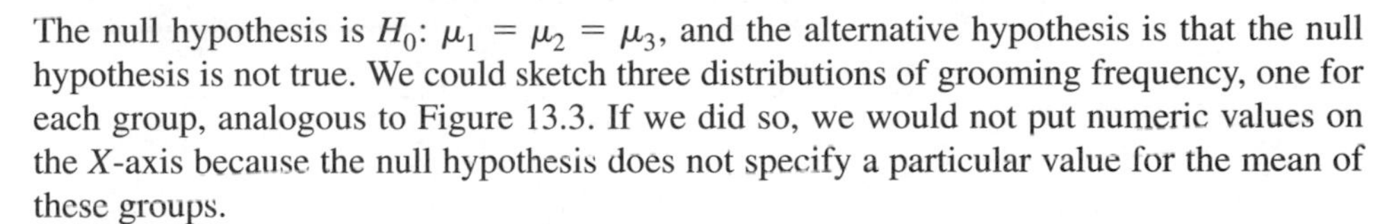

The null hypothesis is H_0: $\mu_1 = \mu_2 = \mu_3$, and the alternative hypothesis is that the null hypothesis is not true. We could sketch three distributions of grooming frequency, one for each group, analogous to Figure 13.3. If we did so, we would not put numeric values on the X-axis because the null hypothesis does not specify a particular value for the mean of these groups.

II. Set the Criterion for Rejecting H_0

Level of significance?
Directional
Shade rejection regions.

In this experiment, there are $k = 3$ groups, so $df_B = k - 1 = 2$. There are 4 degrees of freedom within each group, so $df_W = 4 + 4 + 4 = 12$. We choose $\alpha = .05$, which implies that $F_{cv_{df=2,12}} = 3.89$ (from Table A.4). We sketch the distribution of F and shade the rejection region in Figure 13.5.

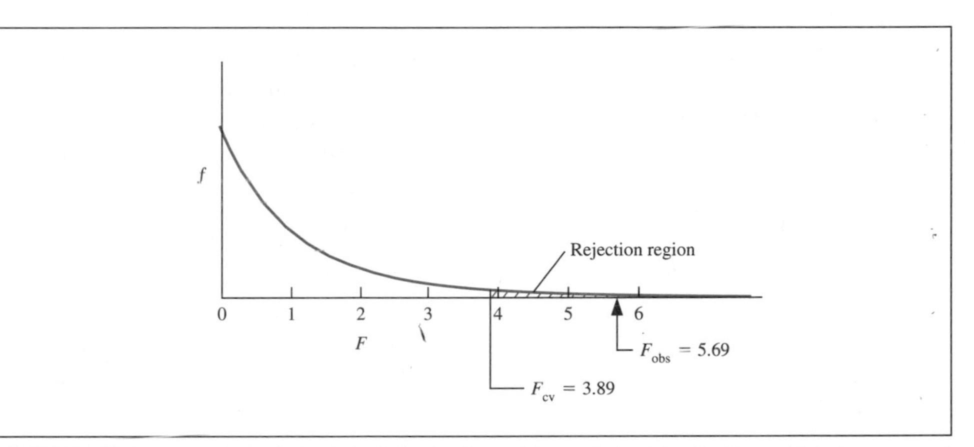

FIGURE 13.5 The ANOVA test statistic for the chimp data (F with 2 and 12 degrees of freedom) with rejection region shaded

III. Collect a Sample and Compute the Observed Value of the Test Statistic

Observed values:
Sample statistic
Test statistic

The data are shown in the top portion of Table 13.6; the summary statistics are shown in the lower portion of that table. Because the n_j are equal, we can use $s_{\overline{X}}^2(n)$ as the numerator of F—that is, as MS_B. $s_{\overline{X}}^2$ is the variance of the three numbers 4.00, 4.20, and 6.00. The mean of those three numbers, because the n_j are equal, is the grand mean $\overline{X}_G = 4.733$, so the variance can be obtained from Equation (13.2) as

$$s_{\overline{X}}^2 = \frac{(4.00 - 4.733)^2 + (4.20 - 4.733)^2 + (6.00 - 4.733)^2}{3 - 1} = 1.213$$

and $MS_B = s_{\overline{X}}^2(n) = 1.213(5) = 6.067$. df_B is the denominator in that variance calculation $(k - 1)$; therefore, $df_B = k - 1 = 2$.

Again because the n_j are equal, we can use Equation (13.4) to pool the within-group variances by simple averaging and then use Equation (13.5) to obtain $MS_W = s_{\text{pooled}}^2$. Thus,

$$MS_W = s_{\text{pooled}}^2 = \frac{.500 + 1.200 + 1.500}{3} = 1.067$$

df_W is the sum of the degrees of freedom within the three groups: $df_W = (5 - 1) + (5 - 1) + (5 - 1) = 12$.

We enter those four values into the ANOVA summary table in Table 13.7. We can complete that table by noting that Equation (13.6) gives $F_{obs} = MS_B/MS_W = 6.067/1.067 = 5.69$. We can complete the table by noting that a mean square equals a sum of squares divided by its degrees of freedom; therefore, a sum of squares equals a mean square *times* its degrees of freedom. Working Equation (13.14) backward gives $SS_B = MS_B(df_B)$; working Equation (13.15) backward gives $SS_W = MS_W(df_W)$. Furthermore, we recall from Equation (13.10b) that $SS_T = SS_W + SS_B$. That completes the ANOVA summary table.

TABLE 13.7 Summary table for the ANOVA of chimpanzee grooming data

Source	SS	*df*	MS	*F*
Between	12.133	2	6.067	5.69*
Within	12.800	12	1.067	
Total	24.933	14		

$*p < .05, F_{cv_{2,12}=3.89}$

We could have begun our computations by first determining the sums of squares: SS_B from Equation (13.12) and SS_W from (13.11). Then we would have built the ANOVA summary table from left to right. That procedure would have produced the identical table.

IV. Interpret the Results

Statistically significant? Practically significant? State results clearly.

Because the observed value $F_{obs} = 5.69$ exceeds the critical value $F_{cv} = 3.89$, we reject the null hypothesis and conclude that there is a statistically significant difference in grooming behaviors among the three groups of chimpanzees.

Because we found a statistically significant difference, we must consider whether this difference is practically significant. The standard deviation ratio effect size $f = s_{\overline{X}}/s_{pooled}$ is computed for our data as follows, $s_{\overline{X}} = \sqrt{s^2_{\overline{X}}} = \sqrt{1.213} = 1.10$ and $s_{pooled} = \sqrt{MS_W} = \sqrt{1.067} = 1.03$, so $f = s_{\overline{X}}/s_{pooled} = 1.10/1.03 = 1.07$, which indicates that the distribution of the sample means is about 1.07 times as wide as the distribution of the original observations. Such an effect size is considered "very large" by Cohen (1988).

We can also compute the maximum mean difference effect size

$$d_M = (\overline{X}_{max} - \overline{X}_{min})/s_{pooled}$$

For our chimpanzee data, $\overline{X}_{max} = \overline{X}_3 = 6.00$, $\overline{X}_{min} = \overline{X}_1 = 4.00$, and $s_{pooled} = 1.03$, so $d_M = (6.00 - 4.00)/1.03 = 1.94$.

We illustrate this effect size in Figure 13.6. There we see that when we take s_{pooled} as the point-estimate of σ, the standard deviation within each group, and the sample means $\overline{X}_1$ and $\overline{X}_3$ as the point-estimates of μ_1 and μ_3, respectively, there is very little overlap between the two distributions: About 97% of group 1 has fewer grooming behaviors than does group 3.[3]

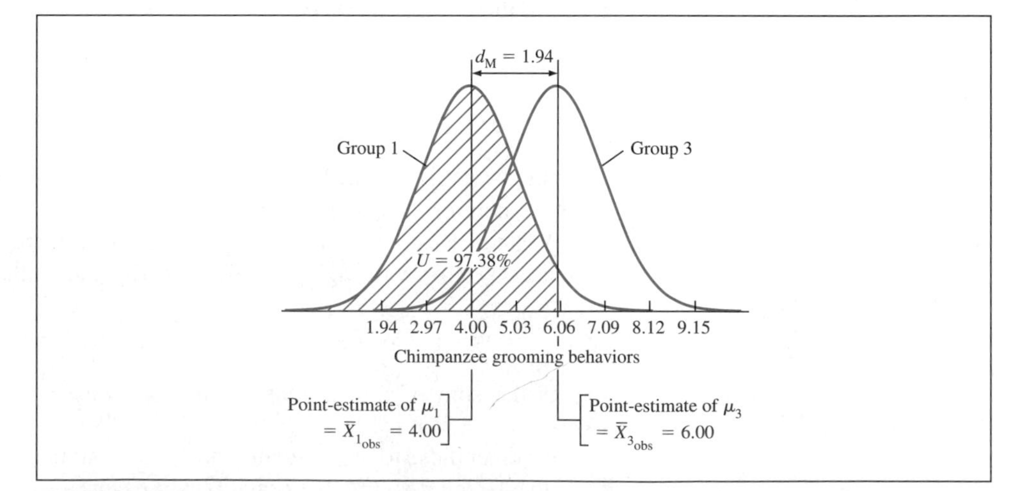

FIGURE 13.6 The largest and smallest population distributions as point-estimated from the chimpanzee grooming data

[3]The proportion of variance accounted for by these data is $R^2 = SS_B/SS_T = 12.13/24.93 = .49$, a "very large" proportion according to Cohen's (1988) suggestion.

Now that we have considered the practical significance, we can state our results in plain English: There is a statistically significant difference in the grooming behaviors of the three chimpanzee groups used in our study. The sample means ranged from 4.00 to 6.00, which (because the pooled standard deviation is 1.03) implies that there is 97% nonoverlap between the smallest and the largest group. This analysis does *not* tell us which groups are in fact significantly different from which other groups.

13.7 ANOVA Eyeball-calibration

We present now a few examples to give you practice in identifying how characteristics of the data affect the ANOVA.

Consider the data shown in Table 13.8. Note that all the columns are identical. What is the value of F for these data? To answer that question, we must consider both the numerator and the denominator of the F ratio.

TABLE 13.8 What is the value of F?

Group 1	*Group 2*	*Group 3*	*Group 4*
3	3	3	3
4	4	4	4
5	5	5	5
6	6	6	6
7	7	7	7

The numerator is $MS_B = s_{\bar{X}}^2(n)$; that is, it is based on the variance (and thus on the standard deviation) of the four group means. Here, the four group means are identical (5.0), so the standard deviation of 5.0, 5.0, 5.0, and 5.0 is 0. Thus, $MS_B = 0$.

The denominator MS_W is the pooled within-group variances, all of which are nonzero, so $MS_W \neq 0$. Therefore, $F = MS_B/MS_W = 0/MS_W = 0$. Regardless of the value of MS_W, the F ratio for these data is 0. F measures the "far-apartness" of the sample means, and these means have zero far-apartness.

Now consider Tables 13.9a and 13.9b. Note that the four sample means in Table 13.9a are 3, 4, 5, and 6, whereas they are 3, 5, 7, and 9 in Table 13.9b. Which F ratio is larger?

TABLE 13.9a Is the F ratio larger for these data . . .

Group 1	*Group 2*	*Group 3*	*Group 4*
1	2	3	4
2	3	4	5
3	4	5	6
4	5	6	7
5	6	7	8

TABLE 13.9b . . . or for these?

Group 1	*Group 2*	*Group 3*	*Group 4*
1	3	5	7
2	4	6	8
3	5	7	9
4	6	8	10
5	7	9	11

The numerator (MS_B) is based on the variance of the means. The means in Table 13.9b are farther apart, so the standard deviation of those means (and thus their variance and thus MS_B) is larger.

The denominator (MS_W) is the pooled within-group variance. Consider the standard deviations of the four groups in Table 13.9a. Note that the deviations from the group mean are identical in all four groups (−2, −1, 0, 1, and 2, respectively). If the deviations are identical, the sum of the squared deviations must be identical, so $s_1 = s_2 = s_3 = s_4$. The data in Table 13.9b also have the same deviations from the mean, so all eight within-group standard deviations in these two tables must be identical. MS_W for Table 13.9a is the pooled variance, the average of the four identical variances in the table. MS_W for Table 13.9b must equal MS_W for Table 13.9a because all the standard deviations (and thus the variances) are identical.

If the numerator is larger in Table 13.9b while the denominators are identical, F must be larger in that table. Computation shows that to be true: F in Table 13.9b is 13.33, whereas F in Table 13.9a is 3.33.

Now consider Tables 13.10a and 13.10b. Note that the mean of group 1 in both tables is 3, the mean of group 2 in both tables is 4, and so on. Which table has the larger F ratio?

TABLE 13.10a Is the *F* ratio larger for these data . . .

Group 1	*Group 2*	*Group 3*	*Group 4*
1	2	3	4
2	3	4	5
3	4	5	6
4	5	6	7
5	6	7	8

TABLE 13.10b . . . or for these?

Group 1	*Group 2*	*Group 3*	*Group 4*
2.8	3.8	4.8	5.8
2.9	3.9	4.9	5.9
3.0	4.0	5.0	6.0
3.1	4.1	5.1	6.1
3.2	4.2	5.2	6.2

The numerator of F (MS_B) is based on the variance of the four means. The four means are identical in both tables, so MS_B must be identical in both tables.

The denominator of F (MS_W) is the pooled within-group variance. Is the standard deviation (and thus the variance) within group 1 larger in either table? We see that s_1 is ten times larger in Table 13.10a than in Table 13.10b (the range is also ten times as large), and the same holds true for s_2, s_3, and s_4. If s_1 is ten times as large, s_1^2 must be 100 times as large in Table 13.10a. The same is true for the other variances. To pool is to average variances, so the pooled variance (MS_W) must be 100 times as large in Table 13.10a.

MS_W appears in the denominator of F, so if MS_W is 100 times as large in Table 13.10a and the numerators are identical, F must be 100 times as *small* in Table 13.10a and that is in fact true. Computation shows that $F = 3.33$ in Table 13.10a and $F = 333$ in Table 13.10b.

Consider the two data sets shown in Tables 13.11a and 13.11b. Note that they are identical except that Table 13.11b has an extra group that is identical to group 3 (same mean and standard deviation). Which table has the larger F ratio?

The numerator (MS_B) of F is based on the variance of the sample means. Because the duplicated mean is not far from the grand mean (that is, it is not an extreme mean), the standard deviation (and thus the variance) of the five means (3, 4, 5, 5, 6) is slightly smaller than the standard deviation of the four means (3, 4, 5, 6). Thus, the numerator in Table 13.11a is slightly larger.

TABLE 13.11a Is the *F* ratio larger for these data . . .

Group 1	*Group 2*	*Group 3*	*Group 4*
1	2	3	4
2	3	4	5
3	4	5	6
4	5	6	7
5	6	7	8

TABLE 13.11b . . . or for these?

Group 1	*Group 2*	*Group 3*	*Group 4*	*Group 5*
1	2	3	3	4
2	3	4	4	5
3	4	5	5	6
4	5	6	6	7
5	6	7	7	8

The denominator (MS_W) of F is the pooled within-group variance. All nine of these groups have identical variance because the deviations from the group mean are identical for all groups. Thus, the pooled variances (and therefore the MS_W values) are identical for the two tables. [There are more degrees of freedom for the denominator in Table 13.11b (20 instead of 16), but MS_W remains the same.]

Because the numerator in Table 13.11a is slightly larger and the denominators are identical, F is slightly larger there. By computation, $F = 3.33$ in Table 13.11a and $F = 2.60$ in Table 13.11b.

Now consider one last pair of tables: Tables 13.12a and 13.12b. Note that Table 13.12b is simply a shuffling of groups; the values within groups stay the same. Which F is larger?

TABLE 13.12a Is the *F* ratio larger for these data . . .

Group 1	*Group 2*	*Group 3*	*Group 4*
1	2	3	4
2	3	4	5
3	4	5	6
4	5	6	7
5	6	7	8

TABLE 13.12b . . . or for these?

Group 1	*Group 2*	*Group 3*	*Group 4*
2	1	4	3
3	2	5	4
4	3	6	5
5	4	7	6
6	5	8	7

The numerator (MS_B) is based on the variance of the sample means. The sample means themselves have not changed, and when we compute a variance of four numbers, it does not matter in which order we take those four numbers. Therefore, MS_B is identical in both tables.

The denominator (MS_W) is the pooled within-group variance. The variances within all eight groups are still all identical, so the pooled variances are identical in both tables.

Because the numerator and the denominator are both identical, F is identical (3.33) in both tables. The individual values within each group could also have been shuffled without changing F.

We now consider much the same material but present it graphically. The next series of figures present two sets of distributions, an upper set and a lower set. Each set represents four sample distributions, one for each of the four groups in our ANOVA. In actuality, sample distributions are rarely perfectly normal as these are drawn, but these idealized sample distributions illustrate the important characteristics of the analysis of variance. In Figures 13.7–13.11, your task is to decide, if we were to calculate two F ratios, one for the upper set of data and one for the lower set, which F ratio would be *larger*, and why. Cover the right-hand side of this and the following pages until you have determined for yourself which F is larger.

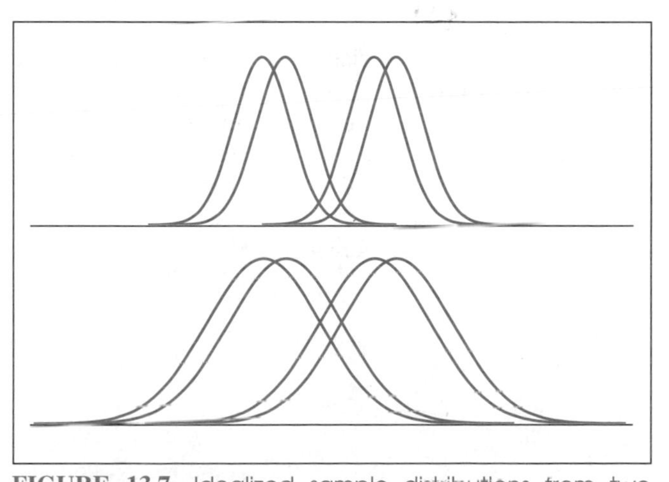

FIGURE 13.7 Idealized sample distributions from two ANOVAs. Would the *F* ratio from the upper ANOVA be larger or smaller than the *F* ratio from the lower ANOVA?

In the distributions of Figure 13.7, the four sample *means* of the upper distributions are identical to the four means below, so the magnitudes of MS_B (the numerator) for the two F ratios are the same. However, the variances of the four groups below are larger than those above, so the pooled within-group variance (MS_W, the denominator) is larger below.

Thus, the F ratio for the upper ANOVA is larger.

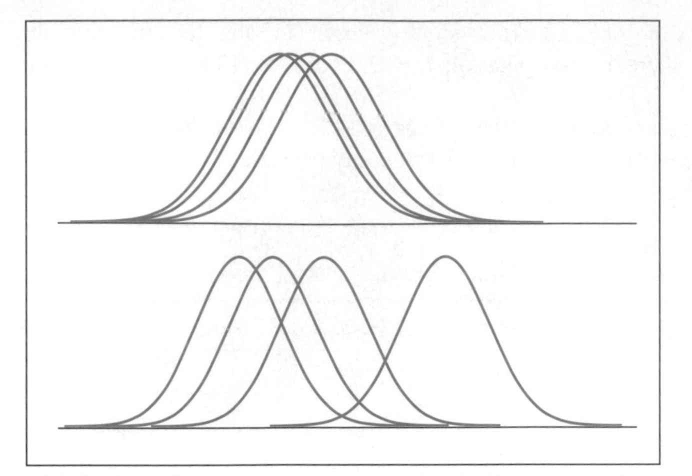

FIGURE 13.8 Idealized sample distributions from two ANOVAs. Which F is larger?

In the distributions of Figure 13.8, the means of the lower distributions are farther apart than those of the upper distributions, so MS_B (the numerator of F) is larger below. The variances above are the same as the variances below, so MS_W (the denominator of F) is the same.

The result is that the F ratio for the lower distributions is larger.

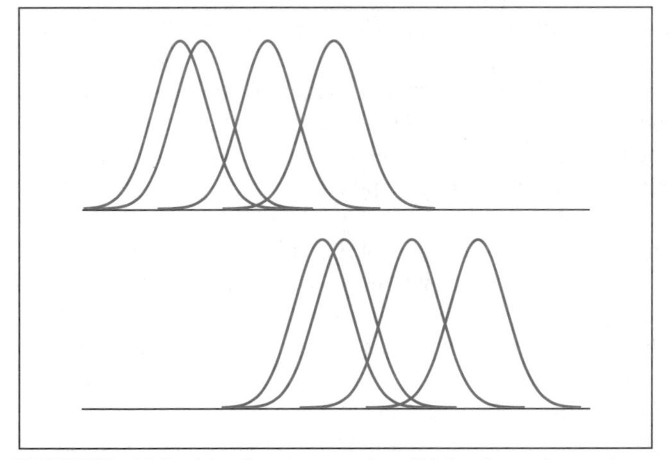

FIGURE 13.9 Idealized sample distributions from two ANOVAs. Which F is larger?

In the distributions of Figure 13.9, the distances *between* the means are identical above and below, so the MS_B values are identical. Furthermore, the variances within each sample are identical above and below, so the MS_W values are identical.

The result is that the F ratios are identical; shifting all the samples to the right or left does not change the ANOVA.

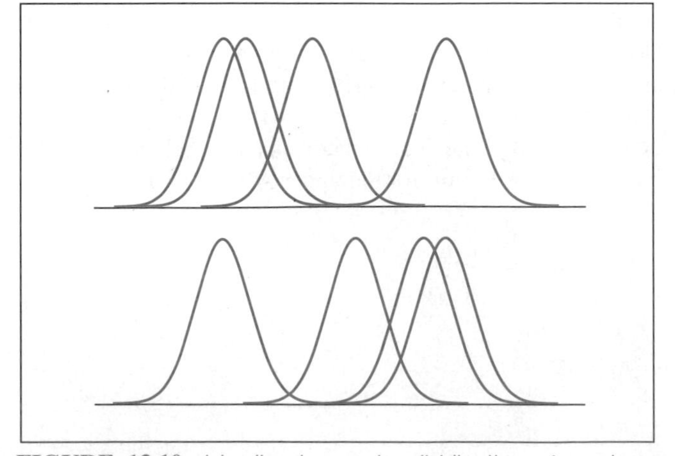

FIGURE 13.10 Idealized sample distributions from two ANOVAs. Which F is larger?

In Figure 13.10, the distances between the means are identical except reversed right to left. Reversing does not change the distances between the means, so the numerator stays the same; that is, the MS_B values are identical. The variances above are identical to the variances below, so the MS_W values are identical.

The F ratios are therefore identical.

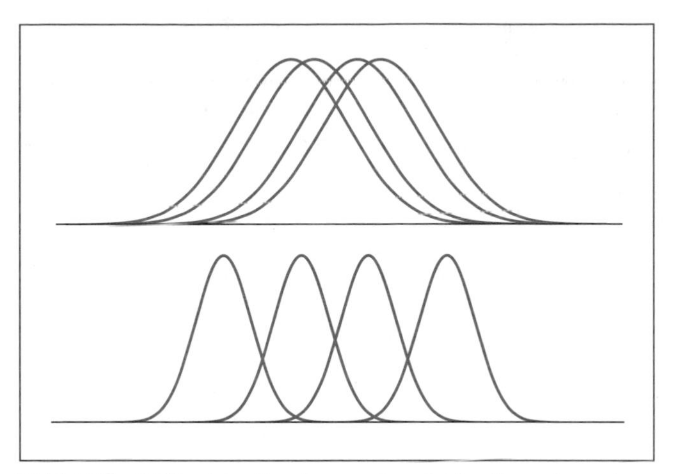

FIGURE 13.11 Idealized sample distributions from two ANOVAs. Which F is larger?

In Figure 13.11, the distances between the means are greater below, so MS_B is greater below. The sample variances are smaller below, so the pooled variance (MS_W) is smaller below.

Both effects work in the *same* direction: The F ratio for the lower distributions is larger.

You may wish to use the ESTAT exercise "anova" at this time. Use the eyeball-calibration paths (version 2.0 or later).

13.8 Number of Subjects Required for Adequate Power

You may recall from Chapter 12 that power is the probability of correctly rejecting the null hypothesis, and that the factors that have the strongest effect on power are the effect size index in the population and the number of subjects. Table 12.4 in Chapter 12 gave the number of subjects required for a t test to have power of .8 or .4 as a function of the effect size index $d_{\text{population}}$. Before we provide a similar table for ANOVA, we must consider the ANOVA population effect size index.

In this chapter, we have been using two *sample* effect size indexes: the standard deviation ratio effect size $f = s_{\overline{X}}/s_{\text{pooled}}$ and the maximum mean difference $d_M = (\overline{X}_{\max} - \overline{X}_{\min})/s_{\text{pooled}}$. Two corresponding *population* effect size indexes are $f_{\text{population}} = \sigma_{\overline{X}}/\sigma$ and $d_{M_{\text{population}}} = (\mu_{\max} - \mu_{\min})/\sigma$, where the population standard deviation within each group is assumed to be equal to the same value σ. Either one can be used to allow the desired power to determine the required sample size in an experiment.

Cohen (1988) calls $f_{\text{population}} = .10$ a "small" effect size, $f_{\text{population}} = .25$ a "medium" effect size, and $f_{\text{population}} = .40$ a "large" effect size. It can be shown that $d_{M_{\text{population}}}$ is approximately twice as large as $F_{\text{population}}$ [see Cohen (1988, p. 278) for a table that compares $d_{M_{\text{population}}}$ and $f_{\text{population}}$], so Cohen would call $d_{M_{\text{population}}} = .20$ "small," $d_{M_{\text{population}}} = .50$ "medium," and $d_{M_{\text{population}}} = .80$ "large." The alert reader will recognize that those are precisely the values Cohen assigned for small, medium, and large $d_{\text{population}}$ values for t tests.

Table 13.13 gives the sample sizes required for ANOVA experiments to have power = .8 or power = .4 for small, medium, and large effect size indexes, given that the level of significance $\alpha = .05$. Note that Table 13.13 is identical in format to Table 12.4 and simply extends that table to the analysis of variance. Table 13.13 gives sample sizes for experiments that have two groups, four groups, or six groups; see Cohen (1988) for other designs. Note that the required sample sizes for the ANOVA two-group design are identical to the independent-sample t test sample sizes in Table 12.4; we have noted that when there are two groups, the ANOVA $F = t^2$ and the ANOVA and independent-sample t test are equivalent.

For example, suppose we are interested in the performance of student singers, and we wish to know whether the length of time that singers can sustain a note is the same among sopranos, altos, tenors, and basses. We know from previous research that the standard deviation of sustain time for sopranos is $\sigma = 8$ seconds, and we are willing to assume

TABLE 13.13 Number of subjects in each group necessary to obtain the required power for an ANOVA with level of significance $\alpha = .05$ as a function of effect size index f.[a]

Test type	Null hypothesis	Effect size index ($f_{population}$)	Power	Magnitude of $f_{population}$ Small	Medium	Large
ANOVA with two independent groups	H_0: $\mu_1 = \mu_2$	$\frac{\sigma_{\bar{X}}}{\sigma}$	.8	393	64	26
			.4	145	24	10
ANOVA with four independent groups	H_0: $\mu_1 = \mu_2 = \mu_3 = \mu_4$	$\frac{\sigma_{\bar{X}}}{\sigma}$	.8	274	44	18
			.4	112	19	8
ANOVA with six independent groups	H_0: $\mu_1 = \mu_2 = \mu_3 = \mu_4 = \mu_5 = \mu_6$	$\frac{\sigma_{\bar{X}}}{\sigma}$	.8	215	35	14
			.4	92	16	7

[a]The number of subjects in the table will be approximately correct if small, medium, and large values of the effect size index $d_{M_{population}}$ are used instead of $f_{population}$. This table is a continuation of Table 12.4 and is derived from Cohen (1988, pp. 289–354).

that the standard deviation within the other groups is the same $\sigma = 8$ seconds. If the population with the largest average sustain time has a mean that is 4 seconds longer than the population with the shortest sustain time, how many subjects do we need for an experiment to have power of .8? Here $\mu_{max} - \mu_{min} = 4$ seconds and $\sigma = 8$ seconds, so $d_{M_{population}} = 4/8 = .5$, which, as we have seen, Cohen would call a "medium" effect size index. We enter Table 13.13 for a four-group experiment and find that the number of subjects required in each group is 44. Cohen (1988) gives more complete tables.

13.9 Connections

Cumulative Review

We continue the discrimination practice begun in Chapter 10. You may recall the two steps:

Step 1: Determine whether the problem asks for (a) the area under a normal distribution, (b) a confidence interval, or (c) the evaluation of a hypothesis. See the Cumulative Review in Section 10.5 for hints about this step.

Step 2: If the problem asks for the evaluation of a hypothesis, consult the summary of hypothesis testing in Table 13.14 to determine which kind of test is appropriate.

Table 13.14 shows that so far we have considered four basic kinds of hypothesis tests (the one-sample test for the mean, two-independent-samples test for means, two-dependent-samples tests for means, and now three-or-more-independent-samples tests). Note that this table is identical to Table 12.6, with the addition of the last line appropriate for this chapter. We will continue to expand this table progressively throughout the remainder of the textbook.

Journals

The typical journal article, to conserve space, does not present the complete ANOVA table but instead provides only a summary. Thus, the report on the antidepressant drug study in a journal article might read, "The four sample means were statistically significantly different from each other (Prozac: $M = 28.0$, $SD = 2.45$; Elavil: $M = 33.5$, $SD = 3.15$; Zoloft: $M = 24.0$, $SD = 2.68$; Tofranil: $M = 32.5$, $SD = 1.97$; $F(3, 20) = 17.04$, $p < .05$)." M and SD refer to the sample mean and standard deviation, respectively. $F(3, 20)$ indicates that the F ratio has 3 and 20 degrees of freedom. $p < .05$ indicates that the test rejected the null hypothesis at the level of significance $\alpha = .05$.

TABLE 13.14 Summary of hypothesis-testing procedures in Chapters 10–13

Design	*Chapter*	*Null hypothesis*	*Sample statistic*	*Test statistic*	*Effect size index*
One sample					
σ known	10	$\mu = a$	$\overline{X}$	z	$d = \frac{\overline{X}_{obs} - \mu}{\sigma}$
σ unknown	10	$\mu = a$	$\overline{X}$	t	$d = \frac{\overline{X}_{obs} - \mu}{s}$
Proportion	Exercise 7, Chapter 10	$\Pi = a$	p	t	—[a]
Two independent samples					
For means	11	$\mu_1 = \mu_2$	$\overline{X}_1 - \overline{X}_2$	t	$d = \frac{\overline{X}_{1_{obs}} - \overline{X}_{2_{obs}}}{s_{pooled}}$
Two dependent samples					
For means	12	$\mu_D = 0$	$\overline{D}$	t	$d = \frac{\overline{D}_{obs}}{s_D}$
k independent samples					
For means	13	$\mu_1 = \mu_2 = \cdots = \mu_k$	n/a	F	$d_M = \frac{\overline{X}_{max} - \overline{X}_{min}}{s_{pooled}}$ or $f = \frac{s_{\overline{X}}}{s_{pooled}}$ or $R^2 = \frac{SS_B}{SS_T}$

[a] Effect size index is not discussed in this textbook.

A good journal report gives you enough information so that you can at least approximately reconstruct the ANOVA summary table. Here, for example, if you assume that the group sizes are equal (the Method section of the journal article should be explicit about that), then $df_B = 3$ implies that there were four groups, and $df_W = 20$ implies that there were five *df* or six observations per group. Therefore, you could compute the variance of the four means 28.0, 33.5, 24.0, and 32.5, and multiply that number by 6 to obtain MS_B. (If the group sizes are not equal, that calculation will be only approximate.) Furthermore, you can obtain MS_W by recognizing that it is the pooled variance, and the pooled variance can be obtained (still assuming equal group sizes) by summing the squared *SD* values $[(2.45)^2 + (3.15)^2 + (2.68)^2 + (1.97)^2]$ and then dividing that sum by 4.

Journal articles often report so-called exact probabilities. In this case, the journal might report "$p = .0005$." This can be interpreted as meaning that α was originally set at .05, but the null hypothesis would also have been rejected if α had been set at any value greater than or equal to .0005.

Computers

Use ESTAT's datagen to compute the ANOVA for the antidepressant drug data. The data are shown in Table 13.1, with H_0: $\mu_1 = \mu_2 = \mu_3 = \mu_4$.

1. Enter the six group 1 values from the first column of Table 13.1 into the first column of the datagen spreadsheet.
2. Enter the six group 2 values from the second column of Table 13.1 into the second column of the datagen spreadsheet.
3. In a similar manner, enter the remaining two columns.
4. Datagen automatically computes the ANOVA, shown in the `datagen Statistics` window.
5. See Homework Tip 6.

Use SPSS to compute the ANOVA for the antidepressant drug data. The data are shown in Table 13.1, with H_0: $\mu_1 = \mu_2 = \mu_3 = \mu_4$.

1. Enter the six group 1 values from the first column of Table 13.1 into the first column of the SPSS spreadsheet. For each value, enter a "1" into the second column of the SPSS spreadsheet. (These "1" values will serve as the independent variable, group membership.)
2. Enter the six group 2 values into the first column of the SPSS spreadsheet beginning with row 7. For each value, enter a "2" into the second column of the SPSS spreadsheet.
3. Enter the six group 3 values into the first column of the SPSS spreadsheet beginning with row 13, and similarly the group 4 values. For each value, enter a "3" or "4" into the second column of the SPSS spreadsheet, as appropriate.
4. Click `Statistics`, then `Compare Means...`, and then `One-Way ANOVA....`
5. Click `var00001`; then click the upper ▶ to move it into the `Dependent List:` window.
6. Click `var00002`; then click the lower ▶ to move it into the `Factor:` window.
7. Click `Define Range...`; enter "1" in `Minimum` and "4" in `Maximum`. Then click `Continue`.
8. Click `OK`.
9. The ANOVA summary table shown in the accompanying printout appears in the `!Output1` window. Note that SPSS does not automatically supply means and standard deviations for each group. Whenever you input data, you should ask to see those statistics and make sure they are correct. Note also that SPSS reports that the exact probability is .0000. This is simply the actual exact probability rounded to four decimal places, which indicates that the actual exact probability is some value less than .00005.

```
SPSS for MS WINDOWS Release 6.1

                  - - - - -  O N E W A Y  - - - - -

     Variable  VAR00001
  By Variable  VAR00002

                        Analysis of Variance

                                  Sum of        Mean          F        F
       Source            D.F.     Squares       Squares       Ratio    Prob.

Between Groups             3     345.0000      115.0000     17.0370    .0000
Within Groups             20     135.0000        6.7500
Total                     23     480.0000
```

Exact probability

10. You may wish to save these data because we will use them in Chapter 14.
11. See Homework Tip 6.

Homework Tips

1. Check the list of learning objectives at the beginning of this chapter. Do you understand each one?

2. The key to understanding ANOVA is to recognize that the F ratio has as its numerator a variance of means (times n) and as its denominator the pooled within-group variance.

3. Be sure you understand that if there are five groups, then the variance of the means is the variance of five numbers computed just as in Chapter 5. Those five numbers happen to be the five sample means. That variance has $5 - 1 = 4$ degrees of freedom.

4. Be sure you understand that if there are five equal-sized groups, then the pooled within-group variance is simply the average of the variance of the first group (computed just as in Chapter 5), the variance of the second group, . . . , and the variance of the fifth group.

5. Memorize the list of ANOVA summary table relationships shown on page 296.

6. Whenever you ask a computer to perform a task, you should (a) make sure the number of cases is correct and (b) eyeball-estimate enough statistics to convince yourself that the computer has interpreted the data in the way you expected.

Exercises for Chapter 13

Section A: Basic exercises (answers in Appendix D, page 520)

1. A marketing person wishes to know whether college students prefer one beer to another. She takes four different beers, Budweiser, Miller, Coors, and Strohs, and removes the labels so that her subjects cannot identify the brand. She finds 20 volunteers and at random gives one beer to each to drink. Then she has the subjects rate their liking for the beer on a 10-point scale (10 being the highest rating). The ratings are shown in the table. At $\alpha = .05$, can she conclude that college students have a differential preference for beer?

   *Omit the parts marked with an asterisk (*) unless you have read the optional eyeball-estimation section at the end of the chapter.

Budweiser	Miller	Coors	Strohs
8	4	9	6
8	4	5	7
7	3	6	6
8	10	9	7
9	4	6	4

I. State the null and alternative hypotheses

(a) What is the null hypothesis? The alternative hypothesis?

(b) Is this test directional or nondirectional? Why?

(c) What is the test statistic? Sketch the distribution of the test statistic assuming the null hypothesis is true.

II. Set the criterion for rejecting H_0

(d) What is the level of significance?

(e) How many degrees of freedom are there?

(f) (Eyeball-estimate and then)* look up the critical value(s) of the test statistic. Enter it (them) on the appropriate distribution and shade the rejection region(s).

III. Collect a sample and compute statistics

(g) (Eyeball-estimate and then)* compute the observed value of the test statistic. Enter it on the appropriate distribution. Show the complete ANOVA summary table.

IV. Interpret the results

(h) Is this result statistically significant (that is, should we reject H_0)? Decide (by eyeball and)* by computation.

(i) If the result was statistically significant, then determine the effect size indexes f and d_M and Cohen's nonoverlap statistic U. Illustrate d_M and U as shown in Figure 13.4.

(j) Describe the results, including both statistical and practical significance, in plain English.

2. A little league baseball bat manufacturer can build aluminum bats using three different manufacturing processes. He wishes to know whether the processes affect the power of the bat. He builds a bat-swinging machine that is designed to simulate a human batter, and he has the machine use each of the three bats to hit eight baseballs. He records the distance each ball travels (in feet) before it hits the ground, as shown in the table. Does manufacturing process affect bat power? Use $\alpha = .05$ and the steps of Exercise 1.

	Process A	Process B	Process C
	213.7	199.8	222.7
	219.3	207.2	245.1
	235.0	191.4	225.0
	217.3	180.1	216.7
	193.8	194.2	223.2
	200.2	209.0	225.1
	222.7	227.1	261.8
	178.8	185.6	236.4
Labor savings:			
$\sum X_i$	1680.8	1594.4	1856.0
$\sum X_i^2$	355,407.681	319,338.264	432,165.632

3. (*Exercise 2 continued*) Suppose that the bat manufacturer suspects that his technicians have made a mistake in the measurements and have mismeasured all the hits in process C by 30 feet. The surmised distances are shown in the table.

	Process A	Process B	Process C
	213.7	199.8	192.7
	219.3	207.2	215.1
	235.0	191.4	195.0
	217.3	180.1	186.7
	193.8	194.2	193.2
	200.2	209.0	195.1
	222.7	227.1	231.8
	178.8	185.6	206.4
Labor savings:			
$\sum X_i$	1680.8	1594.4	1616.0
$\sum X_i^2$	355,407.681	319,338.264	328,005.640

(a) Predict how this mistake will affect the size of F in the ANOVA. Do you expect an increase, a decrease, or no change in F?

(b) Recalculate the ANOVA following the steps of Exercise 1, noting the values that have changed.

4. Exercise 2 in Chapter 11 provided data (reproduced here) that represented WISC-R Vocabulary Subtest scores from random samples of size eight from two special education schools. In that exercise, you used an independent-sample t test to determine whether the two groups were different. Perform an analysis of variance on the same data, again using $\alpha = .05$.

(a) Demonstrate that $F_{obs} = (t_{obs})^2$ from Exercise 2 in Chapter 11.

(b) Demonstrate that $F_{cv} = (t_{cv})^2$ from Exercise 2 in Chapter 11.

(c) Do your results imply that, in general, if a nondirectional independent-sample t test rejects the null hypothesis, an F test on the same data will also reject the null hypothesis, and vice versa?

Sample 1	Sample 2
4	8
1	6
9	2
6	6
2	3
5	9
7	7
7	1

5. A perfume manufacturer is ready to market a new perfume, and she wishes to know whether the shape of the bottle will affect perfume sales. She is trying to decide among five possible bottle shapes. She sends one case of perfume to each of 28 retailers; each retailer receives only one shape of bottle. At the end of one month, she obtains a report of sales from each retailer. The numbers of bottles sold are shown in the table. Does shape of bottle make a difference? Use $\alpha = .05$ and the steps of Exercise 1.

Shape 1	Shape 2	Shape 3	Shape 4	Shape 5
21	30	21	19	26
22	23	23	21	20
27	26	20	18	20
20	31	20	23	23
	23	15	19	31
	23	23		24
		18		

6. (*Exercise 5 continued*) Suppose the perfume manufacturer had decided to use $\alpha = .01$ instead of $\alpha = .05$. What changes does that make to each part of the answer?

7. (a) Fill in the blanks in the ANOVA summary table:

Source	SS	df	MS	F
Between		5	43.00	
Within	435.55	___	14.52	
Total	650.55	35		

(b) How many groups are in this experiment?

(c) Assuming equal n_j, how many subjects are in each group?

8. (a) Fill in the blanks in the ANOVA summary table:

Source	SS	df	MS	F
Between	81.35		27.12	9.19
Within	___	36		
Total	187.57	39		

(b) How many groups are in this experiment?

(c) Assuming equal n_j, how many subjects are in each group?

9. This ANOVA summary table represents the results of an experiment that compares four treatments with nine subjects in each treatment condition. Complete the table.

Source	SS	df	MS	F
Between	60			7.0
Within	___	___		
Total				

10. Suppose you wish to test three methods of improving GRE scores and you are designing an experiment that will use four groups: computer presentation of GRE items, printed presentation of GRE items, live classroom discussion of GRE items, and a no-treatment control group. You know that GRE scores are normally distributed with $\sigma = 100$,

and prior research leads you to believe that the best of the training methods will improve GRE scores by about 20 points over no treatment at all. Approximately how many subjects in each group should you use if you desire the probability of obtaining a significant result to be .8?

Section B: Supplementary exercises

11. A Navy gunnery officer is charged with the task of determining which gun-loading procedure enables his gun teams to fire the fastest. There are five possible procedures. He trains six teams to use each of the five procedures (a total of 30 teams) and then counts the number of shots fired in a 30-minute interval. The numbers of shots are shown in the table. Does loading procedure make a difference? Use $\alpha = .05$ and the steps of Exercise 1.

A	B	C	D	E
18	21	20	18	22
19	18	22	19	23
16	19	23	16	21
19	19	22	19	18
21	20	23	21	24
17	15	18	16	21

12. (*Exercise 1 continued*) Reconsider the beer data from Exercise 1. Suppose we add 1 point to each of the Budweiser ratings, so they become 9, 9, 8, 9, 10. How will that change (increase, decrease, or no change) each measure?
(a) Group means ($\overline{X}_j$)
(b) Standard deviation based on the group means
(c) MS_B
(d) Within-group variances (s_j^2)
(e) Pooled within-group variance (s^2_{pooled})
(f) MS_W
(g) Overall F ratio

13. (*Exercise 1 continued*) Suppose we *subtract* 1 point from each of the Budweiser ratings, so they become 7, 7, 6, 7, 8. How will that change (increase, decrease, or no change) each measure?
(a) Group means ($\overline{X}_j$)
(b) Standard deviation based on the group means
(c) MS_B
(d) Within-group variances (s_j^2)
(e) Pooled within-group variance (s^2_{pooled})
(f) MS_W
(g) Overall F ratio

14. (*Exercise 1 continued*) Suppose we repeat the adding and subtracting of 1 point procedure we just did with the Budweiser data, but this time with the Strohs data instead of the Budweiser data.
(a) Will the Strohs manipulations have a larger effect, a smaller effect, or the same effect as the Budweiser manipulations? [*Hint:* They will *not* have the same effect, even though that may seem intuitively correct.]
(b) Explain why such a manipulation has different effects on the two different groups.
(c) A similar manipulation on one of the remaining groups will have an effect approximately as large as the larger of the Budweiser/Strohs effects. Which one? Why?
(d) State the general principle involved here. [*Hint:* It should have a form something like (choose one of each slashed option): "Increasing/decreasing the largest/smallest/middle mean of an ANOVA data set will have a large/small effect on the F ratio, while altering the largest/smallest/middle mean of the data set will have a small effect on F."]

15. (*Exercise 1 continued*) Suppose we add a fifth beer, Pabst, to our taste test data. Its ratings are 10, 10, 9, 10, 10. How will that change (increase, decrease, otherwise alter, or no change) each measure?
(a) Group means ($\overline{X}_j$)
(b) Standard deviation based on the group means
(c) MS_B
(d) Within-group variances (s_j^2)
(e) Pooled within-group variance (s^2_{pooled})
(f) MS_W
(g) Overall F ratio

16. (*Exercise 1 continued*) Suppose we add a fifth beer, Pabst, to our taste test data, and now its ratings are 6, 4, 7, 6, 7. How will that change (increase, decrease, otherwise alter, or no change) each measure?
(a) Group means ($\overline{X}_j$)
(b) Standard deviation based on the group means
(c) MS_B
(d) Within-group variances (s_j^2)
(e) Pooled within-group variance (s^2_{pooled})
(f) MS_W
(g) Overall F ratio

17. State the general principle involved in Exercises 15 and 16. [*Hint:* Its form should be something like (choose one of each set of slashed choices): "Adding a new group to an ANOVA data set, whose group mean/within-group variance is larger than/smaller than/in the middle of the remaining group means/within-group variances will increase/decrease the overall F ratio, whereas adding a new group whose mean/within-group variance is larger than/smaller than/in the middle of the remaining group means/within-group variances will have little effect on F."]

18. The Acme Nut and Bolt Company wishes to buy a new machine that makes screws. Three such machines are on the market, and all are about equally priced, so Acme decides to determine whether the rate of defective screws differs from machine to machine. It makes ten trial runs on each of the three machines; each trial run involves the production of 1000 screws. The product engineers inspect each batch, counting the number of defective screws in

each batch. The numbers are shown in the table. Is there a difference in the quality of the three screw machines? Use the steps of Exercise 1.

Machine A	Machine B	Machine C
10	11	9
10	9	7
15	10	8
9	10	6
10	10	10
10	11	9
11	10	9
14	10	7
10	12	8
10	9	7

19. A woman decides to take seriously the claim that some shampoos will give hair more body. She buys five different shampoos and uses a different one each day. Each day, she randomly chooses six co-workers and asks them to rate her hair on a 10-point scale (10 being high). The ratings are shown in the table. Does kind of shampoo make a difference? Use the steps of Exercise 1.

A	B	C	D	E
7	6	8	7	6
7	6	7	8	8
7	6	7	8	6
6	7	7	6	7
8	5	7	7	7
7	6	6	5	6

20. Suppose you wish to test three kinds of wood glue: Bondex, SuperBond, and BonzAll. Your experiment will glue n pairs of blocks of wood together using Bondex; then you will measure the force necessary to split each glue joint apart. You will repeat the process for n joints using SuperBond and n joints using BonzAll. You know from previous research that the strength of Bondex joints is typically normally distributed with mean 200 pounds and standard deviation 40 pounds, that SuperBond joints are typically normally distributed with mean 220 pounds and standard deviation 40 pounds, and BonzAll joints are typically normally distributed with mean 188 pounds and standard deviation 40 pounds. Approximately what sample size n should be in each group if you desire the probability of obtaining a significant result to be .8?

Section C: Extending your comprehension (answers in Appendix D, page 523)

21. Suppose that in an ANOVA the groups are all identical (that is, $X_{ij} = X_{iJ}$ for all i, j, and J)? What value does F have?

22. Suppose that in an ANOVA all the members of group 1 have value 1 (that is, $X_{i1} = 1$ for all i), all the members of group 2 have value 2, all the members of group 3 have value 3, and all the members of group 4 have value 4. What value does F have?

23. Suppose that in an ANOVA with four groups, all the members of every group have value 2 (that is, $X_{ij} = 2$ for all i and j). What value does F have?

24. Create a data set with four groups, four members in each group, such that $F = 1.0$.

Section D: Cumulative review (answers in Appendix D, page 523)

For Exercises 25–31, choose the correct answer or fill in the blank. (**Note that computations are *not* required**; use $\alpha = .05$ unless otherwise instructed.)

(a) This problem requires
  (1) Finding the area under a normal distribution [Skip parts (b)–(g) and proceed to (h).]
  (2) Creating a confidence interval
  (3) Testing a hypothesis about the mean of one group
  (4) Testing a hypothesis about the means of two independent groups
  (5) Testing a hypothesis about the means of two dependent groups
  (6) Testing a hypothesis about the means of three or more independent groups
  (7) Performing a power analysis (involving finding the area under a normal distribution). [Skip parts (b)–(h).]

(b) The null hypothesis is of the form:
  (1) $\mu = a$ (2) $\mu_1 = \mu_2$
  (3) $\mu_D = 0$ (4) $\mu_1 = \mu_2 = \cdots = \mu_k$
  (5) There is no null hypothesis.

(c) The appropriate sample statistic(s) is (are):
  (1) $\overline{X}$ (2) p
  (3) $\overline{X}_1 - \overline{X}_2$ (4) $\overline{D}$
  (5) Not applicable for ANOVA

(d) The appropriate test statistic(s) is (are):
  (1) z (2) t
  (3) F

(e) The number of degrees of freedom for this test statistic is ________. (State the number or *not applicable*.)

(f) The hypothesis test (or confidence interval) is:
  (1) Directional (2) Nondirectional

(g) The critical value of the test statistic is ________. (Give the value.)

(h) The appropriate formula for the test statistic (or confidence interval) is given by Equation (________).

Note that computations are *not* required for any of the problems in Section D. See the instructions.

25. A physician wishes to determine a patient's mean systolic blood pressure, so she teaches him to take his own blood pressure, gives him a telephone page beeper to carry with him, and has his nurse beep him at 25 random times, when he is to measure his blood pressure. The blood pressures

are: 131, 121, 138, 147, 143, 132, 129, 130, 134, 126, 143, 134, 126, 121, 143, 128, 121, 123, 132, 135, 136, 132, 132, 130, 128. What can be said about his mean blood pressure?

26. The suitcases that a particular airline carries have weights that are normally distributed with mean 26 pounds and standard deviation 5 pounds. Suppose that the suitcases are carried two at a time by porters, and that each of the two can be considered a random sample from the suitcase population. What percentage of porter's loads (that is, two suitcases) have mean suitcase weight of 30 pounds or more?

27. A psychologist is studying speech patterns. He wishes to know whether the frequency of interruptions (abrupt pauses, abrupt changes of topic, etc.) differs among schizophrenic patients, anxiety patients, and nonpatients. He identifies 12 schizophrenics, 10 anxiety patients, and 15 nonpatients and engages them in a standard interview, which is tape recorded. Later, he counts the number of speech interruptions per 1 minute of speaking. The numbers are shown in the table. Does the number of speech interruptions depend on diagnosis?

Schiz.	*Anxiety*	*Nonpat.*
11	12	8
5	4	7
15	8	6
13	6	4
15	9	6
11	6	7
10	9	4
10	5	8
13	6	4
12	3	6
14		5
11		5
		5
		3
		5

28. I wish to know whether Duracell alkaline batteries have a different capacity than EverReady batteries. How many batteries of each type should I test if I make the following two assumptions?
    (a) I believe that the standard deviation of tested battery capacity is .3 mAH (milliAmp-hours).
    (b) I will consider a difference practically significant if it exceeds .1 mAH.

29. A sociologist wishes to know whether exposure to a natural disaster changes an individual's religiosity. She happens to have given a religiosity scale to 100 students the week before a dormitory fire caused grave damage and injury. She finds that 24 of her subjects were present in the dormitory during the disaster. She locates those 24 subjects and, two weeks later, gives them the religiosity test again. The scores for those individuals are shown in the table. Has the fire changed religiosity?

Table for Exercise 29

Student	*Before*	*After*
1	54	58
2	51	52
3	44	61
4	50	63
5	38	50
6	42	48
7	42	49
8	53	61
9	54	69
10	34	58
11	36	48
12	40	46
13	41	45
14	49	54
15	49	60
16	53	49
17	40	59
18	55	52
19	47	66
20	52	53
21	53	70
22	58	64
23	20	48
24	55	60

30. A glue manufacturer is considering including an additive in its glue to make glued joints stronger. She prepares 20 chains, each of which has one link that is broken. She randomly divides the chains into two groups, with ten chains per group, and repairs the broken link with either regular glue or glue with additives. She then hangs the chains from the ceiling and gradually adds weight to the end of each chain until it breaks. She records the breaking weights (in pounds), as shown here. Is the glue with the additive superior to the regular glue?

    Regular glue: 41, 50, 43, 39, 39, 40, 45, 42, 43, 36
    Glue with additive: 47, 42, 42, 39, 59, 43, 38, 55, 40, 53

31. Suppose that of all the people who purchase soft drinks at Acme Supermarket, 29% purchase Coke. Now Acme lowers the price of Coke by 20 cents per six pack. Of the next 300 customers who purchase soft drinks, 121 purchase Coke. Has lowering the price caused more people to purchase Coke?

Section E: From the journals (answers in Appendix D, page 523)

32. Hall and Honey (1990) investigated "the conditions that influence the transfer of a conditioned emotional response from one context to another" (p. 271). In particular, they predicted that "conditioned responding should be less vigorous when those contextual cues that were present during acquisition are absent at the time of testing." They divided 32 rats into four groups that were to acquire the behavior of pressing a flap to gain access to food pellets. The exper-

imenters created two kinds of salient contextual cues by introducing distinctive odors (either eucalyptus oil or iso-amyl acetate) or distinctive visual scenes (with or without black-and-white checkered wallpaper on the sides of the operant chambers). Their basic design was to have each of the four groups of rats acquire the behaviors in one of these four contexts (eucalyptus oil/checkered wall paper, eucalyptus oil/plain wallpaper, iso-amyl acetate/checkered wallpaper, and iso-amyl acetate/plain wallpaper) and then subsequently to test the behaviors in identical or different contexts. The details of this subsequent testing need not concern us at the moment.

Even though Hall and Honey randomly assigned the rats to the four groups, it was possible that the groups differed before training began. In order to rule out this hypothesis, the researchers measured the response rates of the four groups before they presented any of the contextual stimuli; the mean rates of response for the four groups were $\overline{X} = 27.62$, 26.00, 20.28, and 20.25 responses per minute. If these means were significantly different from one another, then any subsequent finding could be the result of initial differences between the groups, not the result of experimental manipulation.

Hall and Honey concluded that in this preliminary test, "an... ANOVA conducted on... the rates of response... showed that these rates did not differ, $F(3, 27) = 1.76$, $p > .17$" (p. 274).

(a) Are Hall and Honey's degrees of freedom computed correctly?
(b) Is enough information presented to allow you to verify that F was computed correctly? If so, verify the computation of F.
(c) Assuming equal group sizes (which was apparently not actually the case in Hall and Honey's study, although they don't say what the actual group sizes were), what was MS_B?
(d) Assuming that F was computed correctly, approximately what was MS_W?
(e) Create a complete ANOVA summary table for this test.
(f) State the conclusion of this test in plain English.

33. (*Exercise 32 continued*) Suppose we replicate Hall and Honey's (1990) study and, like them, seek to know whether the four groups of rats have significantly different response rates before training begins. We use 24 rats, assigning six to each group. The response rates are shown in the table.

Group 1	*Group 2*	*Group 3*	*Group 4*
21.1	22.0	27.2	28.5
30.1	22.6	21.6	23.5
28.6	26.0	21.7	36.2
25.4	11.1	27.1	19.2
23.6	26.0	18.5	24.7
27.3	20.8	26.3	15.4

(a) How many degrees of freedom are there for our test?
(b) Compute the analysis of variance, showing the complete ANOVA summary table.
(c) State your conclusion in plain English.

Section F: Computer explorations

34. Ask the computer to perform an ANOVA on the following data:

Sample 1: 1, 1, 1, 1, 1
Sample 2: 2, 2, 2, 2, 2
Sample 3: 3, 3, 3, 3, 3

(a) Explain why the computer tells you that the ANOVA F is undefined for these data.
(b) Change the last point in sample 3 from "3" to "3.01." Now what happens to F? Explain why.
(c) Recall that in Exercise 31 in Chapter 11, you performed an independent-sample t test for similar data. Explain the similarities in your results.

35. (a) Have the computer create a random sample of size five from the population of IQ scores ($\mu = 100$, $\sigma = 15$). [*Hint:* We described how to do that in Section 7.3.]
(b) Have the computer create a second random sample of size five from the population of IQ scores ($\mu = 100$, $\sigma = 15$).
(c) Have the computer create a third random sample of size five from the population of IQ scores ($\mu = 100$, $\sigma = 15$).
(d) If these are the data in a three-group ANOVA, what can you say about the truth of the null hypothesis?
(e) Approximately what would you predict the ANOVA F to be?
(f) Why is it reasonable to predict that MS_W will be approximately $(15)^2 = 225$? Why is it reasonable to predict that MS_B will also be approximately 225?
(g) Have the computer determine the ANOVA. How does the actual ANOVA summary compare to your predictions in parts (e) and (f)? [*Hint:* You may be a bit surprised here.]

36. Repeat Exercise 32 several times to get a feel for the magnitude of F when the null hypothesis is true, and for how far the two mean squares can vary from 225. Then try varying the sample sizes (try $n = 200$), the number of groups, and other factors to extend your comprehension.

Eyeball-estimating the Analysis of Variance (Optional)

This section may be omitted without loss of continuity.

Remember that the philosophy of eyeball-estimation is to perform all eyeball computations "in your head," *not* using a calculator. The object is to get a quick approximation, not an exact result.

It is sometimes impossible to get an accurate approximation when eyeball-estimating ANOVA. Eyeball-estimating ANOVA is useful as a way of quickly practicing the ANOVA procedure, however, even if the numerical result is rather imprecise.

General Eyeball Procedure

The eyeball-estimation of ANOVA follows the same general procedure that we have been using since Chapter 9.

I. State the Null and Alternative Hypotheses

Use the same procedure we used when computing.

II. Set the Criterion for Rejecting H_0

$F_{\text{cv}_{\text{by eyeball}}}$ is about 3 or 4. The more degrees of freedom, the smaller F_{cv}, but specifying it more precisely is difficult.

III. Collect a Sample and Eyeball-estimate the Observed Values of the Sample Statistic and the Test Statistic

Eyeballing the test statistic follows the same procedures we used when computing ANOVA in equal n_j situations: First we eyeball-estimate MS_B (the numerator), then MS_W (the denominator), and then the F ratio: $\text{MS}_\text{B}/\text{MS}_\text{W}$.

a. Eyeball-estimate $\text{MS}_{\text{B}_{\text{by eyeball}}} \approx s^2_{\bar{X}_{\text{by eyeball}}}(n)$ according to the following steps:
   i. Eyeball-estimate the mean of the first group (use either the range method or the balancing-point method, whichever is more convenient).
   ii. Eyeball-estimate the means of the remaining groups in the same way.
   iii. Eyeball-estimate the standard deviation of those k means (use the range method with a conversion factor of about 2 or 2.5). That gives $s_{\bar{X}_{\text{by eyeball}}}$.
   iv. Eyeball-square the result of the preceding step and eyeball-multiply it by n, the sample size of each group. If the sample sizes are not all equal, use an approximate mean of the sample sizes (close enough for eyeball purposes). That gives $\text{MS}_{\text{B}_{\text{by eyeball}}} \approx s^2_{\bar{X}_{\text{by eyeball}}}(n)$, the numerator of F. Do not use a calculator; the object is just to get "in the ballpark."

b. Eyeball-estimate $\text{MS}_{\text{W}_{\text{by eyeball}}} \approx s^2_{\text{pooled}_{\text{by eyeball}}}$ according to the following steps:
   i. Eyeball-estimate the standard deviation of the first group (use either the range method or the inflection-point method, whichever is more convenient).
   ii. Eyeball-estimate the standard deviation of the each of the remaining groups in the same way.
   iii. Eyeball-square each standard deviation to obtain the sample variances. Do not use a calculator; the object is just to get "in the ballpark."
   iv. Eyeball-pool those variances: The pooled variance is a value that is somewhere in the middle of the variances from the preceding step: $\text{MS}_{\text{W}_{\text{by eyeball}}} \approx s_{\text{pooled}_{\text{by eyeball}}}$.

c. Obtain $F_{\text{by eyeball}}$ by eyeball-dividing MS_B by MS_W. Do not use a calculator.

IV. Interpret the Results

If your eyeballed test statistic ($F_{\text{by eyeball}}$ from step c) is greater than your eyeballed critical value (from part II), then $\text{reject}_{\text{by eyeball}}$ H_0.

The Antidepressant Drug Example

This eyeball-estimation procedure parallels the computation in the text on page 297.

We now apply that procedure to the eyeball-estimation of the antidepressant data example in the text. You may recall that there were four different groups and each received a different antidepressant drug. The null hypothesis was H_0: $\mu_1 = \mu_2 = \mu_3 = \mu_4$.

I. State the Null and Alternative Hypotheses

Use the same procedure we used when computing.

II. Set the Criterion for Rejecting H_0

These are relatively small samples, so we suspect that $F_{\text{cv}_{\text{by eyeball}}}$ will be rather large, perhaps about 4.

III. Collect a Sample and Eyeball-estimate the Observed Values of the Sample Statistic and the Test Statistic

a. Eyeball-estimate $MS_{B_{\text{by eyeball}}} \approx s^2_{\overline{X}_{\text{by eyeball}}}(n)$.
   i. We can most conveniently use the range method to eyeball-estimate the first sample mean:

$$\overline{X}_{1_{\text{by eyeball}}} \approx \frac{\text{highest } X_{i1} + \text{lowest } X_{i1}}{2} = \frac{32 + 25}{2} = 28.5$$

   ii. The remaining means are obtained the same way:

$$\overline{X}_{2_{\text{by eyeball}}} \approx \frac{\text{highest } X_{i2} + \text{lowest } X_{i2}}{2} = \frac{37 + 28}{2} = 32.5$$

$$\overline{X}_{3_{\text{by eyeball}}} \approx \frac{\text{highest } X_{i3} + \text{lowest } X_{i3}}{2} = \frac{26 + 19}{2} = 22.5$$

$$\overline{X}_{4_{\text{by eyeball}}} \approx \frac{\text{highest } X_{i4} + \text{lowest } X_{i4}}{2} = \frac{36 + 30}{2} = 33.0$$

   iii. The standard deviation of those four values, by the range method, is (32.5 − 22.5) divided by about 2 or 2.5, or 10 divided by a bit more than 2, so $s_{\overline{X}_{\text{by eyeball}}} \approx 5$ (actually a bit less, but 5 is close enough).
   iv. $MS_{B_{\text{by eyeball}}} \approx s^2_{\overline{X}_{\text{by eyeball}}}(n) = 5^2(6) = 150$. That is the numerator$_{\text{by eyeball}}$ of F.

b. Eyeball-estimate $MS_{W_{\text{by eyeball}}} \approx s^2_{\text{pooled}_{\text{by eyeball}}}$.
   i. We can most conveniently use the range method to eyeball-estimate the first sample standard deviation. The conversion factor will be about 2.5 because there are six points:

$$s_{1_{\text{by eyeball}}} \approx \frac{\text{highest } X_{i1} - \text{lowest } X_{i1}}{2.5} = \frac{32 - 25}{2.5} \approx 3$$

   ii. The remaining standard deviations are obtained the same way:

$$s_{2_{\text{by eyeball}}} \approx \frac{\text{highest } X_{i2} - \text{lowest } X_{i2}}{2.5} = \frac{37 - 28}{2.5} \approx 3$$

$$s_{3_{\text{by eyeball}}} \approx \frac{\text{highest } X_{i3} - \text{lowest } X_{i3}}{2.5} = \frac{26 - 19}{2.5} \approx 3$$

$$s_{4_{\text{by eyeball}}} \approx \frac{\text{highest } X_{i4} - \text{lowest } X_{i4}}{2.5} = \frac{36 - 30}{2.5} \approx 2$$

   iii. The four sample variances$_{\text{by eyeball}}$ are therefore about 9, 9, 9, and 4.
   iv. The pooled variance $s^2_{\text{pooled}_{\text{by eyeball}}}$ is the average of those four variances, or $s^2_{\text{pooled}_{\text{by eyeball}}} \approx 7$. Therefore, $MS_{W_{\text{by eyeball}}} = s^2_{\text{pooled}_{\text{by eyeball}}} \approx 7$.

c. $F_{\text{by eyeball}} \approx MS_{B_{\text{by eyeball}}}/MS_{W_{\text{by eyeball}}} = 150/7 \approx 21$.

IV. Interpret the Results

Because $F_{\text{by eyeball}} = 21$ is in fact greater than $F_{\text{cv}_{\text{by eyeball}}} \approx 4$, we do reject$_{\text{by eyeball}}$ the null hypothesis.

As you may recall, we also rejected the null hypothesis by computation in the text: The computed values were $\text{MS}_\text{B} = 115.00$, $\text{MS}_\text{W} = 6.75$, and $F = 17.04$, quite close to our eyeballed values.

The Chimpanzee Example

This eyeball-estimation procedure parallels the computation in the text on page 301.

We now apply our procedure to the eyeball-estimation of the chimpanzee grooming example in the text. You may recall that there were three different groups of chimpanzees, and we counted the number of grooming behaviors in each. The null hypothesis was H_0: $\mu_1 = \mu_2 = \mu_3$.

I. State the Null and Alternative Hypotheses

Use the same procedure we used when computing.

II. Set the Criterion for Rejecting H_0

These are relatively small samples, so we suspect that $F_{\text{cv}_{\text{by eyeball}}}$ will be rather large, perhaps about 4.

III. Collect a Sample and Eyeball-estimate the Observed Values of the Sample Statistic and the Test Statistic

a. Eyeball-estimate $\text{MS}_{\text{B}_{\text{by eyeball}}} \approx s^2_{\overline{X}_{\text{by eyeball}}}(n)$.
   i. We can most conveniently use the range method to eyeball-estimate the first sample mean:

$$\overline{X}_{1_{\text{by eyeball}}} \approx \frac{\text{highest } X_{i1} + \text{lowest } X_{i1}}{2} = \frac{5+3}{2} = 4$$

   ii. The remaining means are obtained the same way:

$$\overline{X}_{2_{\text{by eyeball}}} \approx \frac{\text{highest } X_{i2} + \text{lowest } X_{i2}}{2} = \frac{6+3}{2} = 4.5$$

$$\overline{X}_{3_{\text{by eyeball}}} \approx \frac{\text{highest } X_{i3} + \text{lowest } X_{i3}}{2} = \frac{7+4}{2} = 5.5$$

   iii. The standard deviation of those four values, by the range method, is (5.5 − 4.0) divided by about 2, or 1.5 divided by 2, so $s_{\overline{X}_{\text{by eyeball}}} \approx .75$.
   iv. $\text{MS}_{\text{B}_{\text{by eyeball}}} \approx s^2_{\overline{X}_{\text{by eyeball}}}(n) = .75^2(5) \approx 3$. That is the numerator$_{\text{by eyeball}}$ of F.

b. Eyeball-estimate $\text{MS}_{\text{W}_{\text{by eyeball}}} \approx s^2_{\text{pooled}_{\text{by eyeball}}}$.
   i. We can most conveniently use the range method to eyeball-estimate the first sample standard deviation. The conversion factor will be about 2.5 because there are five points:

$$s_{1_{\text{by eyeball}}} \approx \frac{\text{highest } X_{i1} - \text{lowest } X_{i1}}{2.5} = \frac{5-3}{2.5} \approx 1$$

   ii. The remaining standard deviations are obtained the same way:

$$s_{2_{\text{by eyeball}}} \approx \frac{\text{highest } X_{i2} - \text{lowest } X_{i2}}{2.5} = \frac{6-3}{2.5} \approx 1$$

$$s_{3_{\text{by eyeball}}} \approx \frac{\text{highest } X_{i3} - \text{lowest } X_{i3}}{2.5} = \frac{7-4}{2.5} \approx 1$$

   iii. The three sample variances$_{\text{by eyeball}}$ are therefore about 1, 1, and 1.

iv. The pooled variance $s^2_{\text{pooled}_{\text{by eyeball}}}$ is the average of those three variances, or $s^2_{\text{pooled}_{\text{by eyeball}}} \approx 1$. $\text{MS}_{\text{W}_{\text{by eyeball}}} = s^2_{\text{pooled}_{\text{by eyeball}}} \approx 1$.

c. $F_{\text{by eyeball}} \approx \text{MS}_{\text{B}_{\text{by eyeball}}}/\text{MS}_{\text{W}_{\text{by eyeball}}} = 3/1 \approx 3$.

IV. Interpret the Results

Because $F_{\text{by eyeball}} \approx 3$ is not greater than $F_{\text{cv}_{\text{by eyeball}}} \approx 4$, we do not reject$_{\text{by eyeball}}$ the null hypothesis.

As you may recall, we did reject the null hypothesis by computation in the text: The computed values were $\text{MS}_\text{B} = 6.067$, $\text{MS}_\text{W} = 1.067$, and $F = 5.69$. The discrepancy is in our estimation of MS_B: The actual group 3 mean is 6, but we eyeball-estimated 5.5, and that made our eyeball-estimate of the variance of the means too low. However, our aim here is primarily to understand the process, and even though our estimate was not terribly accurate, it did give us practice in comparing the variance based on means to the within-group variance, which is the essence of ANOVA.

You may wish to use the ESTAT exercise "anova" at this time. In version 2.0 or later, take the eyeball-estimation path.

Computational Formulas for ANOVA (Optional)

This section may be omitted without loss of continuity.

The deviation formulas for the sums of squares—Equations (13.11) and (13.12)—are perfectly adequate methods for computing the analysis of variance, whether or not the n_j are equal. However, the ANOVA deviation formulas can be somewhat cumbersome to compute when data sets are large. We present here computational formulas that are somewhat easier because they do not require the calculation of deviations for each point. The alert reader will recognize that the present situation is entirely analogous to that encountered in Chapter 5, when for the same reason we presented mean square deviation and computational formulas for the standard deviation and variance.

These computational formulas for the sums of squares are applicable whether or not the n_j are equal. The total sum of squares is

total sum of squares, computational formula

$$\text{SS}_\text{T} = \sum X_{ij}^2 - \frac{\left(\sum X_{ij}\right)^2}{n_\text{G}} \tag{13.18}$$

where n_G is the total number of observations: $n_\text{G} = n_1 + n_2 + \cdots + n_k$.

The sum of squares between groups is

sum of squares between groups, computational formula

$$\text{SS}_\text{B} = \left[\frac{\left(\sum X_{i1}\right)^2}{n_1} + \frac{\left(\sum X_{i2}\right)^2}{n_2} + \cdots + \frac{\left(\sum X_{ik}\right)^2}{n_k}\right] - \frac{\left(\sum X_{ij}\right)^2}{n_\text{G}} \tag{13.19}$$

Note that the last terms in Equations (13.18) and (13.19) are identical and so need to be computed only once.

The sum of squares within groups is

sum of squares within groups, computational formula

$$\text{SS}_\text{W} = \sum X_{ij}^2 - \left[\frac{\left(\sum X_{i1}\right)^2}{n_1} + \frac{\left(\sum X_{i2}\right)^2}{n_2} + \cdots + \frac{\left(\sum X_{ik}\right)^2}{n_k}\right] \tag{13.20}$$

Note that the terms in brackets in Equations (13.19) and (13.20) are identical, so they need to be computed only once. Similarly, the first terms in Equations (13.18) and (13.20) are identical.

We use Equation (13.10b) to verify our computations:

$$\text{SS}_\text{T} = \text{SS}_\text{W} + \text{SS}_\text{B}$$

Table 13.15 summarizes the computation using the given formulas for the antidepressant drug data from Table 13.1. Note that we get identical results to those we obtained in the text.

TABLE 13.15 Using the computational formulas for ANOVA on the antidepressant drug data

X_{i1}	X_{i1}^2	X_{i2}	X_{i2}^2	X_{i3}	X_{i3}^2	X_{i4}	X_{i4}^2
29	841	35	1225	23	529	33	1089
32	1024	34	1156	25	625	36	1296
26	676	28	784	26	676	30	900
28	784	32	1024	26	676	32	1024
25	625	37	1369	25	625	32	1024
28	784	35	1225	19	361	32	1024
168	4734	201	6783	144	3492	195	6357

$$\sum X_{ij} = 168 + 201 + 144 + 195 = 708$$

$$\sum X_{ij}^2 = 4734 + 6783 + 3492 + 6357 = 21{,}366$$

$$SS_T = \sum X_{ij}^2 - \frac{\left(\sum X_{ij}\right)^2}{n_G} = 21{,}366 - \frac{(708)^2}{24} = 21{,}366 - 20{,}886 = 480$$

$$SS_B = \left[\frac{(\sum X_{i1})^2}{n_1} + \frac{(\sum X_{i2})^2}{n_2} + \cdots + \frac{(\sum X_{ik})^2}{n_k}\right] - \frac{\left(\sum X_{ij}\right)^2}{n_G}$$

$$= \left[\frac{(168)^2}{6} + \frac{(201)^2}{6} + \frac{(144)^2}{6} + \frac{(195)^2}{6}\right] - \frac{(708)^2}{24}$$

$$= (4704.0 + 6733.5 + 3456.0 + 6337.5) - 20{,}886.0$$

$$= 21{,}231.0 - 20{,}886.0 = 345$$

$$SS_W = -\sum X_{ij}^2 - \left[\frac{(\sum X_{i1})^2}{n_1} + \frac{(\sum X_{i2})^2}{n_2} + \cdots + \frac{(\sum X_{ik})^2}{n_k}\right]$$

$$= 21{,}366 - 21{,}231.0 = 135$$

Check: $SS_T = 480 = SS_W + SS_B = 135 + 345$

Exercise for Optional Section

1. Use the computational formulas to compute the ANOVA for the chimpanzee grooming data of Table 13.6. Show that the computational formulas give the same results shown in Table 13.7.

CHAPTER 14

Post Hoc Tests, A Priori Tests, and Repeated-Measures Analysis of Variance

LEARNING OBJECTIVES

1. What are post hoc tests and when can they be performed?
2. Distinguish between pairwise and complex null hypotheses.
3. What is the Tukey HSD test?
4. What is the *Q* test statistic?
5. What is an a priori test? What are its advantages and disadvantages compared with an omnibus *F* and subsequent post hoc tests?
6. What is a repeated-measures analysis of variance?
7. How do the ANOVA summary tables for independent-sample and repeated-measures tests differ?

This chapter covers three topics related to the ANOVA of Chapter 13: (1) post hoc tests, (2) planned comparisons, and (3) repeated-measures ANOVA.

As we saw in Chapter 13, when we reject an ANOVA null hypothesis, we support one or more of the alternative hypotheses: $\mu_1 \neq \mu_2$ or $\mu_1 \neq \mu_3$ or $\mu_2 \neq \mu_4$ or some other pairwise or complex alternative hypothesis. However, the ANOVA results do not tell us *which one* (or *ones*) of these alternatives to support. Post hoc tests allow us to decide, after rejecting the ANOVA null hypothesis, which of the possible alternative hypotheses to support.

There are many cases when it is preferable to skip the ANOVA altogether, however, and test directly some combination of population means. These tests are called "a priori tests" or "planned comparisons"; they are substantially more powerful than the ANOVA and its subsequent post hoc tests.

The ANOVA of Chapter 13 assumed that the samples in several groups were independent of one another, but that is not always true or desirable. The repeated-measures ANOVA is the analysis of variance analogy to the dependent-sample *t* test of Chapter 12.

14.1 Interpreting ANOVA: Post Hoc Tests

You may recall that the null hypothesis for the independent-sample analysis of variance in Chapter 13 is H_0: $\mu_1 = \mu_2 = \cdots = \mu_k$. We called this an "omnibus" null hypothesis because if H_0 is true, then *all* the possible *pairwise* null hypotheses [e.g., H_{0_1}: $\mu_1 = \mu_2$] and *complex* null hypotheses [e.g., H_{0_7}: $(\mu_1 + \mu_2)/2 = (\mu_3 + \mu_4)/2$] are also true. (See the list of null hypotheses on page 284.) If we reject the ANOVA null hypothesis, then we must reject one or more of these pairwise or complex null hypotheses, *but we do not know which one(s).*

An omnibus test like ANOVA is sometimes compared to a shotgun, which fires many pellets simultaneously in a pattern that covers a broad area. The advantage of a shotgun over a rifle is that it does not have to be aimed precisely; the shooter counts on the spread of the pellets to hit the target. The omnibus (shotgun) ANOVA aims one pellet at each possible (pairwise or complex) null hypothesis. Rejection of the omnibus null hypothesis indicates that one or more of the pellets hit its target (that is, that one or more of the possible null hypotheses should be rejected), but it does not tell us which one(s). The post hoc tests in this section will help us to determine which of the possible (pairwise or complex) null hypotheses to reject.

post hoc tests: hypothesis tests performed after a significant ANOVA to explore which means or combination of means differ from one another

Post hoc is Latin for "after the fact." It is permissible to perform post hoc tests only *after* an omnibus ANOVA null hypothesis is rejected.

Recall from Chapter 13 that we performed an analysis of variance on data that represented scores on the Beck Depression Inventory (BDI) for groups of depressed patients who received Prozac, Elavil, Zoloft, or Tofranil. We reprint the data from Table 13.1 and the ANOVA summary table from Table 13.4 in Table 14.1 for convenience. We rejected the omnibus null hypothesis H_0: $\mu_1 = \mu_2 = \mu_3 = \mu_4$ because $F_{obs} = 17.04$ was greater than the critical value $F_{cv} = 3.10$.

TABLE 14.1 Patients' Beck Depression Inventory scores and analysis of variance

	Group 1 (Prozac)		Group 2 (Elavil)	Group 3 (Zoloft)	Group 4 (Tofranil)
	29		35	23	33
	32		34	25	36
	26		28	26	30
	28		32	26	32
	25		37	25	32
	28		35	19	32
n_j	6		6	6	6
$\overline{X}_j$	28.0		33.5	24.0	32.5
s_j^2	6.000		9.900	7.200	3.900
Source	SS	*df*	MS	*F*	
Between	345	3	115.00	17.04*	
Within	135	20	6.75		
Total	480	23			

$^*p < .05, F_{cv_{df=3,20}} = 3.10$

Rejecting the omnibus null hypothesis implies that we should reject *one or more* of the possible null hypotheses. Perhaps a pairwise null hypothesis such as H_{0_1}: $\mu_1 = \mu_2$ ("The effect of Prozac equals the effect of Elavil") should be rejected; perhaps a complex null hypothesis such as H_{0_7}: $(\mu_1 + \mu_2)/2 = (\mu_3 + \mu_4)/2$ ("The average of the effects of Prozac and Elavil equals the average of the effects of Zoloft and Tofranil") should be rejected. Perhaps both. Perhaps neither one of those but some other pairwise or complex null hypothesis should be rejected.

It may seem that we could simply inspect the sample means: Table 14.1 shows that group 3 (Zoloft) had a mean BDI score ($\overline{X}_3 = 24.0$) lower than that of any other drug (low BDI scores are good—associated with less depression). Does rejecting the ANOVA null hypothesis therefore imply that Zoloft had *significantly* lower BDI scores than the other three drugs? *No, it does not*. The rejection of the omnibus ANOVA H_0 allows us to conclude that *some* drug (or combination thereof) had a significantly lower mean BDI score than *some other* drug (or combination thereof), but it does *not* allow us to conclude *anything* about the significance of the differences among any *particular* groups. Group 3's (Zoloft's) mean BDI score (24.0) is certainly lower than group 1's (Prozac's) mean (28.0), but to determine whether it is *significantly* lower, we need to conduct a statistical test.

Furthermore, it may seem that we could simply conduct a *t* test between a particular pair of means. For example, we might observe that in our data, Zoloft produced the lowest BDI scores ($\overline{X}_3 = 24.0$) and Elavil produced the highest ($\overline{X}_2 = 33.5$). Can we simply perform a *t* test with those two means? No, because that would violate the basic philosophy of the *t* test. The *t* test is a way of limiting (to the probability α) the likelihood that an observed difference is due to random sampling fluctuations. But probabilities must be specified *before* any characteristics of the particular sample data are known. We must decide to perform a *t* test *before* we collect the data.

This analogy may help to clarify why the decision to perform the *t* test must be made in advance. Suppose I roll a fair die. The probability of the die showing "3" is 1/6. But if I roll the die and then peek at the result, the probability that the die shows "3" is *not* 1/6 but is either 1 (if my peek revealed a "3") or 0 (if my peek revealed a face other than "3"). Similarly, if I collect antidepressant drug data and peek at them, then the probability that a particular *t* test will be significant (even if the null hypothesis is true) is no longer α: It is greater than α if my peek revealed that the means are in fact far apart, but less than α if my peek revealed that the means are close together.

Having performed the ANOVA (or otherwise inspecting the data) prohibits us from subsequently performing *t* tests on our antidepressant drug data. We would still like to explore the pattern of significances in our ANOVA data set, however. We would still like to know which drugs are significantly different from which other drugs. Statisticians have developed several methods that allow this kind of exploration. Although it is beyond the scope of this textbook to explain their rationales in detail, all are based on some variation of the following premise: The fact that we rejected the omnibus ANOVA null hypothesis gives us the right to test *all possible* null hypotheses (even though, as we just saw, it does *not* give us the right selectively to choose to test just *one*), as long as the probability of a Type I error is distributed appropriately across *all those possible* null hypotheses. These methods are collectively known as *post hoc* ("after the fact") tests and may be performed only *after the fact* that we have rejected the omnibus ANOVA null hypothesis.

[*Note:* If we perform an ANOVA and *do not* obtain a significant F ratio—that is, if we fail to reject the omnibus null hypothesis—we do *not* have the right to perform *any* subsequent test, because failing to reject the omnibus null hypothesis implies the failure to reject *all* possible null hypotheses for the data. However, we may wish to do a bit of *informal* "data snooping" and use the results to help plan a new, more powerful study (perhaps using planned comparisons, the topic of Section 14.3).]

Tukey's HSD (honestly significant difference) test: a particular kind of post hoc test appropriate for testing pairwise null hypotheses

We present only one of several post hoc methods, *Tukey's HSD* (**honestly significant difference**) method, which is used to test all possible pairwise null hypotheses.[1] Recall that for our antidepressant data, the six possible *pairwise* null hypotheses are:

null hypotheses for pairwise post hoc tests following analysis of variance with four groups

$$\begin{aligned} H_{0_1}&: \mu_1 = \mu_2 & &[\text{drug A} = \text{drug B}]\\ H_{0_2}&: \mu_1 = \mu_3 & &[\text{drug A} = \text{drug C}]\\ H_{0_3}&: \mu_1 = \mu_4 & &[\text{drug A} = \text{drug D}]\\ H_{0_4}&: \mu_2 = \mu_3 & &[\text{drug B} = \text{drug C}]\\ H_{0_5}&: \mu_2 = \mu_4 & &[\text{drug B} = \text{drug D}]\\ H_{0_6}&: \mu_3 = \mu_4 & &[\text{drug C} = \text{drug D}] \end{aligned} \tag{14.1}$$

Studentized range statistic (Q): the test statistic used in the Tukey HSD procedure

Each one of these six null hypotheses is evaluated using a test statistic called Q, the *Studentized range statistic*. Q is the fourth test statistic we have encountered so far (the others are z, t, and F). We compute Q for the Tukey test as follows:

Studentized range statistic for Tukey's post hoc tests

$$Q = \frac{|\bar{X}_i - \bar{X}_j|}{\sqrt{\text{MS}_\text{W}/\tilde{n}}} \tag{14.2}$$

where MS_W is the mean square within groups from the ANOVA and $\tilde{n}$ ("n tilde," rhymes with "Hilda") is the harmonic mean of the number of subjects in each group given by

harmonic mean of sample sizes

$$\tilde{n} = \frac{k}{\dfrac{1}{n_1} + \dfrac{1}{n_2} + \dfrac{1}{n_3} + \cdots + \dfrac{1}{n_k}} \tag{14.3}$$

where k is the number of groups. If all the n_j are equal, then Equation (14.3) simplifies to give

harmonic mean of sample sizes, equal n_j

$$\tilde{n} = n_j \qquad [\text{only if equal } n_j] \tag{14.4}$$

For our antidepressant data, the Tukey procedure will result in six Q test statistics, one for each of the pairwise null hypotheses. All six Q statistics have the same denominator—namely, $\sqrt{\text{MS}_\text{W}/\tilde{n}}$. MS_W, from the original ANOVA, is 6.75, and Equation (14.4) says that, because the n_j are equal, $\tilde{n} = n_j = 6$. Therefore, $\sqrt{\text{MS}_\text{W}/\tilde{n}} = \sqrt{6.75/6} = 1.061$. Table 14.2 summarizes the computations and results.

To find the critical value of Q, we enter Table A.5 in Appendix A with $k = 4$, the total number of groups, and $df_\text{W} = 20$, the degrees of freedom of MS_W. Note that we enter Table A.5 with the number of groups (k), *not* with the degrees of freedom between groups ($k - 1$). We find that if $\alpha = .05$, $Q_\text{cv} = 3.96$. Because the first observed $Q_1 = 5.18 > Q_\text{cv} = 3.96$, we reject H_{0_1} and conclude that $\mu_1 \neq \mu_2$. We indicate that by placing an asterisk next to the Q_1 value.

[1] Other post hoc methods test both pairwise and complex null hypotheses (the most common are the Newman–Keuls test and the Scheffé test), but we leave their discussion to more advanced textbooks.

TABLE 14.2 Calculation of the six Q test statistics for the Tukey post hoc tests

Null hypothesis	$\overline{X}_i$	$\overline{X}_j$	$Q = \dfrac{\lvert\overline{X}_i - \overline{X}_j\rvert}{\sqrt{MS_W/\tilde{n}}}$
H_{0_1}: $\mu_1 = \mu_2$	28.0	33.5	$Q_1 = \dfrac{\lvert 28.0 - 33.5\rvert}{1.061} = 5.18^*$
H_{0_2}: $\mu_1 = \mu_3$	28.0	24.0	$Q_2 = \dfrac{\lvert 28.0 - 24.0\rvert}{1.061} = 3.77$
H_{0_3}: $\mu_1 = \mu_4$	28.0	32.5	$Q_3 = \dfrac{\lvert 28.0 - 32.5\rvert}{1.061} = 4.24^*$
H_{0_4}: $\mu_2 = \mu_3$	33.5	24.0	$Q_4 = \dfrac{\lvert 33.5 - 24.0\rvert}{1.061} = 8.95^*$
H_{0_5}: $\mu_2 = \mu_4$	33.5	32.5	$Q_5 = \dfrac{\lvert 33.5 - 32.5\rvert}{1.061} = .94$
H_{0_6}: $\mu_3 = \mu_4$	24.0	32.5	$Q_6 = \dfrac{\lvert 24.0 - 32.5\rvert}{1.061} = 8.01^*$

$^*p < .05$, $Q_{cv_{k=4, df_W=20}} = 3.96$

We then compute the five Q values corresponding to the five remaining null hypotheses and find that three of them exceed Q_{cv}, so we reject the corresponding null hypotheses (H_{0_3}, H_{0_4}, and H_{0_6}). When $i = 1$ and $j = 3$, and when $i = 2$ and $j = 4$, however, Q is not significant, and we *do not* reject the corresponding null hypotheses: $H_{0_2}: \mu_1 = \mu_3$ and $H_{0_5}: \mu_2 = \mu_4$.

Our conclusions, then, are that Prozac is significantly different from Elavil and Tofranil, and Zoloft is significantly different from Elavil and Tofranil, but Prozac is *not* significantly different from Zoloft, and Elavil is *not* significantly different from Tofranil. In other words, it seems that we have two sets of effects: Prozac and Zoloft are similar and Elavil and Tofranil are similar, but the two sets are different from each other.

Note that all Tukey null hypotheses are nondirectional. The absolute value in the numerator of Q makes Q positive whether the first mean is larger or smaller than the second.

14.2 Instead of ANOVA: A Priori Tests

Post hoc tests are performed only *after* the omnibus ANOVA null hypothesis is rejected. You may recall that we likened an omnibus test to a shotgun and noted the advantage of the omnibus shotgun: It does not have to be aimed precisely because it aims at all possible null hypotheses simultaneously.

The *disadvantage* of a shotgun, compared with a rifle, is that it is not so powerful: Shotgun pellets do not travel as far or as accurately, or penetrate as deep, as a single rifle bullet. The same disadvantage applies to the omnibus F test and its subsequent post hoc tests: They are *not* as powerful as a single test aimed at only one null hypothesis. A test aimed at a single null hypothesis is called an "a priori test" or a "planned comparison." A priori tests are like rifles: They aim all their power at a single target (null hypothesis), ignoring completely all other possible targets.

a priori test: the test of a (pairwise or complex) null hypothesis selected before the data are collected; sometimes called "planned comparison" or "planned contrast"

A priori is Latin for "from before," and the procedure involves selecting, *before data collection*, one of the many possible pairwise or complex null hypotheses and ignoring all the rest.[2] Then we collect data and test just that one null hypothesis.

---

[2]Actually it is generally considered permissible to test *a few* a priori null hypotheses simultaneously, but how many (and by what procedure) is the subject of some debate, which we leave to more advanced textbooks.

A Priori Tests When There Are Two Groups

The z and t tests of Chapters 10, 11, and 12 are actually a priori tests, although we did not explicitly refer to them as such because we had not yet introduced the concept of post hoc tests. For example, the independent-sample t test from Chapter 11 is an a priori rifle aimed at one null hypothesis (H_0: $\mu_1 = \mu_2$). Box 14.1 gives an example to show that the a priori independent-sample t test is more powerful than the shotgun ANOVA and its subsequent post hoc tests.

BOX 14.1 The independent-sample (a priori) t test is more powerful than Tukey's post hoc Q

This box may be omitted without loss of continuity.

Recall from the preceding section that we performed post hoc tests on our antidepressant drug data. One of the null hypotheses we tested was H_{0_2}: $\mu_1 = \mu_3$. Q_2 turned out to be 3.77, and because $Q_{\text{cv}_{k=4, df_W=20}} = 3.96$, we failed to reject the post hoc H_{0_2}, concluding that Prozac ($\overline{X}_1 = 28.0$) and Zoloft ($\overline{X}_3 = 24.0$) were *not* significantly different from each other.

Now let's turn the clock back to before we collected the data and proceed as if our *only* interest was comparing Prozac and Zoloft. Then we would form one (a priori) null hypothesis H_0: $\mu_1 = \mu_3$ (we call the mean of the second group μ_3 to make it clear that we are referring to Zoloft), and we would collect only the data shown in the first and third columns of Table 14.1. The appropriate test of that null hypothesis would be an independent-sample t test just as in Chapter 11. We would compute t_{obs} from $\overline{X}_1 = 28.0$, $\overline{X}_3 = 24.0$, $s_1^2 = 6.000$, $s_3^2 = 7.200$, $s_{\text{pooled}}^2 = 6.600$, and $s_{\overline{X}_1 - \overline{X}_2} = 1.483$, so $t_{\text{obs}} = 2.70$. Because $t_{\text{cv}_{df=10}} = 2.228$, we would reject the a priori null hypothesis (by a substantial margin) and conclude that Prozac and Zoloft *were* different.

Note that the null hypothesis for the post hoc Q is identical to the null hypothesis for the a priori t, and the data for both tests are identical. However, Q fails to reject the null hypothesis, whereas t rejects it by a substantial margin. This example shows that the a priori t test has substantially more power than the post hoc Q test.

Comparisons

The t tests of Chapter 11 were a priori tests, but they apply to only two groups. We would also like to be able to conduct a priori tests when we have three or more groups, and to do that we generalize the t-test procedure.

Suppose we wish to study weight loss, so we recruit 20 subjects and randomly assign each to one of four groups, with five subjects per group. Each group meets three times, 30 minutes each time, for 16 weeks. Group 1 is instructed in Dr. J's diet plan, and the meetings involve a weigh-in and social support. Group 2 performs aerobic ski machine exercises during their meetings, group 3 performs aerobic dance exercises during their meetings, and group 4 performs aerobic treadmill exercises. The outcome measure (dependent variable) is the amount of weight each subject loses over the course of the 16 weeks. The data are shown in Table 14.3.

Table 14.4 shows some of the many possible null hypotheses that we could test for these data.[3] All of them (and many others) are perfectly acceptable null hypotheses, and it is quite possible that we are equally interested in all of them, in which case we would prefer the omnibus analysis of variance procedure described in Chapter 13, followed by a set of post hoc tests.

[3] Planned comparisons can be directional also; in fact, most are. We leave that slight complexity for other textbooks.

TABLE 14.3 Weight lost (pounds) by subjects in four conditions

	Group 1 Dr. J's diet	Group 2 Ski aerobic	Group 3 Dance aerobic	Group 4 Treadmill aerobic
	4	3	4	7
	6	7	7	13
	0	8	8	2
	1	12	10	11
	3	11	9	6
n_j	5	5	5	5
$\overline{X}_j$	2.8	8.2	7.6	7.8
s_j^2	5.700	12.700	5.300	18.700

TABLE 14.4 Selected null hypotheses for the data of Table 14.3

Null hypothesis	Interpretation	Coefficients of the comparison
Some pairwise null hypotheses:		
$H_0\colon\ \mu_1 = \mu_2$	There is no difference between Dr. J's diet and aerobic ski machine exercise.	$1, -1, 0, 0$
$H_0\colon\ \mu_1 = \mu_3$	There is no difference between Dr. J's diet and aerobic dance exercise.	$1, 0, -1, 0$
$H_0\colon\ \mu_2 = \mu_4$	There is no difference between aerobic ski machine exercise and aerobic treadmill exercise.	$0, 1, 0, -1$
Some complex null hypotheses:		
$H_0\colon\ \mu_1 = \frac{\mu_2 + \mu_3 + \mu_4}{3}$	There is no difference between Dr. J's diet and the average of the three aerobic exercises.	$1, -\frac{1}{3}, -\frac{1}{3}, -\frac{1}{3}$
$H_0\colon\ \frac{\mu_1 + \mu_2}{2} = \frac{\mu_3 + \mu_4}{2}$	The average of Dr. J's diet and aerobic ski machine exercise equals the average of aerobic dance exercise and aerobic treadmill exercise.	$\frac{1}{2}, \frac{1}{2}, -\frac{1}{2}, -\frac{1}{2}$
$H_0\colon\ \mu_3 = \frac{\mu_1 + \mu_2 + \mu_4}{3}$	There is no difference between aerobic dance exercise and the average of Dr. J's diet, aerobic ski machine exercise, and aerobic treadmill exercise.	$-\frac{1}{3}, -\frac{1}{3}, 1, -\frac{1}{3}$

Suppose that our interest focuses on only one of these hypotheses, however, and the remainder are of no particular concern to us. Then we create a new statistical construct called a *comparison*, which is a linear combination of means (that is, means multiplied by constants and added together). We use a generalization of the t test to determine whether that comparison is significantly different from 0, just as we did in Chapter 11. Thus, the only new concept to master in this section is the definition and use of the comparison.

comparison: a linear combination of means; sometimes called a "contrast"

coefficients of a comparison: the constants that multiply the means in a comparison; sometimes called "weights"

To define a comparison is simply to specify a series of coefficients that multiply means. We call those coefficients C_j, and the comparison in a population is given by

population comparison

$$\sum C_j\mu_j \quad \text{or} \quad C_1\mu_1 + C_2\mu_2 + C_3\mu_3 + C_4\mu_4 \tag{14.5}$$

(in general) (when there are four groups)

The equation for a comparison in a sample is obtained by substituting $\overline{X}$ for μ:

sample comparison

$$\sum C_j\overline{X}_j \quad \text{or} \quad C_1\overline{X}_1 + C_2\overline{X}_2 + C_3\overline{X}_3 + C_4\overline{X}_4 \tag{14.6}$$

(in general) (when there are four groups)

The comparison constants themselves must sum to 0; that is,

comparison coefficients must sum to 0

$$\sum C_j = 0 \quad \text{or} \quad C_1 + C_2 + C_3 + C_4 = 0 \tag{14.7}$$

(in general) (when there are four groups)

The standard error of a sample comparison is given by

standard error of a comparison

$$s_{\text{sample comparison}} = \sqrt{s^2_{\text{pooled}} = \left(\frac{C_1^2}{n_1} + \frac{C_2^2}{n_2} + \cdots + \frac{C_k^2}{n_k}\right)} \tag{14.8}$$

where n_j is the number of observations in the jth group and s^2_{pooled} is the pooled variance, pooled from all k groups. Recall that Equation (13.4) gave the pooled variance for k equal-sized groups as

$$s^2_{\text{pooled}} = \frac{s_1^2 + s_2^2 + \cdots + s_k^2}{k} \quad [\text{only if equal } n_j] \tag{13.4}$$

If the n_j are not equal, we obtain SS_W from Equation (13.11) and MS_W from Equation (13.15), and then we recall that $s^2_{\text{pooled}} = MS_W$.

The null hypothesis for a comparison is

general form of the null hypothesis

$$H_0\text{: Population comparison} = 0 \tag{14.9}$$

and we test this null hypothesis using the test statistic, Equation (9.1), that we have been using since Chapter 9.

Let's begin with a pairwise example (similar to Chapter 11). We wish to know whether the weight loss due to Dr. J's diet (group 1) differs from the weight loss due to aerobic ski machine exercise (group 2). Following the logic of Chapter 11, the null hypothesis would be $H_0: \mu_1 = \mu_2$ and we would perform a t test given by Equation (11.2): $t = (\overline{X}_1 - \overline{X}_2)/s_{\overline{X}_1 - \overline{X}_2}$. We will demonstrate that that is almost identical to the equation for an a priori (pairwise) comparison that tests the same null hypothesis.

When we performed such t tests in Chapter 11, we rewrote the null hypothesis as H_0: $\mu_1 - \mu_2 = 0$ so that we could substitute $(\mu_1 - \mu_2)$ for the population parameter in the general form of the test statistic:

$$\text{test statistic} = \frac{\text{sample statistic} - \text{population parameter}}{\text{standard error of the sample statistic}} \tag{9.1}$$

Now we observe that "$\mu_1 - \mu_2$" is simply an example of a population comparison with coefficients $C_1 = 1$, $C_2 = -1$, $C_3 = 0$, and $C_4 = 0$, as shown by Equation (14.5):

$$\begin{aligned}\sum C_j\mu_j &= C_1\mu_1 + C_2\mu_2 + C_3\mu_3 + C_4\mu_4 \\ &= (1)\mu_1 + (-1)\mu_2 + (0)\mu_3 + (0)\mu_4 \\ &= \mu_1 - \mu_2\end{aligned}$$

We check to see that the coefficients themselves sum to 0 as Equation (14.7) requires:

$$\begin{aligned}\sum C_j &= C_1 + C_2 + C_3 + C_4 \\ &= (1) + (-1) + (0) + (0) \\ &= 0\end{aligned}$$

so $\mu_1 - \mu_2$ is indeed a legal comparison.

The general form of the test statistic, Equation (9.1), can be used to test hypotheses about *any* legal comparison by recognizing (1) that any population comparison is simply a particular kind of "population parameter," (2) that the sample comparison that point-estimates that population comparison (that is, the sample comparison with the same coefficients as the population comparison) is the "sample statistic," and (3) that the standard error of that sample comparison is then the "standard error of the sample statistic." With those substitutions, Equation (9.1) becomes

general form of the test statistic

$$\text{test statistic} = \frac{\text{sample comparison} - \text{population comparison}}{\text{standard error of the sample comparison}} \tag{14.10}$$

We have seen that the population comparison in the pairwise case is $\mu_1 - \mu_2$, so the numerator of Equation (14.10) is $(\overline{X}_1 - \overline{X}_2) - (\mu_1 - \mu_2)$. However, the null hypothesis for this comparison is the assumption that $\mu_1 - \mu_2 = 0$, so that term disappears, which leaves the numerator of Equation (14.10) to be simply $\overline{X}_1 - \overline{X}_2$.

Now we need the denominator: the standard error of the sample comparison. Equation (14.8) tells us, again because $C_1 = 1$, $C_2 = -1$, $C_3 = 0$, and $C_4 = 0$, that

$$s_{\text{sample comparison}} = \sqrt{s^2_{\text{pooled}}\left(\frac{C_1^2}{n_1} + \frac{C_2^2}{n_2} + \cdots + \frac{C_k^2}{n_k}\right)} = \sqrt{s^2_{\text{pooled}}\left(\frac{1}{n_1} + \frac{1}{n_2}\right)}$$

The test statistic for comparisons is t, so for this pairwise comparison we can write

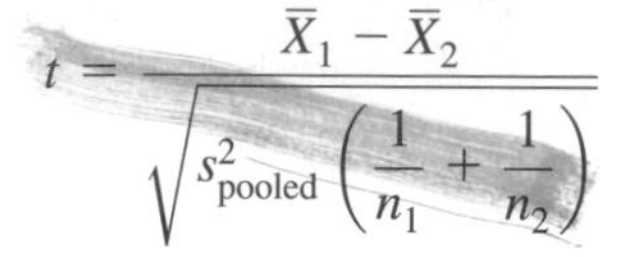

$$t = \frac{\overline{X}_1 - \overline{X}_2}{\sqrt{s^2_{\text{pooled}}\left(\frac{1}{n_1} + \frac{1}{n_2}\right)}}$$

Now let's compare that equation to Equation (11.2): $t = (\overline{X}_1 - \overline{X}_2)/s_{\overline{X}_1 - \overline{X}_2}$. Equation (11.3) gives this expression for the standard error of the difference between two means:

$$s_{\overline{X}_1 - \overline{X}_2} = \sqrt{s^2_{\text{pooled}}\left(\frac{1}{n_1} + \frac{1}{n_2}\right)}$$

Substituting that into Equation (11.2) gives

$$t = \frac{\overline{X}_1 - \overline{X}_2}{\sqrt{s^2_{\text{pooled}}\left(\frac{1}{n_1} + \frac{1}{n_2}\right)}}$$

the identical formula that the a priori (pairwise) comparison produced.

There is one slight difference between the two formulas: s^2_{pooled} is pooled from only two groups in the Chapter 11 test but from four groups in the a priori (pairwise) test, so the magnitude of the two t values may be slightly different. Why do we pool variances from all four groups rather than from just the two under consideration? The object is to get the best point-estimate possible of σ^2, and usually (though not always) that is the result of pooling *all* group variances. Note that the degrees of freedom depend on the number of subjects that make up the pooled variance, so the a priori t based on pooling all groups will have more degrees of freedom (which is desirable) than did the Chapter 11 t.

The main point should be clear: The a priori (pairwise) test is extremely similar to the t tests discussed in Chapter 11.

Complex A Priori Tests

Now let's turn to a complex comparison. Suppose that our actual interest before the data are collected is whether Dr. J's diet is better (or worse) than aerobic exercise in general. In that case we would compare the dieting group mean with the mean of the three aerobic groups—that is, H_0: $\mu_1 = (\mu_2 + \mu_3 + \mu_4)/3$. This is the one null hypothesis that we will test, which implies that we do not particularly care whether any other possible null hypotheses are true or false (whether weight loss from aerobic treadmill exercise is different from that of aerobic dance exercise, and so on).

Just as we did above and in Chapter 11, we rewrite the null hypothesis to put all the parameters on the same side of the equal sign:

$$H_0\text{:}\quad \mu_1 - \frac{\mu_2 + \mu_3 + \mu_4}{3} = \mu_1 + \left(-\frac{1}{3}\right)\mu_2 + \left(-\frac{1}{3}\right)\mu_3 + \left(-\frac{1}{3}\right)\mu_4 = 0$$

The population comparison specified by this null hypothesis thus has the coefficients C_j: 1, $-\frac{1}{3}$, $-\frac{1}{3}$, and $-\frac{1}{3}$. We check to ensure that those coefficients sum to 0 as required by Equation (14.7), and they do. Then we can use the general form of the test statistic, Equation (14.10), to test whether this comparison is significant.

I. State the Null and Alternative Hypotheses

Three sketches illustrate H_0:
Variable
Sample statistic
Test statistic

$$H_0: \quad \mu_1 = \frac{\mu_2 + \mu_3 + \mu_4}{3} \quad \text{or} \quad H_0: \quad \mu_1 + \left(-\frac{1}{3}\right)\mu_2 + \left(-\frac{1}{3}\right)\mu_3 + \left(-\frac{1}{3}\right)\mu_4 = 0$$

$$H_1: \quad \mu_1 \neq \frac{\mu_2 + \mu_3 + \mu_4}{3} \quad \text{or} \quad H_1: \quad \mu_1 + \left(-\frac{1}{3}\right)\mu_2 + \left(-\frac{1}{3}\right)\mu_3 + \left(-\frac{1}{3}\right)\mu_4 \neq 0$$

ⓘ Sketching the variable and the sample statistic for complex comparisons is not informative.

The complex comparisons do not in general have easily visualizable distributions, so we omit the sketches of the variable. The sample statistic is the sample comparison, but its sketch is also not particularly informative. Therefore, we sketch the test statistic t in Figure 14.1.

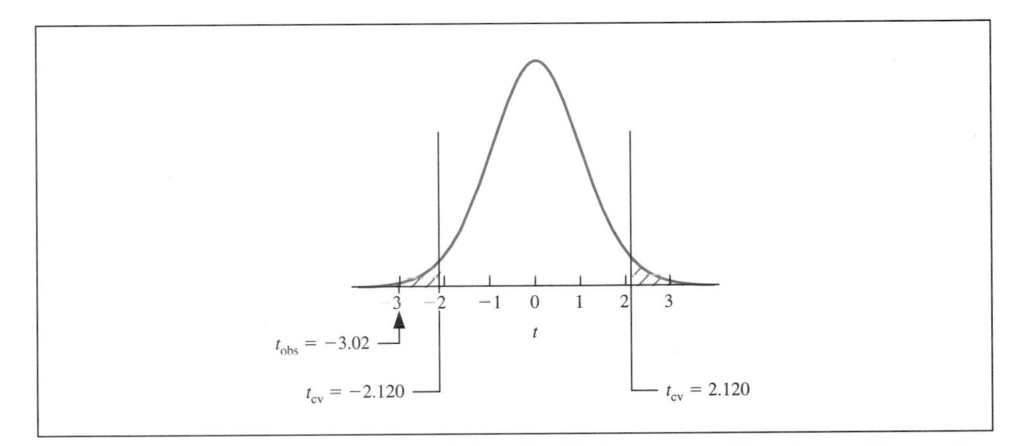

FIGURE 14.1 Distribution of the test statistic (t with 16 degrees of freedom) with rejection regions shaded and observed value indicated

II. Set the Criterion for Rejecting H_0

Level of significance?
Directional?
Shade rejection regions.

Because we have not stated otherwise, we adopt the conventional $\alpha = .05$. The test statistic is t, and the degrees of freedom are those obtained by pooling the variances:[4] $(n_1 - 1) + (n_2 - 1) + \cdots + (n_k - 1)$. In our example, t has $20 - 4 = 16$ degrees of freedom, so because the null hypothesis is nondirectional, Table A.2 shows that the critical value of t with 16 degrees of freedom is $t_{cv_{df=16}} = 2.120$.

We illustrate the critical values and the corresponding rejection regions on the sketch of the test statistic t in Figure 14.1.

III. Collect a Sample and Compute the Observed Value(s) of (the Sample Statistic and) the Test Statistic

Observed values:
Sample statistic
Test statistic

The "sample statistic" here is the sample comparison. Using the data from Table 14.3, we find this sample comparison:

$$\overline{X}_1 + \left(-\frac{1}{3}\right)\overline{X}_2 + \left(-\frac{1}{3}\right)\overline{X}_3 + \left(-\frac{1}{3}\right)\overline{X}_4 = 2.8 - \frac{8.2}{3} - \frac{7.6}{3} - \frac{7.8}{3} = -5.067$$

We compute the test statistic t_{obs} according to Equation (14.10):

$$\text{test statistic} = \frac{\text{sample comparison} - \text{population comparison}}{\text{standard error of the sample comparison}}$$

The population comparison is $\mu_1 + \left(-\frac{1}{3}\right)\mu_2 + \left(-\frac{1}{3}\right)\mu_3 + \left(-\frac{1}{3}\right)\mu_4$, but the null hypothesis specified that this population comparison is 0.

To obtain the standard error of the sample comparison, we need to know the pooled variance s^2_{pooled}, which Equation (13.4) tells us (because the n_j are equal) is just the average of

[4] Note that because $s^2_{pooled} = MS_W$, the degrees of freedom for t are the same as would be used for MS_W had we computed a standard (independent-sample) ANOVA.

the four variances shown in Table 14.3: $s^2_{\text{pooled}} = (5.700 + 12.700 + 5.300 + 18.700)/4 = 10.600$. [If the n_j were not equal, we would have computed s^2_{pooled} by noting that it is identical to MS_W in an independent-sample ANOVA and using Equation (13.15) to obtain MS_W.] Then Equation (14.8) gives

$$\begin{aligned} s_{\text{sample comparison}} &= \sqrt{s^2_{\text{pooled}}\left(\frac{C_1^2}{n_1} + \frac{C_2^2}{n_2} + \cdots + \frac{C_k^2}{n_k}\right)} \\ &= \sqrt{10.6\left[\frac{(1)^2}{5} + \frac{\left(-\frac{1}{3}\right)^2}{5} + \frac{\left(-\frac{1}{3}\right)^2}{5} + \frac{\left(-\frac{1}{3}\right)^2}{5}\right]} \\ &= \sqrt{10.6(.200 + .022 + 0.22 + 0.22)} \\ &= 1.68 \end{aligned}$$

Substituting those values into Equation (14.10) gives

$$t_{\text{obs}} = \frac{-5.067 - 0}{1.68} = -3.02$$

IV. Interpret the Results

Statistically significant?
Practically significant?
State results clearly.

The observed value of the test statistic $t_{\text{obs}} = -3.02$ exceeds (in absolute value) the critical value $t_{\text{cv}_{df=16}} = 2.120$, so we reject the null hypothesis and conclude that Dr. J's diet group, with its mean weight loss of 2.8 pounds, is statistically significantly different from the average of the weight losses in the three aerobic groups, $(8.2 + 7.6 + 7.8)/3 = 7.87$ pounds.

Because the result is statistically significant, we consider its practical significance. Recall that what we would like to know is the magnitude of the population comparison $C_1\mu_1 + C_2\mu_2 + \cdots + C_k\mu_k$, which, in our case, is $(1)\mu_1 + (-\frac{1}{3})\mu_2 + (-\frac{1}{3})\mu_3 + (-\frac{1}{3})\mu_4$. We concluded in the preceding paragraph that this magnitude is different from 0 (unless we made a Type I error); now we wish to know *how much* different from 0. The best point-estimate of the population comparison is the sample comparison—in our case, $(1)\overline{X}_1 + (-\frac{1}{3})\overline{X}_2 + (-\frac{1}{3})\overline{X}_3 + (-\frac{1}{3})\overline{X}_4$. The effect size index d is the sample comparison divided by s_{pooled}, the square root of s^2_{pooled}, the pooled variance. Thus,

$$\begin{aligned} d &= \frac{(1)\overline{X}_1 + \left(-\frac{1}{3}\right)\overline{X}_2 + \left(-\frac{1}{3}\right)\overline{X}_3 + \left(-\frac{1}{3}\right)\overline{X}_4}{s_{\text{pooled}}} \\ &= \frac{(1)2.8 + (-.333)8.2 + (-.333)7.6 + (-.333)7.8}{\sqrt{10.6}} \\ &= -1.56 \end{aligned}$$

Recall that the comparison was created to contrast the results of Dr. J's diet group with the mean of the three aerobic groups. Thus, this value $d = -1.56$ indicates that the mean of Dr. J's diet group is 1.56 standard deviations away from the mean of the three aerobic groups, where we use for the standard deviation its best point-estimate, the square root of the pooled variance. Cohen's nonoverlap U is found by entering Table A.1 at $z = d = 1.56$ (note that we ignore the sign of d) and finding $U = .5 + .4406$, or 94.06%. Thus, we conclude that approximately 94.06% of dieters are expected to lose less weight than the average aerobic exerciser.

14.3 Planned Comparisons Are More Powerful Than ANOVA

The result of the weight loss example was to reject the dieting-aerobics null hypothesis: The observed value of t (−3.02) was well beyond the critical value of t (2.120). What would have happened if, when we designed this experiment, we had chosen to analyze the results using the omnibus F ("shotgun") method of Chapter 13?

The omnibus F analysis of variance summary table is shown in Table 14.5, and the observed value of F is 3.06. The critical value of F with 3 and 16 degrees of freedom is found in Table A.4 to be 3.24. Thus, the omnibus F test *fails* to reject its null hypothesis, whereas the planned comparison easily rejects its null hypothesis. This is an example of the power increase by using the planned comparison bullet over the omnibus F shotgun, and it is the main reason planned comparisons are preferred over omnibus F analyses if the research design permits.

TABLE 14.5 Independent-sample analysis of variance (as in Chapter 13) of weight lost

Source	SS	*df*	MS	*F*
Between	97.2	3	32.4	3.06
Within	169.6	16	10.6	
Total	266.8	19		

F is not significant: $F_{cv_{df=3,16}} = 3.24$

We have seen that planned comparison tests are substantially more powerful than the omnibus ANOVA and its subsequent post hoc tests, which are among the least powerful tests in statistics. The null hypothesis in a post hoc test can be identical to the null hypothesis in a planned comparison, but the probability of rejecting that null hypothesis is far greater in a planned comparison.

In this introductory textbook, we are unfortunately able to give only this one example of complex planned comparisons. There are as many different examples of planned comparisons as there are ways of selecting the coefficients of the comparison, many of them among the most useful and most powerful tests in the statistician's arsenal.

14.4 Repeated-Measures Analysis of Variance

repeated-measures analysis of variance: the statistical procedure for testing hypotheses about two or more means when the samples are repeated observations on the same subjects or are otherwise statistically related to each other; also called "dependent-sample ANOVA."

ⓘ Independent-samples ANOVA is discussed in Chapter 13.

In Chapter 11, we discussed the evaluation of hypotheses regarding two independent samples (the independent-sample t test), and we extended that discussion to three or more samples in Chapter 13 (the independent-sample analysis of variance). In Chapter 12, we considered hypotheses regarding two *dependent* samples (the dependent-sample t test). Now we discuss the extension of that dependent-sample logic to three or more samples, a topic usually called *repeated-measures analysis of variance*.

Suppose ΨΦΠ conducts study sessions Monday through Thursday and takes attendance every 15 minutes, thus keeping track of the number of 15-minute time blocks its members use these sessions. ΨΦΠ wishes to know whether attendance at the sessions differs on the different days of the week. It selects six members at random and determines the number of time blocks used by each of the six members averaged across the last four Mondays, Tuesdays, Wednesdays, and Thursdays. The data are shown in Table 14.6.

14.5 Logic of Repeated-Measures Analysis of Variance

ⓘ In this paragraph and the next four we are pretending that the data are independent samples (like the data in Chapter 13) so that you can see how the repeated-measures ANOVA is similar to and different from the independent-sample ANOVA.

Let us begin our discussion of the repeated-measures analysis of variance by covering up the "Person" column in Table 14.6 and pretending for the moment that the data represent four independent groups. We could then perform an independent-sample ANOVA with four groups (see Chapter 13). As you may recall, Section 13.3 described the logic of the analysis of variance; the results of the procedures described in that section are summarized at the bottom of Table 14.6. We find the numerator of the F ratio (MS_B) by first obtaining the four day-of-week means $\bar{X}_j$: 9.167, 9.833, 6.500, and 5.167. The variance of those four

TABLE 14.6 Time spent in study sessions

	Time blocks (15-minute periods) spent studying				
Person	*Monday*	*Tuesday*	*Wednesday*	*Thursday*	*Person mean*
Pat	15	10	8	7	10
Bobby	10	11	4	7	8
Riki	4	9	7	0	5
Jean	10	10	7	1	7
Lynn	4	2	4	2	3
Jo	12	17	9	14	13
Means	9.167	9.833	6.500	5.167	
Variance of those four means: 4.8510					
Variance of those means times n = $\text{MS}_{\text{between groups}}$ = (4.8510)(6) = 29.11					
Standard deviations	4.401	4.792	2.074	5.269	
Variances	19.367	22.967	4.300	27.767	
Pooled variance (MS_W) = 18.60					

means is $s_{\bar{X}}^2 = 4.8510$. Because the n_j are equal, the numerator of the F ratio is, according to Equation (13.3), $\text{MS}_\text{B} = s_{\bar{X}}^2(n) = (4.8510)(6) = 29.11$.

Again using the independent-sample procedure, we obtain the denominator of the F ratio (MS_W) by pooling the variances within each group. Because the group sizes are equal, pooling is simply averaging the four within-group variances—19.367, 22.967, 4.300, and 27.767—to obtain the denominator of F according to Equation (13.5): $\text{MS}_\text{W} = s_{\text{pooled}}^2 = 18.60$.

Then the observed value of the independent-sample F is simply $F_{\text{obs}} = \text{MS}_\text{B}/\text{MS}_\text{W} = 29.11/18.60 = 1.57$, as shown in Table 14.7. The critical value of F with 3 and 20 degrees of freedom (assuming $\alpha = .05$) is found from Table A.4 to be 3.10, so F is not significant. All this follows the independent-sample logic of Chapter 13, and we would conclude that the study times are not significantly different from one day to the next.

TABLE 14.7 Analysis of variance summary table of study session data pretending that the samples are independent (as in Chapter 13)

Source	*SS*	*df*	*MS*	*F*
Between groups	87.33	3	29.11	1.57
Within groups	372.00	20	18.60	
Total	459.33	23		

F is not significant: $F_{\text{cv}_{df=3,20}} = 3.10$

Why is this F ratio small? There are two possibilities (still reviewing Chapter 13): Either the numerator of F is small (that is, there is a small difference between the means for the days of the week) or the denominator of F is large (that is, the standard deviation within each day is large). Of most interest here is the second possibility: There are in fact large standard deviations within each day.

For example, consider Tuesday in Table 14.6. The observations range from 2 to 17 time blocks: a large range and therefore a large standard deviation. Why is the range so large? To a great extent because the ΨΦΠ members are quite different from one another: Jo studies a lot, and this is reflected in Jo's Tuesday value of 17 time blocks. On the other hand, Lynn studies only a little, and this is also reflected in Lynn's Tuesday value of only 2 time blocks. Thus, a large portion of the standard deviation within each day is accounted for by the *between-subjects effect*. If we could somehow eliminate the between-subjects effect—if Jo and Lynn studied the same amount—then the standard deviation within Tuesday (and within the other days as well) would be smaller, so F might well be significant. But how can we eliminate the between-subjects effect?

between-subjects effect: the portion of the total sum of squares in a repeated-measures design that is due to individual differences between subjects

The remainder of this section provides the deviation method of understanding repeated-measures ANOVA. The sum of squares formulas in the next section are easier to compute.

Let's subtract each person's mean number of time blocks from each data point. Thus, for example, Pat's mean is 10 time blocks so we subtract 10 from her Monday value (15) to obtain an adjusted value of 15 − 10 = 5, as shown in Table 14.8. We call these differences between the day value and the person's own mean *deviations*, using that term in the same way we did in Chapter 5.

TABLE 14.8 Deviations (original observation minus person's own mean) for study session data

	Deviations from person's own mean				
Person	*Monday*	*Tuesday*	*Wednesday*	*Thursday*	*Mean*
Pat	5	0	−2	−3	0
Bobby	2	3	−4	−1	0
Riki	−1	4	2	−5	0
Jean	3	3	0	−6	0
Lynn	1	−1	1	−1	0
Jo	−1	4	−4	1	0
Mean	1.500	2.167	−1.167	−2.500	
Variance of those four means: 4.8510					
n times variance of those means = $MS_{\text{between occasions}}$ = 6(4.8510) = 29.11					
Standard deviation	2.345	2.137	2.563	2.665	

Computing the deviations (as in Table 14.8) clearly eliminates the between-subjects effect from the data: The mean of the deviations is now identical (0) for each person, as can be seen in the right-hand column of Table 14.8. Note at the bottom of Table 14.8 that removing the between-subjects effect has markedly reduced the within-day standard deviations, from 4.401, 4.792, 2.074, and 5.269 in the original data (Table 14.6) to 2.345, 2.137, 2.563, and 2.665 in the deviation data (Table 14.8).

Now we wish to know whether there are significant differences between days *for the deviations*. It might seem that we could simply perform another analysis of variance on the deviations of Table 14.8. After all, we no longer need to *pretend* that the observations are independent because they *are in fact* independent now that we have removed the between-subjects effect. Indeed we *can* now simply perform an analysis of variance on the deviations, *as long as we adjust the degrees of freedom within groups* to account for forming the deviations. Recall that we lose a degree of freedom whenever we put a constraint on the data, and now we have applied six "person constraints": We required that the mean for each person be 0. One of those constraints is already accounted for by the grand mean, so we lose 6 − 1 = 5 additional degrees of freedom as the price to be paid for removing the between-subjects effect. With that adjustment, we can perform an analysis of variance on the deviation data using the sum-of-squares Equations (13.9), (13.11), and (13.12), just as we would have in Chapter 13. This is shown in Table 14.9, where the X_{ij} values in the first column are the deviations from the subjects' own means. [Note that we *cannot* use Equation (13.5), the pooled-variance formula for the mean square, because of the adjustment to the degrees of freedom.]

occasions: the columns in a repeated-measures design

Now we can compare the summary table of Table 14.9 (deviations ANOVA) with Table 14.7 (independent-sample ANOVA). First, we have substituted the term *occasion* for "group." When we pretended that these data were independent samples, we called the day-of-week effect "between groups," which implied measuring *different groups* of people on the several days. That is misleading, however, when we are repeatedly observing the *same individuals* on the several days of the week. Therefore, in repeated-measures logic, we call the day-of-week effect "between *occasions*." Note that "between groups" and "between occasions" are two different names for identical computations; thus, $MS_{\text{between groups}} = MS_{\text{between occasions}} = 29.11$.

Second, the mean square that will form the denominator of F is now called "residual" instead of "within groups." Perhaps it seems that "within occasions" would have sufficed

TABLE 14.9 Analysis of variance of deviations

(Note: grand mean $\bar{X}_G = 0.0$)	Deviation from subject's own mean	Total: Deviation from grand mean		Within occasions: Deviation from occasion mean		Between occasions: deviation of occasion mean from grand mean	
	X_{ij}	$X_{ij} - \bar{X}_G$	$(X_{ij} - \bar{X}_G)^2$	$X_{ij} - \bar{X}_j$	$(X_{ij} - \bar{X}_j)^2$	$\bar{X}_j - \bar{X}_G$	$(\bar{X}_j - \bar{X}_G)^2$
	5	5	25	3.5	12.25	1.5	2.25
	2	2	4	.5	.25	1.5	2.25
Monday	−1	−1	1	−2.5	6.25	1.5	2.25
($\bar{X}_1 = 1.500$)	3	3	9	1.5	2.25	1.5	2.25
	1	1	1	−.5	.25	1.5	2.25
	−1	−1	1	−2.5	6.25	1.5	2.25
	0	0	0	−2.167	4.696	2.167	4.696
	3	3	9	.833	.694	2.167	4.696
Tuesday	4	4	16	1.833	3.360	2.167	4.696
($\bar{X}_2 = 2.167$)	3	3	9	.833	.694	2.167	4.696
	−1	−1	1	−3.167	10.030	2.167	4.696
	4	4	16	1.833	3.360	2.167	4.696
	−2	−2	4	−.833	.694	−1.167	1.362
	−4	−4	16	−2.833	8.026	−1.167	1.362
Wednesday	2	2	4	3.167	10.030	−1.167	1.362
($\bar{X}_3 = -1.167$)	0	0	0	1.167	1.362	−1.167	1.362
	1	1	1	2.167	4.696	−1.167	1.362
	−4	−4	16	−2.833	8.026	−1.167	1.362
	−3	−3	9	−.5	.25	−2.5	6.25
	−1	−1	1	1.5	2.25	−2.5	6.25
Thursday	−5	−5	25	−2.5	6.25	−2.5	6.25
($\bar{X}_4 = -2.500$)	−6	−6	36	−3.5	12.25	−2.5	6.25
	−1	−1	1	1.5	2.25	−2.5	6.25
	1	1	1	3.5	12.25	−2.5	6.25
		$SS_{deviations}$	= 206	$SS_{residual}$	= 118.67	$SS_{between\ occasions}$	= 87.33

Source	SS	*df*	MS	*F*
Between occasions	87.33	3	29.11	3.68*
Residual	118.67	15	7.91	
Total of deviations	206.00	18		

$^*p < .05$, $F_{cv_{df=3,15}} = 3.29$

residual: what remains after both the subject effect and the occasion effect have been subtracted

ⓘ A simple way of determining the degrees of freedom for the residual is $df_{residual} = (k - 1)(n - 1)$

because we are substituting "occasion" for "groups." However, because we have also removed the between-subjects effect (by taking deviations), we use the term *residual* to emphasize that this is what remains after *both* the between-occasion *and* the between-subject effects have been removed.

Third, the degrees of freedom for the denominator of F (the residual effect) is now $20 - 5 = 15$ instead of $df_W = 20$. This reflects the adjustment for forming the deviations that we described above.

Finally, the observed value of F (3.68) for the deviations data is larger than it was for the original data (1.57), and this F is now significant: The critical value of F with 3 and 15 degrees of freedom is 3.29, as shown in Table A.4.

To recapitulate: A repeated-measures analysis of variance can be obtained by computing the analysis of variance after the "between-subjects effect" is removed from the data by computing the deviations. The degrees of freedom for the residual must be adjusted to account for removing the between-subjects effect.

14.6 Partitioning the Sum of Squares in Repeated-Measures ANOVA

Another way of understanding the repeated-measures analysis of variance is to observe how the sums of squares are partitioned. We saw in Section 13.4 that in the independent-sample ANOVA we partition the total sum of squares into two components—the sum of squares between groups and the sum of squares within groups, or $SS_T = SS_B + SS_W$. In repeated-measures ANOVA, we partition the total sum of squares into *three* components instead of two:

partition of the total sum of squares in repeated-measures ANOVA

$$SS_{total} = SS_{between\ occasions} + SS_{between\ subjects} + SS_{residual} \tag{14.11}$$

Recall "occasions" and "groups" refer to the same mathematical procedures, so

$$SS_{between\ occasions} = SS_{between\ groups}$$

This partitioning is illustrated in Figure 14.2.

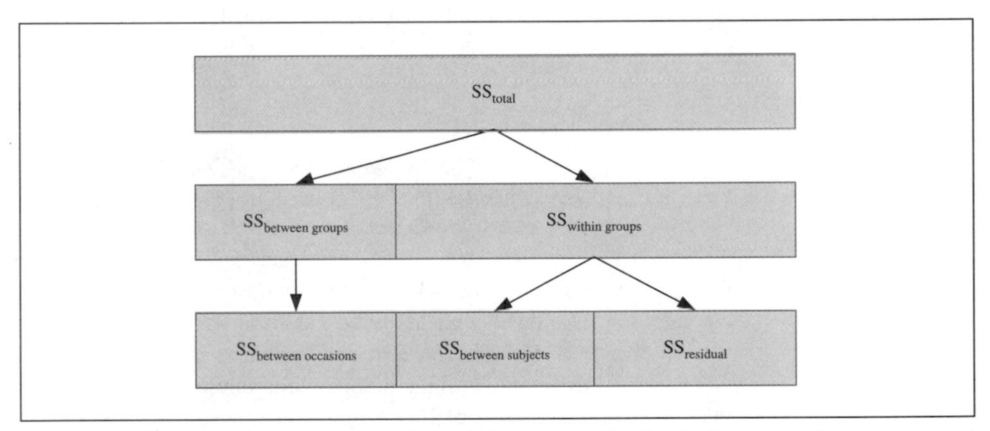

FIGURE 14.2 Partitioning the total sum of squares, first for independent samples and then for dependent samples

The degrees of freedom are partitioned in the same way as the sums of squares. Thus, in a repeated-measures ANOVA,

partition of the total degrees of freedom in repeated-measures ANOVA

$$\begin{aligned} df_{total} &= df_{between\ occasions} + df_{between\ subjects} + df_{residual} \\ &= nk - 1 \end{aligned} \tag{14.12}$$

where n is the number of subjects and k is the number of occasions. Note that in all repeated-measures designs, there are *always* the same number of subjects at each occasion [all $n_j = n$ because they are the same subjects (or related subjects) on each occasion].

We can determine the mean square between subjects in a manner exactly analogous to the way we obtained the mean square between occasions above, where $MS_{between\ occasions} = ns_{\bar{X}}^2$. Thus,

mean square between subjects in repeated-measures ANOVA, definitional formula

$$MS_{between\ subjects} = ks_{\bar{X}}^2 \tag{14.13}$$

where k is the number of occasions (instead of the number of subjects) and $s_{\bar{X}}^2$ is the variance of the six subject means (instead of the variance of the four day means). Table 14.6 showed, in the right-hand column, the six subject means 10, 8, 5, 7, 3, and 13. The variance of those six subject means is 12.667, so $MS_{between\ subjects} = ks_{\bar{X}}^2 = 4(12.667) = 50.667$. Because there are six means, $df_{between\ subjects} = 6 - 1 = 5$. Because in general $MS = SS/df$ or $SS = df(MS)$, we have $SS_{between\ subjects} = (df_{between\ subjects})(MS_{between\ subjects}) = 5(50.667) = 253.333$.

SS_{total} for our original data was 459.333, and that remains the same regardless of whether we perform an independent-sample or a repeated-measures ANOVA.

We compute the test statistic F by dividing the between-occasions mean square by the residual mean square:

test statistic (F) for repeated-measures ANOVA

$$F = \frac{MS_{\text{between occasions}}}{MS_{\text{residual}}} \tag{14.14}$$

The complete repeated-measures ANOVA summary table is shown in Table 14.10. Note that in the repeated-measures ANOVA we partition SS_{total}, 459.33, into three components: $SS_{\text{between occasions}} = 87.33$, $SS_{\text{between subjects}} = 253.333$, and $SS_{\text{residual}} = 118.667$.

ⓘ Repeated-measures ANOVA computation summary

- compute $SS_{\text{total}} = \sum(X_{ij} - X_G)^2$ as in Chapter 13
- compute SS_{residual} and $SS_{\text{between occasions}}$ as in Table 14.9
- compute $MS_{\text{between subjects}}$ from Equation (14.13)
- $df_{\text{between occasions}} = k - 1$
- $df_{\text{between subjects}} = n - 1$
- $df_{\text{residual}} = (k - 1)(n - 1)$
- complete the summary table (Table 14.10)

TABLE 14.10 Repeated-measures analysis of variance summary table for study session data

Source	SS	*df*	MS	*F*
Between occasions	87.33	3	29.11	3.68*
Between subjects	253.33	5		
Residual	118.67	15	7.91	
Total	459.33	23		

$*p < .05, F_{cv_{df=3,15}} = 3.29$

Let's recapitulate again: A repeated-measures analysis of variance can be obtained by removing the between-subjects sum of squares from the total sum of squares. The between-subjects degrees of freedom within groups must also be removed.

Note that the repeated-measures ANOVA summary table (Table 14.10) is identical to the deviation-score ANOVA summary table (Table 14.9) except that the "Between subjects" line has been added and the Total sum of squares and *df* adjusted accordingly.[5]

How is the repeated-measures analysis of variance (Table 14.10) the same as and different from the independent-sample analysis of variance (Table 14.7)?

- The total sums of squares are identical.
- The sums of squares for the column effect ("groups" or "occasions") are identical.
- The within groups SS (372.00) in Table 14.7 has been partitioned into the between-subjects SS (253.33) and the residual SS (118.67) and the *df* is adjusted accordingly in Table 14.10.
- The total sum of squares in Table 14.10 is partitioned into three components instead of two.
- The denominator of the F ratio is MS_{residual} (Table 14.10) instead of MS_{within} (Table 14.7). MS_{residual} has 15 degrees of freedom, whereas MS_{within} has 20.

14.7 Evaluating Repeated-Measures Hypotheses

The procedure for evaluating repeated-measures hypotheses is identical to that of independent-sample ANOVA.

[5]We note in passing that it is possible to compute an F ratio for the between-subjects effect as follows: The mean square between subjects is obtained by dividing the sum of squares between subjects by its degrees of freedom: $MS_{\text{between subjects}} = SS_{\text{between subjects}}/df_{\text{between subjects}} = 253.33/5 = 50.67$. Then the between-subjects F ratio is obtained by dividing $MS_{\text{between subjects}}$ by MS_{residual}: $F = MS_{\text{between subjects}}/MS_{\text{residual}} = 50.67/7.91 = 6.41$. The critical value of F with 5 and 15 degrees of freedom is 2.90, so this between-subjects F is statistically significant. The between-subjects differences are usually (as here) not the main interest of the study, so we omitted those values from Table 14.10.

I. State the Null and Alternative Hypotheses

Three sketches illustrate H_0:
Variable
Sample statistic
Test statistic

It is difficult to provide meaningful illustrations of the null hypothesis because we would need to sketch the distributions of the study time with the between-subjects effect removed, so we illustrate only the test statistic.

$$H_0: \mu_1 = \mu_2 = \mu_3 = \mu_4$$
$$H_1: \text{The null hypothesis is false.}$$

We sketch the test statistic F in Figure 14.3.

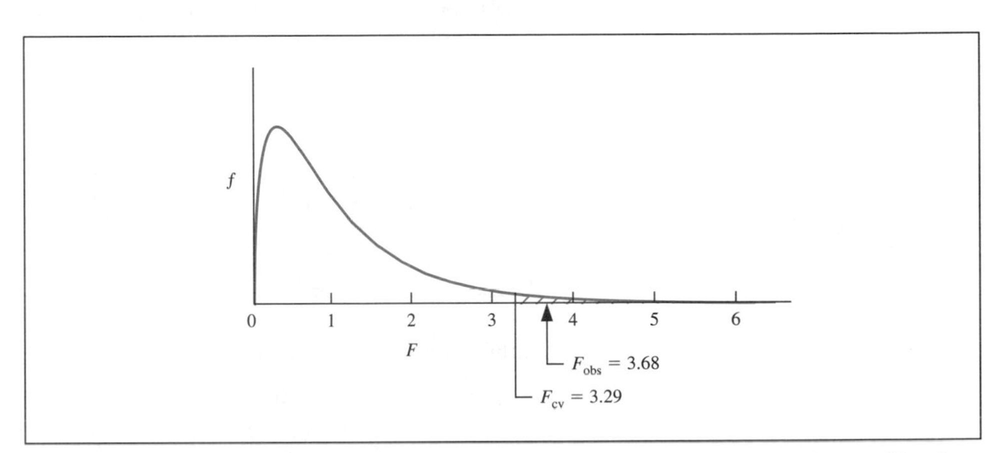

FIGURE 14.3 The ANOVA test statistic (F with 3 and 15 degrees of freedom) with rejection region shaded and observed value indicated

II. Set the Criterion for Rejecting H_0

Level of significance?
Directional?
Shade rejection regions.

We choose $\alpha = .05$ as is customary, and that will specify the critical value of the test statistic F. As we have seen, F will have 3 and 15 degrees of freedom, so Table A.4 gives the critical value $F_{cv_{df=3,15}} = 3.29$. We plot the critical value of the test statistic in Figure 14.3. Note that, as in Chapter 13, the F test is always nondirectional, even though the rejection region appears in only the right-hand tail of the F distribution.

III. Collect a Sample and Compute the Observed Value(s) of (the Sample Statistic and) the Test Statistic

Observed values:
Sample statistic
Test statistic

As with independent-sample ANOVA, there is no sample statistic of particular interest.

We computed $F_{obs} = 3.68$, and we show that value on the distribution of the test statistic in Figure 14.3.

IV. Interpret the Results

Statistically significant?
Practically significant?
State results clearly.

Because $F_{obs} = 3.68$ is greater than $F_{cv} = 3.29$, we reject H_0 and conclude that the use of the study sessions is different on different days. Because the results are statistically significant, we must consider the practical significance. There is no universally agreed best measure of effect size for repeated-measures ANOVA, so I recommend the following procedure, analogous to one we used in Chapter 13, as being the most clearly visualizable. Recall from Chapter 11 that the effect size index for two-independent-samples tests was $d = (\bar{X}_1 - \bar{X}_2)/s_{pooled}$; then in Chapter 13 we generalized that formula to the k-independent-samples ANOVA case, transforming d into the *maximum* effect size index $d_M = (\bar{X}_{max} - \bar{X}_{min})/s_{pooled}$, where s_{pooled} is the result of pooling k variances in Chapter 13 rather than two variances in Chapter 11. Now we perform the analogous generalization for the dependent-sample situation. Recall from Chapter 12 that the effect size index for two-dependent-samples tests was $d = \bar{D}/s_D$; now we generalize that formula to the k-dependent-samples ANOVA case, transforming d into the the maximum effect size index $d_M = \bar{D}_{max}/\sqrt{2MS_{residual}}$, where $\bar{D}_{max} = (\bar{X}_{max} - \bar{X}_{min})$ is the maximum difference between occasion means and $\sqrt{2MS_{residual}}$ is the k-occasion equivalent of s_D (that is, the best point-estimate of σ_D; see Exercise 13). In our example, the day means are 9.167, 9.833, 6.500, and 5.167 time blocks, so $\bar{D}_{max} = (\bar{X}_{max} - \bar{X}_{min}) =$

$(\overline{X}_{\text{Tuesday}} - \overline{X}_{\text{Thursday}}) = (9.833 - 5.167) = 4.666$. $\sqrt{2\text{MS}_{\text{residual}}} = \sqrt{2(7.91)} = 3.977$, so $d_{\text{M}} = \overline{D}_{\text{max}}/\sqrt{2\text{MS}_{\text{residual}}} = 4.666/3.977 = 1.17$. If we assume that the population is normally distributed, then entering Table A.1 with the effect size index of 1.17 implies that for about $U = 100(.5 + .3790) = 87.9\%$ of people, Tuesday study times would be expected to be longer than Thursday study times. Inspection of the means shows that usage is greater on Monday and Tuesday than on Wednesday and Thursday, although post hoc tests would be necessary to be confident in that conclusion.[6] We would need to consult the ΨΦΠ members to see whether the differences in usage (e.g., Tuesday's 9.833 versus Thursday's 5.167 time blocks) are of practical significance.

14.8 Connections

Cumulative Review

We continue the discrimination practice begun in Chapter 10. You may recall the two steps:

Step 1: Determine whether the problem asks for (a) the area under a normal distribution, (b) a confidence interval, or (c) the evaluation of a hypothesis. See the Cumulative Review in Section 10.5 for hints about this step.

Step 2: If the problem asks for the evaluation of a hypothesis, consult the summary of hypothesis testing in Table 14.11 to determine which kind of test is appropriate.

Table 14.11 shows that so far we have used seven basic kinds of hypothesis tests (the one-sample test for the mean, two-independent-samples test for means, two-dependent-samples tests for means, three-or-more-independent-samples tests, and now post hoc tests, a priori tests, and three-or-more-repeated-measures tests). This table is identical to Table 13.14 with the addition of the last three lines appropriate for this chapter. We will continue to expand this table progressively throughout the remainder of the book.

Journals

Post hoc tests: Here is an example of how the post hoc tests of the antidepressant drug study might be reported in a journal article: "The four sample means were statistically significantly different from each other (Prozac: $M = 28.0$, $SD = 2.45$; Elavil: $M = 33.5$, $SD = 3.15$; Zoloft: $M = 24.0$, $SD = 2.68$; Tofranil: $M = 32.5$, $SD = 1.97$; $F(3, 20) = 17.04$, $p < .05$. Tukey post hoc tests (again with $p < .05$) revealed that Prozac was significantly different from Elavil [$Q(4, 20) = 5.18$] and from Tofranil [$Q(4, 20) = 4.24$], and Zoloft was significantly different from Elavil [$Q(4, 20) = 8.95$] and from Tofranil [$Q(4, 20) = 8.01$]."

A priori tests: Commentators (e.g., Cohen, 1994) have often observed (but to no avail) that a priori tests are markedly underrepresented in the research literature. Many researchers use an omnibus F test when they should have used a more powerful a priori test.

A priori tests are often reported using the F statistic with one degree of freedom in the numerator, instead of t. For example, a journal might report, "An a priori test showed that mean weight loss with Dr. J's diet (2.8 pounds) is significantly different from the average weight loss in the aerobic conditions (7.87 pounds), $F(1, 16) = 9.08$." Recall from Chapter 13 that F with 1 and 16 degrees of freedom is identical to the *square* of t with 16 degrees of freedom, so our $(t_{\text{obs}})^2 = (-3.02)^2 = 9.08$, which is F as reported in the journal. F and t are equivalent ways of reporting a priori tests except that F does not allow directional hypotheses. Note also that the words "An a priori test showed..." are often replaced by "A test of a planned comparison showed..." or "Planned contrast analysis showed...."

[6] We leave post hoc tests for repeated measures to more advanced textbooks.

TABLE 14.11 Summary of hypothesis-testing procedures in Chapters 10–14

Design	Chapter	Null hypothesis	Sample statistic	Test statistic	Effect size index
One sample					
σ known	10	$\mu = a$	$\overline{X}$	z	$d = \frac{\overline{X}_{\text{obs}} - \mu}{\sigma}$
σ unknown	10	$\mu = a$	$\overline{X}$	t	$d = \frac{\overline{X}_{\text{obs}} - \mu}{s}$
Proportion	Exercise 7, Chapter 10	$\Pi = a$	p	t	—[a]
Two independent samples					
For means	11	$\mu_1 = \mu_2$	$\overline{X}_1 - \overline{X}_2$	t	$d = \frac{\overline{X}_{1_{\text{obs}}} - \overline{X}_{2_{\text{obs}}}}{s_{\text{pooled}}}$
Two dependent samples					
For means	12	$\mu_D = 0$	$\overline{D}$	t	$d = \frac{\overline{D}_{\text{obs}}}{s_D}$
k independent samples					
For means (omnibus)	13	$\mu_1 = \mu_2 = \cdots = \mu_k$	n/a	F	$d_M = \frac{\overline{X}_{\max} - \overline{X}_{\min}}{s_{\text{pooled}}}$ or $f = \frac{s_{\overline{X}}}{s_{\text{pooled}}}$ or $R^2 = \frac{\text{SS}_B}{\text{SS}_T}$
Post hoc	14	$\mu_i = \mu_j$	$\overline{X}_i - \overline{X}_j$	Q	—
A priori	14	$C_1\mu_1 + \cdots + C_k\mu_k = 0$	$C_1\overline{X}_1 + \cdots + C_k\overline{X}_k$	t	$d = \frac{C_1\overline{X}_1 + \cdots + C_k\overline{X}_k}{s_{\text{pooled}}}$
k dependent samples					
For means	14	$\mu_1 = \mu_2 = \cdots = \mu_k$	n/a	F	$d_M = \frac{\overline{D}_{\max}}{\sqrt{2\text{MS}_{\text{Residual}}}}$

[a] Effect size index is not discussed in this textbook.

Repeated-measures ANOVA: Repeated-measures ANOVA is often called "dependent-samples ANOVA" or "related-samples ANOVA" in the journals. A typical journal report might read: "Repeated-measures ANOVA revealed that students' study hall usage depended significantly on day of week. Means were 9.167 time blocks for Mondays, 9.833 for Tuesdays, 6.500 for Wednesdays, and 5.167 for Thursdays." It would be desirable for the journal report also to provide an effect size index, but that is not yet the norm.

Computers

Note that ESTAT does not perform post hoc tests or a priori tests. Versions earlier than 2.0 do not perform repeated-measures ANOVA. ESTAT does provide the means and sums of squares to facilitate the computation of tests that it does not itself directly support.

Note that the student version of SPSS does not perform repeated-measures ANOVA. This test is available in the complete SPSS.

Use SPSS to compute the Tukey HSD post hoc tests for the antidepressant drug data in Table 14.1.

1 Follow the first eight steps of the SPSS instructions in Section 13.9 or open the file you saved there and begin the ANOVA procedure.

2. Click `Post Hoc....` Then check the `Tukey's honestly significant difference` check box. Then click the `Harmonic average of all groups` button. Click `Continue` and then click `OK`.
3. The ANOVA summary table appears in the `!Output1` window, exactly as in Section 13.9. The Tukey HSD test is presented as in the accompanying display. Note that SPSS calls the Studentized range statistic Q "RANGE."

```
SPSS for MS WINDOWS Release 6.1

                     - - - - -  O N E W A Y  - - - - -

     Variable  VAR00001
  By Variable  VAR00002

Multiple Range Tests:  Tukey-HSD test with significance level .050
Harmonic Mean Cell size = 6.0000

  the actual range used is the listed RANGE * 1.0607
  with the following value(s) for RANGE: 3.95

   (*) Indicates significant differences which are shown in the lower triangle

                              G G G G
                              r r r r
                              p p p p

                              3 1 4 2
   Mean        VAR00002

   24.0000     Grp 3
   28.0000     Grp 1
   32.5000     Grp 4          * *
   33.5000     Grp 2          * *
```

Annotations: $\tilde{n}$ (for 6.0000); $\sqrt{MS_W/\tilde{n}}$ (for 1.0607); Critical value of the Studentized range statistic Q_{cv} (for 3.95); $\bar{X}_j$ (for Mean); Note that groups are arranged in increasing order; Asterisks indicate which pairs of means are significantly different from each other

4. Consider the asterisks circled at the bottom of the sample output. SPSS identifies which pairs of groups are significantly different from other pairs simply by placing an asterisk at the intersection of the group means. Thus, the upper left asterisk indicates that group 3 is significantly different from group 4.
5. SPSS does not report the magnitudes of the individual Q statistics. SPSS does supply the quantity $\sqrt{MS_W/\tilde{n}}$, although it is not clearly labeled as such (see sample output). Then you can compute the individual Q statistics from Equation (14.2).
6. SPSS provides some additional information about "homogeneous subsets" (not shown in our sample output). This is a somewhat more advanced concept regarding post hoc tests not covered in this textbook.
7. If you failed to click the `Harmonic average of all groups` check box (and thus performed the default Tukey HSD test), SPSS would perform a test that is similar but not identical to Tukey's HSD test, sometimes called the Tukey/Kramer test.

Use SPSS to compute the a priori test for the Dr. J's diet example given in the text (data in Table 14.3):

1. Enter the data from Table 14.3 as you did for the ANOVA example in Section 13.9. Put the dependent variable in the first column and the independent (group) variable in the second column.

2. Click Statistics, then Compare Means, and then One-Way ANOVA. . . . Set up the ANOVA just as you did in Section 13.9: Put var00001 in the Dependent List: box, put var00002 in the Factor: box, and define the range from 1 to 4.
3. Click Contrasts. . . . Now enter the contrast coefficients, which, as you recall, are $+1$, $-\frac{1}{3}$, $-\frac{1}{3}$, and $-\frac{1}{3}$. However, SPSS requires that we use integers as coefficients. Therefore, we convert our coefficients to integers by multiplying by the least common denominator of the coefficients, which in this case is 3. The new coefficients are 3, -1, -1, and -1. Note that the new coefficients still sum to 0 as required. Therefore, enter "3" in the cell labeled Coefficients: and click Add; then enter "-1" in the Coefficients: cell and click Add; and repeat this "-1" entry two more times. Check to make sure that the Coefficient Total: equals 0. Click Continue and then click OK.
4. The ANOVA summary table appears in the !Output1 window, exactly as in Section 13.9; the a priori test follows. The test that corresponds to the one presented in this textbook is called the "Pooled Variance Estimate" test. Note that SPSS gives t_{obs} as -3.014, whereas the value shown in this text is $+3.02$. The sign is entirely arbitrary (depending on whether you subtract the right side of the null hypothesis from the left, or vice versa); the difference between 3.014 and 3.02 is due to roundoff.
5. SPSS gives the value of the comparison (which it calls a "contrast") and its standard error as -15.2000 and 5.0438, respectively, whereas the text gives those values as -5.067 and 1.68. The difference is due to the scaling of the coefficients: Recall that we multiplied the coefficients by 3 to make them integers for SPSS. You can verify for yourself that -15.2000 is $3(-5.067)$ and 5.0438 is $3(1.68)$.

Homework Tips

1. Check the list of learning objectives at the beginning of this chapter. Do you understand each one?

2. Post hoc tests: Remember that there are as many Tukey Q statistics as there are pairs of means.

3. Post hoc tests: Note that the denominator of Tukey's Q is the same for all Q statistics in any given experiment.

4. Post hoc tests: The most common mistake in post hoc tests is the incorrect computation of the harmonic mean. Check to make sure that your computed $\tilde{n}$ is approximately equal (exactly equal if the n_j are equal) to n_j.

5. A priori tests: Note that an a priori hypothesis must be decided on *before* you look at the data.

6. A priori tests: The notation for comparisons may appear to be difficult, but it really is not. Make sure you understand all the comparison coefficients in Table 14.4.

7. Repeated-measures ANOVA: Note that Section 14.5 explains how repeated-measures ANOVA works, but Section 14.6 provides formulas that are much easier to compute.

Exercises for Chapter 14

Section A: Basic exercises (answers in Appendix D, page 524)

1. The little league baseball bat manufacturer of Exercise 2 in Chapter 13 collected data on the distance a baseball traveled when hit with bats built by three different manufacturing processes. The distances (in feet) each ball traveled and the subsequent analysis of variance summary table are reproduced here.

Process A	Process B	Process C
213.7	199.8	222.7
219.3	207.2	245.1
235.0	191.4	225.0
217.3	180.1	216.7
193.8	194.2	223.2
200.2	209.0	225.1
222.7	227.1	261.8
178.8	185.6	236.4

Source	SS	df	MS	F
Between	4441.44	2	2220.72	8.60*
Within	5419.58	21	258.08	
Total	9861.02	23		

$^*F_{cv} = 3.47, p < .05$

(a) Is it permissible to perform a Tukey HSD test in this situation? Why or why not?

I. State the null and alternative hypotheses

(b) How many null hypotheses will there be? State them. State the alternative hypotheses.
(c) Are these tests directional or nondirectional? Why?
(d) How many test statistics are there in this problem? What is the statistic?

II. Set the criterion for rejecting H_0

(e) What is the level of significance?
(f) How many degrees of freedom are there?
(g) What is (are) the critical value(s) of the test statistic?

III. Collect a sample and compute statistics

(h) Compute the harmonic mean of the sample sizes ($\tilde{n}$).
(i) Compute the observed value(s) of the test statistics.

IV. Interpret the results

(j) Decide which null hypotheses should be rejected.
(k) Interpret your result(s).

2. You are interested in whether life satisfaction depends on marital status. You plan to randomly sample ten individuals in each of five classifications: single male, married male, single female, married female, and divorced individual (whether male or female). None of these 50 individuals is to be related to any other individual in the sample. You will administer the Life Satisfaction Scale (LSS) to each person and have five groups with ten scores in each group.

For parts (a) and (b), suppose you simply wish to know whether life satisfaction depends on marital status.
(a) State the null hypothesis.
(b) Describe (but do not perform) the appropriate analysis: What kind of test is appropriate? What is the test statistic? How many degrees of freedom does it have?

For the remaining portions of this exercise, suppose instead that you wish to know whether the life satisfaction of married individuals is different from that of nonmarried (single or divorced) individuals.
(c) State the null hypothesis.
(d) If you have not done so, state this null hypothesis in the form H_0: $\sum C_j \mu_j = 0$.
(e) Demonstrate for this comparison that $\sum C_j = 0$.
(f) Describe the appropriate analysis: What kind of test is appropriate? What is the test statistic? How many degrees of freedom does it have?

Now you collect the LSS data shown here (higher scores = more satisfaction).

Single Male	Married Male	Single Female	Married Female	Divorced
49	49	55	51	46
47	54	50	47	48
49	47	44	47	48
38	56	41	51	47
54	52	46	52	41
44	50	46	48	54
47	46	47	50	44
50	51	51	42	45
47	44	41	59	50
46	55	42	58	46

(g) What is the numerical value of the sample comparison?
(h) What is the numerical value of the standard error of the sample comparison?
(i) What is the observed value of the test statistic?
(j) What is the critical value of the test statistic?
(k) Should you reject H_0?
(l) If you rejected H_0, what is the magnitude of the raw effect size?
(m) If you rejected H_0, what is the magnitude of the effect size index d?
(n) If you rejected H_0, what is the magnitude of Cohen's nonoverlap statistic U?
(o) State your results in plain English.

3. *(Exercise 2 continued)* If you had performed the omnibus analysis of variance in part (b) of Exercise 2, you would have reached this ANOVA summary table:

Source	SS	df	MS	F
Between groups	166.32	4	41.5800	2.2360
Within groups	836.80	45	18.5956	
Total	1003.12	49		

(a) Would you have rejected the null hypothesis?
(b) In view of the fact that you did reject the planned comparison H_0 in part (k) of Exercise 2, what general principle does this illustrate?

4. Suppose you are investigating the levels of glucose in the blood of diabetic individuals, and you wish to know whether the glucose level depends on the time of day. You select five patients, and for each patient, measure the glucose level three times: at 8:00 A.M., 5:00 P.M., and 10:00 P.M. The data are shown here.

Patient	8:00 A.M.	5:00 P.M.	10:00 P.M.
1	143	138	155
2	122	108	137
3	186	188	196
4	146	140	142
5	159	151	157

(a) Is this an independent-sample or a repeated-measures situation?
(b) What is the test statistic? How many degrees of freedom does it have? [*Hint:* You may wish to lay out the *df* column of the analysis of variance summary table before you answer this question.]
(c) What is the critical value of the test statistic, assuming $\alpha = .05$.
(d) Fill in the complete analysis of variance summary table.
(e) Should you reject H_0?
(f) If you rejected H_0, what is the magnitude of the maximum effect size index d_M.
(g) If you rejected H_0, what is the magnitude of Cohen's nonoverlap statistic U?
(h) State your results in plain English.

5. *(Exercise 4 continued)* If you had performed an *independent-sample* analysis of variance for the data in Exercise 4, you would have produced this ANOVA summary table:

Source	*SS*	*df*	*MS*	*F*
Between groups	384.4	2	192.20	.300
Within groups	7696.0	12	641.33	
Total	8080.4	14		

(a) Would you have rejected the null hypothesis?
(b) In view of the fact that you did reject the dependent-sample H_0 in part (e) of Exercise 4, what general principle does this illustrate?

Section B: Supplementary exercises

6. In Exercise 5 of Chapter 13, a perfume manufacturer tested five bottle shapes to see whether shape affected perfume sales. She sent one type of bottle to 28 retailers and recorded sales, and then she performed an analysis of variance on the resulting data. The data and the analysis of variance summary table are given here. She used the .05 level of significance. Perform the steps in Exercise 1.

Shape 1	*Shape 2*	*Shape 3*	*Shape 4*	*Shape 5*
21	31	21	19	26
22	23	23	21	20
27	26	18	18	20
20	31	20	23	24
	23	15	20	31
	23	23		24
		18		

Source	*SS*	*df*	*MS*	*F*
Between groups	170.07	4	44.52	3.99*
Within groups	256.90	23	11.17	
Total	434.96	27		

$*F_{cv} = 2.80, p < .05$

7. Dr. Teachwell teaches statistics, and she wishes to use a computer statistical package that facilitates the acquisition of concepts. She has a class of 24 students, and she assigns them at random to one of three groups. Group 1 will use ESTAT, group 2 will use Multi-Stat, and group 3 will use Psy-Lab. The dependent variable will be the score (out of 100) on the final exam. Previous research has shown that ESTAT is substantially different from the other packages. Is that true? The data are shown here. Perform the steps in Exercise 2.

Group 1	*Group 2*	*Group 3*
67	79	82
96	65	86
96	84	72
69	86	72
93	75	48
97	73	64
81	94	87
99	56	83

8. *(Exercise 7 continued)* If you had performed the omnibus analysis of variance in part (b) of Exercise 7, you would have produced this ANOVA summary table:

Source	*SS*	*df*	*MS*	*F*
Between groups	772.33	2	386.167	2.3377
Within groups	3469.00	21	165.190	
Total	4241.33	23		

(a) Would you have rejected the null hypothesis?
(b) In view of the fact that you did reject the planned comparison H_0 in part (k) of Exercise 7, what general principle does this illustrate?

9. Suppose you are a restaurant owner and you wish to determine whether brand of coffee makes any difference to your customers. You randomly select ten customers and have each one rate the overall taste of four brands of coffee: Columbian Premium, Santa Rosa Blend, Leader's Instant, and Mike's Special. The rating scale is from 1 to 10, with 10 being highest. The data are shown here. You recognize that some of your raters like coffee more than

Person	*Columbian*	*Santa Rosa*	*Leader's*	*Mike's*
1	9	8	10	8
2	7	8	8	9
3	5	5	7	4
4	9	6	7	6
5	8	7	9	6
6	5	6	6	6
7	5	3	7	4
8	7	7	6	6
9	8	7	9	8
10	8	7	9	9

others, so you expect that some will give higher ratings than others. You also recognize that these data are ordinal, but previous research has shown that they can be treated as if they were interval for purposes of analysis of variance. Does type of coffee make a difference? Perform the steps in Exercise 4.

10. *(Exercise 9 continued)* If you had performed an *independent-sample* analysis of variance for the data in Exercise 9, you would have produced this ANOVA summary table:

Source	SS	df	MS	F
Between groups	11.675	3	3.892	1.53
Within groups	91.300	36	2.536	
Total	102.975	39		

(a) Would you have rejected the null hypothesis?
(b) In view of the fact that you did reject the dependent-samples H_0 in part (e) of Exercise 4, what general principle does this illustrate?

Section C: Extending your comprehension (answers in Appendix D, page 524)

11. Is it possible to create a data set for which an omnibus F is significant but *no* post hoc test is significant? Why?
12. Is it possible to create a data set for which an omnibus F is significant but *no* planned comparison is significant? Why?
13. Assume that the data shown here are for dependent samples.

15	12
20	17
17	14
11	7
11	11

(a) Demonstrate that the observed value of F as computed from the dependent-sample ANOVA is equal to the square of the observed value of t as computed from the dependent-sample t test.
(b) Demonstrate that the critical value of F for the dependent-sample ANOVA is equal to the square of the critical value of t for the dependent-sample t test. [Parts (a) and (b) imply that F and t are equivalent when $k = 2$.]
(c) Demonstrate that d_M as computed from the dependent-sample ANOVA is identical to d as computed from the dependent-sample t test.

Section D: Cumulative review (answers in Appendix D, page 524)

For Exercises 14–21, choose the correct answer or fill in the blank. [**Note that computations are *not* required**; use $\alpha = .05$ unless otherwise instructed.]

(a) This problem requires:
(1) Finding the area under a normal distribution [Skip parts (b)–(g) and proceed to (h).]
(2) Creating a confidence interval
(3) Testing a hypothesis about the mean of one group
(4) Testing a hypothesis about the means of two independent groups
(5) Testing a hypothesis about the means of two dependent groups
(6) Testing a hypothesis about the means of three or more independent groups
(7) Test hypotheses about means using post hoc tests
(8) Testing an a priori hypothesis about means (planned comparison)
(9) Testing a hypothesis about the means of three or more dependent groups
(10) Peforming a power analysis (involving finding the area under a normal distribution). [Skip parts (b)–(h).]

(b) The null hypothesis is of the form:
(1) $\mu = a$ (2) $\mu_1 = \mu_2$
(3) $\mu_D = 0$ (4) $\mu_1 = \mu_2 = \cdots = \mu_k$
(5) A series of hypotheses of the form $\mu_i = \mu_j$ (State how many such hypotheses.)
(6) One null hypothesis of the form $C_1\mu_1 + C_2\mu_2 + \cdots + C_k\mu_k = 0$
(7) There is no null hypothesis.

(c) The appropriate sample statistic(s) is (are):
(1) $\overline{X}$ (2) p
(3) $\overline{X}_1 - \overline{X}_2$ (4) $\overline{D}$
(5) $\overline{X}_i - \overline{X}_j$
(6) $C_1\overline{X}_1 + C_2\overline{X}_2 + \cdots + C_k\overline{X}_k$
(7) Not applicable for ANOVA

(d) The appropriate test statistic(s) is (are):
(1) z (2) t
(3) F (4) Q

(e) The number of degrees of freedom for this test statistic is ______. (State the number or *not applicable*: if there are multiple null hypotheses, state the number of degrees of freedom for each.)

(f) The hypothesis test (or confidence interval) is:
(1) Directional (2) Nondirectional

(g) The critical value(s) of the test statistic(s) is(are) ______. (Give the value.) (If there are multiple null hypotheses, state the critical value for each.)

(h) The appropriate formula(s) for the test statistic(s) (or confidence interval) is (are) given by Equation (______). (If there are multiple null hypotheses, state the equation number for each.)

Note that computations are *not* required for any of the problems in Section D. See the instructions.

14. A breakfast cereal manufacturer wants its machines to dispense on average 17 ounces of corn flakes into its "16-oz" boxes. It knows the standard deviation of the weights of

corn flakes in its boxes is .5 ounce. If the machines are operating correctly, what percentage of box contents weigh less than 16 ounces?

15. (*Exercise 14 continued*) The breakfast cereal manufacturer wonders whether one of its corn flake packaging machines has slipped out of adjustment. It knows from past experience that when adjustment problems occur, the mean amount of corn flakes per box shifts but the standard deviation stays the same. To find out whether the adjustment has shifted, the manufacturer proposes to take a random sample of 20 boxes from this machine's production and weigh them. Suppose that the machine has slipped out of adjustment and the true mean is now 16 ounces. What is the probability that a 20-sample experiment will conclude that the adjustment has shifted?

16. (*Exercises 14 and 15 continued*) The breakfast cereal manufacturer conducts the experiment described in Exercise 15. It takes a random sample of 20 boxes from this machine's production; the weights (in ounces) are: 15.5, 15.9, 15.9, 16.5, 15.6, 15.9, 15.8, 16.2, 15.1, 15.9, 16.5, 16.4, 16.2, 16.2, 16.2, 16.3, 16.3, 15.7, 14.9, 16.6. Has this machine slipped out of adjustment?

17. The "autokinetic effect" refers to the fact that if a subject is placed in a totally dark room and views a stationary point of light, the light appears to move with random, jerky movements. Dr. Smith wishes to know whether the distance the light is perceived to move depends on the color of the light. She selects nine subjects and presents each color of light to them in random order. The apparent distances moved (in inches) are shown in the table. Does the size of the autokinetic effect depend on color?

Subject	*Red*	*Green*	*Blue*	*White*
1	11	14	15	14
2	13	15	10	15
3	10	14	17	12
4	10	19	12	22
5	15	14	12	23
6	10	18	9	23
7	13	14	9	20
8	12	12	20	16
9	16	16	13	16

18. (*Exercise 17 continued*) Dr. Jones uses an experiment (different from the one in Exercise 17) to examine the effect of color on autokinetic perception. He has reason to believe, from previous research, that green and white affect autokinetic perception differently from red and blue. He obtains 32 subjects and assigns each to one of four groups; each group sees a different color. The apparent distances moved (in inches) are shown in the table. Do green and white affect the size of the autokinetic effect differently from red and blue?

Red	*Green*	*Blue*	*White*
23	19	7	17
5	11	7	25
17	24	15	19
9	20	13	14
6	28	21	28
12	11	18	19
12	20	14	20
18	24	15	24

19. A restaurant owner wishes to increase his clientele, so he issues two-for-one passes in hopes that customers will like his restaurant and come back to pay full price. He obtains the names and telephone numbers of the 200 patrons who use his coupons. After one month, he contacts a random sample of 25 of these individuals and asks whether they have made a return visit to his restaurant. Of these 25, 11 say they have indeed returned. What can the restaurateur say about the percentage of repeat business spawned by his coupons?

20. An experimenter wishes to know whether hospital admissions are related to the phase of the moon. She obtains data on the number of daily admissions for the last year and randomly selects 24 days, organizing them according to moon phase and recording whether the day in question occurred when the moon was "full" (at least 75%), "partial" (25% to 75%), or "new" (less than 25%). The numbers of admissions are shown here. Did moon phase affect hospital admissions?

New: 15, 10, 21, 12, 8, 19
Partial: 11, 31, 15, 14, 16, 21, 20, 14, 21, 12, 21
Full: 14, 4, 16, 10, 11, 10, 8

21. A researcher has a pigeon laboratory and has five different grains available to feed to the pigeons. She wishes to know whether there is a difference in the birds' rate of responding as a function of the different grains. She randomly assigns ten pigeons to each of the five grains and then measures the average rate of responding. The means for the five different grains (in responses per minute) are 14.6, 13.9, 20.0, 19.4, and 16.8. An analysis of variance shows that the means are not all the same; MS_W was 8.4 in this ANOVA. Now she wishes to know which of the grains are significantly different from which other grains. How does she proceed?

Section E: From the journals (answers in Appendix D, page 524)

22. Erinosho (1990) "examined the effects of teacher-directed and student-oriented remediation methods in high school physics classes in Nigeria. Students ($N = 121$) from three Nigerian high schools were assigned randomly to three conditions: teacher-directed remediation [where the researcher worked through all the items on a formative test], student-oriented remediation [where the students were encouraged by the researcher to find answers to items they missed on the formative test using individual or group efforts], or no treatment" (p. 177). She measured the effect of the interaction using three instruments, but we will

focus on only one, an achievement test based on the Ordinary Level West African School Certificate examination. This achievement test has 45 multiple-choice items and was administered two weeks after the end of the treatment. Erinosho's results are given in the tables.

Achievement test scores by subgroup

Subgroup	*n*	$\overline{X}$	*s*
Teacher-directed	40	59.31	8.55
Student-oriented	40	46.23	7.88
No treatment	41	44.09	9.53

ANOVA summary table

Source	SS	*df*	MS	*F*
Between	2,900.3	2	1450.2	10.98*
Within	15,540.4	117	132.8	
Total	18,340.6	119		

*$p < .05$

"Post hoc analyses of the means ... indicate that [the teacher-directed subgroup] attained a higher performance than [the student-oriented group] ($t = 2.88$, $p < .05$) and [the no treatment group] ($t = 7.59, p < .05$). [The student-oriented group] did not score significantly higher than [the no treatment group] ($t = 1.14, p > .05$)" (p. 180).

(a) Did Erinosho compute the degrees of freedom in the ANOVA correctly?
(b) Is the ANOVA table internally correct (for example, are the totals correct and the mean squares computed correctly)?
(c) Does MS_W correspond to the standard deviations listed for each group?
(d) Is Erinosho's conclusion regarding the ANOVA correct?
(e) Are the t tests reported in the post hoc analysis simple two-independent-samples t tests? Verify that by performing the t tests yourself.
(f) Are such t tests the appropriate way to test the post hoc hypotheses? If not, why not?
(g) Perform the Tukey post hoc procedure. Use the value of MS_W reported by Erinosho, even though it may not be correct.
(h) How do your Tukey results compare with Erinosho's multiple t test results?

23. (*Exercise 22 continued*) Suppose you replicate Erinosho's (1990) study using ten subjects in each group, with the test scores shown in the table.

Teacher directed	*Student oriented*	*No treatment*
66	32	50
53	45	41
40	62	36
47	41	55
66	44	48
62	44	47
57	34	37
64	46	46
70	41	39
61	43	44

(a) Compute the analysis of variance.
(b) State your results in plain English.
(c) If you have the right to perform post hoc tests, perform them.
(d) State your results in plain English.

Section F: Computer explorations

24. Create a data set with three groups and five values per group so that the omnibus F test is significant but no Tukey post hoc test is significant.
25. Create a data set with three groups and five values per group so that a planned comparison is significant but the omnibus F test is not significant.
26. Create a data set with three groups and five values per group so that the omnibus F test is significant but the repeated-measures F is not significant.
27. Create a data set with three groups and five values per group so that the repeated-measures F is significant but the omnibus F test is not significant.

CHAPTER 15

Two-way Analysis of Variance: Factorial Design

LEARNING OBJECTIVES

1. What is a two-way design?
2. What is a factorial design?
3. How are factorial designs graphically displayed?
4. What are main effects?
5. What is an interaction?
6. What are the steps in testing hypotheses in two-way ANOVAs?
7. Why must one examine a table of residuals to interpret interaction?

This chapter describes two-way analyses of variance, the tests appropriate for designs that have two independent (or treatment) variables and one dependent (outcome) variable. A two-way analysis of variance results in the computation of three F ratios and three corresponding tests of significance: two tests for main effects (one for each of the independent variables, analogous to the one-way analyses of variance described in Chapter 13) and a test for interaction, which measures the extent that the effect of one treatment variable depends on the level of the other treatment variable.

levels: the possible values of an independent variable

In Chapter 13, we considered the analysis of variance, which we used to test the hypothesis that three or more groups have the same population mean. Our example was a test of whether four different antidepressant drugs all had the same effect on depression. We called the factor that distinguishes these four groups from one another (the type of drug, in our example) the "independent variable" (or "treatment variable"), and we called the values of the treatment variable *levels*. Thus, in the depression example, there was one treatment variable (type of drug) with four levels (Prozac, Elavil, Zoloft, and Tofranil).

15.1 A Two-way Factorial Design

two-way design: an experiment in which outcomes corresponding to several levels of two independent variables are observed

It is possible to design experiments that have two or more treatment variables, each of which has two or more levels. A study that has two treatment variables (regardless of the number of levels within each) is called a *two-way design*, a study that has three treatment variables is called a *three-way design*, and so on. Higher-order designs are possible but relatively rare in practice. This textbook considers two-way analyses only.

Let us take an example of a two-way design that has three levels of the first variable and two levels of the second. Suppose we are administrators of a large mental health clinic, and we are interested in the outcome of the psychotherapy our therapists provide. Is one kind of therapy better than another? We call our variable of interest "therapy outcome," and we decide to measure it by using a Mental Health Scale (MHS) that has a maximum score of 99, indicating excellent mental health, and a minimum score of 0, indicating severe psychological problems.[1]

Suppose three kinds of psychotherapy are practiced at our clinic: rational emotive therapy (RET), behavior therapy (BT), and person-centered therapy (PCT). Therapy type is thus the first treatment variable, and it has three levels: RET, BT, and PCT.

Suppose also that we can divide our therapists into two groups: those with much experience in performing psychotherapy and those with relatively little experience. We therefore make therapist's experience the second treatment variable, and it has two levels: low and high.

[1] In practice, such a scale is difficult to construct because mental health is a multidimensional construct. Here we will ignore that difficulty and treat mental health as a unitary phenomenon because it provides a useful example of how the two-way analysis can proceed. The data in this example are fictitious and do not necessarily represent the outcomes of actual therapy outcome research.

If all possible combinations of the two treatment variables are represented in the study, we call this a *factorial design*. In our example, this requires that high-experience therapists conduct RET, BT, and PCT, and also that low-experience therapists conduct all three kinds of therapy. This is the case in our experiment, as shown by the data in Table 15.1. A study that has one treatment variable with three levels and another treatment variable with two levels is called a "3 by 2" or "3 × 2" factorial study.

factorial design: a two-way (or higher-order) design in which outcomes corresponding to all possible combinations of the treatment variables are observed

TABLE 15.1 Mental Health Scale scores for patients at the conclusion of therapy

		Therapist's experience	
		Low	*High*
Therapy type	RET	60	60
		50	65
		55	70
		62	51
		48	67
	BT	70	55
		58	64
		63	60
		65	60
		62	58
	PCT	59	66
		41	80
		38	71
		45	64
		49	66

We emphasize that even though this design has two independent variables (therapy type and therapist's experience), it has only *one* dependent variable—namely, each patient's therapy outcome score as quantified by the Mental Health Scale.

15.2 Displaying the Outcome of a Two-way Design

There are six combinations of levels of the independent variables in a 3 × 2 design: RET therapists who have low experience, RET therapists who have high experience, BT therapists who have low experience, and so on. In a display, the data from each combination of levels are placed in a *cell*; there are thus six cells in Table 15.1. The data in the first-row–first-column cell are considered to be random samples from the population of all RET therapists who have low experience. This population is assumed to be normally distributed with (unknown) mean $\mu_{\text{RET,low}}$ and (unknown) standard deviation σ. We assume that the population standard deviations in all the cells are identical, so we do not need to write the "RET, low" subscript on σ. Likewise, the data in the third-row–second-column cell are considered to be random samples from the population of all PCT therapists who have high experience. This population is assumed to be normal with (unknown) mean $\mu_{\text{PCT,high}}$ and standard deviation σ.

cell: the intersection of a row and a column

We would like to know the parameters $\mu_{\text{RET,low}}$, $\mu_{\text{PCT,high}}$, and so on, but we do not; that is why we are conducting the research. The sample mean (usually called the "cell" mean) $\overline{X}_{\text{RET,low}}$ of the first-row–first-column cell is the best point-estimate of $\mu_{\text{RET,low}}$; $\overline{X}_{\text{PCT,high}}$ is the best point-estimate of the third-row–second-column $\mu_{\text{PCT,high}}$; and so on. As in Chapters 9–13, we will draw inferences about the six population μ values based on the six sample (cell) means.

We display the cell means in Table 15.2. For example, in the upper left cell $\overline{X}_{\text{RET,low}} = (60 + 50 + 55 + 62 + 48)/5 = 55.0$. Also shown in Table 15.2 are the row means [for example, $\overline{X}_{\text{RET}} = 58.8$ is the mean of all ten observations in the first (RET) row of Table 15.1]; the column means [for example, $\overline{X}_{\text{high}} = 63.8$ is the mean of all 15 observations in the second (high experience) column of Table 15.1]; and the grand mean $\overline{X}_{\text{G}} = 59.4$, the mean of all 30 observations in the entire Table 15.1.

TABLE 15.2 Cell, row, and column means for the therapy outcome data of Table 15.1

		Therapist's experience		*Row*
		Low	*High*	*mean*
	RET	55.0	62.6	58.8
Therapy type	BT	63.6	59.4	61.5
	PCT	46.4	69.4	57.9
Column mean		55.0	63.8	59.4

We can display these results graphically in either of the two equivalent ways shown in Figure 15.1. On the left side of Figure 15.1, we place the three levels of therapy type on the X-axis and show two superimposed outcomes, one for low and one for high experience. On the right side, we place the two levels of therapist's experience on the X-axis and show three superimposed outcomes, one each for BT, RET, and PCT. For example, the upper left point on the left-hand graph is the cell mean for high-experience RET therapists ($\overline{X}_{\text{RET,high}} = 62.6$), and the upper left point on the right-hand graph is the cell mean for low-experience BT therapists ($\overline{X}_{\text{BT,low}} = 63.6$).

The row and column means can be inferred from either graphical presentation. For example, the row mean for the first (RET) row of the Table 15.1 data must be halfway between the two points on the left side of the left-hand graph in Figure 15.1: $\overline{X}_{\text{RET}} = 58.8$ is halfway between $\overline{X}_{\text{RET,low}} = 55.0$ and $\overline{X}_{\text{RET,high}} = 62.6$. Note that the row mean will be *exactly* halfway between the two points *only* if the cell sizes are equal; all of our examples in this chapter will assume that the cell sizes are in fact equal. As another example, the column mean for the first (low) column of Table 15.1 must be midway between the three solid points on the left side of Figure 15.1: $\overline{X}_{\text{low}} = 55.0$ is midway between $\overline{X}_{\text{RET,low}} = 55.0$, $\overline{X}_{\text{BT,low}} = 63.6$, and $\overline{X}_{\text{PCT,low}} = 46.4$ (again exactly midway only if the cell sizes are equal). The same conclusions can be drawn from the right-hand graph of Figure 15.1.

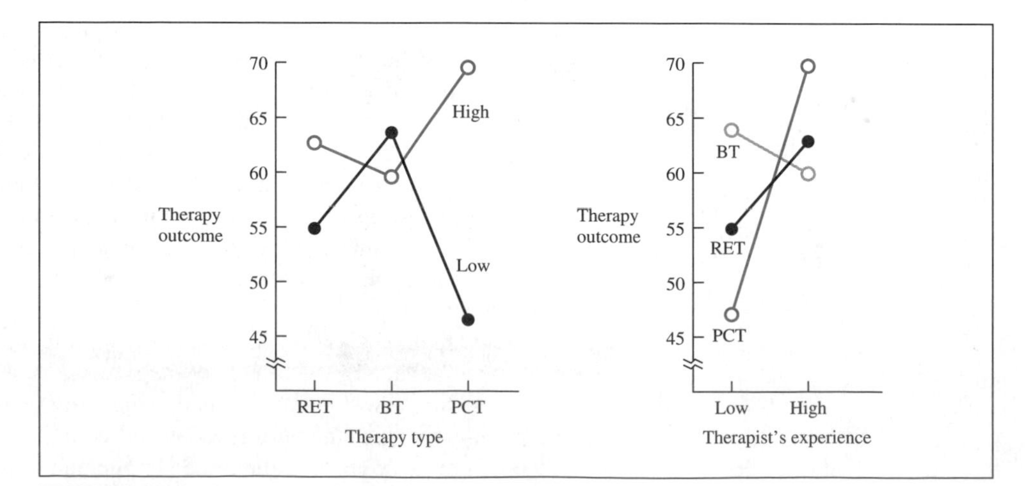

FIGURE 15.1 Two equivalent displays of the therapy outcome data

The two displays, though equivalent, emphasize different aspects of the data. For example, it is easier to note in the left-hand display that high-experience therapists seem in general to have better outcomes than low-experience therapists, but it is easier to see in the right-hand display that the therapy outcome for PCT therapists seems to depend greatly on the level of the therapist's experience.

15.3 Main Effects

The displays in Table 15.2 and Figure 15.1 are of the *sample* data. High-experience therapists do have on average better *sample* outcomes than do low-experience therapists ($\bar{X}_{\text{high}} = 63.8$, whereas $\bar{X}_{\text{low}} = 55.0$, for example). But does that imply that the population mean of all high-experience therapists (μ_{high}) is greater than the population mean of all low-experience therapists (μ_{low})? That is exactly the kind of question we have been asking since Chapter 9: One column (sample) mean is larger than the other, but is it *significantly* larger? To answer that question requires testing a null hypothesis that we will call the *main effect for columns*, or the *main effect for experience*.

main effect: the extent to which the dependent (outcome) variable differs from one level of one independent variable to another level of the same independent variable

In a two-way factorial study, there are actually *three* null hypotheses that can be tested: the main effect for experience that we have just been discussing, the main effect for therapy type (or equivalently, the main effect for rows), and the interaction between therapy type and experience. We will discuss each in turn.

The null hypothesis for a main effect is identical to the null hypothesis for a one-way ANOVA as described in Chapter 13. Thus, when we consider the effect of therapy type on therapy outcome, we could ignore the fact that we have low- and high-experience therapists and perform a one-way ANOVA on the data in Table 15.1, exactly as we did in Chapter 13. If we were to do so, the null hypothesis would be

$$H_0\colon \quad \mu_{\text{RET}} = \mu_{\text{BT}} = \mu_{\text{PCT}}$$

that is, that it makes no difference which therapy type is used.

Similarly, if we are interested only in the effect of a therapist's level of experience on therapy outcome, we could ignore the fact that we have three different therapy types and perform another one-way ANOVA on the data of Table 15.1, again exactly as we discussed in Chapter 13. In this case, the null hypothesis would be

$$H_0\colon \quad \mu_{\text{low}} = \mu_{\text{high}}$$

which states that the level of a therapist's experience has no effect on the therapy outcome.

A two-way ANOVA tests *both* of those null hypotheses at the same time. It is customary to identify these two main-effect null hypotheses as the row null hypothesis H_{0_R} and the column null hypothesis H_{0_C}, referring to the way the data are arranged in the table. In our example, there are three separate rows, one for each of the therapy types, so we refer to therapy type as the row variable. There are two separate columns, one for each level of experience, so we refer to experience as the column variable. Thus, we have two main-effect null hypotheses:

$$H_{0_R}\colon \quad \mu_{\text{RET}} = \mu_{\text{BT}} = \mu_{\text{PCT}}$$
$$H_{0_C}\colon \quad \mu_{\text{low}} = \mu_{\text{high}}$$

In general terms, these null hypotheses are written

null hypothesis for rows

null hypothesis for columns

$$H_{0_R}\colon \quad \mu_{\text{row }1} = \mu_{\text{row }2} = \cdots = \mu_{\text{row }R} \tag{15.1}$$
$$H_{0_C}\colon \quad \mu_{\text{col }1} = \mu_{\text{col }2} = \cdots = \mu_{\text{col }C} \tag{15.2}$$

We say there is a "significant main effect for rows" (or for therapy type) if the null hypothesis for therapy type (or for rows) is rejected, and there is a "significant main effect for columns" (or for therapist's experience) if the null hypothesis for therapist's experience (or for columns) is rejected.

We will return to the actual data in Table 15.1 (and the means displayed in Table 15.2 and Figure 15.1) in a moment, but first let us explore some different possible results of this kind of study. Figures 15.2 through 15.8 illustrate how main effects and interactions appear in differing sets of outcomes.

The two graphs in Figure 15.2 both represent the same data, just like the two graphs in Figure 15.1. The left-hand graph of Figure 15.2 shows that the means of the high-experience cells are all greater than the means of the corresponding low-experience cells, regardless of the type of therapy performed. Thus, it appears that this graph illustrates a main effect for therapist's experience: High-experience therapists have better results than low-experience therapists. These same data show that BT therapists have higher cell means than the other therapists, regardless of level of experience, so it appears that these data illustrate a main effect for therapy type also. Thus, Figure 15.2 shows main effects for *both* independent variables; there is no interaction (to be discussed in the next section).

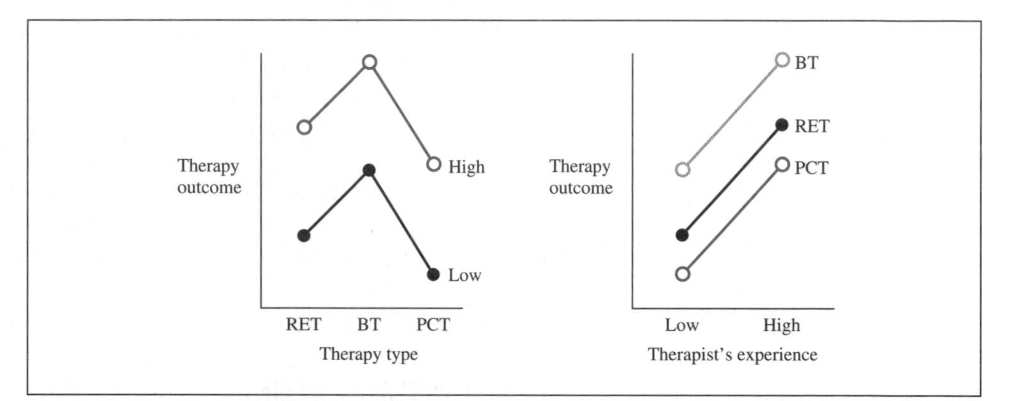

FIGURE 15.2 Possible result of a two-way ANOVA: significant main effect for both therapy type and experience; no interaction. Displayed in two equivalent ways.

It should be emphasized that a statistical test (namely, the ANOVA) is required to establish that the differences illustrated in the figures are significant. It is quite possible to construct data with cell means that have the pattern of Figure 15.2 but where *no* differences are significant (as when within-cell variances are large).

Figure 15.3 shows a result that has a significant main effect for only one of the independent variables: for therapy type but not for therapist's experience. PCT produces better outcomes than RET or BT whether the therapist has high or low experience, but experience by itself has no effect on outcome.

Figure 15.4 shows a result where only the main effect for therapist's experience is significant. High-experience therapists are better with each therapy, but therapy type does not make much difference with respect to outcome.

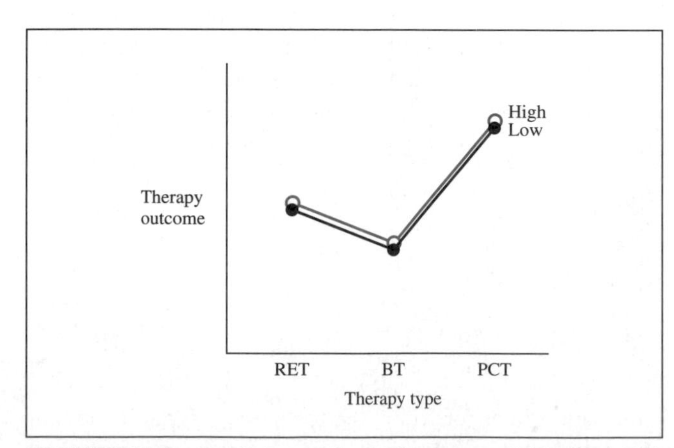

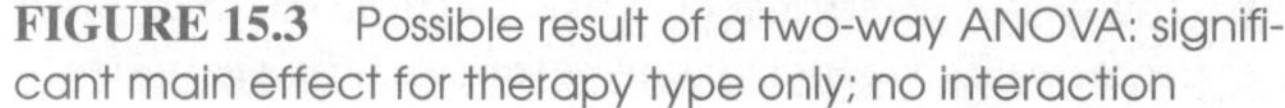

FIGURE 15.3 Possible result of a two-way ANOVA: significant main effect for therapy type only; no interaction

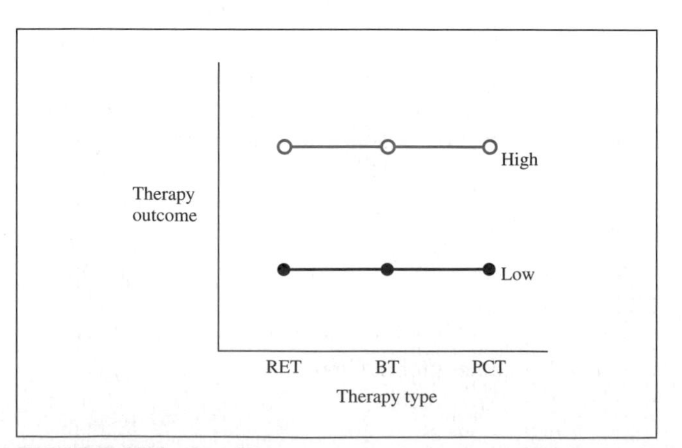

FIGURE 15.4 Possible result of a two-way ANOVA: significant main effect for therapist's experience only; no interaction

Figure 15.5 shows a possible result if *neither* main effect is significant. The outcome is the same regardless of therapy type or therapist's experience.

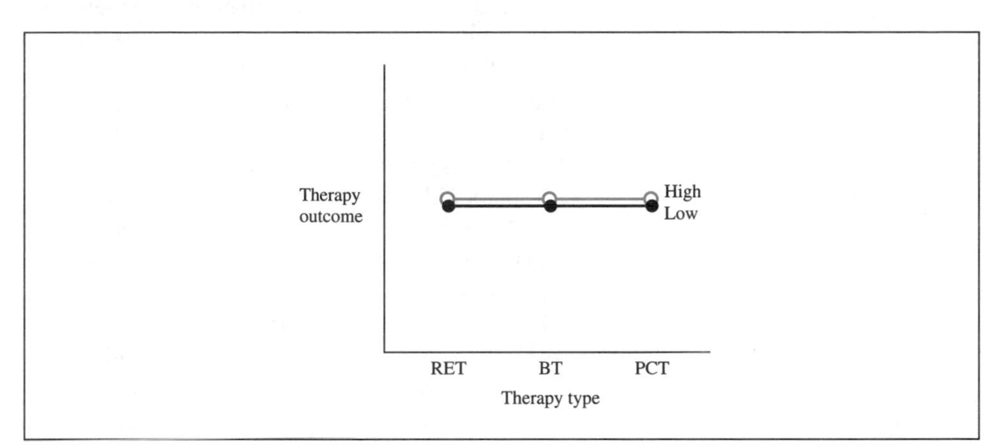

FIGURE 15.5 Possible result of a two-way ANOVA: neither main effect significant; no interaction

15.4 Interaction

interaction: the extent to which the outcome associated with one independent variable depends on the level of the other independent variable

A two-way ANOVA not only tests both main-effect null hypotheses at the same time, but also tests the null hypothesis that there is no *interaction* between the two effects—in our example, that there is no interaction between therapy type and therapist's experience. What is meant by *interaction*? We say there is an interaction when the effectiveness of one treatment variable *depends on the level* of the other treatment variable. Interaction is present when the lines on the outcome graphs are (significantly) not parallel to each other.

Kinds of Interaction

Figure 15.6 illustrates four of the many possible kinds of interaction when there is no main effect. In the upper left corner of Figure 15.6 verify for yourself that there are no main effects: The mean of the two RET cells equals the mean of the two BT cells equals the

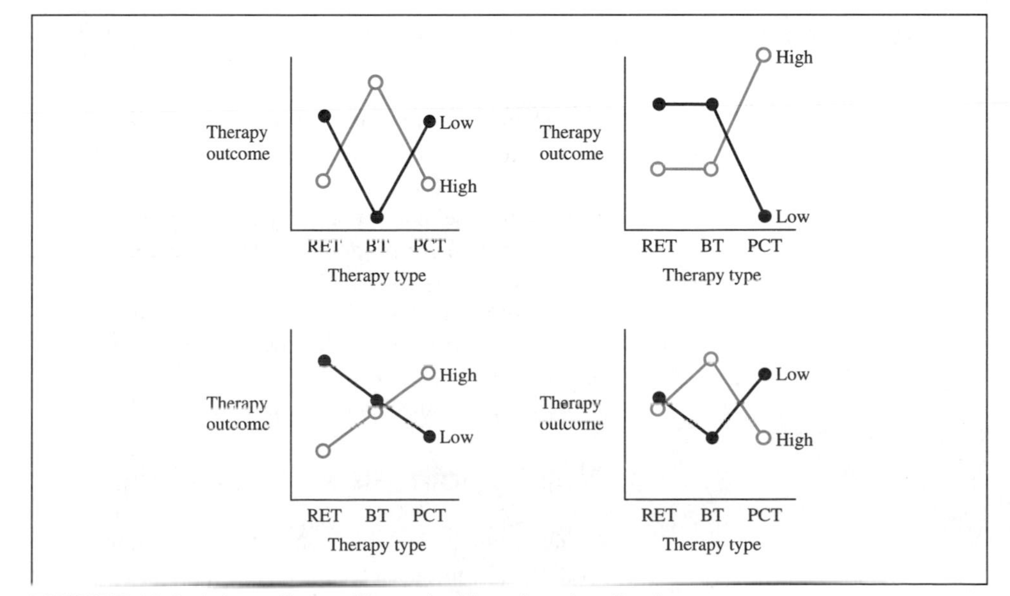

FIGURE 15.6 Interactions with *no* significant main effects

mean of the two PCT cells, and the mean of the three high-experience cells equals the mean of the three low-experience cells.

Because there is no main effect for experience in the upper left graph, we *cannot* say that low-experience therapists perform better (or worse) than high-experience therapists in general. However, the graph shows that low-experience therapists perform better than high-experience therapists *if they are using RET or PCT* but *not* if they are using BT. The effectiveness of experience thus *depends* on which type of therapy the therapist is using; that is, there is a significant interaction.

You should demonstrate for yourself that all of the graphs in Figure 15.6 show *no* main effect for *either* treatment variable but appear to show a *significant* interaction. Here again, a statistical test (the two-way ANOVA) is required to demonstrate that the apparent interaction is indeed significant.

Figures 15.7 and 15.8 illustrate that both main effects and interactions can occur at the same time. In Figure 15.7, we illustrate a significant main effect for therapist's experience (but *not* for therapy type) *and* a significant interaction. The main effect *means* for therapy type are shown on that graph as plus signs. For example, the leftmost plus sign is halfway between the two RET cell means and thus indicates the mean between the two (low and high experience) RET conditions. The three plus signs all have about the same outcome value, which implies that there is no main effect for therapy type. On the other hand, the main effect means for experience are shown as two-ended arrows. The upper arrow is the mean of the three (RET, BT, and PCT) low-experience conditions, and the lower arrow is the mean of the three high-experience conditions. Because those arrows *do* have much different values (as an ANOVA would show), we say that there is a significant main effect for experience.

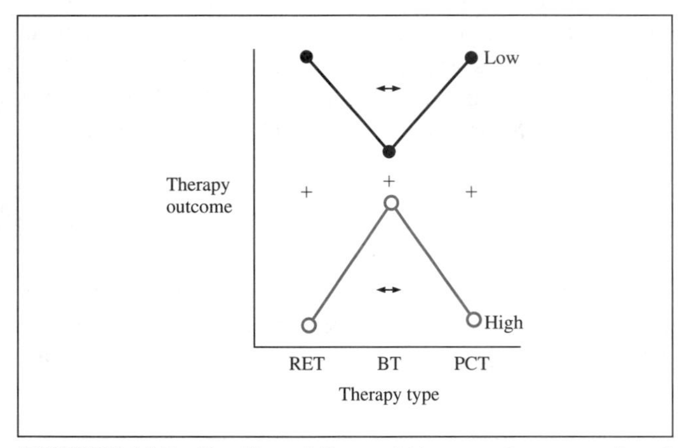

FIGURE 15.7 Significant therapist's experience main effect and interaction

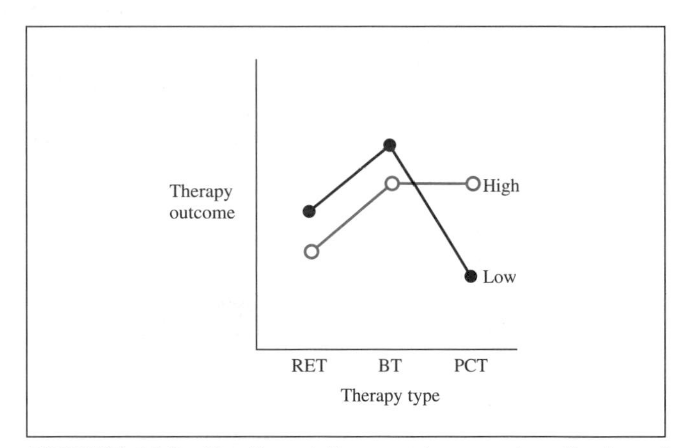

FIGURE 15.8 Significant therapy type main effect and interaction

There is also an interaction effect in these data because therapist's experience has a small effect when BT is used but a large effect when either RET or PCT is used. Thus, we can say that, according to the Figure 15.7 results, low experience is in general better than high experience (that is, the main effect of therapist's experience is significant) *and* that effect is more pronounced when therapists are using RET or PCT (the interaction is significant).

Figure 15.8 shows a main effect for therapy type (BT is better than either RET or PCT) but *no* main effect for therapist's experience. It is left as an exercise for the reader to show the main-effect means and verify those claims. There is also an interaction: High experience is better for PCT but *not* for RET or BT.

Null Hypothesis for Interaction

If there is *no* interaction, we can think of each main-effect level as adding its own increment to all the cell means at its level. For example, let μ_G be the grand mean of all the populations from which the six therapy outcome cells come. If there is no interaction, the therapy type main effect should shift both of the cell population means in the RET row (that is, the RET/high and the RET/low cells) away from the grand mean by the same increment, whose value is $(\mu_{RET} - \mu_G)$. Similarly, the main effect for therapist's experience should shift all the cell population means in the low-experience column (that is, the RET/low, BT/low, and PCT/low cells) away from the grand mean by an increment whose value is $(\mu_{low} - \mu_G)$. If that is true (that is, there is no interaction), the population mean for the RET/low cell (the upper left cell of Table 15.2) should be shifted away from the grand mean by *two* main effects, one for therapy type and one for therapist's experience, so the population mean for the RET/low cell should be $\mu_{RET,low} = \mu_G + (\mu_{RET} - \mu_G) + (\mu_{low} - \mu_G)$. Thus, the upper left cell mean should equal the grand mean plus an increment due to rows plus an increment due to columns. The same is true for all six cells.

There are a total of six equations like this one (one for each cell). For example, the cell population mean for the second row, second column of Table 15.2 should be $\mu_{BT,high} = \mu_G + (\mu_{BT} - \mu_G) + (\mu_{high} - \mu_G)$ if there is no interaction. Because those six equations are true *only* if there is *no* interaction, they can serve as a statement of the null hypothesis for interaction: *If* it is true that $\mu_{RET,low} = \mu_G + (\mu_{RET} - \mu_G) + (\mu_{low} - \mu_G)$, *then* we can conclude that there is no interaction effect applying to the RET/low cell.

We can simplify these six equations by removing the parentheses and collecting all the μ_G terms together (for example, obtaining for the RET/low cell $\mu_{RET,low} = \mu_{RET} + \mu_{low} - \mu_G$ or, moving all terms to the same side of the equal sign, $\mu_{RET,low} - \mu_{RET} - \mu_{low} + \mu_G = 0$). When we substitute the general term "row" for RET, BT, or PCT and the general term "column" for high or low, we get the standard form of the null hypothesis for interaction:

null hypothesis for interaction

$$H_{0_{RC}}: \text{ all } \mu_{row,col} - \mu_{row} - \mu_{col} + \mu_G = 0 \qquad (15.3)$$

We call this null hypothesis $H_{0_{RC}}$ to indicate that it involves the interaction between rows and columns. Furthermore, the word *all* in this expression implies that there is one such equation for each cell of the design, a total of six such equations in our therapy outcome example. We have written out two of them (for the RET/low and BT/high cells); it is left as an exercise for the reader to produce the remaining four. If all six of these equations are in fact true, then we can say that there is no interaction in the data.

15.5 Testing Hypotheses

There are thus three null hypotheses that can be stated in a two-way ANOVA: the main effect for rows $H_{0_R}: \mu_{RET} = \mu_{BT} = \mu_{PCT}$ given in Equation (15.1), the main effect for columns $H_{0_C}: \mu_{low} = \mu_{high}$ from Equation (15.2), and the interaction of rows with columns $H_{0_{RC}}$: all $\mu_{row,col} - \mu_{row} - \mu_{col} + \mu_G = 0$ from Equation (15.3). Each of these null hypotheses will be tested using its own F ratio. Because, as we saw in Chapter 13, every F is the ratio of one mean square to another, the three F ratios are $F_R = MS_R/MS_W$, $F_C = MS_C/MS_W$, and $F_{RC} = MS_{RC}/MS_W$.

To determine the critical values of these three F ratios, we need to determine the degrees of freedom for each. To do so, we first let R equal the number of rows (three in our example). That is the number of levels of the first treatment variable (we called that k in the one-way ANOVAs of Chapter 13). Let C be the number of columns (two in our example). Then there are $RC = 3(2) = 6$ cells in the design.

The *denominator* for all three F ratios is MS_W, which is obtained by pooling the variances within each of the $RC = 6$ cells of Table 15.2. Each of those cells in our example has $n = 5$ data points, and therefore there are $n - 1 = 4$ degrees of freedom within each cell. Because pooling adds degrees of freedom, $df_W = 4 + 4 + 4 + 4 + 4 + 4 = RC(n - 1) = 6(4) = 24$.

The degrees of freedom for the row treatment variable can be obtained in the same way as in a one-way ANOVA: It is the number of levels of the row treatment variable minus one; that is, $df_R = R - 1 = 3 - 1 = 2$. Similarly, $df_C = C - 1 = 2 - 1 = 1$.

So far we have "used up" 24 + 2 + 1 = 27 degrees of freedom. Now we know that the total degrees of freedom are the total number of points in the data set (disregarding row and column divisions) minus one. That is, $df_T = N - 1 = 30 - 1 = 29$. What has happened to the remaining 2 degrees of freedom?

The answer is that the interaction mean square MS_{RC} has $(R-1)(C-1) = (3-1)(2-1) = 2$ degrees of freedom in our example. Now we can see that $df_T = df_R + df_C + df_{RC} + df_W = 2 + 1 + 2 + 24 = 29$. When we recall that any mean square is its sum of squares divided by its degrees of freedom, we can lay out the general form of the two-way summary ANOVA table as shown in Table 15.3.

We can now illustrate our standard procedure for testing hypotheses using the data from the therapy outcome experiment (see Table 15.1).

TABLE 15.3 General form of the two-way ANOVA summary table

Source of variation	*Sum of squares (SS)*	*Degrees of freedom (df)*	*Mean square (MS)*	*F*
Row	SS_R	$R-1$	SS_R/df_R	MS_R/MS_W
Column	SS_C	$C-1$	SS_C/df_C	MS_C/MS_W
RC interaction	SS_{RC}	$(R-1)(C-1)$	SS_{RC}/df_{RC}	MS_{RC}/MS_W
Within cell	SS_W	$RC(n-1)$	SS_W/df_W	
Total	SS_T	$N-1$		

I. State the Null and Alternative Hypotheses

Three sketches illustrate H_0:
Variable
Sample statistic
Test statistic

ℹ Note that there are three separate null hypotheses. The sketches that we customarily use to illustrate H_0 are not particularly informative, so we omit them.

We have seen that there are three separate null hypotheses, one for each main effect and one for the interaction between the two:

$$\text{Row main effect:} \quad H_{0_R}: \quad \mu_{RET} = \mu_{BT} = \mu_{PCT} \tag{15.1}$$

$$\text{Column main effect:} \quad H_{0_C}: \quad \mu_{low} = \mu_{high} \tag{15.2}$$

$$\text{Interaction:} \quad H_{0_{RC}}: \quad \text{all } \mu_{row,col} - \mu_{row} - \mu_{col} + \mu_G = 0 \tag{15.3}$$

The alternative hypothesis for each is analogous to the case for one-way ANOVAs; we simply state that the null hypothesis is not true. Note that although we could perform two separate one-way ANOVAs to test the null hypotheses for the two main effects, it is impossible to test for interaction in one-way ANOVAs.

II. Set the Criterion for Rejecting H_0

Level of significance?
Directional?
Shade rejection regions.

If we set $\alpha = .05$, then we can determine the critical values of the three F ratios. For the test for therapy type (the row main effect), $F_R = MS_R/MS_W$; the numerator MS_R has $R - 1 = 2$ degrees of freedom, and the denominator MS_W has $RC(n-1) = 3(2)(5-1) = 24$ degrees of freedom, so F_R has 2 and 24 degrees of freedom. From Table A.4, the table of critical values of the F statistic in Appendix A, $F_{R_{cv_{2,24}}} = 3.40$. The test for therapist's experience (the column main effect) is $F_C = MS_C/MS_W$ and has $C - 1 = 1$ and $RC(n-1) = 24$ degrees of freedom, so, from Table A.4, $F_{C_{cv_{1,24}}} = 4.26$. The F for the interaction between therapy type and therapist's experience has $(R-1)(C-1) = (3-1)(2-1) = 2$ and $RC(n-1) = 24$ degrees of freedom, so, from Table A.4, $F_{RC_{cv_{2,24}}} = 3.40$. We could, if we liked, sketch three F distributions, one for each null hypothesis, and shade the rejection regions in the right-hand tail of each. In all three cases, if the observed F_{obs} exceeds the appropriate critical value F_{cv}, we reject the corresponding null hypothesis.

III. Collect a Sample and Compute the Observed Values of the Test Statistics

Observed values:
Sample statistic
Test statistic

As in the one-way ANOVA in Chapter 13, no sample statistic is informative here, so we deal with only the test statistics—that is, the three F_{obs} values. We showed the data in Table 15.1 and the cell means in Table 15.2. We now turn to the computation of the three F ratios.

There is no separate sample statistic for ANOVA.

Recall from Chapter 13 that the total sum of the squared deviations from the grand mean (usually called the total sum of squares) is given by Equation (13.9):

$$SS_T = \sum(X_{ij} - \overline{X}_G)^2$$

where X_{ij} is the ith observation in the jth group and $\overline{X}_G$ is the grand mean. That notation can be extended to the two-way ANOVA situation:

total sum of squares

$$SS_T = \sum(X_{irc} - \overline{X}_G)^2 \qquad (15.4)$$

where X_{irc} is now the ith observation in the cell that is in the rth row and the cth column, and $\overline{X}_G$ is the grand mean. Just as in Chapter 13, the total sum of squares, SS_T, is the numerator in the computation of the variance of all the observations, ignoring the fact that the observations are divided into rows and columns.

Recall further that in Chapter 13 we "partitioned" the total sum of squares into two parts, the sum of squares within groups (SS_W) and the sum of squares between groups (SS_B): $SS_T = SS_W + SS_B$ [Equation (13.10b)]. Now in two-way ANOVA we partition the total sum of squares into *four* parts, the sum of squares within cells, the sum of squares for rows, the sum of squares for columns, and the sum of squares for interaction:

partitioning the sum of squares in two-way ANOVA

$$SS_T = SS_W + SS_R + SS_C + SS_{RC} \qquad (15.5)$$

Just as we did in Chapter 13, to create this partition we transform SS_T by adding and subtracting the same quantity $(\overline{X}_r + \overline{X}_c - 2\overline{X}_G)$ to Equation (15.4):

$$\begin{aligned} SS_T &= \sum(X_{irc} - \overline{X}_G)^2 \\ &= \sum\left(X_{irc} - \overline{X}_G + [\overline{X}_r + \overline{X}_c - 2\overline{X}_G] - [\overline{X}_r + \overline{X}_c - 2\overline{X}_G]\right)^2 \end{aligned}$$

A little algebra gives this fundamental result:

partitioning the sum of squares in two-way ANOVA

$$\underset{\text{(total)}}{SS_T} = \underset{\text{(within cells)}}{\sum(X_{irc} - \overline{X}_{rc})^2} + \underset{\text{(between rows)}}{\sum(\overline{X}_r - \overline{X}_G)^2} + \underset{\text{(between columns)}}{\sum(\overline{X}_c - \overline{X}_G)^2} + \underset{\text{(interaction)}}{\sum(\overline{X}_{rc} - \overline{X}_r - \overline{X}_c + \overline{X}_G)^2} \qquad (15.6)$$

Table 15.4 shows this partition for the data of Table 15.1. The deviations of individual points from the grand mean are given by $(X_{irc} - \overline{X}_G)$, shown in the second data column of Table 15.4. The sum of the squared deviations is, by Equation (15.4), SS_T, which is found by summing down the third data column: $SS_T = 2505.20$.

A deviation of an individual point from its own cell mean is $(X_{irc} - \overline{X}_{rc})$. The sum of the squared within-cell deviation is SS_W:

sum of squares within cells

$$SS_W = \sum(X_{irc} - \overline{X}_{rc})^2 \qquad (15.7)$$

For our data, SS_W is obtained by summing down the fifth data column: $SS_W = 924.00$.

row effect: deviation of a row mean from the grand mean

A *row effect* is the deviation of the mean of a particular row from the grand mean $(\overline{X}_r - \overline{X}_G)$. The sum of the squared row effects for each point is the sum of squares for rows:

sum of squares for rows

$$SS_R = \sum(\overline{X}_r - \overline{X}_G)^2 \qquad (15.8)$$

For our data, $SS_R = 70.20$.

column effect: deviation of a column mean from the grand mean

Similarly, a *column effect* is the deviation of the mean of a particular column from the grand mean $(\overline{X}_c - \overline{X}_G)$. The sum of the squared column effects for each point is the sum of squares for columns:

sum of squares for columns

$$SS_C = \sum(\overline{X}_c - \overline{X}_G)^2 \qquad (15.9)$$

For our data, $SS_C = 580.80$.

TABLE 15.4 Partitioning the sum of squares in the two-way ANOVA

		Total		*Within cell*		*Row effect*		*Column effect*		*Interaction*	
	X_{irc}	$X_{irc} - \bar{X}_G$	$(X_{irc} - \bar{X}_G)^2$	$X_{irc} - \bar{X}_{rc}$	$(X_{irc} - \bar{X}_{rc})^2$	$\bar{X}_r - \bar{X}_G$	$(\bar{X}_r - \bar{X}_G)^2$	$\bar{X}_c - \bar{X}_G$	$(\bar{X}_c - \bar{X}_G)^2$	$\bar{X}_{rc} - \bar{X}_r - \bar{X}_c + \bar{X}_G$	$(\bar{X}_{rc} - \bar{X}_r - \bar{X}_c + \bar{X}_G)^2$
	60	.6	.36	5.0	25.00	−.6	.36	−4.4	19.36	.6	.36
Row 1	50	−9.4	88.36	−5.0	25.00	−.6	.36	−4.4	19.36	.6	.36
Column 1	55	−4.4	19.36	.0	.00	−.6	.36	−4.4	19.36	.6	.36
$\bar{X}_{11} = 55.0$	62	2.6	6.76	7.0	49.00	−.6	.36	−4.4	19.36	.6	.36
	48	−11.4	129.96	−7.0	49.00	−.6	.36	−4.4	19.36	.6	.36
	70	10.6	112.36	6.4	40.96	2.1	4.41	−4.4	19.36	6.5	42.25
Row 2	58	−1.4	1.96	−5.6	31.36	2.1	4.41	−4.4	19.36	6.5	42.25
Column 1	63	3.6	12.96	−.6	.36	2.1	4.41	−4.4	19.36	6.5	42.25
$\bar{X}_{21} = 63.6$	65	5.6	31.36	1.4	1.96	2.1	4.41	−4.4	19.36	6.5	42.25
	62	2.6	6.76	−1.6	2.56	2.1	4.41	−4.4	19.36	6.5	42.25
	59	−.4	.16	12.6	158.76	−1.5	2.25	−4.4	19.36	−7.1	50.41
Row 3	41	−18.4	338.56	−5.4	29.16	−1.5	2.25	−4.4	19.36	−7.1	50.41
Column 1	38	−21.4	457.96	−8.4	70.56	−1.5	2.25	−4.4	19.36	−7.1	50.41
$\bar{X}_{31} = 46.4$	45	−14.4	207.36	−1.4	1.96	−1.5	2.25	−4.4	19.36	−7.1	50.41
	49	−10.4	108.16	2.6	6.76	−1.5	2.25	−4.4	19.36	−7.1	50.41
	60	.6	.36	−2.6	6.76	−.6	.36	4.4	19.36	−.6	.36
Row 1	65	5.6	31.36	2.4	5.76	−.6	.36	4.4	19.36	−.6	.36
Column 2	70	10.6	112.36	7.4	54.76	−.6	.36	4.4	19.36	−.6	.36
$\bar{X}_{12} = 62.6$	51	−8.4	70.56	−11.6	134.56	−.6	.36	4.4	19.36	−.6	.36
	67	7.6	57.76	4.4	19.36	−.6	.36	4.4	19.36	−.6	.36
	55	−4.4	19.36	−4.4	19.36	2.1	4.41	4.4	19.36	−6.5	42.25
Row 2	64	4.6	21.16	4.6	21.16	2.1	4.41	4.4	19.36	−6.5	42.25
Column 2	60	.6	.36	.6	.36	2.1	4.41	4.4	19.36	−6.5	42.25
$\bar{X}_{22} = 59.4$	60	.6	.36	.6	.36	2.1	4.41	4.4	19.36	−6.5	42.25
	58	−1.4	1.96	−1.4	1.96	2.1	4.41	4.4	19.36	−6.5	42.25
	66	6.6	43.56	−3.4	11.56	−1.5	2.25	4.4	19.36	7.1	50.41
Row 3	80	20.6	424.36	10.6	112.36	−1.5	2.25	4.4	19.36	7.1	50.41
Column 2	71	11.6	134.56	1.6	2.56	−1.5	2.25	4.4	19.36	7.1	50.41
$\bar{X}_{32} = 69.4$	64	4.6	21.16	−5.4	29.16	−1.5	2.25	4.4	19.36	7.1	50.41
	66	6.6	43.56	−3.4	11.56	−1.5	2.25	4.4	19.36	7.1	50.41
Σ	1782	.0	$2505.20 = SS_T$	.0	$924.00 = SS_W$	.0	$70.20 = SS_R$	.0	$580.80 = SS_C$	.0	$930.20 = SS_{RC}$

interaction effect: the deviation of a cell mean from the grand mean that is not accounted for by either the row effect or the column effect

An *interaction effect* is the deviation of a cell mean that is not accounted for by either the row effect or the column effect $[\bar{X}_{rc} - (\bar{X}_r - \bar{X}_G) - (\bar{X}_c - \bar{X}_G) - \bar{X}_G = \bar{X}_{rc} - \bar{X}_r - \bar{X}_c + \bar{X}_G]$. The sum of the squared interaction effects for each point is the sum of squares for interaction:

sum of squares for interaction

$$SS_{RC} = \sum(\bar{X}_{rc} - \bar{X}_r - \bar{X}_c + \bar{X}_G)^2 \tag{15.10}$$

For our data, $SS_{RC} = 930.20$.

Mean squares for rows, columns, interaction, and within cells are obtained by dividing the sums of squares by their appropriate degrees of freedom, as shown in Table 15.5. Then the three observed F ratios are computed (also shown in Table 15.5):

test statistic (F) for rows

$$F_{R_{obs}} = \frac{MS_R}{MS_W} \tag{15.11}$$

test statistic (F) for columns

$$F_{C_{obs}} = \frac{MS_C}{MS_W} \tag{15.12}$$

test statistic (F) for interaction

$$F_{RC_{obs}} = \frac{MS_{RC}}{MS_W} \tag{15.13}$$

The ANOVA summary table in Table 15.5 is the accepted way of summarizing the two-way ANOVA computations.

TABLE 15.5 Summary table for two-way ANOVA of therapy outcome by therapy type and therapist's experience

Source of variation	*Sum of squares (SS)*	*Degrees of freedom (df)*	*Mean square (MS)*	*F*
Rows (therapy type)	70.20	2	35.10	.91[a]
Columns (experience)	580.80	1	580.80	15.09[b]
RC interaction (type by experience)	930.20	2	465.10	12.08[c]
Within cell	924.00	24	38.50	
Total	2505.20	29		

[a] $p > .05, F_{cv_{2,24}} = 3.40$
[b] $p < .05, F_{cv_{1,24}} = 4.26$
[c] $p < .05, F_{cv_{2,24}} = 3.40$

IV. Interpret the Results

Statistically significant? Practically significant? State results clearly.

There are three null hypotheses in this experiment. For the row main effect, $F_{R_{obs}} = 35.10/38.50 = .91$, which we see at a glance is not significant because F_{cv} is always greater than 1 (in this case $F_{cv_{2,24}} = 3.40$), so we conclude that there is no significant difference between the three therapy types. For the column main effect, $F_{C_{obs}} = 580.80/38.50 = 15.09$, which, with $df = 1$ and 24, has a critical value of 4.26, so we do reject the column null hypothesis, and we conclude that therapist's experience *is* an important factor. For the interaction effect, $F_{RC_{obs}} = 465.10/38.50 = 12.08$ is significant ($F_{cv} = 3.40$ with $df = 2, 24$), so we conclude that the effectiveness of a particular therapy type *does* depend on whether the therapist has low or high experience.

Recall that the means for these data were displayed in Figure 15.1. It can be observed there that high-experience therapists in general have better outcomes than low-experience therapists; that is the significant main effect for therapist's experience. From that figure it is not at all clear that one therapy type is better than another, illustrating that the therapy type main effect is not significant. The distinct nonparallelism of Figure 15.1 illustrates the significant interaction, a subject we turn to next.

15.6 Interpreting Interaction

Because $F_{RC_{obs}} = 12.08$ was greater than the critical value $F_{cv_{2,24}} = 3.40$, we concluded that therapy outcome has a significant experience-by-type interaction. The existence of an interaction implies in general that the extent to which therapist's experience influences outcome depends on the type of therapy being performed. But the only way to examine a significant interaction in detail is to create a table of residuals like Table 15.6.

TABLE 15.6 Residuals[a] of therapy outcome data

		Therapist's experience		
		Low	*High*	*Row mean*
	RET	.60	−.60	0
Therapy type	BT	6.50	−6.50	0
	PCT	−7.10	7.10	0
Column mean		0	0	0

[a]Entries are the values from Table 15.2 after subtracting the row mean, subtracting the column mean, and adding the grand mean.

Recall that the null hypothesis for the interaction effect is $H_{0_{RC}}$: all $\mu_{row,col} - \mu_{row} - \mu_{col} + \mu_G = 0$. For each combination of experience and type, the best point-estimate of the interaction effect is the "residual": $\overline{X}_{row,col} - \overline{X}_{row} - \overline{X}_{col} + \overline{X}_G$. Thus, for example, the residual for the upper left (low experience/RET) cell in Table 15.2 is the mean of the low experience/RET cell minus the mean of all RET (top row) values minus the mean of all low-experience (first column) values plus the grand mean: $\text{residual}_{low,RET} = 55.0 - 58.8 - 55.0 + 59.40 = .60$. This value is shown in the upper left corner of Table 15.6. The other residuals in Table 15.6 are computed in the same way.

The residual in each cell is a measure of the extent to which the mean of that cell *cannot* be accounted for by the main effect for its row and the main effect for its column; that is, the residual is the measure of the magnitude of the interaction that affects that cell. Let us consider, for example, the upper left (RET/low) cell and what contributes to its magnitude. The grand mean for all the data is $\overline{X}_G = 59.4$, as shown in the lower right-hand corner of Table 15.2. The main effect for the top (RET) row, as estimated from the data, is $\overline{X}_{RET} - \overline{X}_G = 58.8 - 59.4 = -.60$, which implies that all the values in the RET row (both low and high experience) are, on average, .60 Mental Health Scale point below the grand mean.

Similarly, the main effect for the first (Low) column is $\overline{X}_{low} - \overline{X}_G = 55.0 - 59.4 = -4.40$, which implies that all the values in the Low column (including RET, BT, and PCT) are, on average, 4.40 Mental Health Scale points below the grand mean.

Combining those two main effects gives the value for the cell that is in both the RET row and the Low column. Thus, $\overline{X}_{RET,low}$ as accounted for by the main effects is $\overline{X}_G + (\overline{X}_{RET} - \overline{X}_G) + (\overline{X}_{low} - \overline{X}_G) = 59.4 + (-.60) + (-4.40) = 54.4$. Now the actual observed mean in that cell is $\overline{X}_{RET,low} = 55.0$, and the difference between the actual mean and the value accounted for by the main effects is the residual: $\text{residual}_{RET,low} = \overline{X}_{RET,low} - [\overline{X}_G + (\overline{X}_{RET} - \overline{X}_G) + (\overline{X}_{low} - \overline{X}_G)] = 55.0 - 54.40 = .60$, as shown in the upper left cell of Table 15.6. That value, .60, is the extent to which interaction (and experimental error) affects the RET/low cell. That is, $\overline{X}_{RET,low}$ is higher than is accounted for by the two main effects by .60 Mental Health Scale point.

Let us rearrange the equation for the residual:

$$\begin{aligned}\text{residual}_{RET,low} &= \overline{X}_{RET,low} - [\overline{X}_G + (\overline{X}_{RET} - \overline{X}_G) + (\overline{X}_{low} - \overline{X}_G)] \\ &= \overline{X}_{RET,low} - \overline{X}_{RET} - \overline{X}_{low} + \overline{X}_G\end{aligned}$$

That parallels for our sample the null hypothesis statement about the parameters of that cell: H_0: $\mu_{RET,low} - \mu_{RET} - \mu_{low} + \mu_G = 0$. Similar observations can be made for the five other cells in the data. Therefore, if the null hypothesis for interaction is true, then the cell means will for the most part be accounted for by the row and column effects, and so the residuals will be approximately 0. If the residuals are significantly different from 0, we reject the null hypothesis for interaction.

If we inspect Table 15.6, we see that the interaction effect is large for therapy type BT: Outcome in the BT/low cell is 6.50 points higher than would be accounted for by the BT row mean and the low-experience column mean alone, and outcome in the BT/high cell is 6.50 points lower than would be accounted for by the row and column means. Furthermore, the interaction effect is large for therapy type PCT: Outcome in the PCT/low cell is 7.10 points lower than would be accounted for by the main (row and column) effects, whereas outcome in the PCT/high cell is 7.10 points higher than expected. However, the interaction is small for RET: The residuals are only plus or minus .60 points.

Thus, we see that the interaction of therapist's level of experience with therapy type is important for BT and PCT but not for RET. This conclusion is *not* immediately apparent from an examination of Figure 15.1: The RET line crosses the other two lines and is not parallel to the x-axis. Why then does the residual matrix show that there is almost no interaction effect for RET? Because the increase in RET outcome from low- to high-experience therapists is almost entirely accounted for by the main effect for experience: RET outcome is better for more experienced therapists because experienced therapists produce better outcome regardless of therapy type.

To recapitulate: The general conclusion about interaction—that the extent to which therapist's experience influences outcome depends on the type of therapy being performed—actually applies primarily to only BT and PCT therapists. The only way we could see that refinement was to examine a table of residuals. (See Rosnow & Rosenthal, 1989.)

15.7 Comparing Two-way and One-way ANOVAs

As we noted in the opening of this chapter, we could perform two separate one-way ANOVAs on the data shown in Table 15.1. The first of these tests the null hypothesis that therapy types are equally effective. That null hypothesis was originally stated as Equation (15.1), where we called it the null hypothesis for the row main effect: H_{0_R}: $\mu_{RET} = \mu_{BT} = \mu_{PCT}$. The results of the one-way ANOVA to analyze that hypothesis are shown in Table 15.7.

TABLE 15.7 One-way ANOVA summary for therapy type

Source	SS	*df*	MS	F
Between	70.20	2	35.10	.39[a]
Within	2435.00	27	90.19	
Total	2505.20	29		

[a] $p > .05$, $F_{cv_{2,27}} = 3.35$

TABLE 15.8 One-way ANOVA summary for therapist's experience

Source	SS	*df*	MS	F
Between	580.80	1	580.80	8.45[a]
Within	1924.40	28	68.73	
Total	2505.20	29		

[a] $p < .05$, $F_{cv_{1,28}} = 4.20$

The second one-way ANOVA tests the null hypothesis that therapist's experience is not important. That null hypothesis was originally stated as Equation (15.2), where we called it the null hypothesis for column main effect: H_{0_C}: $\mu_{low} = \mu_{high}$. The summary table for the one-way ANOVA to analyze that hypothesis is shown in Table 15.8.

How do those one-way ANOVAs compare with each other and with the two-way ANOVA of Table 15.5? First, we note that SS_T and df_T are identical in all three ANOVAs. That should not surprise us because SS_T is the numerator of the variance calculation when any division into groups is ignored. The SS_T is a characteristic of the 30 data points, regardless of their order or the way they are displayed.

Second, the sum of squares, degrees of freedom, and mean square for therapy type are identical wherever they appear, which is in the "Rows" portion of the two-way ANOVA and the "Between" section of the one-way ANOVA for therapy type (Table 15.7). That is because that sum of squares depends on only the differences between the RET, BT, and PCT means, averaged across *all* the subjects who receive that respective type of therapy.

Similarly, the sum of squares, degrees of freedom, and mean square for therapist's experience are identical wherever they appear, which is in the "Columns" portion of the two-way ANOVA and the "Between" section of the one-way ANOVA for therapist's experience (Table 15.8).

Consider the therapy type one-way ANOVA in Table 15.7. Transforming that one-way analysis into the two-way analysis of Table 15.5 would involve including the main effect for experience and the interaction. Where do the sums of squares and degrees of freedom come from for this analysis? The answer is that the within groups sum of squares in the one-way design gets partitioned into the experience sum of squares, the interaction sum of squares, and the new (two-way) within cells sum of squares as shown in Figure 15.9, where we note that the "Groups" of Table 15.7 (therapy type) are the "Rows" of Table 15.5 (also therapy type). That is, the SS_W for the one-way ANOVA of Table 15.7 (2435.00) gets divided into the three pieces SS_C, SS_{RC}, and SS_W of Table 15.5; thus, $2435.00 = 580.80 + 930.20 + 924.00$. The degrees of freedom also partition in the same way: df_W of Table 15.7 (27) equals $df_C + df_{RC} + df_W$ of Table 15.5: $27 = 1 + 2 + 24$.

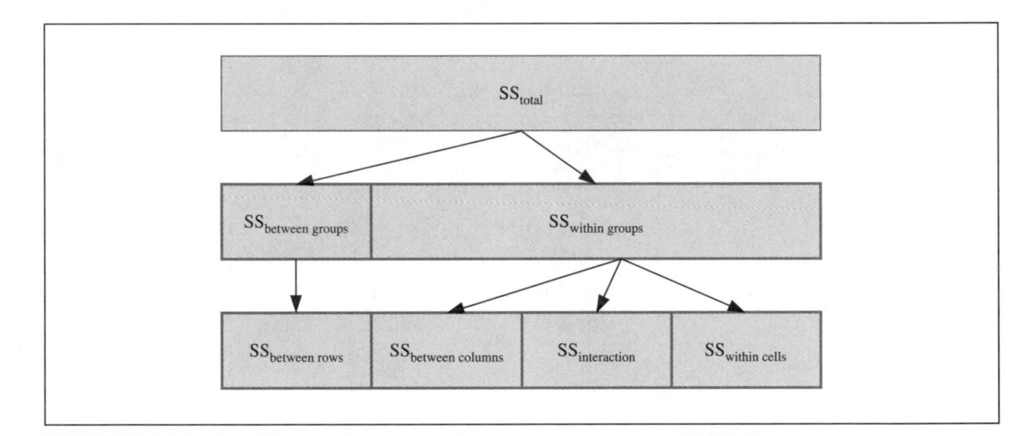

FIGURE 15.9 Partitioning the sum of squares in two-way ANOVA

The same is true if we begin with the one-way ANOVA for therapist's experience (Table 15.8). There, the SS_W (1924.40) is partitioned into the therapy type main effect sum of squares, the interaction, and the new within sum of squares of Table 15.5: $SS_{\text{within groups}} = 1924.40 = SS_R + SS_{RC} + SS_{\text{within cells}} = 70.20 + 930.20 + 924.00$. The degrees of freedom also partition: df_W of Table 15.8 (28) equals $df_R + df_{RC} + df_W$ of Table 15.5 ($28 = 2 + 2 + 24$).

There are two reasons to prefer a two-way ANOVA to two one-way ANOVAs. First, the two-way ANOVA allows the testing for an interaction, which is impossible with two one-way ANOVAs. Second, the SS_W is smaller in the two-way case (Table 15.5) than it is in either one-way ANOVA (Table 15.7 or 15.8). The decrease from SS_W (one-way) to SS_W (two-way) will be large if either or both of the added main effect and the interaction sum of squares are large. Because SS_W (when divided by df_W) becomes the denominator of the F ratios, it is desirable to have SS_W small. Thus, the power of the row effect test (likelihood of rejecting the row main effect null hypothesis) *increases* if the column effect (and/or interaction) is large, and vice versa.

15.8 Interpreting the Two-way ANOVA

Recall that MS_W is the pooled within-cell variance; that is true whether the ANOVA is one-way or two-way. Also recall that MS_W is the best point-estimate of σ^2. Now MS_W is almost always, as in our example, smaller in a two-way design than in either of the related one-way designs; in our example, $MS_{W_{\text{two-way}}} = 38.50$, whereas $MS_{W_{\text{one-way–therapy type}}} = 90.19$ and $MS_{W_{\text{one-way–experience}}} = 68.73$. How can the two-way estimate of σ^2 be so much smaller than either of the one-way estimates of the same thing? The answer is that the two-way MS_W is a *better*, or one might say *purer*, estimate of σ^2 than is either of the one-way MS_W values. MS_W in the two-way case has the effects of both therapy type *and* therapist's experience removed from it, whereas the two one-way MS_W have one or the other but not both removed. Thus, for example, MS_W in the experience one-way design is contaminated as an estimate of σ^2 by the inclusion of the experience effect.

There are eight possible outcomes of a two-way design: Each of three null hypotheses can be independently accepted and/or rejected. In Figures 15.2 through 15.8 we illustrated seven of these possibilities. The remaining possibility is the case where both main effects are significant and also the interaction is significant. The illustration for this case is left as an exercise.

Box 15.1 is a comment about eyeball-estimating the two-way ANOVA.

BOX 15.1 Eyeball-estimating the two-way ANOVA

The same techniques that we used to eyeball-estimate the one-way ANOVAs of Chapter 13 can be used to eyeball-estimate the main effects in two-way ANOVAs, but they are quite cumbersome to use, so we will not present them. The interested reader can follow the Chapter 13 steps to estimate the row and column main effects and the within-cell statistics and can expect accuracies of about $\pm 30\%$. The absence or presence of interaction can be eyeballed from the parallelism or lack of parallelism in plots such as those shown in Figures 15.4 through 15.8. The magnitude of interaction, and whether it is significant, is difficult, if not impossible, to eyeball directly.

15.9 Connections

Cumulative Review

We continue the discrimination practice begun in Chapter 10. You may recall the two steps:

Step 1: Determine whether the problem asks for (a) the area under a normal distribution, (b) a confidence interval, or (c) the evaluation of a hypothesis. See the Cumulative Review in Section 10.5 for hints about this step.

Step 2: If the problem asks for the evaluation of a hypothesis, consult the summary of hypothesis testing in Table 15.9 to determine which kind of test is appropriate.

Table 15.9 shows that so far we have eight basic kinds of hypothesis tests (the one-sample test for the mean, two-independent-samples test for means, two-dependent-samples tests for means, three-or-more-independent-samples tests, post hoc tests, a priori tests, three-or-more-repeated-measures tests, and now two-way ANOVA). This table is identical to Table 14.11 with the addition of the last three lines appropriate for this chapter. We will continue to expand this table progressively throughout the remainder of the textbook.

TABLE 15.9 Summary of hypothesis-testing procedures in Chapters 10–15

Design	*Chapter*	*Null hypothesis*	*Sample statistic*	*Test statistic*	*Effect size index*
One sample					
σ known	10	$\mu = a$	$\overline{X}$	z	$d = \frac{\overline{X}_{obs} - \mu}{\sigma}$
σ unknown	10	$\mu = a$	$\overline{X}$	t	$d = \frac{\overline{X}_{obs} - \mu}{s}$
Proportion	Exercise 7, Chapter 10	$\Pi = a$	p	t	—[a]
Two independent samples					
For means	11	$\mu_1 = \mu_2$	$\overline{X}_1 - \overline{X}_2$	t	$d = \frac{\overline{X}_{1_{obs}} - \overline{X}_{2_{obs}}}{s_{pooled}}$
Two dependent samples					
For means	12	$\mu_D = 0$	$\overline{D}$	t	$d = \frac{\overline{D}_{obs}}{s_D}$
k independent samples					
For means (omnibus)	13	$\mu_1 = \mu_2 = \cdots = \mu_k$	n/a	F	$d_M = \frac{\overline{X}_{max} - \overline{X}_{min}}{s_{pooled}}$ or $f = \frac{s_{\overline{X}}}{s_{pooled}}$ or $R^2 = \frac{SS_B}{SS_T}$
Post hoc	14	$\mu_i = \mu_j$	$\overline{X}_i - \overline{X}_j$	Q	—
A priori	14	$C_1\mu_1 + \cdots + C_k\mu_k = 0$	$C_1\overline{X}_1 + \cdots + C_k\overline{X}_k$	t	$d = \frac{C_1\overline{X}_1 + \cdots + C_k\overline{X}_k}{s_{pooled}}$
k dependent samples					
For means	14	$\mu_1 = \mu_2 = \cdots = \mu_k$	n/a	F	$d_M = \frac{\overline{D}_{max}}{\sqrt{2MS_{residual}}}$
RC independent samples					
For row means	15	$\mu_{r1} = \mu_{r2} = \cdots = \mu_R$	n/a	F	—
For column means	15	$\mu_{c1} = \mu_{c2} = \cdots = \mu_C$	n/a	F	—
Interaction	15	$\mu_{rc} - \mu_r - \mu_c + \mu_G = 0$	n/a	F	—

[a] Effect size index is not discussed in this textbook.

Journals

The interpretation of interaction is often misreported in the literature. Rosnow and Rosenthal (1989) report that only about 1% of all journal articles that discuss interaction make explicit reference to the examination of residuals, even though such examination is necessary to the proper interpretation of the interaction effect. The correct residual-examination procedure is described in Section 15.6.

A typical journal report of a two-way ANOVA would involve presenting a table of means, such as our Table 15.2, and then stating, "Mean therapy outcome scores are presented in Table 15.2. There was no significant effect for therapy type, $F(2, 24) = .91$, *ns*, but therapist's experience, $F(1, 24) = 15.09$, $p < .05$, and the interaction between therapy type and therapist's experience, $F(2, 24) = 12.08$, $p < .05$, were statistically significant.

Inspection revealed that more experienced therapists in general produced better outcomes. Inspection of the table of residuals revealed that high experience improved PCT therapists' outcomes but worsened BT therapists' outcomes."

Computers

ESTAT will not compute a two-way ANOVA. Use ESTAT to perform the two one-way ANOVAs, and use the observations of Section 15.7 to get the correct two-way sums of squares and mean squares.

Use SPSS to compute the two-way ANOVA for the therapy outcome data. The data are shown in Table 15.1.

1. Enter the five RET/Low values from the upper left cell of Table 15.1 into the first column of the SPSS spreadsheet. For each value, enter a "1" into the second column of the SPSS spreadsheet (the second column will hold the independent variable that defines the row). Also for each value, enter a "1" into the third column of the SPSS spreadsheet (the third column will hold the independent variable that defines the column).
2. Enter the five values from the BT/Low cell into the first column of the SPSS spreadsheet beginning in row 6. For each value, enter a "2" into the second column of the SPSS spreadsheet (indicating that these values are from the second "BT" row). Also for each value, enter a "1" into the third column of the SPSS spreadsheet (indicating that these values are from the first "Low" column).
3. Enter the five values from the PCT/Low cell into the first column of the SPSS spreadsheet beginning in row 11. For each value, enter a "3" into the second column of the SPSS spreadsheet (indicating that these values are from the third "PCT" row). Also for each value, enter a "1" into the third column of the SPSS spreadsheet (indicating that these values are from the first "Low" column).
4. Enter the five values from the RET/High cell into the first column of the SPSS spreadsheet beginning in row 16. For each value, enter a "1" into the second column of the SPSS spreadsheet (indicating that these values are from the first "RET" row). Also for each value, enter a "2" into the third column of the SPSS spreadsheet (indicating that these values are from the second "High" column).
5. Enter the five values from the BT/High cell in the first column beginning in row 21, with "2" and "2" in the second and third columns, respectively.
6. Enter the five values from the PCT/High cell in the first column beginning in row 26, with "3" and "2" in the second and third columns, respectively.
7. Click `Statistics`, then `ANOVA Models`, and then `Simple Factorial....`
8. Click the upper ▶ to move `var00001` into the `Dependent:` window.
9. Click `var00002`; then click the middle ▶ to move it into the `Factor(s):` window.
10. Click `var00003`; then click the middle ▶ to move it into the `Factor(s):` window.
11. Click `var00002(? ?)`; then click `Define Range...`; then enter "1" in `Minimum` and "3" in `Maximum`. Click `Continue`.
12. Click `var00003(? ?)`; then click `Define Range...`; then enter "1" in `Minimum` and "2" in `Maximum`. Click `Continue`.
13. Click `OK`.
14. The ANOVA summary table shown on the following page appears in the `!Output1` window. With my annotations, it is self-explanatory.

Homework Tips

1. Check the list of learning objectives at the beginning of this chapter. Do you understand each one?

2. Staying organized and keeping track of which is the row variable and which is the column variable are the keys to easy computation.

```
SPSS for MS WINDOWS Release 6.1

            * * *  A N A L Y S I S   O F   V A R I A N C E  * * *

                 VAR00001    Dependent variable
            by   VAR00002    Row independent variable
                 VAR00003    Column independent variable

                 UNIQUE sums of squares                 } Not covered in this text
                 All effects entered simultaneously     }

                                                  Sum of            Mean                 Sig
Source of Variation                               Squares    DF     Square        F      of F

Main Effects   R + C (not covered in this text)   651.000     3    217.000    5.636     .005
   VAR00002             Rows                       70.200     2     35.100     .912     .415
   VAR00003             Columns                   580.800     1    580.800   15.086     .001

2-Way Interactions      Interaction               930.200     2    465.100   12.081     .000
   VAR00002 VAR00003                              930.200     2    465.100   12.081     .000

Explained R + C + RC (not covered in this text)  1581.200     5    316.240    8.214     .000

Residual                Within cells              924.000    24     38.500

Total                   Total                    2505.200    29     86.386
```

Exercises for Chapter 15

Section A: Basic exercises (answers in Appendix D, page 525)

1. Suppose you are interested in convincing high school students to avoid taking drugs, and you prepare three films to be shown to the students. You wish to know whether there is any difference in the films' effectiveness in persuading students to avoid drugs. You select a group of 30 students, 15 male and 15 female, and randomly assign each to one of three groups. Each group will then see one of the films, and students will rate themselves on the drug attitude scale (DAS) you have designed. Previous research has shown that a film might differentially affect males and females, so you note each subject's gender. Thus, five males and five females view each of the films. The DAS cell means for the data you collect are given in the table (high DAS scores indicate a high tendency to avoid drugs).

		Male	*Female*
	A	88.0	68.6
Film	B	81.4	78.4
	C	80.2	83.4

   (a) Prepare a plot of mean DAS score by film and another plot of mean DAS score by gender (analogous to Figure 15.1).
   (b) From these plots, does it appear that there is a main effect for film? (In Exercise 4, we will perform the computation.) If so, describe it in plain English. Which plot shows this more clearly, the one with film on the X-axis or the one with gender on the X-axis?
   (c) From these plots, does it appear that there is a main effect for gender? (In Exercise 4, we will perform the computation.) If so, describe it in plain English. Which plot shows this more clearly, the one with film on the X-axis or the one with gender on the X-axis?
   (d) From these plots, does it appear that there is an interaction? (In Exercise 4, we will perform the computation.) If so, describe it in plain English. Which plot shows this more clearly, the one with film on the X-axis or the one with gender on the X-axis?

2. (*Exercise 1 continued*)
   (a) State in plain English and in symbols the null hypothesis for the main effect of film.
   (b) State in plain English and in symbols the null hypothesis for the main effect of gender.

(c) State in plain English and in symbols the null hypothesis for interaction in the male/film A cell.
(d) State in symbols the general statement for the null hypothesis for interaction.

3. (*Exercise 1 continued*)
(a) How many total degrees of freedom are there in this problem?
(b) How many degrees of freedom are there for the main effect of film?
(c) How many degrees of freedom are there for the main effect of gender?
(d) How many degrees of freedom are there for interaction?
(e) How many degrees of freedom are there within each cell?
(f) How many degrees of freedom are there within all cells combined?
(g) Does the sum of parts (b), (c), (d), and (f) equal (a)? Should it?

4. (*Exercise 1 continued*) The DAS scores are shown in the accompanying table. (You should verify that these data do in fact have cell means as shown in the table of Exercise 1.)

	Males			Females		
	Film A	Film B	Film C	Film A	Film B	Film C
	80	82	78	75	77	86
	86	79	91	68	77	82
	90	84	71	71	80	84
	93	76	83	71	82	79
	91	86	78	58	76	86
Labor savings:						
$\sum X$	440	407	401	343	392	417

(a) Provide tabulations of the data similar to Table 15.4.
(b) Provide the complete ANOVA summary table.
(c) State your main effect conclusions in plain English.
(d) Is there a significant interaction? If so, prepare a table of residuals (like Table 15.6) and describe the interaction effect in plain English.

5. Suppose we have a 4 × 3 factorial design with one dependent variable (called Y) and two independent variables A (with four levels a_1, a_2, a_3, and a_4) and B (with three levels b_1, b_2, and b_3). We have three observations in each cell, as shown in the accompanying table.

		A			
		a_1	a_2	a_3	a_4
	b_1	35	40	45	50
		55	60	65	70
		15	20	25	30
B	b_2	37.5	42.5	47.5	52.5
		57.5	62.5	67.5	72.5
		17.5	22.5	27.5	32.5
	b_3	40	45	50	55
		60	65	70	75
		20	25	30	35

(a) Prepare a plot of mean Y score by A, and another plot of mean Y score by B (analogous to Figure 15.1).
(b) From these plots, does it appear that there is a main effect for A? (In Exercise 8, we will perform the computation.) If so, describe it in plain English.
(c) From these plots, does it appear that there is a main effect for B? (In Exercise 8, we will perform the computation.) If so, describe it in plain English.
(d) Consider carefully the representation of interaction in these plots. Would you expect the interaction sum of squares (and mean square and F) to be small, or would you expect it to be exactly 0? Why?

6. (*Exercise 5 continued*)
(a) State in plain English and in symbols the null hypothesis for the main effect of A.
(b) State in plain English and in symbols the null hypothesis for the main effect of B.
(c) State in plain English and in symbols the null hypothesis for interaction for the cell in the a_2 column and the b_3 row.
(d) State in symbols the general statement for the null hypothesis for interaction.

7. (*Exercise 5 continued*)
(a) How many total degrees of freedom are there in this problem?
(b) How many degrees of freedom are there for the main effect of A?
(c) How many degrees of freedom are there for the main effect of B?
(d) How many degrees of freedom are there for interaction?
(e) How many degrees of freedom are there within each cell?
(f) How many degrees of freedom are there within all cells combined?
(g) Does the sum of parts (b), (c), (d), and (f) equal (a)? Should it?

8. (*Exercise 5 continued*) Compute the ANOVA for the data of Exercise 5.
(a) Provide the complete ANOVA summary table.
(b) If your interpretation in Exercise 5(b) was that there is a main effect for A, how do you account for your results here?
(c) If your interpretation in Exercise 5(c) was that there is a main effect for B, how do you account for your results here?
(d) Reconsider why the interaction sum of squares turned out to be exactly 0.

9. (*Exercise 5 continued*) Note that the data in each cell in Exercise 5 were created according these rules: The first value is the cell mean, the second value is the cell mean plus 20, and the third value is the cell mean minus 20. The data shown in the accompanying table have cell means identical to those in the table of Exercise 5. However, now the second and third values are plus and minus 1 (instead of plus and minus 20). If you were to plot mean Y

score by A and also mean Y score by B, how would these plots be similar to or different from those you prepared in Exercise 5?

		A			
		a_1	a_2	a_3	a_4
	b_1	35	40	45	50
		36	41	46	51
		34	39	44	49
B	b_2	37.5	42.5	47.5	52.5
		38.5	43.5	48.5	53.5
		36.5	41.5	46.5	51.5
	b_3	40	45	50	55
		41	46	51	56
		39	44	49	54

10. For the data of Exercise 9, which of the following values would you expect to change from the results you computed in Exercise 5?
    (a) Sum of squares for main effect of A
    (b) Sum of squares for main effect of B
    (c) Sum of squares for interaction between A and B
    (d) Sum of squares within cells
    (e) Degrees of freedom for main effect of A
    (f) Degrees of freedom for main effect of B
    (g) Degrees of freedom for interaction
    (h) Degrees of freedom within cells
    (i) Mean square for main effect of A
    (j) Mean square for main effect of B
    (k) Mean square for interaction
    (l) Mean square within cells
    (m) F for main effect of A
    (n) F for main effect of B
    (o) F for interaction

Section B: Supplementary exercises

11. A factory is about to begin production of a new widget. The industrial psychologist wishes to investigate the effect of three methods of training on workers' widget production. He takes 45 factory workers and randomly assigns each to one of three groups. Group A receives videotaped instruction on how to assemble widgets, group B receives a live lecture/demonstration on how to assemble widgets, and group C receives no formal training but is given time with the machines and allowed to experiment for themselves. Then widget production is begun and the numbers of widgets each worker produces is recorded. The numbers of widgets are shown in the accompanying table. Does method of training make a difference?

    I. State the null and alternative hypotheses
    (a) What is the null hypothesis? The alternative hypothesis?
    (b) Is this test directional or nondirectional? Why?
    (c) What is the test statistic? Sketch the distribution of the test statistic assuming the null hypothesis is true.

	Group A	Group B	Group C
	27	22	19
	21	32	14
	33	24	24
	29	30	21
	27	27	17
	21	24	26
	29	27	24
	25	30	22
	20	22	20
	30	32	28
	23	25	36
	20	29	22
	26	23	27
	19	31	29
	27	27	31
Labor savings:			
$\sum X$	377	405	360

II. Set the criterion for rejecting H_0
(d) What is the level of significance?
(e) How many degrees of freedom are there?
(f) Look up the critical value(s) of the test statistic. Enter it (them) on the appropriate distribution and shade the rejection region(s).

III. Collect a sample and compute statistics
(g) Compute the observed value of the test statistic. Enter it on the appropriate distribution.

IV. Interpret the results
(h) Should we reject H_0?
(i) Interpret your result.

12. (*Exercise 11 continued*) Plot mean widget production as a function of type of training. From this plot, does it appear that there is a significant effect of training type? Explain in light of your results in Exercise 11.
13. (*Exercise 11 continued*) Suppose the psychologist varied not only the type of training but also the *length* of training, so that of each group's 15 workers, 5 received training for 2 hours, 5 received training for 4 hours, and 5 received training for 6 hours. Note that the numbers of widgets in the accompanying table are identical to those from Exercise 11 except that each group is now divided into three subgroups. Do training type and training duration affect widget production?
    (a) Compute the cell means and prepare a plot of widget production as a function of training type for each duration. Also prepare a plot of widget production as a function of duration of training for each training type.
    (b) From these plots, does it appear that there is a main effect for training type? If so, describe it in plain English.
    (c) From these plots, does it appear that there is a main effect for duration of training? If so, describe it in plain English.
    (d) From these plots, does it appear that there is an interaction? If so, describe it in plain English.

Table for Exercise 13

Length	Group A	Group B	Group C
2 hours	27	22	19
	21	32	14
	33	24	24
	29	30	21
	27	27	17
4 hours	21	24	26
	29	27	24
	25	30	22
	20	22	20
	30	32	28
6 hours	23	25	36
	20	29	22
	26	23	27
	19	31	29
	27	27	31

14. (*Exercises 11 and 13 continued*)
    (a) State the main and alternative hypotheses for the main effect of training type.
    (b) State the main and alternative hypotheses for the main effect of training duration.
    (c) State the main and alternative hypotheses for the interaction. State both in symbols and in plain English.
    (d) How many total degrees of freedom are there in this problem? How does that compare to the total degrees of freedom in Exercise 11?
    (e) How many degrees of freedom are there for the main effect of training type in this problem? How does that compare to the degrees of freedom "between groups" in Exercise 11?
    (f) How many degrees of freedom are there within cells in this problem? How does that compare to the degrees of freedom within groups in Exercise 11? How do you account for the difference?
    (g) How many degrees of freedom are there for the main effect of training duration?
    (h) How many degrees of freedom are there for interaction?
    (i) Does the sum of the degrees of freedom for training type, training duration, interaction, and within cells equal the total degrees of freedom? Should it?

15. (*Exercises 11, 13, and 14 continued*)
    (a) Compute the two-way ANOVA for these data and show the complete ANOVA summary table.
    (b) How do the sum of squares and mean square for training type compare to the results of Exercise 11?
    (c) Show that the within-cell sum of squares of Exercise 11 has been partitioned into the sums of squares for duration, interaction, and within cells.
    (d) What is your conclusion regarding training type? State it in plain English. How does that compare to your conclusion in Exercise 11? Account for any difference.
    (e) What is your conclusion regarding training duration? State it in plain English.
    (f) What is your conclusion regarding interaction? If there is a significant interaction, prepare a table of residuals (like Table 15.6) and describe the interaction effect in plain English.

16. A farmer grows corn. He can use any of four kinds of hybrid corn seed (A, B, C, and D) and either of two kinds of fertilizer (X and Y). He divides his farm into 32 plots and randomly assigns corn type and fertilizer to each plot. At harvest, he records the output in bushels from each plot as shown in the table. Are there differences between the corn and fertilizer types?

Fertilizer	Corn type A	B	C	D
X	104	118	98	126
	114	109	108	115
	118	136	99	131
	125	138	97	115
Y	108	125	93	113
	105	115	114	123
	104	112	97	116
	128	108	93	121
Partial labor savings: $\sum X$	906	961	799	960

    (a) State the null and alternative hypotheses for both main effects and for interaction.
    (b) Provide the complete ANOVA summary table.
    (c) State your results in plain English.

Section C: Extending your comprehension (answers in Appendix D, page 527)

17. Is it possible to create a data set for a two-way ANOVA where $SS_R < 0$ or $SS_C < 0$ or $SS_{RC} < 0$?

18. Is it possible to create a data set with three columns and six members per column such that: if it is subjected to a one-way analysis of variance at $\alpha = .05$, the null hypothesis will *not* be rejected; but if each column is divided into two (row) groups of three members each and subjected to a two-way ANOVA, the main effect for columns *will* be rejected at $\alpha = .05$? If so, create such a data set and explain its characteristics.

19. Is it possible to create a data set with three columns and six members per column such that: if it is subjected to a one-way analysis of variance at $\alpha = .05$, the null hypothesis *will* be rejected; but if each column is divided into two (row) groups of three members each and subjected to a two-way ANOVA, the main effect for columns will *not* be rejected at $\alpha = .05$? If so, create such a data set and explain its characteristics.

20. Is it possible to create a data set with three columns and two rows, three members per cell, such that both $SS_R = 0$ and $SS_C = 0$, but with a significant interaction? If so, create such a data set and explain its characteristics.

Section D: Cumulative review (answers in Appendix D, page 527)

For Exercises 21–29, choose the correct answer or fill in the blank. (**Note that computations are *not* required;** use $\alpha = .05$ unless otherwise instructed.)

(a) This problem requires:
(1) Finding the area under a normal distribution [Skip parts (b)–(g) and proceed to (h).]
(2) Creating a confidence interval
(3) Testing a hypothesis about the mean of one group
(4) Testing a hypothesis about the means of two independent groups
(5) Testing a hypothesis about the means of two dependent groups
(6) Testing a hypothesis about the means of three or more independent groups
(7) Testing hypotheses about means using post hoc tests
(8) Testing an a priori hypothesis about means (planned comparison)
(9) Testing a hypothesis about the means of three or more dependent groups
(10) Testing hypotheses for main effects and interaction
(11) Performing a power analysis (involving finding the area under a normal distribution) [Skip parts (b)–(h).]

(b) The null hypothesis is of the form:
(1) $\mu = a$ (2) $\mu_1 = \mu_2$
(3) $\mu_D = 0$ (4) $\mu_1 = \mu_2 = \cdots = \mu_k$
(5) A series of null hypotheses of the form $\mu_i = \mu_j$. (State how many such hypotheses.)
(6) One null hypothesis of the form $C_1\mu_1 + C_2\mu_2 + \cdots + C_k\mu_k = 0$
(7) Three null hypotheses:
rows: $\mu_{\text{row }1} = \mu_{\text{row }2} = \cdots = \mu_{\text{row }R}$;
columns: $\mu_{\text{col }1} = \mu_{\text{col }2} = \cdots = \mu_{\text{col }C}$; and
interaction: all $\mu_{\text{row,col}} - \mu_{\text{row}} - \mu_{\text{col}} + \mu = 0$
(8) There is no null hypothesis.

(c) The appropriate sample statistic(s) is (are):
(1) $\overline{X}$ (2) p
(3) $\overline{X}_1 - \overline{X}_2$ (4) $\overline{D}$
(5) $\overline{X}_i - \overline{X}_j$
(6) $C_1\overline{X}_1 + C_2\overline{X}_2 + \cdots + C_k\overline{X}_k$
(7) Not applicable for ANOVA

(d) The appropriate test statistic(s) is (are):
(1) z (2) t
(3) F (4) Q

(e) The number of degrees of freedom for this test statistic is ________. (State the number or *not applicable*; if there are multiple null hypotheses, state the number of degrees of freedom for each.)

(f) The hypothesis test (or confidence interval) is:
(1) Directional (2) Nondirectional

(g) The critical value(s) of the test statistic(s) is (are) ________. (Give the value. If there are multiple null hypotheses, state the critical value for each.)

(h) The appropriate formula(s) for the test statistic(s) (or confidence interval) is (are) given by Equation (________). (If there are multiple null hypotheses, state the equation number for each.)

Computations are *not* required for any of the problems in Section D. See the instructions.

21. To estimate the number of fish in a particular river, a biologist throws a net into the river, pulls it up, and counts the number of fish caught in the net. She then empties the net and repeats the procedure several more times. The biologist knows that the mean number of fish caught in such nettings in the East Fork River is 37 with standard deviation 6.5. Last year, an oil tank truck crashed on a bridge and spilled part of its contents into the river. The biologist wishes to know whether this year's counts are lower than those in previous years. She selects ten places on the river at random and casts her net in each place. The counts are: 28, 30, 35, 25, 42, 35, 28, 35, 33, 33. Has the number of fish in the East Fork River decreased?

22. Physicians are investigating the treatment of a particular virus. They have three analgesics available: aspirin, acetaminophen, and ibuprofen. Furthermore, they wish to know whether to recommend eating or fasting during the course of this illness. They prepare a 10-point rating scale to measure distress, with 10 being high distress and 0 being no distress. They assign 30 patients randomly to these conditions. The distress ratings are shown in the table. Do the treatments have any differential effect?

	Aspirin	Aceto.	Ibupr.
Eat	4	2	1
	4	4	3
	8	3	4
	5	6	1
	5	1	8
Fast	9	0	4
	1	2	0
	10	1	8
	3	4	10
	6	0	5

23. Dr. Dixon wishes to compare three memory encoding strategies: deep processing encoding, verbal encoding, and visual encoding. She enlists eight subjects and trains all of them in all three strategies. Each subject reads three complicated stories with instructions to use deep processing encoding for one, verbal encoding for another, and visual encoding for the third. Then the subject is tested about the details in the story. The percentages of correct

responses for each subject in each of the three encoding conditions are shown in the table. Is there a difference in the effectiveness of the encoding strategies?

Subject	Deep processing	Verbal	Visual
1	66	58	58
2	51	56	55
3	68	47	56
4	47	66	62
5	67	51	72
6	51	78	71
7	64	48	64
8	58	73	67

24. A milk wholesaler purchases milk from four dairy farms. She wishes to know whether the fat content in the milk differs from farm to farm. She selects ten days at random and measures the fat content in samples from each of the four farms (40 samples in all). She conducts an analysis of variance and finds a significant F. Now she wishes to know which farms are significantly different from which other farms. How should she proceed?

25. The telephone company is going to install telephone booths in airports. It has two kinds of booths: those with seats and those where the user stands. The company wishes to know whether long-distance calls made from sit-down booths have longer or shorter durations than those made from stand-up booths. It installs a sit-down and a stand-up booth side by side and records the durations (in minutes) of 15 long-distance calls from each booth. Is the duration of sit-down calls different from that of stand-up calls?

    Sitting down: 21.4, 6.0, 27.1, 6.2, 15.8, 14.3, 6.9, 1.0, 13.6, 28.2, 16.8, 14.1, 15.8, 26.1, 30.9

    Standing up: 4.4, 13.3, 10.0, 20.9, 22.3, 17.7, 14.1, 4.4, 13.6, 21.3, 14.1, 2.0, 22.6, 9.5, 4.3

26. A pollster wishes to know the proportion of voters who favor candidate Smith. She contacts 300 voters chosen at random and asks whether they will vote for Smith. Of these 300, 174 say yes. What can she say about the percentage of voters who favor Smith?

27. Dr. Skan studies the relationship of schizophrenia to the brain. Of particular interest is the size of the ventricles (the fluid-filled regions). She obtains three groups of eight individuals: schizophrenic, depressed, and normal. The sizes of the ventricles (in millimeters) are listed in the table.

Schizophrenic	Depressed	Normal
8	12	11
11	9	17
12	12	10
9	7	7
13	12	8
14	13	12
9	13	12
14	15	12

Previous research has indicated that the size of ventricles of schizophrenics is different from the size in depressed and normal individuals. Is that true in this sample?

28. A dietician wishes to know whether his lecture on increasing protein intake at breakfast time has an effect. He has a group of 20 patients record their breakfast intake for a week. Then he gives the lecture and asks the same patients to record their breakfast intake for the next week. The dietician determines the total protein intake in both conditions. The data on protein intake (in grams) are given in the table. Has the lecture increased protein intake?

Patient	Before	After
1	21	37
2	30	36
3	36	50
4	26	51
5	39	56
6	36	28
7	30	47
8	38	41
9	54	57
10	17	38
11	38	42
12	31	48
13	28	37
14	40	51
15	43	60
16	43	53
17	39	59
18	24	49
19	33	36
20	39	41

29. Life expectancy in a particular culture has mean 75 years and standard deviation 9.5 years. What percentage of individuals live 90 years or longer?

Section E: From the journals (answers in Appendix D, page 527)

30. Rich and Dahlheimer (1989) were interested in the thinking patterns of irrational versus rational persons, and they tested the prediction that "negative thinking is often used as a strategy for protecting self-esteem and that this strategy often facilitates performance in evaluative [that is, success or failure] situations" (p. 15). They administered the Rational Behavior Inventory to 146 undergraduate students; the 16 students who scored highest formed the "high irrational" group, and the 16 students who scored lowest formed the "low irrational" group. These subjects were given a word construction task that included six stimulus words. The task for each stimulus word was to form as many new words as possible (e.g., rearrange the letters of the word TURKEY to form KEY, RUT, etc.). Subjects were given 2 minutes to generate new words for each of the six stimulus words.

"In order to manipulate success–failure, subjects were told that the purpose of the task was to see if they could construct two special words out of the letters of each stimulus word that they worked on. Specifically, they were asked to construct the most common word listed by others and also a creative word, a word listed by only 15% of the people who worked on the task" (p. 19). Subjects in the "success" condition were told that they had succeeded in five of the six stimulus words, performance that was well above the average; subjects in the "failure" condition were told they had succeeded only once, which was well below the average. Half the subjects were in the success condition; the other half were in the failure condition.

Did such a manipulation alter individuals' self-esteem? To measure self-esteem, Rich and Dahlheimer used the Adjective Generation Technique (AGT): "Subjects are directed to spontaneously generate five adjectives that were descriptive of them at the present moment" (page 19). Each adjective was later assigned a numerical rating of favorability based on norms that have been established for over 1300 words. Self-esteem scores were derived by summing the favorability ratings over the five words. They administered the AGT twice, once before and once after the success–failure feedback, and they computed the difference between the two scores for each subject. This difference formed the dependent variable in their design, but we will consider here only a preliminary test that Rich and Dahlheimer performed on the "pre" scores.

Although Rich and Dahlheimer randomly assigned subjects to the success and the failure feedback groups, it was possible that the four groups (high irrational/success, low irrational/success, high irrational/failure, low irrational/failure) differed before training began. In order to rule out this hypothesis, they subjected the self-esteem (AGT) "pre" scores to a two-way analysis of variance. The four group means on the AGT were $\overline{X}$ = 307, 332, 303, and 312 (these values are interpolated from a graphical presentation). If these means were significantly different from one another, then any subsequent finding could be the result of initial self-esteem differences between the groups, not the result of experimental manipulation.

Rich and Dahlheimer reported that "a 2×2 (irrationality $\times$ feedback) analysis of variance of the pretest scores indicated no significant differences between the groups" (p. 23).

(a) Do Rich and Dahlheimer provide their degrees of freedom for you to check? If not, how many degrees of freedom are there?
(b) Is enough information presented to allow you to know whether F was close to 1 or close to the critical value of F? Might that distinction be important?
(c) Do you have enough information to compute MS_R, MS_C, MS_{RC}, or MS_W?

31. (*Exercise 30 continued*) Suppose you replicate Rich and Dahlheimer's (1989) study and obtain the pretest data shown in the accompanying table.

	Irrationality	
	High	*Low*
Success	184	434
	266	362
	329	276
	332	361
	321	284
	557	340
	192	252
	388	332
Failure	202	346
	215	397
	328	294
	328	374
	286	438
	258	331
	223	411
	201	368

(a) Perform the two-way analysis of variance and provide a complete ANOVA summary table.
(b) State your results in plain English.

32. Swoope and Johnson (1988) examined "the effects of text-based (text structure) and reader-based (prior knowledge) factors on judgments of importance of information in expository text" (p. 5). They presented subjects with an expository passage that described an imaginary insect that they called the "E-bug." Subjects were randomly divided into groups that differed in two ways: in subjects' prior knowledge of insects (trained vs. untrained) and in the text structure of the expository passage (elaborated vs. unelaborated) that they would read. There were thus four groups: trained/elaborated, trained/unelaborated, untrained/elaborated, and untrained/unelaborated. The researchers forced group size to be 31 by randomly discarding subjects in groups that exceeded 31 subjects. Following the reading of the expository text, students rated the importance of each item in a 12-item list. The sum of these importance ratings provided the dependent measure in this experiment. The mean and standard deviation for the importance rating scores for the unelaborated/untrained group were $\overline{X} = .26$ and $s = .58$; for the elaborated/untrained group, $\overline{X} = .61$ and $s = .84$; for the unelaborated/trained group, $\overline{X} = .71$ and $s = .69$; and for the elaborated/trained group, $\overline{X} = 1.03$ and $s = 1.14$.

Swoope and Johnson reported that "the [2×2] analysis [of variance] showed significant main effects for both prior knowledge training [$F(1, 120) = 8.33$, $p = .005$] and text elaboration [$F(1, 120) = 5.04$, $p = .027$]. The training by elaboration interaction [$F(1, 120) = .01$, $p = .915$] was not significant" (p. 8).

(a) Are the degrees of freedom computed correctly for the F tests?
(b) Do you have enough information to perform the 2×2 analysis of variance? If so, compute it.
(c) How do your results compare with Swoope and Johnson's?

(d) Are the standard deviations within each group reported correctly?

33. (*Exercise 32 continued*) Suppose we replicate Swoope and Johnson's (1988) study with ten subjects in each group; the data are shown in the table.

	Trained	*Untrained*
Unelab.	.1	.6
	1.0	.9
	1.1	.0
	1.2	1.9
	1.3	.1
	.1	.1
	.4	.1
	.2	1.2
	.3	.7
	.2	1.0
Elab.	.8	.1
	.3	1.8
	1.3	.1
	.8	.7
	1.2	.9
	.1	.9
	1.1	.8
	.1	1.5
	1.9	.1
	.4	.8

(a) Compute the two-way ANOVA and show the summary table.
(b) State your results in plain English.

Section F: Computer explorations

34. Enter the data of Table 15.1 and perform the two-way ANOVA described in the text. Verify that the computations in the text are correct.

35. Consider adding the constant 1.0 to all the values in the first (Low) column of Table 15.1. Predict how that will change the cell, row, and column means that are displayed in Table 15.2. Predict how it will change the F ratios for rows, for columns, and for interaction. Then have the computer perform the additions and the two-way ANOVA and verify your predictions.

36. What constant would you have to add to each value in the first (Low) column of Table 15.1 to make the F ratio for the column effect equal 0? Have the computer verify your prediction by actually adding that constant and performing the two-way ANOVA.

37. How would you alter the data to make the F ratio for rows equal 0? Have the computer verify your prediction by actually altering the data and performing the two-way ANOVA.

38. How would you alter the data to make the F ratio for interaction equal 0? Have the computer verify your prediction by actually altering the data and performing the two-way ANOVA.

Measures of the Relationship Between Two Variables: Correlation

CHAPTER

LEARNING OBJECTIVES

1. What type of information is provided by a correlation coefficient?
2. What is a scatter diagram or scatterplot?
3. How can correlation be inferred from a scatter diagram?
4. What are the possible values of a correlation coefficient?
5. What correlation coefficient is used with interval/ratio data?
6. How is the Pearson r correlation coefficient calculated?
7. How is a correlation coefficient eyeball-estimated?
8. What two factors affect the size of r?
9. Why doesn't correlation imply causation?
10. How are hypotheses tested about Pearson's r?
11. What correlation coefficient is used with ordinal data?
12. How are hypotheses tested about Spearman's r_S?

This chapter describes measuring the strength of the relationship between two variables. The Pearson product-moment correlation coefficient is used when the data are interval/ratio, and the Spearman rank-order correlation coefficient is used when the data are ordinal. The chapter introduces the scatter diagram, a plot of two variables that shows their degree of relationship. Correlation coefficients are descriptive statistics, and hypotheses regarding them can be tested using procedures similar to those outlined in previous chapters.

Up to now we have been interested in single dependent variables. We described the distributions of single variables (specified their mean and standard deviation, for example, in Chapters 4 and 5) and asked whether the distribution of the single variable in one condition is the same as the distribution of that same variable in another condition (in the two-sample t test of Chapter 11, for example). Even in the two-way ANOVA, our interest was in a single dependent variable (the therapy outcome in our example from Chapter 15), although it took two treatment variables (therapy type and therapist's experience) to specify the six conditions under which that single dependent variable was measured.

We turn now to consider variables taken *two* at a time, and we ask to what extent one variable of this pair is related to the other. If a subject has a high score on one variable, does that imply that the same subject is likely to have a high score on the other variable? Or does a high score on one variable imply a *low* score on the other? Or does a high score on one variable have no implication at all for the same subject's score on the other variable?

16.1 Correlation Coefficient

correlation coefficient: the descriptive statistic that measures the degree of the relationship between two variables

The statistic that allows us to quantify the answers to questions such as these is called the *correlation coefficient*. Just as the mean, median, range, and standard deviation (among others) are descriptive statistics useful for characterizing the properties of single distributions, correlation coefficients are descriptive statistics that characterize a property of the relationship between two variables.

Let's begin with an example. It is a common observation that many human traits "run in families": height, weight, intelligence, facial features, eye color, and so on. When we say that "height runs in the family," we mean, for example, that if George and Sam are brothers and we happen to know that George is short, then we expect that Sam is also short. If Sally and Mary are sisters and we happen to know that Sally is tall, then we expect that Mary is also tall.

The correlation coefficient is the statistic we use to describe the extent to which such measurements are related. Table 16.1 lists the correlation coefficients for a variety of

TABLE 16.1 Approximate correlation coefficients for human traits

	Identical twins reared together	Identical twins reared apart	Siblings reared together	Siblings reared apart	Unrelated children reared together
Intelligence	.9	.7	.5	.4	.3
Educational achievement	.9	.6	.8	.5	.5
Height	.9	.9	.5	.5	.0
Weight	.9	.9	.6	.4	.2
Head length	.9	.9	.5	.5	.1
Head breadth	.9	.9	.5	.5	.1
Eye color	1.0	1.0	.5	.5	.1

Based on Matarazzo (1972, p. 305).

measurements. As we shall see later in this chapter, a correlation coefficient of 1.0 is the strongest possible degree of relationship: If the correlation for height among brothers was 1.0 and we happened to know that George is short, we would know *for a fact (without doubt)* that his brother Sam is also short. Table 16.1 shows that the correlation for height among brothers is actually about .5 (not 1.0), however, which indicates that if we know that George is short, then we know that it is *likely (but not necessarily the case)* that Sam is also short.

Table 16.1 also shows that the correlation for height among identical twins is .9. If George and Sam are identical twins and we happen to know that George is short, then it is *very likely (but not absolutely necessarily the case)* that Sam is also short. A correlation of .9 is a stronger degree of relationship than a correlation of .5, so it is fair to say that height is more strongly related among identical twins than among siblings in general.

A correlation coefficient of .0 indicates no statistical relationship. Table 16.1 shows that the correlation for height among unrelated children who happen to be reared together is about .0. If George and Sam are unrelated children and we happen to know that George is short, then we know *nothing at all* about the height of Sam.

TABLE 16.2 Heights and weights of ten college men

Man	Height (in.)	Weight (lb)
George	72	160
Sam	66	144
Bill	68	154
Don	74	210
Jim	68	182
Ron	64	159
Chuck	63	199
Adam	73	198
Walt	61	132
Kyle	62	126

A correlation coefficient is always based on pairs of data: pairs of identical twins, or pairs of siblings, or pairs of unrelated children reared together. Sometimes these pairs are the same measurement on different (but paired) subjects (as in all the correlations in Table 16.1), but sometimes these pairs are different measurements on the same subjects. For example, consider the relationship between height and weight in college men. We can observe that as a general rule college men who are relatively tall weigh somewhat more than college men who are relatively short;[1] that is, the two variables height and weight are related (or "covary") to some extent. The correlation coefficient between height and weight gives us a measure that specifies precisely what we mean when we say that tall men are "somewhat" heavier and that height and weight are related "to some extent."

To ascertain the magnitude of this correlation coefficient, we begin by taking a random sample of college men and measuring each man's height and weight. Table 16.2 shows the (hypothetical) height and weight data for a sample of size ten. Note that for each man we have *two* observations, his height and his weight.

Scatter Diagrams

scatter diagram: a graphical method of representing the relationship between two variables where one variable is plotted on the *X*-axis and the other is plotted on the *Y*-axis

To explore the concept of correlation, it is convenient to illustrate our data with a kind of display called a *scatter diagram* or *scatterplot*, in which one variable is plotted on the *X*-axis and the other variable is plotted on the *Y*-axis. For our data, we choose to put height

[1] This does not imply that tall men are overweight. It simply means, for example, that although 130 pounds might be a common weight for men who are 5 feet, 6 inches tall, 130 pounds would be unusually light for men who are 6 feet, 2 inches tall.

on the X-axis, although it would be equally satisfactory to put weight there. Figure 16.1 shows the plotting of the first two points. George's data are displayed by plotting a point at the intersection of lines drawn at 72 inches (George's height) and 160 pounds (his weight). Similarly, Sam's data point is at the intersection of 66 inches and 144 pounds.

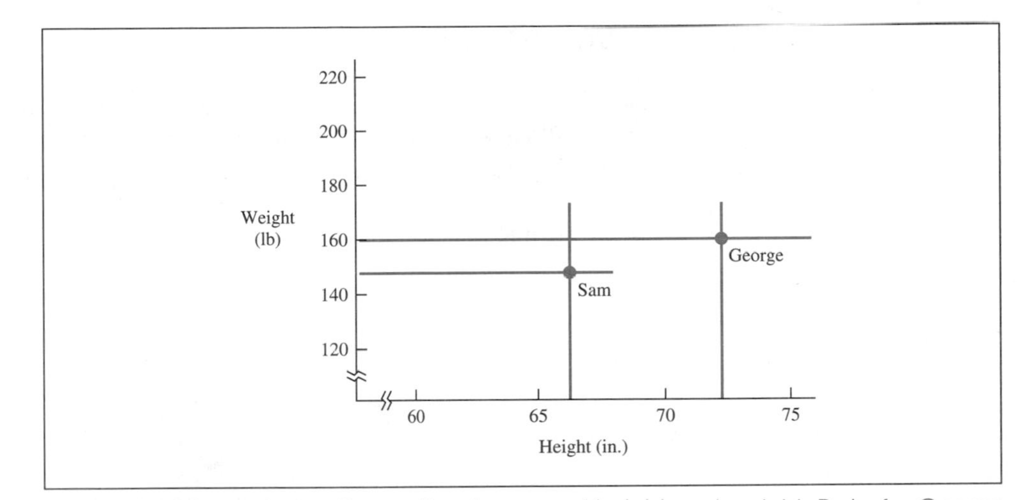

FIGURE 16.1 Beginning the scatter diagram of height and weight: Data for George and Sam are plotted

Figure 16.2 is the completed scatter diagram for all ten men. Let us explore what kind of information the scatter diagram gives us. First, we can see that the individual who *weighs the most* is the one whose point is *highest* on the scatter diagram. The highest weight is 210 pounds, and inspection of Table 16.2 shows that that weight belongs to Don, who is the heaviest subject.

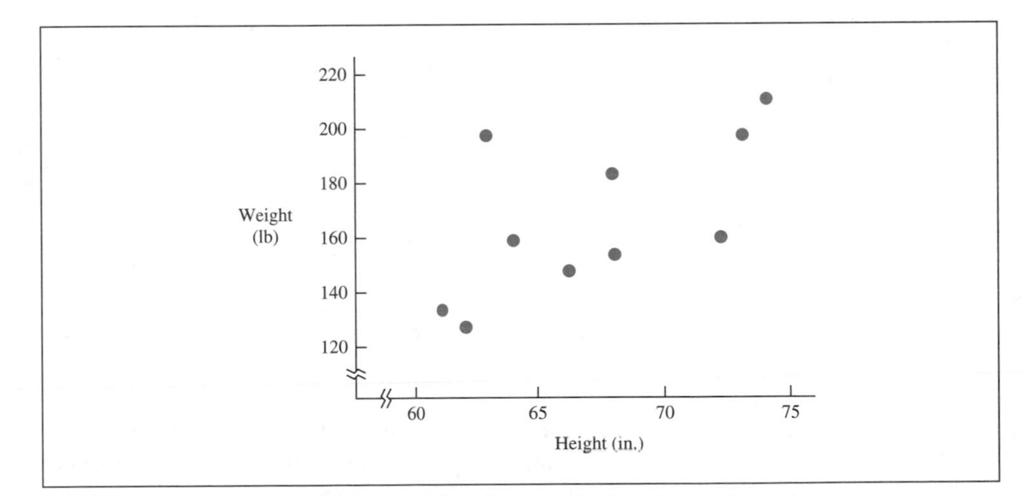

FIGURE 16.2 Completing the scatter diagram for height and weight

Second, if we define "portly" as meaning short and heavy, we can see that the most *portly* individual is the man in the *farthest upper left* (heavy and short) corner of the scatter diagram. That point has coordinates 63 inches (5 feet, 3 inches) and 199 pounds, and Table 16.2 shows the individual to be Chuck.

Third, the *tallest* individual is the one whose point lies *farthest to the right* in the scatter diagram. We can see that the rightmost height is 74 inches and belongs to Don.

Fourth, if we define "skinny" as meaning tall and light, then the *skinniest* individual is the man whose point is in the *farthest lower right* (light and tall) corner of the scatter diagram. Those coordinates are 72 inches and 160 pounds, and the skinniest subject is George.

We can also see that there is a general trend for taller men to be somewhat heavier because the points in Figure 16.2 generally lie in a pattern that goes up and to the right. It is the strength of this trend that we seek to measure with the correlation coefficient.

Values of *r*

The correlation coefficient, which we call r when we are describing a sample and ρ ("rho," Greek r) when we describe a population, is a number that lies between -1 and $+1$. Figure 16.3 shows how the value of r (or ρ) is related to the scatter diagram. Two characteristics of scatter diagrams should be noted: the *slope* and the *width* of the imaginary ellipse that can be drawn around most of the points in the scatter diagram.

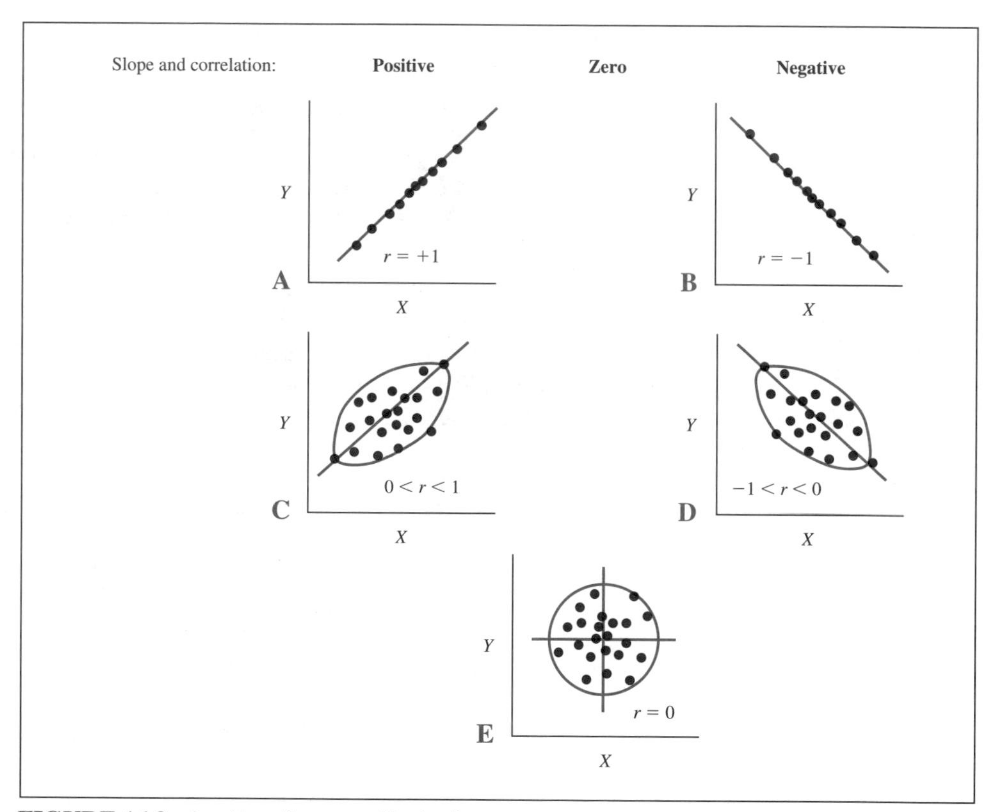

FIGURE 16.3 Scatter diagrams illustrating various values of the correlation coefficient

The slope determines the *sign* of r. If the slope is from lower left to upper right, as in diagrams A and C in Figure 16.3, then the correlation coefficient is *positive*. If the slope is from upper left to lower right, as in diagrams B and D, the correlation coefficient is *negative*. If it is impossible to determine whether the slope goes from upper left to lower right or from lower left to upper right, as in E, the correlation coefficient is 0.

The width of the imaginary ellipse that can be drawn around most of the points of the scatter diagram corresponds to the *magnitude* of the correlation coefficient. If the ellipse is extremely narrow, as in diagrams A and B, the magnitude is 1. If the ellipse is as wide as it can be, thus making it a circle as in diagram E, the magnitude is 0. If the width of the ellipse lies somewhere between as wide as possible and extremely narrow, as in diagrams C and D, the magnitude is somewhere between 0 and 1. Thus, scatter diagram A shows a perfect positive correlation, $r = +1.0$, whereas scatter diagram B shows a perfect negative correlation, $r = -1.0$. The correlation in scatter diagram C is somewhere between 0 and $+1$, perhaps .4 or .5, and the correlation in scatter diagram D is somewhere between -1 and 0, perhaps $-.4$ or $-.5$. The correlation coefficient in diagram E is 0.

Let us now refer back to our scatter diagram of height and weight in Figure 16.2. The slope is from lower left to upper right, so the sign of the correlation coefficient must be positive, and the ellipse that we might draw to include most of the points is somewhere between narrow and wide, so the magnitude must be between 0 and 1. Later we will compute the correlation coefficient for these data and find that $r = +.63$.

What would college men be like if the correlation between weight and height were $+1.0$ (instead of .63)? There would then be a perfect relationship between height and weight; for every increase in height, there would be a proportional increase in weight, and vice versa. If we knew the height of a man, we would also know his exact weight, and vice versa. There would be no such thing as a "portly" or a "skinny" man. The proportion of height to weight would be exactly the same in every man.

What if the correlation between height and weight were perfectly negative: $r = -1.0$? Then there would be a proportional *decrease* in weight for every increase in height; the taller a man was, the lighter he would be. The world would be populated by men ranging from short and wide to tall and narrow.

If the correlation between height and weight were 0, then knowing a man's height would tell us *nothing* about his weight, and vice versa. There would be the same percentage of 5 feet tall men who weigh 200 pounds as of 6 feet tall men who weigh 200 pounds.

Positive correlation implies that relatively large values of one variable are in general associated with relatively large values of the other variable, and that small values of one are associated with small values of the other. All the variables in Table 16.1 (intelligence among siblings, height among siblings, etc.) are positively correlated, as are the following: IQ and school grades (people with high IQ generally earn somewhat better grades), age and income (people's income generally increases with age), reading ability and math ability, number of runs and number of hits in baseball games, amount of sugar in breakfast cereal and the number of calories, and amount of alcohol consumed and reaction time (measured in seconds). Examples of perfect or nearly perfect positive correlations are temperature in degrees Fahrenheit and temperature in degrees Centigrade, number of nickels in a pile and the weight of that pile, and speed of an automobile and distance traveled.

Negative correlation implies that relatively large values of one variable are in general associated with relatively *small* values of the other variable. Examples of negative correlations in the real world are distance to the archery target and number of arrows in the bull's-eye (the farther the target, the less the accuracy), number of flaws in a diamond and its price (the more flaws, the lower the price), the size of an automobile and the number of miles per gallon it gets, and a baby's weight at birth and the number of birth defects. Examples of perfect negative correlation are speed and time taken to travel from point A to point B and the distance an airplane is from New York and its distance from Paris, assuming it flies in a perfectly straight line between New York and Paris.

Examples of 0 or near 0 correlation are head size and IQ, visual acuity and running speed, and elevation of the center of a city and average income of city residents.

16.2 Pearson's *r*

Just as there were several measures of central tendency depending on the level of measurement of the data (mode for nominal data, median for ordinal data, and mean for interval/ratio data), there are several measures of correlation, also depending on the level of measurement of the data. We will discuss two of these correlation coefficients: Pearson's product-moment correlation coefficient r, which requires that data be measured at the interval/ratio level, and Spearman's rank-order correlation coefficient r_S, which requires that data be measured at the ordinal (or interval/ratio) level.[2] We will discuss the Pearson r first because it is the most frequently used correlation coefficient.

[2] Other correlation coefficients measure the strength of relationship between nominal variables, between one nominal and one ordinal variable, between one nominal and one interval/ratio variable, and so on. We leave their discussion to more advanced textbooks.

z Score Formulas

Pearson *r*: the correlation coefficient appropriate for interval/ratio data

The *Pearson r* correlation coefficient (sometimes called the Pearson product-moment correlation coefficient) between any two variables X and Y is defined by

Pearson product-moment correlation coefficient for a sample, standard score formula

Equation (16.1a) on page 386 is identical but easier to compute.

$$r = \frac{\sum z_X z_Y}{n-1} \tag{16.1}$$

where z_X is the z score associated with the value X, z_Y is the z score associated with the value Y, and n is the number of *pairs* of data in the sample. The corresponding formula for the population correlation coefficient is

Pearson product-moment correlation coefficient for a population, standard score formula

Equation (16.2a) on page 386 is identical but easier to compute.

$$\rho = \frac{\sum z_X z_Y}{N} \tag{16.2}$$

where N is the number of *pairs* of data in the population.

Recall from Chapter 6 that a z score is formed by subtracting the mean from each raw score and dividing by the standard deviation. Referring back to our height and weight data from Table 16.2, we compute the mean and standard deviation of height to be $\overline{X}_H = 67.1$ inches and $s_H = 4.7$ inches. Thus, the z scores for our height data are given by $z_H = (\text{height} - \overline{X}_H)/s_H = (\text{height} - 67.1)/4.7$. Similarly, we can compute the mean and standard deviation of the weights from Table 16.2, obtaining $\overline{X}_W = 166.4$ pounds and $s_W = 29.4$ pounds. The z scores for weight are then obtained from $z_W = (\text{weight} - 166.4)/29.4$.

Figure 16.4 superimposes the z axes on our scatterplot of height and weight. Recall that $\overline{X}_H = 67.1$ inches and $s_H = 4.7$ inches. From Chapter 6, that implies that $z_H = 0$ when height $= \overline{X}_H = 67.1$ inches, and $z_H = 1$ when height is one standard deviation above $\overline{X}_H$; that is, $z_H = 1$ when height $= 67.1 + 4.7 = 71.8$ inches. These relationships are shown on the two horizontal axes of Figure 16.4.

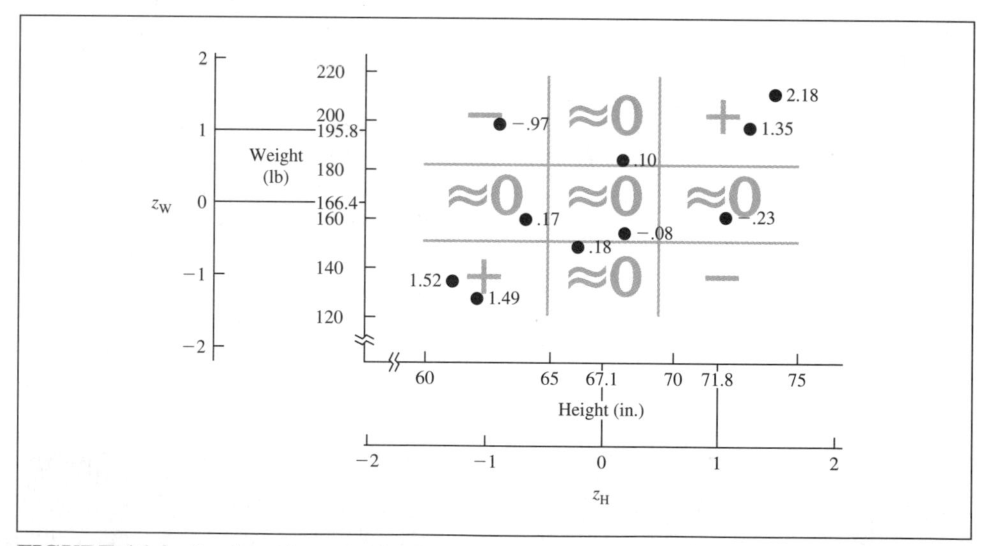

FIGURE 16.4 Scatter diagram from Figure 16.2 with z axes superimposed and values of $z_H z_W$ indicated. Positive, negative, and approximately 0 contributions to Pearson r are shown

Similarly, $\overline{X}_W = 166.4$ pounds and $s_W = 29.4$ pounds. That implies that $z_W = 0$ when weight $= 166.4$ pounds, and $z_W = 1$ when weight $= 166.4 + 29.4 = 195.8$ pounds. These relationships are shown on the two vertical axes of Figure 16.4.

Each of the points in Figure 16.4 has its corresponding value of $z_H z_W$ next to it. Thus, for example, the upper right point in this scatterplot represents Don, whose height is 74 inches and whose weight is 210 pounds. The z score corresponding to Don's height is therefore $z_H = (74 - 67.1)/4.7 = 1.468$, and the z score corresponding to his weight is $z_W = (210 - 166.4)/29.4 = 1.483$. The product of the two z scores is $z_H z_W = 1.468(1.483) = 2.18$, as shown next to Don's point at the upper right.

Equation (16.1), $r = \sum z_H z_W/(n-1)$, tells us that the sample correlation coefficient is the sum of all these $z_H z_W$ values divided by the number of points minus 1, and that gives us a clue to eyeball-estimating the correlation coefficient.

Pearson's r will be large when $\sum z_H z_W$ is large, small when that sum is small, negative when that sum is negative, and so on. Thus, to understand the operation of Pearson's r, we must answer the question: Under what conditions will $\sum z_H z_W$ be large?

Recall that a z score is positive when it is above the mean and negative when it is below the mean. Thus, the z_H's are positive for relatively tall men and negative for relatively short men. Similarly, the z_W's are positive for relatively heavy men and negative for relatively light men. Values of $z_H z_W$ are positive, then, for men who are both relatively tall and relatively heavy (because both z scores are positive) *and* for men who are both relatively short and relatively light (because both z scores are negative and their product is therefore positive).

If we superimpose a tic-tac-toe grid on the Figure 16.2 scatter diagram for the height and weight data, as shown in Figure 16.4, we can see that $z_H z_W$ is positive for points in the upper right (z_H and z_W both positive) and the lower left (z_H and z_W both negative) sections of the scatter diagram. These portions of Figure 16.4 are indicated with a + sign.

On the other hand, $z_H z_W$ is negative when one of the z scores is positive and the other is negative—that is, for men who are tall and light or short and heavy. These are the points in the upper left (z_H negative but z_W positive) and lower right (z_H positive, z_W negative) corners of the scatter diagram, indicated with a − sign.

If either (or both) z_H or z_W is approximately 0, then $z_H z_W$ is also approximately 0. The middle column of the tic-tac-toe grid includes points whose z_H values are approximately 0 (because they are close to the mean height); the middle row of the grid includes points whose z_W values are approximately 0. Thus, in both the middle column and the middle row, $z_H z_W \approx 0$. Those sections of the tic-tac-toe diagram are indicated with ≈ 0.

The numerator of the correlation coefficient is the sum of the $z_H z_W$ values for all the points in the scatter diagram, but only the four corners make large contributions. Thus, for our data, the five most extreme points (with $z_H z_W$ values 2.18, 1.35, 1.52, 1.49, and −.97) contribute by far the most to the total $\sum z_H z_W$: The sum of $z_H z_W$ for those five points is 5.57, whereas the sum of $z_H z_W$ for all the other points is only .14. If a scatter diagram has a preponderance of + points, the correlation coefficient will be strongly positive, and if it has a preponderance of − points, the coefficient will be negative.

The scatter diagram for the height and weight data depicts relatively strongly positively correlated variables because there are four points in the + regions of the scatter diagram (that is, four contributions that increase the numerator sum) but only one point in the − regions (only one point that decreases that sum). Because the data appear to be strongly but not perfectly positively correlated, our eyeball-estimate is that $r \approx .7$. The actual value is .634, as shown in Box 16.1.

You should practice eyeball-estimating correlation coefficients for the scatter diagrams in Box 16.2.

Computational Formulas

There are two methods of computing the Pearson r: the one that uses z scores [described previously; see Equations (16.1) and (16.2) and Box 16.1] and the algebraically equivalent computational forms described in this section that use raw scores [see Equations (16.1a) and (16.1b)]. The z score formula was useful in understanding how the correlation coefficient measures the strength of a relationship, but it is computationally clumsy because one must first compute the means and standard deviations of both variables and then compute z scores for every data point, a laborious process. On the other hand, the computational formula is easier to use, but it obscures the logic behind the correlation coefficient.

You may wish to use the ESTAT exercise "scatter" at this time.

BOX 16.1 Computing the correlation coefficient from the z score formula

This box may be omitted without loss of continuity.

Table 16.3 shows the z_H and z_W scores for each of the subjects in Table 16.2. The last column of Table 16.3 shows the product $z_H z_W$ for each subject. Then Equation (16.1) gives $r = \sum z_H z_W/(n - 1) = 5.71/(10 - 1) = .634$. It is quite cumbersome to compute z scores for each data point, so the computational (raw scores) formulas, Equations (16.1a) and (16.2a), are easier to compute.

TABLE 16.3 Height and weight data with subcomputations necessary for the z score formula for Pearson's r

	Height	Weight	z_H	z_W	$z_H z_W$
George	72	160	1.04	−.22	−.23
Sam	66	144	−.23	−.76	.18
Bill	68	154	.19	−.42	−.08
Don	74	210	1.47	1.48	2.18
Jim	68	182	.19	.53	.10
Ron	64	159	−.66	−.25	.17
Chuck	63	199	−.87	1.11	−.97
Adam	73	198	1.26	1.07	1.35
Walt	61	132	−1.30	−1.17	1.52
Kyle	62	126	−1.08	−1.37	1.49
$\sum$	671	1664	.01	.00	5.71

BOX 16.2 The correlation coefficient: Eyeball-calibration

We present in Figures 16.5 through 16.17 a series of scatter diagrams for you to develop the skills of eyeball-estimating Pearson's r. The general procedure is to superimpose a tic-tac-toe grid that puts the center of the scatter diagram's points in the center cell of the grid, as we did in Figure 16.4. Pearson's r is highly positive if many points are in the + cells and very few points are in the − cells, r is highly negative if many points are in the − cells and few are in the + cells, and r is 0 if points are equally distributed in the + and − cells. Inspect a few scatter diagrams and you will obtain a feel for the intermediate-sized coefficients.

We have superimposed a tic-tac-toe grid over the scatter diagram in Figure 16.5. Because many points lie in both + cells of the grid and none lies in the − cells, r must be very strongly positive. Because the points do not lie on a perfectly straight line, r cannot be +1.0. Our eyeball-estimate is therefore $r \approx +.9$. The computed value is +.88.

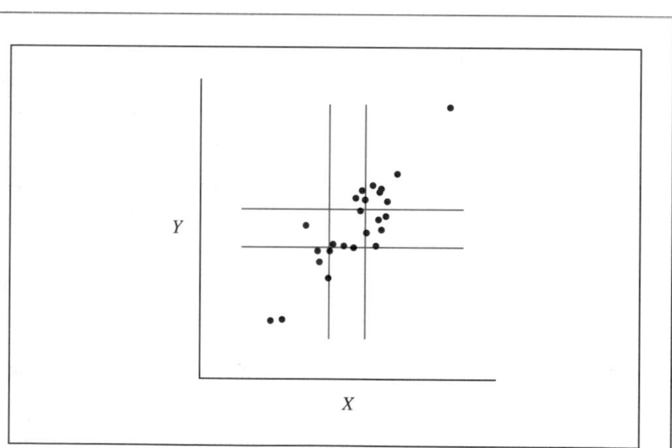

FIGURE 16.5 Scatter diagram to illustrate the eyeball-estimation procedure

Note that it does not particularly matter how wide the tic-tac-toe diagram is, as long as the vertical lines are approximately centered on the mean of X and the horizontal lines are approximately centered on the mean of Y. Furthermore, it does not matter at all what magnitude the values on the X- and Y-axes have (that is why we have omitted them entirely). It is the shape of the scatterplot that determines the correlation coefficient, *not* the values on the axes.

Twelve additional scatter diagrams are shown on the next page; the correlation coefficient is given in the caption for each figure. *Cover the caption of each figure* and eyeball-estimate the correlation coefficient, paying attention to the slope (giving the sign) and the tic-tac-toe grid and the width of the imaginary ellipse (giving the magnitude). Then uncover the caption to find the actual correlation coefficient.

(continued)

BOX 16.2 *(continued)*

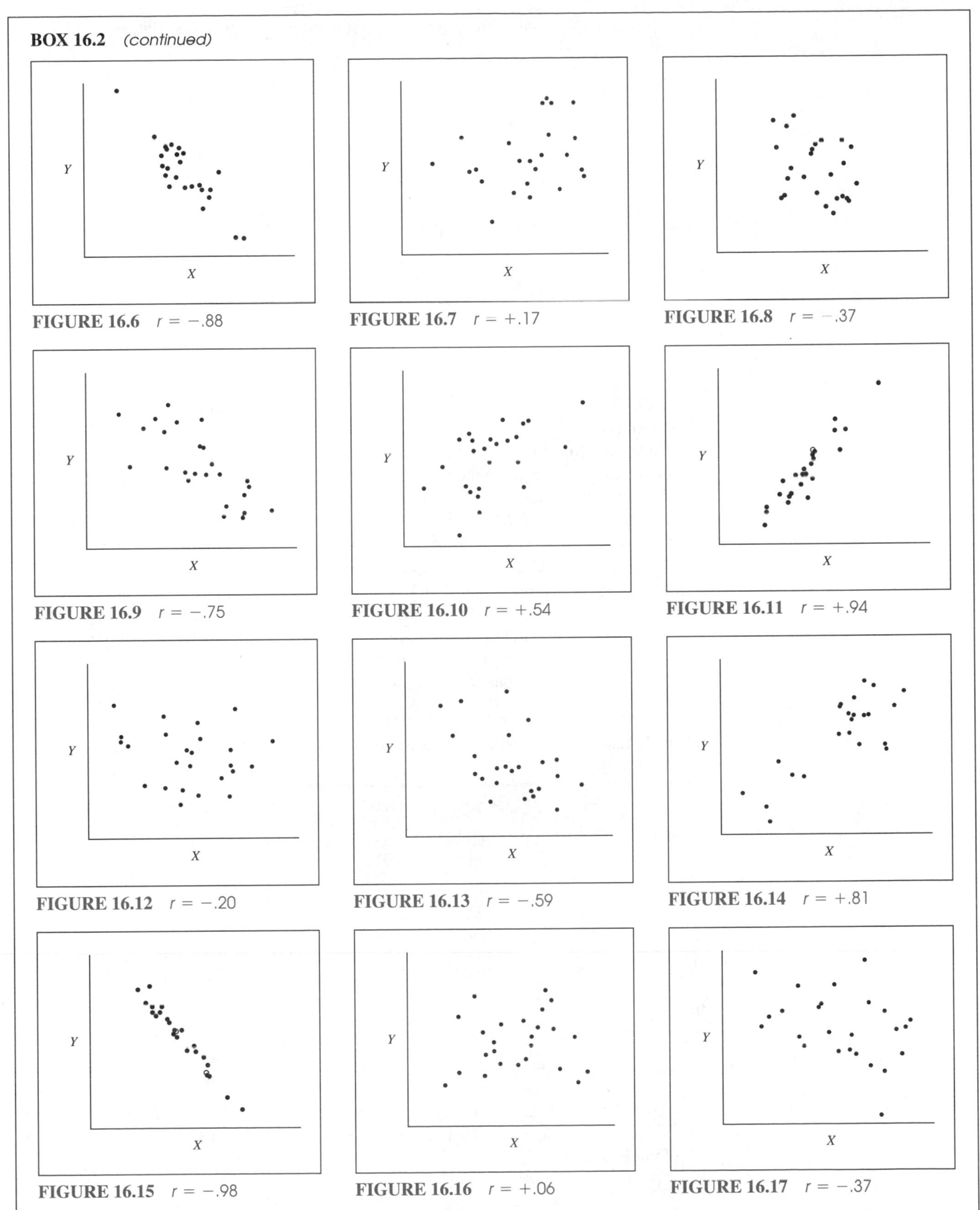

FIGURE 16.6 $r = -.88$

FIGURE 16.7 $r = +.17$

FIGURE 16.8 $r = -.37$

FIGURE 16.9 $r = -.75$

FIGURE 16.10 $r = +.54$

FIGURE 16.11 $r = +.94$

FIGURE 16.12 $r = -.20$

FIGURE 16.13 $r = -.59$

FIGURE 16.14 $r = +.81$

FIGURE 16.15 $r = -.98$

FIGURE 16.16 $r = +.06$

FIGURE 16.17 $r = -.37$

The computational formula for the Pearson product-moment correlation coefficient in a sample is

Pearson product-moment correlation coefficient for a sample, computational formula

$$r = \frac{\sum XY - \dfrac{\sum X \sum Y}{n}}{\sqrt{\left[\sum X^2 - \dfrac{(\sum X)^2}{n}\right]\left[\sum Y^2 - \dfrac{(\sum Y)^2}{n}\right]}} \tag{16.1a}$$

The corresponding formula for a population is

Pearson product-moment correlation coefficient for a population, computational formula

$$\rho = \frac{\sum XY - \dfrac{\sum X \sum Y}{N}}{\sqrt{\left[\sum X^2 - \dfrac{(\sum X)^2}{N}\right]\left[\sum Y^2 - \dfrac{(\sum Y)^2}{N}\right]}} \tag{16.2a}$$

Note that these two equations are identical except that ρ replaces r and N replaces n in Equation (16.2a).[3]

The two terms in the denominator of both Equations (16.1a) and (16.2a) should look familiar: They are the numerators of the raw score formulas, Equation (5.4a) for a sample or (5.5a) for a population, for the variance of X and Y, respectively. The numerators of Equations (16.1a) and (16.2a) have the same form as each term in the denominators, except that where a variable is squared in the denominator, a cross-product of variables appears in the numerator; that is, $\sum X^2$ is replaced by $\sum XY$. And where the sum of a variable is squared in the denominator, the sum of one variable is multiplied by the sum of the other variable in the numerator; that is, $(\sum X)^2$ is replaced by $(\sum X)(\sum Y)$.

The computational formula requires fewer computations because z scores, each of which requires a subtraction and a division, do not have to be calculated. The subcomputations for the computational formula are shown in Table 16.4.

TABLE 16.4 Height and weight data with subcomputations for raw score formula computation

Man	*Height*	*Weight*	*Height²*	*Weight²*	*(Height)(Weight)*
George	72	160	5,184	25,600	11,520
Sam	66	144	4,356	20,736	9,504
Bill	68	154	4,624	23,716	10,472
Don	74	210	5,476	44,100	15,540
Jim	68	182	4,624	33,124	12,376
Ron	64	159	4,096	25,281	10,176
Chuck	63	199	3,969	39,601	12,537
Adam	73	198	5,329	39,204	14,454
Walt	61	132	3,721	17,424	8,052
Kyle	62	126	3,844	15,876	7,812
$\sum$	671	1,664	45,223	284,662	112,443

For our data, we can enter the sums shown at the bottom of Table 16.4 into the sample raw score formula, Equation (16.1a), to obtain the same result as when we applied Equation (16.1) in Box 16.1:

[3] It may seem, because the z score formula, Equation (16.1), has $(n - 1)$ as its denominator, that there should be an $(n - 1)$ term in the denominator of Equation (16.1a). That is *not* the case, however, because each z in the numerator of Equation (16.1) is a deviation divided by a standard deviation. Each of those standard deviations has $\sqrt{n - 1}$ as its denominator, and all the $(n - 1)$ terms together cancel out.

$$r = \frac{112{,}443 - \dfrac{(671)(1664)}{10}}{\sqrt{\left[45{,}223 - \dfrac{(671)^2}{10}\right]\left[284{,}662 - \dfrac{(1664)^2}{10}\right]}}$$

$$= \frac{788.6}{\sqrt{[198.9][7772.4]}}$$

$$= \frac{788.6}{1243.354} = .634$$

Factors That Affect the Size of r

Perhaps you have come to the conclusion that a low correlation coefficient means that there is little or no relationship between the two variables being measured. Although that is usually true, there are two important exceptions.

Nonlinear Relationships

Pearson's r measures only the degree of the *linear* relationship between two variables. It can happen (though not often in practice) that variables have a strong *nonlinear* relationship and yet Pearson's r is quite small or 0.

Figure 16.18 shows such a scatter diagram. The points are strongly, one might even say perfectly, related to one another, but low values of X are related to high values of Y *and also* high values of X are related to high values of Y. Pearson's r for these data is 0, which indicates that these data have no linear trend even though the curvilinear relationship is strong.

Figure 16.19 shows another scatter diagram that has a clear curvilinear relationship, but there is a measurable linear component. Here, Pearson's $r = -.82$, a strong negative relationship.

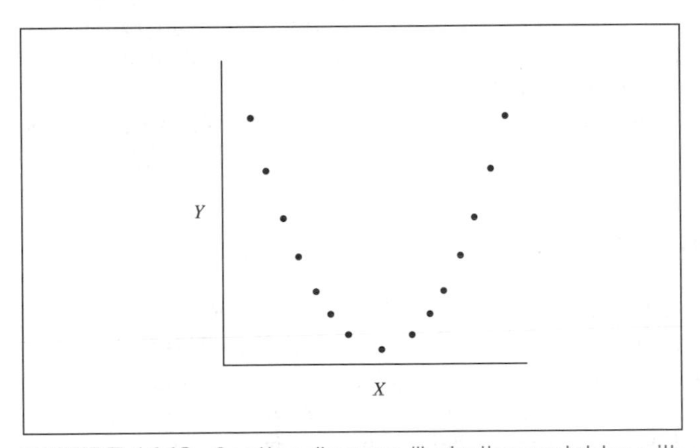

FIGURE 16.18 Scatter diagram illustrating variables with a strong but nonlinear relationship

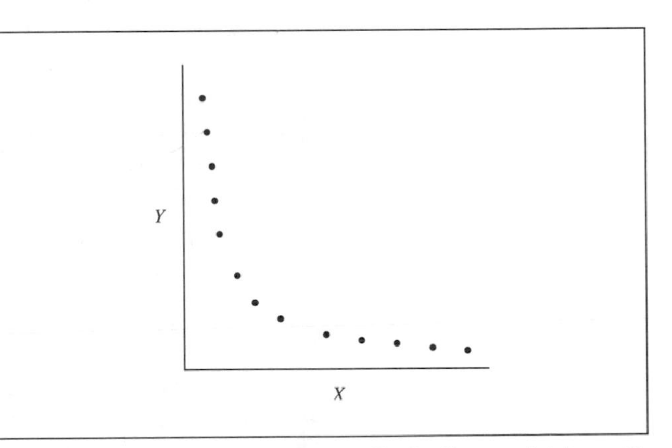

FIGURE 16.19 Scatter diagram illustrating variables with a strong nonlinear relationship that has a linear component

Thus, it can be seen that Pearson's r measures only the linear portion of the relationship between variables.

Restrictions of the Range

If the total range of one (or both) of the variables we are measuring is truncated or diminished in any way, the correlation between the two variables is likely to be reduced. Figure 16.20 shows such a hypothetical situation. Suppose the top of Figure 16.20 represents a scatter diagram for family income and the price of the automobile driven by the family. The

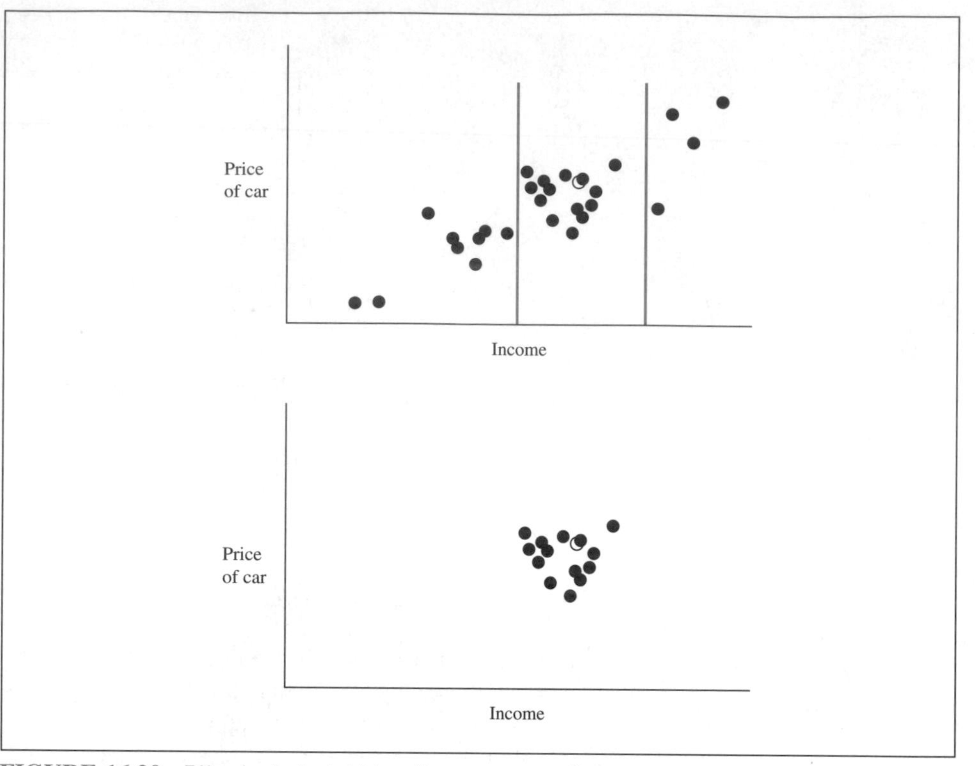

FIGURE 16.20 Effect of diminishing the range on the correlation coefficient. *Top:* r is strongly positive. *Bottom:* r is quite small.

correlation between income and price is seen to be quite high (actually, $r = .88$ in this scatter diagram).

Now suppose we had collected data in only one upper-middle-class neighborhood, as shown between the two vertical lines of the top scatter diagram; that is, we exclude the very wealthy and the very poor individuals. The remaining data are shown as the scatter diagram at the bottom of Figure 16.20. Note that the points in the bottom diagram are exactly the same as those in the middle of the top diagram. However, r has dropped dramatically (to about .1) when the extreme points are excluded.

In order for a correlation coefficient to be large, there must be sufficient variation of both of the two variables being measured.

Correlation Does Not Imply Causation

It is frequently tempting to conclude that if a correlation coefficient is large, then one of the variables (usually the one displayed on the X-axis) *causes* the other. This is, however, not always true. In the income/price of automobile example, it *is* probably true: Increasing one's income probably does cause an increase in the price of the automobile one drives. On the other hand, it is probably *not* true in the height and weight example: People are not heavy *because* they are tall, and they are not tall *because* they are heavy. Rather, they are both tall and heavy because of some third factor or factors (such as genetic makeup or infant nutrition).

The mere fact of a strong correlation does not prove anything about the nature of cause and effect. Inferences about cause and effect must be made on the basis of other evidence.

Testing Hypotheses About r

Thus far we have been considering Pearson's r as a *descriptive* statistic; r describes the relationship between a sample of two variables in much the same way as $\overline{X}$ describes the central tendency of a sample or s describes its variation. We turn now to the use of r as an inferential statistic. Just as we used $\overline{X}$ to test inferences about the population mean μ, we can also use r to test inferences about the population correlation coefficient ρ.

The most frequently used test regarding correlation coefficients asks whether any significant relationship exists between two variables. The null hypothesis is then that no such relationship exists; that is,

null hypothesis, Pearson correlation

$$H_0: \quad \rho = 0 \tag{16.3}$$

r is used as the test statistic to test this hypothesis, and if r is greater (in absolute value) than its critical value, we reject H_0.

Let us consider our height and weight data and ask: Is there in the population of college men any relationship (that is, any significant correlation) between height and weight? We use the data of Table 16.2 to test the hypothesis. The question we ask is whether the obtained value of r is significantly different from 0.

BOX 16.3 Critical values of Pearson's r as a function of degrees of freedom

> Students are often surprised at how large a correlation coefficient must be before it is judged to be significantly greater than 0. In our example, with ten subjects, the correlation coefficient critical value is $r_{\text{cv}_{\alpha=.05,\, df=8,\, \text{nondirectional}}} = .6319$, quite a large r_{cv}. If the hypotheses had been tested at the .01 level of significance, the critical values would be larger still: $r_{\text{cv}_{\alpha=.01,\, df=8,\, \text{nondirectional}}} = .765$.
>
> Additional critical values are listed in Table A.8 in Appendix A.

I. State the Null and Alternative Hypotheses

Three sketches illustrate H_0:
Variable
Sample statistic
Test statistic

ⓘ The sketches are not particularly informative, so we omit them here.

The hypotheses are

$$H_0: \quad \rho = 0$$
$$H_1: \quad \rho \neq 0 \qquad \text{(nondirectional)}$$

where ρ is the correlation coefficient in the population from which our sample is drawn. We show the alternative hypothesis as being nondirectional.

II. Set the Criterion for Rejecting H_0

Level of significance?
Directional?
Shade rejection regions.

If we choose a .05 level of significance, we can find directly from Table A.8 in Appendix A the critical value r_{cv}. A correlation coefficient has $n - 2$ degrees of freedom. For our data, $n = 10$ (*pairs* of data), so we enter Table A.8 with $df = 10 - 2 = 8$, nondirectional, and find $r_{\text{cv}} = .6319$. For a comment on the critical values of correlation coefficients, see Box 16.3.

III. Collect a Sample and Compute the Observed Value of the Statistic

Observed values:
Sample statistic
Test statistic

In our problem, we randomly sampled ten individuals, measured their heights and weights, and computed $r_{\text{obs}} = .634$. r_{obs} is the sample statistic, but because critical values of this statistic are provided in Table A.8, r_{obs} can also be considered to be its own test statistic. The test statistic from which Table A.8 is derived is actually t, which follows the general pattern for test statistics shown in Equation (9.1). See Box 16.4.

BOX 16.4 The significance test for r is derived from the test statistic t

ⓘ This box may be omitted without loss of continuity.

> This box shows that testing a hypothesis about r is another application of the Equation (9.1) general formula for the test statistic: test statistic = (sample statistic − population parameter)/(standard error of the sample statistic).
>
> The population parameter (specified by the null hypothesis) is ρ, the population correlation coefficient; the sample statistic that point-estimates ρ is r; and the denominator of the test statistic must therefore be s_r, the standard error of the correlation coefficient.
>
> *(continued)*

BOX 16.4 *(continued)*

standard error of *r*: the standard deviation of the distribution of sample correlation coefficients

The *standard error of r* can be defined analogously to the standard error of the mean $s_{\overline{X}}$. Suppose we take a sample of size n, measure each subject on variables X and Y, and compute r for those data. Then we take another sample of size n and again compute r, and another sample (and another r), and so on, indefinitely often. The standard deviation of the resulting distribution of all the r values is s_r, the standard error of r. If the null hypothesis is true, then it can be shown that

standard error of a Pearson correlation

$$s_r = \sqrt{\frac{1 - r^2}{n - 2}} \tag{16.4}$$

For our height and weight data, $s_r = \sqrt{(1 - .634^2)/(10 - 2)} = \sqrt{.075} = .274$.

The test statistic can be shown to be t, so following the Equation (9.1) model, $t = (r - \rho)/s_r$, but because $\rho = 0$ by the null hypothesis, the test statistic becomes simply

test statistic (t), Pearson correlation

$$t = \frac{r}{s_r} \qquad [df = n - 2] \tag{16.5}$$

We can sketch the distribution of the sample statistic r and the test statistic t as shown in Figure 16.21. The distribution on the upper axis shows the values we might expect r to take if we computed the correlation coefficients of a long series of independent samples of size ten drawn from a population in which the null hypothesis $\rho = 0$ is true. The lower axis shows t as defined by Equation (16.5).

Choosing a .05 level of significance, we find from Table A.2 that the critical value of t with $df = 10 - 2 = 8$ is $t_{\text{cv}_{\alpha=.05,\text{ nondirectional}}} = 2.306$. The critical value of r is then $r_{\text{cv}} = \rho \pm t_{\text{cv}}(s_r) = 0 \pm 2.306(.274) = \pm.632$. We show these values on the appropriate axes in Figure 16.21 and shade the rejection regions that lie beyond them.

We show the observed value of the statistic $r_{\text{obs}} = .634$ on the upper axis of Figure 16.21, and we compute $t_{\text{obs}} = r_{\text{obs}}/s_r = .634/.274 = 2.314$ and show it on the lower axis. Because the observed values exceed the critical values (just barely!), we reject H_0, just as when using Table A.8.

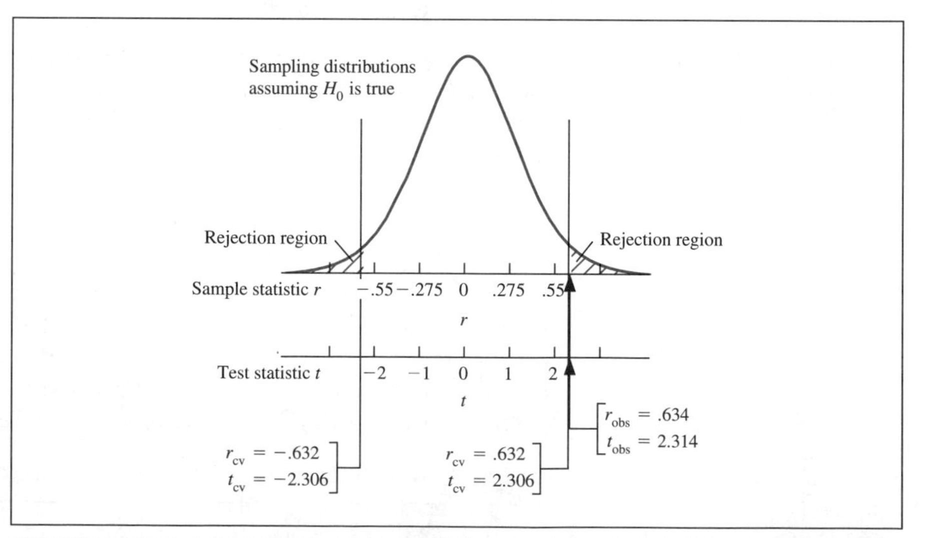

FIGURE 16.21 Distributions of the statistic r and the test statistic t with rejection regions shaded and observed values shown

Note that if you try this process with a different correlation coefficient, you may arrive at a different critical value r_{cv}. For example, if r_{obs} had been .4, s_r would have been $\sqrt{(1 - .4^2)/(10 - 2)} = .324$, so the critical values would have

(continued)

BOX 16.4 *(continued)*
been $\pm 2.306(.324) = \pm .747$ instead of $\pm .632$. The larger the r_{obs}, the smaller the s_r. As r_{obs} moves closer and closer to r_{cv}, r_{cv} moves closer and closer to r_{obs}, with the result that the hypothesis test based on t will reject H_0 in exactly those instances where the test based on Table A.8 rejects H_0. (See Exercise 20 at the end of the chapter.)

IV. Interpret the Results

Statistically significant? Practically significant? State results clearly.

ℹ See Section 17.6 for a discussion of practical significance.

Because the observed value $r_{obs} = .634$ exceeds its critical value $r_{cv} = .6319$, we reject the null hypothesis and conclude that there is indeed a positive correlation between height and weight in college men.

Power

We have seen since Chapter 12 that statistical power is the probability of correctly rejecting the null hypothesis, and that one of the most important uses of power analysis is in determining the sample size necessary for an experiment to have the desired power. This requires specifying the effect size index in the population from which we expect the sample subjects to be drawn. In the case of a correlation experiment, the appropriate effect size index is ρ, the correlation coefficient in the population. Cohen (1988) has called $\rho = .1$ a "small" effect size, $\rho = .3$ a "medium" effect size, and $\rho = .5$ a "large" effect size, and with those conventions, Table 16.5 shows the number of subjects necessary for an experiment to have power of .8 or .4. Table 16.5 is thus a continuation of Tables 12.4 and 13.13, which showed the same information for t tests and ANOVA, respectively.

TABLE 16.5 Number of pairs of observations necessary to obtain the required power for a nondirectional (two-tailed) test of the correlation coefficient with level of significance $\alpha = .05$

	Effect size index			*Magnitude of ρ*		
Test type	*Null hypothesis*	ρ	*Power*	*Small*	*Medium*	*Large*
Correlation	H_0: $\rho = 0$	ρ	.8	625	66	23
			.4	195	22	9

Source: Cohen (1988, pp. 86–87).

For example, if we assume that the correlation of height with weight in the population is actually $\rho = .3$ (a "medium" correlation), then an experiment must have 66 subjects to have an 80% chance of (correctly) rejecting the null hypothesis. Cohen's (1988, p. 86) tables show that the experiment in this chapter, with ten subjects (and still assuming that $\rho = .3$), has power .22—that is, only a 22% chance of (correctly) rejecting H_0.

16.3 Spearman's r_S

ℹ Some reports use "ρ" to symbolize r_s and call it "Spearman's rho." We avoid that terminology because of its confusion with the correlation coefficient in a population.

The Pearson product-moment correlation coefficient that we have been considering is applicable only if the variables we are correlating are both measured at the interval/ratio level of measurement. How do we proceed if both of the variables that we wish to correlate are measured at only the ordinal level (or if one is ordinal and the other is interval/ratio)?

For example, suppose we wish to know how strongly teachers' impressions of their students' intelligence are related to the students' actual intelligence as measured by a standard IQ test. We ask a teacher to rate each of his students on a scale ranging from

1 (not smart) to 10 (very smart). There are 15 students in the teacher's classroom. The data (hypothetical) for our study are in Table 16.6 and displayed in the scatter diagram in Figure 16.22.

TABLE 16.6 Data for teacher rating example and subcomputations required for computing Spearman's r_S

	Teacher's rating		Student's IQ		Difference, $D_i = R_{X_i} - R_{Y_i}$	
	Rating	Rank	IQ	Rank		
Student	X_i	R_{X_i}	Y_i	R_{Y_i}	D_i	D_i^2
1	3	5.5	99	7	−1.5	2.25
2	4	7	103	8	−1.0	1.00
3	6	10.5	106	10	.5	.25
4	1	2	72	1	1.0	1.00
5	5	8.5	98	6	2.5	6.25
6	1	2	105	9	−7.0	49.00
7	5	8.5	91	5	3.5	12.25
8	3	5.5	81	2	3.5	12.25
9	7	12	120	13	−1.0	1.00
10	10	15	127	14	1.0	1.00
11	8	13.5	135	15	−1.5	2.25
12	1	2	83	3	−1.0	1.00
13	8	13.5	110	11.5	2.0	4.00
14	6	10.5	110	11.5	−1.0	1.00
15	2	4	85	4	.0	.00
Σ					.0	94.50

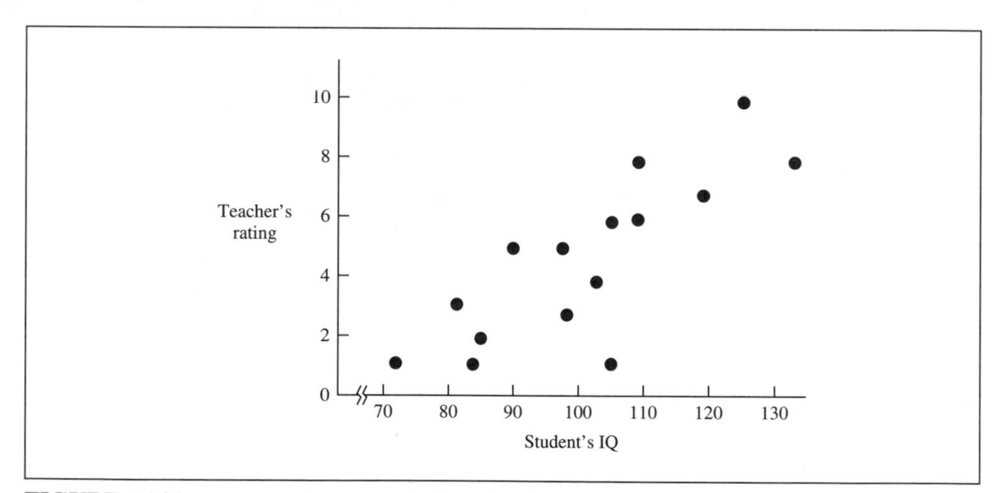

FIGURE 16.22 Scatter diagram of teacher's rating versus student's IQ

The teacher's rating cannot be considered to be at the interval/ratio level of measurement because there is no reason to believe that the difference in intelligence between, say, a teacher's rating of 2 and a rating of 3 is the same as the difference between, say, the ratings of 9 and 10. Thus, the scale is only ordinal, so we cannot use Pearson's r.

Spearman r_S: the correlation coefficient appropriate for ordinal data

The correlation coefficient designed to measure the degree of the relationship between ordinal variables or between a pair of variables where one is ordinal and the other is interval/ratio is called *Spearman's* r_S, or sometimes Spearman's rank-order correlation coefficient, because it depends on only the *ordering* of the data within each variable, *not* on the values of the data themselves.

Spearman's r_S is eyeball-estimated in Box 16.5 and computed from

Spearman rank correlation coefficient

$$r_S = 1 - \frac{6 \sum D_i^2}{n(n^2 - 1)} \tag{16.6}$$

where D_i is the difference *between the ranks* of the variables and n is the number of pairs. [Equation (16.6) can be shown to be identical to the raw score formula for Pearson's r (Equation 16.1a) when ranks are substituted for the original data.]

BOX 16.5 Eyeball-estimating r_S

Because in most cases Spearman's r_S is approximately equal to Pearson's r, we eyeball-estimate Spearman's r_S using the identical procedure as for Pearson's r. In the scatter diagram in Figure 16.22, a narrow ellipse would contain most of the points, as superimposing the tic-tac-toe grid would show. Thus, our eyeball-estimate is strongly positive: $r_{S_{\text{by eyeball}}} \approx .85$.

Ranking

The first step in computing Spearman's r_S is to assign an ordered rank to each data point. Let us call the teacher's ratings X_i, and we assign a rank, called R_{X_i}, to each. The rank of the *lowest* value of X_i is 1, the rank of the second lowest is 2, the rank of the third lowest is 3, . . . , and the rank of the highest value is 15 (because $n = 15$).

Inspection of the teacher's ratings column in Table 16.6 shows that the teacher has given *three* students the lowest rating (that is, $X_4 = X_6 = X_{12} = 1$). We would like to assign ranks of 1, 2, and 3 to these students, but how? This is an example of *tied ranks*, and the problem is resolved by averaging the ranks we would like to assign and then assigning the average to each subject who is tied. The average of 1, 2, and 3 is 2, so we assign that rank to each (that is, $R_{X_4} = R_{X_6} = R_{X_{12}} = 2$). Note that such an assignment "uses up" the three ranks 1, 2, and 3, so the next available rank is 4, which we assign to the fourth lowest value of X_i (which is $X_{15} = 2$, so $R_{X_{15}} = 4$). Note also that we do not assign the ranks of 1 and 3 because we assigned 2 three times.

We continue to assign ranks until all n values of X_i have been ranked. As a check to ensure that the ranks have been assigned correctly, we see that the highest rank is n (15 in our example) unless that itself is a tied rank.

In a similar fashion, we assign ranks (which we call R_{Y_i}) to each value of Y_i, again being careful to treat tied ranks correctly and checking to see that the highest rank is n.

We then compute the differences *between the ranks*:

difference between ranks for the *i*th individual

$$D_i = R_{X_i} - R_{Y_i} \tag{16.7}$$

as shown in the second to last column of Table 16.6. Note that we subtract the ranks, *not* the scores themselves. As a check of both our rank assignments and our subtractions, we compute $\sum D_i$, which should be exactly 0 if we have proceeded correctly.

We then square each of the D_i values to form the last column of Table 16.6 and sum down that column, obtaining, for our data, $\sum D_i^2 = 94.50$. We can compute Spearman's r_S from Equation (16.6):

$$r_S = 1 - \frac{6 \sum D_i^2}{n(n^2 - 1)}$$

$$= \frac{6(94.5)}{15(15^2 - 1)} = .83$$

We interpret r_S in the same way we would interpret r (it lies between -1 and $+1$), so we conclude that there is a strong correlation between this teacher's ratings of intelligence and his students' IQ scores.

Testing Hypotheses About r_S

As with Pearson's r, the most frequent hypothesis we wish to test regarding Spearman's r_S is whether it is significantly different from 0—that is, whether our data are related at all. Thus, the null hypothesis is H_0: $\rho_S = 0$. In our example, we have reason to expect the teacher's ratings to be positively related to actual IQ, so the test is directional; the alternative hypothesis is thus H_1: $\rho_S > 0$. As with Pearson's r, r_S serves as its own test statistic.

Note that we enter Table A.9 at n, not $n - 2$.

This test is straightforwardly performed by consulting Table A.9, which shows critical values of r_S as a function of the number of pairs of data. If the absolute value of r_S exceeds the critical value, we reject the null hypothesis.

The teacher's ratings example has 15 pairs of data. If, as usual, we assume that $\alpha = .05$, directional, then with $n = 15$ the critical value $r_{S_{cv}} = .441$ from Table A.9. Because our observed value $r_{S_{obs}} = .83$ is greater than $r_{S_{cv}} = .441$, we reject the null hypothesis and conclude that there is a significant correlation between our teacher's ratings and students' IQ scores.

The Pearson correlation coefficient is appropriate when both variables are measured at the interval/ratio level of measurement, and the Spearman correlation coefficient is appropriate when both variables are measured at either the ordinal or interval/ratio level. Other correlation coefficients exist for data measured at the nominal level, but we leave their discussion to other textbooks.

16.4 Connections

Cumulative Review

We continue the discrimination practice begun in Chapter 10. You may recall the two steps:

Step 1: Determine whether the problem asks for (a) the area under a normal distribution, (b) a confidence interval, or (c) the evaluation of a hypothesis. See the Cumulative Review in Section 10.5 for hints about this step.

Step 2: If the problem asks for the evaluation of a hypothesis, consult the summary of hypothesis testing in Table 16.7 to determine which kind of test is appropriate.

Table 16.7 shows that so far we have ten basic kinds of hypothesis tests (the one-sample test for the mean, two-independent-samples test for means, two-dependent-samples tests for means, three-or-more-independent-samples tests, post hoc tests, a priori tests, three-or-more-repeated-measures tests, the two-way ANOVA, and now two tests for correlation). Note that this table is identical to Table 15.9 with the addition of the last two lines appropriate for this chapter. We will continue to expand this table progressively throughout the remainder of the textbook.

Journals

A typical journal report of a correlation might be, "Height was statistically significantly related to weight, $r(8) = .63$, $p < .05$."

The correlations we have discussed in this chapter have described the degree of relationship between two variables. Researchers often make observations about more than two variables at a time. For example, a study that uses 100 subjects may gather several pieces of demographic information such as age and income, as well as several personality variables such as scores on tests of ego strength, aggressiveness, and affiliation. We could then compute a correlation coefficient for all the pairs of observations: age with income, age with ego strength, income with affiliation, aggressiveness with affiliation, and so on. Such results are often reported in a correlation matrix such as the one shown in Table 16.8, which shows that the correlation between Age and Income was .47, and between Aggression and Ego strength was −.23, and so on. Note that this table has *four* rows and columns, even though *five* variables are being intercorrelated.

To understand how this table is arranged, it is easiest to look first at the bottom row of the table, which shows the correlations between Affiliation and the four remaining variables. The second to last row would also show the four correlations between Aggression and the four remaining variables, but one of these (Aggression with Affiliation) has already been

TABLE 16.7 Summary of hypothesis-testing procedures in Chapters 10–16

Design	*Chapter*	*Null hypothesis*	*Sample statistic*	*Test statistic*	*Effect size index*
One sample					
σ known	10	$\mu = a$	$\overline{X}$	z	$d = \dfrac{\overline{X}_{obs} - \mu}{\sigma}$
σ unknown	10	$\mu = a$	$\overline{X}$	t	$d = \dfrac{\overline{X}_{obs} - \mu}{s}$
Proportion	Exercise 7, Chapter 10	$\Pi = a$	p	t	—[a]
Two independent samples					
For means	11	$\mu_1 = \mu_2$	$\overline{X}_1 - \overline{X}_2$	t	$d = \dfrac{\overline{X}_{1_{obs}} - \overline{X}_{2_{obs}}}{s_{pooled}}$
Two dependent samples					
For means	12	$\mu_D = 0$	$\overline{D}$	t	$d = \dfrac{\overline{D}_{obs}}{s_D}$
k independent samples					
For means (omnibus)	13	$\mu_1 = \mu_2 = \cdots = \mu_k$	n/a	F	$d_M = \dfrac{\overline{X}_{max} - \overline{X}_{min}}{s_{pooled}}$ or $f = \dfrac{s_{\overline{X}}}{s_{pooled}}$ or $R^2 = \dfrac{SS_B}{SS_T}$
Post hoc	14	$\mu_i = \mu_j$	$\overline{X}_i - \overline{X}_j$	Q	—
A priori	14	$C_1\mu_1 + \cdots + C_k\mu_k = 0$	$C_1\overline{X}_1 + \cdots + C_k\overline{X}_k$	t	$d = \dfrac{C_1\overline{X}_1 + \cdots + C_k\overline{X}_k}{s_{pooled}}$
k dependent samples					
For means	14	$\mu_1 = \mu_2 = \cdots = \mu_k$	n/a	F	$d_M = \dfrac{\overline{D}_{max}}{\sqrt{2MS_{residual}}}$
RC independent samples					
For row means	15	$\mu_{r1} = \mu_{r2} = \cdots = \mu_R$	n/a	F	—
For column means	15	$\mu_{c1} = \mu_{c2} = \cdots = \mu_C$	n/a	F	—
Interaction	15	$\mu_{rc} - \mu r - \mu_c + \mu_G = 0$	n/a	F	—
Correlation					
Parametric	16	$\rho = 0$	r	r (or t)	r^2
Ordinal	16	$\rho_s = 0$	r_s	r_s	—

[a] Effect size index is not discussed in this textbook.

TABLE 16.8 Pearson product-moment correlations between two demographic and three personality variables (illustrative data)

	Age	*Income*	*Ego strength*	*Aggression*
Income	.47***			
Ego strength	.22*	.15		
Aggression	−.14	.21*	−.23*	
Affiliation	.09	.11	.31**	−.15

$df = 98$. *$p < .05$; **$p < .01$; ***$p < .001$

displayed in the last row, and duplicating it would be redundant. Thus, the upper right half of this correlation matrix would duplicate the lower left half and so is conventionally left blank (or used for some other purpose). In a matrix that illustrates the correlations between k variables, there are $k(k-1)/2$ nonduplicated correlations, or $5(4)/2 = 10$ nonduplicated correlations in our example.

Note also that an asterisk is customarily used to indicate which of the correlations are significant at $p < .05$; two asterisks indicate $p < .01$, and so on. Ten hypothesis tests are described in Table 16.8, so if the level of significance $\alpha = .05$ and even if all the null hypotheses are in fact true, we should not be surprised if one or more of these correlation coefficients exceed the critical value by chance alone. Researchers often ignore this fact. For example, if our study had measured ten variables instead of five, there would have been $10(9)/2 = 45$ correlation coefficients, and 5% of them (or between two and three of them) should be expected to exceed the critical value by chance alone. (There are ways of avoiding this risk. For example, the "Bonferroni procedure" specifies how to reduce α when many hypotheses are being tested simultaneously. We leave the discussion of that and similar procedures to other textbooks.)

Note that some journal reports refer to the Spearman r_S correlation coefficient as "Spearman's ρ" or "Spearman's rho."

Computers

Use ESTAT's datagen to compute the correlation coefficient for the height and weight data in Table 16.2:

1. Enter the first subject's height and weight values from the first row of Table 16.2 into the first row (first two columns) of the datagen spreadsheet.
2. Enter the remaining nine height and weight values into the second through tenth rows of the datagen spreadsheet.
3. Datagen automatically computes the correlation coefficient, shown in the `datagen Statistics` window in the section headed "assuming the two variables are paired."
4. See Homework Tip 8. ESTAT provides the means and standard deviations automatically in the `datagen Descriptive Statistics` window.
5. You may wish to save this data set because we will use it again in Chapter 17.

Use SPSS to compute the correlation coefficient for the height and weight data in Table 16.2:

1. Enter the first subject's height and weight values from the first row of Table 16.2 into the first row (first two columns) of the SPSS spreadsheet.
2. Enter the remaining nine height and weight values into the second through tenth rows of the SPSS spreadsheet.
3. Click `Statistics`, then `Correlate`, and then `Bivariate....`
4. Click ▶ to move var00001 into the `Variables:` window.
5. Click `var00002`; then click ▶ to move it into the `Variables:` window.
6. Click `OK`. The accompanying correlation matrix appears in the `!Output1` window. This matrix has the general form shown in Table 16.8, with two exceptions: (a) Table 16.8 has one fewer row and column than the original number of variables, so this case would display only one row and one column. By having the extra row and column, SPSS can show the correlations between var00001 and itself and var00002 and itself on the main diagonal of its output. Those correlations are always 1.00. (b) SPSS prints the duplicated correlations in the upper right half of the matrix. Thus, .6343 appears as the correlation both between var00001 and var00002 and between var00002 and var00001, which are of course the same.
7. See Homework Tip 8. SPSS will provide means and standard deviations for you: Click the `Options...` button on the `Bivariate correlations` dialog window.
8. You may wish to save this data set because we will use it again in Chapter 17.

```
SPSS for MS WINDOWS Release 6.1

                    - -  Correlation Coefficients  - -

              VAR00001    VAR00002

VAR00001       1.0000       .6343
              (   10)     (   10)
              P= .        P= .049

VAR00002        .6343      1.0000
              (   10)     (   10)
              P= .049     P= .

(Coefficient / (Cases) / 2-tailed Significance)

" . " is printed if a coefficient cannot be computed
```

Pearson correlation coefficient, followed by number of cases and exact significance

Homework Tips

1. Check the list of learning objectives at the beginning of this chapter. Do you understand each one?

2. Note that the raw score formulas—Equations (16.1a) and (16.2a)—are easier to compute than their z score counterparts—Equations (16.1) and (16.2).

3. Equations (16.1a) and (16.2a) are not as difficult to compute as they may look. Only three new columns of computations are required: X^2, Y^2, and XY, as shown in Table 16.4.

4. Consider Equation (16.1a): The denominator is the square root of two bracketed expressions. Both those bracketed expressions must be positive; if your computation produces a negative value, you have made a mistake. The same applies to Equation (16.1b).

5. The correlation coefficient must be between -1 and $+1$. If your computation produces a value outside that range, you have made a mistake.

6. The most frequent single mistake in this chapter is made in determining tied ranks when computing Spearman's correlation coefficient. For example, the values 12, 13, 13, and 17 would receive the ranks 1, 2.5, 2.5, and 4 (*not* 3), and the values 132, 136, 140, 140, 140, and 163 would receive the ranks 1, 2, 4, 4, 4, and 6 (*not* 5), respectively.

7. When ranks are determined for n subjects, the last rank assigned must be n (unless it itself is tied). If the highest assigned rank is less than n, you have probably made the mistake of Homework Tip 6.

8. As always, if you use a computer to compute statistics for you, you should verify that the computer has interpreted your data correctly by checking the number of subjects and eyeball-estimating (or verifying by some other method) the means and standard deviations of your data.

Exercises for Chapter 16

Section A: Basic exercises (answers in Appendix D, page 528)

1. Use the data shown in the table.
   (a) Prepare a scatterplot.
   (b) Eyeball-estimate the correlation coefficient.
   (c) Compute the Pearson correlation coefficient.

X	Y
1	12
4	14
2	12
5	15
3	13

2. The data in the table represent family income (in thousands of dollars) and the size of house (in hundreds of square feet) that the family lives in.

Family	Income (× \$1000)	House size (× 100 sq ft)
1	23.5	16.4
2	11.8	11.2
3	28.1	19.4
4	26.0	16.9
5	18.3	13.3
6	27.5	11.9
7	38.8	13.4
8	35.7	17.4
9	28.5	17.3
Partial labor savings:		
$\sum X$	238.2	137.2
$\sum X^2$	6840.42	2156.48

(a) Prepare a scatterplot, placing family income on the X-axis.
(b) Eyeball-estimate the correlation coefficient.
(c) Compute the Pearson correlation coefficient. Why is it permissible to use Pearson's coefficient?
(d) Describe your results in plain English.

3. (*Exercise 2 continued*) Is the correlation coefficient you computed in Exercise 2 significantly different from 0?

I. State the null and alternative hypotheses

(a) What is the null hypothesis? The alternative hypothesis?
(b) Is this test directional or nondirectional? Why?
(c) What is the sample statistic?

II. Set the criterion for rejecting H_0

(d) What is the level of significance?
(e) How many degrees of freedom are there?
(f) What is (are) the critical value(s) of the test statistic from Table A.8 in Appendix A?

III. Collect a sample and compute statistics

(g) What is r_{obs}? [You eyeball-estimated it in Exercise 2(b) and then computed it in Exercise 2(c).]

IV. Interpret your results

(h) Should we reject H_0? Decide by eyeball and by computation.
(i) Interpret your result.

If you read Box 16.4, continue on. Otherwise, skip the remaining steps.

(j) As shown in Box 16.4, sketch the distribution of the sample statistic assuming the null hypothesis is true.
(k) As shown in Box 16.4, sketch the distribution of the test statistic assuming the null hypothesis is true.
(l) As shown in Box 16.4, what are the critical values of the test statistic (t_{cv}) and the sample statistic (r_{cv}) derived from it? Enter them on the appropriate distribution and shade the rejection region. Explain why r_{cv} is somewhat different here than it was in part (f).
(m) As in Box 16.4, show the observed value of the sample statistic on the appropriate distribution.
(n) As in Box 16.4, compute the observed value of the test statistic and show it on the appropriate distribution.
(o) What is your conclusion?

4. The data in the table represent the number of times a rat has run a maze and the time it took the rat to run the maze on its last trial.

(a) Prepare a scatterplot, placing number of trials on the X-axis.
(b) Eyeball-estimate the correlation coefficient.
(c) Compute the Pearson correlation coefficient. Why is it permissible to use Pearson's coefficient?
(d) Describe your results in plain English.

Rat	Trials	Time (sec)
1	8	10.9
2	9	8.6
3	6	11.4
4	5	13.6
5	3	10.3
6	6	11.7
7	3	10.7
8	2	14.8
Partial labor savings:		
$\sum X$	42	92.0
$\sum X^2$	264	1084.2

5. (*Exercise 4 continued*) Assume that the experimenter had reason to believe that the rats would learn to run the maze faster as they gained experience in the maze. Is the correlation coefficient you computed in Exercise 4 significant? Use parts (a)–(o) of Exercise 3.

6. (*Exercise 2 continued*) Suppose that each of the family incomes is increased by \$5000.

(a) How would that change the scatterplot of Exercise 2(a)?
(b) How would that change the eyeball-estimation of Exercise 2(b)?
(c) How would that change the correlation coefficient calculated in Exercise 2(c)? Verify your answer by computing the new correlation coefficient.

7. (*Exercise 2 continued*) Suppose that each of the family incomes is doubled.

(a) How would that change the scatterplot of Exercise 2(a)?
(b) How would that change the eyeball-estimation of Exercise 2(b)?
(c) How would that change the correlation coefficient calculated in Exercise 2(c)? Verify your answer by computing the new correlation coefficient.

8. Use the data of Exercise 1.

(a) Compute the Spearman r_S.
(b) How do the Pearson and Spearman correlations compare?

9. A teabag producer wishes to investigate whether the perceived pleasantness of the color of his brewed tea is related to the perceived pleasantness of its taste. He asks 15 subjects to rate his tea's color on a seven-point scale that ranges from 1 = very unpleasant to 7 = very pleasant, and also to rate its taste on a ten-point scale that ranges from 1 = very unpleasant to 10 = very pleasant. The data are shown in the table.

Rater	Color rating	Taste rating
1	5	7
2	4	6
3	5	10
4	2	1
5	5	8
6	3	6
7	4	6
8	6	5
9	6	6
10	5	9
11	4	8
12	5	5
13	5	5
14	1	3
15	4	4

(a) Prepare a scatterplot.
(b) Eyeball-estimate the correlation coefficient.
(c) What is the appropriate correlation coefficient here? Why? Compute it.
(d) Interpret this correlation in plain English.
(e) Is this correlation significantly different from 0?

Section B: Supplementary exercises

10. Consider the sample data in the table.

i	X_i	Y_i
1	6	11
2	3	8
3	5	7
4	8	13
5	2	7
6	2	9

(a) Prepare a scatterplot and eyeball-estimate the correlation coefficient.
(b) Compute z scores for each X_i.
(c) Compute z scores for each Y_i.
(d) Compute the Pearson correlation coefficient using the z score formula, Equation (16.1).
(e) Compute the Pearson correlation coefficient using the raw score formula, Equation (16.1a).
(f) Demonstrate that both z score and raw score formulas give identical results.

11. The data in the table represent the ages of students and the times they require to learn to perform a particular task.

Student	Age	Time (min)
1	18	12
2	12	18
3	12	15
4	10	27
5	4	24
6	15	17
7	9	16
8	17	24
9	11	25
10	11	26
11	10	16
12	19	14
13	5	29
14	12	19
15	13	23

(a) Prepare a scatterplot and eyeball-estimate the correlation coefficient.
(b) Compute the Pearson correlation coefficient.

12. (*Exercise 11 continued*)
(a) On your scatterplot, shade away all students who are younger than 7 or older than 14.
(b) Eyeball-estimate the correlation coefficient of the nine points that remain.
(c) Compute the Pearson correlation coefficient for the nine points that remain.
(d) Is the correlation computed here higher or lower than the one in Exercise 11? Is that what you would expect in general? Why?

13. Joe thinks that it takes him longer to mow his lawn when the grass is longer. Suppose he measures the time required and the grass length for the next ten times he mows his grass and then computes the Pearson correlation coefficient. Should he expect the coefficient to be positive, negative, or 0? Why?

14. (*Exercise 13 continued*) Joe's lawn mowing data are shown in the table.

Occasion	Length (in.)	Time (min)
1	2.4	68
2	3.1	57
3	2.1	66
4	3.7	54
5	3.3	62
6	1.6	63
7	.6	60
8	3.2	61
9	2.7	61
10	3.8	62
Partial labor savings:		
$\sum X$	26.5	614
$\sum X^2$	79.25	37,844

(a) Prepare a scatterplot and eyeball-estimate the correlation coefficient.
(b) Compute the Pearson correlation coefficient.

15. (*Exercise 14 continued*) Is Joe's correlation coefficient significantly different from 0? Do parts (a)–(o) of Exercise 3.

16. A psychologist wishes to know to what degree verbal skills and mathematical skills are interrelated. She obtains the verbal and quantitative scores on the Graduate Record Examination for a sample of 13 students. The scores are shown in the table.

Student	Verbal	Quantitative
1	440	390
2	401	501
3	388	443
4	430	496
5	359	512
6	671	692
7	669	480
8	386	491
9	592	665
10	583	622
11	543	407
12	586	611
13	420	505
Partial labor savings:		
$\sum X$	6,468	6,815
$\sum X^2$	3,370,522	3,680,959

(a) Prepare a scatterplot and eyeball-estimate the correlation coefficient.
(b) Compute the Pearson correlation coefficient.

17. (*Exercise 16 continued*) Is the correlation you obtained in Exercise 16 significant? Do parts (a)–(o) of Exercise 3.

18. A social researcher is aware of research that shows that physical attractiveness has important effects on many interpersonal behaviors. Critics of this research maintain that there is no general agreement on what is physically attractive: Some prefer thin, some prefer blonde, and so on. To test this, the researcher takes photographs of 20 individuals and asks two raters to rate each photograph on a 10-point attractiveness scale (10 = highly attractive). If the two raters have a strong correlation, then at least these two raters *do* agree on what is attractive. The ratings are listed in the table.

Photo	Rater 1	Rater 2	Photo	Rater 1	Rater 2
1	7	3	11	5	5
2	4	3	12	2	2
3	5	3	13	3	8
4	3	4	14	10	8
5	1	1	15	1	1
6	6	7	16	1	2
7	6	2	17	4	7
8	3	5	18	3	3
9	4	7	19	2	6
10	6	4	20	8	10

(a) Prepare a scatterplot.
(b) Eyeball-estimate the correlation coefficient.
(c) What is the appropriate correlation coefficient here? Why? Compute it.
(d) Is this correlation coefficient significant?
(e) Interpret your results in plain English.

19. A wine connoisseur claims to be able to tell the value of a wine by its taste. To test her claim, we purchase ten different cabernet sauvignon wines, with varying prices per bottle, and pour a glass from each. We identify each glass by a code number so that the connoisseur cannot identify the bottle from which the wine came. The connoisseur tastes each wine and rates it on a five-point scale: 1 = inexpensive, 2 = modest, 3 = moderate, 4 = expensive, and 5 = very expensive. The wine costs and ratings are listed in the table. Does the connoisseur know her wines? Use the same procedure as in Exercise 18.

Wine	Cost ($)	Rating
1	9.56	1
2	18.42	5
3	11.10	2
4	14.21	4
5	20.13	5
6	7.15	3
7	7.95	2
8	15.06	4
9	13.89	3
10	15.02	1

Section C: Extending your comprehension (answers in Appendix D, page 530)

20. Use the t test procedure described in Box 16.4 to determine the smallest positive correlation coefficient that would be significantly different from 0 (nondirectional test) in these cases. [*Hint:* Use an iterative method: Make a guess of r and see whether it is significant; use the results of the first guess to generate a second (better) guess of r; and continue until you have a guess of r that is just barely significant (accurate to within .01).]
(a) $n = 10$ and $\alpha = .05$
(b) $n = 30$ and $\alpha = .05$
(c) $n = 120$ and $\alpha = .05$
(d) $n = 30$ and $\alpha = .01$
(e) Compare your answers with the critical values of Pearson's r in Table A.8. Could Table A.8 have been generated using the procedure of this problem?

21. (a) Show algebraically that Equation (16.1) is equivalent to Equation (16.1a).
(b) Show algebraically that Equation (16.2) is equivalent to Equation (16.2a).
(c) Explain why there is no $(n - 1)$ in the denominator of Equation (16.1a).

22. What is the value of Pearson's r when $X_i = C$ for some constant C and all i?

23. What is the value of Pearson's r when $X_i = Y_i$ for all i?

Section D: Cumulative review (answers in Appendix D, page 530)

For Exercises 24–32, choose the correct answer or fill in the blank. (**Note that computations are *not* required;** use $\alpha = .05$ unless otherwise instructed.)

(a) This problem requires:
  (1) Finding the area under a normal distribution [Skip parts (b)–(g) and proceed to (h).]
  (2) Creating a confidence interval
  (3) Testing a hypothesis about the mean of one group
  (4) Testing a hypothesis about the means of two independent groups
  (5) Testing a hypothesis about the means of two dependent groups
  (6) Testing a hypothesis about the means of three or more independent groups
  (7) Testing hypotheses about means using post hoc tests
  (8) Testing an a priori hypothesis about means (planned comparison)
  (9) Testing a hypothesis about the means of three or more dependent groups
  (10) Testing hypotheses for main effects and interaction
  (11) Testing a hypothesis about the strength of a relationship
  (12) Performing a power analysis (involving finding the area under a normal distribution) [Skip parts (b)–(h).]

(b) The null hypothesis is of the form:
  (1) $\mu = a$
  (2) $\mu_1 = \mu_2$
  (3) $\mu_D = 0$
  (4) $\mu_1 = \mu_2 = \cdots = \mu_k$
  (5) A series of null hypotheses of the form $\mu_i = \mu_j$ (State how many such hypotheses.)
  (6) One null hypothesis of the form $C_1\mu_1 + C_2\mu_2 + \cdots + C_k\mu_k = 0$
  (7) Three null hypotheses:
  rows: $\mu_{\text{row }1} = \mu_{\text{row }2} = \cdots = \mu_{\text{row }R}$;
  columns: $\mu_{\text{col }1} = \mu_{\text{col }2} = \cdots = \mu_{\text{col }C}$; and
  interaction: all $\mu_{\text{row,col}} - \mu_{\text{row}} - \mu_{\text{col}} + \mu = 0$
  (8) $\rho = 0$
  (9) $\rho_S = 0$
  (10) There is no null hypothesis.

(c) The appropriate sample statistic(s) is (are):
  (1) $\bar{X}$
  (2) p
  (3) $\bar{X}_1 - \bar{X}_2$
  (4) $\bar{D}$
  (5) $\bar{X}_i - \bar{X}_j$
  (6) $C_1\bar{X}_1 + C_2\bar{X}_2 + \cdots + C_k\bar{X}_k$
  (7) r
  (8) r_S
  (9) Not applicable for ANOVA

(d) The appropriate test statistic(s) is (are):
  (1) z
  (2) t
  (3) F
  (4) Q
  (5) r
  (6) r_S
  (7) Not applicable

(e) The number of degrees of freedom for this test statistic is ________. (State the number or *not applicable*; if there are multiple null hypotheses, state the number of degrees of freedom for each.)

(f) The hypothesis test (or confidence interval) is
  (1) Directional
  (2) Nondirectional
  (3) Not applicable

(g) The critical value(s) of the test statistic(s) is (are) ________. (Give the value. If there are multiple null hypotheses, state the critical value for each; if not applicable, so state.)

(h) The appropriate formula(s) for the test statistic(s) (or confidence interval or strength of relationship) is (are) given by Equation (________). (If there are multiple null hypotheses, state the equation number for each.)

Computations are *not* required for any of the problems in Section D. See the instructions.

24. It is claimed that the British Broadcasting Corporation (BBC) news programs devote more time to each news story than do their American network news counterparts. To check this hypothesis, a random sample of 20 news stories are timed in Britain, and 20 stories are timed in the United States. The times (in minutes) are listed here. Are the BBC segments longer?

BBC: 40, 42, 66, 55, 56, 40, 44, 25, 40, 56, 43, 31, 21, 43, 35, 43, 53, 48, 40, 40

USA: 36, 32, 32, 34, 41, 42, 29, 39, 40, 29, 42, 31, 46, 30, 18, 20, 40, 32, 46, 35

25. A researcher wishes to know whether men who are judged to be more attractive marry women who are also judged to be more attractive, and men who are less attractive marry women who are less attractive. He photographs 20 couples, separates the husband from the wife, and has a

Couple	*Husband*	*Wife*
1	10	10
2	7	10
3	3	3
4	5	9
5	10	7
6	5	6
7	3	4
8	8	6
9	4	2
10	4	6
11	8	6
12	10	9
13	5	4
14	4	3
15	9	10
16	3	5
17	4	6
18	6	6
19	6	5
20	2	1

"blind" rater rate each photograph on a 10-point attractiveness scale (10 = very attractive). The ratings are listed in the table. Is there a significant positive relationship between husband and wife attractiveness?

26. An experimenter puts ten overweight husband–wife couples on a diet and wonders whether men will lose significantly more weight than women, or vice versa. The weight losses (in pounds) are listed in the table. What is the conclusion?

Couple	Husband	Wife
1	19	17
2	13	14
3	13	11
4	16	12
5	17	15
6	10	12
7	9	8
8	12	13
9	19	18
10	13	11

27. A pediatrician is studying weight gain in infants. He divides the infants into two groups: those who are primarily breast fed and those who are primarily bottle fed. Furthermore, he divides the infants into those whose mothers feed them primarily on a timed schedule and those whose mothers feed them when they cry ("on demand"). Weight gain is defined as weight at age 2 months minus weight at birth. The weight gains (in pounds) are in the table. Do method of feeding and feeding schedule affect weight gain?

	Breast	Bottle
	.6	7.1
	4.8	3.3
Schedule	1.9	3.0
	4.0	5.8
	2.2	7.6
	5.5	6.0
	8.2	4.7
Demand	4.5	4.3
	6.3	4.0
	3.2	7.3

28. A researcher wishes to know the speed of sound in pure nitrogen at a particular pressure and temperature. She measures the speed ten separate times. The speeds (in feet per second) are: 1127.1, 1123.1, 1101.2, 1102.2, 1112.4, 1100.6, 1098.5, 1102.8, 1096.8, 1097.1. What can she say about the speed of sound in nitrogen?

29. A production line produces widgets. The mean number of widgets produced per hour on this line has been 29.5 with standard deviation 3.4. The company has just installed better lighting on the production line. It has reason to believe that the lighting may change the mean but that the standard deviation will remain 3.4. It proposes to randomly sample 20 hours and record the number of widgets produced. Suppose that the new rate of widget production is actually 31.1. What is the probability that an experiment with this design will in fact conclude that the new lighting has increased productivity?

30. (*Exercise 29 continued*) The widget producer now actually performs the experiment and finds these numbers of widgets: 36, 30, 32, 35, 32, 26, 28, 29, 31, 28, 29, 35, 31, 32, 29, 31, 30, 39, 27, 29. Has the productivity of the line increased?

31. Dr. Warner studies the effectiveness of warning labels by showing a product with a warning label to a subject and asking the subject to rate the dangerousness of the product on a scale where 100 is extreme danger. This experiment explores the impact of the color of the label's text. Dr. Warner randomly assigns 28 subjects to one of four groups; each group sees one product and rates its dangerousness. The labels in the four groups are printed in one of the following fonts: red-roman, red-bold, black-italic, or black-inverted. Previous research indicates that the label color is important to the perception of danger, regardless of the font. Is that true in this study?

Red-roman	Red-bold	Black-italic	Black-inverted
37	47	26	34
33	34	36	41
42	39	27	30
49	43	26	32
37	43	26	27
34	47	27	28
36	43	18	36

32. Dr. T. Cher gives her ten-student class four 50-point multiple choice examinations during the semester. Do T's exams differ in degree of difficulty?

Student	Exam 1	Exam 2	Exam 3	Exam 4
1	40	39	46	43
2	35	38	28	35
3	43	38	43	36
4	26	36	34	25
5	39	36	37	32
6	33	31	30	22
7	39	35	36	32
8	36	32	38	35
9	38	37	33	39
10	28	21	40	37

Section E: From the journals (answers in Appendix D, page 530)

33. Riggio, Lippa, and Salinas (1990) investigated whether expressive movements (expansiveness or emphasis in hand gestures, for example) were related to personality traits. Among other things, they administered to 149 subjects the 16-PF Extraversion Scale (a personality trait mea-

sure) and collected a handwriting sample on which they measured loop width (the width of the loops in the letters a, p, and d, measured in millimeters at their widest point). Loop width served as a measure of expressive movement. They found that extraversion and loop width were correlated at .30 and stated that this value was significant ($p < .01$).

(a) What is the null hypothesis being tested here?

(b) How many degrees of freedom are there for this correlation?

(c) Verify that $r = .30$ is indeed significant.

34. (*Exercise 33 continued*) Suppose you replicate Riggio's (1990) study with 12 subjects. The results are shown in the table.

Extraversion	*Loop width*
25	3
26	1
17	2
20	4
10	3
18	4
22	3
33	8
19	3
20	4
25	5
23	3

(a) How many degrees of freedom do you have?

(b) Compute the correlation coefficient.

(c) How does your correlation coefficient compare to Riggio's?

(d) Is your correlation coefficient significant at $\alpha = .01$? How does that compare with Riggio's result?

35. Brown (1990) investigated the organization of public events in long-term memory. In particular, he was interested in the accuracy with which a person recalls the dates of past events and whether the extent of a person's knowledge about the events in question was related to his or her ability to date the events correctly. Brown presented a series of 32 events, such as "Sadat and Begin win the Nobel Prize," "Iraq initiates Iranian war," and "first space shuttle launched," and had each of 15 subjects rate their level of knowledge about each of these events on a ten-point scale ranging from 0 (no knowledge of the event) to 9 (a great deal of knowledge). Subjects were also asked to give their best estimates of the 32 dates in questions. Dating errors were computed by subtracting the actual date from the estimated date and then taking the absolute value of that error. The correlation between the knowledge ratings and dating errors was $-.56$ ($p < .001$). Although Brown apparently did not say so explicitly, this value seems to be a Pearson product-moment correlation coefficient taken across $15(32) = 480$ data points.

(a) How many degrees of freedom are there for this correlation? (Brown does not state this value.)

(b) Is this correlation in fact significant, $p < .001$?

(c) State in plain English your interpretation of Brown's result.

36. (*Exercise 35 continued*) Brown's (1990) study apparently computed the correlation coefficient both across and within subjects (statisticians would say that the study confounded between-subject and within-subject variation). Suppose you wish to replicate this study in such a way that avoids between-subject variability, and you do so by performing the study on *only one* subject. The data within this single subject are shown in the table.

Knowledge rating	*Dating errors*	*Knowledge rating*	*Dating errors*
1	2.5	4	1.9
9	1.2	5	7.5
5	3.9	7	1.4
7	2.8	7	1.3
8	3.7	2	8.2
3	4.3	7	5.9
7	3.5	9	5.3
8	.8	8	1.9
9	1.4	2	1.1
8	.0	7	2.3
8	.0	6	3.0
5	3.0	2	3.1
1	3.1	5	3.3
9	.4	5	3.0
7	3.6	7	.3
3	6.4	5	1.1

(a) How many degrees of freedom are there for this correlation?

(b) What is the Pearson product-moment correlation coefficient for these data?

(c) Is this correlation significant, $p < .01$?

(d) State in plain English your interpretation of this result.

37. Caddisfly larvae create silk "retreats" and defend those retreats against intruding larvae. Englund and Olsson (1990) noted that these defensive fights are sometimes won and sometimes lost, and they wished to investigate the factors that determined the outcome of the fights. One of the factors they considered was the relative sizes (which they called "size asymmetry") of the two larvae involved in the fight. They measured the weights of the larvae and computed the size asymmetry for each fight as follows: Size asymmetry = 0 if the two larvae have equal weights; size asymmetry = .1 if the defender is 10% heavier than the intruder; size asymmetry = $-.1$ if the defender is 10% lighter than the intruder; and so on.

Englund and Olsson observed 130 fights, each of which had two possible outcomes: takeover and non-takeover. They divided the size asymmetry scores into seven classes with approximately the same number of observations in each class. They then computed the frequency of non-takeovers within each class. To ascertain whether size asymmetry was related to outcome, they computed the Spearman rank-order correlation between these seven pairs of data: frequency of non-takeover and size asymmetry (as classified into the seven classes). Their results (as derived from the scatterplot that they presented) are shown in the table. Their computation of the Spearman rank-order correlation coefficient was $r_S = .96$, $p < .01$, with $n = 7$.

Size asymmetry	*Frequency of non-takeover*
−.41	.4
−.30	.4
−.10	.6
.05	.6
.15	.8
.35	.8
.43	1.0

(a) Why did Englund and Olsson use the Spearman correlation coefficient?
(b) Compute the Spearman correlation coefficient for these data. Does your result agree with theirs?
(c) Is this result significant at $\alpha = .01$?

38. (*Exercise 37 continued*) Suppose we replicate Englund and Olsson's (1990) experiment, also dividing size asymmetry into seven categories. We observe 247 fights, and our data are listed in the table.
(a) Compute the Spearman rank-order correlation for these data.
(b) Is this result significant at $\alpha = .01$?

Size asymmetry	*Frequency of non-takeover*
−.47	.39
−.35	.37
−.20	.54
.05	.65
.18	.80
.31	.75
.40	1.00

39. Best and Luckenbill (1990) tested Harris's theory of deviant typescripts, which predicts higher female participation in deviance when male dominance is low. For each of the 50 states, they measured male dominance within each state using the Status of Women Index (SWX). In Best and Luckenbill's use of the SWX, high values indicate high male dominance. Previous researchers had provided a rating of SWX scores for each of the 50 states. Best and Luckenbill computed the male–female homicide rate ratio for each state "by dividing the number of female-offender homicides by the 1980 female population, then performing the analogous calculation for males. The female homicide rate was then divided by the male rate to arrive at the population-based sex ratio for homicides" (p. 75).

Best and Luckenbill computed the Pearson product-moment correlation for the SWX and sex ratio for homicides across the 50 states and reported that $r = .366$, $p < .01$. They concluded that "in general, the results contradict Harris's prediction: male dominance is positively related to the population-based sex ratios for homicides" (p. 75).
(a) How many degrees of freedom are there for this correlation coefficient? (Best and Luckenbill do not specify this.)
(b) Is the reported correlation in fact significant at $p < .01$?
(c) Do you agree with Best and Luckenbill's conclusion? State it in your own words.

Section F: Computer explorations

40. (a) Have the computer create a random sample of size five from the population of IQ scores ($\mu = 100, \sigma = 15$). [*Hint:* We described how to do that in Section 7.3.]
(b) Have the computer create a second random sample of size five from the population of IQ scores ($\mu = 100, \sigma = 15$).
(c) If these are the data for a Pearson correlation coefficient, what can you say about the truth of the null hypothesis?
(d) Approximately what would you predict r to be?
(e) How does the actual r compare to your prediction in part (d)?

41. Repeat Exercise 40 several times to get a feel for the magnitude of r when the null hypothesis is true. Then try varying the sample size (try $n = 10$, 50, or 200, for example) to extend your comprehension.

42. Repeat the procedure of two random samples of size five in Exercise 40 until you get a small r (say, less than .2 and greater than −.2). Now add a single (sixth) point (called an "outlier") at (200, 200). How does that change r? Now add the outlier (1000, 1000). How does that change r? Note that 1000 is just 100.0 with the decimal point omitted. Speculate about the importance of being careful with decimal points when computing a correlation coefficient.

43. Repeat Exercise 42 except use samples of size 100 instead of five. Does that change your conclusion about decimal points in Exercise 42?

44. (a) Create a data set with five pairs of points that has correlation $r = 1.0$.
(b) Now alter one pair of those points until the correlation is $r = .5$.
(c) Alter one pair of those points until the correlation is $r = .0$.
(d) Alter one pair of those points until the correlation is $r = -.5$.
(e) Is it possible to alter one pair of those points until the correlation is $r = -1.0$? Explore and then explain.

CHAPTER

Prediction: Linear Regression

LEARNING OBJECTIVES

1. What does linear regression allow us to do?
2. What is the equation for the regression line?
3. How is the slope of the regression line computed?
4. How is the intercept of the regression line computed?
5. What is an error of prediction?
6. What is the standard error of estimate?
7. What is the least squares criterion?
8. What is the coefficient of determination?
9. How are hypotheses tested in regression?
10. How is the total sum of squares partitioned in regression?

This chapter describes linear regression, the science of using the linear (straight-line) relationship between two variables to predict the value of one variable (usually called Y) from the value of another (called X). The "error of prediction" is the difference between the actual value Y and its predicted value. The "best" regression line is the one that minimizes the sum of the squared errors of prediction. Specifying the equation for the best regression line requires specifying its slope b and its intercept a; this chapter gives formulas for computing those regression constants. The "standard error of estimate"—that is, the standard deviation of the errors of estimate—is a measure of the accuracy of the regression line when used as a predictor. Another measure of the adequacy of the regression line is the "coefficient of determination," which is the square of the Pearson correlation coefficient and the measure of how much variance X and Y share in common.

In Chapter 16, we saw that when two variables are positively correlated, we know that if a subject has a high score on one of the variables, he or she is also likely to have a high score on the other variable. Furthermore, if a subject has a low score on one variable, we can predict that he or she will also have a low score on the other. The higher the correlation coefficient, the more confident we are of our predictions.

In the height and weight example of Chapter 16, we found that the correlation between height and weight was $r = +.634$. Suppose we know that a man is 72 inches tall (which is taller than our average man). How heavy should we expect that man to be? The positive correlation tells us we should expect him to be heavier than average.

Although the positive correlation coefficient says we should predict a high weight, it does not tell us directly *how high* that predicted weight should be. Linear regression gives us the tool to make such a quantitative prediction.

17.1 Regression Lines

linear regression: determining the best straight line through a data set and using it to predict Y from X

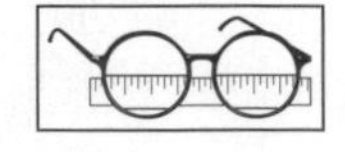

Recall from Chapter 16 that Pearson's r measures the degree of the *linear* relationship between two variables. *Linear regression* plots a straight line through the scatter diagram and uses it to predict the value of one variable from the value of the other. The science of linear regression provides rules for determining which is the *best* line for predicting one variable from the other.

Figure 17.1 shows the height/weight scatter diagram from Figure 16.2 with two possible regression lines (labeled A and B) superimposed. Do you think either of the lines fits the data points well? No; both lines A and B evidently have the wrong slope. The data points in general become higher as we move to the right in Figure 17.1, but lines A and B become lower or stay the same.

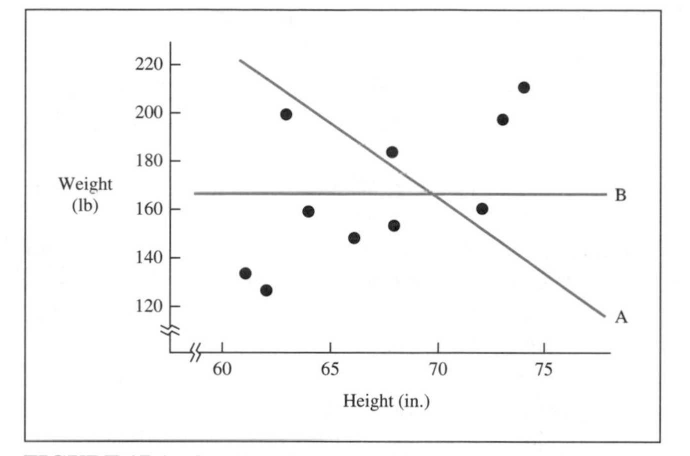

FIGURE 17.1 Scatter diagram of height and weight data with two possible (but incorrect) regression lines superimposed

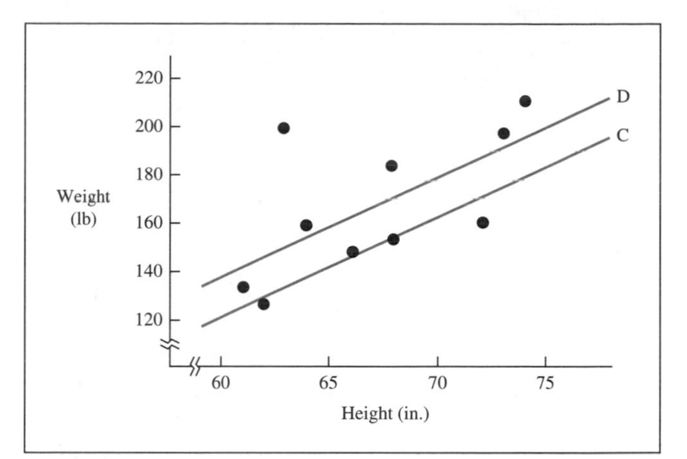

FIGURE 17.2 Scatter diagram with two possible regression lines superimposed. C has the correct slope, but the entire line is too low. D is the best line.

Figure 17.2 shows two more possible regression lines, both with the correct slope: up and to the right. Line C seems to be too low, however, because most of the points are above it. Line D seems by eyeball to be a good line to use to predict weight from height, and as we shall see, it is in fact the best such line for these data.

We have seen that is possible to eyeball-estimate a good regression line. There is a mathematically "best" regression line, and we will return to discuss how "best" is defined later. In the meantime, let's just accept that line D is mathematically the best line.

Figure 17.3 shows how a regression line is used to predict one variable from another. Suppose, for example, we know that George, the first subject in our data, is 72 inches tall, but we do not know anything else about him. We could use our regression line to predict his weight by starting at his height ($X_1 = 72$ inches) on the X-axis, going up to the regression line, then going over to the Y-axis. There we find that his weight, as predicted from the regression line, is expected to be 187 pounds.

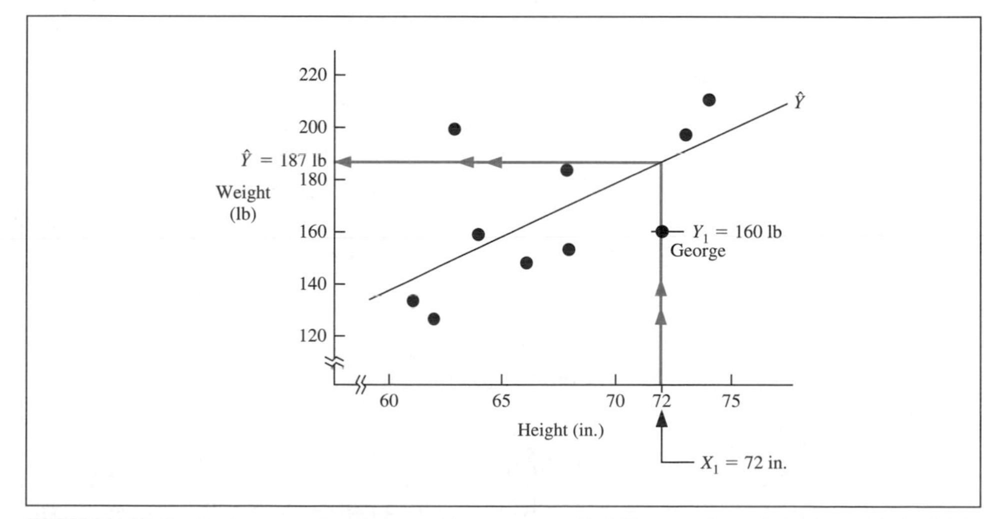

FIGURE 17.3 Best regression line and how to predict George's weight from his height

George's actual weight, as his point on the scatter diagram shows, is 160 pounds, not 187 pounds as we predicted, so our prediction is not perfect. We shall see later that the correlation coefficient can be thought of as a measure of accuracy of prediction: The closer the correlation coefficient is to 1 (or −1), the better the prediction.

Notation

As usual, we call the variable on the horizontal axis (height in our example) X and the individual value of that variable for the ith subject X_i. In Table 16.2, the first subject is George, so $X_1 = 72$ inches, shown in Figure 17.3 as the X coordinate of George's data point. Similarly, Y_i is the value of the Y variable (weight) for the ith subject; thus, $Y_1 = 160$ pounds, George's weight.

It is customary to put the variable we wish to predict on the Y-axis and the predictor variable on the X-axis. In our height and weight example, this choice is rather arbitrary because we could just as easily predict weight from height as height from weight. In other examples, it is quite clear which variable is to be predicted from which. For example, suppose we are interested in the experience of pain. We administer electric shocks of various intensities and measure the strength of avoidant muscle contractions. Here it is apparent that the natural order of prediction is to predict muscle response from the shock intensity, so muscle response would be placed on the Y-axis and shock intensity on the X-axis.

We call the values of Y as predicted from the regression line $\hat{Y}$ (read "Y hat"). The predicted Y value for the ith subject is called $\hat{Y}_i$; thus, $\hat{Y}_1 = 187$ pounds, our prediction for George. The *entire* regression line comprises values of $\hat{Y}$, points that we would predict from values of X.

The Equation

We are continuing to accept on faith that the line in Figure 17.3 (like line D in Figure 17.2) is mathematically the best line. Now we need to see how to write an equation for this line. It can be shown in plane geometry that any straight line can be defined by specifying two constants, called the *slope* of the line and its *intercept*. The general equation for a straight line is: The height of the line at any point X is equal to the slope of the line times X plus the intercept.

slope: the change in Y divided by the change in X
intercept: the value of Y when $X = 0$

The line we are interested in is the one we will use to predict Y values given values of X. Because every point on this line will be a predicted value of Y and because, as we have seen, we are referring to all predicted values of Y as $\hat{Y}$, Figure 17.3 labels the entire line $\hat{Y}$. It is customary in statistics to call the slope of the regression line b and its intercept a. Therefore, the equation for the regression line is

regression line

$$\hat{Y} = bX + a \tag{17.1}$$

[Many geometry books use as the notation for a straight line $Y = mX + b$, where m is the slope of the line and b is the intercept. To avoid confusion, you should keep clearly in mind that b refers to the *slope* in Equation (17.1) but to the *intercept* in those geometry books.]

When we know the X value for a particular individual—say, the ith subject—and we wish to predict the $\hat{Y}$ value, we write

predicted value of Y for the ith individual

$$\hat{Y}_i = bX_i + a \tag{17.2}$$

Understanding the concepts of regression requires that we be clear about the geometry of straight lines; we turn now to a review of that topic.

The Equation for a Straight Line

In order to specify a particular straight line, we specify its slope and intercept. Conversely, if we know the slope and intercept, we know the straight line.

The slope b of a straight line is calculated by the expression "rise over run." Choose any two points on the straight line shown in Figure 17.4. The *rise* is the change in Y as you travel the line from the left-hand point (point 1) to the right-hand point (point 2); that is, the rise is $Y_2 - Y_1$. Let us take as our two points the places on the line where $X_1 = 3$ and where $X_2 = 7$. As we traverse this line from point 1 to point 2, Y increases from $Y_1 = 3$ to $Y_2 = 5$, so the rise $= Y_2 - Y_1 = 5 - 3 = 2$.

The *run* is the increase in the X value as we make the same traverse; that is, the run is $X_2 - X_1$. Here the run $= X_2 - X_1 = 7 - 3 = 4$. The slope is the rise divided by the run; thus, b = slope = rise/run = $2/4 = .5$. You should verify for yourself that we can choose any two points on this line and the slope will always be .5.

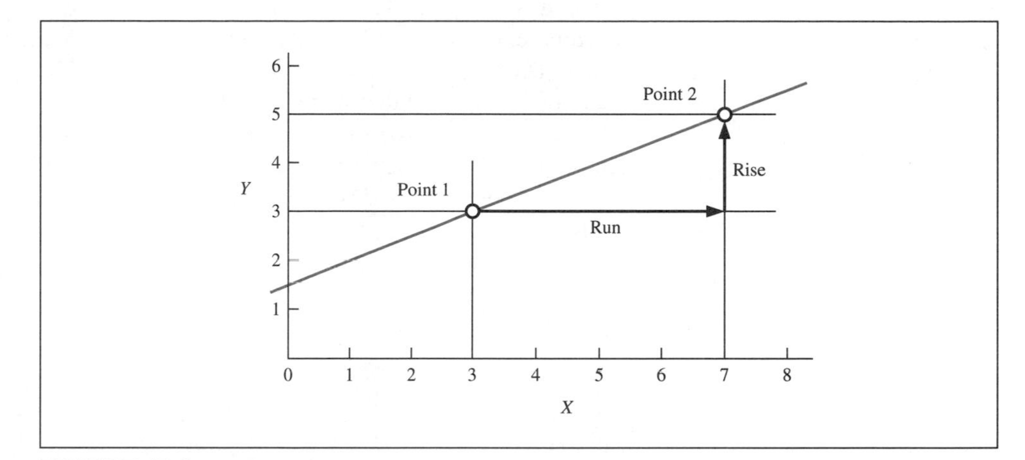

FIGURE 17.4 A straight line showing rise and run

Note that the *units of the slope* are the units of the rise (that is, the units of the Y-axis) divided by the units of the run (that is, the units of the X-axis). Thus, in our height and weight example, the units of b are pounds divided by inches, commonly called "pounds per inch."

The intercept, a, is defined to be the value of Y when $X = 0$. In Figure 17.4, when $X = 0$, $Y = 1.5$, so $a = 1.5$. Because the intercept is a value on the Y-axis, the *units of the intercept* are the same as the units of Y. In our height and weight example, the units of a are thus pounds.

The equation of the line in Figure 17.4 is $Y = bX + a = .5X + 1.5$.

Figure 17.5 shows that both slope and intercept can be either positive or negative. Notice that if the slope is positive, as in Figure 17.5C, the line slants from lower left to upper right. If the slope is negative, as in Figures 17.5A and D, the slant is from upper left to lower right. If the slope is 0, as in Figure 17.5B, the line is horizontal. A vertical line has infinite slope. Calculate each slope in Figure 17.5 by the rise over run method, making sure that your answers agree with those shown in the four equations of Figure 17.5 (that is, -1, 0, $+2$, and $-\frac{1}{2}$).

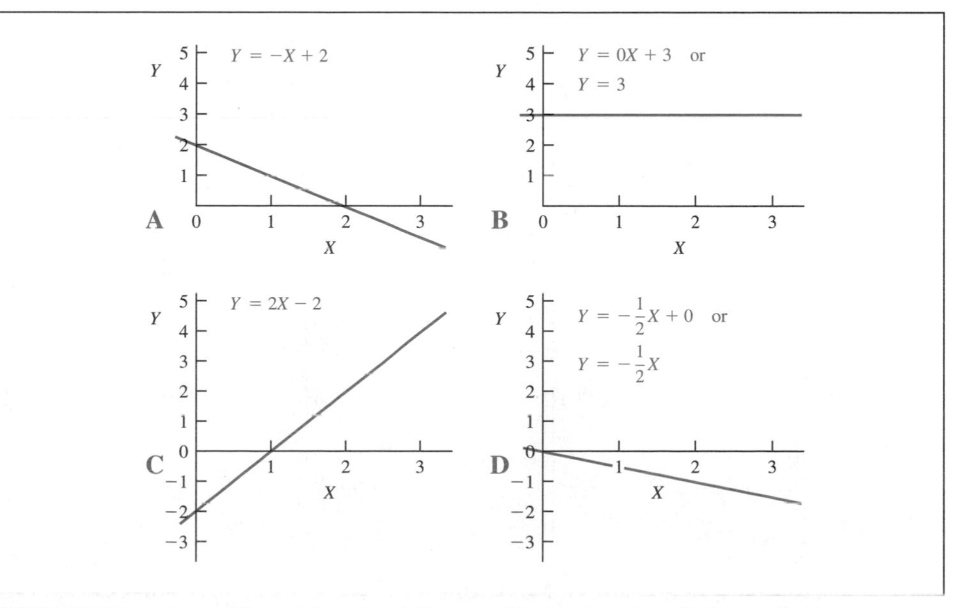

FIGURE 17.5 Lines with positive, negative, and zero slope and intercept

Note that the intercept is positive in Figures 17.5A and B, negative in C, and 0 in D. Again, make sure that your understanding of the intercept agrees with the values shown in the equations in the figure (+2, +3, −2, and 0).

We have seen how we can obtain the slope and intercept if we are given the regression line. It is also possible to plot the regression line if we are given the slope and intercept. For example, suppose we know that $b = +.5$ and $a = +.25$. How do we plot the regression line? The equation for the line is $Y = bX + a = .5X + .25$. We plot that line by identifying any two points and then connecting them.

For example, when $X = 1$, $Y = .5(1) + .25 = .75$, so we know the line must pass through the point $(X, Y) = (1, .75)$. We plot that point in Figure 17.6. For our second point, let us choose $X = 3$. Then $Y = .5(3) + .25 = 1.75$, and we plot the point (3, 1.75) also. All that remains is to connect the dots. As a check, we can see that the line does in fact pass through the intercept (0, .25).

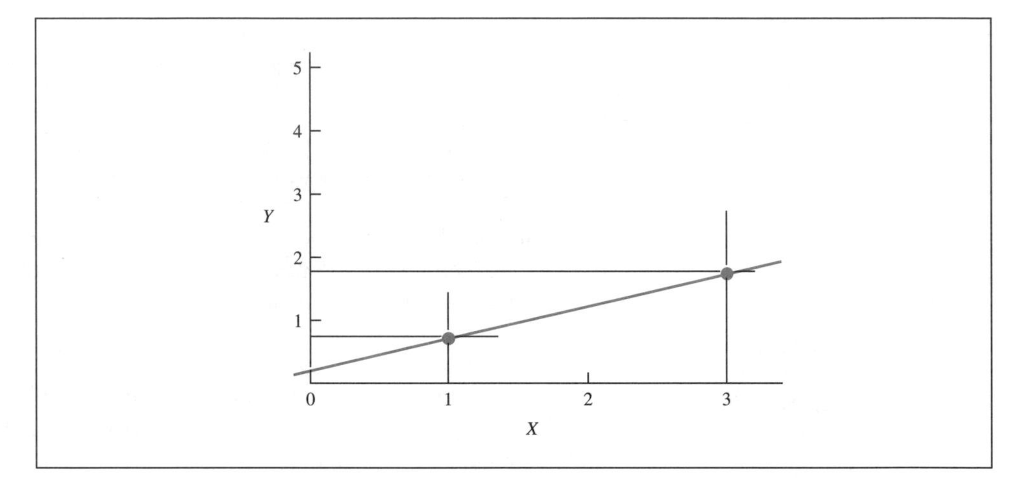

FIGURE 17.6 Constructing the line $Y = .5X + .25$

Eyeball-estimating the Constants

We saw in Figure 17.2 how to fit by eyeball a regression line through a scatter diagram. We redraw that figure here as Figure 17.7. Now we show how to eyeball-estimate the regression constants b and a from that line.

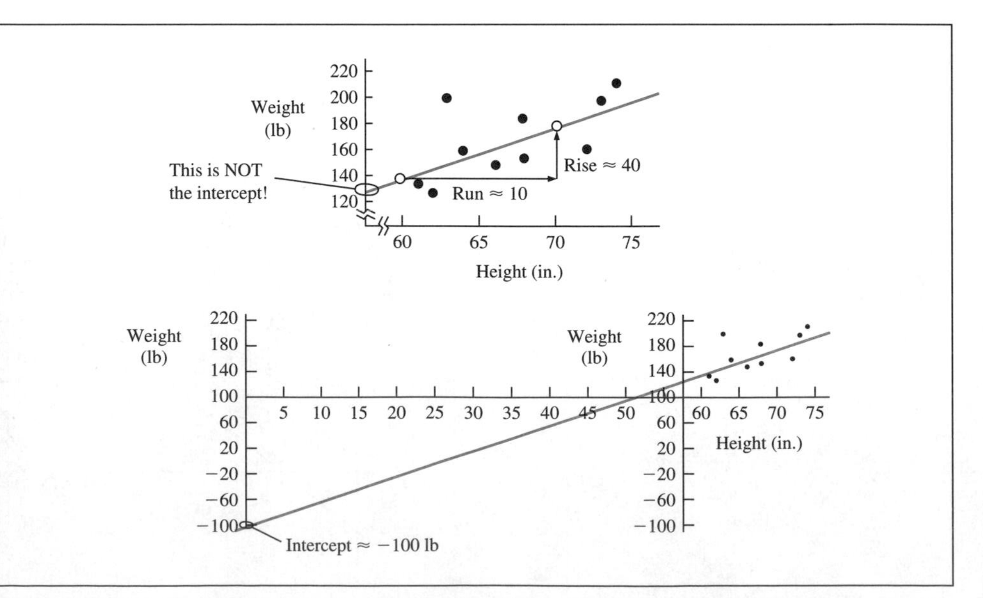

FIGURE 17.7 Eyeball-estimating slope and intercept. *Top:* Finding the slope by the rise over run method. *Bottom:* Extending the *X*-axis to find the intercept.

Slope: We eyeball-estimate the slope of that line using the rise over run method. We choose any two points on that line. Any two will do, although it is easiest if we choose round numbers on the X-axis. Our choices here are $X_1 = 60$ and $X_2 = 70$ (see the top of Figure 17.7). The run is therefore $X_2 - X_1 = 70 - 60 = 10$ inches. The rise is $\hat{Y}_2 - \hat{Y}_1$, which appears to be about $180 - 140 = 40$ pounds. Then the slope is rise over run, or $b_{\text{by eyeball}} \approx 40/10 = 4$ pounds per inch.

Intercept: The intercept is the value of $\hat{Y}$ when $X = 0$. It may seem that the intercept should be the point where the regression line crosses the Y-axis in the top portion of the figure, but that is *not* the case because the Y-axis is *not* drawn at $X = 0$. In order to find the intercept, we must extend the X-axis out to its zero point, as shown in the bottom of the figure. Note that the right-hand end of the bottom figure is a scaled-down version of the original scatter diagram. We can see that the intercept $a_{\text{by eyeball}} \approx -100$ pounds.

We shall see that the actual values are $b = 3.965$ and $a = -99.65$, very close to our eyeball-estimates.

Computing the Constants

We compute the regression constants as follows: We will provide two equivalent formulas for computing b and one for computing a.

The slope of the regression line is given by

slope of the regression line computed from the correlation coefficient

$$b = r\frac{s_Y}{s_X} \tag{17.3}$$

where r is the Pearson correlation coefficient between X and Y, s_Y is the standard deviation of all the Y values, and s_X is the standard deviation of all the X values. An equivalent formula is

slope of the regression line computed from raw scores

$$b = \frac{\sum XY - \dfrac{\sum X \sum Y}{n}}{\sum X^2 - \dfrac{(\sum X)^2}{n}} \tag{17.4}$$

The two formulas for b are entirely equivalent. Equation (17.3) is more convenient to use if you already know the correlation coefficient r and the standard deviations of X and Y (s_X and s_Y), whereas Equation (17.4) is more convenient if you have only the raw scores for X and Y.

Equation (17.3) shows how closely the slope of the regression line b is related to the correlation coefficient r: b is the correlation coefficient scaled by the ratio of the two standard deviations.

The intercept of the regression line is given by

intercept of the regression line

$$a = \overline{Y} - b\overline{X} \tag{17.5}$$

Another way of expressing Equation (17.5) is $\overline{Y} = b\overline{X} + a$, which implies that *all* regression lines pass through $(X, \overline{Y})$, the point that is both the mean of X and the mean of Y.

The statistics we computed in Chapter 16 for our height and weight example are shown in Table 17.1. For these data, we let height be the X variable and weight be Y. Then, from Equation (17.3):

$$\begin{aligned} b &= r\frac{s_Y}{s_X} \\ &= .634\frac{29.387}{4.701} \\ &= 3.963 \text{ pounds per inch} \end{aligned}$$

TABLE 17.1 Summary of statistics computed in Chapter 16 for the height and weight example

n	10
$\sum X$	671 inches
$\sum Y$	1,664 pounds
$\sum X^2$	45,223 inches^2
$\sum Y^2$	284,662 pounds^2
$\sum XY$	112,443 inch-pounds
$\overline{X}$	67.1 inches
s_X	4.701 inches
$\overline{Y}$	166.4 pounds
s_Y	29.387 pounds
r	.634

Alternatively, we could have computed b from Equation (17.4):

$$b = \frac{\sum XY - \dfrac{\sum X \sum Y}{n}}{\sum X^2 - \dfrac{(\sum X)^2}{n}}$$

$$= \frac{112{,}443 - \dfrac{(671)(1664)}{10}}{45{,}223 - \dfrac{(671)^2}{10}}$$

$$= 3.965 \text{ pounds per inch}$$

This is the same value except for rounding error. Rounding errors can be important in regression computations, so it is wise to keep as many decimal places as possible in each subcomputation. Thus, when r is used as an intermediate value, as above, we prefer at least three decimal places (.634) in the correlation coefficient.

We can then compute a from Equation (17.5):

$$a = \overline{Y} - b\overline{X}$$
$$= 166.4 - 3.965(67.1)$$
$$= -99.65 \text{ pounds}$$

This value of a is physically meaningless because there is no such thing as negative weight.

The formula for the regression of weight ($\hat{Y}$) on height (X) is therefore given by Equation (17.2) to be:

$$\hat{Y} = bX + a$$
$$= 3.965X - 99.65 \text{ pounds}$$

This is in fact the line that is plotted in Figure 17.3 (and Figure 17.2 D). To verify that, let us choose two values of X and compute their respective $\hat{Y}$ values from the preceding regression equation. For example, let $X_1 = 60$ inches, so $\hat{Y}_1 = 3.965(60) - 99.65 = 138.25$ pounds. Then one of the points on the regression line should be $(X_1, \hat{Y}_1) = (60, 138.25)$. For the second point, choose $X_2 = 75$ inches, so $\hat{Y}_2 = 3.965(75) - 99.65 = 197.72$ pounds. Thus, $(X_2, \hat{Y}_2) = (75, 197.72)$ should also lie on the line. Inspection of Figure 17.2 shows that these two points do in fact lie on the regression line that is drawn there. Because two points determine a line, the drawn regression line is in fact the line represented by the equation $\hat{Y} = 3.965X - 99.65$.

Regression Line Eyeball-calibration

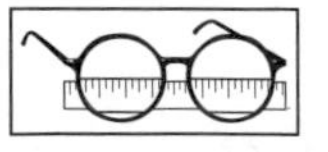

We provide a series of scatter diagrams for you to practice eyeball-estimating the regression lines and constants. We will go through the first one step by step. The scatter diagram in Figure 17.8 depicts the rate of return on an investment (in percent) as a function of the size of the investment (in thousands of dollars). Before we go further, sketch on this scatterplot the line that your eyeball tells you is the best regression line. Note that the letters "A, B, C,..." and "a, b, c,..." are located on the sides of the scatterplot. If you extend your line so that it intersects these letters, the letters will allow you to compare your line to my eyeball-estimate, which is shown in Figure 17.9.

My eyeballed line intersects "G" and "d." Your line should be close, but not necessarily through exactly the same letters as mine. Now that we have eyeballed the line itself, we must eyeball-estimate the slope and intercept of this line.

We choose any two points to estimate the slope. The points $X_1 = 20$ and $X_2 = 30$ produce the run $\approx 30 - 20 = 10$ thousand dollars. The corresponding $\hat{Y}$ values appear to be $\hat{Y}_1 \approx 8.9\%$ and $\hat{Y}_2 \approx 10.3\%$, so the rise $\approx 10.3 - 8.9 = 1.4\%$. The slope $b_{\text{by eyeball}} =$ rise/run $\approx 1.4/10 = .14\%$ per thousand dollars. Check to make sure the slope has the

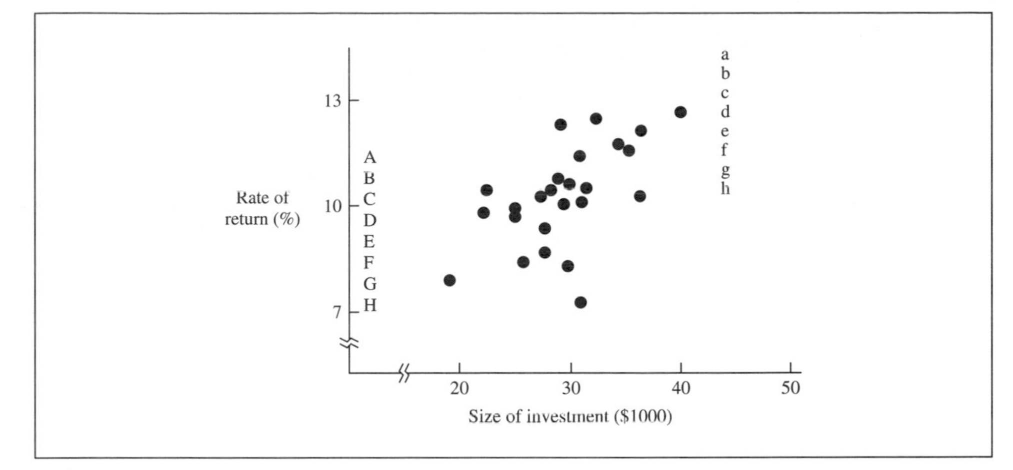

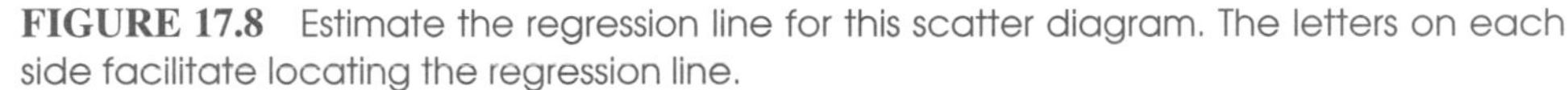
FIGURE 17.8 Estimate the regression line for this scatter diagram. The letters on each side facilitate locating the regression line.

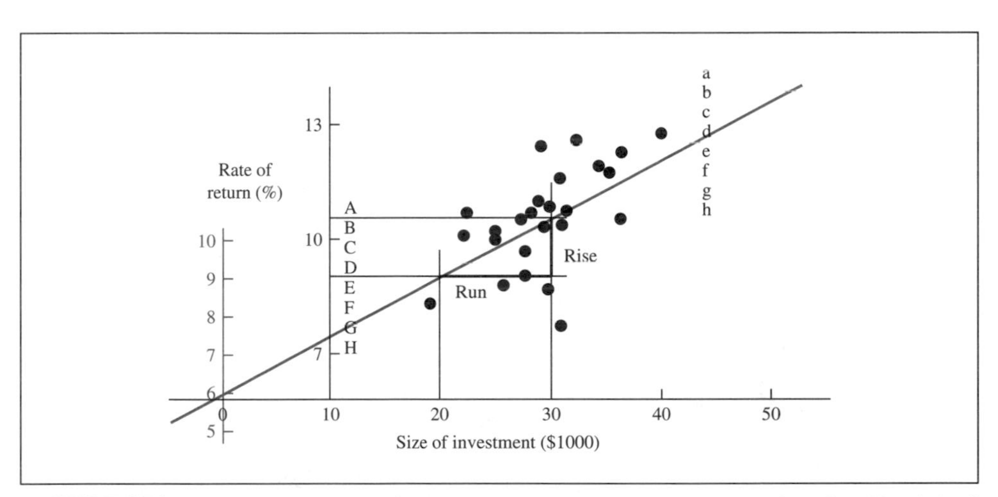

FIGURE 17.9 Eyeball-estimating the regression line, including extending the X-axis to 0

correct sign: b is positive if the line goes from lower left to upper right, and negative if the line goes from upper left to lower right (just like the correlation coefficient).

The intercept is the value of $\hat{Y}$ when $X = 0$. Because the vertical axis is *not* drawn at $X = 0$ in Figure 17.9, but at $X = 10$ thousand dollars, we must extend the axes until the point corresponding to $X = 0$ is found. Thus, the intercept $a_{\text{by eyeball}} \approx 6\%$.

The eyeballed equation for the regression line is thus $\hat{Y}_{\text{by eyeball}} \approx .14X + 6\%$.

The computed values for this scatterplot are $b = .15\%$ per thousand dollars and $a = 5.86\%$, so the actual regression equation is $\hat{Y} = .15X + 5.86\%$. The actual regression line passes through the letters G and d, approximately as we have drawn it.

Figures 17.10 through 17.15 are six scatter diagrams for you to practice eyeball-estimating the regression lines and eyeball-estimating b and a. The actual values of b and a, as well as the letters through which the actual regression line passes, are given in Table 17.2.

You may wish to use the ESTAT exercise "regtry" at this time.

You will find that you can be considerably more accurate in eyeball-estimating b than in estimating a, particularly in those cases that require extending the X-axis to its zero point. We shall see later that b is the more important regression constant. You may also find that the closer the correlation coefficient is to 0, the more difficult it is to accurately fit the regression line.

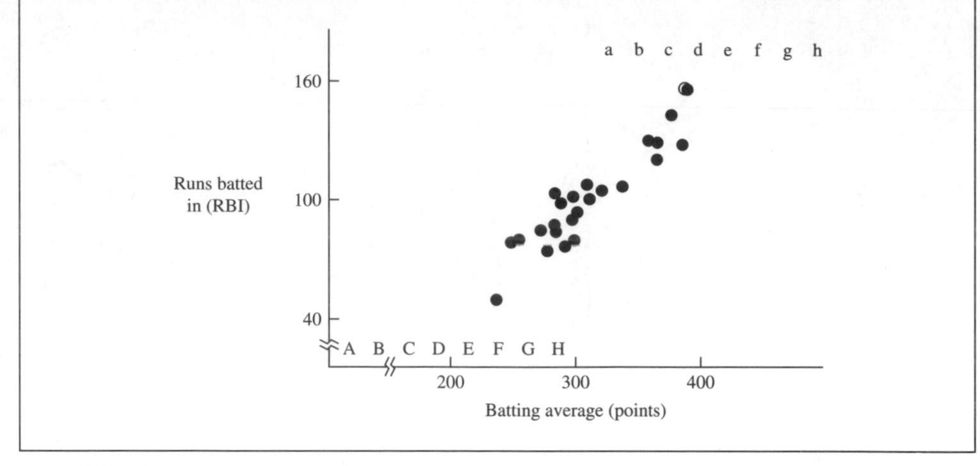

FIGURE 17.10

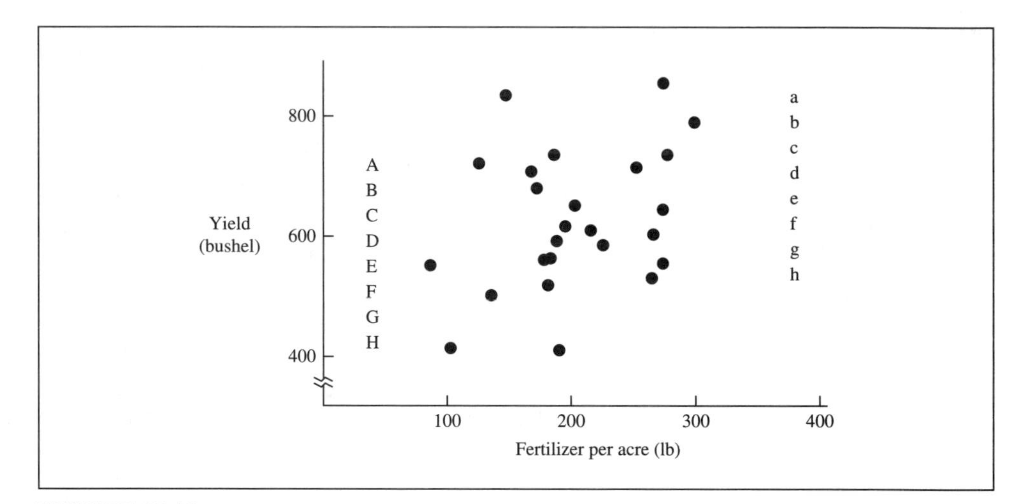

FIGURE 17.11

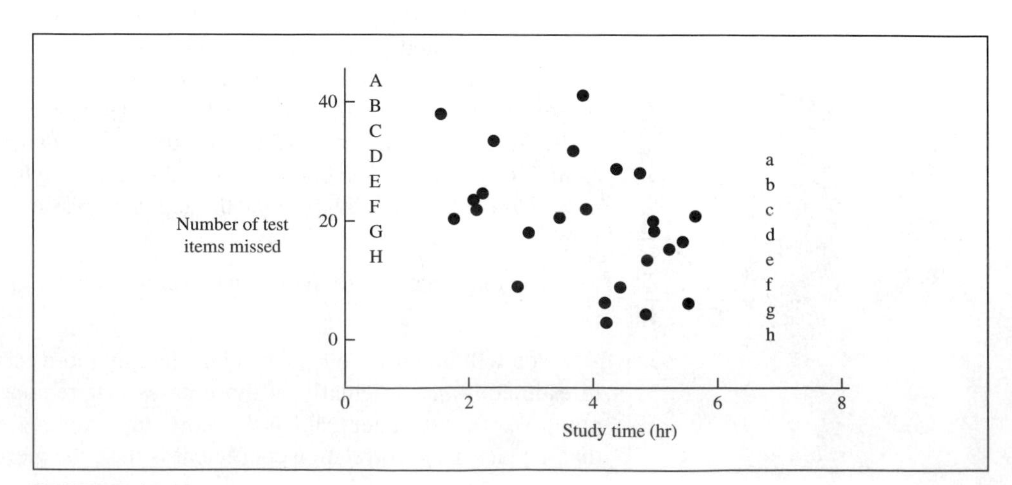

FIGURE 17.12

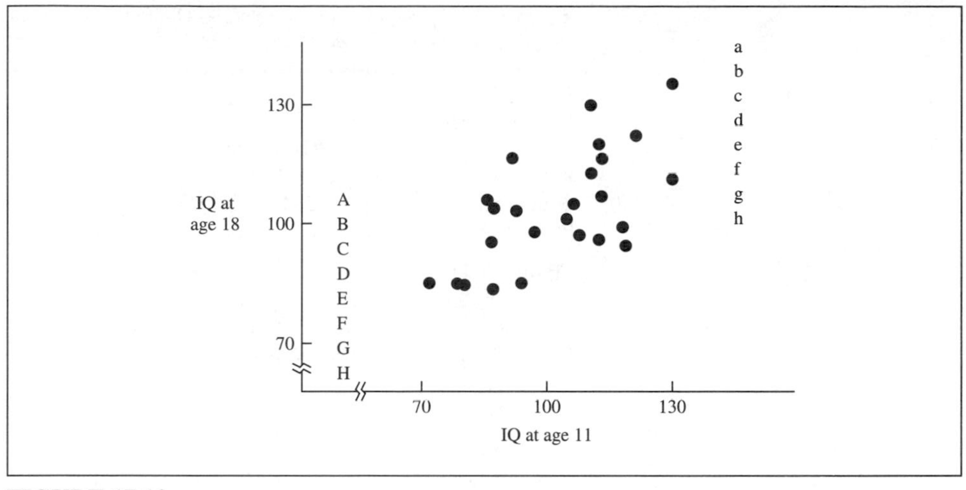

FIGURE 17.13

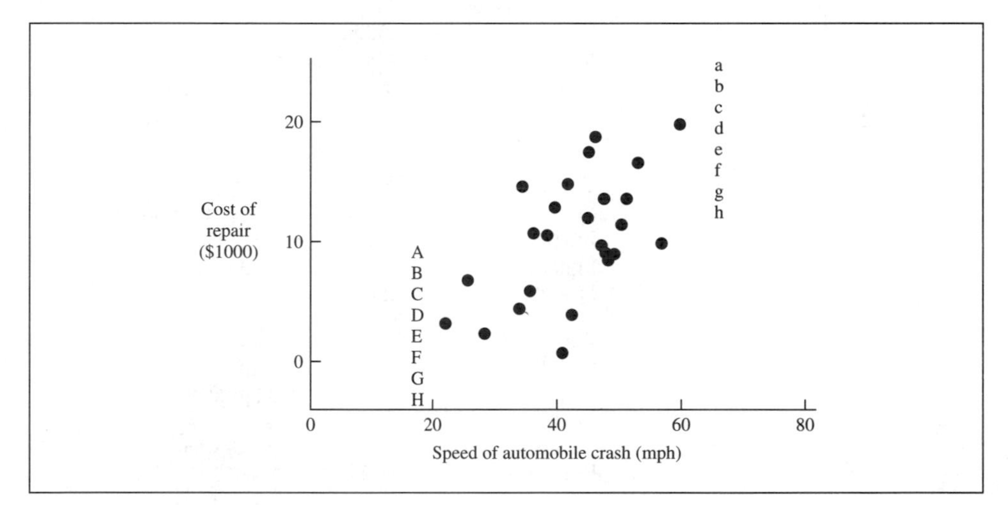

FIGURE 17.14

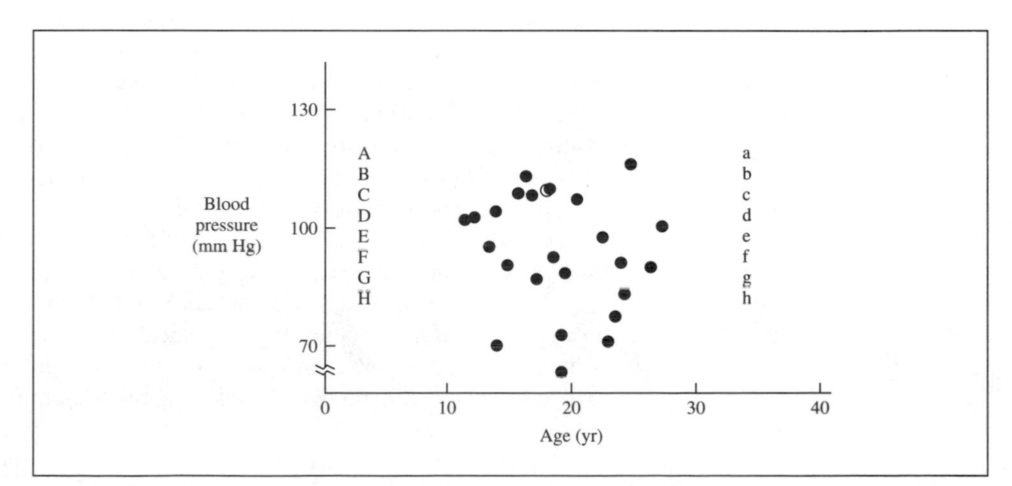

FIGURE 17.15

TABLE 17.2 Answers to eyeball-estimation of regression lines (Figures 17.10–17.15)

Figure 17.10	
Line intersects	Cf
Slope b	.534 RBI per point
Intercept a	−64 RBIs
Equation	$\hat{Y} = .534X - 64$ RBIs
Pearson's r	.93
Figure 17.11	
Line intersects	Fc
Slope b	.67 bushel per pound
Intercept a	486.7 bushels
Equation	$\hat{Y} = .67X + 486.7$ bushels
Pearson's r	.37
Figure 17.12	
Line intersects	Df
Slope b	−3.3 items per hour
Intercept a	32.47 items
Equation	$\hat{Y} = -3.3X + 32.47$ items
Pearson's r	−.42
Figure 17.13	
Line intersects	Fd
Slope b	.53 point per point
Intercept a	49.2 points
Equation	$\hat{Y} = .53X + 49.2$ points
Pearson's r	.65
Figure 17.14	
Line intersects	Ee
Slope b	.32 thousand dollars per mph
Intercept a	−3.3 thousand dollars
Equation	$\hat{Y} = .32X - 3.3$ thousand dollars
Pearson's r	.58
Figure 17.15	
Line intersects	Dg
Slope b	−.53 mm Hg per year
Intercept a	104.1 mm Hg
Equation	$\hat{Y} = -.53X + 104.1$ mm Hg
Pearson's r	−.17

Interpreting the Constants

The slope of the regression line b gives the increase in the predicted value of Y for a unit increase in X. Thus, in our height and weight example, because $b = 3.965$ pounds per inch, we see that for every 1 inch taller a man is, we should predict that he will be 3.965 pounds heavier. Additional examples of the importance of b are found in Figures 17.10–17.15. In Figure 17.10, for example, the slope $b = .534$ RBI per point indicates that, on average, a batting average (X) increase of one point is associated with an additional half (actually, .534) an RBI (Y). The data of Figure 17.11 show that, on average, each additional pound of fertilizer is associated with an increase of .67 bushel of produce. A statistically sophisticated farmer would know the cost per pound of fertilizer and the price he can receive per bushel of produce; if the price exceeds the cost, he would use that fertilizer.

The intercept a is the $\hat{Y}$ value when $X = 0$. In many cases (as in our height and weight example), a has no real meaning. Technically it is the weight we would predict for a man whose height is 0; however, there are no zero-height men. Furthermore, the weight we would predict is −99.65 pounds, and negative weight is absurd. For another example, the intercept a in Figure 17.14 is a repair cost of −3.3 thousand dollars. A negative repair cost is an impossible result. The intercept is a value that allows us to write the regression line equation in standard form, but it often does not have real significance.

Note that b always has the same *sign* as Pearson's r. For our height and weight data, both are positive because the scatter diagram slopes from lower left to upper right. This fact can be derived from Equation (17.3). Standard deviations are always positive, so s_Y/s_X must be positive and $b = r(s_Y/s_X)$ must have the same sign as r.

17.2 Regression Line in Standard Form

As you may recall from Chapter 16, the equations for the correlation coefficient were substantially simpler when expressed in standard scores (z scores) than when expressed in raw scores. The same is true in the regression situation. Just as we did in Figure 16.4, we superimpose z_X and z_Y axes on our height/weight scatterplot in Figure 17.16. The points in the scatterplot stay the same when we superimpose these axes; therefore, the correlation coefficient stays the same and the best regression line stays the same. However, we now have to write an equation for the best regression line in terms of $\hat{z}_Y$ and z_X instead of $\hat{Y}$ and X.

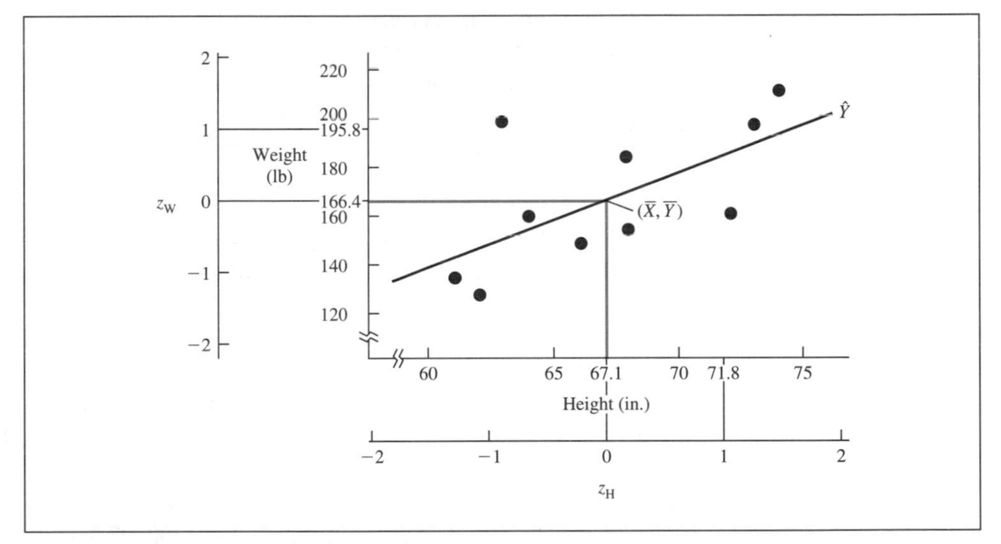

FIGURE 17.16 Regression line with z score axes added

Recall from Equation (17.1) that the general formula for the regression line is $\hat{Y} = bX + a$. If $\hat{z}_Y$ and z_X are z scores for $\hat{Y}$ and X, respectively, then $\hat{z}_Y = bz_X + a$. Recall that Equation (17.3) told us how to obtain b: $b = r(s_Y/s_X)$. Now, however, we wish to use that equation with z scores, so $b = r(s_{z_Y}/s_{z_X})$. Recall from Chapter 6 that z scores always have standard deviation 1, so $s_{z_Y} = 1$ and $s_{z_X} = 1$ and our equation for b becomes $b = r(1/1)$, or simply $b = r$.

Now recall that Equation (17.5) told us how to obtain a: $a = \overline{Y} - b\overline{X}$, which for our z scores becomes $a = \overline{z}_Y - b\overline{z}_X$. Because $b = r$ and because z scores always have mean 0, we see that $a = \overline{z}_Y - b\overline{z}_X = 0 - r0 = 0$.

Now we perform our substitutions to obtain the equation for the regression line for standard scores: $\hat{z}_Y = bz_X + a = rz_X + 0$, or simply

regression line expressed as standard scores

$$\hat{z}_Y = rz_X \tag{17.6}$$

where $\hat{z}_Y$ (read "z sub Y hat") is the predicted value of z_Y (that is, the z score associated with $\hat{Y}$), z_X is the z score associated with X, and r, the correlation coefficient, is the slope of the regression line when expressed as standard scores.

Equation (17.6), the best regression line expressed in z scores, defines the *same line* as does Equation (17.1), $\hat{Y} = bX + a$, the best regression line expressed in raw scores. The points in the scatterplot do not care how we transform the axes. Furthermore, the slope of the standard-score regression line is simply r, the correlation coefficient.

Equation (17.6) tells us that the intercept of the standard-score regression line is always 0. Recall that the definition of the intercept is the value of $\hat{z}_Y$ when $z_X = 0$. Thus, for all standard-score regression lines, $\hat{z}_Y = 0$ when $z_X = 0$. But $z_X = 0$ is the mean of the X values and $\hat{z}_Y$ is the mean of the Y values, which implies that the best regression line always passes through the point that is the mean of both variables.

The standard-score form of the regression line, Equation (17.6), can give us important insights into the nature of regression. First, consider the case where $r = 0$. Equation (17.6) tells us that if $r = 0$, we will *always* predict $\hat{z}_Y = 0$ (that is, we will always predict $\overline{Y}$) regardless of the value of z_X (and therefore regardless of the value of X).

Second, consider the case where $r = 1$. Equation (17.6) tells us that if $r = 1$, we will *always* predict $\hat{z}_Y = z_X$. For example, if X is two standard deviations above its mean ($z_X = 2$), then the predicted value of Y will be two standard deviations above its mean ($\hat{z}_Y = 2$).

regression to the mean: the predicted variable lies closer to its mean than does the predictor variable

Third, and perhaps most surprising in its ramifications, for intermediate values of r (that is, not 0 and not ± 1) we should always predict a Y value *closer to its mean* than was the predicting X value, closer by a factor of r. This phenomenon is called *regression to the mean*. Let's consider three examples.

Example 1 (a surprise): IQ is relatively (but not absolutely) stable over time; for example, the correlation between IQ measured at age 5 years and IQ measured again at adulthood is .5. Suppose that Sally is 5 years old, you administer an IQ test, and her score is 120. What should you expect Sally's adult IQ to be? The best prediction is *not* 120 but rather 110! Here's why: The predictor is Sally's current IQ score, $X = 120$. Because for IQ scores, $\mu = 100$ and $\sigma = 15$, Equation (6.1) gives $z_X = (X - \mu)/\sigma = (120 - 100)/15 = 1.33$. This tells us that Sally's IQ of 120 is 1.33 standard deviations above the mean of 100. Because $r = .5$, our best prediction of Sally's adult IQ would have a z score of $\hat{z}_Y = rz_X = .5(1.33) = .667$, which implies that Sally's adult IQ should be only .667 (instead of 1.33) standard deviation above its mean. Translating that back to IQ scores using Equation (6.3) gives $\hat{Y} = \hat{z}_Y\sigma + \mu = .667(15) + 100 = 110$, the best prediction of Sally's adult IQ.

Example 2 (nature regresses to the mean): The heights of parent and child are positively correlated with r approximately .5. If John is exceptionally tall [say, 2 standard deviations above the population mean height ($z_X = 2$)], then his son Jack is also expected to be taller than the average man [$z_Y = .5(2)$, or 1 standard deviation above the population mean)], but that is *shorter* than John himself. Furthermore, if Sally is exceptionally short [say, 3 standard deviations below the population mean ($z_X = -3$)], then her daughter Jill is also expected to be shorter than the average woman [$z_Y = .5(-3) = -1.5$, or 1.5 standard deviations below the population mean), but that is *taller* than Sally herself. In both cases, the offspring are expected to be closer to the mean than were the parents.

Example 3 (a frequent mistake): Dr. Nostat believes he has a program that increases IQ among low-IQ children. He gives 1000 children a standard IQ test that has test–retest reliability of .9. ("Test–retest reliability" is the correlation coefficient between the same IQ test administered to the same children on two different occasions.) Dr. Nostat selects the lowest 100 scorers on this test and finds that their mean IQ is 80. He enrolls these 100 children in his IQ-raising program and at its conclusion measures their IQs again. The mean IQ for these children is now 82. He concludes that his program raised IQ by 2 points. Is he correct? No. *An increase of 2 points is exactly what would be expected by regression to the mean alone*. The average child in his group has at the beginning of his study IQ $X = 80$. The corresponding z score from Equation (6.1) is thus $z_X = (X - \mu)/\sigma = (80 - 100)/15 = -1.33$, or 1.33 standard deviations below the mean. Then because $r = .9$, $\hat{z}_Y = rz_X = .9(-1.33) = -1.197$, which implies that on second testing the typical student should be 1.197 standard deviations below the mean, which, from Equation (6.3), gives $\hat{Y} = \hat{z}_Y\sigma + \mu = -1.197(15) + 100 = 82$. Thus, regression to the mean *alone* (with *no* help from an IQ-raising program) predicts that this group will, on average, gain 2 points in IQ.

As these considerations show, the correlation coefficient r is a measure of how well the regression line predicts Y from X. If $r = 0$, the regression line is worthless: We always predict $\overline{Y}$ regardless of X. If $r = 1$ (or -1), the regression line is a perfect predictor. The larger (in absolute value) the r, the better the prediction (that is, the less the expected regression to the mean).

You may wish to use ESTAT exercise "corest" (particularly Path 2) at this time.

17.3 The Best (Least Squares) Regression Line

We have said that the equations we have been using in this chapter are for the "best" regression line, but until now we have not questioned what "best" actually means. We turn to that topic now.

In Figure 17.3 we predicted the weight of the first subject George given his height of 72 inches. Our prediction was that he would weigh $\hat{Y}_1 = 187$ pounds; however, his actual weight is $Y_1 = 160$ pounds. Because the correlation is not perfect, our prediction is not perfect.

Error of Prediction

error of prediction: the difference between the actual value of Y and the value predicted from X

We call the difference between the actual value of Y and its predicted value $\hat{Y}$ the *error of prediction* (or sometimes the *residual*) and denote it by e. Thus, the error of prediction for the ith individual is e_i, which is obtained from

error of prediction for the ith individual

$$e_i = Y_i - \hat{Y}_i \qquad (17.7)$$

That is, the error of prediction e_i is the difference between the ith subject's actual Y value (Y_i) and the value we would predict from the regression line ($\hat{Y}_i$). The error of prediction for George's weight is $e_1 = Y_1 - \hat{Y}_1 = 160 - 187 = -27$ pounds.

The errors of prediction are shown in Figure 17.17 as arrows *from* the regression line *to* the data point. This emphasizes that the values of e_i are positive for points that lie above the line and negative for points that lie below the line. Because George's actual weight is less than the predicted weight, the error of prediction for him is negative.

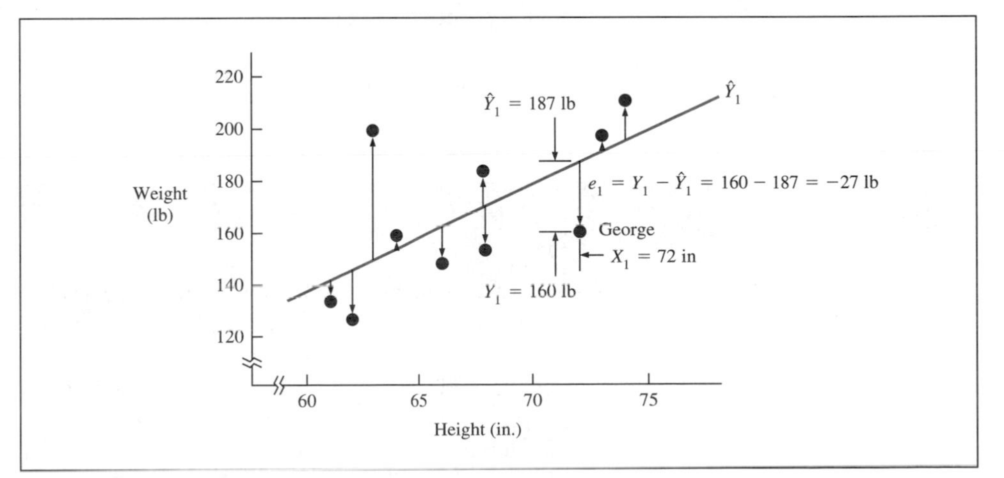

FIGURE 17.17 Errors of prediction

Note that the error of prediction is a *vertical* distance between the subject's data point and the regression line. (Students frequently hold the *mistaken* impression that the error of prediction is a distance *perpendicular* to the regression line. That is *not* the case.)

The Best Line

It might seem that a line that minimizes the sum of the errors of prediction (that is, for which $\sum e_i$ is a minimum) would be the best regression line. Unfortunately, this is not adequate as a criterion because there are many lines for which $\sum e_i = 0$.[1] However, any line that passes through the point $(\overline{X}, \overline{Y})$ has $\sum e_i = 0$. For example, Figure 17.18 shows a line that passes through the point ($\overline{X} = 67.1$, $\overline{Y} = 166.4$). This line has $\sum e_i = 0$ (verify for yourself that the sum of the lengths of the upward-pointing arrows appears to equal the sum of the lengths of the downward-pointing arrows). However, the errors of prediction themselves (the e_i) are much longer than in Figure 17.17, so this line is clearly *not* as good a predictor as the line in Figure 17.17.

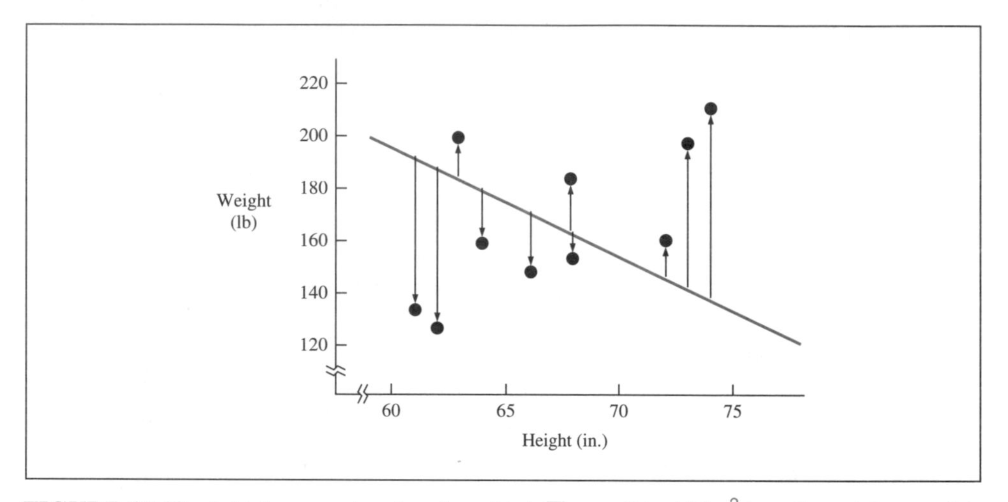

FIGURE 17.18 A trial regression line for which $\sum e_i = 0$ but $\sum e_i^2$ is *not* a minimum. It is clearly *not* the best line.

We just saw that there are many lines for which $\sum e_i = 0$, but only one of those lines can be the "best" regression line. We call a regression line "best" if (1) the sum of its *errors* of prediction is 0 (that is, $\sum e_i = 0$) *and* (2) the sum of its *squared errors* of prediction is smaller than the sum of the squared errors for any other possible line. This second criterion, which requires that $\sum e_i^2$ be a minimum, is referred to as the *least squares criterion.* It is possible (but beyond the scope of this textbook) to show that there is one and only one line for any given scatter diagram for which *both* $\sum e_i = 0$ *and* $\sum e_i^2$ is a minimum; that is, there is one and only one "best" regression line. This best least squares regression line has precisely those equations that we have been using in this chapter.

least squares criterion: the rule that states that the best regression line is the one that produces the smallest sum of the squared errors of prediction

17.4 Standard Error of Estimate

We have seen that the correlation coefficient r is one measure of the goodness of a regression line: The closer r is to 1 (or to -1), the better the prediction. That is illustrated in Figure 17.19. The correlation on the left is clearly stronger than the one on the right, so the errors of prediction are clearly smaller on the left.

The size of the errors of prediction provides another measure of the goodness of the estimates made using regression lines. The mean of the e_i values, $\bar{e}$, is 0 for both lines in Figure 17.19 (because one of the requirements of a regression line is that $\sum e_i = 0$, and $\bar{e} = \sum e_i / n$), so $\bar{e}$ does not provide a useful measure of goodness. However, note that the standard deviation of the e_i values is larger in the right-hand graph of Figure 17.19.

[1]We faced a similar dilemma in Chapter 5 when we considered using the mean of the deviations as a measure of the width of a distribution. See Box 5.1.

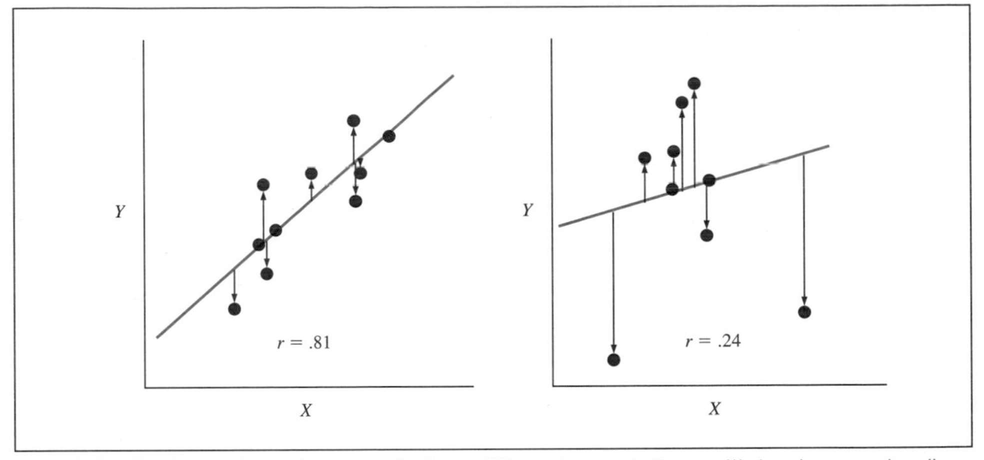

FIGURE 17.19 Scatter diagrams for two different correlations with best regression lines and errors of prediction indicated

[Your eyeball should detect that: Some e_i values in the right-hand graph are quite large and positive, while some are quite negative, so the range (and therefore the standard deviation) is quite large, whereas on the left, the e_i values are all close to 0, so the standard deviation of the e_i values is quite small.] Thus, the standard deviation of the errors of prediction is related to the size of the errors of prediction themselves and can thus serve as a measure of the goodness of regression lines. The smaller the standard deviation of the errors of estimate, the better the regression line.

standard error of estimate: the standard deviation of the errors of prediction

We call the standard deviation of the errors of prediction the *standard error of estimate*, or s_e. We compute it the same way we would compute any standard deviation:

standard error of estimate as the standard deviation of the errors of estimate

$$s_e = \sqrt{\frac{\sum(e_i - \bar{e})^2}{n - 2}} \qquad (17.8)$$

except that we divide by $n - 2$ because we have lost two degrees of freedom, one for the mean of X and one for the mean of Y.

Because for the best regression line the mean of the e_i values (that is, $\bar{e}$) is 0, Equation (17.8) simplifies to

standard error of estimate

$$s_e = \sqrt{\frac{\sum(e_i)^2}{n - 2}} \qquad (17.9)$$

We can also compute s_e from the correlation coefficient:

standard error of estimate computed from the correlation coefficient

$$s_e = s_Y\sqrt{1 - r^2}\sqrt{\frac{n - 1}{n - 2}} \qquad (17.10)$$

where s_Y is the standard deviation of all the Y values. Equation (17.10) will give the same results as Equation (17.9) if errors of rounding are kept small. Equation (17.10) makes it clear that as the correlation coefficient *increases*, the standard error of estimate *decreases*. If $r = 0$, then $s_e = s_Y$, adjusted by the factor $\sqrt{(n - 1)/(n - 2)}$, which will be quite close to 1 for reasonably large sample sizes. Thus, the standard error of estimate can never be larger than the standard deviation of Y, and it is approximately equal to the standard deviation of Y if $r = 0$. On the other hand, if $r = 1$, then Equation (17.10) shows that $s_e = 0$; that is, all the errors of prediction are 0, so their standard deviation is 0.

For our height and weight data of Table 16.2, we could compute s_e from Equation (17.9). That would require determining the predicted $\hat{Y}$ value for each point, subtracting the actual Y value, squaring, and so on. It is easier to use the equivalent Equation (17.10).

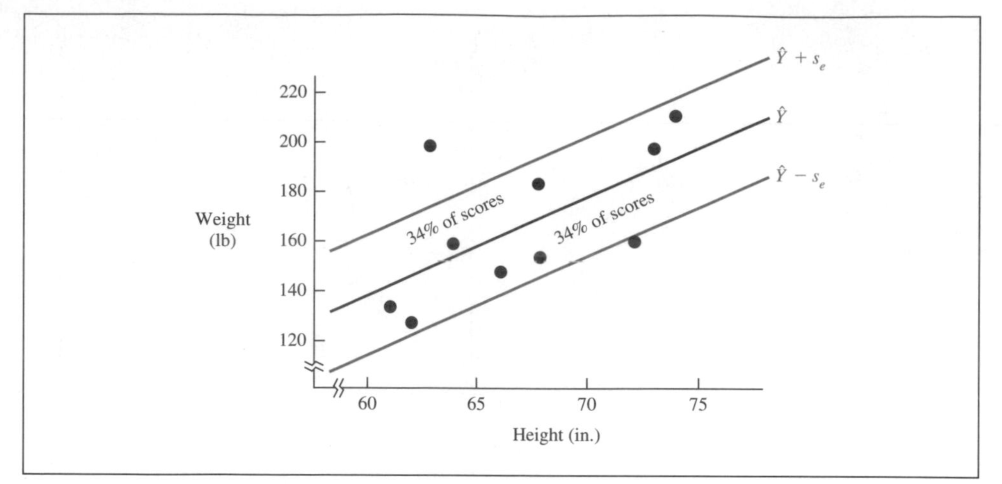

FIGURE 17.20 Regression of weight on height with parallel lines one s_e above and below the regression line

For the data in Table 16.2, $s_Y = 29.4$ pounds, so

$$\begin{aligned} s_e &= s_Y\sqrt{1-r^2}\sqrt{\frac{n-1}{n-2}} \\ &= 29.4\sqrt{1-(.634)^2}\sqrt{\frac{10-1}{10-2}} \\ &= 24.12 \text{ pounds} \end{aligned}$$

The standard error of estimate is the standard deviation of the errors of prediction. Thus, approximately 68% of all points in a scatter diagram will lie within one standard error of the regression line (34% above and 34% below). This is illustrated in Figure 17.20 for our height and weight data. Two blue lines are added to our plot of the regression line. The upper blue line is 24.12 pounds (one s_e) above the regression line and the lower blue line is 24.12 pounds below it (see Box 17.1).

BOX 17.1 Homoscedasticity

One of the assumptions of linear regression, called *homoscedasticity*, is that the size of the errors of prediction does *not* depend on the value of X. For the height and weight data, under this assumption, the band that includes 68% of the points is the same width [2(24.12) pounds] when the height is 65 inches as when it is 70 or 75 inches.

We have seen that the standard error of estimate, s_e, measures the magnitude of the errors of prediction and that $s_e = s_Y\sqrt{1-r^2}\sqrt{(n-1)/(n-2)}$. Now $\sqrt{(n-1)/(n-2)}$ is generally an inconsequential factor, particularly when the number of observations n is moderate or large. For example, when n is 20, $\sqrt{(n-1)/(n-2)} = \sqrt{19/18} = 1.027$, and when n is 100, $\sqrt{(n-1)/(n-2)} = \sqrt{99/98} = 1.005$, so we can safely ignore that factor. Thus, the standard error of estimate is approximately given by $s_e = s_Y\sqrt{1-r^2}$, and we can therefore easily see that s_e depends primarily on two factors: the standard deviation of the Y values (s_Y) and $\sqrt{1-r^2}$.

Coefficient of Determination

coefficient of determination: the square of the Pearson r

The square of the correlation coefficient (r^2) is often called the *coefficient of determination* because it measures the extent to which one variable determines the magnitude of another or, equivalently, because it determines how accurate a prediction is likely to be. Recall that we said that a prediction is "good" when r is close to ± 1. In that case, the coefficient of determination r^2 is close to 1 and $\sqrt{1 - r^2}$ is close to 0. Therefore, the standard error of estimate, s_e, is also close to 0, which implies that the typical error of estimate is very small.

We also said that a prediction is "bad" when r is close to 0. In that case, the coefficient of determination r^2 is close to 0 and $\sqrt{1 - r^2}$ is close to 1. Therefore, the standard error of estimate, s_e, is approximately s_Y, which is as large as it can be, and the typical error of estimate is s_Y.

To review: s_e can take any value between 0 and about s_Y, ignoring the $\sqrt{(n - 1)/(n - 2)}$ factor. The better the prediction, the closer s_e is to 0 (and the closer r^2 is to 1); the worse the prediction, the closer s_e is to s_Y (and the closer r^2 is to 0). A small s_e is thus a desirable thing: A good prediction shrinks s_e.

It is useful to acquire a sense of how much information particular values of the correlation coefficient provide, that is, of how $(1 - r^2)$ and $\sqrt{1 - r^2}$ are related to r. Table 17.3 provides some values of r and the corresponding values of r^2 and $\sqrt{1 - r^2}$. Remembering that, at least approximately, $s_e = s_Y\sqrt{1 - r^2}$, we see from this table that when $r = 0$, the standard error of estimate $s_e = s_Y$, the original standard deviation of the Y values. Note that $\sqrt{1 - r^2}$ does *not* decrease rapidly with r; it takes values of r near .75 before there are large decreases in $\sqrt{1 - r^2}$ (and therefore large decreases in s_e).

TABLE 17.3 r^2, $(1 - r^2)$, and $\sqrt{1 - r^2}$ as a function of r

r	r^2	$1 - r^2$	$\sqrt{1 - r^2}$
.00	.0000	1.0000	1.0000
.25	.0625	.9375	.9682
.50	.2500	.7500	.8660
.75	.5625	.4375	.6614
.90	.8100	.1900	.4359
1.00	1.0000	.0000	.0000

Predicted Distribution

We have seen that the mean weight in our sample is $\overline{Y} = 166.4$ pounds, with the standard deviation $s_Y = 29.4$ pounds. If we assume that our data are typical of men in general, what can we say about the distribution of the weights of all men *who are 65 inches tall*? The mean weight of all 65-inch-tall men will be the weight predicted by the regression equation: $\hat{Y} = 3.965(65) - 99.65 = 158.08$ pounds, lower than $\overline{Y} = 166.4$ pounds. The standard deviation of the weights of all 65-inch-tall men is the standard error of measurement $s_e = 24.12$ pounds. As shown in Figure 17.21, the distribution of weights of 65-inch-tall

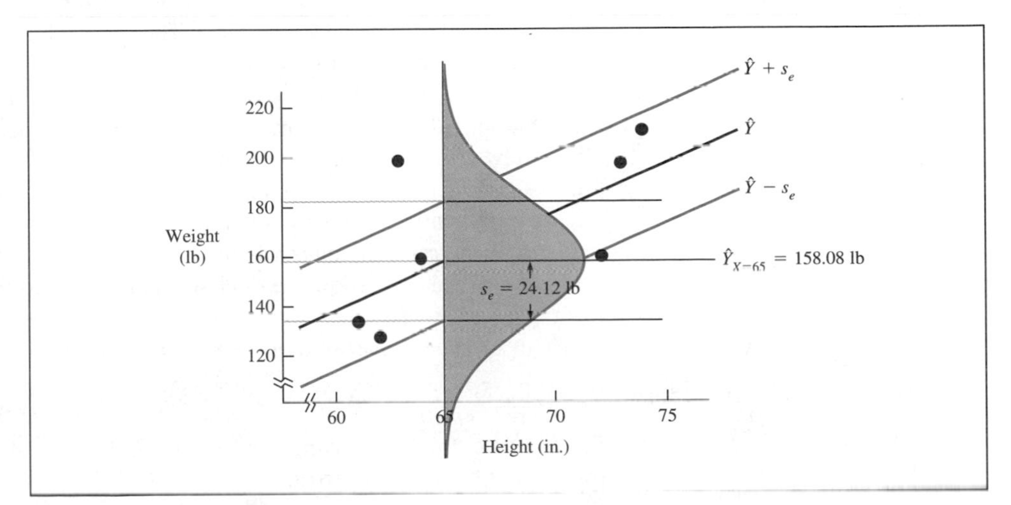

FIGURE 17.21 Distribution of weights of 65-inch-tall men

men (with $\hat{Y} = 158.08$ pounds and $s_e = 24.12$ pounds) has a *lower* mean ($\hat{Y} = 158.08$ instead of $\overline{Y} = 166.4$ pounds) and is *narrower* than the distribution of weights of men in general ($s_e = 24.12$ instead of $s_Y = 29.4$ pounds) because knowing a man's height implies something about his weight. The extent of that implication is measured by the coefficient of determination r^2. The assumption of homoscedasticity implies that the standard deviation of the weights of all men who are 75 inches tall (or 73 inches or any other particular height) is also $s_e = 24.12$ pounds.

17.5 Partitioning the Regression Sum of Squares

Recall that in Chapter 13 (and similarly in the other analysis of variance chapters) it was possible to partition the total sum of squares into the sum of squares between groups and the sum of squares within groups. We can perform the same kind of analysis here, using exactly the same logic we used in deriving Equations (13.9)–(13.12).

The two variables in regression are X and Y. The variable whose sum of squares we will partition is the *predicted* variable Y, so the total sum of squares is the sum of the squared deviations about the grand mean:

total sum of squares

$$\mathrm{SS_T} = \sum(Y_i - \overline{Y})^2 \qquad (17.11)$$

If we transform $\mathrm{SS_T}$ by both adding and subtracting the predicted value $\hat{Y}_i$, we obtain

$$\mathrm{SS_T} = \sum(Y_i - \hat{Y}_i + \hat{Y}_i - \overline{Y})^2$$

and a little algebraic manipulation gives this fundamental result:

partitioning the total sum of squares

$$\underset{\text{(total)}}{\mathrm{SS_T}} = \underset{\text{(error)}}{\sum(Y_i - \hat{Y}_i)^2} + \underset{\text{(predicted)}}{\sum(\hat{Y}_i - \overline{Y})^2} \qquad (17.11a)$$

Now $Y_i - \hat{Y}_i$ is the error of prediction for the ith point, so we call the first term on the right-hand side of Equation (17.11a) the error sum of squares, $\mathrm{SS_{error}}$:

error sum of squares

Sometimes called the "residual" sum of squares.

$$\mathrm{SS_{error}} = \sum(Y_i - \hat{Y}_i)^2 \qquad (17.12)$$

Furthermore, $\hat{Y}_i - \overline{Y}$ is the deviation of the ith point's predicted value from the grand mean—that is, the part of the Y variable that is "explained" by the X variable. Thus, we call the second sum on the right side of Equation (17.11a) $\mathrm{SS_{explained}}$:

explained sum of squares

Sometimes called the "regression" sum of squares.

$$\mathrm{SS_{explained}} = \sum(\hat{Y}_i - \overline{Y})^2 \qquad (17.13)$$

We illustrate this for our height and weight data in Table 17.4. The first two columns are the original height and weight data. The third column contains the predicted values $\hat{Y}_i$, obtained from the regression equation $\hat{Y}_i = 3.965X_i - 99.65$. The fourth and fifth columns show the deviations and squared deviations of the individual points from the grand mean $\overline{Y}$. Thus, the sum of the fifth column is the total sum of squares: $\mathrm{SS_T} = 7772.40$. [If we divided 7772.40 by $(n - 1)$ and then took the square root, we would get the standard deviation of the Y values: $s_Y = \sqrt{7772.40/(10 - 1)} = 29.4$ pounds.]

The sixth and seventh columns show the errors and squared errors of estimate. Thus, the sum of the seventh column is the error sum of squares: $\mathrm{SS_{error}} = 4646.2620$ pounds. The last two columns show the deviations and squared deviations of the predicted values from the grand mean. Thus, the sum of the last column is $\mathrm{SS_{explained}} = 3127.1060$.

This has been a partition of the sum of squares, so

partitioning the sum of squares

$$\mathrm{SS_T} = \mathrm{SS_{error}} + \mathrm{SS_{explained}} \qquad (17.14)$$

For our data, this partitioning implies that $7772.40 = 4646.2620 + 3127.1060$, which is true except for rounding error.

TABLE 17.4 Partitioning the sum of squares for regression

			Total		*Error*		*Explained*	
			Deviation from grand mean		*Deviation from prediction (error of estimate)*		*Deviation of prediction from grand mean*	
X_i	Y_i	$\hat{Y}_i$	$Y_i - \overline{Y}$	$(Y_i - \overline{Y})^2$	$Y_i - \hat{Y}_i$	$(Y_i - \hat{Y}_i)^2$	$\hat{Y}_i \; \overline{Y}$	$(\hat{Y}_i - \overline{Y})^2$
72	160	185.83	−6.4	40.96	−25.83	667.1889	19.43	377.5249
66	144	162.04	−22.4	501.76	−18.04	325.4416	−4.36	19.0096
68	154	169.97	−12.4	153.76	−15.97	255.0409	3.57	12.7449
74	210	193.76	43.6	1900.96	16.24	263.7376	27.36	748.5696
68	182	169.97	15.6	243.36	12.03	144.7209	3.57	12.7449
64	159	154.11	−7.4	54.76	4.89	23.9121	−12.29	151.0441
63	199	150.14	32.6	1062.76	48.86	2387.2990	−16.26	264.3876
73	198	189.80	31.6	998.56	8.20	67.2400	23.40	547.5600
61	132	142.22	−34.4	1183.36	−10.22	104.4484	−24.18	584.6724
62	126	146.18	−40.4	1632.16	−20.18	407.2324	−20.22	408.8484
			SS_T = 7772.40		SS_{error} = 4646.2620		$SS_{explained}$ = 3127.1060	

It can be shown, although it is beyond the scope of this book, that the proportion of the total sum of squares that is explained is equal to the coefficient of determination:

coefficient of determination as the proportion of sum of squares that is explained

$$r^2 = \frac{SS_{explained}}{SS_T} = \frac{SS_T - SS_{error}}{SS_T} \tag{17.15}$$

Thus, for our data,

$$\begin{aligned} r^2 &= \frac{SS_{explained}}{SS_T} \\ &= \frac{3127.1060}{7772.40} \\ &= .4023 \end{aligned}$$

The Pearson correlation coefficient is thus $r = \sqrt{r^2} = \sqrt{.4023} = .634$, as we have seen since Chapter 16.

The coefficient of determination, as the proportion of the sum of squares accounted for ("explained"), is a highly important concept in more advanced statistics. Many of the topics in statistics involve partitioning the sum of squares, often into many more than two pieces. The coefficient of determination as the percentage of explained variability is always useful as a measure of the "success" or "utility" of the partition. Thus r^2 can be considered to be an effect size index for correlation and regression.

17.6 Interpreting Correlation and Regression

Students frequently ask reasonable questions such as, How large must a correlation coefficient be before it is of practical significance? One can spend a lifetime answering that question, but we give two examples that may shed some light on the subject.

IQ

We choose IQ as one example because it is widely discussed and because the correlation coefficients between IQ and other measures are among the highest that a behavioral science researcher encounters. For example, the correlation between IQ and educational attainment (years in school) is about .7.[2]

[2] From Matarazzo (1972, p. 296).

Table 17.5 gives some approximate correlations of intelligence between relatives with different relationships, and of height for the same relatives.[3] The correlation of intelligence between parents and their children (.5) is somewhat higher than the correlation of height between parents and their children (.4).

To review and improve our grasp of the size and real significance of correlation and prediction, we will explore the measurements of IQs for the same individuals on two separate occasions (as a child of various ages and as an adult). Table 17.6 is constructed to answer two questions.[4] First, suppose we measure a group of children's IQs. Then later, when the same children are adults, we measure their IQs again. What is the correlation between these scores? Second, if a child's IQ score is 115, what should we predict that child's adult IQ to be?

TABLE 17.5 Approximate correlations between relatives' intelligence and height

Relationship	*IQ*	*Height*
Parent–child	.5	.4
Brother–sister	.6	.5
Identical twins	.9	.8

TABLE 17.6 Correlation of child's IQ with adult's IQ, best predicted adult IQ for children whose IQs are 115, and 95% range of adult IQs for children whose IQs are 115

Age at first test	*r*	*Predicted adult IQ*	*95% range*
3	.4	106	79–133
5	.5	107.5	82–133
10	.7	110.5	90–131
15	.8	112	94–130

IQ (for most IQ tests) is normally distributed with $\mu = 100$ and $\sigma = 15$ at all ages. Table 17.6 shows that the correlation between IQ_{adult} and IQ_3 is approximately .4. It may surprise you that that correlation is so low because many people believe that IQ is constant for individuals over their lifetime. That simply is not true, as the low correlations demonstrate. Suppose we want to predict adult IQ for a 3-year-old whose $IQ_3 = 115$; we let $IQ_{adult} = \hat{Y}$ and $IQ_3 = X$. Then, from Equation (17.3), $b = r(s_Y/s_X) = .4(15/15) = .4$, and from Equation (17.5), $a = \overline{Y} - b\overline{X} = 100 - .4(100) = 60$. The equation for the regression line that predicts adult IQ from IQ at age 3 is, then, from Equation (17.1), $\widehat{IQ}_{adult} = .4(IQ_3) + 60$. If a child's IQ at age 3 is 115, as we are assuming for this entire exercise, then we should predict that child's IQ as an adult to be $\widehat{IQ}_{adult} = .4(115) + 60 = 106$.

Note, as we saw above, that we do not predict the adult's IQ to be the same as the child's. The fact is that regression equations *always* predict scores *closer to the mean* (by a factor of r) than the original scores, the phenomenon we have called "regression toward the mean."

Now suppose we collect a group of 1000 3-year-olds, *all* of whom have $IQ_3 = 115$. The middle 95% of IQ_{adult} scores can be obtained by noting that these scores should have a normal distribution with mean $\widehat{IQ}_{adult} = 106$ and standard deviation $s_e = s_Y\sqrt{1-r^2}\sqrt{(n-1)/(n-2)} = 15\sqrt{1-.4^2}\sqrt{999/998} = 13.75$ IQ points from Equation (17.9). By the logic of confidence intervals we developed in Chapter 8, the middle 95% of those 1000 now-adults will lie within $\pm 1.96 s_e$ of 106—that is, between $106 - 1.96(13.75) = 79.05$ and $106 + 1.96(13.75) = 132.95$, or between about 79 and 133. [By contrast, the middle 95% of *all* IQs lie between $100 \pm 1.96(15)$, or between 70.6 and 129.4.]

These results imply that a psychologist who has administered an IQ test to a 3-year-old child and obtained a score of 115 should tell the parent something like, "The best bet for your child's adult IQ is 106, and there is a 95% chance that his adult IQ will lie between 79 and 133." If that seems a wide range to you, that is simply the result of predicting when the correlation is .4.

Results for first-test ages of 5, 10, and 15 are also shown in Table 17.6.

---

[3] Compiled from Cattell (1982, pp. 145–146) and Matarazzo (1972, p. 303).

[4] Approximate correlations based on Jensen (1980, pp. 278–280).

You may wish to use the ESTAT exercise "corest" at this time.

Personality

Researchers in the field of personality often report correlation coefficients with magnitudes between about .2 and .3. These coefficients are so common that Mischel (1968, p. 78), in a classic summary of such coefficients, wrote:

personality coefficient: the observation that most correlations between personality measures and behavior are approximately .2 to .3

> Indeed, the phrase "personality coefficient" might be coined to describe the correlation between .20 and .30 which is found persistently when virtually any personality dimension inferred from a questionnaire is related to almost any conceivable external criterion involving responses sampled in a *different* medium—that is, not by another questionnaire.

There is still some debate about why personality correlation coefficients are seldom greater than .3. For whatever reason, Mischel's general observation in 1968 still holds today, and it is therefore important for us to "get a feel" for a correlation coefficient of .3.

We saw in the preceding section that the coefficient of determination (r^2) measures the percentage of the variability of Y (the predicted variable) that is accounted for by X (a predictor variable). If the correlation between X and Y is $r = .3$, then $r^2 = .3^2 = .09$, so about one-tenth (actually closer to one-eleventh, but round numbers will suffice for our discussion here) of the variance of Y will be accounted for by ("predicted by") X.

If X accounts for one-tenth of the variability of Y, there may be *nine other* measures $(X_1, X_2, \ldots, X_9)$ that are totally independent of X and also independent of each other that are at least as good predictors of Y as is X. [Alternatively, there may be 18 other predictors of Y that are half as good as is X, or five other predictors that are twice as good as is X (although that is not likely in the personality arena), or any combination thereof. We examine the case of "nine other measures" because it is easiest to conceptualize.]

I find the following thought experiment helpful: Picture a specially constructed automobile in the middle of the Bonneville salt flats (a wide, flat expanse of smooth desert). This automobile has ten drivers, all with their own steering wheels, all in their own private compartments with no communication between them (that is, the drivers are "independent"). All "drive" according to their own independent desires, turning hard right, turning a bit left, or proceeding straight ahead according to their own whims. The car is constructed so that the direction of the car's front wheels is the average of the ten steering wheel positions.

What is the correlation between any one driver's steering wheel position and the direction that the car actually takes? About $r = .3$, because each driver is accounting for 10% of the variability in the car's direction. What is the correlation between any two drivers' steering wheel positions? Zero, because each driver is steering independent of each other. Thus, each of the ten drivers is fully justified in saying, "My driving had substantial impact on the direction of the car: The correlation between my driving and the car's direction was .3." Yet all of the ten drivers were trying to go, in the long run, in totally unrelated directions.

This analysis has many important implications. If you (or a professional psychologist, for that matter) attempt to predict the behavior of other people, it is highly unlikely that you can do better than a good psychological test (Meehl, 1954; Kleinmuntz, 1990), which means that your predictions will likely account for less than about 10% of the total variability of their behavior. That means that there may be nine other people whose opinions are entirely unrelated to yours and to each other's who are, nonetheless, every bit as good as you at predicting the behavior in question (or five others who are twice as good, etc.). Taking such an observation seriously should evoke a proper sense of humility.

Is it possible for a predictor to account for more than 10% of the variability in behavior? Yes; basing a prediction on two or more *independent* predictors can improve the accuracy of prediction, and knowing the person extremely well (better than could any psychological test) can improve the accuracy of prediction. But the fact remains that it is a common mistake among statistically naive individuals to overestimate greatly the accuracy of their own predictions and to fail to appreciate the likelihood that the predictions of those with whom they disagree are likely to be every bit as good as their own.

17.7 Hypothesis Testing in Regression

As we have seen, the regression line has two constants, b and a, and of these, b is generally the more important to us. There are many situations where we wish to know whether the regression of Y on X is significant, by which we mean, Do we have reason to believe that the slope of the regression line in the population is different from 0?

We know that b is the slope of the regression line that predicts $\hat{Y}$ from X in a sample. The slope in the corresponding population is the parameter that we will call $b_{\text{population}}$. We wish to know whether $b_{\text{population}}$ is nonzero. The null hypothesis, then, is

null hypothesis for slope of regression line

$$H_0\colon\ b_{\text{population}} = 0 \tag{17.16}$$

As has been our custom, we could use the Greek letter β for the population parameter, but we are reserving β for the probability of a Type II error.

Just as the sample correlation r is a point-estimate of the population correlation ρ, the slope of the sample regression line b is a point-estimate of the population slope $b_{\text{population}}$. If the null hypothesis is true, b should be approximately 0, but b is not likely to be *exactly* 0 because of sampling fluctuations. We need a test to find out whether the slope of the regression line b is *significantly* different from 0.

It can be shown that we reject the null hypothesis H_0: $b_{\text{population}} = 0$ *under exactly the same conditions as we reject* H_0: $\rho = 0$, the null hypothesis for Pearson correlation coefficients that we discussed in Chapter 16. Thus, to test the regression null hypothesis, we compute the Pearson r for the same data and ask whether r is significantly different from 0, using the Chapter 16 procedures. If r is significantly different from 0, then we reject the regression null hypothesis (and also the correlation null hypothesis).

Thus, it makes sense to think of b as the sample statistic in the hypothesis-testing procedure, and r as its test statistic. For example, suppose we wish to know whether there is a "significant regression" of weight on height; that is, is b significantly different from 0? Said another way, does an increase in height imply an increase in weight? The null hypothesis is H_0: $b_{\text{population}} = 0$—that is, there is no increase in weight with height.

We compute $r_{\text{obs}} = .634$, just as in Chapter 16. We find the critical value of r from Table A.8, entering with $n - 2$ degrees of freedom: $r_{\text{cv}} = .6319$. Because $r_{\text{obs}} = .634 > r_{\text{cv}} = .6319$, we reject the regression null hypothesis and conclude that in fact an increase in height does imply an increase in weight.

Note that to test whether b is significantly different from 0, we need not compute b; r is all that is required.

17.8 Connections

Cumulative Review

We continue the discrimination practice begun in Chapter 10. You may recall the two steps:

Step 1: Determine whether the problem asks for (a) the area under a normal distribution, (b) a confidence interval, or (c) the evaluation of a hypothesis. See the Cumulative Review in Section 10.5 for hints about this step.

Step 2: If the problem asks for the evaluation of a hypothesis, consult the summary of hypothesis testing in Table 17.7 to determine which kind of test is appropriate.

Table 17.7 shows that so far we have 11 basic kinds of hypothesis tests (the one-sample test for the mean, two-independent-samples test for means, two-dependent-samples tests for means, three-or-more-independent-samples tests, post hoc tests, a priori tests, three-or-more-repeated-measures tests, the two-way ANOVA, two tests for correlation, and now the test for regression). Note that this table is identical to Table 16.7 with the addition of the last line appropriate for this chapter. We will continue to expand this table progressively throughout the remainder of the book.

TABLE 17.7 Summary of hypothesis-testing procedures in Chapters 10–17

Design	Chapter	Null hypothesis	Sample statistic	Test statistic	Effect size index
One sample					
σ known	10	$\mu = a$	$\overline{X}$	z	$d = \frac{\overline{X}_{obs} - \mu}{\sigma}$
σ unknown	10	$\mu = a$	$\overline{X}$	t	$d = \frac{\overline{X}_{obs} - \mu}{s}$
Proportion	Exercise 7, Chapter 10	$\Pi = a$	p	t	—[a]
Two independent samples					
For means	11	$\mu_1 = \mu_2$	$\overline{X}_1 - \overline{X}_2$	t	$d = \frac{\overline{X}_{1_{obs}} - \overline{X}_{2_{obs}}}{s_{pooled}}$
Two dependent samples					
For means	12	$\mu_D = 0$	$\overline{D}$	t	$d = \frac{\overline{D}_{obs}}{s_D}$
k independent samples					
For means (omnibus)	13	$\mu_1 = \mu_2 = \cdots = \mu_k$	n/a	F	$d_M = \frac{\overline{X}_{max} - \overline{X}_{min}}{s_{pooled}}$ or $f = \frac{s_{\overline{X}}}{s_{pooled}}$ or $R^2 = \frac{SS_B}{SS_T}$
Post hoc	14	$\mu_i = \mu_j$	$\overline{X}_i - \overline{X}_j$	Q	—
A priori	14	$C_1\mu_1 + \cdots + C_k\mu_k = 0$	$C_1\overline{X}_1 + \cdots + C_k\overline{X}_k$	t	$d = \frac{C_1\overline{X}_1 + \cdots + C_k\overline{X}_k}{s_{pooled}}$
k dependent samples					
For means	14	$\mu_1 = \mu_2 = \cdots = \mu_k$	n/a	F	$d_M = \frac{\overline{D}_{max}}{\sqrt{2MS_{residual}}}$
RC independent samples					
For row means	15	$\mu_{r1} = \mu_{r2} = \cdots = \mu_R$	n/a	F	—
For column means	15	$\mu_{c1} = \mu_{c2} = \cdots = \mu_C$	n/a	F	—
Interaction	15	$\mu_{rc} - \mu_r - \mu_c + \mu_G = 0$	n/a	F	—
Correlation					
Parametric	16	$\rho = 0$	r	r (or t)	r^2
Ordinal	16	$\rho_s = 0$	r_s	r_s	—
Regression					
Slope	17	$b_{population} = 0$	b	r	r^2

[a] Effect size index is not discussed in this textbook.

Journals

A journal might report the results of our height and weight regression simply as follows: "The regression of weight on height was significant ($df = 8, p < .05$)." Note that the report implies, but does not state, that the hypothesis that is being tested is about the *slope* of the regression line, and the null hypothesis is $H_0\colon b_{population} = 0$. Furthermore, note that the statement "regression of weight *on* height" implies that weight is the predicted (Y) variable and height is the predictor (X) variable.

Multiple regression is a frequently used method of analyzing behavioral science data. Multiple regression is beyond the scope of this text, but we can give a hint about its operation. Equation (17.6) is the regression equation in standard form: $\hat{z}_Y = rz_X$. We can call this a "bivariate" equation because it involves two variables: X and Y (or z_X and z_Y). We use a bivariate regression equation to predict one variable (like weight) from one other variable (like height). Often, however, we wish to predict one variable *from several* other variables, and the equation to make such a prediction is a "multiple regression equation."

Computers

Use ESTAT's datagen to compute the regression line for the height and weight data listed in Table 16.2:

1. Open the data set you saved in Chapter 16 (or follow the instruction in the Computers section of Chapter 16).
2. Datagen automatically computes the regression line, shown in the `datagen Statistics` window in the section headed "assuming the two variables are paired."

Use SPSS to compute the regression line for the height and weight data listed in Table 16.2:

1. Open the data set you saved in Chapter 16 (or follow the instruction in the Computers section of Chapter 16).

```
SPSS for MS WINDOWS Release 6.1

         * * * *  M U L T I P L E   R E G R E S S I O N   * * * *

Listwise Deletion of Missing Data

Equation Number 1    Dependent Variable..   VAR00002

Block Number  1.  Method:  Enter      VAR00001

Variable(s) Entered on Step Number  } Not discussed
   1..     VAR00001                 } in this text

Multiple R               .63425   Pearson r
R Square                 .40228   Coefficient of determination r^2
Adjusted R Square        .32756   Not discussed in this text
Standard Error         24.09812   Standard error of estimate s_e

Analysis of Variance
                   DF      Sum of Squares     Mean Square  }
Regression          1          3126.64635      3126.64635  } Not discussed
Residual            8          4645.75365       580.71921  } in this text
                                                           }
F =        5.38409        Signif F =  .0489                }
------------------ Variables in the Equation ------------------
                                                  t (see Box 16.4)  "Exact"
Variable              B          SE B        Beta        T   Sig T  probability

VAR00001   (b)   3.964806     1.708700    .634252    2.320   .0489
(Constant) (a) -99.638512   114.906709               -.867   .4111
                               Not discussed in this text
End Block Number   1   All requested variables entered.
```

2. Click `Statistics`, then `Regression`, and then `Linear....`
3. Click `var00002`; then click the upper ▶ to move it into the `Dependent:` window.
4. Click `var00001`; then click the second ▶ to move it into the `Independent(s):` window. Click `OK`.
5. The accompanying regression output appears in the `!Output1` window. Bivariate regression (the topic discussed in this chapter) is a special case of multiple regression (see the Journals section), so SPSS simply presents the multiple regression output with one predictor variable.

Homework Tips

1. Check the list of learning objectives at the beginning of this chapter. Do you understand each one?

2. Make sure you follow the custom of putting the variable you wish to predict on the Y-axis and the predictor variable on the X-axis.

3. Many students find the concept of regression to the mean elusive. A prediction is always closer to its mean than the predictor variable is to its own mean unless the correlation is ± 1.

4. Remember that there are *two* criteria for the best regression line: $\sum e_i = 0$ and $\sum e_i^2$ is a minimum.

5. An error of prediction is the *vertical* distance between a point and the regression line, *not* the *perpendicular* distance.

6. If a point is below the regression line, its error of prediction is negative.

7. The standard error of estimate is a measure of the typical length of the errors because the mean error of estimate is 0.

8. If you prepare a table that partitions the sum of squares (like Table 17.4), make sure you use many significant figures in the computation of b and a. Otherwise, roundoff in the computation of $\hat{Y}$ will be extreme.

Exercises for Chapter 17

Section A: Basic exercises (answers in Appendix D, page 531)

1. (a) Using the scatterplot you prepared for Exercise 1 in Chapter 16, plot by eyeball the regression line. The data are duplicated in the table.
   (b) Compute the regression line constants b and a and plot the resulting line on the same scatterplot. How did your eyeball-estimate compare with your computation?

X	Y
1	12
4	14
2	12
5	15
3	13

2. Exercise 2 in Chapter 16 presented data on family income (in thousands of dollars) and the size of the house (in hundreds of square feet) in which the family lives. The data are duplicated in the table.

Family	Income (× $1000)	House size (× 100 sq ft)
1	23.5	16.4
2	11.8	11.2
3	28.1	19.4
4	26.0	16.9
5	18.3	13.3
6	27.5	11.9
7	38.8	13.4
8	35.7	17.4
9	28.5	17.3
Partial labor savings:		
$\sum X$	238.2	137.2
$\sum X^2$	6840.42	2156.48

(a) Using the scatterplot you prepared in Exercise 2 of Chapter 16, plot by eyeball the regression line.

(b) Compute the regression line constants and plot the resulting line on the same scatterplot. How did your eyeball-estimate compare with your computation?
(c) State in plain English what the value of b in this problem tells us.
(d) Given that the Jones family income is \$31,000, what house size would you predict for them? Eyeball-estimate this value from your regression line and then compute it.
(e) Given that the Smith family income is \$13,000, what house size would you predict for them? Eyeball-estimate this value from your regression line and then compute it.

3. (*Exercise 2 continued*) Suppose we find 100 families, each of which (like the Jones family) has a family income of \$31,000.
   (a) What would you expect the mean house size for these 100 families to be? Why?
   (b) What would you expect the standard deviation of these 100 house sizes to be? Why?
   (c) Superimpose on your scatterplot/regression line from Exercise 2(b) the distribution you would expect these 100 house sizes to have.

4. (*Exercise 2 continued*) Suppose we find 1000 families, each of which (like the Smith family) has a family income of \$13,000.
   (a) What would you expect the mean house size for these 1000 families to be? Why?
   (b) What would you expect the standard deviation of these 1000 house sizes to be? Why?
   (c) Superimpose on your scatterplot/regression line from Exercise 2(b) the distribution you would expect these 1000 house sizes to have.

5. What do we call the assumption that led us to believe that the standard deviations in Exercises 3(b) and 4(b) would be identical?

6. (*Exercise 2 continued*) Is the regression constant b you computed in Exercise 2(b) significantly different from 0?

I. State the null and alternative hypotheses
(a) What is the null hypothesis? The alternative hypothesis?
(b) Is this test directional or nondirectional? Why?
(c) What is the sample statistic?
(d) What is the test statistic?

II. Set the criterion for rejecting H_0
(e) What is the level of significance?
(f) How many degrees of freedom are there?
(g) Look up the critical value(s) of the test statistic.

III. Collect a sample and compute statistics
(h) Eyeball-estimate and then compute the observed value of the sample statistic.
(i) Eyeball-estimate and compute the observed value of the test statistic.

IV. Interpret the results
(j) Should we reject H_0? Decide by eyeball and by computation.
(k) Interpret your result.
(l) How does the hypothesis test here compare with that in Exercise 2 in Chapter 16?

7. Exercise 4 in Chapter 16 presented data showing the number of times a rat has run through a maze and the time it took the rat to run the maze on its last trial. The data are reproduced in the table.

Rat	*Trials*	*Time (sec)*
1	8	10.9
2	9	8.6
3	6	11.4
4	5	13.6
5	3	10.3
6	6	11.7
7	3	10.7
8	2	14.8
Partial labor savings:		
$\sum X$	42	92.0
$\sum X^2$	264	1084.2

(a) Using the scatterplot you prepared for Exercise 4 in Chapter 16, plot by eyeball the regression line.
(b) Compute the regression line constants and plot the resulting line on the same scatterplot. How did your eyeball-estimate compare with your computation?
(c) State in plain English what the value of b in this problem tells us.
(d) Given that a rat has run a maze six times, what would you predict to be the time for the rat's last trial? Eyeball-estimate this value from your regression line and then compute it.
(e) Given that a rat has run a maze three times, what would you predict to be the time for the rat's last trial? Eyeball-estimate this value from your regression line and then compute it.

8. In Exercise 7(b), you computed the regression slope b using either Equation (17.3) or Equation (17.4). Now use the other equation and compute b again to verify that Equations (17.3) and (17.4) give identical results.

9. *(Exercises 7 and 8 continued)* Use the data and regression line you computed in Exercises 7 and 8.
   (a) Determine the predicted value $\hat{Y}_i$ for each point.
   (b) Determine the error of prediction e_i for each point.
   (c) Prepare a table that partitions the sum of squares (like Table 17.4).
   (d) Show that $SS_T = SS_{error} + SS_{explained}$.
   (e) The correlation coefficient r for these data was computed in Exercise 4 of Chapter 16 to be $-.57$. Show that r^2 equals the proportion of the total sum of squares that is explained by regression.

10. The accompanying scatterplot represents the relationship between scores on a vocabulary test and IQ. For these data, the mean and standard deviation of vocabulary scores are 9.57 and 2.51 points, respectively; the mean and standard deviation of IQ scores are 97.43 and 13.87 points, respectively; and the Pearson correlation between them is .61.
    (a) Eyeball-estimate the regression line of vocabulary score on IQ.
    (b) Compute the constants for the regression of vocabulary score on IQ.
    (c) Plot the computed regression line. Why doesn't it cross the Y-axis at $\hat{Y} = a$?

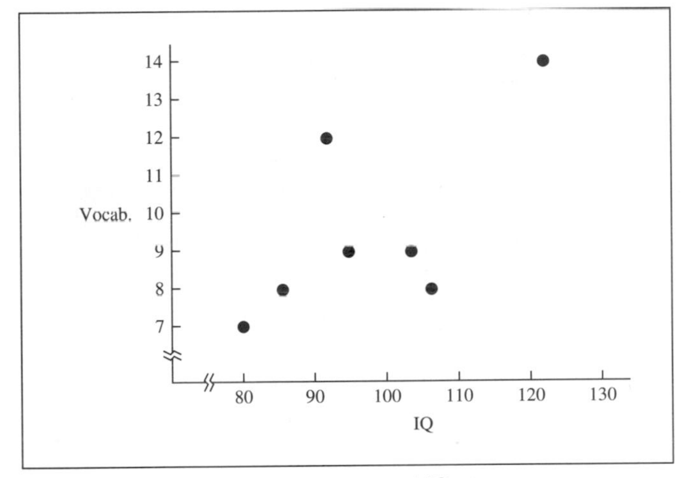

Scatterplot of vocabulary score and IQ

11. (*Exercise 10 continued*)
    (a) If Jack's IQ is 127 and Jill's IQ is 128, how much higher would we expect Jill's vocabulary score to be than Jack's?
    (b) If Bob's IQ is 102 and Ted's IQ is 112, how much higher would we expect Ted's vocabulary score to be than Bob's?

Section B: Supplementary exercises

12. A public swimming pool operator believes that the amount of chlorine necessary to keep his pool clean depends on the air temperature that day. To verify his belief, he selects ten days at random from last summer's pool records. Then

Day	*Temp. (°F)*	*Chlorine (oz)*
1	77	11
2	79	9
3	62	3
4	65	6
5	73	7
6	91	12
7	68	8
8	73	7
9	89	10
10	74	8
Labor savings:		
$\sum Y$	751	81
$\sum X^2$	57,199	717

he looks up the day's temperature in the library's newspapers for those days. The data on temperature and chlorine amounts are shown in the table. What is the relationship between temperature and chlorine use?
    (a) Prepare a scatterplot.
    (b) Eyeball-estimate the regression line.
    (c) Eyeball-estimate the regression constants a and b.
    (d) Compute the regression constants and write the equation for the regression line. Plot it on the scatterplot.

13. (*Exercise 12 continued*) Is the slope of the regression between temperature and chlorine amount significant? Use the steps of Exercise 6.

14. (*Exercise 12 continued*) How much chlorine should the operator expect to use for these air temperatures?
    (a) 75°F (b) 85°F

15. (*Exercise 12 continued*) Suppose the pool operator went back through the records for the last 20 years and identified all those days on which the air temperature was 75°F. There were 150 such days in the past 20 years.
    (a) If the operator computed the mean amount of chlorine used on those 150 days, what value would he expect that mean to have?
    (b) If he computed the standard deviation of chlorine used on those 150 days, what value should he expect that standard deviation to have?

16. (*Exercise 15 continued*) Suppose the pool operator identified all those days on which the air temperature was 85°F. There were 200 such days in the past 20 years.
    (a) Should the operator expect the mean chlorine used on these 200 days to be larger than, smaller than, or the same as the mean chlorine used when the temperature was 75°F? Why?
    (b) Should he expect the standard deviation of chlorine used on those 200 days to be larger than, smaller than, or the same as the standard deviation of chlorine used when the temperature was 75°F? Why?

17. (*Exercise 12 continued*)
    (a) Compute the standard error of the estimate for the chlorine-use data.
    (b) For each of the ten points in the temperature and chlorine-use data, compute $\hat{Y}_i$ and e_i.
    (c) Compute the standard deviation of the e_i's you computed in part (b).
    (d) Should the value you obtained in part (c) be the same as the value you obtained in part (a)? Why?

18. (*Exercises 12–17 continued*) Use the data and regression line you computed in Exercises 12–17.
    (a) Prepare a table that partitions the sum of squares (like Table 17.4).
    (b) Show that $SS_T = SS_{error} + SS_{explained}$.
    (c) Compute the correlation coefficient r for these data from Equation (17.3). Square that value and show that this r^2 equals the proportion of the total sum of squares that is explained by regression.

19. (*Exercise 17 continued*) Suppose you add 1 to the regression constant a and compute a new (incorrect) set of $\hat{Y}_i$ and e_i. Then you compute the standard deviation of these new e_i's. Would you expect this new standard deviation to be larger than, smaller than, or the same as the standard deviation of the e_i's you computed in Exercise 17? Why?

20. The amount of gasoline used (miles per gallon) is related to the size of an automobile's engine. State the linear relationship between engine size and gasoline used. The data relating engine size (in cubic inches) to fuel consumption (in miles per gallon) are shown in the table. Use the steps of Exercise 12.

Auto	Size	Consumption
1	206	20
2	348	12
3	239	6
4	363	10
5	212	11
6	234	12
7	341	9
8	202	19
9	309	7
10	162	24
11	218	16
12	405	6
13	211	22
14	314	5
15	348	12
Labor savings:		
$\sum X$	4,112	191
$\sum X^2$	1,206,710	2,957

21. To what extent is crowding in a neighborhood related to the crime rate in that neighborhood? A city's neighborhoods are rated according to the room occupancy rate (ROR, the number of occupants in a dwelling divided by the number of rooms in that dwelling) and the number of times police were summoned to that neighborhood in the past year. The data are shown in the table. Use the steps of Exercise 12.

Neighborhood	ROR	Crimes
1	1.0	36
2	1.1	39
3	2.4	55
4	.3	26
5	.5	39
6	1.0	40
7	.6	46
8	1.4	39
9	1.6	60
10	1.4	49
11	.3	37
12	.5	46
Labor savings:		
$\sum X$	12.1	512
$\sum X^2$	16.49	22,762

22. A widget production manager knows that as the rate of widget production goes up on her assembly line, the number of defective widgets also goes up. She computes the regression line used to predict the number of defectives from the production rate.
    (a) Is the slope b of this regression line likely to be positive/negative/0/cannot say? Why?
    (b) Is the regression constant a of the regression line likely to be positive/negative/0/cannot say? Why?

23. (*Exercise 22 continued*) The widget production data for nine days selected at random are shown in the table. How is the number of defectives related to the production? Use the steps of Exercise 12.

Day	Production	Defective
1	301	18
2	264	6
3	356	17
4	328	9
5	353	10
6	284	8
7	322	14
8	332	7
9	255	11
Labor savings:		
$\sum X$	2,795	100
$\sum X^2$	878,815	1,260

24. (*Exercise 23 continued*) Is the slope of the regression line between production rate and defective rate significant? Use the steps of Exercise 6.

Section C: Extending your comprehension (answers in Appendix D, page 532)

25. What are the regression constants a and b for a regression line where the Pearson $r = 0$?

26. What must be the characteristics of the data if the slope b of the regression line approaches infinity?

Section D: Cumulative review (answers in Appendix D, page 532)

For Exercises 27–35, choose the correct answer or fill in the blank. (**Note that computations are *not* required;** use $\alpha = .05$ unless otherwise instructed.)

(a) This problem requires:
   (1) Finding the area under a normal distribution [Skip parts (b)–(g) and proceed to (h).]
   (2) Creating a confidence interval
   (3) Testing a hypothesis about the means of one group
   (4) Testing a hypothesis about the means of two independent groups
   (5) Testing a hypothesis about the means of two dependent groups
   (6) Testing a hypothesis about the means of three or more independent groups

(7) Testing hypotheses about means using post hoc tests
(8) Testing an a priori hypothesis about means (planned comparison)
(9) Testing a hypothesis about the means of three or more dependent groups
(10) Testing hypotheses for main effects and interaction
(11) Testing a hypothesis about the strength of a relationship
(12) Testing a hypothesis about the slope of the regression line
(13) Performing a power analysis (involving finding the area under a normal distribution) [Skip parts (b)–(h).]

(b) The null hypothesis is of the form:
(1) $\mu = a$ (2) $\mu_1 = \mu_2$
(3) $\mu_D = 0$ (4) $\mu_1 = \mu_2 = \cdots = \mu_k$
(5) A series of null hypotheses of the form $\mu_i = \mu_j$. How many such hypotheses?
(6) One null hypothesis of the form $C_1\mu_1 + C_2\mu_2 + \cdots + C_k\mu_k = 0$
(7) Three null hypotheses:
rows: $\mu_{\text{row }1} = \mu_{\text{row }2} = \cdots = \mu_{\text{row }R}$;
columns: $\mu_{\text{col }1} = \mu_{\text{col }2} = \cdots = \mu_{\text{col }C}$; and
interaction: all $\mu_{\text{row,col}} - \mu_{\text{row}} - \mu_{\text{col}} + \mu = 0$
(8) $\rho = 0$ (9) $\rho_S = 0$
(10) $b_{\text{population}} = 0$ (equivalent to $\rho = 0$)
(11) There is no null hypothesis.

(c) The appropriate sample statistic(s) is (are):
(1) $\overline{X}$ (2) p
(3) $\overline{X}_1 - \overline{X}_2$ (4) $\overline{D}$
(5) $\overline{X}_i - \overline{X}_j$
(6) $C_1\overline{X}_1 + C_2\overline{X}_2 + \cdots + C_k\overline{X}_k$
(7) r
(8) r_S
(9) a and/or b
(10) Not applicable for ANOVA

(d) The appropriate test statistic(s) is(are):
(1) z (2) t
(3) F (4) Q
(5) r (6) r_S
(7) Not applicable

(e) The number of degrees of freedom for this test statistic is ________. (State the number or *not applicable*; if there are multiple null hypotheses, state the number of degrees of freedom for each.)

(f) The hypothesis test (or confidence interval) is:
(1) Directional (2) Nondirectional
(3) Not applicable

(g) The critical value(s) of the test statistic(s) is (are) ________. (Give the value. If there are multiple null hypotheses, state the critical value for each; if not applicable, so state.)

(h) The appropriate formula(s) for the test statistic(s) (or confidence interval or strength of relationship) is (are) given by Equation (________). (If there are multiple null hypotheses, state the equation number for each.)

Computations are *not* required for any of the problems in Section D. See the instructions.

27. Dr. Smith is a sports psychologist. She wishes to evaluate the effectiveness of three different strategies of shooting basketball free throws: imaginal visualization, concentration on follow-through, and relaxed release. She enlists the ten worst free-throw shooters on the team and trains them for three weeks. During one week, they use imaginal visualization. During another week, they concentrate on the follow-through. During another week, they concentrate on relaxing the release. Each day they shoot 50 free throws, or 300 total in a six-day week. The table shows the number of free throws made. Does strategy make a difference?

Player	*Visualization*	*Follow-through*	*Relaxed*
1	186	185	211
2	222	208	213
3	241	237	277
4	208	230	226
5	250	258	271
6	272	261	291
7	257	252	258
8	204	180	212
9	178	228	212
10	249	251	246

28. A researcher is exploring the features of World Wide Web sites. In particular, she wishes to know whether the characteristics of the screen background influences the time a surfer spends on a particular page. She sets up two identical pages except that the background of one is plain white, whereas the other background is a pastel design. She records the amount of time between the display of the page and the moment the surfer clicks the "Continue" button. The data for one day's surfers on each page are shown here. Does type of background make a difference?

Plain white: 6.2, 9.6, 5.7, 7.4, 4.7, 8.9, 9.0, 10.6, 6.6, 9.2, 9.2, 6.9, 6.7, 5.3, 7.8, 6.8, 8.3, 9.0, 8.0, 6.7

Pastel design: 6.3, 7.3, 3.1, 5.5, 6.1, 4.3, 7.6, 8.2, 5.3, 9.1, 4.7, 7.3, 7.8, 5.9, 5.5

29. An anesthesiologist has her choice of four different anesthetic techniques (A, B, C, D) to conduct a particular kind of operation. She decides to investigate the effec-

A	*B*	*C*	*D*
115	82	127	105
114	99	116	97
120	93	130	85
157	101	114	86
87	81	99	92
123	75	137	87
122	77	142	93
141	97	121	98
157	88	139	99
151	68	144	81

tiveness of these four techniques by randomly assigning a technique to each of 40 patients (ten per technique) and then recording the time the patient spends in the recovery room. The times (in minutes) for each technique are listed in the table. Does the type of anesthesia technique make a difference?

30. (*Exercise 29 continued*) The test required in the anesthesiology experiment had significant results. Which pairs of anesthetics are significantly different from one another.

31. An advertising executive wishes to know how effective television advertising is in selling Smellgood perfume. He chooses eight cities as test marketing sites and buys advertising time in all those cities. He varies randomly the number of advertising spots he buys in each city. For each city, he records the percentage increase in sales as shown in the table and determines a formula to predict the percentage increase in sales from the number of advertising spots. Is there a statistically significant reason to use this formula, or should he just predict by using the mean sales?

City	Spots	Percent increase
1	21	25
2	12	11
3	32	34
4	11	16
5	30	30
6	44	28
7	29	30
8	32	19

32. A zoo curator wishes to know whether playing recordings of animal sounds alters the time visitors spend watching the particular animals. He selects 16 of his exhibits at random and installs the apparatus necessary to play the animal sounds. He sets the apparatus so that sometimes it plays and sometimes it does not play. He follows patrons and records the number of seconds each patron stays in front of the particular exhibit, and he also notes for each patron whether the recording was playing or not. The mean times spent at each exhibit under both conditions are shown in the table. Does the recording influence the time spent viewing the exhibit?

Exhibit	Off	On
1	21	55
2	49	50
3	40	36
4	38	52
5	39	51
6	26	8
7	24	26
8	31	32
9	1	5
10	29	20
11	24	18
12	9	7
13	25	25
14	23	53
15	29	62
16	22	32

33. A highway engineer wishes to know the average speed of automobiles at a particular point on Interstate 90. She randomly measures by radar the speeds of 20 automobiles: 58, 67, 71, 57, 68, 65, 66, 70, 67, 66, 75, 61, 66, 66, 62, 71, 76, 64, 73, 62. What can she say about the mean speed at this point on Interstate 90?

34. A university administrator wishes to know whether a student's family income depends on his or her gender and/or class standing. He takes a stratified random sample of 40 students and ascertains their family incomes. The incomes (in $1000) are shown in the table. Is income related to gender or to class standing?

	Freshman	Sophomore	Junior	Senior
Male	35.3	32.2	31.7	30.3
	31.6	36.6	30.5	29.4
	30.0	31.3	29.4	25.9
	28.9	32.1	31.5	30.5
	32.3	28.7	27.0	30.4
Female	34.3	19.4	28.9	35.7
	34.2	27.8	33.4	30.7
	31.8	22.6	29.8	26.0
	32.4	36.2	30.2	44.8
	33.1	28.8	28.5	26.9

35. Dr. Transen studies the physiology of meditation by measuring oxygen intake. He uses four groups of six subjects each. One group meditates using transcendental meditation (TM), the second group meditates using internalized chanting (IC), the third group simply relaxes (SR), and the fourth group relaxes with imagery (IR). The oxygen intakes (in cc per minute) are shown in the table. Is meditation different from relaxation?

TM	IC	SR	IR
230	208	203	206
207	210	231	205
226	210	179	224
194	207	221	233
211	211	195	192
221	213	190	215

Section E: From the journals (answers in Appendix D, page 532)

36. Shei, Iwakuma, and Fujii (1988) were interested in the ecology of a particular fish-free pond in Japan. "In the pond, a large cladoceran [a kind of crustacean], *Daphnia rosea*, has dominated the zooplankton community since 1983. . . . In order to analyze the factors regulating the population dynamics of *D. rosea*, the seasonal characteristics of the population parameters (sexual reproduction, size distribution of females, and parthenogenetic reproduc-

tion) and the related environmental factors were studied" (pp. 291–292). We will focus here on the size distribution of females.

Shei wished to count the number of mature females but was faced with the difficulty of how to define "mature" in these organisms that are only about 2 millimeters long. Furthermore, the researchers observed that the size of mature females varied with water temperature. They decided to use a regression equation to specify the minimum length of a mature female. Any organism longer than that minimum (which depended on water temperature) would be considered mature (females are much longer than males in this species).

They derived the regression relationship between mature size and water temperature in the following way: They measured the lengths of 1454 egg-bearing (and thus by definition mature) females and the water temperature of the sample from which each female came. They divided the water temperatures (which ranged from 4 to 25°C) into seven groups at intervals of 3°C and inspected each of the seven groups, looking for the smallest egg-bearing female in each group. They thus selected seven females (the smallest egg-bearing female in each of the seven temperature classes) and fit a regression line through those seven data points. The resulting regression line was $\hat{L} = -.02T + 1.72$ ($r = -.92$, $n = 7$, $\hat{L}$ = predicted length in millimeters, T = temperature in °C). Thereafter, "individuals larger than the mature size determined by the above regression equation are referred to as adult females" (p. 296).

(a) Is the slope of this regression line significantly different from 0? (Shei et al. did not report this result.)

(b) What is the minimum length of a mature female when the water temperature is 4°C? 25°C?

37. (*Exercise 36 continued*) Suppose you replicate the Shei et al. (1988) study, using the same methodology, and find the seven minimum egg-bearing female lengths shown in the table.

Temp. (°C)	Length (mm)
5	1.59
8	1.71
11	1.44
14	1.46
17	1.47
20	1.34
23	1.38

(a) Compute the regression line.

(b) Compute the Pearson correlation coefficient.

(c) Is the slope of the regression line significantly different from 0?

38. Alzheimer's disease is a disorder that causes cognitive impairment that generally worsens over time. Ortof and Crystal (1989) sought to specify just how quickly this impairment worsens. "The rate of progression of Alzheimer's disease (AD) is a major concern to families of Alzheimer patients. Determination of rate is useful to predict future disability and to check that the course of the illness is consistent with that of AD" (p. 511). The degree of cognitive impairment was measured by a test called the Blessed Test, which is scored from 0 to 33 points (33 = maximal cognitive impairment). Ortof and Crystal reviewed the records of 54 AD patients who were followed for at least a year and for whom at least three Blessed results were available during that period.

Ortof and Crystal then computed a regression line for *each* of the 54 patients. In each case, the Y variable was the Blessed score and the X variable was the time measured in years (or decimal fractions of years) from that patient's first Blessed testing. Thus, for each patient, Ortof and Crystal computed the regression intercept a (whose value was approximately the first Blessed score) and the regression slope b (the change in the Blessed score per year). It was b, the rate of change of the Blessed score, that was of primary interest.

The mean value of b for all 54 patients was 4.1; that is, the average patient gained 4.1 Blessed points per year (recall that an increase in the Blessed score corresponds to an increase in impairment). "There was, however, considerable variability among the individual rates of our 54 subjects. While a majority of the patients progressed at rates between 2.4 and 7.2 points/year, five subjects had negative rates (suggesting that they improved over time) and 11 subjects had rates as high as 7.2 to 9.6 points/year. We were unable to account for these differences based on age of onset, duration of illness, or family history of dementia" (p. 512).

(a) Does each patient's regression slope b in fact give the increase in the Blessed score for that patient?

(b) In this situation, which is more desirable: a regression slope b that is positive or one that is negative?

39. (*Exercise 38 continued*) Suppose you are a physician and decide to replicate Ortof and Crystal's results with your own patient. The patient's original (at time 0) Blessed score was 17, .5 year later his score was 19, and .5 year after that (that is, 1 year after the initial testing) his score was 23.

(a) Do that patient's scores seem to be typical of the results reported by Ortof and Crystal?

(b) Compute the regression line for this patient. Is the regression slope b typical of those reported by Ortof and Crystal?

(c) When X (time) = 0, Y (Blessed score) = 17. Why, then, isn't the regression intercept a exactly 17?

Section F: Computer explorations

40. (a) Prepare five pairs of IQ-type data points as follows (this is the same procedure we followed in Chapter 16): Have the computer create a random sample of size five from the population of IQ scores ($\mu = 100$, $\sigma = 15$). [*Hint:* We described how to do that in Section 7.3.] Then have the computer create a second

random sample of size five from the population of IQ scores ($\mu = 100$, $\sigma = 15$).

(b) Alter one point in this data set until the correlation coefficient is 0. [*Hint:* If the existing correlation is positive, choose a pair for which both values are relatively large, and alter one of them. If the existing correlation is negative, choose a pair for which one value is relatively large and the other is relatively small, and alter one of those points.]

(c) When you have succeeded in making the correlation 0, what is the value of the regression constant b in the equation Variable2 = b(Variable1) + a?

(d) Alter the point to make r positive. What happens to b?

(e) Alter the point to make r negative. What happens to b?

41. (a) Have the computer create five pairs of IQ-type data points as in Exercise 40(a).

(b) Jot down the correlation coefficient r and the regression constants b and a in the equation Variable2 = b(Variable1) + a.

(c) Predict what will happen to r, b, and a if you add 20 points to each value of Variable 2. Then actually add 20 points to all of the values of Variable 2 and verify your predictions. Add and subtract other values until you thoroughly understand this concept.

(d) Now repeat parts (b) and (c) except alter Variable 1 instead of Variable 2.

42. (a) Have the computer create five pairs of IQ-type data points as in Exercise 40(a) (or continue on with the points you were using in Exercise 41).

(b) Jot down the correlation coefficient r and the regression constants b and a in the equation Variable2 = b(Variable1) + a.

(c) Predict what will happen to r, b, and a if you multiply each value of Variable 2 by 10. Then actually multiply each value of Variable 2 by 10 and verify your predictions. Experiment with other values until you thoroughly understand this concept.

(d) Now repeat parts (b) and (c) except alter Variable 1 instead of Variable 2.

43. Create a data set with five pairs of data that has this regression line: Variable2 = Variable1 (that is, for which $b = 1$ and $a = 0$) and for which $r = 1.0$. [*Hint:* Sketch a scatterplot.]

44. *(Exercise 43 continued)* Create a data set with five pairs of data that has this regression line: Variable2 = 2(Variable1) + 0 and for which $r = 1.0$. [*Hint:* Sketch a scatterplot.]

45. *(Exercise 44 continued)* Create a data set with five pairs of data that has this regression line: Variable2 = 2(Variable1) + 3 and for which $r = 1.0$. [*Hint:* Sketch a scatterplot.]

46. *(Exercise 43 continued)* Add two pairs of points to the data set of Exercise 43 so that b and a stay the same but r becomes .5. [*Hint:* The easiest way is to keep the Variable 1 value the same for both added points.]

47. *(Exercises 43–46 continued)* The procedures you just practiced should make it clear that you can create a regression line with whatever values of r, b, and a that you desire. Demonstrate that fact to yourself by specifying your own values of r, b, and a and then creating a regression line that has those characteristics.

CHAPTER 18

Some Nonparametric Statistical Tests

LEARNING OBJECTIVES

1. What is a parametric test?
2. What effects does violating the assumption of normality have on a parametric test?
3. What is a nonparametric statistical test?
4. What are the two main kinds of nonparametric tests?
5. When should a χ^2 goodness of fit test be used?
6. When should a χ^2 test of independence be used?
7. For what type of samples is the McNemar test for significance of change used?
8. When should a Mann–Whitney U test be used?
9. When should a Wilcoxon matched-pairs signed-rank test be used?
10. When should a Kruskal–Wallis H test be used?
11. What criteria should you use in determining whether to use a parametric or a nonparametric test?

The hypothesis-testing procedures of previous chapters have all assumed that data (or at least the means of data) are normally distributed. Nonparametric tests are used when we cannot make this assumption. This chapter describes the effects of violating the assumption of normality and the impossibility of then determining important probabilities such as the probability of making a Type I error. Nonparametric tests are used for data that are *not* measured at the interval/ratio level (and therefore could not possibly be normally distributed) as well as for nonnormal interval/ratio data. There are two broad classes of nonparametric tests: those that depend only on the classification of the data (and are therefore appropriate for nominal or ordinal or interval/ratio data) and those that depend on the inherent order of the variables (and are therefore appropriate for ordinal or interval/ratio data). Classification-type tests are based on the χ^2 distribution, whereas order-type tests use different distributions (U, T, and H) for each test.

As you may recall, we have been building a review of hypothesis-testing procedures ever since Chapter 10; the most recent table was Table 17.7. Inspection of that table shows that all the hypotheses we have used are statements *about the parameters of normal distributions*.[1] This implies that we have been *assuming* in Chapters 10–17 that our data are samples from normal distributions.[2] The hypothesis tests simply ask particular questions about the means (or correlations) of those normal distributions. Such tests are called *parametric* to emphasize that we are inquiring about the *parameters* of normal distributions.

parametric test: a hypothesis test about the parameters of a normal distribution

Many variables are *not* normally distributed; some populations have markedly skewed distributions, for example. What effect does such a deviation from normality have on the hypothesis tests of Chapters 10–17? This is a rather complicated question to answer precisely, but we will give an example to illustrate the ramifications of using a test that assumes a normal distribution when the population is in fact not normal but skewed.

18.1 Testing with a Nonnormal Distribution

Suppose we wish to test a hypothesis about a variable that actually has a skewed distribution. Suppose further that we perform a one-sample (parametric) test of the kind discussed in

[1] Or about the parameters of bivariate normal distributions in the case of correlation.

[2] If the variable was not normally distributed, the sample size was large enough so that (as the central limit theorem predicts) the *means* of the samples could be considered normally distributed.

Chapter 10 even though the distribution is not normal. Assume we know that the standard deviation is σ so that our test statistic is z. The null hypothesis is H_0: $\mu = 0$. Let us assume that the test is directional, so the alternative hypothesis is H_1: $\mu > 0$.

If the distribution of the variable were normal, we would sketch the distribution of the test statistic z as shown by the broken line in Figure 18.1. z would be normal with mean 0 and standard deviation 1, exactly as in Chapter 10.

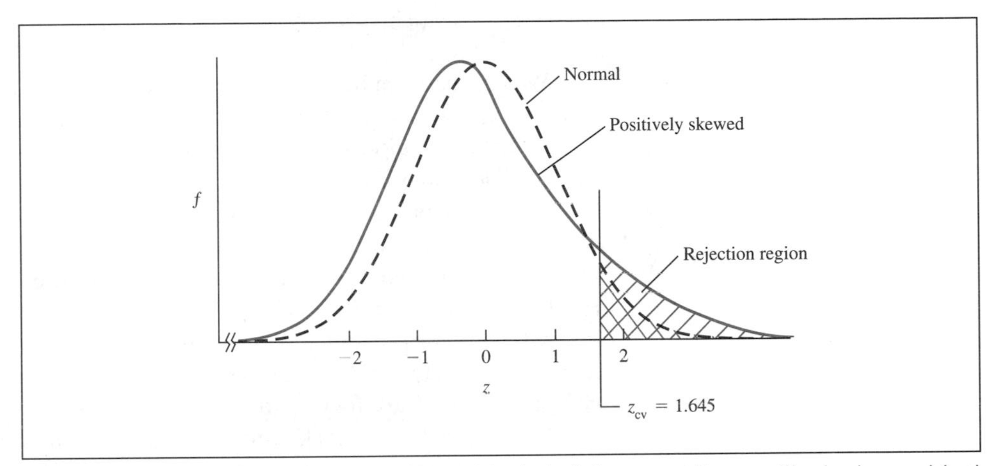

FIGURE 18.1 Distributions of the normal test statistic z and the positively skewed test statistic with the same mean and standard error. Rejection regions are shaded.

However, our variable is actually very positively skewed. Then the distribution of the *means of samples* of that variable will also be somewhat positively skewed.[3] If the distribution of means is positively skewed, the distribution of the test statistic (the mean divided by the standard error of the mean) will also be positively skewed. We show this positively skewed distribution of the test statistic (which also has mean 0 and standard error 1) in Figure 18.1 superimposed over the normal distribution of z.

If we assume that $\alpha = .05$, the parametric hypothesis-testing decision rule from Chapter 10 determines the directional critical value of z to be 1.645. That value is shown in Figure 18.1, and the rejection region beyond it is shaded. The area of the rejection region under the *normal* distribution (shaded with cross-hatching) is therefore .05, the probability of a Type I error assuming the distribution of the test statistic is normal. However, we can see that the area beyond z_{cv} under the *skewed* actual distribution (shaded diagonally) is *greater* than .05, so the probability of a Type I error in the skewed-distribution case is somewhat greater than .05. If we wish to maintain the probability of a Type I error at .05, we would have to choose a critical value somewhat greater than 1.645. How much greater is difficult to determine exactly.

Our general conclusion is that a departure from normality in the variable affects the probability of a Type I error. How much the probability is affected depends on the degree of departure from normality. In actual practice, small departures from normality do not appreciably change the probability of Type I errors, and so they can be disregarded, especially with large sample sizes. The analysis of variance is particularly unaffected (statisticians say the ANOVA is "robust") by departures from normality. However, because one of our primary concerns in inferential statistics is limiting the probability of a Type I error to some clearly specified value, it is desirable to develop procedures that can account for skewed or otherwise nonnormal data without making the probability of a Type I error undeterminable.

---

[3]The means of samples are not as strongly skewed as the original distribution of the variable. Recall from Chapter 7 that the central limit theorem states that distributions of means tend to become normal as the sample size increases regardless of the shape of the original distribution of the variable. If the original distribution is skewed and the sample size is only small or moderate, however, the distribution of means will also be somewhat, though not as strongly, skewed.

18.2 Nonparametric Statistical Tests

We just saw that skewed or otherwise nonnormal distributions can present problems for the kinds of hypotheses we tested in Chapters 10–17. Are there other kinds of data for which those parametric tests are inappropriate? Yes; it is inappropriate to use parametric tests on variables that are measured at only the nominal or ordinal level of measurement because only variables measured at the interval/ratio level of measurement can be normally distributed.

We *can* perform hypothesis tests on data that are measured at the nominal or ordinal level or on data that are measured at the interval/ratio level but are skewed or otherwise nonnormal. Such tests will quite clearly *not* be about the parameters (μ, ρ, or $b_{\text{population}}$) of normal distributions; therefore, we refer to these tests as *nonparametric* tests.

nonparametric test: a hypothesis test that does not depend on the parameters of a normal distribution

There are, broadly speaking, two kinds of nonparametric tests: those that depend on only the categorization of variables (and are thus appropriate for variables that are measured at least at the nominal level) and those that depend on the order inherent in the data (and are thus appropriate for variables that are measured at least at the ordinal level). Table 18.1 lists all the nonparametric tests that we will consider in this chapter (there are others that more advanced texts discuss), as well as the parametric tests appropriate for research with similar designs. Thus, for example, the second row of Table 18.1 shows tests appropriate for two-independent-samples designs. If the data are interval/ratio and normally distributed, we would perform a parametric test, the two-sample t (H_0: $\mu_1 = \mu_2$) from Chapter 11. If the data are nominal (making no assumptions about the order of categories), we would use a nonparametric test, the χ^2 test of independence (χ is Greek "chi," pronounced like the first part of "kite"; the test is called "chi-square test of independence"). If the data are ordinal, we would use another nonparametric test, the Mann–Whitney U test. We will discuss each of the nonparametric tests in turn, beginning with those appropriate for nominal data.

TABLE 18.1 Summary of parametric and nonparametric tests presented in this text

	Nonparametric		
Design	*At least nominal*	*At least ordinal*	*Parametric normal*
One sample	Goodness of fit	—[a]	One-sample t or z test
Two independent samples	χ^2 test of independence	Mann–Whitney U test	Two-independent-samples t test
Two dependent samples	McNemar test for significance of change	Wilcoxon matched-pairs signed-rank test	Two-dependent-samples t test
k independent samples	χ^2 test of independence	Kruskal–Wallis H test	One-way ANOVA
Correlation	—	Test of Spearman r_S	Test of Pearson r

[a] Test not discussed in this textbook

18.3 Tests for Data Measured at the Nominal Level

We will describe three tests that make no assumptions about whether or not data have inherent order. Thus, these tests can be used on data measured at the nominal level. They can also be used with data measured at the ordinal or interval/ratio level by ignoring the orderedness inherent in such data.

One Sample: The χ^2 Goodness of Fit Test

It may not be immediately obvious how a statistical test can be performed without making any assumptions about the normality of a distribution, so let us take a simple example and explore it in relative detail to familiarize ourselves with the general strategy of nonparametric tests for nominal variables.

Assume you have been playing some game involving dice, and you come to suspect that the die you are rolling is not "fair" in the sense that each of the six faces does not come up equally often. Perhaps the die is "loaded" or "shaved," so you decide to conduct a statistical experiment to ascertain whether your hunch is correct. You decide that you will roll the die 120 times and count the number of times each face appears.

If the die is fair, each face should have an equal probability of appearance on any given roll, so a "1" should appear $1/6$ of the time (20 times in 120 rolls), a "2" should appear $1/6$ of the time (20 times), and so on. The sequence of what we will call the "expected" frequencies for the six outcomes in this experiment is [20, 20, 20, 20, 20, 20]. It is unlikely, because of sampling fluctuations, that 1 or any other number would appear *exactly* 20 times in 120 rolls. Thus, if the outcome of our experiment was [19, 21, 20, 22, 18, 20], we would probably conclude that the die is fair even though the "observed" frequencies do not exactly match the "expected" frequencies. On the other hand, if the outcome of our experiment was [5, 35, 20, 20, 5, 35], we would probably conclude that the die is not fair.

Suppose that the actual outcome of our 120-roll experiment was [15, 23, 19, 26, 21, 16]. Is this a sufficient departure from the expected [20, 20, 20, 20, 20, 20] to justify concluding that the die is unfair? The χ^2 *goodness of fit test* furnishes a criterion that will provide an objective answer to that question.

goodness of fit test: a test of the null hypothesis that the sampled data come from a specified distribution

This kind of problem is called a "goodness of fit" problem because we begin with a statement of the expected frequencies [20, 20, 20, 20, 20, 20] and ask how well the observed frequencies "fit" our expectations. We test this goodness of fit by using the χ^2 distribution as a test statistic:

χ^2 test statistic for goodness of fit test and test of independence

$$\chi^2 = \sum \frac{(O_i - E_i)^2}{E_i} \qquad [df = k - 1] \tag{18.1}$$

where k is the number of categories (six in our dice example) and O_i and E_i are the observed and expected frequencies of the ith category, respectively. Thus, $O_1 = 15$ is the actual number of times "1" appears in 120 rolls, and $E_1 = 20$ is the expected number of times "1" would appear in 120 rolls.

χ^2, like t, is a family of distributions whose characteristics depend on the number of degrees of freedom. In general, χ^2 is positively skewed, and markedly so when the number of degrees of freedom is small. χ^2 distributions for selected degrees of freedom ($df = 1, 3, 5$, and 10) are shown in Figure 18.2. Note how the χ^2 distribution becomes more and more normal as the number of degrees of freedom increases. When the degrees of freedom are large—say, greater than 30—the χ^2 distribution becomes very close to a normal distribution whose mean is $df - 1/2$.

Table A.3 in Appendix A gives critical values for the χ^2 distribution when $df < 30$. When $df > 30$, χ^2_{cv} can be obtained from[4]

approximate critical value of χ^2 when $df > 30$

$$\chi^2_{cv} = \frac{z^2_{cv}}{2} + z_{cv}\sqrt{2\,df - 1} + \frac{2\,df - 1}{2} \qquad [\text{only if } df > 30] \tag{18.2}$$

where z_{cv} is the critical value of the normal distribution obtained from Table A.1 (use a directional value of z_{cv} to provide a critical value in one tail of the χ^2 distribution). If the observed value of χ^2 is greater than χ^2_{cv}, then we will reject the null hypothesis.

Let us apply the general hypothesis-testing procedure we have been using since Chapter 9 to the fair-die situation.

[4]Equation (18.2) is sometimes written $z = \sqrt{2\chi^2} - \sqrt{2\,df - 1}$.

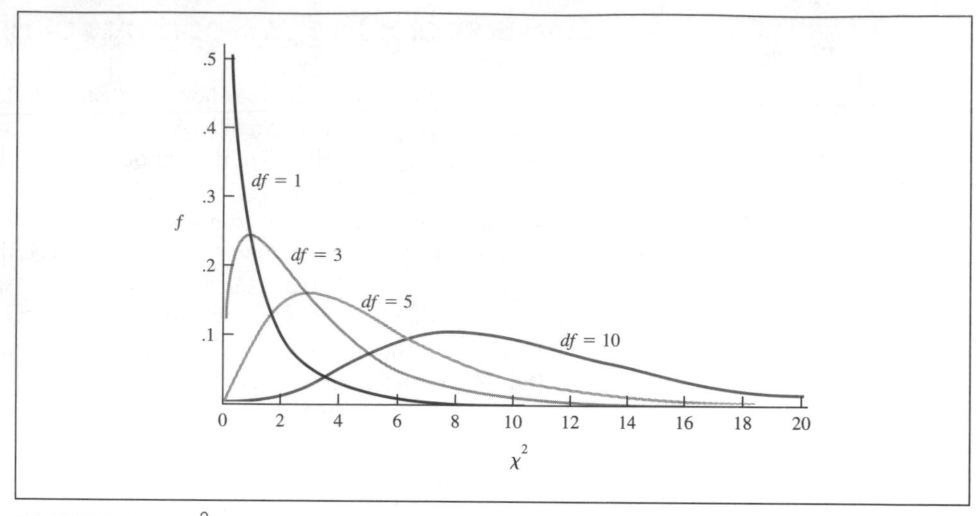

FIGURE 18.2 χ^2 distributions for selected degrees of freedom

Three sketches illustrate H_0:
Variable
Sample statistic
Test statistic

The only sketch that is informative here is the test statistic, so we omit the others.

I. State the Null and Alternative Hypotheses

H_0: The true frequency distribution is [20, 20, 20, 20, 20, 20].

H_1: The true frequency distribution has some other form.

Note that this null hypothesis is different from any we have encountered thus far. Until now the null hypothesis has specified a value for some *parameter* of a *normal* distribution. For example, we might have written H_0: $\mu = 3.5$ because 3.5 is the expected mean value, $(1 + 2 + 3 + 4 + 5 + 6)/6$, of dice rolls. Here, however, we specify the *form* of the *entire* distribution, not just some aspect of it (such as its mean). If the actual outcome of 120 rolls was [50, 10, 0, 0, 10, 50], the observed mean would be exactly 3.500 and yet we would still reject the null hypothesis because the shape of the distribution is not the specified [20, 20, 20, 20, 20, 20].

In the past, we have sketched distributions of variables, statistics, and test statistics, but the distributions of the variable and statistic are not particularly useful in χ^2 problems. Therefore, we show only the distribution of the test statistic χ^2 in Figure 18.3. Note that the test statistic has $df = k - 1 = 6 - 1 = 5$.

II. Set the Criterion for Rejecting H_0

Level of significance?
Directional?
Shade rejection regions.

As we have seen, the appropriate test statistic is

$$\chi^2 = \sum \frac{(O_i - E_i)^2}{E_i} \qquad [df = k - 1] \tag{18.1}$$

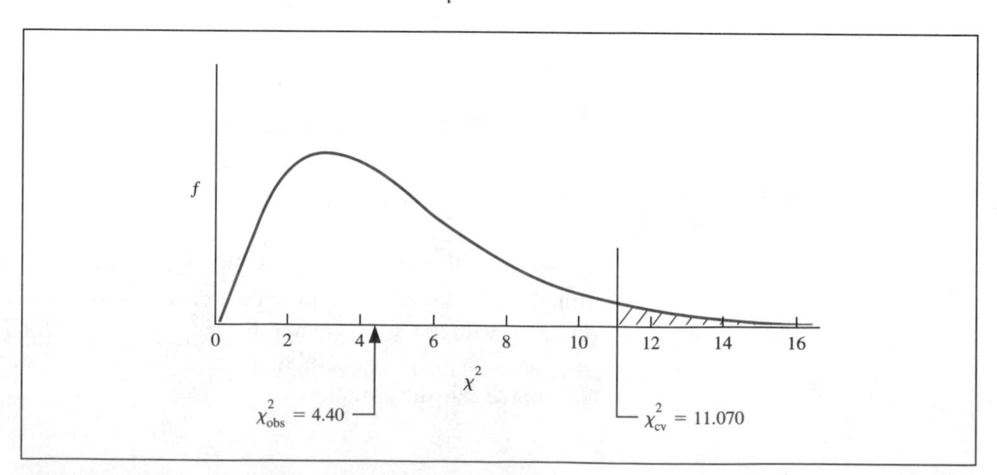

FIGURE 18.3 Dice-rolling example: distribution of the test statistic χ^2 ($df = 5$) with rejection region shaded and observed value indicated

where k is the number of categories (six in our example). Note that any term in this sum gets larger whenever the (squared) discrepancy between the observed and expected frequencies $(O_i - E_i)^2$ gets larger, regardless of whether the observed frequency is larger or smaller than the expected frequency (because the difference is squared). Thus, because χ^2 increases for *any* departure from the expected values, the larger the discrepancies, the larger the χ^2. Therefore, the critical value χ^2_{cv} is a value of χ^2 that leaves a particular α (say, .05) in the *right-hand* tail of the χ^2 distribution. We look up this critical value in Table A.3 and find that for $df = 5$ and $\alpha = .05$, $\chi^2_{cv} = 11.070$. We enter that value on our sketch of the test statistic and shade the rejection region beyond it, as shown in Figure 18.3. If the observed value χ^2_{obs} falls in the rejection region, we will (as usual) reject the null hypothesis.

The χ^2 goodness of fit test is always nondirectional: A departure from the specified distribution in *any* direction will increase χ^2_{obs}. The rejection region happens to be in one tail of the χ^2 distribution, but the test is still *nondirectional*.

III. Collect a Sample and Compute the Observed Value of the Test Statistic

Observed values:
Sample statistic
Test statistic
There is no separate sample statistic for χ^2 tests.

We now roll the die 120 times, each time recording which face is up. Then we count the frequency of occurrence of each face. Our results are that we rolled 15 "1's," 23 "2's," 19 "3's," 26 "4's," 21 "5's," and 16 "6's"; that is, our observed frequencies are (15, 23, 19, 26, 21, 16).

We need, then, to compute the six terms of the sum of Equation (18.1), as shown in Table 18.2. Then $\chi^2_{obs} = \Sigma(O_i - E_i)^2/E_i = 1.25 + .45 + .05 + 1.80 + .05 + .80 = 4.40$. We show that value on our distribution of the test statistic in Figure 18.3.

TABLE 18.2 Computing the χ^2 test statistic for the dice-roll example

Category	O_i	E_i	$O_i - E_i$	$(O_i - E_i)^2$	$(O_i - E_i)^2/E_i$
1	15	20	−5	25	1.25
2	23	20	3	9	.45
3	19	20	−1	1	.05
4	26	20	6	36	1.80
5	21	20	1	1	.05
6	16	20	−4	16	.80
Σ	120	120	0		$\chi^2_{obs} = 4.40$

IV. Interpret the Results

Statistically significant?
Practically significant?
State results clearly.

Because the observed value of the test statistic $\chi^2_{obs} = 4.40$ is less than the critical value of the test statistic $\chi^2_{cv} = 11.070$, we do not reject the null hypothesis. Our conclusion is that the observed frequencies are of the kind to be expected by chance fluctuations in 120 rolls.

Eyeball-calibration

The eyeball-estimation techniques we discussed in previous chapters do not lend themselves easily to nonparametric statistics. However, we can get a sense of what it takes to reject a goodness of fit null hypothesis by asking: If the dice frequencies we have just been considering [15, 23, 19, 26, 21, 16] are not sufficiently different from the expected [20, 20, 20, 20, 20, 20] to reject the null hypothesis, how far must the dice frequencies depart from what we expect so we *can* reject the null hypothesis? As a way of calibrating our eyeball, we need to get a sense of how much variability can be expected by chance. Table 18.3 gives an outcome of 120 rolls that *just barely* leads to the rejection of H_0. When the results are [11, 24, 19, 30, 21, 15], $\chi^2_{obs} = 11.20$. df is still 5, so χ^2_{cv} remains 11.070, and we would reject the null hypothesis. Anything closer to [20, 20, 20, 20, 20, 20] for example, [12, 24, 19, 29, 21, 15]—would *not* result in the rejection of H_0.

TABLE 18.3 Dice-roll outcome that just barely rejects H_0

Category	O_i	E_i	$O_i - E_i$	$(O_i - E_i)^2$	$(O_i - E_i)^2/E_i$
1	11	20	−9	81	4.05
2	24	20	4	16	.80
3	19	20	−1	1	.05
4	30	20	10	100	5.00
5	21	20	1	1	.05
6	15	20	−5	25	1.25
Σ	120	120	0		$\chi^2_{obs} = 11.20$

Another Example

Consider another example of the goodness of fit test. Suppose you are the manager of the American Psychological Association annual convention. This year you make a concerted effort to get young students to attend the convention by creating activities designed especially for them. You would like to know whether your efforts to attract younger students were successful. You know from past years that among students who attend the APA convention, 10% are freshmen, 20% are sophomores, 30% are juniors, and 40% are seniors. Does the student attendance at this year's convention follow that same pattern?

To answer that, we decide to take a random sample of 100 students and ask them to tell us their class standing. On the basis of those results, we will ascertain whether we can reject the null hypothesis that attendance is [10%, 20%, 30%, 40%].

I. State the Null and Alternative Hypotheses

The null and alternative hypotheses are:

H_0: The distribution of this year's student attendance is [10%, 20%, 30%, 40%] freshmen, sophomores, juniors, and seniors, respectively.

H_1: The distribution has some other shape.

II. Set the Criterion for Rejecting H_0

Because we have not otherwise specified it, we assume that $\alpha = .05$. We will be performing a goodness of fit test on a variable with four categories, so the test statistic is χ^2 with 3 degrees of freedom. That distribution is shown in Figure 18.4. We know that any departure from the distribution specified by the null hypothesis will cause χ^2 to increase, so the rejection region will be only in the right-hand tail of this distribution even though the test

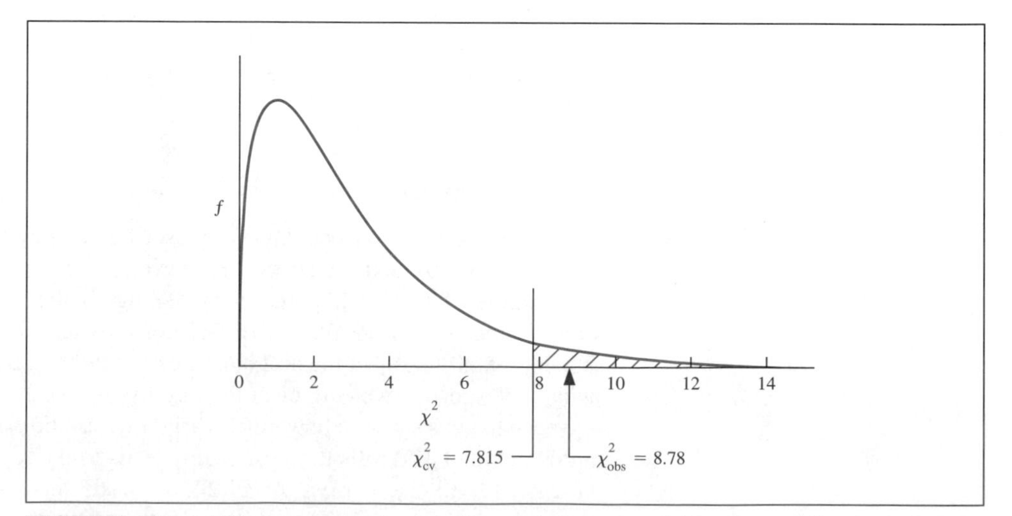

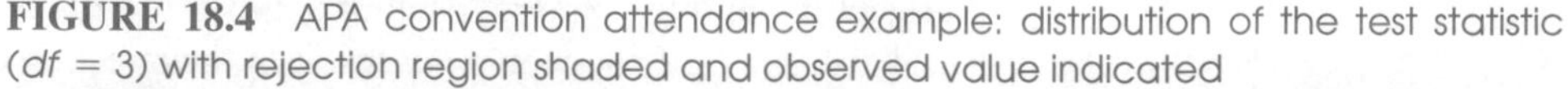
FIGURE 18.4 APA convention attendance example: distribution of the test statistic ($df = 3$) with rejection region shaded and observed value indicated

is nondirectional. Table A.3 shows that $\chi^2_{cv,df=3} = 7.815$. We enter that value on our sketch and shade the rejection region to its right.

III. Collect a Sample and Compute the Observed Value of the Test Statistic

There is no sample statistic that is informative, so we omit discussion of that entirely.

We take our random sample of 100 students and ask their class standing. Our results show that 17 of the 100 respondents were freshmen, 25 were sophomores, 28 were juniors, and 30 were seniors. We compute χ^2 from Equation (18.1) as shown in Table 18.4, obtaining $\chi^2_{obs} = 8.78$.

TABLE 18.4 Computing χ^2_{obs} for students attending the APA convention

Category	O_i	E_i	$O_i - E_i$	$(O_i - E_i)^2$	$(O_i - E_i)^2/E_i$
Freshman	17	10	7	49	4.90
Sophomore	25	20	5	25	1.25
Junior	28	30	−2	4	0.13
Senior	30	40	−10	100	2.50
Σ	100	100	0		$\chi^2_{obs} = 8.78$

IV. Interpret the Results

Because $\chi^2_{obs} = 8.78 > \chi^2_{cv} = 7.815$, we reject the null hypothesis that this year's distribution of class standings fits that of previous years. Apparently the efforts have paid off (or some other factor has led to increased attendance of young students).

Two Independent Samples: The χ^2 Test of Independence

In the examples we have just discussed, we considered one variable at a time: the outcome of the roll of a die in the first case and class standing in the second. The χ^2 distribution can also be used on two variables at a time to test whether the outcome of one variable depends on the outcome of the other.[5]

To explore this, let us assume we are interested in whether a person's attitude toward abortion is related to the size of his or her family of origin. We develop an interview technique designed to classify respondents into one of three abortion-attitude categories: ProChoice (women have the right to decide abortion issues for themselves), Moderate (abortion should be allowed in some circumstances), and ProLife (abortion should not be allowed). Family-of-origin size is categorized as Only (only child), Small (one sibling), Middle-sized (two or three siblings), and Large (four or more siblings).

The two distributions (abortion attitude and family size) are "independent" if the percentages of ProChoice, Moderate, and ProLife people who were Only children are approximately the same as the percentages of ProChoice, Moderate, and ProLife people who came from Small families, and so on. We use the χ^2 *test of independence* to determine whether the two distributions are independent.

test of independence: a nonparametric test of whether the shape of a distribution depends on the value of another variable (or variables)

I. State the Null and Alternative Hypotheses

H_0: The distribution of attitudes toward abortion is independent of family size.

H_1: Attitude toward abortion depends on family size.

There are three things to note about the null hypothesis in this example. First, as in the earlier goodness of fit examples, the null hypothesis is a statement about the *entire* shape of a distribution, *not* about just some aspect ("parameter") of the distribution.

[5] In fact, χ^2 can be used to test three or more variables simultaneously, although we leave such examples to more advanced texts.

Second, just as for correlation coefficients in Chapter 16, the null hypothesis here is *not* actually a statement of cause and effect. From the standpoint of the logic of the χ^2 test, it would have been perfectly legitimate for the null hypothesis to be reversed—H_0: Family size is independent of attitude toward abortion. The first null hypothesis statement is somewhat more natural because family size precedes in time (by many years) a person's attitude toward abortion, so it seems more reasonable to conclude that family size might affect attitude rather than that attitude would affect family size. However, such an argument is based on considerations (our general understanding of nature) entirely separate from statistics. The χ^2 test itself is entirely indifferent to the direction in which the null hypothesis is stated.

Third, although the variables we are considering in this problem are in fact ordinal (for family size, Large > Middle-sized > Small > Only, and for Abortion Attitude, ProChoice > Moderate > ProLife), the χ^2 independence analysis does *not* make use of the order properties; these variables are treated as if they were nominal.

II. Set the Criterion for Rejecting H_0

The χ^2 test for independence requires the computation of a χ^2 test statistic according to Equation (18.1), our general equation for computing χ^2. Here, however, the number of degrees of freedom is $df = (R - 1)(C - 1)$, where R is the number of categories in the "row" variable and C is the number of categories in the "column" variable. It makes absolutely no difference which variable is assigned to rows and which to columns. Let us arbitrarily let abortion attitude be the row variable and family size be the column variable. Then $R = 3$ and $C = 4$, so $df = (R - 1)(C - 1) = (3 - 1)(4 - 1) = 6$. We sketch the χ^2 distribution with 6 degrees of freedom in Figure 18.5.

Because we have not specified otherwise, we let $\alpha = .05$; then because $df = 5$, $\chi^2_{cv} = 12.592$ from Table A.3. We show that value in Figure 18.5 and shade the rejection region as usual. Here again, the test is nondirectional, but we reject H_0 only for large values of χ^2.

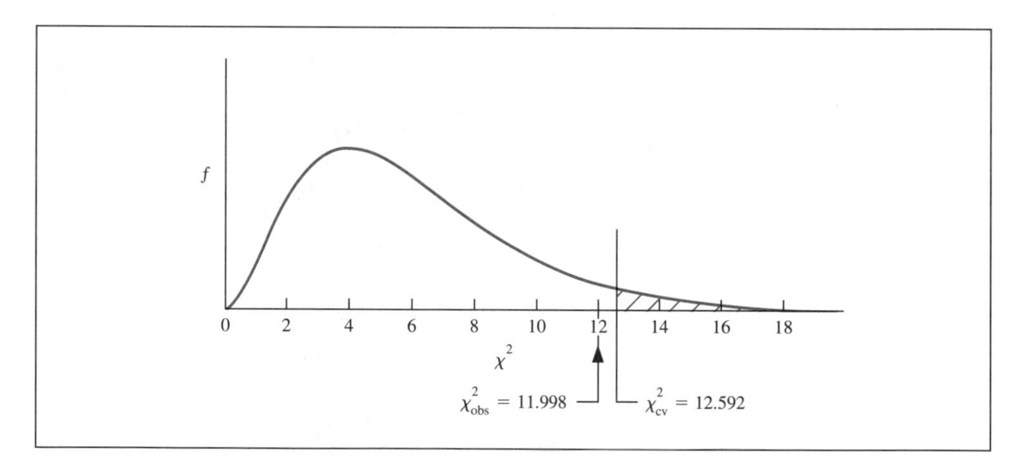

FIGURE 18.5 Abortion attitude example: distribution of the test statistic χ^2 ($df = 6$) with rejection region shaded and observed value indicated

III. Collect a Sample and Compute the Observed Value of the Test Statistic

contingency table: a table presentation of two variables so that their interrelationships can be examined; all combinations of categories of all variables are presented

We take a random sample of 150 people and ascertain each person's abortion attitude and family size. The results of these interviews are shown in Table 18.5. For example, our first subject was ProChoice and came from a Middle-sized family. We summarize these data by preparing a 3 × 4 *contingency table*, as shown in Table 18.6. This table has three rows, one for each level of the abortion attitude variable, and four columns, one for each level of family size. The first entry in each cell is the frequency of occurrence of the particular row and column categories. Thus, the value in the upper left corner of this table is the count of the number of respondents who identified themselves being both ProLife and Only children. There are 11 such subjects (subjects numbered 8, 20, 58, 82, 85, 108, 122, 125,

TABLE 18.5 Abortion attitude example: Results of interviews with 150 individuals

Subject	*Subject's report of abortion attitude/family size*				
1– 5	ProCho/Midl	ProCho/Larg	Modrat/Smal	ProCho/Smal	ProCho/Only
6– 10	ProCho/Midl	ProLif/Smal	ProLif/Only	Modrat/Larg	ProLif/Smal
11– 15	Modrat/Larg	Modrat/Midl	Modrat/Midl	ProCho/Only	Modrat/Midl
16– 20	Modrat/Only	Modrat/Only	ProLif/Midl	Modrat/Midl	ProLif/Only
21– 25	Modrat/Midl	ProCho/Smal	ProLif/Smal	ProLif/Larg	ProCho/Smal
26– 30	Modrat/Smal	ProCho/Midl	ProLif/Smal	Modrat/Midl	ProCho/Smal
31– 35	ProCho/Midl	ProCho/Larg	ProCho/Midl	ProCho/Larg	ProCho/Midl
36– 40	ProCho/Midl	ProCho/Only	Modrat/Smal	ProCho/Midl	ProCho/Larg
41– 45	ProCho/Midl	ProLif/Smal	Modrat/Only	Modrat/Larg	ProLif/Smal
46– 50	Modrat/Larg	Modrat/Midl	Modrat/Midl	ProCho/Only	Modrat/Midl
51– 55	ProLif/Midl	ProCho/Only	Modrat/Smal	ProCho/Midl	ProCho/Larg
56– 60	ProCho/Midl	ProLif/Smal	ProLif/Only	Modrat/Larg	ProLif/Smal
61– 65	Modrat/Larg	Modrat/Midl	Modrat/Midl	ProCho/Only	Modrat/Midl
66– 70	Modrat/Smal	ProLif/Larg	ProCho/Midl	Modrat/Midl	ProCho/Smal
71– 75	Modrat/Smal	ProCho/Smal	ProLif/Smal	Modrat/Midl	ProCho/Smal
76– 80	Modrat/Smal	Modrat/Only	ProLif/Smal	Modrat/Midl	Modrat/Only
81– 85	Modrat/Only	ProLif/Only	ProCho/Midl	Modrat/Midl	ProLif/Only
86– 90	Modrat/Midl	ProCho/Smal	Modrat/Smal	ProLif/Larg	ProCho/Midl
91– 95	Modrat/Smal	ProCho/Smal	Modrat/Smal	Modrat/Midl	ProCho/Midl
96–100	Modrat/Smal	Modrat/Only	ProLif/Smal	Modrat/Midl	Modrat/Only
101–105	ProCho/Midl	ProCho/Only	Modrat/Smal	ProCho/Smal	ProCho/Larg
106–110	ProCho/Midl	ProLif/Smal	ProLif/Only	Modrat/Larg	ProLif/Smal
111–115	Modrat/Smal	ProCho/Smal	ProLif/Smal	Modrat/Midl	ProCho/Smal
116–120	Modrat/Smal	Modrat/Only	ProLif/Smal	Modrat/Midl	Modrat/Only
121–125	Modrat/Only	ProLif/Only	ProLif/Midl	Modrat/Midl	ProLif/Only
126–130	Modrat/Midl	ProCho/Smal	ProLif/Smal	ProLif/Larg	ProCho/Smal
131–135	Modrat/Smal	ProCho/Smal	ProLif/Smal	Modrat/Midl	ProCho/Smal
136–140	Modrat/Smal	Modrat/Only	ProLif/Smal	Modrat/Midl	Modrat/Only
141–145	Modrat/Only	ProLif/Only	ProLif/Midl	Modrat/Midl	ProLif/Only
146–150	Modrat/Only	Modrat/Only	ProLif/Midl	Modrat/Midl	ProLif/Only

TABLE 18.6 A 3 × 4 contingency table showing frequencies of abortion attitude and family size (expected frequencies in parentheses)

		Family size				
		Only	*Small*	*Middle*	*Large*	*Row totals*
Abortion attitude	ProLife	11 (8.61)	18 (12.67)	5 (12.41)	4 (4.31)	38
	Moderate	16 (14.96)	16 (22.00)	27 (21.56)	7 (7.48)	66
	ProChoice	7 (10.43)	16 (15.33)	17 (15.03)	6 (5.21)	46
	Column totals	34	50	49	17	150

142, 145, and 150 in Table 18.5), so we enter the value 11 in the "ProLife" row and "Only" column. We determine the remaining frequencies in a similar manner: There are 18 ProLife people who came from Small families, and so on.

The right-hand column of the contingency table shows the total for each row. Thus, for the first row, $11 + 18 + 5 + 4 = 38$ individuals in our sample identified themselves as being ProLife. The bottom row represents the totals for each column. Thus, for the "Middle" column, $5 + 27 + 17 = 49$ individuals came from Middle-sized families. The grand total (150 respondents) is shown in the bottom right-hand corner. As a computational check, the grand total should equal both the sum of the row totals $(38 + 66 + 46)$ and the sum of the column totals $(34 + 50 + 49 + 17)$.

Next we compute the expected frequency of each cell and show it in parentheses in the appropriate cell of Table 18.6. The expected frequency of the cell in the rth row and cth column is given by

expected frequency of cell in contingency table

$$\text{expected frequency of cell } rc = \frac{f_r f_c}{N} \tag{18.3}$$

where f_r is the total for row r, f_c is the total for column c, and N is the total number of points. For example, the expected frequency of the ProLife/Middle-sized family cell ($r = 1, c = 3$) is $f_{\text{row 1}} f_{\text{col 3}}/N = 38(49)/150 = 12.41$.

Understanding Equation (18.3) is the key to understanding why these χ^2 tests are called tests of *independence*. As we just saw, $f_{\text{row 1}} = 38$ is the total number of ProLife individuals in our sample. Therefore, $f_{\text{row 1}}/N = 38/150 = .253$ (or 25.3%) is the proportion of ProLife individuals in our total sample. If the abortion attitude and family size variables are *independent*, then about 25.3% of Only children would be expected to be ProLife, 25.3% of Small-family children would be expected to be ProLife, and also 25.3% of Middle-sized and 25.3% of Large-family children. If, on the other hand, the two variables are *not independent*, then we would expect those cell percentages to *differ* from one another.

The same logic holds true for the column variables. For example, $f_{\text{col 3}}/N = 49/150 = .327$ implies that 32.7% of all our respondents came from Middle-sized families. If the null hypothesis is true (that is, if the two variables are independent), about 32.7% of ProLife individuals would be expected to come from Middle-sized families, and 32.7% of Moderate and 32.7% of ProChoice individuals would be expected to come from Middle-sized families as well.

If all that is true, *how many* individuals would be expected to be both ProLife and from Middle-sized families? There are $f_{\text{row 1}} = 38$ ProLife individuals, and $f_{\text{col 3}}/N = 49/150 = .327$ of them would be expected (assuming independence of the variables) to come from Middle-sized families. Therefore, we would expect that $f_{\text{row 1}}(f_{\text{col 3}}/N) = 38(.327) = 12.41$ individuals to be both ProLife and from Middle-sized families. Removing the parentheses shows that $f_{\text{row 1}}(f_{\text{col 3}}/N) = f_{\text{row 1}} f_{\text{col 3}}/N$ exactly as shown in Equation (18.3). We show the expected value, 12.41, in parentheses in the first-row, third-column cell of Table 18.6.

We calculate the remaining expected values and show them in Table 18.6 also. Now we are ready to calculate χ^2_{obs} according to Equation (18.1), just as we have done in previous situations. The computations are shown in Table 18.7, with the result that $\chi^2_{\text{obs}} = 11.998$.

TABLE 18.7 Computing χ^2_{obs} for abortion attitude and family size data

Category	O_i	E_i	$O_i - E_i$	$(O_i - E_i)^2$	$(O_i - E_i)^2/E_i$
ProLife/Only	11	8.61	2.39	5.712	.663
ProLife/Smal	18	12.67	5.33	28.409	2.242
ProLife/Midl	5	12.41	−7.41	54.908	4.424
ProLife/Larg	4	4.31	−.31	.096	.022
Moderat/Only	16	14.96	1.04	1.082	.072
Moderat/Smal	16	22.00	−6.00	36.000	1.636
Moderat/Midl	27	21.56	5.44	29.594	1.373
Moderat/Larg	7	7.48	−.48	.230	.031
ProChoi/Only	7	10.43	−3.43	11.765	1.128
ProChoi/Smal	16	15.33	.67	.449	.029
ProChoi/Midl	17	15.03	1.97	3.881	.258
ProChoi/Larg	6	5.21	.79	.624	.120
Σ	150	150.00	.00		$\chi^2_{\text{obs}} = 11.998$

IV. Interpret the Results

Because $\chi^2_{\text{obs}} = 11.998$ is not greater than $\chi^2_{\text{cv}} = 12.592$, we do not reject H_0. Our conclusion is that abortion attitude and family size can be considered to be independent of each other.

Two Dependent Samples: The McNemar Test for Significance of Change

Suppose that in August 1990, we took a random sample of 100 individuals in the town of Oakville and asked them, "Do you favor requiring the Pledge of Allegiance to be said in elementary schools? Yes or no?" In August 1991, we located those same 100 individuals and asked them the same question again. Between those two Augusts there occurred a very quick, popular, dominant American/Allied victory over forces in Iraq, and we wish to ascertain whether such a political event changed people's opinions about saying the pledge.

This problem differs from the test of independence situation we just considered because we use the *same* people in the two conditions (before and after). For example, the three rows of the abortion-attitude example referred to three separate, unrelated groups of individuals; the first was ProLife, the second was Moderate, and the third was ProChoice. The same can be said about the four column groups; one was composed of Only children, the second of children from Small families, and so on. It would have violated the assumptions of the test of independence model to use the same person in any two cells. For example, if a person changed his attitude about abortion, he could not be used in a second cell.

Recall that in Chapters 11 and 12 we distinguished between two-sample t tests that were applicable to groups whose members were independent of members in the other groups (as in Chapter 11) and tests for groups whose members were in some way related to members of the other sample (as in Chapter 12). Recall further that one of the most common types of relatedness described in Chapter 12 was in the repeated-measures design, where the same individuals appear in both groups. It is that same distinction between independent and dependent groups that we are considering here.

McNemar test for significance of change: a nonparametric test of dependent-sample proportions

The *McNemar test for significance of change* provides a way to use χ^2 to analyze dependent samples if each of the samples is measured on one dichotomous variable—that is, in situations where we have a 2 × 2 contingency table. Here again, the most frequent use of the McNemar test (like the dependent-sample t tests of Chapter 12) is in repeated-measures situations, although other relationships between the individuals in the two groups are possible.

I. State the Null and Alternative Hypotheses

H_0: The proportion of individuals who said yes in 1990 is the same as the proportion of those same individuals who said yes in 1991.

H_1: The proportions differ.

II. Set the Criterion for Rejecting H_0

Because we have not specified otherwise, we use $\alpha = .05$. The test is nondirectional in the same sense as the other χ^2 tests, so the rejection region is in the right-hand tail. All McNemar tests have $df = 1$, so $\chi^2_{cv} = 3.841$ from Table A.3. We show this distribution with the rejection region shaded in Figure 18.6.

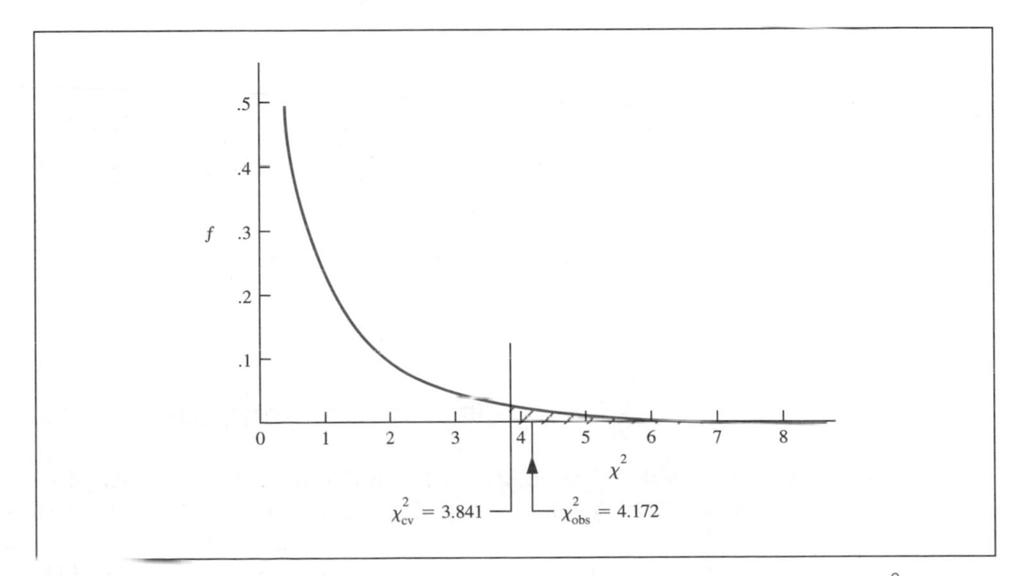

FIGURE 18.6 McNemar test pledge example: distribution of the test statistic χ^2 ($df = 1$) with rejection region shaded and observed value indicated

III. Collect a Sample and Compute the Observed Value of the Test Statistic

The results of our two samples are shown in Table 18.8. Note that the column total in the lower left corner shows that before the war, 42 of the 100 individuals were in favor of requiring the pledge. The upper left cell indicates that of those 42, 33 were still in favor of the pledge a year later. The middle left cell shows that 9 of those 42 had changed their minds from yes to no.

TABLE 18.8 Pledge of Allegiance example: 2 × 2 contingency table showing attitudes toward the pledge before and after the war in Iraq (Expected frequencies of cells where attitudes change in parentheses)

		Before the war		
		Yes	*No*	*Row totals*
After the war	Yes	33	20 (14.5)	53
	No	9 (14.5)	38	47
Column totals		42	58	100

The McNemar test always proceeds by considering only those two cells where individuals change. In our example, those are the cells that switch from yes to no ($f_{\text{yes}\rightarrow\text{no}} = 9$) and from no to yes ($f_{\text{no}\rightarrow\text{yes}} = 20$). As in the other χ^2 tests, we need to compute expected frequencies for those cells. Here, the expected frequency is the average of the observed frequencies in those two cells, because if the null hypothesis is true, the same number of individuals would be expected to switch their attitudes in each direction. Thus, the expected frequency should be the same for both cells and equal to $(f_{\text{yes}\rightarrow\text{no}} + f_{\text{no}\rightarrow\text{yes}})/2 = (9 + 20)/2 = 14.5$. These expected frequencies are shown in parentheses in Table 18.8.

We calculate χ^2_{obs} according to Equation (18.1), except here $df = 1$ and we consider *only* those two cells where responses differ from one condition to the other (in our example, those that have changed attitudes from before to after). Thus, $\chi^2_{\text{obs}} = \sum(O_i - E_i)^2/E_i = (9 - 14.5)^2/14.5 + (20 - 14.5)^2/14.5 = 4.172$. We show that value in Figure 18.6.

IV. Interpret the Results

Because $\chi^2_{\text{obs}} = 4.172 > \chi^2_{\text{cv}} = 3.841$, we reject the null hypothesis and conclude that the war in Iraq did in fact change the attitudes about requiring the pledge among the citizens of Oakville.

18.4 Tests for Data Measured at the Ordinal Level

The tests we have been considering in this chapter so far require only that data be divided into categories—that is, be measured at the nominal level of measurement. We now turn to consider tests used for data that are measured at the ordinal level of measurement, or for interval/ratio data that are not normally distributed.

Two Independent Samples: The Mann–Whitney *U* Test

Mann–Whitney *U* test: a nonparametric (ordinal data) test for two independent samples

The ordinal analog of the two-independent-samples *t* test (see Table 18.1) is the Mann–Whitney *U* test. Suppose you are called upon by the operators of an assertiveness training program to determine whether the trainers should invest in videotape equipment. The trainers wonder whether videotaped feedback could help their clients become assertive; on the other hand, they worry that the videotape might prove to be a distraction that undermines training efforts. You create two assertiveness training groups: one that will operate with and one without videotape feedback. You randomly assign clients to the two groups. At the end of the training, you have each client respond in a standard assertiveness situation. This response is observed by a "blind" observer (that is, one who does not know the purpose of the experiment and who does not know to which group each client belonged) who rates the response on a ten-point assertiveness scale (10 = very high assertiveness).

If these ratings could be considered interval/ratio data, we should use a parametric test, the two-independent-samples t test discussed in Chapter 11. However, ratings on a ten-point scale are not likely to be interval/ratio data because there is no guarantee that the "distance" from, say, 2 to 3 on the scale is the same as the distance from 9 to 10. Because the ratings are ordered but do not have equal intervals, we conclude that the level of measurement is ordinal. The appropriate test in this situation is thus nonparametric, the Mann–Whitney U test.

I. State the Null and Alternative Hypotheses

H_0: The distribution of assertiveness ratings in the population who receive videotaped feedback is the same as that in the population who do not receive such feedback.

H_1: The distributions are not the same.

Note that here, as in the χ^2 tests, the hypotheses are statements about entire distributions, not merely about parameters.

II. Set the Criterion for Rejecting H_0

Because we have not specified otherwise, we take the level of significance α to be .05. The Mann–Whitney U test will result in the calculation of a test statistic U, and the critical value of U depends on the number of observations n_1 and n_2 in the two groups. Our group sizes in this example turn out to be $n_1 = 10$ and $n_2 = 8$. Table A.6 in Appendix A shows that $U_{\text{cv}_{n_1=10,n_2=8}} = 17$. Note that the Mann–Whitney test is an exception to the general rule about critical values: We will reject the null hypothesis if U_{obs} is *less than* U_{cv}.[6]

III. Collect a Sample and Compute the Observed Value of the Test Statistic

The observer's ratings are shown in Table 18.9. We rank the original data as shown in the same table. Note carefully that tied ranks are handled just as they were in the case of Spearman's r_S in Chapter 16.[7] Here, for example, the lowest observer's rating (3) was given three times; the rank assigned to each is $(1 + 2 + 3)/3 = 2$. This has "used up" ranks 1, 2, and 3, so the next available rank is 4.

Note also that, *unlike* the case of the Spearman r_S, the rankings are *not* performed within each group separately but in the *combined* groups. For example, the lowest observer rating (3) appears once in the With group and twice in the Without group, and the rank of 2 is assigned to all three observer ratings.

The remaining calculations all are performed on the *ranks*, not on the original observations. First, we compute two statistics U_1 and U_2, given by

Mann–Whitney U statistic, first

$$U_1 = n_1 n_2 + \frac{n_1(n_1 + 1)}{2} - R_1 \qquad (18.4)$$

Mann–Whitney U statistic, second

$$U_2 = n_1 n_2 + \frac{n_2(n_2 + 1)}{2} - R_2 \qquad (18.5)$$

[6] There is only one other similar exception that we discuss in this text: the Wilcoxon matched-pairs signed-rank test discussed in the next section.

[7] The occurrence of too many tied ranks violates assumptions that underlie the Mann–Whitney test. See, for example, Conover (1980).

TABLE 18.9 Observer's ratings of assertiveness in videotape and no-videotape groups, and rankings of those ratings

With video		*Without video*	
Rating	*Rank*	*Rating*	*Rank*
8	15.5	4	5
7	13	7	13
4	5	5	8.5
7	13	8	15.5
9	17.5	3	2
3	2	3	2
5	8.5	5	8.5
6	11	4	5
9	17.5		
5	8.5		
Σ	111.5		59.5

where n_1 = the number of observations in group 1, n_2 = the number of observations in group 2, R_1 = the sum of the ranks assigned to group 1, and R_2 = the sum of the ranks assigned to group 2. For the data as shown in Table 18.9, $n_1 = 10$, $n_2 = 8$, $R_1 = 111.5$, and $R_2 = 59.5$, so $U_1 = 10(8) + 10(10 + 1)/2 - 111.5 = 23.5$ and $U_2 = 10(8) + 8(8 + 1)/2 - 59.5 = 56.5$.

The observed value of the test statistic U_{obs} is the *smaller* of U_1 and U_2. Thus, in our example, $U_1 = 23.5$ and $U_2 = 56.5$, so $U_{obs} = 23.5$.

Note that Table A.6 shows critical values of U up to n_1 and n_2 values of 20. If n_1 and n_2 are greater than 20, then U is approximately normally distributed with mean

mean of Mann–Whitney U if $n_1 > 20$ and $n_2 > 20$

$$\mu_U = \frac{n_1 n_2}{2} \qquad [\text{if } n_1 \text{ and } n_2 > 20] \qquad (18.6)$$

and standard deviation

standard deviation of Mann–Whitney U if $n_1 > 20$ and $n_2 > 20$

$$\sigma_U = \sqrt{\frac{n_1 n_2 (n_1 + n_2 + 1)}{12}} \qquad [\text{if } n_1 \text{ and } n_2 > 20] \qquad (18.7)$$

Then we compute

normal approximation to Mann–Whitney U if $n_1 > 20$ and $n_2 > 20$

$$z_{obs} = \frac{U_{obs} - \mu_U}{\sigma_U} \qquad [\text{if } n_1 \text{ and } n_2 > 20] \qquad (18.8)$$

and use the critical values of z in Table A.1 of Appendix A.

IV. Interpret the Results

Remember in the Mann-Whitney test we reject H_0 if the observed value is *smaller* than the critical value.

Because $U_{obs} = 23.5$ is greater than $U_{cv} = 17$, we *do not* reject the null hypothesis (remember, in the Mann–Whitney U test we reject when U_{obs} is *smaller* than U_{cv}). We conclude that we do not have enough evidence to conclude that videotaped feedback affects assertiveness training results.

Two Dependent Samples: The Wilcoxon Matched-pairs Signed-rank Test

Wilcoxon matched-pairs signed-rank test: a nonparametric (ordinal data) test for two dependent samples

The ordinal analog of Chapter 12's two-dependent-samples t test (see Table 18.1) is the Wilcoxon matched-pairs signed-rank test. Suppose we are wine marketers interested in the effects of the style of the bottle on the public's perception of the quality of the wine. We enlist subjects for an evaluation of different kinds of foods and give them a series of foods to rate on a 100-point scale (from 1 = "terrible" to 100 = "excellent"). In this series of foods we present the same wine twice, once poured (in sight of the subject) from a "jug"-wine-type bottle with a screw-on cap, and once poured from a "fine" wine bottle with a cork wrapped in heavy foil. We expect that our subjects will prefer wine poured from a fine bottle. The order of servings is counterbalanced so that some subjects get the "jug" bottle first and some get the "fine" bottle first. There are sufficient other foods to taste between wine servings so as to make it unlikely that the subject will directly remember the first wine. Each time, the subject is asked to rate the wine on the 100-point scale.

This is a repeated-measures design because each subject rates both wine presentations. If the data could be considered interval/ratio, we would use a parametric test, the dependent-sample t test described in Chapter 12. However, because the data are only ordinal, we use a nonparametric test, the Wilcoxon matched-pairs signed-rank test.

I. State the Null and Alternative Hypotheses

H_0: The distribution of the population of ratings of wine poured from a jug is the same as the distribution of the ratings of wine poured from a fine bottle.

H_1: The ratings in the "fine" bottle situation are higher than those in the "jug" situation.

Note that this alternative hypothesis is directional: Fine wine bottles are expected to produce *higher* ratings.

II. Set the Criterion for Rejecting H_0

We use $\alpha = .05$ because we have not specified otherwise. The test statistic that we will compute is called T, and the critical values of T are shown in Table A.7. Note that the statistic is a *capital* T, sometimes called "Wilcoxon's T" to distinguish it from Student's (lowercase) t that we have used since Chapter 8. In all of our other tests, we can ascertain the critical value of a test statistic as soon as we know how many groups there are and how many subjects there are in each group. The Wilcoxon test is an exception for reasons we shall see in the next section.

III. Collect a Sample and Compute the Observed Value of the Test Statistic

Twelve subjects participate in this procedure; the results are shown in Table 18.10. The steps for calculating the test statistic T are as follows:

steps for determining the Wilcoxon T

Step 1. Compute the difference scores $D_i = X_i - Y_i$ (exactly as we did in Chapter 12). These are shown in the fourth column of Table 18.10.

TABLE 18.10 Ratings of wine poured from jug and fine bottles, and computations necessary for Wilcoxon's test

Subject	*Jug*	*Fine*	*Diff.*	*Sign and rank of diff.*
1	71	66	5	5.5
2	50	66	−16	(−) 9
3	70	68	2	1
4	78	75	3	2.5
5	60	66	−6	(−) 7
6	86	83	3	2.5
7	58	62	−4	(−) 4
8	49	78	−29	(−) 11
9	58	63	−5	(−) 5.5
10	56	76	−20	(−) 10
11	81	81	0	—
12	50	58	−8	(−) 8

Step 2. Omit any subject for whom the difference score is 0. In our data, we omit subject #11, thus reducing the total number of subjects we are considering from 12 to 11.

Step 3. Rank the *absolute value of the difference scores* in ascending order (that is, ignore the signs when making this ranking) and handle tied ranks as you did with the Mann–Whitney U test.[8] Indicate which difference scores are negative by writing $(-)$ next to them.

Step 4. Sum the ranks whose differences are positive: $T_{\text{positive}} = 5.5 + 1 + 2.5 + 2.5 = 11.5$. Then sum the ranks whose signs are negative: $T_{\text{negative}} = 9 + 7 + 4 + 11 + 5.5 + 10 + 8 = 54.5$. T_{obs} is the *smaller* of T_{positive} and T_{negative}; in our case, $T_{\text{obs}} = 11.5$.

IV. Interpret the Results

Remember in the Wilcoxon test we reject H_0 if the observed value is smaller than the critical value.

There are 12 pairs of data in our example; we omitted one because its difference score was 0. Thus, the critical value of T is based on 11 pairs. Table A.7 in Appendix A shows that for directional tests at $\alpha = .05$, $T_{\text{cv}} = 13$. We reject the null hypothesis in a Wilcoxon test, as in a Mann–Whitney U test, when the observed value is *less than* the critical value. Because $T_{\text{obs}} = 11.5$ is less than $T_{\text{cv}} = 13$, we do reject H_0. Therefore, we conclude that type of wine bottle does make a difference in subjects' ratings of wine quality.

k Independent Samples: The Kruskal–Wallis *H* Test

Kruskal–Wallis *H* test: a nonparametric (ordinal data) test for three or more independent samples

The ordinal analog of Chapter 13's analysis of variance (see Table 18.1) is the Kruskal–Wallis H test. Suppose we are interested in the effect that type of music has on judges' ratings in an ice skating competition. Skaters can choreograph their routines to any kind of music they wish; the question is whether their choice affects their score. Suppose that we divide music into three categories: classical, rock, and standards. Unknown to the judges, we contact skaters prior to a contest and randomly assign them one of the three music types. The skaters then perform in the contest to the assigned type of music, and we record the skaters' scores as shown in Table 18.11.

TABLE 18.11 Judges' ratings for skaters using three kinds of music, and computations necessary for Kruskal–Wallis *H* test

Classical		*Rock*		*Standards*	
Rating	*Rank*	*Rating*	*Rank*	*Rating*	*Rank*
9.4	2	9.6	7.5	9.9	14
9.3	1	9.5	4.5	9.8	12
9.5	4.5	9.5	4.5	9.7	9.5
9.6	7.5	9.8	12	9.8	12
		9.7	9.5	9.5	4.5
Σ	15.0		38.0		52.0

Judges' ratings, like the observer's ratings in the Mann–Whitney assertiveness example, are ordinal data. If the data were interval/ratio, we would analyze this case with a parametric test, the analysis of variance. However, because the data are ordinal, we use the nonparametric analog of ANOVA, the Kruskal–Wallis H test.

I. State the Null and Alternative Hypotheses

H_0: The distributions of scores in the three music populations are identical.

H_1: The distributions are not identical.

[8] The occurrence of too many tied ranks violates assumptions that underlie the Wilcoxon test. See, for example, Conover (1980).

II. Set the Criterion for Rejecting H_0

It can be shown that H has the χ^2 distribution with $df = k - 1$, where k is the number of groups.[9] Here, $k = 3$, so if $\alpha = .05$, $H_{cv} = \chi^2_{cv,df=2} = 5.991$ from Table A.3. If our observed value H_{obs} is greater than H_{cv}, we will reject the null hypothesis.

III. Collect a Sample and Compute the Observed Value of the Test Statistic

The data are shown in Table 18.11. Computation proceeds just as for the Mann–Whitney U test. We first rank all data and find the R_j values by summing the ranks. The test statistic H is computed from the following equation:[10]

Kruskal–Wallis H statistic

$$H = \frac{12}{N(N+1)} \sum \frac{R_j^2}{n_j} - 3(N+1) \tag{18.9}$$

where n_j = number of observations in the jth group, R_j = sum of rankings in the jth group, and N = total number of observations ($\sum n_j$). In our example, $n_1 = 4$, $n_2 = 5$, $n_3 = 5$, $R_1 = 15.0$, $R_2 = 38.0$, $R_3 = 52.0$, and $N = 4 + 5 + 5 = 14$. Thus,

$$\begin{aligned} H_{obs} &= \frac{12}{14(14+1)}\left(\frac{15^2}{4} + \frac{38^2}{5} + \frac{52^2}{5}\right) - 3(14+1) \\ &= \left(\frac{12}{210}\right)(885.85) - 45 \\ &= 5.62 \end{aligned}$$

IV. Interpret the Results

Because $H_{obs} = 5.62$ is not greater than $H_{cv} = 5.991$, we fail to reject the null hypothesis. We cannot say that type of music affects judges' ratings.

18.5 Choosing Between Parametric and Nonparametric Tests

In some cases—for example, when data are nominal—the only possible statistical test is nonparametric. But in other cases, as we have seen, it is often mathematically possible to perform two different tests on the same data: one parametric and the other nonparametric—the two-independent-samples t test and the Mann–Whitney U test, for example. How does one determine which test is more appropriate? This decision is often a judgment call based on experience or common practice in the research field, but a few general rules apply.

First, the parametric tests are usually more powerful than their nonparametric analogs. Thus, if we have a choice, we would in general prefer a parametric test.

Second, it is always satisfactory to perform a nonparametric test even if a parametric test is also appropriate. The only penalty for such a choice is the loss of power, which makes the nonparametric test more conservative (less likely to result in a Type I error).

Third, the larger the sample size, the less critical are violations of assumptions about normality, and so the less reason to prefer an ordinal test to a parametric one. This follows directly from the central limit theorem, which states that the distribution of means of

[9] Actually, H is only approximately distributed as χ^2. Exact probability levels can be found in Iman, Quade, and Alexander (1975).

[10] This formula for H actually assumes that there are no tied ranks. The value of the exact formula including ties is most often extremely close to the value of the formula given. For the exact formula and a discussion, see Conover (1980).

samples of size n tends to become normal as n increases, regardless of the shape of the original distribution.

Fourth, the analysis of variance has been shown to be robust to violations of the assumption of normality. *Robust* means that small departures from normality do not appreciably affect the probability distribution of the F ratio.

Weighing these advantages and disadvantages is one of the important topics of research design.

18.6 Connections

Cumulative Review

We continue the discrimination practice begun in Chapter 10. You may recall the two steps:

Step 1: Determine whether the problem asks for (a) the area under a normal distribution, (b) a confidence interval, or (c) the evaluation of a hypothesis. See the Cumulative Review in Section 10.5 for hints about this step.

Step 2: If the problem asks for the evaluation of a hypothesis, consult the summary of hypothesis testing in Table 18.12 to determine which kind of test is appropriate.

TABLE 18.12 Summary of hypothesis-testing procedures in Chapters 10–18

Design	*Chapter*	*Null hypothesis*	*Sample statistic*	*Test statistic*	*Effect size index*
One sample					
σ known	10	$\mu = a$	$\overline{X}$	z	$d = \frac{\overline{X}_{obs} - \mu}{\sigma}$
σ unknown	10	$\mu = a$	$\overline{X}$	t	$d = \frac{\overline{X}_{obs} - \mu}{s}$
Proportion	Exercise 7, Chapter 10	$\Pi = a$	p	t	—[a]
No parameter (goodness of fit)	18	Distribution has specified shape	n/a	χ^2	—
Two independent samples					
For means	11	$\mu_1 = \mu_2$	$\overline{X}_1 - \overline{X}_2$	t	$d = \frac{\overline{X}_{1_{obs}} - \overline{X}_{2_{obs}}}{s_{pooled}}$
No parameter (test of independence)	18	Distributions are independent	n/a	χ^2	—
No parameter (Mann–Whitney)	18	Distributions are the same	n/a	U	—
Two dependent samples					
For means	12	$\mu_D = 0$	$\overline{D}$	t	$d = \frac{\overline{D}_{obs}}{s_D}$
No parameter (Wilcoxon)	18	Distributions are the same	n/a	T	—
Proportion (McNemar)	18	Proportion has not changed	n/a	χ^2	—

[a] Effect size index was not discussed in this textbook.

(continued)

TABLE 18.12 Summary of hypothesis-testing procedures in Chapters 10–18 *(continued)*

Design	Chapter	Null hypothesis	Sample statistic	Test statistic	Effect size index
k independent samples					
For means (omnibus)	13	$\mu_1 = \mu_2 = \cdots = \mu_k$	n/a	F	$d_M = \frac{\overline{X}_{max} - \overline{X}_{min}}{s_{pooled}}$ or $f = \frac{s_{\overline{X}}}{s_{pooled}}$ or $R^2 = \frac{SS_B}{SS_T}$
Post hoc	14	$\mu_i = \mu_j$	$\overline{X}_i - \overline{X}_j$	Q	—
A priori	14	$C_1\mu_1 + \cdots + C_k\mu_k = 0$	$C_1\overline{X}_1 + \cdots + C_k\overline{X}_k$	t	$d = \frac{C_1\overline{X}_1 + \cdots + C_k\overline{X}_k}{s_{pooled}}$
No parameter (test of independence)	18	Distributions are independent	n/a	χ^2	—
No parameter (Kruskal–Wallis)	18	Distributions are the same	n/a	H	—
k dependent samples					
For means	14	$\mu_1 = \mu_2 = \cdots = \mu_k$	n/a	F	$d_M = \frac{\overline{D}_{max}}{\sqrt{2MS_{residual}}}$
RC independent samples					
For row means	15	$\mu_{r1} = \mu_{r2} = \cdots = \mu_R$	n/a	F	—
For column means	15	$\mu_{c1} = \mu_{c2} = \cdots = \mu_C$	n/a	F	—
Interaction	15	$\mu_{rc} - \mu_r - \mu_c + \mu_G = 0$	n/a	F	—
Correlation					
Parametric	16	$\rho = 0$	r	r (or t)	r^2
Ordinal	16	$\rho_s = 0$	r_s	r_s	—
Regression					
Slope	17	$b_{population} = 0$	b	r	r^2

Table 18.12 shows that this text has covered 18 basic kinds of hypothesis tests (the one-sample test for the mean, two-independent-samples test for means, two-dependent-samples tests for means, three-or-more-independent-samples tests, post hoc tests, a priori tests, three-or-more-repeated-measures tests, the two-way ANOVA, two tests for correlation, and now seven nonparametric tests). Note that this table is identical to Table 17.7 with the addition of the seven lines appropriate for this chapter.

Journals

Of all the nonparametric tests, the χ^2 test of independence is the most commonly used. A typical journal report might present a table of frequencies such as our Table 18.6, although the expected values (in parentheses) are typically omitted. Furthermore, percentages rather than actual frequencies are often presented. The text of the article might then read, "Percentages of abortion attitude are presented in Table 18.6. There was no statistically significant relationship between family size and abortion attitude, $\chi^2(6, N = 150) = 11.998$, *ns*."

There is no universal agreement about the criteria for deciding whether to use a nonparametric test, such as the Kruskal–Wallis H test, or ANOVA. The common practice is to rely on the central limit theorem (to make the distribution of means approximately normal even if the original distribution is not normal) and the robustness of ANOVA, and present an ANOVA rather than H.

Computers

ESTAT does not compute nonparametric statistics. Version 2.0 and later will determine the ranks of variables and thus facilitate computation of statistics based on ranks.

Use SPSS to compute the χ^2 for the abortion attitude data listed in Table 18.5:

1. Enter codes for abortion attitude into the first column of the SPSS spreadsheet (use 1 = ProLif, 2 = Modrat, 3 = ProCho) and codes for family size into the second column of the SPSS spreadsheet (use 1 = Only, 2 = Small, 3 = Midl, 4 = Larg). Thus, the first subject, "ProCho/Midl," would be coded "3" in column one and "3" in column two, and so on. Note that any numerical codes for these values (e.g., 3 = ProLif, 2 = Modrat, 1 = ProCho) would be satisfactory but might rearrange the rows or columns of the SPSS output.
2. Click `Statistics`, then `Summarize`, and then `Crosstabs....`
3. Click the upper ▶ to move var00001 into the `Row(s):` window. Then click `var00002` and then the second ▶ to move var00002 into the `Column(s):` window.
4. Click `Statistics....` Then click the `Chi-square` check box. Click `Continue`.
5. Click `Cells....` Then click the `Expected` check box and then `Continue`.
6. Click `OK`. The contingency table (similar to the text's Table 18.6) appears in the `!Output1` window as shown here. The SPSS output shows that χ^2_{obs} = 12.00114, which is different from the text's χ^2obs = 11.998 only because of roundoff.

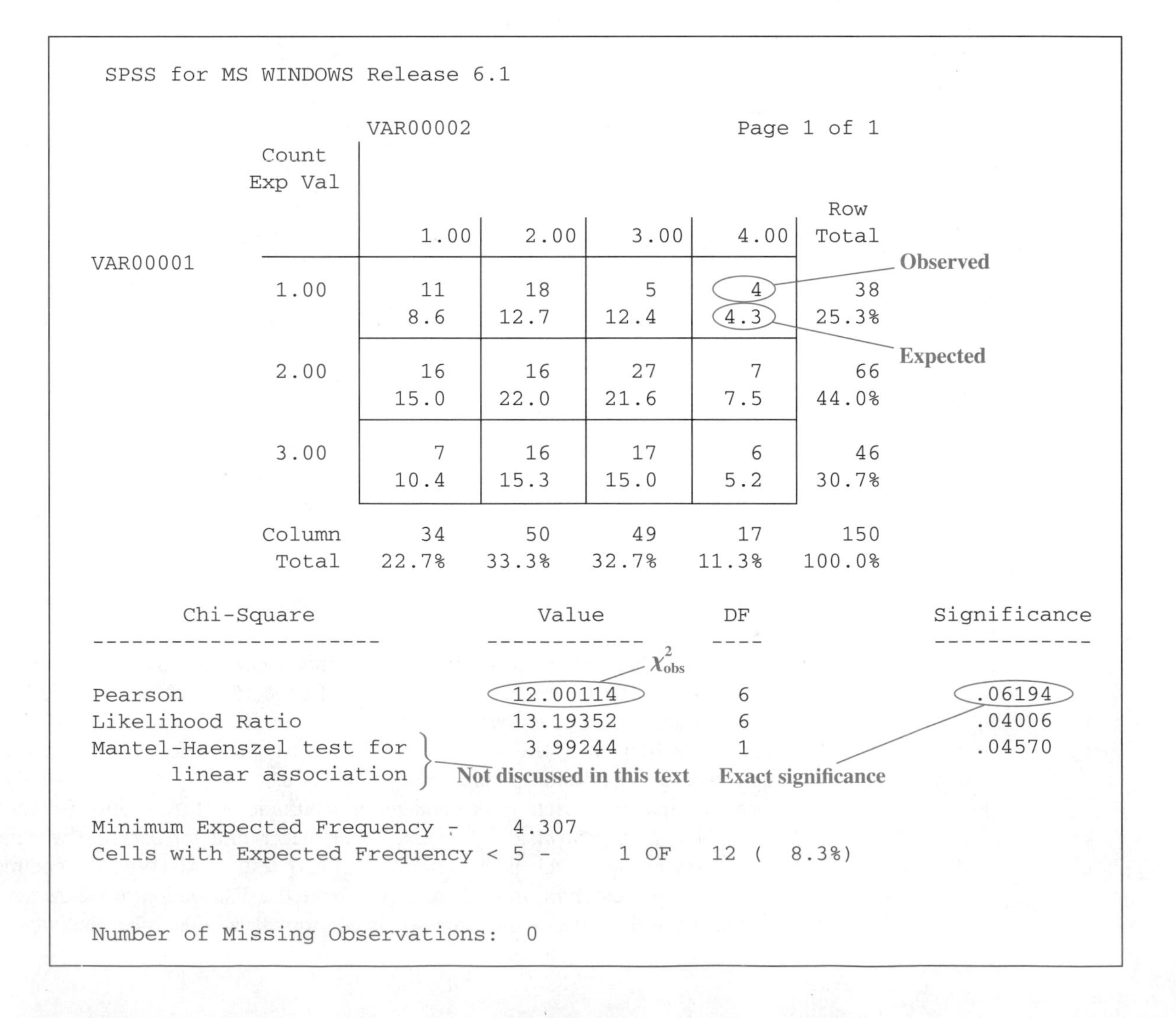

SPSS for MS WINDOWS Release 6.1

VAR00002 Page 1 of 1

VAR00001	Count Exp Val	1.00	2.00	3.00	4.00	Row Total
	1.00	11 8.6	18 12.7	5 12.4	4 4.3	38 25.3%
	2.00	16 15.0	16 22.0	27 21.6	7 7.5	66 44.0%
	3.00	7 10.4	16 15.3	17 15.0	6 5.2	46 30.7%
	Column Total	34 22.7%	50 33.3%	49 32.7%	17 11.3%	150 100.0%

Chi-Square	Value	DF	Significance
Pearson	12.00114	6	.06194
Likelihood Ratio	13.19352	6	.04006
Mantel-Haenszel test for linear association	3.99244	1	.04570

Minimum Expected Frequency - 4.307
Cells with Expected Frequency < 5 - 1 OF 12 (8.3%)

Number of Missing Observations: 0

Homework Tips

1. Check the list of learning objectives at the beginning of this chapter. Do you understand each one?

2. Entering the data for a large data set such as the one in Table 18.5 entails a fair amount of work and careful attention to detail. Though seemingly a thankless task, the ability to enter data correctly is the fundamental basis of all statistical computation. Too many incorrect analyses are the result of improper data entry.

3. Remember when determining ranks for the tests in this chapter that you determine the ranks *for all the groups combined*, not within each group.

4. When determining the ranks of data, recall this tip from Chapter 16: The most frequent single mistake is made in determining tied ranks. For example, the values 12, 13, 13, and 17 receive the ranks 1, 2.5, 2.5, and 4 (*not* 3), and the values 132, 136, 140, 140, 140, and 163 receive the ranks 1, 2, 4, 4, 4, and 6 (*not* 5), respectively.

5. Remember that unlike all the other tests in this textbook, the Mann–Whitney U and the Wilcoxon T are rejected when the observed value is *less* than the critical value.

Exercises for Chapter 18

Section A: Basic exercises (answers in Appendix D, page 532)

1. Suppose that Hazel's Nut Company sells mixed nuts and claims that 30% of the nuts in its mix are cashews, 20% are pecans, 20% are Brazil nuts, and 30% are peanuts. You wonder whether this claim is true, so you take a random sample of 200 nuts and count the number of each kind. Your sample has 45 cashews, 35 pecans, 41 Brazil nuts, and 79 peanuts. Do you have a right to complain to the company? Use $\alpha = .05$.
   (a) What is the appropriate test in this situation?

   I. State the null and alternative hypotheses
   (b) What is the null hypothesis? The alternative hypothesis?
   (c) Is this test directional or nondirectional? Why?
   (d) What is the test statistic? How many degrees of freedom does it have? Sketch the distribution of the test statistic assuming the null hypothesis is true.

   II. Set the criterion for rejecting H_0
   (e) What is the level of significance?
   (f) What is (are) the critical value(s) of the test statistic? Enter it (them) on the appropriate distribution and shade the rejection region(s).

   III. Collect a sample and compute statistics
   (g) Compute the observed value of the test statistic. Show this value on the appropriate distribution.

   IV. Interpret the results
   (h) Should we reject H_0?
   (i) Interpret your result in plain English.

2. A wine distributor wishes to know whether the patrons at different supermarkets purchase the same mix of wines. He stations an observer for one day at each of three supermarket wine departments (FoodKing, Safeway, and Lucky) to watch patrons make their selections. The observer records whether the patron selects white wine (W), red wine (R), or blush wine (B). The selections are shown in the table. Does the pattern of wine selection depend on the supermarket? Use $\alpha = .05$ and the steps of Exercise 1.

FoodKing: W B R B B W W B R R R B
B W W W W B R B R W B R B W B R
B R B W B W R B R B R R W B R B
R R R W W R B R B R B W W R B R
B R B R R R B W W B W B R B R B
R W B R B W W R B R B W B R B R
B R B R W W B R B R B W B R R B
Labor savings: There are 26 Ws, 40 Rs, and 42 Bs

Safeway: W B R B R B R R R R W W
W B R B W B R W W R B W R B W R
B W R B W R W B R W B R W B R R
W B R W B R B W R W B R W B R W
B R W B R W R W B R B W R B W R
B W R B W R B W B R B W R
Labor savings: There are 29 Ws, 32 Rs, and 28 Bs

Lucky: W B B W B R W B R W B R B
W R B W R B W R B W B R W B R R
B W R B W B R B W R B W B R B R
B W B R B W R B W B B B R B B W
B R B W R B W W B W R B W B R W
B B W B B W R B B R W B R W B W
R B W B B
Labor savings: There are 29 Ws, 25 Rs, and 44 Bs

3. (*Exercise 2 continued*) Suppose that at each of the three supermarkets we observe exactly *twice as many* selections as we reported in Exercise 2. Furthermore, suppose that the

percentages of white, red, and blush wines remain identical for each supermarket. How does that change these measures?

(a) The null and alternative hypotheses
(b) The number of degrees of freedom
(c) The critical value of the test statistic
(d) The observed entries in the contingency table (and in the table for the computation of χ^2)
(e) The expected entries in the contingency table (and in the table for the computation of χ^2)
(f) The entries in the column headed $(O_i - E_i)$
(g) The entries in the column headed $(O_i - E_i)^2$
(h) The entries in the column headed $(O_i - E_i)^2/E_i$
(i) χ^2_{obs}
(j) The conclusion

4. In the early autumn of 1991, a researcher administered to 100 workers a questionnaire that included the question, "Do you consider a supervisor's discussion of pornography with a subordinate in the workplace to be sexual harassment?" In late autumn, following the U.S. Senate's confirmation hearings for a Supreme Court justice where sexual harassment was a major issue, he located the same 100 employees and administered the questionnaire again. On the original questionnaire, 38 answered yes and 62 answered no. Of the 38 who originally answered yes, 36 also answered yes on the second administration, while 2 switched to no. Of the 62 who originally answered no, 50 also answered no on the second administration, while 12 switched to yes. Has there been a significant change in responding to this question? Use the steps of Exercise 1.

5. The text presented an example of the McNemar test on repeated measurements of attitudes about requiring the Pledge of Allegiance. Suppose that the data presented in Table 18.8 were for *independent* samples; that is, we sampled 100 people before the war and 42 said yes. Then we sampled a *different* 100 people after the war, and 53 said yes. Did the war affect the attitudes among the Oakville population? Use the steps of Exercise 1.

6. A professor wishes to know whether studying with the aid of visual imagery is more effective than nonvisual memorization. She randomly divides her class into two equal groups. She teaches one group the techniques of visualization, and she teaches the other group a rote memorization technique. She assigns letter grades (A, B, C, D, and F) on the next exam. The grades with the two techniques are listed here. Did the visual imagery group perform better than the nonvisual group? Use the steps of Exercise 1 except do not sketch distributions. [*Hint:* Convert letter grades to numbers, A = 4, B = 3, etc.]

Visual imagery group: B, B, A, C, D, A, A, C, A, B, A
Nonvisual group: A, C, F, D, B, B, C, A, B, A, C

7. The example in the text about videotaped feedback in assertiveness training was analyzed using the Mann–Whitney U test because the ratings were assumed to be ordinal data. Now assume that the observer has been adequately trained so that his ratings can be assumed to be interval/ratio data that are normally distributed.
(a) What test will you use now?
(b) What is its null hypothesis?
(c) How is its null hypothesis different from that used in the Mann–Whitney example?
(d) Perform the test. What is the observed value of the test statistic?
(e) What is your conclusion? How does your conclusion compare to the Mann–Whitney conclusion?

8. In the development of a textbook, the authors wish to know whether the use of color photos alters the perceived readability of the text. Two text passages are identified that past research has shown to have equal readability. Now one of these is illustrated in color and the other in black and white. Both passages are given to a group of 13 students who rate each passage on a readability scale (10 = high readability; 1 = difficult to read). The readability ratings are given in the table. Does the use of color alter readability? Use the steps of Exercise 1 except do not sketch the distributions.

Student	*Color*	*B/W*
1	7	4
2	4	3
3	4	2
4	3	5
5	10	6
6	8	6
7	1	1
8	8	6
9	8	4
10	10	9
11	6	1
12	4	8
13	5	4

9. The example in the text on the fine wine/jug wine bottle effect was analyzed using the Wilcoxon matched-pairs signed-rank test because the ratings were assumed to be ordinal data. Now assume that the judges have been adequately trained so that their ratings can be assumed to be interval/ratio data that are normally distributed.
(a) What test will you use now?
(b) What is its null hypothesis?
(c) How is its null hypothesis different from that in the Wilcoxon example?
(d) Perform the test. What is the observed value of the test statistic?
(e) What is your conclusion? How does your conclusion compare to the Wilcoxon conclusion?

10. A researcher is interested in whether different fraternities have different levels of political awareness. There are five fraternities on campus. He randomly selects six members of each and interviews them. At the conclusion of each interview, he rates each student on a nine-point political awareness scale (9 = very politically aware). The awareness ratings are given in the table. Do the fraternities

Table for Exercise 10

ABΓ	ΔEZ	HΘI	KΛM	NΞO
8	6	2	9	9
5	5	3	8	6
7	6	3	9	5
8	2	2	8	4
5	8	5	8	6
5	5	3	8	9

differ? Use the steps of Exercise 1 except do not sketch the distributions.

11. The example in the text about the music choice of ice skaters was analyzed using the Kruskal–Wallis H test because the ratings were assumed to be ordinal data. Now assume that the judges have been adequately trained so that their ratings can be assumed to be interval/ratio data that are normally distributed.
    (a) What test will you use now?
    (b) What is its null hypothesis?
    (c) How is its null hypothesis different from that in the Kruskal–Wallis example?
    (d) Perform the test. What is the observed value of the test statistic?
    (e) What is your conclusion? How does your conclusion compare to the Kruskal–Wallis conclusion?

Section B: Supplementary exercises

12. A video poker machine is a television screen driven by a computer programmed to simulate the deal of a hand of poker. The machine says that it uses a 52-card deck. You wonder whether that is in fact true, so you count the number of times each value (ace, king, . . . , 2) occurs in the next 348 cards. The data are shown in the table. Is the actual distribution significantly different from what would be expected if the machine correctly simulated a 52-card deck? Use the steps of Exercise 1.

Value	Frequency
Ace	21
King	27
Queen	29
Jack	23
10	30
9	24
8	24
7	32
6	31
5	27
4	29
3	26
2	25

13. Suppose you wish to know whether the symptomatology in schizophrenia is related to age at onset. You go back through the records of hospitalized schizophrenics and classify their symptoms into three major classes: primarily linguistic, primarily perceptual, and primarily motor. You classify the age at onset of these symptoms into four classes: under 20, 20–25, 26–30, and over 30. A contingency table for the hypothetical data is provided here. Is age at onset independent of symptomatology? Use the steps of Exercise 1.

Symptom	Age at onset			
	Under 20	20–25	26–30	Over 30
Primarily linguistic	23	42	32	25
Primarily perceptual	36	50	51	40
Primarily motor	23	21	13	9

14. Candidate Smith will address a meeting of 100 voters. His staff gives each voter a questionnaire to fill out both before Smith speaks and afterward. One of the questions is, "Will you vote for candidate Smith?" Before the speech, 35 answer yes, and 65 answer no. After the speech, 10 of those who originally said no now say yes (the result candidate Smith hoped for), but 5 of those who originally said yes now say no. Has the speech significantly increased the percentage of voters who favor Smith? Use the steps of Exercise 1.

15. A chef knows there are two kinds of olive oil, regular and "extra virgin," which is made from select olives and cold-processed. Extra virgin oil is therefore more expensive than regular olive oil, but it also tastes better. He wishes to know whether to use extra virgin oil in his salad dressing. To decide, he recruits 28 volunteers. He prepares identical salads and salad dressing for all, but for a randomly selected 14 he uses extra virgin oil; the remainder get regular. He asks all 28 to rate his salad dressing on a ten-point scale (10 = excellent). The ratings are listed here. Does the extra virgin oil improve the salad? Use the steps of Exercise 1 except do not sketch the distributions.

    Regular: 2, 8, 8, 8, 3, 6, 9, 10, 2, 4, 4, 5, 7, 6
    Extra virgin: 2, 7, 10, 8, 7, 7, 10, 10, 5, 6, 3, 6, 7, 8

16. (*Exercise 15 continued*) Suppose that the 28 raters are so well trained that their ratings can be considered to be interval/ratio data. What test is now appropriate? How is the null hypothesis different from the null hypothesis in Exercise 15? Perform the analysis. How are the results similar to or different from those in Exercise 15?

17. (*Exercise 15 continued*) Suppose that the olive oil data represent *two* measurements on 14 individuals; that is, 14 individuals tasted two salads, one made with regular and one with extra virgin oil. Does extra virgin oil improve the salad? Use the steps of Exercise 1 except do not sketch the distributions.

18. (*Exercise 17 continued*) Suppose that the 14 raters are so well trained that their ratings can be considered to be interval/ratio data. What test is now appropriate? How is the null hypothesis different from the null hypothesis in Exercise 17? Perform the analysis. How are the results similar or different from those in Exercise 17?

19. Researchers wish to know whether the warmth shown by a mother to her first-born child is related to the age she was when that child was born. They observe a standard interaction between mother and child, and rate maternal warmth on a seven-point scale. They divide the mothers' ages at first child's birth into four categories as shown in the table. Does warmth depend on the mother's age at birth of the child? Use the steps of Exercise 1 except do not sketch the distributions.

Maternal warmth ratings			
0–18	*19–22*	*23–26*	*27+*
1	4	7	5
3	3	6	5
3	3	3	5
4	2	5	3
5	2	5	4
3	6	5	7
4	4	7	1

20. (*Exercise 19 continued*) Suppose that the maternal warmth raters are so well trained that their ratings can be considered to be interval/ratio data. What test is now appropriate? How is the null hypothesis different from the null hypothesis in Exercise 19? Perform the analysis. How are the results similar to or different from those in Exercise 19?

Section C: Extending your comprehension (answers in Appendix D, page 534)

21. Is it possible to create a data set with two groups, five points in each group, so that an independent-sample t test is significant but a Mann–Whitney U test is not significant at $\alpha = .05$? If so, show such a data set.
22. Is it possible to create a data set with two groups, five points in each group, so that a t test is not significant but a Mann–Whitney U test is significant at $\alpha = .05$? If so, show such a data set.
23. Is it possible for two data sets, five points in each, to have identical means and yet have a significant independent-sample t test? If so, show such a data set.
24. Is it possible for two data sets, five points in each, to have identical means and yet have a significant Mann–Whitney U test? If so, show such a data set.
25. Is it possible for two data sets, five points in each, to have identical medians and yet have a significant independent-sample t test? If so, show such a data set.
26. Is it possible for two data sets, five points in each, to have identical medians and yet have a significant Mann–Whitney U test? If so, show such a data set.

Section D: Cumulative review (answers in Appendix D, page 534)

For Exercises 27–38, choose the correct answer or fill in the blank. (**Note that computations are *not* required;** use $\alpha = .05$ unless otherwise instructed.)

(a) This problem requires:
 (1) Finding the area under a normal distribution [Skip parts (b)–(g) and proceed to (h).]
 (2) Creating a confidence interval
 (3) Testing a hypothesis about the mean of one group
 (4) Testing a hypothesis about the means of two independent groups
 (5) Testing a hypothesis about the means of two dependent groups
 (6) Testing a hypothesis about the means of three or more independent groups
 (7) Testing hypotheses about means using post hoc tests
 (8) Testing an a priori hypothesis about means (planned comparison)
 (9) Testing a hypothesis about the means of three or more dependent groups
 (10) Testing hypotheses for main effects and interaction
 (11) Testing a hypothesis about the strength of a relationship
 (12) Testing a hypothesis about the slope of the regression line
 (13) Testing a hypothesis about one nominal, ordinal, or interval/ratio distribution
 (14) Testing a hypothesis about the independence of two or more nominal, ordinal, or interval/ratio distributions
 (15) Testing a hypothesis about the change of a nominal, ordinal, or interval/ratio distribution
 (16) Testing a hypothesis about two independent, ordinal distributions
 (17) Testing a hypothesis about two dependent, ordinal distributions
 (18) Testing a hypothesis about three or more independent, ordinal distributions
 (19) Performing a power analysis (involving finding the area under a normal distribution) [Skip parts (b)–(h).]

(b) The null hypothesis is of the form:
 (1) $\mu = a$
 (2) $\mu_1 = \mu_2$
 (3) $\mu_D = 0$
 (4) $\mu_1 = \mu_2 = \cdots = \mu_k$
 (5) A series of null hypotheses of the form $\mu_i = \mu_j$ (State how many such hypotheses.)
 (6) One null hypothesis of the form $C_1\mu_1 + C_2\mu_2 + \cdots + C_k\mu_k = 0$
 (7) Three null hypotheses:
 rows: $\mu_{\text{row }1} = \mu_{\text{row }2} = \cdots = \mu_{\text{row }R}$;
 columns: $\mu_{\text{col }1} = \mu_{\text{col }2} = \cdots = \mu_{\text{col }C}$; and
 interaction: all $\mu_{\text{row,col}} - \mu_{\text{row}} - \mu_{\text{col}} + \mu = 0$
 (8) $\rho = 0$
 (9) $\rho_S = 0$
 (10) $b_{\text{population}} = 0$ (equivalent to $\rho = 0$)
 (11) The distribution has the form specified.
 (12) The distribution of one variable is independent of the other(s).
 (13) The distribution has not changed.
 (14) The two distributions are the same.
 (15) The three or more distributions are the same.
 (16) There is no null hypothesis.

(c) The appropriate statistic(s) is (are):
(1) $\overline{X}$ (2) p
(3) $\overline{X}_1 - \overline{X}_2$ (4) $\overline{D}$
(5) $\overline{X}_i - \overline{X}_j$
(6) $C_1\overline{X}_1 + C_2\overline{X}_2 + \cdots + C_k\overline{X}_k$
(7) r
(8) r_S
(9) a and/or b
(10) Not applicable for ANOVA or nonparametric tests

(d) The appropriate test statistic(s) is (are):
(1) z (2) t
(3) F (4) Q
(5) r (6) r_S
(7) χ^2 (8) U
(9) T (10) Not applicable

(e) The number of degrees of freedom for this test statistic is ________. (State the number or *not applicable*. If there are multiple null hypotheses, state the number of degrees of freedom for each.)

(f) The hypothesis test (confidence interval) is:
(1) Directional
(2) Nondirectional
(3) Not applicable

(g) The critical value(s) of the test statistic(s) is (are) ________. (Give the value. If there are multiple null hypotheses, state the critical value for each; if not applicable, so state.)

(h) The appropriate formula(s) for the test statistic(s) (or confidence interval or strength of relationship) is (are) given by Equation (________). (If there are multiple null hypotheses, state the equation number for each. If the test is nonparametric, give its name.)

Computations are *not* required for any of the problems in Section D. See the instructions.

27. The coach of a ski team wishes to know whether different waxes affect times in downhill racing. There are four waxes: A, B, C, and D. He suspects that the effectiveness of a type of wax depends on the snow conditions, which he divides into two groups (wet and dry). The coach randomly assigns five skiers to each of the eight (4 waxes × 2 snow types) conditions and has them run the standard downhill course. He records the times for each as shown in the table. Do type of wax and snow condition affect times?

	Wax			
	A	B	C	D
Wet snow	73.7	52.6	77.3	55.1
	76.0	52.4	66.5	48.0
	77.6	54.8	72.6	48.0
	70.1	53.3	58.5	69.9
	70.0	56.8	73.3	64.0
Dry snow	64.7	66.1	69.3	78.4
	65.8	64.2	68.1	60.7
	66.4	62.3	76.5	63.1
	63.3	65.2	75.0	61.5
	63.0	67.0	70.9	54.8

28. An instructor wishes to know whether there is a relationship between the grades students receive on exams and the order in which they turn their exams in. Is there? The data are shown in the table.

Sequence number	Grade
1	C
2	A
3	A
4	B
5	A
6	C
7	F
8	C
9	C
10	D
11	D
12	B
13	C

29. A researcher wishes to know whether cause of death is related to place of birth. He consults death records and divides the causes of death into four categories: heart, cancer, accident, and other. He divides places of birth into three categories: East, Midwest, and West. The data are given in the table. Is cause of death related to place of birth?

	Cause of death			
Origin	Heart	Cancer	Accident	Other
East	34	46	23	26
Midwest	46	29	16	23
West	72	89	49	80

30. Candidate Jones's campaign manager conducts a random sample of 225 voters and asks whether they will vote for Jones. Of these 225, 145 say yes. What can she say about the percentage of the population who favor Jones?

31. A rifle markswoman knows that at a particular distance on the average she hits 40% bull's-eyes, 25% in the first ring, 25% in the second ring, and 10% outside that. She decides to try a new kind of shell-loading technique, and she wants to know whether it affects her shooting. She fires 100 rounds using the new shells. Of these 100 rounds, 43 are bull's-eyes, 29 are in the first ring, 27 in the second ring, and 1 outside that. Has the new technique affected her shooting?

32. The publishers of a novel wish to know whether placing a colored picture of a scantily clad couple on the cover of the book will increase its sales. They select 20 bookstores to test-market the book. They put the scantily clad covers on the books they consign to ten (randomly chosen) stores; the other ten get more conservative covers.

The stores record the sales in the first week they have the book. Does the choice of cover make a difference?

Scantily clad: 78, 77, 71, 79, 82, 52, 65, 96, 85, 98
Conservative: 63, 56, 74, 58, 79, 60, 57, 62, 93, 69

33. A photograph processor knows that he processes on the average 120 rolls of film a day, with standard deviation 15 rolls. He starts a radio advertisement, and for the next n days, he plans to record the number of rolls he processes. On the assumption that the advertisement actually increases developing requests by ten rolls per day, what value should he assign to n (the number of days over which the sales will be recorded) so that the probability is .8 that his study will demonstrate an increase?

34. (*Exercise 33 continued*) The processor places the advertisement and for ten days records these daily numbers of rolls he develops: 110, 137, 141, 131, 127, 158, 141, 122, 118, 155. Has the advertisement increased sales?

35. A modeling agency is considering teaching its models a technique designed to enhance their individuality, and it wants to know whether this affects the overall ratings of the models' attractiveness. The agency has 24 models and randomly assigns 12 to each of two groups. One group learns the new technique; the other does not. Then all 24 are rated by a "blind" judge on attractiveness using a ten-point scale. These ratings are listed here. Does the technique affect attractiveness?

Enhanced: 6, 8, 10, 3, 5, 6, 6, 4, 9, 6, 6, 7
Not enhanced: 6, 10, 10, 9, 10, 6, 9, 8, 10, 6, 8, 5

36. Ergodynamics, Inc., makes three kinds of computer keyboards: "standard," "split," and "featherlight." They would like to know whether typing speed is affected by the kind of keyboard. They enlist six subjects and have them type a standard passage on each of the three keyboards. They then measure how long (in seconds, with penalties for mistakes made) it takes to type the passage. Does the kind of keyboard make a difference?

Typist	*Standard*	*Split*	*Featherlight*
1	121	115	124
2	122	117	127
3	106	116	100
4	113	102	120
5	108	86	112
6	98	94	96

37. Sandra runs a weight-lifting gym. She wishes to know how fast her clients' strength increases. She measures strength increase by taking the weight a client can bench press now and subtracting the weight he could press when he first came to the gym. She measures the strength increase for 20 clients, and notes how many months they have been training in her gym. Her data are listed in the table. She would like to be able to predict strength increase from length of time in training. Is that prediction significant?

Table for Exercise 37

Client	*Training (months)*	*Strength increase*
1	16	34
2	10	68
3	28	75
4	31	87
5	17	53
6	14	51
7	24	49
8	9	56
9	13	56
10	15	31
11	24	53
12	26	86
13	15	64
14	23	46
15	33	78
16	26	46
17	32	50
18	13	73
19	22	67
20	10	34

38. Dr. Cognit is interested in the memory for faces. She shows the picture of a face to a subject and then has the subject perform a task. Then she shows the subject five new pictures (one of which is of the same person) and asks the subject to identify the correct picture. This process is repeated 20 times, so each subject can obtain a score between 0 and 20 correct identifications. Dr. Cognit's interest is in the interference effect of the task. Each subject uses only one kind of task. The experiment uses two verbal tasks (task A: describe the picture you just saw, and task B: describe your mother's face) and two imaginal tasks (task C: visualize the picture you just saw, and task D: visualize your mother's face). Is there a difference in performance between subjects who use verbal tasks and those who use visual tasks?

Task A	*Task B*	*Task C*	*Task D*
16	11	16	19
12	13	12	11
8	8	13	13
10	7	12	12
14	12	15	15
10	13	16	17
11	11	10	15
16	11	14	11

Section E: From the journals (answers in Appendix D, page 534)

39. Finkelhor and colleagues (1990) reported the results of a national survey of adults concerning their history of childhood sexual abuse. They concluded that "separation from a natural parent for a major portion of one's childhood

... was a risk factor...: girls showed markedly higher risk under all family circumstances except that of living with two natural parents" (pp. 24–25). The overall rate of sexual abuse in their sample of 1485 women was 27%. The rates of abuse broken down by predominant family structure are shown in the table. To determine whether the different family structures did in fact have a significant effect on rate of abuse, Finkelhor performed a χ^2 test of independence, finding that $\chi^2 = 24.91, p < .0001$.

Predominant family structure	*Percent victimized*	*n*
Both natural parents	26	1209
Mother alone	35	98
Father alone	50	22
Both nonnatural parents	48	29
Natural mother/stepfather	40	53
Natural father/stepmother	44	16
Other	37	52

(a) Does the table correspond to Finkelhor's statement that there were 1485 subjects?
(b) How many degrees of freedom are there for this test? (Finkelhor does not state this.)
(c) Compute χ^2. [*Hint:* You will have to use the percentage column to create new columns corresponding to the number victimized and the number not victimized.] Does your computation agree with Finkelhor's? If not, why not?
(d) Is your χ^2 significant at $\alpha = .01$?
(e) State your conclusion in plain English.

40. Goldkamp and colleagues (1990) were interested in predicting drug use among 1820 felony defendants in Dade County, Florida. They developed a model that included 14 variables such as type of offense, whether a weapon was used, age, whether drug charges had been filed, and whether the subject reported using cocaine. On the basis of this model, they classified defendants into four groups. Individuals in group 4 were predicted to have the highest risk of using cocaine, whereas individuals in group 1 had the lowest risk. They also conducted a urinalysis with either positive or negative results for cocaine use. The observed percentages of felony defendants with positive tests among the four groups are shown in the table.

Goldkamp et al. reported that "$\chi^2 = 361.95$, 3 *d.f.*, $p \leq 0.00$" (p. 652).

Relative risk of positive test	*Number of defendants*	*Observed percentages with positive tests*
Group 1 (lowest)	469	44.1
Group 2	431	74.0
Group 3	449	87.1
Group 4 (highest)	471	94.1

(a) Are the degrees of freedom stated correctly by Goldkamp?
(b) Perform the χ^2 test. [*Hint:* You will need to create two columns headed "Number of positive tests" and "Number of negative tests."] Do your results correspond with Goldkamp's?
(c) What can be said about the statement "$p \leq 0.00$"?

41. Weintraub (1986) analyzed the verbal behavior of Presidents Jimmy Carter and Ronald Reagan during their news conferences and used his analysis to draw conclusions about the personalities of those presidents. For example, he counted the number of times the pronoun "I" occurred in samples of 1000 words taken from news conferences held during the first year of each presidency. "In general, clinical observers agree that the infrequent use of 'I' indicates an avoidance of intimacy, commitment, and candor; whereas a very high frequency suggests inordinate self-preoccupation" (p. 288). Is there a significant difference in the number of times Carter and Reagan used "I"?

Weintraub could identify 20 1000-word samples for Carter and 16 such samples for Reagan. The median number of "I" references for Carter was 31 and for Reagan 25. He performed a Mann–Whitney U test on these data, finding that $U = 77$, which he reported was significant at $p < .01$.
(a) What are n_1 and n_2 in this study?
(b) What is U_{cv} if $\alpha = .01$? Is the reported $U_{obs} = 77$ in fact significant at $p < .01$?
(c) Is there enough information to verify that U was computed correctly?

42. Siddall and Conway (1988) investigated whether the type of admission to a residential drug treatment program affected the outcome of the program for the person. They recorded three types of admission: voluntary, involuntary (criminal justice referral), and involuntary (criminal charges pending). Each person's outcome ("discharge disposition") was rated on a seven-point scale that was rank-ordered from least successful to most successful: D1. Absence without authorization; D2. Drug abuse; D3. Criminal charge; D4. Sexual misconduct; D5. Withdrawal against advice; D6. Reached maximum therapeutic benefit; and D7. Completed all program requirements.

The records of 100 patients were examined and subjected to a Kruskal–Wallis analysis of variance where the discharge disposition was the dependent variable and the type of admission was the independent variable. Siddall and Conway reported that $\chi^2 = 3.78, p < .05$, but they did not specify the degrees of freedom used in this test.
(a) How many degrees of freedom are there for a Kruskal–Wallis test with three groups (three types of admission)? What is the critical value of χ^2 in that case? Is Siddall and Conway's reported observed value of χ^2 significant at $p < .05$?
(b) Perhaps Siddall and Conway collapsed the two kinds of involuntary admission into one category, so that type of admission was divided into two groups (voluntary and involuntary) instead of three. Now how many degrees of freedom are there for the Kruskal–

Wallis test with two groups? What is the critical value of χ^2 in that case? Is Siddall and Conway's reported observed value of χ^2 significant at $p < .05$?

(c) Is there any other way to interpret Siddall and Conway's results so that their observed value of χ^2 is in fact correctly reported to be significant at $p < .05$?

43. Del Cerro and Borrell (1987) studied the effect of β-endorphin on memory processes in Wistar rats because much research had shown that such neurohumors can interfere with memory. They trained 56 rats as follows: On day 1, rats were given an injection (described below), and then 10 minutes later, they were placed in a cage that had a door on one side that led to a dark compartment. Rats were allowed to enter and explore freely the dark compartment. Day 2 was a repeat of day 1. Day 3 was a repeat of days 1 and 2 except that when the rat entered the dark compartment, it received an inescapable electric shock of 2 seconds duration. On day 4, each rat was given the injection, and then 10 minutes later, it was placed directly in the dark compartment for 5 minutes, during which time no shock was administered; this procedure is called "forced extinction." Day 5 was the last day. The rat was given the injection and then placed in the original compartment that had the door leading into the dark compartment. The time ("latency") was measured before the rat entered the dark compartment. This latency was the dependent variable in the study.

The 56 rats were divided into four groups, each of which received a different series of injections. Group I received saline solution (no β-endorphin); group II received .1 microgram per kilogram of body weight (μg/kg) β-endorphin; group III received 1.0 μg/kg; and group IV received 10 μg/kg. Did the different doses of β-endorphin affect the latencies?

Because the latencies are interval/ratio data, it would seem appropriate to use an analysis of variance to answer this question. However, the latencies varied greatly, from a few seconds to more than 300 seconds. Even though these latencies are interval/ratio data, a very long latency would have a great effect on the mean for that group, overshadowing all the other measurements, so the latencies were converted to ranks and a Kruskal–Wallis analysis of variance was performed. Del Cerro and Borrell reported that $H = 9.62$, $df = 3$, $p < .025$.

Del Cerro and Borrell also reported median latencies for each group presented as a graph. This graph can be interpreted to show that the median latency for group I (saline) = 20 seconds, the median latency for group II (.1 μg/kg) = 150 seconds, the median latency for group III (1 μg/kg) = 300 seconds, and the median latency for group IV (10 μg/kg) = 85 seconds.

(a) Did del Cerro and Borrell report the degrees of freedom correctly?

(b) What is the critical value of H (that is, of χ^2) with 3 degrees of freedom when $\alpha = .05$? When $\alpha = .01$?

(c) Is del Cerro and Borrell's claim that their observed $H = 9.62$ is significant with $p < .025$ reasonable?

(d) Do the median latency data provide enough information to verify the computation of H? Does that information make it seem reasonable that H was computed correctly?

Section F: Computer explorations

44. (a) Verify that the text's computation of U for Table 18.9 is correct. [*Hint:* If using SPSS, you will need to define an independent variable (similar to the t tests of Chapter 11). Then click `Statistics`, then `Nonparametric Tests`, and then `2 Independent Samples....`]

(b) Use t to analyze the same data. Can you conclude anything about the comparative power of U and t?

45. Verify that the text's computation of T for Table 18.10 is correct. [*Hint:* If using SPSS, click `Statistics`, then `Nonparametric Tests`, and then `2 Related Samples....` Note that SPSS uses a z approximation for the Wilcoxon test, but the results are the same (remember that the text's example is directional).]

46. (a) Verify that the text's computation of H for Table 18.11 is correct. [*Hint:* If using SPSS, click `Statistics`, then `Nonparametric Tests`, and then `k Related Samples....` Note that SPSS uses the χ^2 approximation for H.]

(b) Use ANOVA to analyze the same data. Can you conclude anything about the comparative power of H and the ANOVA's F?

REFERENCES

Best, J., & Luckenbill, D. F. (1990). Male dominance and female criminality: A test of Harris' theory of deviant type-scripts. *Sociological Inquiry, 60,* 71–86.

Brown, N. R. (1990). Organization of public events in long-term memory. *Journal of Experimental Psychology: General, 119,* 297–314.

Burger, J. M. (1990). Desire for control and interpersonal interaction style. *Journal of Research in Personality, 24,* 32–44.

Cattell, R. B. (1982). *The inheritance of personality and ability.* New York: Academic Press.

Christensen, L. B. (1991). *Experimental methodology,* 5th ed. Boston: Allyn & Bacon.

Cohen, J. (1988). *Statistical power analysis for the behavioral sciences,* 2nd ed. Hillsdale, NJ: Erlbaum.

———. (1990). Things I have learned (so far). *American Psychologist, 45,* 1304–1312.

———. (1992). A power primer. *Psychological Bulletin, 112,* 155–159.

———. (1994). The earth is round ($p < .05$). *American Psychologist, 49,* 997–1003.

Columbino, U., & Del Boca, D. (1990). The effect of taxes on labor supply in Italy. *The Journal of Human Resources, 25,* 390–414.

Conover, W. J. (1980). *Practical nonparametric statistics,* 2nd ed. New York: Wiley.

del Cerro, S., & Borrell, J. (1987). β-endorphin impairs forced extinction of an inhibitory avoidance response in rats. *Life Sciences, 41,* 579–584.

Drotar, D., Eckerle, D., Satola, J., Pallotta, J., & Wyatt, B. (1990). Maternal interactional behavior with nonorganic failure-to-thrive infants: A case comparison study. *Child Abuse and Neglect, 14,* 41–51.

Englund, G., & Olsson, T. I. (1990). Fighting and assessment in the net-spinning caddis larva *Arctopsyche ladogensis*: A test of the sequential assessment game. *Animal Behavior, 39,* 55–62.

Erinosho, S. V. (1990). The effect of two remediation methods in high school physics classes in Nigeria. *The Journal of Experimental Education, 58,* 177–183.

Fagley, N. S. (1985). Applied statistical power analysis and the interpretation of nonsignificant results by research consumers. *Journal of Counseling Psychology, 32,* 391–396.

Finkelhor, D., Hotaling, G., Lewis, I. A., & Smith, C. (1990). Sexual abuse in a national survey of adult men and women: Prevalence, characteristics, and risk factors. *Child Abuse and Neglect, 14,* 19–28.

Folnegovic-Smalc, V., Folnegovic, Z., & Kulcar, Z. (1990). Age of disease onset in Croatia's hospitalized schizophrenics. *British Journal of Psychiatry, 156,* 368–372.

Gigerenzer, G. (1993). The superego, the ego, and the id in statistical reasoning. In G. Keren & C. Lewis (Eds.), *A handbook for data analysis in the behavioral sciences: Methodological issues* (pp. 311–339). Hillsdale, NJ: Erlbaum.

Gigerenzer, G., Swijtink, A., Porter, T., Daston, L., Beatty, J., & Krüger, L. (1989). *The empire of chance: How probability changed science and everyday life.* Cambridge, England: Cambridge University Press.

Goldkamp, J. S., Gottfredson, M. R., & Weiland, D. (1990). Pretrial drug testing and defendant risk. *The Journal of Criminal Law and Criminology, 81,* 585–652.

Hall, G., & Honey, R. C. (1990). Context-specific conditioning in the conditioned-emotional response procedure. *Journal of Experimental Psychology: Animal Behavior Processes, 16,* 271–278.

Hearne, K. M. T. (1989). A questionnaire and personality study of self-styled psychics and mediums. *Journal of the Society for Psychical Research, 55,* 404–411.

Hurlburt, R. T. (1993). Developing estimation skills to increase students' comprehension of the mean and standard deviation. *Teaching Sociology,* 21, 177–181.

Iman, R. L., Quade, D., & Alexander, D. A. (1975). Exact probability levels for the Kruskal-Wallis test. *Selected Tables in Mathematical Statistics, 3,* 329–384.

Jensen, A. R. (1980). *Bias in mental testing.* New York: Free Press.

Kleinmuntz, B. (1990). Why we still use our heads instead of formulas: Toward an integrative approach. *Psychological Bulletin, 107,* 296–310.

Landon, 1,293,669; Roosevelt, 972,897. (1936). *Literary Digest, 122,* 5.

Matarazzo, J. D. (1972). *Wechsler's measurement and appraisal of adult intelligence.* Baltimore: Williams & Wilkins.

Meehl, P. (1954). *Clinical versus statistical prediction: A theoretical analysis and a review of the evidence.* Minneapolis: University of Minnesota Press.

Mischel, W. (1968). *Personality and assessment.* New York: Wiley.

Ortof, E., & Crystal, H. A. (1989). Rate of progession of Alzheimer's disease. *Journal of the American Geriatrics Society, 37,* 511–514.

Pearson, E. S. (1966). *Biometrika tables for statisticians.* Cambridge, MA: Cambridge University Press.

Rich, A. R., & Dahlheimer, D. (1989). The power of negative thinking: A new perspective on "irrational" cognitions. *Journal of Cognitive Psychotherapy: An International Quarterly, 3*, 15–30.

Riggio, R. E., Lippa, R., & Salinas, C. (1990). The display of personality in expressive movement. *Journal of Research in Personality, 24*, 16–31.

Rosenthal, R., & Fode, K. (1963). The effect of experimenter bias on the performance of the albino rat. *Behavioral Science, 8*, 183–189.

Rosenthal, R., & Jacobson, L. (1968). *Pygmalion in the classroom: Teacher expectation and pupils' intellectual development*. New York: Holt, Rinehart & Winston.

Rosenthal, R., & Rosnow, R. L. (1991). *Essentials of behavioral research: Methods and data analysis* (2nd ed.). New York: McGraw-Hill.

Rosnow, R. L., & Rosenthal, R. (1989). Definition and interpretation of interaction effects. *Psychological Bulletin, 105*, 143–146.

Schmidt, F. L. (1992). What do data really mean? Research findings, meta-analysis, and cumulative knowledge in psychology. *American Psychologist, 47*, 1173–1181.

Sedlmeier, P., & Gigerenzer, G. (1989). Do studies of statistical power have an effect on the power of studies? *Psychological Bulletin, 195*, 309–316.

Shei, P., Iwakuma, T., & Fujii, K. (1988). Population dynamics of *Daphnia rosea* in a small eutrophic pond. *Ecological Research, 3*, 291–304.

Siddall, J. W., & Conway, G. L. (1988). Interactional variables associated with retention and success in residential drug treatment. *International Journal of the Addictions, 23*, 1241–1254.

Simons, C. W., & Birkimer, J. C. (1988). An exploration of factors predicting the effects of aerobic conditioning on mood state. *Journal of Psychosomatic Research, 32*, 63–75.

Sobolski, J. C., Kolesar, J. J., Kornitzer, M. D., de Backer, G. G., Mikes, Z., Dramaix, M. M., Degre, S. G., & Denolin, H. F. (1988). Physical fitness does not reflect physical activity patterns in middle-aged workers. *Medicine and Science in Sports and Exercise, 20*, 6–13.

Swoope, K. F., & Johnson, C. S. (1988). A reexamination of the effects of reader- and text-based factors on priority judgments in expository prose. *The Journal of Educational Research, 82*, 5–9.

Thompson, B. (1994). The concept of statistical significance testing. *Measurement Update, 4*, 5–6. ERIC/AE Digest, EDO-TM-94-1: Tracking No. TM 021 079.

Weintraub, W. (1986). Personality profiles of American presidents as revealed in their public statements: The presidential news conferences of Jimmy Carter and Ronald Reagan. *Political Psychology, 7*, 285–295.

Wu, T., Tashkin, D. P., Rose, J. E., & Benham, D. (1988). Influence of marijuana potency and amount of cigarette consumed on marijuana smoking pattern. *Journal of Psychoactive Drugs, 20*, 43–46.

Zeidner, M. (1990). Some demographic and health correlates of trait anger in Israeli adults. *Journal of Research in Personality, 24*, 1–15.

Statistical Tables

APPENDIX

TABLE A.1 Proportions of areas under the normal curve

For a description in the text, see page 106.

AN EXAMPLE USING TABLE A.1

If $z = .80$, then the area between $z = 0$ and $z = .80$ is shown in the region marked "A"; that area is .2881, or 28.81% of the total area under the curve.

The area beyond z is shown in the region marked "B"; that area is .2119, or 21.19% of the total area under the curve.

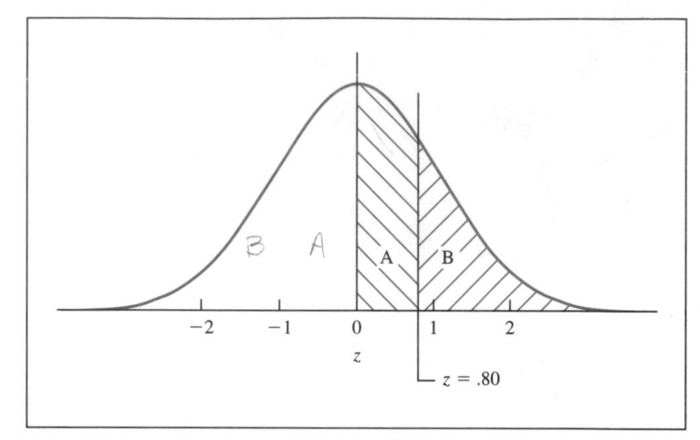

$z = .80$; the area in region A is .2881; the area in region B is .2119

z	Area between mean and z (A)	Area beyond z (B)	z	Area between mean and z (A)	Area beyond z (B)
.00	.0000	.5000	.30	.1179	.3821
.01	.0040	.4960	.31	.1217	.3783
.02	.0080	.4920	.32	.1255	.3745
.03	.0120	.4880	.33	.1293	.3707
.04	.0160	.4840	.34	.1331	.3669
.05	.0199	.4801	.35	.1368	.3632
.06	.0239	.4761	.36	.1406	.3594
.07	.0279	.4721	.37	.1443	.3557
.08	.0319	.4681	.38	.1480	.3520
.09	.0359	.4641	.39	.1517	.3483
.10	.0398	.4602	.40	.1554	.3446
.11	.0438	.4562	.41	.1591	.3409
.12	.0478	.4522	.42	.1628	.3372
.13	.0517	.4483	.43	.1664	.3336
.14	.0557	.4443	.44	.1700	.3300
.15	.0596	.4404	.45	.1736	.3264
.16	.0636	.4364	.46	.1772	.3228
.17	.0675	.4325	.47	.1808	.3192
.18	.0714	.4286	.48	.1844	.3156
.19	.0753	.4247	.49	.1879	.3121
.20	.0793	.4207	.50	.1915	.3085
.21	.0832	.4168	.51	.1950	.3050
.22	.0871	.4129	.52	.1985	.3015
.23	.0910	.4090	.53	.2019	.2981
.24	.0948	.4052	.54	.2054	.2946
.25	.0987	.4013	.55	.2088	.2912
.26	.1026	.3974	.56	.2123	.2877
.27	.1064	.3936	.57	.2157	.2843
.28	.1103	.3897	.58	.2190	.2810
.29	.1141	.3859	.59	.2224	.2776

Adapted from Table IIi of R. A. Fisher & F. Yates, *Statistical Tables for Biological, Agricultural, and Medical Research*, Sixth Edition, published by Longman Group UK, Ltd., (1974). Adapted with permission.

TABLE A.1 *(continued)* Proportions of areas under the normal curve

z	*Area between mean and z (A)*	*Area beyond z (B)*	*z*	*Area between mean and z (A)*	*Area beyond z (B)*
.60	.2257	.2743	1.10	.3643	.1357
.61	.2291	.2709	1.11	.3665	.1335
.62	.2324	.2676	1.12	.3686	.1314
.63	.2357	.2643	1.13	.3708	.1292
.64	.2389	.2611	1.14	.3729	.1271
.65	.2422	.2578	1.15	.3749	.1251
.66	.2454	.2546	1.16	.3770	.1230
.67	.2486	.2514	1.17	.3790	.1210
.68	.2517	.2483	1.18	.3810	.1190
.69	.2549	.2451	1.19	.3830	.1170
.70	.2580	.2420	1.20	.3849	.1151
.71	.2611	.2389	1.21	.3869	.1131
.72	.2642	.2358	1.22	.3888	.1112
.73	.2673	.2327	1.23	.3907	.1093
.74	.2704	.2296	1.24	.3925	.1075
.75	.2734	.2266	1.25	.3944	.1056
.76	.2764	.2236	1.26	.3962	.1038
.77	.2794	.2206	1.27	.3980	.1020
.78	.2823	.2177	1.28	.3997	.1003
.79	.2852	.2148	1.29	.4015	.0985
.80	.2881	.2119	1.30	.4032	.0968
.81	.2910	.2090	1.31	.4049	.0951
.82	.2939	.2061	1.32	.4066	.0934
.83	.2967	.2033	1.33	.4082	.0918
.84	.2995	.2005	1.34	.4099	.0901
.85	.3023	.1977	1.35	.4115	.0885
.86	.3051	.1949	1.36	.4131	.0869
.87	.3078	.1922	1.37	.4147	.0853
.88	.3106	.1894	1.38	.4162	.0838
.89	.3133	.1867	1.39	.4177	.0823
.90	.3159	.1841	1.40	.4192	.0808
.91	.3186	.1814	1.41	.4207	.0793
.92	.3212	.1788	1.42	.4222	.0778
.93	.3238	.1762	1.43	.4236	.0764
.94	.3264	.1736	1.44	.4251	.0749
.95	.3289	.1711	1.45	.4265	.0735
.96	.3315	.1685	1.46	.4279	.0721
.97	.3340	.1660	1.47	.4292	.0708
.98	.3365	.1635	1.48	.4306	.0694
.99	.3389	.1611	1.49	.4319	.0681
1.00	.3413	.1587	1.50	.4332	.0668
1.01	.3438	.1562	1.51	.4345	.0655
1.02	.3461	.1539	1.52	.4357	.0643
1.03	.3485	.1515	1.53	.4370	.0630
1.04	.3508	.1492	1.54	.4382	.0618
1.05	.3531	.1469	1.55	.4394	.0606
1.06	.3554	.1446	1.56	.4406	.0594
1.07	.3577	.1423	1.57	.4418	.0582
1.08	.3599	.1401	1.58	.4429	.0571
1.09	.3621	.1379	1.59	.4441	.0559

TABLE A.1 *(continued)* Proportions of areas under the normal curve

z	*Area between mean and z (A)*	*Area beyond z (B)*	*z*	*Area between mean and z (A)*	*Area beyond z (B)*
1.60	.4452	.0548	2.10	.4821	.0179
1.61	.4463	.0537	2.11	.4826	.0174
1.62	.4474	.0526	2.12	.4830	.0170
1.63	.4484	.0516	2.13	.4834	.0166
1.64	.4495	.0505	2.14	.4838	.0162
1.65	.4505	.0495	2.15	.4842	.0158
1.66	.4515	.0485	2.16	.4846	.0154
1.67	.4525	.0475	2.17	.4850	.0150
1.68	.4535	.0465	2.18	.4854	.0146
1.69	.4545	.0455	2.19	.4857	.0143
1.70	.4554	.0446	2.20	.4861	.0139
1.71	.4564	.0436	2.21	.4864	.0136
1.72	.4573	.0427	2.22	.4868	.0132
1.73	.4582	.0418	2.23	.4871	.0129
1.74	.4591	.0409	2.24	.4875	.0125
1.75	.4599	.0401	2.25	.4878	.0122
1.76	.4608	.0392	2.26	.4881	.0119
1.77	.4616	.0384	2.27	.4884	.0116
1.78	.4625	.0375	2.28	.4887	.0113
1.79	.4633	.0367	2.29	.4890	.0110
1.80	.4641	.0359	2.30	.4893	.0107
1.81	.4649	.0351	2.31	.4896	.0104
1.82	.4656	.0344	2.32	.4898	.0102
1.83	.4664	.0336	2.33	.4901	.0099
1.84	.4671	.0329	2.34	.4904	.0096
1.85	.4678	.0322	2.35	.4906	.0094
1.86	.4686	.0314	2.36	.4909	.0091
1.87	.4693	.0307	2.37	.4911	.0089
1.88	.4699	.0301	2.38	.4913	.0087
1.89	.4706	.0294	2.39	.4916	.0084
1.90	.4713	.0287	2.40	.4918	.0082
1.91	.4719	.0281	2.41	.4920	.0080
1.92	.4726	.0274	2.42	.4922	.0078
1.93	.4732	.0268	2.43	.4925	.0075
1.94	.4738	.0262	2.44	.4927	.0073
1.95	.4744	.0256	2.45	.4929	.0071
1.96	.4750	.0250	2.46	.4931	.0069
1.97	.4756	.0244	2.47	.4932	.0068
1.98	.4761	.0239	2.48	.4934	.0066
1.99	.4767	.0233	2.49	.4936	.0064
2.00	.4772	.0228	2.50	.4938	.0062
2.01	.4778	.0222	2.51	.4940	.0060
2.02	.4783	.0217	2.52	.4941	.0059
2.03	.4788	.0212	2.53	.4943	.0057
2.04	.4793	.0207	2.54	.4945	.0055
2.05	.4798	.0202	2.55	.4946	.0054
2.06	.4803	.0197	2.56	.4948	.0052
2.07	.4808	.0192	2.57	.4949	.0051
2.08	.4812	.0188	2.58	.4951	.0049
2.09	.4817	.0183	2.59	.4952	.0048

TABLE A.1 *(concluded)* Proportions of areas under the normal curve

z	Area between mean and z (A)	Area beyond z (B)	z	Area between mean and z (A)	Area beyond z (B)
2.60	.4953	.0047	3.00	.4987	.0013
2.61	.4955	.0045	3.01	.4987	.0013
2.62	.4956	.0044	3.02	.4987	.0013
2.63	.4957	.0043	3.03	.4988	.0012
2.64	.4959	.0041	3.04	.4988	.0012
2.65	.4960	.0040	3.05	.4989	.0011
2.66	.4961	.0039	3.06	.4989	.0011
2.67	.4962	.0038	3.07	.4989	.0011
2.68	.4963	.0037	3.08	.4990	.0010
2.69	.4964	.0036	3.09	.4990	.0010
2.70	.4965	.0035	3.10	.4990	.0010
2.71	.4966	.0034	3.11	.4991	.0009
2.72	.4967	.0033	3.12	.4991	.0009
2.73	.4968	.0032	3.13	.4991	.0009
2.74	.4969	.0031	3.14	.4992	.0008
2.75	.4970	.0030	3.15	.4992	.0008
2.76	.4971	.0029	3.16	.4992	.0008
2.77	.4972	.0028	3.17	.4992	.0008
2.78	.4973	.0027	3.18	.4993	.0007
2.79	.4974	.0026	3.19	.4993	.0007
2.80	.4974	.0026	3.20	.4993	.0007
2.81	.4975	.0025	3.21	.4993	.0007
2.82	.4976	.0024	3.22	.4994	.0006
2.83	.4977	.0023	3.23	.4994	.0006
2.84	.4977	.0023	3.24	.4994	.0006
2.85	.4978	.0022	3.25	.4994	.0006
2.86	.4979	.0021	3.30	.4995	.0005
2.87	.4979	.0021	3.35	.4996	.0004
2.88	.4980	.0020	3.40	.4997	.0003
2.89	.4981	.0019	3.45	.4997	.0003
2.90	.4981	.0019	3.50	.4998	.0002
2.91	.4982	.0018	3.60	.4998	.0002
2.92	.4982	.0018	3.70	.4999	.0001
2.93	.4983	.0017	3.80	.4999	.0001
2.94	.4984	.0016	3.90	.49995	.00005
2.95	.4984	.0016	4.00	.49997	.00003
2.96	.4985	.0015			
2.97	.4985	.0015			
2.98	.4986	.0014			
2.99	.4986	.0014			

TABLE A.2 Critical values of t

For a description in the text, see page 154.

AN EXAMPLE USING TABLE A.2

Assume $df = 12$, the level of significance $(\alpha) = .05$, and the test is nondirectional. If t_{obs} is greater than or equal to $t_{cv} = 2.179$, then t_{obs} is statistically significant.

	Level of significance (α) for directional (one-tailed) test			
	.05	.025	.01	.005
	Level of significance (α) for nondirectional (two-tailed) test			
df	.10	.05	.02	.01
1	6.314	12.706	31.821	63.657
2	2.920	4.303	6.965	9.925
3	2.353	3.182	4.541	5.841
4	2.132	2.776	3.747	4.604
5	2.015	2.571	3.365	4.032
6	1.943	2.447	3.143	3.707
7	1.895	2.365	2.998	3.499
8	1.860	2.306	2.896	3.355
9	1.833	2.262	2.821	3.250
10	1.812	2.228	2.764	3.169
11	1.796	2.201	2.718	3.106
12	1.782	2.179	2.681	3.055
13	1.771	2.160	2.650	3.012
14	1.761	2.145	2.624	2.977
15	1.753	2.131	2.602	2.947
16	1.746	2.120	2.583	2.921
17	1.740	2.110	2.567	2.898
18	1.734	2.101	2.552	2.878
19	1.729	2.093	2.539	2.861
20	1.725	2.086	2.528	2.845
21	1.721	2.080	2.518	2.831
22	1.717	2.074	2.508	2.819
23	1.714	2.069	2.500	2.807
24	1.711	2.064	2.492	2.797
25	1.708	2.060	2.485	2.787
26	1.706	2.056	2.479	2.779
27	1.703	2.052	2.473	2.771
28	1.701	2.048	2.467	2.763
29	1.699	2.045	2.462	2.756
30	1.697	2.042	2.457	2.750
40	1.684	2.021	2.423	2.704
60	1.671	2.000	2.390	2.660
120	1.658	1.980	2.358	2.617
∞	1.645	1.960	2.326	2.576

Adapted from Table III of R. A. Fisher & F. Yates, *Statistical Tables for Biological, Agricultural, and Medical Research*, Sixth Edition, published by Longman Group UK, Ltd., (1974). Adapted with permission.

TABLE A.3 Critical values of χ^2

For a description in the text, see page 443.

AN EXAMPLE USING TABLE A.3

Assume $df = 3$ and $\alpha = .05$. The critical value of χ^2 that leaves .05 in the right-hand tail is $\chi^2_{cv} = 7.815$. χ^2_{obs} is statistically significant if it is greater than or equal to χ^2_{cv}.

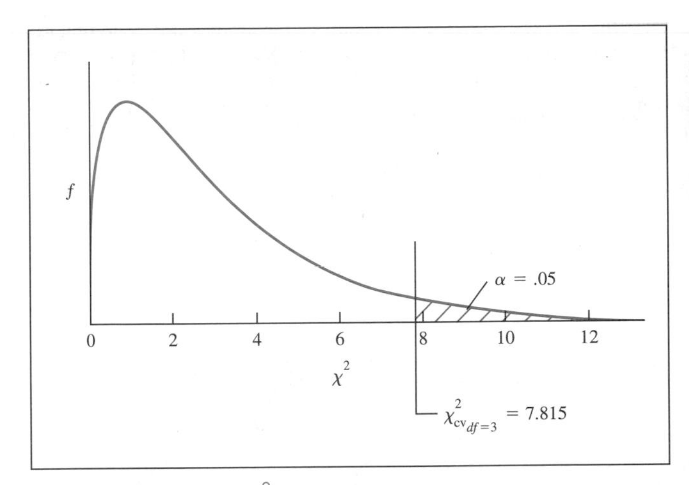

The critical value of χ^2 with $df = 3$ is 7.815

df	Level of significance (α) .05	.01
1	3.841	6.635
2	5.991	9.210
3	7.815	11.345
4	9.488	13.277
5	11.070	15.086
6	12.592	16.812
7	14.067	18.475
8	15.507	20.090
9	16.919	21.666
10	18.307	23.209
11	19.675	24.725
12	21.026	26.217
13	22.362	27.688
14	23.685	29.141
15	24.996	30.578
16	26.296	32.000
17	27.587	33.409
18	28.869	34.805
19	30.144	36.191
20	31.410	37.566
21	32.671	38.932
22	33.924	40.289
23	35.172	41.638
24	36.415	42.980
25	37.652	44.314
26	38.885	45.642
27	40.113	46.963
28	41.337	48.278
29	42.557	49.588
30	43.773	50.892
> 30 See note at left		

Adapted from Table IV of R. A. Fisher & F. Yates, *Statistical Tables for Biological, Agricultural, and Medical Research*, Sixth Edition, published by Longman Group UK, Ltd., (1974). Adapted with permission.

If $df > 30$, then χ^2_{cv} can be approximated by using the normal distribution:

$$\chi^2_{cv} = \frac{z^2_{cv}}{2} + z_{cv}\sqrt{2df - 1} + \frac{2df - 1}{2}$$

where z_{cv} is obtained from Table A.1. Note that in order to obtain directional χ^2_{cv} values (as are shown in Table A.3), one must also use directional z_{cv} values from Table A.1.

Some algebra will show that the preceding expression is equivalent to the more frequently seen expression

$$z = \sqrt{2\chi^2} - \sqrt{2df - 1}$$

TABLE A.4 Critical values of F (α = .05)

For a description in the text, see page 291.

AN EXAMPLE USING TABLE A.4

Assume that the numerator of F has df = 3 and the denominator has df = 20, and α = .05. The critical value of F that leaves .05 in the right-hand tail is $F_{cv_{3,20}} = 3.10$. F_{obs} is statistically significant if it is greater than or equal to F_{cv}.

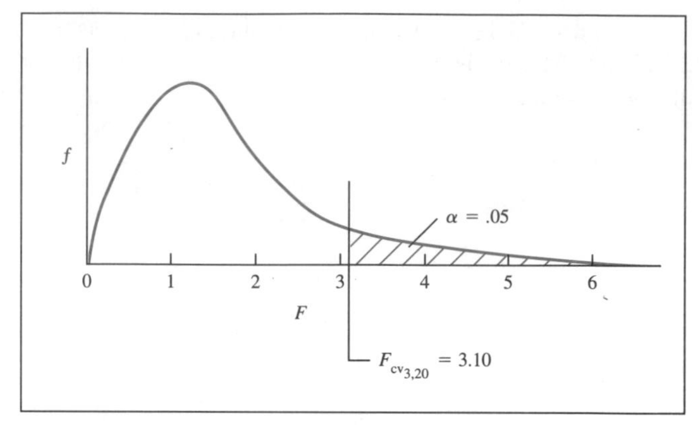

Numerator df = 3; denominator df = 20; α = .05; $F_{CV_{3,20}}$ = 3.10

df for denominator	Level of significance (α) = .05 df for numerator									
	1	2	3	4	5	6	10	30	120	∞
1	161.4	199.5	215.7	224.6	230.2	234.0	241.9	250.1	253.3	254.3
2	18.51	19.00	19.16	19.25	19.30	19.33	19.40	19.46	19.49	19.50
3	10.13	9.55	9.28	9.12	9.01	8.94	8.79	8.62	8.55	8.53
4	7.71	6.94	6.59	6.39	6.26	6.16	5.96	5.75	5.66	5.63
5	6.61	5.79	5.41	5.19	5.05	4.95	4.74	4.50	4.40	4.36
6	5.99	5.14	4.76	4.53	4.39	4.28	4.06	3.81	3.70	3.67
7	5.59	4.74	4.35	4.12	3.97	3.87	3.64	3.38	3.27	3.23
8	5.32	4.46	4.07	3.84	3.69	3.58	3.35	3.08	2.97	2.93
9	5.12	4.26	3.86	3.63	3.48	3.37	3.14	2.86	2.75	2.71
10	4.96	4.10	3.71	3.48	3.33	3.22	2.98	2.70	2.58	2.54
11	4.84	3.98	3.59	3.36	3.20	3.09	2.85	2.57	2.45	2.40
12	4.75	3.89	3.49	3.26	3.11	3.00	2.75	2.47	2.34	2.30
13	4.67	3.81	3.41	3.18	3.03	2.92	2.67	2.38	2.25	2.21
14	4.60	3.74	3.34	3.11	2.96	2.85	2.60	2.31	2.18	2.13
15	4.54	3.68	3.29	3.06	2.90	2.79	2.54	2.25	2.11	2.07
16	4.49	3.63	3.24	3.01	2.85	2.74	2.49	2.19	2.06	2.01
17	4.45	3.59	3.20	2.96	2.81	2.70	2.45	2.15	2.01	1.96
18	4.41	3.55	3.16	2.93	2.77	2.66	2.41	2.11	1.97	1.92
19	4.38	3.52	3.13	2.90	2.74	2.63	2.38	2.07	1.93	1.88
20	4.35	3.49	3.10	2.87	2.71	2.60	2.35	2.04	1.90	1.84
21	4.32	3.47	3.07	2.84	2.68	2.57	2.32	2.01	1.87	1.81
22	4.30	3.44	3.05	2.82	2.66	2.55	2.30	1.98	1.84	1.78
23	4.28	3.42	3.03	2.80	2.64	2.53	2.27	1.96	1.81	1.76
24	4.26	3.40	3.01	2.78	2.62	2.51	2.25	1.94	1.79	1.73
25	4.24	3.39	2.99	2.76	2.60	2.49	2.24	1.92	1.77	1.71
26	4.23	3.37	2.98	2.74	2.59	2.47	2.22	1.90	1.75	1.69
27	4.21	3.35	2.96	2.73	2.57	2.46	2.20	1.88	1.73	1.67
28	4.20	3.34	2.95	2.71	2.56	2.45	2.19	1.87	1.71	1.65
29	4.18	3.33	2.93	2.70	2.55	2.43	2.18	1.85	1.70	1.64
30	4.17	3.32	2.92	2.69	2.53	2.42	2.16	1.84	1.68	1.62
40	4.08	3.23	2.84	2.61	2.45	2.34	2.08	1.74	1.58	1.51
60	4.00	3.15	2.76	2.53	2.37	2.25	1.99	1.65	1.47	1.39
120	3.92	3.07	2.68	2.45	2.29	2.17	1.91	1.55	1.35	1.25
∞	3.84	3.00	2.60	2.37	2.21	2.10	1.83	1.46	1.22	1.00

TABLE A.4 *(continued)* Critical values of F (α = .01)

df for denominator	Level of significance (α) = .01 df for numerator									
	1	*2*	*3*	*4*	*5*	*6*	*10*	*30*	*120*	∞
1	4052	4999.5	5403	5625	5764	5859	6056	6261	6339	6366
2	98.50	99.00	99.17	99.25	99.30	99.33	99.40	99.47	99.49	99.50
3	34.12	30.82	29.46	28.71	28.24	27.91	27.23	26.50	26.22	26.13
4	21.20	18.00	16.69	15.98	15.52	15.21	14.55	13.84	13.56	13.46
5	16.26	13.27	12.06	11.39	10.97	10.67	10.05	9.38	9.11	9.02
6	13.75	10.92	9.78	9.15	8.75	8.47	7.87	7.23	6.97	6.88
7	12.25	9.55	8.45	7.85	7.46	7.19	6.62	5.99	5.74	5.65
8	11.26	8.65	7.59	7.01	6.63	6.37	5.81	5.20	4.95	4.86
9	10.56	8.02	6.99	6.42	6.06	5.80	5.26	4.65	4.40	4.31
10	10.04	7.56	6.55	5.99	5.64	5.39	4.85	4.25	4.00	3.91
11	9.65	7.21	6.22	5.67	5.32	5.07	4.54	3.94	3.69	3.60
12	9.33	6.93	5.95	5.41	5.06	4.82	4.30	3.70	3.45	3.36
13	9.07	6.70	5.74	5.21	4.86	4.62	4.10	3.51	3.25	3.17
14	8.86	6.51	5.56	5.04	4.69	4.46	3.94	3.35	3.09	3.00
15	8.68	6.36	5.42	4.89	4.56	4.32	3.80	3.21	2.96	2.87
16	8.53	6.23	5.29	4.77	4.44	4.20	3.69	3.10	2.84	2.75
17	8.40	6.11	5.18	4.67	4.34	4.10	3.59	3.00	2.75	2.65
18	8.29	6.01	5.09	4.58	4.25	4.01	3.51	2.92	2.66	2.57
19	8.18	5.93	5.01	4.50	4.17	3.94	3.43	2.84	2.58	2.49
20	8.10	5.85	4.94	4.43	4.10	3.87	3.37	2.78	2.52	2.42
21	8.02	5.78	4.87	4.37	4.04	3.81	3.31	2.72	2.46	2.36
22	7.95	5.72	4.82	4.31	3.99	3.76	3.26	2.67	2.40	2.31
23	7.88	5.66	4.76	4.26	3.94	3.71	3.21	2.62	2.35	2.26
24	7.82	5.61	4.72	4.22	3.90	3.67	3.17	2.58	2.31	2.21
25	7.77	5.57	4.68	4.18	3.85	3.63	3.13	2.54	2.27	2.17
26	7.72	5.53	4.64	4.14	3.82	3.59	3.09	2.50	2.23	2.13
27	7.68	5.49	4.60	4.11	3.78	3.56	3.06	2.47	2.20	2.10
28	7.64	5.45	4.57	4.07	3.75	3.53	3.03	2.44	2.17	2.06
29	7.60	5.42	4.54	4.04	3.73	3.50	3.00	2.41	2.14	2.03
30	7.56	5.39	4.51	4.02	3.70	3.47	2.98	2.39	2.11	2.01
40	7.31	5.18	4.31	3.83	3.51	3.29	2.80	2.20	1.92	1.80
60	7.08	4.98	4.13	3.65	3.34	3.12	2.63	2.03	1.73	1.60
120	6.85	4.79	3.95	3.48	3.17	2.96	2.47	1.86	1.53	1.38
∞	6.63	4.61	3.78	3.32	3.02	2.80	2.32	1.70	1.32	1.00

TABLE A.5 Critical values of Q, the Studentized range statistic ($\alpha = .05$)

For a description in the text, see page 325.

AN EXAMPLE USING TABLE A.5

If there are $k = 4$ groups, the df within cells is 24, and $\alpha = .05$, then Q_{obs} is significant if it is greater than or equal to $Q_{cv} = 3.90$.

	Level of significance (α) = .05 k = the number of means or number of steps between ordered means											
df_W	2	3	4	5	6	7	8	9	10	12	15	20
1	17.97	26.98	32.82	37.08	40.41	43.12	45.40	47.36	49.07	51.96	55.36	59.56
2	6.08	8.33	9.80	10.88	11.74	12.44	13.03	13.54	13.99	14.75	15.65	16.77
3	4.50	5.91	6.82	7.50	8.04	8.48	8.85	9.18	9.46	9.95	10.52	11.24
4	3.93	5.04	5.76	6.29	6.71	7.05	7.35	7.60	7.83	8.21	8.66	9.23
5	3.64	4.60	5.22	5.67	6.03	6.33	6.58	6.80	6.99	7.32	7.72	8.21
6	3.46	4.34	4.90	5.30	5.63	5.90	6.12	6.32	6.49	6.79	7.14	7.59
7	3.34	4.16	4.68	5.06	5.36	5.61	5.82	6.00	6.16	6.43	6.76	7.17
8	3.26	4.04	4.53	4.89	5.17	5.40	5.60	5.77	5.92	6.18	6.48	6.87
9	3.20	3.95	4.41	4.76	5.02	5.24	5.43	5.59	5.74	5.98	6.28	6.64
10	3.15	3.88	4.33	4.65	4.91	5.12	5.30	5.46	5.60	5.83	6.11	6.47
11	3.11	3.82	4.26	4.57	4.82	5.03	5.20	5.35	5.49	5.71	5.98	6.33
12	3.08	3.77	4.20	4.51	4.75	4.95	5.12	5.27	5.39	5.61	5.88	6.21
13	3.06	3.73	4.15	4.45	4.69	4.88	5.05	5.19	5.32	5.53	5.79	6.11
14	3.03	3.70	4.11	4.41	4.64	4.83	4.99	5.13	5.25	5.46	5.71	6.03
15	3.01	3.67	4.08	4.37	4.59	4.78	4.94	5.08	5.20	5.40	5.65	5.96
16	3.00	3.65	4.05	4.33	4.56	4.74	4.90	5.03	5.15	5.35	5.59	5.90
17	2.98	3.63	4.02	4.30	4.52	4.70	4.86	4.99	5.11	5.31	5.54	5.84
18	2.97	3.61	4.00	4.28	4.49	4.67	4.82	4.96	5.07	5.27	5.50	5.79
19	2.96	3.59	3.98	4.25	4.47	4.65	4.79	4.92	5.04	5.23	5.46	5.75
20	2.95	3.58	3.96	4.23	4.45	4.62	4.77	4.90	5.01	5.20	5.43	5.71
24	2.92	3.53	3.90	4.17	4.37	4.54	4.68	4.81	4.92	5.10	5.32	5.59
30	2.89	3.49	3.85	4.10	4.30	4.46	4.60	4.72	4.82	5.00	5.21	5.47
40	2.86	3.44	3.79	4.04	4.23	4.39	4.52	4.63	4.73	4.90	5.11	5.36
60	2.83	3.40	3.74	3.98	4.16	4.31	4.44	4.55	4.65	4.81	5.00	5.24
120	2.80	3.36	3.68	3.92	4.10	4.24	4.36	4.47	4.56	4.71	4.90	5.13
∞	2.77	3.31	3.63	3.86	4.03	4.17	4.29	4.39	4.47	4.62	4.80	5.01

TABLE A.5 *(continued)* Critical values of Q, the Studentized range statistic (α = .01)

Level of significance (α) = .01
k = the number of means or number of steps between ordered means

df_W	2	3	4	5	6	7	8	9	10	12	15	20
1	90.03	135.0	164.3	185.6	202.2	215.8	227.2	237.0	245.6	260.0	277.0	298.0
2	14.04	19.02	22.29	24.72	26.63	28.20	29.53	30.68	31.69	33.40	35.43	37.95
3	8.26	10.62	12.17	13.33	14.24	15.00	15.64	16.20	16.69	17.53	18.52	19.77
4	6.51	8.12	9.17	9.96	10.58	11.10	11.55	11.93	12.27	12.84	13.53	14.40
5	5.70	6.98	7.80	8.42	8.91	9.32	9.67	9.97	10.24	10.70	11.24	11.93
6	5.24	6.33	7.03	7.56	7.97	8.32	8.61	8.87	9.10	9.48	9.95	10.54
7	4.95	5.92	6.54	7.01	7.37	7.68	7.94	8.17	8.37	8.71	9.12	9.65
8	4.75	5.64	6.20	6.62	6.96	7.24	7.47	7.68	7.86	8.18	8.55	9.03
9	4.60	5.43	5.96	6.35	6.66	6.91	7.13	7.33	7.49	7.78	8.13	8.57
10	4.48	5.27	5.77	6.14	6.43	6.67	6.87	7.05	7.21	7.49	7.81	8.23
11	4.39	5.15	5.62	5.97	6.25	6.48	6.67	6.84	6.99	7.25	7.56	7.95
12	4.32	5.05	5.50	5.84	6.10	6.32	6.51	6.67	6.81	7.06	7.36	7.73
13	4.26	4.96	5.40	5.73	5.98	6.19	6.37	6.53	6.67	6.90	7.19	7.55
14	4.21	4.89	5.32	5.63	5.88	6.08	6.26	6.41	6.54	6.77	7.05	7.39
15	4.17	4.84	5.25	5.56	5.80	5.99	6.16	6.31	6.44	6.66	6.93	7.26
16	4.13	4.79	5.19	5.49	5.72	5.92	6.08	6.22	6.35	6.56	6.82	7.15
17	4.10	4.74	5.14	5.43	5.66	5.85	6.01	6.15	6.27	6.48	6.73	7.05
18	4.07	4.70	5.09	5.38	5.60	5.79	5.94	6.08	6.20	6.41	6.65	6.97
19	4.05	4.67	5.05	5.33	5.55	5.73	5.89	6.02	6.14	6.34	6.58	6.89
20	4.02	4.64	5.02	5.29	5.51	5.69	5.84	5.97	6.09	6.28	6.52	6.82
24	3.96	4.55	4.91	5.17	5.37	5.54	5.69	5.81	5.92	6.11	6.33	6.61
30	3.89	4.45	4.80	5.05	5.24	5.40	5.54	5.65	5.76	5.93	6.14	6.41
40	3.82	4.37	4.70	4.93	5.11	5.26	5.39	5.50	5.60	5.76	5.96	6.21
60	3.76	4.28	4.59	4.82	4.99	5.13	5.25	5.36	5.45	5.60	5.78	6.01
120	3.70	4.20	4.50	4.71	4.87	5.01	5.12	5.21	5.30	5.44	5.61	5.83
∞	3.64	4.12	4.40	4.60	4.76	4.88	4.99	5.08	5.16	5.29	5.45	5.65

TABLE A.6 Critical values of the Mann–Whitney U ($\alpha = .05$)

Directional (one-tailed) values are shown in lightface type; nondirectional (two-tailed) values are shown in **boldface** type.

For a description in the text see page 453.

AN EXAMPLE USING TABLE A.6

If there are 8 observations in the first group ($n_1 = 8$) and 12 observations in the second group ($n_2 = 12$), and $\alpha = .05$, nondirectional, then $U_{cv} = 26$. U_{obs} is statistically significant if it is *less than* or equal to U_{cv}.

n_2, number of observations in second group	*n_1, number of observations in first group*																			
	1	*2*	*3*	*4*	*5*	*6*	*7*	*8*	*9*	*10*	*11*	*12*	*13*	*14*	*15*	*16*	*17*	*18*	*19*	*20*
1	—	—	—	—	—	—	—	—	—	—	—	—	—	—	—	—	—	—	0	0
																			—	**—**
2	—	—	—	—	0	0	0	1	1	1	1	2	2	2	3	3	3	4	4	4
					—	**—**	**—**	**0**	**0**	**0**	**0**	**1**	**1**	**1**	**1**	**1**	**2**	**2**	**2**	**2**
3	—	—	0	0	1	2	2	3	3	4	5	5	6	7	7	8	9	9	10	11
	—	**—**	**—**	**—**	**0**	**1**	**1**	**2**	**2**	**3**	**3**	**4**	**4**	**5**	**5**	**6**	**6**	**7**	**7**	**8**
4	—	—	0	1	2	3	4	5	6	7	8	9	10	11	12	14	15	16	17	18
	—	**—**	**—**	**0**	**1**	**2**	**3**	**4**	**4**	**5**	**6**	**7**	**8**	**9**	**10**	**11**	**11**	**12**	**13**	**13**
5	—	0	1	2	4	5	6	8	9	11	12	13	15	16	18	19	20	22	23	25
	—	**—**	**0**	**1**	**2**	**3**	**5**	**6**	**7**	**8**	**9**	**11**	**12**	**13**	**14**	**15**	**17**	**18**	**19**	**20**
6	—	0	2	3	5	7	8	10	12	14	16	17	19	21	23	25	26	28	30	32
	—	**—**	**1**	**2**	**3**	**5**	**6**	**8**	**10**	**11**	**13**	**14**	**16**	**17**	**19**	**21**	**22**	**24**	**25**	**27**
7	—	0	2	4	6	8	11	13	15	17	19	21	24	26	28	30	33	35	37	39
	—	**—**	**1**	**3**	**5**	**6**	**8**	**10**	**12**	**14**	**16**	**18**	**20**	**22**	**24**	**26**	**28**	**30**	**32**	**34**
8	—	1	3	5	8	10	13	15	18	20	23	26	28	31	33	36	39	41	44	47
	—	**0**	**2**	**4**	**6**	**8**	**10**	**13**	**15**	**17**	**19**	**22**	**24**	**26**	**29**	**31**	**34**	**36**	**38**	**41**
9	—	1	3	6	9	12	15	18	21	24	27	30	33	36	39	42	45	48	51	54
	—	**0**	**2**	**4**	**7**	**10**	**12**	**15**	**17**	**20**	**23**	**26**	**28**	**31**	**34**	**37**	**39**	**42**	**45**	**48**
10	—	1	4	7	11	14	17	20	24	27	31	34	37	41	44	48	51	55	58	62
	—	**0**	**3**	**5**	**8**	**11**	**14**	**17**	**20**	**23**	**26**	**29**	**33**	**36**	**39**	**42**	**45**	**48**	**52**	**55**
11	—	1	5	8	12	16	19	23	27	31	34	38	42	46	50	54	57	61	65	69
	—	**0**	**3**	**6**	**9**	**13**	**16**	**19**	**23**	**26**	**30**	**33**	**37**	**40**	**44**	**47**	**51**	**55**	**58**	**62**
12	—	2	5	9	13	17	21	26	30	34	38	42	47	51	55	60	64	68	72	77
	—	**1**	**4**	**7**	**11**	**14**	**18**	**22**	**26**	**29**	**33**	**37**	**41**	**45**	**49**	**53**	**57**	**61**	**65**	**69**
13	—	2	6	10	15	19	24	28	33	37	42	47	51	56	61	65	70	75	80	84
	—	**1**	**4**	**8**	**12**	**16**	**20**	**24**	**28**	**33**	**37**	**41**	**45**	**50**	**54**	**59**	**63**	**67**	**72**	**76**
14	—	2	7	11	16	21	26	31	36	41	46	51	56	61	66	71	77	82	87	92
	—	**1**	**5**	**9**	**13**	**17**	**22**	**26**	**31**	**36**	**40**	**45**	**50**	**55**	**59**	**64**	**67**	**74**	**78**	**83**
15	—	3	7	12	18	23	28	33	39	44	50	55	61	66	72	77	83	88	94	100
	—	**1**	**5**	**10**	**14**	**19**	**24**	**29**	**34**	**39**	**44**	**49**	**54**	**59**	**64**	**70**	**75**	**80**	**85**	**90**
16	—	3	8	14	19	25	30	36	42	48	54	60	65	71	77	83	89	95	101	107
	—	**1**	**6**	**11**	**15**	**21**	**26**	**31**	**37**	**42**	**47**	**53**	**59**	**64**	**70**	**75**	**81**	**86**	**92**	**98**
17	—	3	9	15	20	26	33	39	45	51	57	64	70	77	83	89	96	102	109	115
	—	**2**	**6**	**11**	**17**	**22**	**28**	**34**	**39**	**45**	**51**	**57**	**63**	**67**	**75**	**81**	**87**	**93**	**99**	**105**
18	—	4	9	16	22	28	35	41	48	55	61	68	75	82	88	95	102	109	116	123
	—	**2**	**7**	**12**	**18**	**24**	**30**	**36**	**42**	**48**	**55**	**61**	**67**	**74**	**80**	**86**	**93**	**99**	**106**	**112**
19	0	4	10	17	23	30	37	44	51	58	65	72	80	87	94	101	109	116	123	130
	—	**2**	**7**	**13**	**19**	**25**	**32**	**38**	**45**	**52**	**58**	**65**	**72**	**78**	**85**	**92**	**99**	**106**	**113**	**119**
20	0	4	11	18	25	32	39	47	54	62	69	77	84	92	100	107	115	123	130	138
	—	**2**	**8**	**13**	**20**	**27**	**34**	**41**	**48**	**55**	**62**	**69**	**76**	**83**	**90**	**98**	**105**	**112**	**119**	**127**

Based on *Introductory Statistics*, by R. E. Kirk, Brooks/Cole Publishing Company, 1984.

TABLE A.6 *(continued)* Critical values of the Mann–Whitney U ($\alpha = .01$)

Directional (one-tailed) values are shown in lightface type; nondirectional (two-tailed) values are shown in **boldface** type.

n_2, number of observations in second group	*n_1, number of observations in first group*																			
	1	*2*	*3*	*4*	*5*	*6*	*7*	*8*	*9*	*10*	*11*	*12*	*13*	*14*	*15*	*16*	*17*	*18*	*19*	*20*
1	—	—	—	—	—	—	—	—	—	—	—	—	—	—	—	—	—	—	—	—
2	—	—	—	—	—	—	—	—	—	—	—	—	0	0	0	0	0	0	1	1
													—	—	—	—	—	—	**0**	**0**
3	—	—	—	—	—	—	0	0	1	1	1	2	2	2	3	3	4	4	4	5
							—	—	**0**	**0**	**0**	**1**	**1**	**1**	**2**	**2**	**2**	**2**	**3**	**3**
4	—	—	—	—	0	1	1	2	3	3	4	5	5	6	7	7	8	9	9	10
					—	**0**	**0**	**1**	**1**	**2**	**2**	**3**	**3**	**4**	**5**	**5**	**6**	**6**	**7**	**8**
5	—	—	—	0	1	2	3	4	5	6	7	8	9	10	11	12	13	14	15	16
				—	**0**	**1**	**1**	**2**	**3**	**4**	**5**	**6**	**7**	**7**	**8**	**9**	**10**	**11**	**12**	**13**
6	—	—	—	1	2	3	4	6	7	8	9	11	12	13	15	16	18	19	20	22
				0	**1**	**2**	**3**	**4**	**5**	**6**	**7**	**9**	**10**	**11**	**12**	**13**	**15**	**16**	**17**	**18**
7	—	—	0	1	3	4	6	7	9	11	12	14	16	17	19	21	23	24	26	28
			—	**0**	**1**	**3**	**4**	**6**	**7**	**9**	**10**	**12**	**13**	**15**	**16**	**18**	**19**	**21**	**22**	**24**
8	—	—	0	2	4	6	7	9	11	13	15	17	20	22	24	26	28	30	32	34
			—	**1**	**2**	**4**	**6**	**7**	**9**	**11**	**13**	**15**	**17**	**18**	**20**	**22**	**24**	**26**	**28**	**30**
9	—	—	1	3	5	7	9	11	14	16	18	21	23	26	28	31	33	36	38	40
			0	**1**	**3**	**5**	**7**	**9**	**11**	**13**	**16**	**18**	**20**	**22**	**24**	**27**	**29**	**31**	**33**	**36**
10	—	—	1	3	6	8	11	13	16	19	22	24	27	30	33	36	38	41	44	47
			0	**2**	**4**	**6**	**9**	**11**	**13**	**16**	**18**	**21**	**24**	**26**	**29**	**31**	**34**	**37**	**39**	**42**
11	—	—	1	4	7	9	12	15	18	22	25	28	31	34	37	41	44	47	50	53
			0	**2**	**5**	**7**	**10**	**13**	**16**	**18**	**21**	**24**	**27**	**30**	**33**	**36**	**39**	**42**	**45**	**48**
12	—	—	2	5	8	11	14	17	21	24	28	31	35	38	42	46	49	53	56	60
			1	**3**	**6**	**9**	**12**	**15**	**18**	**21**	**24**	**27**	**31**	**34**	**37**	**41**	**44**	**47**	**51**	**54**
13	—	0	2	5	9	12	16	20	23	27	31	35	39	43	47	51	55	59	63	67
		—	**1**	**3**	**7**	**10**	**13**	**17**	**20**	**24**	**27**	**31**	**34**	**38**	**42**	**45**	**49**	**53**	**56**	**60**
14	—	0	2	6	10	13	17	22	26	30	34	38	43	47	51	56	60	65	69	73
		—	**1**	**4**	**7**	**11**	**15**	**18**	**22**	**26**	**30**	**34**	**38**	**42**	**46**	**50**	**54**	**58**	**63**	**67**
15	—	0	3	7	11	15	19	24	28	33	37	42	47	51	56	61	66	70	75	80
		—	**2**	**5**	**8**	**12**	**16**	**20**	**24**	**29**	**33**	**37**	**42**	**46**	**51**	**55**	**60**	**64**	**69**	**73**
16	—	0	3	7	12	16	21	26	31	36	41	46	51	56	61	66	71	76	82	87
		—	**2**	**5**	**9**	**13**	**18**	**22**	**27**	**31**	**36**	**41**	**45**	**50**	**55**	**60**	**65**	**70**	**74**	**79**
17	—	0	4	8	13	18	23	28	33	38	44	49	55	60	66	71	77	82	88	93
		—	**2**	**6**	**10**	**15**	**19**	**24**	**29**	**34**	**39**	**44**	**49**	**54**	**60**	**65**	**70**	**75**	**81**	**86**
18	—	0	4	9	14	19	24	30	36	41	47	53	59	65	70	76	82	88	94	100
		—	**2**	**6**	**11**	**16**	**21**	**26**	**31**	**37**	**42**	**47**	**53**	**58**	**64**	**70**	**75**	**81**	**87**	**92**
19	—	1	4	9	15	20	26	32	38	44	50	56	63	69	75	82	88	94	101	107
		0	**3**	**7**	**12**	**17**	**22**	**28**	**33**	**39**	**45**	**51**	**56**	**63**	**69**	**74**	**81**	**87**	**93**	**99**
20	—	1	5	10	16	22	28	34	40	47	53	60	67	73	80	87	93	100	107	114
		0	**3**	**8**	**13**	**18**	**24**	**30**	**36**	**42**	**48**	**54**	**60**	**67**	**73**	**79**	**86**	**92**	**99**	**105**

TABLE A.7 Critical values of Wilcoxon's T

For a description in the text, see page 455.

AN EXAMPLE USING TABLE A.7

T is the smaller sum of ranks associated with differences that all have the same sign. If there are 20 pairs of differences (n = 20), α = .05, and the test is nondirectional, then T_{cv} = 52. T_{obs} is statistically significant if it is *less than* or equal to T_{cv}.

	Level of significance (α) for directional (one-tailed) test			
	.05	.025	.01	.005
	Level of significance (α) for nondirectional (two-tailed) test			
n	**.10**	**.05**	**.02**	**.01**
5	0	—	—	—
6	2	0	—	—
7	3	2	0	—
8	5	3	1	0
9	8	5	3	1
10	10	8	5	3
11	13	10	7	5
12	17	13	9	7
13	21	17	12	9
14	25	21	15	12
15	30	25	19	15
16	35	29	23	19
17	41	34	27	23
18	47	40	32	27
19	53	46	37	32
20	60	52	43	37
21	67	58	49	42
22	75	65	55	48
23	83	73	62	54
24	91	81	69	61
25	100	89	76	68
26	110	98	84	75
27	119	107	92	83
28	130	116	101	91
29	140	126	110	100
30	151	137	120	109
31	163	147	130	118
32	175	159	140	128
33	187	170	151	138
34	200	182	162	148
35	213	195	173	159
36	227	208	185	171
37	241	221	198	182
38	256	235	211	194
39	271	249	224	207
40	286	264	238	220
41	302	279	252	233
42	319	294	266	247
43	336	310	281	261
44	353	327	296	276
45	371	343	312	291
46	389	361	328	307
47	407	378	345	322
48	426	396	362	339
49	446	415	379	355
50	466	434	397	373

TABLE A.8 Critical values of r, the Pearson product-moment correlation coefficient

For a description in the text, see page 389.

AN EXAMPLE USING TABLE A.8

If we let $\alpha = .05$, the test is nondirectional, and there are $n = 20$ pairs of scores, then $df = n - 2 = 18$. Then the critical value of Pearson's $r_{cv} = \pm .4438$. If r_{obs} is greater than or equal to (in absolute value) r_{cv}, then r_{obs} is statistically significant.

	Level of significance (α) for directional (one-tailed) test			
	.05	.025	.01	.005
	Level of significance (α) for nondirectional (two-tailed) test			
$df = n - 2$	.10	.05	.02	.01
1	.98769	.99692	.999507	.999877
2	.90000	.95000	.98000	.990000
3	.8054	.8783	.93433	.95873
4	.7293	.8114	.8822	.91720
5	.6694	.7545	.8329	.8745
6	.6215	.7067	.7887	.8343
7	.5822	.6664	.7498	.7977
8	.5494	.6319	.7155	.7646
9	.5214	.6021	.6851	.7348
10	.4973	.5760	.6581	.7079
11	.4762	.5529	.6339	.6835
12	.4575	.5324	.6120	.6614
13	.4409	.5139	.5923	.6411
14	.4259	.4973	.5742	.6226
15	.4124	.4821	.5577	.6055
16	.4000	.4683	.5425	.5897
17	.3887	.4555	.5285	.5751
18	.3783	.4438	.5155	.5614
19	.3687	.4329	.5034	.5487
20	.3598	.4227	.4921	.5368
25	.3233	.3809	.4451	.4869
30	.2960	.3494	.4093	.4487
35	.2746	.3246	.3810	.4182
40	.2573	.3044	.3578	.3932
45	.2428	.2875	.3384	.3721
50	.2306	.2732	.3218	.3541
60	.2108	.2500	.2948	.3248
70	.1954	.2319	.2737	.3017
80	.1829	.2172	.2565	.2830
90	.1726	.2050	.2422	.2673
100	.1638	.1946	.2301	.2540

Adapted from Table VII of R. A. Fisher & F. Yates, *Statistical Tables for Biological, Agricultural, and Medical Research*, Sixth Edition, published by Longman Group UK, Ltd., (1974). Adapted with permission.

TABLE A.9 Critical values of r_S, the Spearman rank-order correlation coefficient

For a description in the text, see page 394.

AN EXAMPLE USING TABLE A.9

If we let $\alpha = .05$, the test is nondirectional, and there are $n = 20$ pairs of scores, then $r_{S_{cv}} = \pm.450$. If $r_{S_{obs}}$ is greater than or equal to (in absolute value) $r_{S_{cv}}$, then $r_{S_{obs}}$ is statistically significant.

	Level of significance (α) for directional (one-tailed) test			
	.05	.025	.01	.005
	Level of significance (α) for nondirectional (two-tailed) test			
n	.10	.05	.02	.01
5	.900	—	—	—
6	.829	.886	.943	—
7	.714	.786	.893	.929
8	.643	.738	.833	.881
9	.600	.700	.783	.833
10	.564	.648	.745	.794
11	.536	.618	.709	.818
12	.497	.591	.703	.780
13	.475	.566	.673	.745
14	.457	.545	.646	.716
15	.441	.525	.623	.689
16	.425	.507	.601	.666
17	.412	.490	.582	.645
18	.399	.476	.564	.625
19	.388	.462	.549	.608
20	.377	.450	.534	.591
21	.368	.438	.521	.576
22	.359	.428	.508	.562
23	.351	.418	.496	.549
24	.343	.409	.485	.537
25	.336	.400	.475	.526
26	.329	.392	.465	.515
27	.323	.385	.456	.505
28	.317	.377	.448	.496
29	.311	.370	.440	.487
30	.305	.364	.432	.478

TABLE A.10 2000 Random digits

For a description in the text, see page 128.

AN EXAMPLE USING TABLE A.10

A "seed" digit is chosen by some quasi-random procedure (such as closing your eyes and pointing at the page). The digits immediately following the seed are a random sequence.

	1	*6*	*11*	*16*	*21*	*26*	*31*	*36*	*41*	*46*
1	10480	15011	01536	02011	81647	91646	69179	14194	62590	36207
2	22368	46573	25595	85393	30995	89198	27982	53402	93965	34095
3	24130	48360	22527	97265	76393	64809	15179	24830	49340	32081
4	42167	93093	06243	61680	07856	16376	39440	53537	71341	57004
5	37570	39975	81837	16656	06121	91782	60468	81305	49684	60672
6	77921	06907	11008	42751	27756	53498	18602	70659	90655	15053
7	99562	72905	56420	69994	98872	31016	71194	18738	44013	48840
8	93601	91977	05463	07972	18876	20922	94595	56869	69014	60045
9	89579	14342	63661	10281	17453	18103	57740	84378	25331	12566
10	85475	36857	43342	53988	53060	59533	38867	62300	08158	17983
11	28918	69578	88231	33276	70997	79936	56865	05859	90106	31595
12	63553	40961	48235	03427	49626	69445	18663	72695	52180	20847
13	09429	93969	52636	92737	88974	33488	36320	17617	30015	08272
14	10365	61129	87529	85689	48237	52267	67689	93394	01511	26358
15	07119	97336	71048	08178	77233	13916	47564	81056	97735	85977
16	51085	12765	51821	51259	77452	16308	60756	92144	49442	53900
17	02368	21382	52404	60268	89368	19885	55322	44819	01188	65255
18	01011	54092	33362	94904	31273	04146	18594	29852	71585	85030
19	52162	53916	46369	58586	23216	14513	83149	98736	23495	64350
20	07056	97628	33787	09998	42698	06691	76988	13602	51851	46104
21	48663	91245	85828	14346	09172	30168	90229	04734	59193	22178
22	54164	58492	22421	74103	47070	25306	76468	26384	58151	06646
23	32639	32363	05597	24200	13363	38005	94342	28728	35806	06912
24	29334	27001	87637	87308	58731	00256	45834	15398	46557	41135
25	02488	33062	28834	07351	19731	92420	60952	61280	50001	67658
26	81525	72295	04839	96423	24878	82651	66566	14788	76797	14780
27	29676	20591	68086	26432	46901	20849	89768	81536	86645	12659
28	00742	57392	39064	66432	84673	40027	32832	61362	98947	96067
29	05366	04213	25669	26422	44407	44048	37937	63904	45766	66134
30	91921	26418	64117	94305	26766	25940	39972	22209	71500	64568
31	00582	04711	87917	77341	42206	35126	74087	99547	81817	42607
32	00725	69884	62797	56170	86324	88072	76222	36086	84637	93161
33	69011	65797	95876	55293	18988	27354	26575	08625	40801	59920
34	25976	57948	29888	88604	67917	48708	18912	82271	65424	69774
35	09763	83473	73577	12908	30883	18317	28290	35797	05998	41688
36	91567	42595	37958	30134	04024	86385	29880	99730	55536	84855
37	17955	56349	90999	49127	20044	59931	06115	20542	18059	02008
38	46503	18584	18845	49618	02304	51038	20655	58727	28168	15475
39	92157	89634	94824	78171	84610	82834	09922	25417	44137	48413
40	14577	62765	35605	81263	39667	47358	56873	56307	61607	49518

APPENDIX

Review of Basic Arithmetic

Equality

$=$	Is equal to
$3 = 3$	Three equals three.
$3 = +3$	Three equals positive three.
$-4 = -4$	Negative four equals negative four.
$-4 = -(4)$	Negative four equals negative the quantity four.
$X = X$	X equals X.
$X = 4$	X equals 4.
$\approx$	Is approximately equal to
$3.1 \approx 3.2$	3.1 is approximately equal to 3.2.

Not Equality

$\neq$	Is not equal to
$4 \neq 3$	4 is not equal to 3.
$4 \neq -4$	4 is not equal to negative 4.
$X \neq -X$	X is not equal to negative X.

Inequality

$>$	Is greater than
$4 > 3$	4 is greater than 3.
$4 > -4$	4 is greater than negative 4.
$a > b$	a is greater than b.
$<$	Is less than
$5 < 7$	5 is less than 7.
$-1 < 0$	Negative 1 is less than 0.
$X < 14$	X is less than 14.
$\geq$	Is greater than or equal to
$4 \geq 3$	4 is greater than or equal to 3.
$4 \geq 4$	4 is greater than or equal to 4.
$7 \geq 0$	7 is greater than or equal to 0.
$\leq$	Is less than or equal to

$2 \leq 5$	2 is less than or equal to 5.
$X \leq 0$	X is less than or equal to 0.
$-7 \leq +7$	Negative 7 is less than or equal to positive 7.
$0 < 3 < 5$	3 is greater than 0 and less than 5.
$5 > 3 > 0$	3 is less than 5 and greater than 0 (a true statement, but we prefer the order $0 < 3 < 5$).
$4 < X \leq 10$	X is greater than 4 and less than or equal to 10.

Absolute Value

$\lvert X \rvert$	The absolute value of X is the value of the number with the negative sign (if any) removed.
$\lvert 7 \rvert = 7$	The absolute value of 7 is 7.
$\lvert -2 \rvert = 2$	The absolute value of negative 2 is 2.
$\lvert +6 \rvert = 6$	The absolute value of positive 6 is 6.

Addition and Subtraction

$3 + 5 = 8$	3 plus 5 equals 8.
$5 - 4 = 1$	5 minus 4 equals 1.
$3 + 5 = 5 + 3$	The order in which numbers are added does not change the result.
$3 + 7 + 2 = 7 + 2 + 3$	The order in which numbers are added does not change the result.
$5 + (-3) = 5 - 3 = 2$	Adding a negative number is the same as subtracting the same positive number.
$5 + 0 = 5$	Anything plus 0 equals itself.
$4 - 0 = 4$	Anything minus 0 equals itself.

Multiplication and Division

$2(3) = 6$	2 times 3 equals 6.
$3(2) = 2(3) = 6$	The order in which terms are multiplied is not important.
$7(1) = 7$	Anything times 1 equals itself.
$5(0) = 0$	Anything times 0 equals 0.
$2(-3) = -6$	A positive number times a negative number is always negative.
$\frac{6}{2} = 6/2 = 3$	6 divided by 2 equals 3.
$\frac{6}{1} = 6/1 = 6$	Anything divided by 1 equals itself.
$\frac{-6}{2} = (-6)/2 = -3$	The result of a negative number divided by a positive number is negative.
$\frac{a}{b}$	The upper portion of a fraction (here, a) is called the *numerator*; the lower portion (here, b) is called the *denominator*.
$\frac{6}{-3} = 6/(-3) = -2$	The result of a positive number divided by a negative number is negative; if *either* the numerator *or* the denominator (but not both) is negative, the result is negative.
$\frac{-6}{-3} = (-6)/(-3) = 2$	If both numerator and denominator are negative, the result is positive.

$$\frac{0}{6} = 0/6 = 0$$

Zero divided by anything is 0.

$$\frac{6}{0} = 6/0 = \infty$$

Anything divided by 0 is infinite.

$$\frac{0}{0} = 0/0 \text{ (undefined)}$$

The result of dividing 0 by 0 is undefined.

$$(-6)(-2)(-3) = -36$$

When a series of numbers are multiplied, if the total number of negative terms is *odd*, the result is negative.

$$(-6)(2)(-3) = 36$$

When a series of numbers are multiplied, if the total number of negative terms is *even*, the result is positive.

$$(-a)(-b)(-c) = -abc$$

$$a(-b)(-c) = abc$$

The previous two rules hold for both variables and numbers.

$$\frac{(-6)(-3)}{-2} = (-6)(-3)/(-2) = -9$$

When a series of numbers are divided, the same rules apply: If the total number of negative signs is *odd*, the result is negative; if the number of negative signs is *even*, the result is positive.

Exponentiation

$$4^2 = 4(4) = 16$$

4 squared (or 4 to the second power) equals 16.

$$a^2 = a(a)$$

Any number squared is that number times itself.

$$5^1 = 5$$

Any number to the first power is that number itself.

$$5^0 = 1$$

Any number to the zeroth power equals 1.

$$(-a)^2 = (-a)(-a) = a^2$$

A negative number squared is positive.

$$(10^3)(10^2) = 10^{3+2} = 10^5$$

If two exponential quantities that have the same base are multiplied, the result can be obtained by adding the exponents.

$$\frac{10^3}{10^2} = 10^{3-2} = 10^1 = 10$$

If two exponential quantities that have the same base are divided, the result can be obtained by subtracting the exponents.

$$10^3 = 1000$$

10 raised to the third power is 1 followed by three zeros.

$$10^5 = 100{,}000$$

10 raised to the fifth power is 1 followed by five zeros.

$$\sqrt{}$$

The square root of

$$\sqrt{4} = \pm 2$$

The square root of 4 is either +2 or −2.

$$\sqrt{4}(\sqrt{4}) = 4$$

The definition of the square root: The square root of a number is that value that multiplied by itself gives the original number.

$$4^{1/2} = \sqrt{4} = \pm 2$$

The square root can be written as an exponent of 1/2.

$$4^{1/2}(4^{1/2}) = 4^1 = 4$$

This follows from the rule that states that multiplication is the addition of exponents, and also from the definition of the square root.

$$(10^3)^2 = 10^{3(2)} = 10^6 = 1{,}000{,}000$$

$$(\sqrt{3})^2 = (3^{1/2})^2 = 3^{1/2(1)} = 3^1 = 3$$

When a value with an exponent is itself raised to a power, the exponents are multiplied.

$\sqrt{-4}$ is undefined

The square root of any negative number is not defined (in the real number system); if your computation results in taking the square root of a negative number, you have made a mistake somewhere.

Fractions

$$\frac{a}{b}$$

The upper portion of a fraction (here, a) is called the *numerator*; the lower portion (here, b) is called the *denominator*.

$$\frac{2}{5} + \frac{3}{5} = \frac{5}{5} = 1$$

When fractions that are to be added have the same denominator, add the numerators and divide the sum by the denominator.

$\frac{1}{3} + \frac{1}{2} = \frac{2}{6} + \frac{3}{6} = \frac{5}{6}$

When fractions that are to be added do not have the same denominator, they must be converted to fractions that do have the same ("common") denominator and then added.

$\frac{2}{7}\left(\frac{3}{5}\right) = \frac{6}{35}$

When fractions are to be multiplied, multiply the numerators together and then divide the product by the product of the denominators.

$\frac{4}{7} = .57143$

Fractions may be converted to their decimal equivalents by dividing.

$.57143 = .57$

Sometimes numbers are rounded (here, to two decimal places) by procedures described in Chapter 2.

$.57$ is $.57(100) = 57\%$

To convert a decimal fraction to a percent, multiply by 100.

$\frac{2}{3}\left(\frac{3}{5}\right) = \frac{2}{5}$

When the same factor appears in both the numerator and the denominator of two fractions that are being multiplied, they may be canceled. Here, the 3's disappear.

Order of Operations

$2 + 3 = 3 + 2$

The order of addition does not matter.

$2(3) = 3(2)$

The order of multiplication does not matter.

$2(3 + 4) = 2(7) = 14$

When parentheses (or brackets) are indicated, operations within the parentheses must be performed first.

$2(3) + 4 = 6 + 4 = 10$

$2^3(4) = 8(4) = 32$

$4/2 - 5^2 = 4/2 - 25 = 2 - 25 = -23$

$2 + 3(4) = 2 + 12 = 14$

When the order of operations is ambiguous, the following sequence must be followed: (1) exponentiation, (2) multiplication or division, (3) addition or subtraction.

Parentheses

$2(3 + 4) = 2(7) = 14$

When parentheses (or brackets) are indicated, operations within the parentheses must be performed first.

$2(3 + 4 + 5) = 2(3) + 2(4) + 2(5) = 24$

$a(b + c + d) = ab + ac + ad$

When a sum contained in parentheses is multiplied by a factor, the factor must be multiplied by all the terms in the sum.

$2[(3)(4)(5)] = (2)(3)(4)(5) = 120$

When a product contained in parentheses is multiplied by a factor, the factor multiplies the product.

$a(b - c) = ab - ac$

$(a + b)^2 = (a + b)(a + b) = a^2 + 2ab + b^2$

$(a + b)(a - b) = a^2 - b^2$

Equations: Solving for *Y*

$$\begin{aligned} Y - 3 &= 4 \\ Y &= 4 + 3 \\ Y &= 7 \end{aligned}$$

We can add the same value (here, 3) to both sides of an equation without altering the equality.

$$\begin{aligned} Y + 6 &= 12 \\ Y &= 12 - 6 \\ Y &= 6 \end{aligned}$$

We can subtract the same value (here, 6) from both sides of an equation without altering the equality.

$$\begin{aligned} \frac{Y}{3} &= 5 \\ Y &= 5(3) \\ Y &= 15 \end{aligned}$$

We can multiply both sides of an equation by the same value (here, 3) without altering the equality.

$$2Y = 10$$
$$Y = \frac{10}{2}$$
$$Y = 5$$

We can divide both sides of an equation by the same value (here, 2) without altering the equality.

$$2Y - 4 = 6$$
$$2Y = 6 + 4$$
$$2Y = 10$$
$$Y = \frac{10}{2}$$
$$Y = 5$$

We can perform any sequence of the above four rules without altering the equality as long as we perform the same operation (addition, subtraction, multiplication, or division) on both sides of the equal sign.

$$\frac{5}{3Y + 2} = 4$$
$$5 = 4(3Y + 2)$$
$$1.25 = 3Y + 2$$
$$-.75 = 3Y$$
$$-.25 = Y$$
or $Y = -.25$

Self-test for Arithmetic (answers follow)

In questions 1–20, answer true or false.

Equality

1. $7 = 7$
2. $31 = +31$
3. $+(27) = (+27)$
4. $+2 = -2$
5. $-X = -(X)$

Not Equality

6. $+3 \neq +(3)$
7. $-5 \neq 5$
8. $247 \neq 248$
9. $a \neq a$

Inequality

10. $10 > 9$
11. $237 > 1$
12. $3 < 2$
13. $-4 > -3$
14. $-21 < 21$
15. $5 \leq 5$
16. $10 \geq -9$
17. $-412 < -399 \leq -4$

Absolute Value

18. $|4| = 4$
19. $|-21| = -|21|$
20. $|-39| = +39$

What is the value of X?

Addition and Subtraction

21. $X = 214 + 5$
22. $214 + (-3) = X$
23. $X = -47 + 4$
24. $33 + 0 = X$

Multiplication and Division

25. $X = 3(4)$
26. $14(1) = X$
27. $14(0) = X$
28. $X = 6(-3)$
29. $X = (5)(-5)$
30. $X = 229/229$
31. $X = 25/2$
32. $\frac{-4}{2} = X$

33. $X = \frac{12}{-3}$

34. $X = 234/1$

35. $X = \frac{234}{0}$

36. $X = (-a)(-b)(c)$

37. $X = \frac{-2(24)}{2}$

Exponentiation

38. $3^2 = X$

39. $X = 10^3$

40. $X = 7^1$

41. $X = 29^0$

42. $X = (-4)^2$

43. $X = \sqrt{9}$

44. $\sqrt{-3}$

45. $X = 16^{1/2}$

46. $X = 4^{-2}$

47. $X = (10^2)(10^3)$

48. $X = (64^{1/2})(64^{1/2})$

49. $X = (211^{1/2})^2$

50. $10^3(10^3) = X$

51. $10^3 + 10^3 = X$

52. $X = \sqrt{-9}$

Fractions

53. $X = \frac{12}{6}$

54. $X = \frac{1}{2} + \frac{1}{4}$

55. $X = \frac{4}{9}\left(\frac{9}{4}\right)$

56. $X = \frac{1}{3} + \frac{1}{9} - \frac{1}{27}$

Order of Operations

57. $X = 5 + 2(4)$

58. $X = 5(2) + 4$

59. $X = 4^2/2 + 3$

60. $X = (3 + 2)^2 + 2$

Parentheses

61. $X = 14(3 - 2)$

62. $X = (3 + 2)(3 - 2)$

63. $X = (1 + 21)(2 + 2)^2$

In questions 64–66, answer true or false.

64. $(5 + 4)^2 = (5 + 4)(5 + 4) = 25 + 2(20) + 16$

65. $(5 - 4)^2 = (5 - 4)(5 - 4) = 25 - 2(20) + 16$

66. $(5 + 4)(5 - 4) = 25 - 16$

Equations: Solve for *Y*

67. $Y + 1 = 5$

68. $2Y = 16$

69. $3Y + 12 = 24$

70. $\frac{Y}{5} = 30$

71. $\frac{10}{4Y - 3} = 5$

72. $Y^2 = 9$

73. $2Y + \frac{3Y - 2}{2} = 10$

74. $Y^2(Y^{-1}) - 5 = 0$

75. $3Y - 8 = 1$

Answers to Self-test for Arithmetic

1. True	2. True	3. True	4. False
5. True	6. False	7. True	8. True
9. False	10. True	11. True	12. False
13. False	14. True	15. True	16. True
17. True	18. True	19. False	20. True
21. 219	22. 211	23. -43	24. 33
25. 12	26. 14	27. 0	28. -18
29. -25	30. 1	31. 12.5	32. -2
33. -4	34. 234	35. ∞	36. abc
37. -24	38. 9	39. 1000	40. 7

41. 1
42. 16
43. ± 3
44. Undefined
45. ± 4
46. $1/16$
47. 100,000
48. 64
49. 211
50. 1,000,000
51. 2000
52. Undefined
53. 2
54. $3/4 = .75$
55. 1
56. $11/27 = .407$
57. 13
58. 14
59. 11
60. 27
61. 14
62. 5
63. 352
64. True
65. True
66. True
67. 4
68. 8
69. 4
70. 150
71. $5/4 = 1.25$
72. $+3$ or -3
73. $22/7 = 3.143$
74. 5
75. 3

APPENDIX

Summary of Statistical Formulas Used in This Text

This appendix lists all statistical formulas used in this text, along with the equation number and page number where the formula first appeared.

Description	Formula	Equation number	Page
Probability of an event	$P(E) = \frac{\text{number of outcomes favorable to } E}{\text{total number of possible outcomes}}$	(1.1)	9
Sum of values of a variable	$\sum X_i = X_1 + X_2 + X_3 + X_4 + X_5 + X_6$	(2.1)	18
Mean of a sample	$\overline{X} = \frac{\sum X_i}{n}$	(4.1)	56
Mean of a sample computed from a frequency distribution	$\overline{X} = \frac{\sum fX}{n}$	(4.1a)	66
Mean of a population	$\mu = \frac{\sum X_i}{N}$	(4.2)	59
Mean of a population computed from a frequency distribution	$\mu = \frac{\sum fX}{N}$	(4.2a)	66
Approximate mean of a sample computed from a grouped frequency distribution	$\overline{X} \approx \frac{\sum fX_{\text{mid}}}{n}$	(4.3)	67
Approximate mean of a population computed from a grouped frequency distribution	$\mu \approx \frac{\sum fX_{\text{mid}}}{N}$	(4.4)	67
Range	Range = highest value − lowest value	(5.1)	71
Deviation in a sample	$\text{deviation}_i = X_i - \overline{X}$	(5.2)	73
Deviation in a population	$\text{deviation}_i = X_i - \mu$	(5.3)	73
Standard deviation of a sample (mean squared deviation formula)	$s = \sqrt{\frac{\sum(X_i - \overline{X})^2}{n-1}}$	(5.4)	73
Standard deviation of a sample (computational formula)	$s = \sqrt{\frac{\sum X_i^2 - \frac{(\sum X_i)^2}{n}}{n-1}}$	(5.4a)	82
Standard deviation of a sample (computational formula for data in a frequency distribution)	$s = \sqrt{\frac{\sum fX^2 - \frac{(\sum fX)^2}{n}}{n-1}}$	(5.4b)	92
Standard deviation of a population (mean squared deviation formula)	$\sigma = \sqrt{\frac{\sum(X_i - \mu)^2}{N}}$	(5.5)	73

Description	Formula	Equation number	Page
Standard deviation of a population (computational formula)	$\sigma = \sqrt{\dfrac{\sum X_i^2 - \dfrac{(\sum X_i)^2}{N}}{N}}$	(5.5a)	82
Standard deviation of a population (computational formula for data in a frequency distribution)	$\sigma = \sqrt{\dfrac{\sum fX^2 - \dfrac{(\sum fX)^2}{N}}{N}}$	(5.5b)	92
Variance of a sample (mean squared deviation formula)	$s^2 = \dfrac{\sum(X_i - \bar{X})^2}{n-1}$	(5.6)	83
Variance of a sample (computational formula)	$s^2 = \dfrac{\sum X_i^2 - \dfrac{(\sum X_i)^2}{n}}{n-1}$	(5.6a)	83
Variance of a sample (computational formula for data in a frequency distribution)	$s^2 = \dfrac{\sum fX^2 - \dfrac{(\sum fX)^2}{n}}{n-1}$	(5.6b)	92
Variance of a population (mean squared deviation formula)	$\sigma^2 = \dfrac{\sum(X_i - \mu)^2}{N}$	(5.7)	83
Variance of a population (computational formula)	$\sigma^2 = \dfrac{\sum X_i^2 - \dfrac{(\sum X_i)^2}{N}}{N}$	(5.7a)	83
Variance of a population (computational formula for data in a frequency distribution)	$\sigma^2 = \dfrac{\sum fX^2 - \dfrac{(\sum fX)^2}{N}}{N}$	(5.7b)	92
Approximate standard deviation of a sample (computational formula for data in a grouped frequency distribution)	$s \approx \sqrt{\dfrac{\sum fX_{\text{mid}}^2 - \dfrac{(\sum fX_{\text{mid}})^2}{n}}{n-1}}$	(5.8)	93
Approximate standard deviation of a population (computational formula for data in a grouped frequency distribution)	$\sigma \approx \sqrt{\dfrac{\sum fX_{\text{mid}}^2 - \dfrac{(\sum fX_{\text{mid}})^2}{N}}{N}}$	(5.9)	93
Approximate variance of a sample (computational formula for data in a grouped frequency distribution)	$s \approx \dfrac{\sum fX_{\text{mid}}^2 - \dfrac{(\sum fX_{\text{mid}})^2}{n}}{n-1}$	(5.10)	93
Approximate variance of a population (computational formula for data in a grouped frequency distribution)	$\sigma^2 \approx \dfrac{\sum fX_{\text{mid}}^2 - \dfrac{(\sum fX_{\text{mid}})^2}{N}}{N}$	(5.11)	93
Normal distribution with mean μ and standard deviation σ	$f = \dfrac{1}{\sigma\sqrt{2\pi}} e^{-(X-\mu)^2/2\sigma^2}$	Box 6.1	102
Transformation from a raw score to a standard score in a population	$z = \dfrac{X - \mu}{\sigma}$	(6.1)	97
Transformation from a raw score to a standard score in a sample	$z = \dfrac{X - \bar{X}}{s}$	(6.2)	97
Transformation from a z score to a standardized score in a population	$X = z\sigma + \mu$	(6.3)	111
Transformation from a z score to a standardized score in a sample	$X = zs + \bar{X}$	(6.4)	113
Transformation from a standardized score to a z score in a sample	$z = \dfrac{X - \bar{X}}{s}$	(6.5)	113
Standard error of the mean in a population	$\sigma_{\bar{X}} = \dfrac{\sigma}{\sqrt{n}}$	(7.1)	134
Standard score in the distribution of means in a population	$z = \dfrac{\bar{X} - \mu}{\sigma_{\bar{X}}}$	(7.2)	136

Description	Formula	Equation number	Page
Confidence interval for the population mean when σ is known	$\overline{X} - z_{cv}(\sigma_{\overline{X}}) < \mu < \overline{X} + z_{cv}(\sigma_{\overline{X}})$	(8.1)	147
Standard crror of the mean in a sample	$s_{\overline{X}} = \dfrac{s}{\sqrt{n}}$	(8.2)	153
Confidence interval for the population mean when σ is unknown	$\overline{X} - t_{cv}(s_{\overline{X}}) < \mu < \overline{X} + t_{cv}(s_{\overline{X}})$ $[df = n - 1]$	(8.3)	155
Standard error of a proportion	$s_p = \sqrt{\dfrac{pq}{n-1}}$	(8.4)	157
Confidence interval for a population proportion	$p - t_{cv}(s_p) < \Pi < p + t_{cv}(s_p)$	(8.5)	157
General formula for the test statistic	$\text{Test statistic} = \dfrac{\text{sample statistic} - \text{population parameter}}{\text{standard error of the sample statistic}}$	(9.1)	175
Test statistic (z), one-sample test for the mean when σ is known	$z = \dfrac{\overline{X} - \mu}{\sigma_{\overline{X}}}$	(10.1)	185
Critical value of the sample mean when σ is known	$\overline{X}_{cv} = \mu \pm z_{cv}(\sigma_{\overline{X}})$	(10.2)	187
Effect size index d, one-sample test for the mean when σ is known	$d = \dfrac{\lvert\overline{X}_{obs} - \mu\rvert}{\sigma}$ (nondirectional test) or $d = \dfrac{\overline{X}_{obs} - \mu}{\sigma}$ (directional test)	(10.3)	192
Test statistic (t), one-sample test for the mean when σ is unknown	$t = \dfrac{\overline{X} - \mu}{s_{\overline{X}}}$ $[df = n - 1]$	(10.4)	194
Critical value of the sample mean when σ is unknown	$\overline{X}_{cv} = \mu \pm t_{cv}(s_{\overline{X}})$	(10.5)	194
Effect size index d, one-sample test for the mean when σ is unknown	$d = \dfrac{\lvert\overline{X}_{obs} - \mu\rvert}{s}$ (nondirectional test) or $d = \dfrac{\overline{X}_{obs} - \mu}{s}$ (directional test)	(10.6)	194
Null hypothesis, two-independent-samples test for means (nondirectional test)	$H_0\text{: } \mu_1 = \mu_2$	(11.1)	215
Null hypothesis, two-independent-samples test for means (nondirectional test), alternative form	$H_0\text{: } \mu_1 - \mu_2 = 0$	(11.1a)	217
Test statistic (t), two-independent-samples test for means	$t = \dfrac{\overline{X}_1 - \overline{X}_2}{s_{\overline{X}_1 - \overline{X}_2}}$ $[df = n_1 + n_2 - 2]$	(11.2)	218
Standard error of the difference between two means	$s_{\overline{X}_1 - \overline{X}_2} = \sqrt{s^2_{pooled}\left(\dfrac{1}{n_1} + \dfrac{1}{n_2}\right)}$	(11.3)	218
Pooled variance when sample sizes are equal	$s^2_{pooled} = \dfrac{s_1^2 + s_2^2}{2}$ [only when $n_1 = n_2$]	(11.4)	219
Pooled variance for cqual or unequal sample sizes	$s^2_{pooled} = \dfrac{df_1 s_1^2 + df_2 s_2^2}{df_1 + df_2}$	(11.5)	219
Pooled variance for equal or unequal sample sizes	$s^2_{pooled} = \dfrac{(n_1 - 1)s_1^2 + (n_2 - 1)s_2^2}{n_1 + n_2 - 2}$	(11.5a)	220
Standard error of the difference between two means when sample sizes are equal	$s_{\overline{X}_1 - \overline{X}_2} = \sqrt{s^2_{pooled}\dfrac{2}{n}}$ [only when $n_1 = n_2 = n$]	(11.6)	220
Standard error of the difference between two means when sample sizes are equal, alternative form	$s_{\overline{X}_1 - \overline{X}_2} = \left(\dfrac{s_{pooled}}{\sqrt{n}}\right)\sqrt{2}$ [only when $n_1 = n_2 = n$]	(11.6a)	220

Description	Formula	Equation number	Page
Effect size index d, two-independent-samples test for means	$d = \frac{\lvert(\bar{X}_1 - \bar{X}_2)_{\text{obs}}\rvert}{s_{\text{pooled}}}$ (nondirectional test) or $d = \frac{(\bar{X}_1 - \bar{X}_2)_{\text{obs}}}{s_{\text{pooled}}}$ (directional test)	(11.7)	224
Difference score for the ith individual	$D_i = X_{i1} - X_{i2}$	(12.1)	246
Null hypothesis, two-dependent-samples test for means	$H_0\colon \mu_D = 0$ (nondirectional test) or $H_0\colon \mu_D \leq 0$ or $H_0\colon \mu_D \geq 0$ (directional test)	(12.2)	249
Test statistic (t), two-dependent-samples test for means	$t = \frac{\bar{D}}{s_{\bar{D}}}$ $[df = n - 1]$	(12.3)	249
Standard error of the mean differences	$s_{\bar{D}} = \frac{s_D}{\sqrt{n}}$ [where n is number of *pairs*]	(12.4)	249
Standard deviation of the differences, mean squared deviation and computational forms	$s_D = \sqrt{\frac{\sum(D_i - \bar{D})^2}{n-1}} = \sqrt{\frac{\sum D_i^2 - \frac{(\sum D_i)^2}{n}}{n-1}}$	(12.5)	249
Effect size index d, two-dependent-samples test for means	$d = \frac{\lvert\bar{D}_{\text{obs}}\rvert}{s_D}$ (nondirectional test) or $d = \frac{\bar{D}_{\text{obs}}}{s_D}$ (directional test)	(12.6)	253
Power	$\text{Power} = 1 - \beta$	(12.7)	258
Population effect size index, one-sample test	$d_{\text{population}} = \frac{\mu_{\text{real}} - \mu_{H_0}}{\sigma}$	(12.8)	266
Population effect size index, two-independent-samples test	$d_{\text{population}} = \frac{\mu_1 - \mu_2}{\sigma}$	(12.9)	267
Population effect size index, two-dependent-samples test	$d_{\text{population}} = \frac{\mu_{D_{\text{real}}}}{\sigma_D}$	(12.10)	267
Null hypothesis, analysis of variance, four groups	$H_0\colon \mu_1 = \mu_2 = \mu_3 = \mu_4$	(13.1)	284
Variance of the sample means, definitional formula	$s_{\bar{X}}^2 = \frac{\sum(\bar{X}_j - \bar{X}_G)^2}{k-1}$	(13.2)	288
Mean square between groups, definitional formula	$MS_B = s_{\bar{X}}^2(n)$ [only if $n_1 = n_2 = \cdots = n_k = n$]	(13.3)	289
Pooled variance, k equal-sized groups	$s_{\text{pooled}}^2 = \frac{s_1^2 + s_2^2 + s_3^2 + \cdots + s_k^2}{k}$ [only if equal n_j]	(13.4)	289
Mean square within groups, definitional formula	$MS_W = s_{\text{pooled}}^2$	(13.5)	290
Test statistic (F), analysis of variance	$F = \frac{\text{between-group point-estimate of } \sigma^2}{\text{within-group point-estimate of } \sigma^2} = \frac{MS_B}{MS_W}$	(13.6)	290
Degrees of freedom between groups (degrees of freedom for the numerator of F)	$df_B = k - 1$	(13.7)	291
Degrees of freedom within groups (degrees of freedom for the denominator of F)	$df_W = df_1 + df_2 + \cdots + df_k$	(13.8a)	291
Degrees of freedom within groups, alternative form	$df_W = (n_1 - 1) + (n_2 - 1) + \cdots + (n_k - 1)$	(13.8b)	291
Degrees of freedom within groups, alternative form	$df_W = n_G - k$	(13.8c)	291
Total sum of squares, deviation formula	$SS_T = \sum(X_{ij} - \bar{X}_G)^2$	(13.9)	294
Partitioning the total sum of squares	$SS_T = \sum(X_{ij} - \bar{X}_j)^2 + \sum(\bar{X}_j - \bar{X}_G)^2$	(13.10a)	294

Description	Formula	Equation number	Page
Partitioning the total sum of squares	$SS_T = SS_B + SS_W$	(13.10b)	295
Sum of squares within groups, deviation formula	$SS_W = \sum(X_{ij} - \bar{X}_j)^2$	(13.11)	295
Sum of squares between groups, deviation formula	$SS_B = \sum(\bar{X}_j - \bar{X}_G)^2$	(13.12)	295
Total degrees of freedom	$df_T = n_G - 1$	(13.13)	295
Partitioning the total degrees of freedom	$df_T = df_W + df_B$	(13.13a)	295
Mean square between groups	$MS_B = SS_B/df_B$	(13.14)	296
Mean square within groups	$MS_W = SS_W/df_W$	(13.15)	296
ANOVA effect size as the ratio of standard deviations	$f = \frac{s_{\bar{X}}}{s_{\text{pooled}}}$	(13.16)	299
ANOVA effect size (the ratio of standard deviations) when sample sizes are equal	$f = \sqrt{\frac{F}{n}}$ [only if all n_j equal n]	(13.16a)	299
ANOVA effect size by analogy to previous chapters	$d_M = \frac{\bar{X}_{\text{max}} - \bar{X}_{\text{min}}}{s_{\text{pooled}}}$	(13.17)	299
Total sum of squares, computational formula	$SS_T = \sum X_{ij}^2 - \frac{\left(\sum X_{ij}\right)^2}{n_G}$	(13.18)	320
Sum of squares between groups, computational formula	$SS_B = \left[\frac{(\sum X_{i1})^2}{n_1} + \frac{(\sum X_{i2})^2}{n_2} + \cdots + \frac{(\sum X_{ik})^2}{n_k}\right] - \frac{\left(\sum X_{ij}\right)^2}{n_G}$	(13.19)	320
Sum of squares within groups, computational formula	$SS_W = \sum X_{ij}^2 - \left[\frac{(\sum X_{i1})^2}{n_1} + \frac{(\sum X_{i2})^2}{n_2} + \cdots + \frac{(\sum X_{ik})^2}{n_k}\right]$	(13.20)	320
Null hypotheses for pairwise post hoc tests following analysis of variance with four groups	$H_{0_1}: \mu_1 = \mu_2$ $H_{0_2}: \mu_1 = \mu_3$ $H_{0_3}: \mu_1 = \mu_4$ $H_{0_4}: \mu_2 = \mu_3$ $H_{0_5}: \mu_2 = \mu_4$ $H_{0_6}: \mu_3 = \mu_4$	(14.1)	325
Studentized range statistic for Tukey's post hoc tests	$Q = \frac{\lvert\bar{X}_i - \bar{X}_j\rvert}{\sqrt{MS_W/\tilde{n}}}$	(14.2)	325
Harmonic mean of sample sizes	$\tilde{n} = \frac{k}{\frac{1}{n_1} + \frac{1}{n_2} + \frac{1}{n_3} + \cdots + \frac{1}{n_k}}$	(14.3)	325
Harmonic mean of sample sizes, equal n_j	$\tilde{n} = n_j$ [equal n_j]	(14.4)	325
Population comparison	$\sum C_j\mu_j$ (in general) or $C_1\mu_1 + C_2\mu_2 + C_3\mu_3 + C_4\mu_4$ (when there are four groups)	(14.5)	328
Sample comparison	$\sum C_j\bar{X}_j$ (in general) or $C_1\bar{X}_1 + C_2\bar{X}_2 + C_3\bar{X}_3 + C_4\bar{X}_4$ (when there are four groups)	(14.6)	328
Comparison coefficients must sum to 0	$\sum C_j = 0$ (in general) or $C_1 + C_2 + C_3 + C_4 = 0$ (when there are four groups)	(14.7)	329
Standard error of a comparison	$s_{\text{sample comparison}} = \sqrt{s_{\text{pooled}}^2\left(\frac{C_1^2}{n_1} + \frac{C_2^2}{n_2} + \cdots + \frac{C_k^2}{n_k}\right)}$	(14.8)	329
General form of the null hypothesis	H_0: population comparison = 0	(14.9)	329

Description	Formula	Equation number	Page
General form of the test statistic	$\text{test statistic} = \dfrac{\text{sample comparison} - \text{population comparison}}{\text{standard error of the sample comparison}}$	(14.10)	329
Partition of the total sum of squares in repeated-measures ANOVA	$SS_{total} = SS_{between\ occasions} + SS_{between\ subjects} + SS_{residual}$	(14.11)	337
Partition of the total degrees of freedom in repeated-measures ANOVA	$df_{total} = df_{between\ occasions} + df_{between\ subjects} + df_{residual} = nk - 1$	(14.12)	337
Mean square between subjects in repeated-measures ANOVA, definitional formula	$MS_{between\ subjects} = ks^2_{\bar{X}}$	(14.13)	337
Test statistic (F) for repeated-measures ANOVA	$F = \dfrac{MS_{between\ occasions}}{MS_{residual}}$	(14.14)	338
Null hypothesis for rows	$H_{0_R}: \mu_{row\ 1} = \mu_{row\ 2} = \cdots = \mu_{row\ R}$	(15.1)	353
Null hypothesis for columns	$H_{0_C}: \mu_{col\ 1} = \mu_{col\ 2} = \cdots = \mu_{col\ C}$	(15.2)	353
Null hypothesis for interaction	$H_{0_{RC}}: \text{all } \mu_{row,col} - \mu_{row} - \mu_{col} + \mu_G = 0$	(15.3)	357
Total sum of squares	$SS_T = \sum(X_{irc} - \bar{X}_G)^2$	(15.4)	359
Partitioning the sum of squares in two-way ANOVA	$SS_T = SS_W + SS_R + SS_C + SS_{RC}$	(15.5)	359
Partitioning the sum of squares in two-way ANOVA	$SS_T = \sum(X_{irc} - \bar{X}_{rc})^2 + \sum(\bar{X}_r - \bar{X}_G)^2 + \sum(\bar{X}_c - \bar{X}_G)^2 + \sum(\bar{X}_{rc} - \bar{X}_r - \bar{X}_c + \bar{X}_G)^2$	(15.6)	359
Sum of squares within cells	$SS_W = \sum(X_{irc} - \bar{X}_{rc})^2$	(15.7)	359
Sum of squares for rows	$SS_R = \sum(\bar{X}_r - \bar{X}_G)^2$	(15.8)	359
Sum of squares for columns	$SS_C = \sum(\bar{X}_c - \bar{X}_G)^2$	(15.9)	359
Sum of squares for interaction	$SS_{RC} = \sum(\bar{X}_{rc} - \bar{X}_r - \bar{X}_c + \bar{X}_G)^2$	(15.10)	361
Test statistic (F) for rows	$F_R = MS_R/MS_W$	(15.11)	361
Test statistic (F) for columns	$F_C = MS_C/MS_W$	(15.12)	361
Test statistic (F) for interaction	$F_{RC} = MS_{RC}/MS_W$	(15.13)	361
Pearson product-moment correlation coefficient for a sample, standard score formula	$r = \dfrac{\sum z_X z_Y}{n - 1}$	(16.1)	382
Pearson product-moment correlation coefficient for a sample, computational formula	$r = \dfrac{\sum XY - \dfrac{\sum X \sum Y}{n}}{\sqrt{\left[\sum X^2 - \dfrac{(\sum X)^2}{n}\right]\left[\sum Y^2 - \dfrac{(\sum Y)^2}{n}\right]}}$	(16.1a)	386
Pearson product-moment correlation coefficient for a population, standard score formula	$\rho = \dfrac{\sum z_X z_y}{N}$	(16.2)	382
Pearson product-moment correlation coefficient for a population, computational formula	$\rho = \dfrac{\sum XY - \dfrac{\sum X \sum Y}{N}}{\sqrt{\left[\sum X^2 - \dfrac{(\sum X)^2}{N}\right]\left[\sum Y^2 - \dfrac{(\sum Y)^2}{N}\right]}}$	(16.2a)	386
Null hypothesis, Pearson correlation	$H_0: \rho = 0$	(16.3)	389
Standard error of a Pearson correlation	$s_r = \sqrt{\dfrac{1 - r^2}{n - 2}}$	(16.4)	390

Description	Formula	Equation number	Page
Test statistic (t), Pearson correlation	$t = \frac{r}{s_r} \quad [df = n - 2]$	(16.5)	390
Spearman rank correlation coefficient	$r_s = 1 - \frac{6\sum D_i^2}{n(n^2 - 1)}$	(16.6)	392
Difference between ranks for the ith individual	$D_i = R_{X_i} - R_{Y_i}$	(16.7)	393
Regression line	$\hat{Y} = bX + a$	(17.1)	408
Predicted value of Y for the ith individual	$\hat{Y}_i = bX_i + a$	(17.2)	408
Slope of the regression line computed from the correlation coefficient	$b = r\frac{s_Y}{s_X}$	(17.3)	411
Slope of the regression line computed from raw scores	$b = \frac{\sum XY - \frac{\sum X \sum Y}{n}}{\sum X^2 - \frac{(\sum X)^2}{n}}$	(17.4)	411
Intercept of the regression line	$a = \overline{Y} - b\overline{X}$	(17.5)	411
Regression line expressed as standard scores	$\hat{z}_Y = rz_X$	(17.6)	417
Error of prediction for the ith individual	$e_i = Y_i - \hat{Y}_i$	(17.7)	419
Standard error of estimate as the standard deviation of the errors of estimate	$s_e = \sqrt{\frac{\sum(e_i - \bar{e})^2}{n - 2}}$	(17.8)	421
Standard error of estimate	$s_e = \sqrt{\frac{\sum(e_i)^2}{n - 2}}$	(17.9)	421
Standard error of estimate computed from the correlation coefficient	$s_e = s_Y\sqrt{1 - r^2}\sqrt{\frac{n - 1}{n - 2}}$	(17.10)	421
Total sum of squares	$SS_T = \sum(Y_i - \overline{Y})^2$	(17.11)	424
Partitioning the total sum of squares	$SS_T = \sum(Y_i - \hat{Y}_i)^2 + \sum(\hat{Y}_i - \overline{Y})^2$	(17.11a)	424
Error sum of squares	$SS_{error} = \sum(Y_i - \hat{Y}_i)^2$	(17.12)	424
Explained sum of squares	$SS_{explained} = \sum(\hat{Y}_i - \overline{Y})^2$	(17.13)	424
Partitioning the sum of squares	$SS_T = SS_{error} + SS_{explained}$	(17.14)	424
Coefficient of determination as the proportion of sum of squares that is explained	$r^2 = \frac{SS_{explained}}{SS_T} = \frac{SS_T - SS_{error}}{SS_T}$	(17.15)	425
Null hypothesis for slope of regression line	$H_0\colon b_{population} = 0$	(17.16)	428
χ^2 test statistic for goodness of fit test and test of independence	$\chi^2 = \sum\frac{(O_i - E_i)^2}{E_i} \quad [df = k - 1]$	(18.1)	444
Approximate critical value of χ^2 when $df > 30$	$\chi^2_{cv} = \frac{z^2_{cv}}{2} + z_{cv}\sqrt{2df - 1} + \frac{2df - 1}{2}$ [only if $df > 30$]	(18.2)	443
Expected frequency of cell in contingency table	Expected frequency of cell $rc = \frac{f_r f_c}{N}$	(18.3)	449
Mann–Whitney U statistic, first	$U_1 = n_1n_2 + \frac{n_1(n_1 + 1)}{2} - R_1$	(18.4)	453
Mann–Whitney U statistic, second	$U_2 = n_1n_2 + \frac{n_2(n_2 + 1)}{2} - R_2$	(18.5)	453

Description	Formula	Equation number	Page
Mean of Mann–Whitney U if $n_1 > 20$ and $n_2 > 20$	$\mu_U = \frac{n_1 n_2}{2}$ [if $n_1 > 20$ and $n_2 > 20$]	(18.6)	454
Standard deviation of Mann–Whitney U if $n_1 > 20$ and $n_2 > 20$	$\sigma_U = \sqrt{\frac{n_1 n_2 (n_1 + n_2 + 1)}{12}}$ [if $n_1 > 20$ and $n_2 > 20$]	(18.7)	454
Normal approximation to Mann–Whitney U if $n_1 > 20$ and $n_2 > 20$	$z_{obs} = \frac{U_{obs} - \mu_U}{\sigma_U}$ [if $n_1 > 20$ and $n_2 > 20$]	(18.8)	454
Kruskal–Wallis H statistic	$H = \frac{12}{N(N + 1)} \sum \frac{R_j^2}{n_j} - 3(N + 1)$	(18.9)	457

Answers to Selected Exercises

APPENDIX D

Chapter 1, Section A: Basic exercises

1. A statement whose truth can be assessed by collecting and analyzing data
2. (a) and (c)
3. False
4. The distribution of a variable, the sampling distribution of means, and the test statistic
5. People act in accordance with others' expectations. Experiments are based on pre-experimental observations, we must distinguish between samples and populations, and we must be able to deal with inconsistent data.
6. A sample is a subset of the population of interest. A population is all the members of the group that we are interested in.
7. Sample; my interest is in all UNLV students.
8. Population; my interest is in this class itself.
9. **(a)** $1/52 = .02$ **(b)** $1/52 = .02$ **(c)** $5/52 = .10$ **(d)** $20/52 = 5/13 = .38$
10. **(a)** $8/36 = .22$ **(b)** $4/36 = .11$
11. **(a)** $1/365 = .003$ **(b)** $31/365 = .085$ **(c)** $28/365 = .077$ **(d)** $52/365 = .142$ **(e)** $53/365 = .145$
12. **(a)** $4/52 = 1/13 = .08$ **(b)** $4/20 = 1/5 = .20$ **(c)** $1/13 = .08$

Chapter 2, Section A: Basic exercises

1. **(a)** Yes
   (b) No; the measurement operation must be the same for all subjects.
2. **(a)** Interval/ratio **(b)** Ordinal **(c)** Nominal **(d)** Ordinal **(e)** Nominal **(f)** Interval/ratio
3. **(a)** Continuous **(b)** Discrete **(c)** Discrete **(d)** Continuous
4. **(a)** −5 **(b)** 7 **(c)** 14 **(d)** 188 **(e)** 196 **(f)** No **(g)** 84 **(h)** No **(i)** 153

4.

i	X_i	X_i^2	$X_i + 10$	$X_i^2 - 5$
1	6	36	16	31
2	1	1	11	−4
3	9	81	19	76
4	−5	25	5	20
5	6	36	16	31
6	0	0	10	−5
7	−3	9	7	4
	$\sum X_i = 14$	$\sum X_i^2 = 188$	$\sum(X_i + 10) = 84$	$\sum(X_i^2 - 5) = 153$

5. **(a)** Five

(b)

X_i	X_i^2	$3X_i^2$	$3X_i^2 - 6$	$(3X_i^2 - 6)/4$
6	36	108	102	25.50
1	1	3	−3	−.75
9	81	243	237	59.25
−5	25	75	69	17.25
6	36	108	102	25.50
0	0	0	−6	−1.50
−3	9	27	21	5.25
				$\sum[(3X_i^2 - 6)/4] = 130.50$

6. **(a)** 6

(b)

X	Y	XY
3	2	6
4	4	16
7	1	7
3	11	33
		$\sum XY = 62$

7. **(a)** Seven

(b)

X	Y	$X - Y$	$(X - Y)^2$	$2(X - Y)^2$	$X + Y$	$\frac{2(X - Y)^2}{(X + Y)}$
3	2	1	1	2	5	.4
4	4	0	0	0	8	0
7	1	6	36	72	8	9
3	11	−8	64	128	14	9.1429
						$\sum[2(X - Y)^2/(X + Y)] = 18.5429$

Rounding, we report $\sum[2(X - Y)^2/(X + Y)] = 18.54$.

8. **(a)** $83 \neq 289$ **(b)** $37 \neq 22$
   (c) $353 \neq 484$ **(d)** $62 \neq 306$

9. **(a)** 2 **(b)** 5.67

10. **(a)** 14.64 **(b)** −14.64 **(c)** 2.15
    (d) 3.40 **(e)** 3.41 **(f)** 3.32
    (g) 3.32 **(h)** 3.34 **(i)** 27.43

Chapter 2, Section C: Extending comprehension

19. **(a)** 2.90 **(b)** 7.47
    (c) The square root of a sum is not equal to the sum of the square roots.

20. **(a)** 158 **(b)** 394 **(c)** 5264
    (d) 31,334 **(e)** 12,693
    (f) The numerator is $12{,}693 - [(158)(394)]/5 = 242.600$; the denominator is
    $\sqrt{[5264 - (158)^2/5][31{,}334 - (394)^2/5]} =$
    $\sqrt{[271.200][286.800]} = \sqrt{77{,}780.160} = 278.891.$
    The answer is therefore .87.

Chapter 2, Section D: Cumulative review

21. **(a)** .01 **(b)** 0 **(c)** .50 **(d)** .50
    (e) .26 **(f)** .09 **(g)** .19 **(h)** .09

22. .10

23. .02

Chapter 3, Section A: Basic exercises

1. **(a)** Frequency distribution

Distance ridden (mi)	f
22	2
21	1
20	1
17	2
16	3
15	5
14	1
13	1
12	1
11	1
9	1
8	1
	20

(b) Grouped frequency distribution

Distance ridden (mi)	f
22–23	2
20–21	2
18–19	0
16–17	5
14–15	6
12–13	2
10–11	1
8–9	2
	20

1. **(c)** Histogram

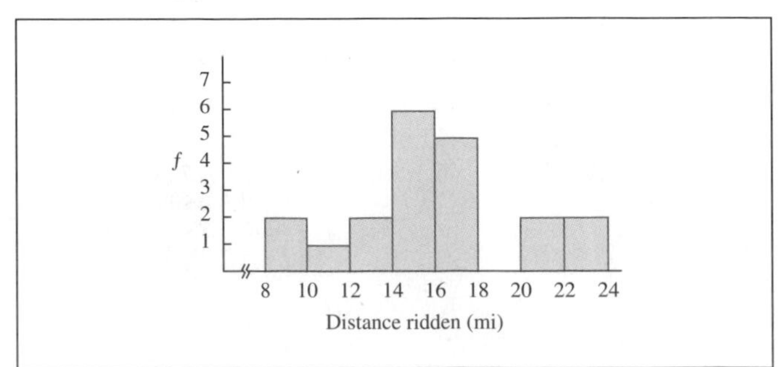

1. **(d)** Frequency polygon

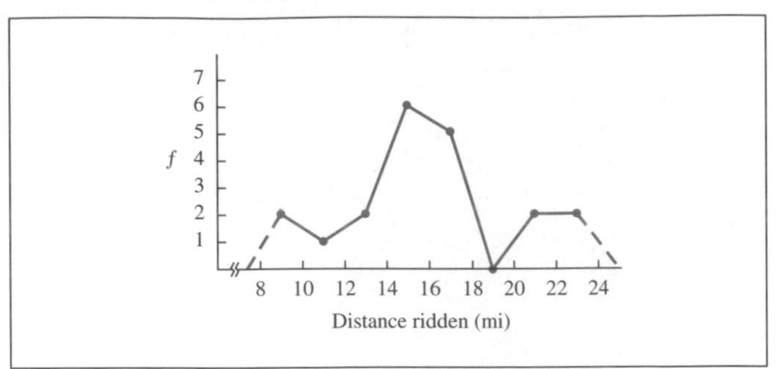

2. **(a)**

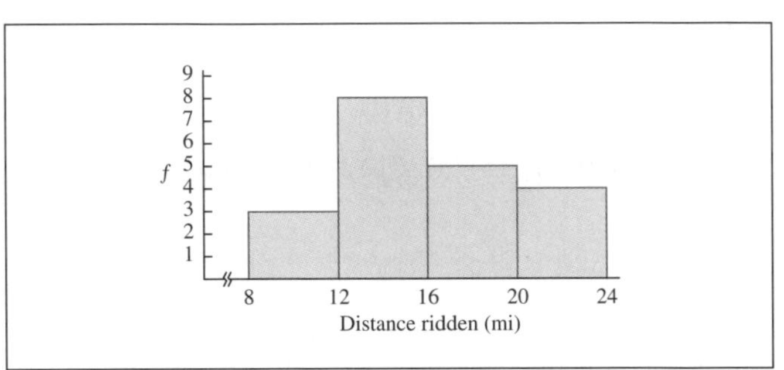

(b) Perhaps the figure with four groups is preferred because, given the small number of data points, the histogram with wider intervals makes the distribution appear smoother.

3. **(a)** Frequency distribution

Score on first exam	f
96	1
95	1
91	3
90	1
89	1
88	2
86	2
84	2
83	2
82	1
81	1
75	1
72	1
71	1
68	1
61	1
	22

(b) Grouped frequency distribution

Score on first exam	f
96–99	1
92–95	1
88–91	7
84–87	4
80–83	4
76–79	0
72–75	2
68–71	2
64–67	0
60–63	1
	22

(b) (*continued*) With 60 as the lowest lower limit, the groups break on even numbers, which is desirable. Furthermore, the groups correspond to letter grades if one uses a 92–84–76–68 system. Those choices are made on the basis of judgment and knowledge about the data, not on purely statistical grounds, so you may have chosen a different (equally defensible) grouping.

3. (c) Histogram

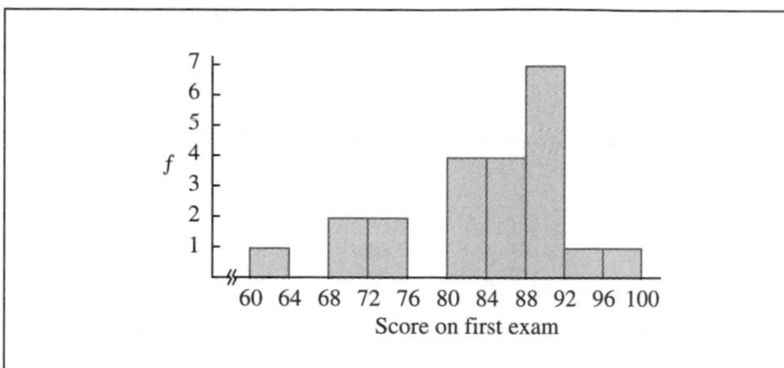

3. (d) Frequency polygon

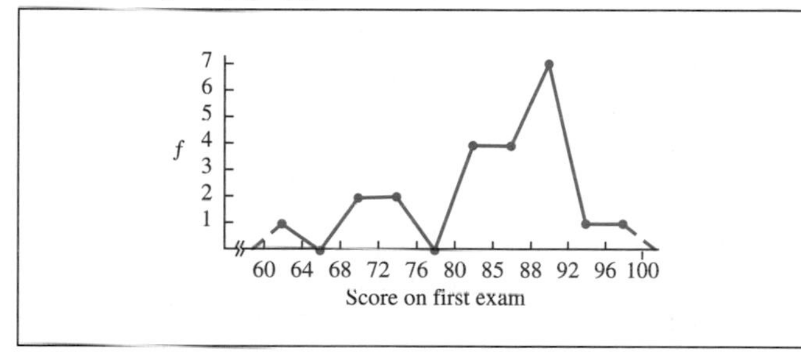

4.

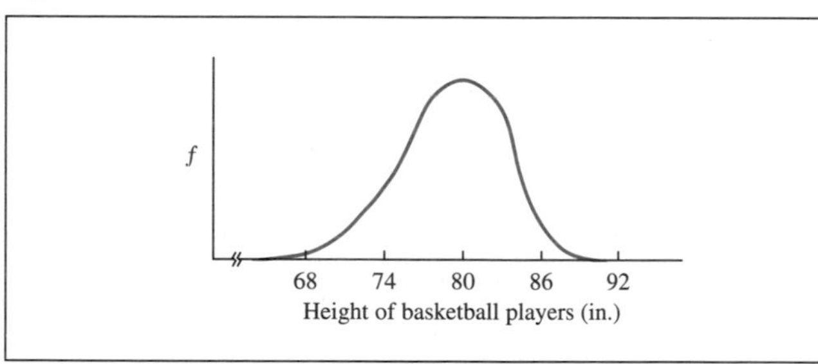

Distribution of heights of basketball players (unimodal, slight negative skew), according to my eyeball

5.

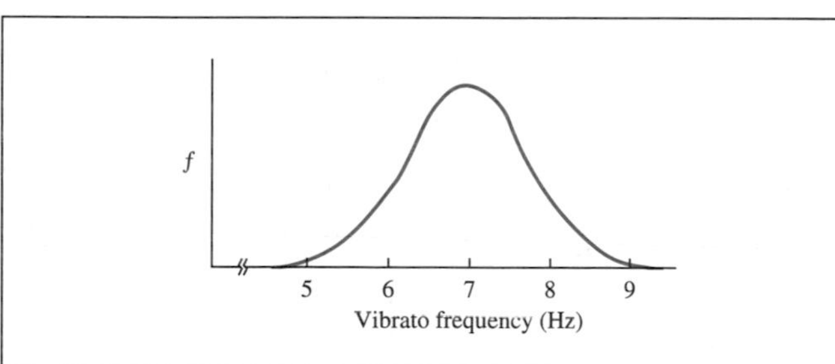

Distribution of froquencies of musicians' vibrato (symmetric, approximately normal), according to my eyeball

6.

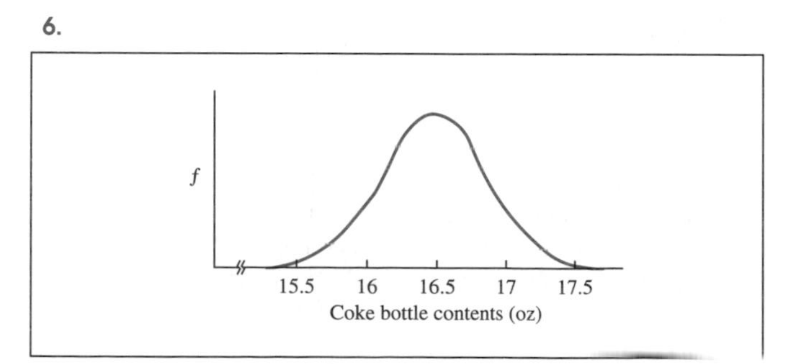

Distribution of contents of Coke bottles (symmetric, approximately normal), according to my eyeball

7.

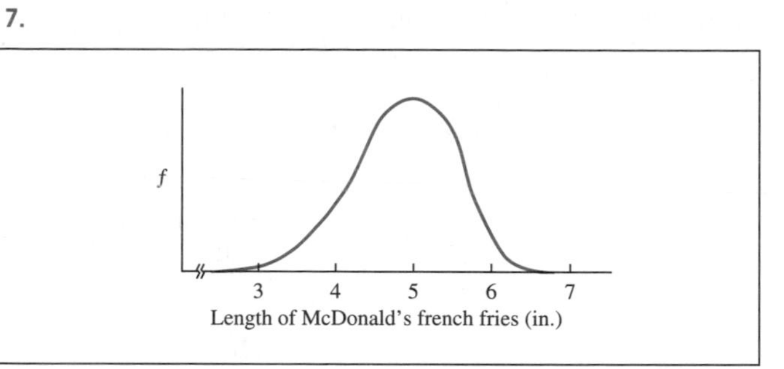

Distribution of lengths of McDonald's french fries (slight negative skew), according to my eyeball

8.

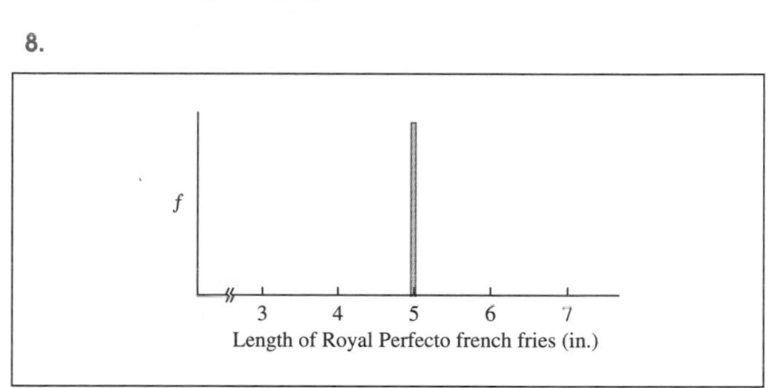

Distribution of lengths of Royal Perfecto french fries (unimodal, symmetric)

9.

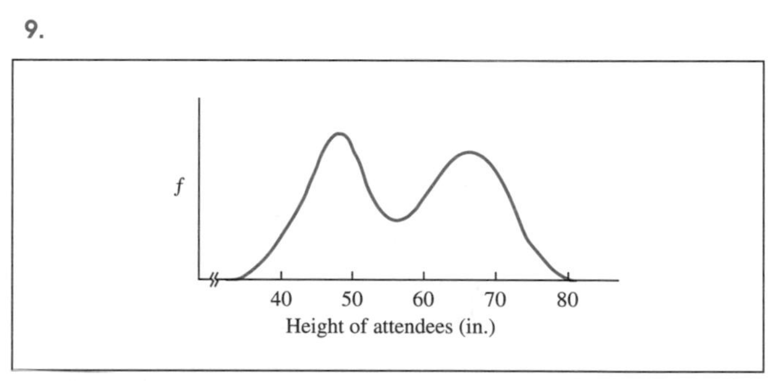

Heights of attendees at Big Brothers Club luncheon (bimodal, slightly asymmetric), according to my eyeball

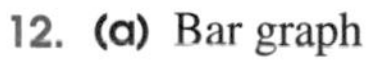

12. (a) Bar graph
(b)

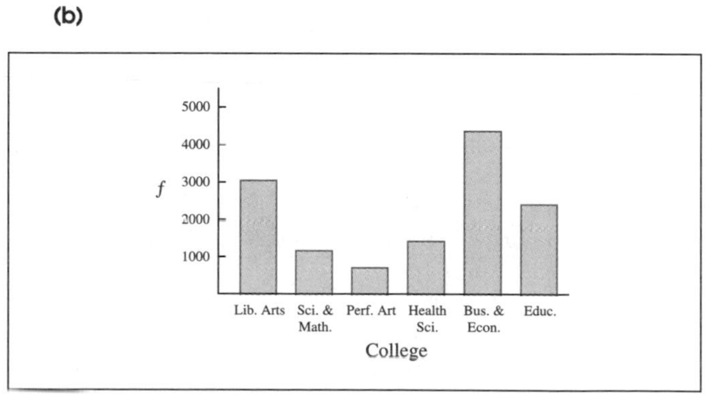

Chapter 3, Section D: Cumulative review

23. (a) 1/4 **(b)** 1/4 **(c)** 1/2

24. (a) 17 **(b)** 4722

26. (a) Interval/ratio **(b)** Ordinal **(c)** Nominal

Chapter 3, Section E: From the journals

27. Unimodal and positively skewed

Chapter 3, Exercises for optional section

1. **(a)** 25
   (b) Weights in pounds: 123, 122, 135, 131, 143, 147, 148, 149, 152, 152, 154, 154, 154, 159, 159, 160, 161, 165, 169, 173, 173, 176, 184, 189, 191

2.

Stem	Leaf
13	1
12	13
11	7
10	000123346
9	3458
8	59
7	9

3. **(a)** It is difficult to tell.
   (b)

Mr. Smith Leaf	Stem	Mrs. Jones Leaf
0	13	1
61	12	13
875321	11	7
65430	10	000123346
983	9	3458
996	8	59
	7	9

   (c) Mr. Smith's class is brighter on average.

4.

Stem	Leaf	Stem	Leaf
11	2234469	11	69
10	35667789	11	22344
		10	5667789
		10	3

The one with .5-second stems is more informative because it shows that most times are in the low 11s or high 10s.

Chapter 4, Section A: Basic exercises

1. **(a)** 3.5 **(b)** 3.00 **(c)** 3.00
   (d) Sample and population mean formulas always give identical results.
   (e) 2 **(f)** 2.5
2. **(a)** 2.0 **(b)** 2.89 **(c)** 2.89
   (d) Sample and population mean formulas always give identical results.
   (e) There is no mode. **(f)** 4
3. **(a)** 78.5°F
   (b) Equation (4.2); this is a population because your interest is in only that week.
   (c) 80.43°F
   (d) Estimate was close enough.
   (e) The method for an odd number of data points
   (f) 82°F **(g)** 82°F
4. **(a)** 63.5°F
   (b) Equation (4.1); this is a sample of all July days.
   (c) 64.0°F
   (d) Estimate was close enough.
   (e) The method for an even number of data points
   (f) 64.5°F **(g)** 66°F
5. **(a)** 15 miles
   (d) Equation (4.1); this is a sample.
   (e) 15.45 miles
   (g) The method for an even number of data points
   (h) 15 miles **(j)** 15 miles
6. **(a)** 78.5 points
   (d) Equation (4.2), assuming that the question implies that we are interested only in that class
   (e) 83.41 points
   (g) The method for an even number of data points
   (h) 85 points **(j)** 91 points **(k)** 89.5 points
   (l) The lowest point (61) is only a single, unusually low point (often called an "outlier"). This single point affects the eyeball-estimate more than it affects the computed mean.
7. Actual mean is 52.8.
8. **(a)** 7 Hz **(b)** 7 Hz **(c)** 7 Hz
9. **(a)** 16.5 oz **(b)** 16.5 oz **(c)** 16.5 oz
10. **(a)** 4.8 in. **(b)** 4.9 in. **(c)** 5.0 in.
    (d) The mean (and to a lesser extent the median) is pulled down by the negative skew.
11. **(a)** 5.0 in. **(b)** 5.0 in. **(c)** 5.0 in.
12. **(a)** Ordinal; mode or median
    (b) Nominal; mode
    (c) Interval/ratio; mode, median, or mean
    (d) Nominal; mode
    (e) Ordinal; mode or median
    (f) Interval/ratio; mode, median, or mean
13. **(a)** Sample **(b)** Population **(c)** Sample

Chapter 4, Section C: Extending comprehension

23. **(a)** Yes; it is even. **(b)** Cannot be determined
24. 8
25. **(a)** Increases by 2 **(b)** Multiplied by 2
    (c) Cannot be specified in general

Chapter 4, Section D: Cumulative review

27. **(a)** .64 **(b)** .46 **(c)** .17 **(d)** .47
    (e) $P(\text{Republican female}) \neq P(\text{Republican} \mid \text{Female})$
28. **(a)** 148 **(b)** 900
    (c) The sum of the squares is not equal to the square of the sum.
    (d) 31

Chapter 4, Section E: From the journals

29. **(a)** Positively skewed; mode 800 kJ per day
    (b) Positively skewed; mode about 1100 kJ per day
    (c) Higher in Slovakia
    (d) Higher in Slovakia; yes
    (e) About 800 kJ per day in Belgium (approximately as Sobolski reported); 2000 kJ per day in Slovakia (much higher than Sobolski reported); no. Perhaps Sobolski et al. really were reporting the modes, not the medians, or

perhaps the correct Slovakia median is 1975, not 975, but those are just guesses.
(f) Higher in Slovakia

Chapter 4, Exercises for optional section

1. **(a)** 83.4 points
   (b) Exactly the same; the two methods are identical.
2. **(a)** 83.32 points
   (b) Approximately the same, but not identical; the values don't lie exactly on the midpoints.

Score on first exam	X_{mid}	f	fX_{mid}
96–99	97.5	1	97.5
92–95	93.5	1	93.5
88–91	89.5	7	626.5
84–87	85.5	4	342.0
80–83	81.5	4	236.0
76–79	77.5	0	.0
72–75	73.5	2	147.0
68–71	69.5	2	139.0
64–67	65.5	0	.0
60–63	61.5	1	61.5
		22	1833.0

Chapter 5, Section A: Basic exercises

1. **(a)** 2 **(b)** .8 **(c)** .89
   (d) Close enough **(e)** .89
   (f) Yes; two formulas for the same thing
   (g) .80
2. **(a)** 7 **(b)** 2.8 **(c)** 2.58
   (d) Close enough **(e)** 2.58
   (f) Yes; two formulas for the same thing
   (g) 6.64
3. **(a)** 9°F **(b)** 3°F
   (c) Computational formula for the standard deviation of a population
   (d) 2.87°F **(e)** No discrepancy
   (g) Computational formula for the variance of a population
   (h) $8.245[°F]^2$
4. **(a)** 11°F **(b)** 4.4°F
   (c) The computational formula for the standard deviation of a sample
   (d) 3.85°F
   (f) The one relatively extreme point (58°F) makes the range large and has a larger effect on the eyeball-estimated standard deviation than it does on the computed standard deviation.
   (g) The computational formula for the variance of a sample
   (h) $14.80[°F]^2$
5. **(a)** 3.5 miles
   (d) The computational formula for the standard deviation of a sample
   (e) 3.85 miles
   (g) The computational formula for the variance of a sample
   (h) 14.79 miles^2
6. **(a)** 8.75 points
   (d) The computational formula for the standard deviation of a population
   (e) 8.79 points **(f)** Close enough
   (g) The computational formula for the variance of a population
   (h) 77.24 points^2
7. Actual standard deviation is 7.0.
8. About .6 Hz
9. About .3 ounce
10. About .7 inch
11. About .0 inch

Chapter 5, Section D: Cumulative review

27. **(a)** Interval/ratio **(b)** Ordinal **(c)** Nominal
    (d) Interval/ratio **(e)** Nominal
28. **(a)** 47.5 years **(b)** ~43 years
    (c) ~42.5 years **(d)** ~8.5 years
29. **(a)** 27 **(b)** 29,079 **(c)** 461,041
30. **(a)** 5 **(b)** 27.5
    (c) 28.29 **(d)** 20.71

Chapter 5, Section E: From the journals

31. **(a)** Men: unimodal with the mode at 1792–1976 hours and with a slight positive skew. Women: bimodal with modes at 0 and at 1872–2059 hours.
    (b) Yes; the intervals overlap (the upper limit of one is the lower limit of the next); furthermore, in the women's distribution, the intervals are not all the same size (most are 187, but some are 188). The single point at 0 can be accepted as an exception to the general rule because such a heavy concentration of points occurs precisely there.
    (c) Yes
    (d) Yes; no, because the distribution is bimodal. Very few members of this sample lie near the mean.
    (e) No; the range method gives values of about 570 hours in both cases; furthermore, the standard deviation for men is likely to be considerably smaller than that because there is such a strong peak at 1792–1976 hours and very little left in the tails. The standard deviation for women is likely to be somewhat greater than 600 hours because of the concentration at the left-hand tail.
    (g) The reported values seem much too large to be possible. [My computations from the tables (using the midpoint of the class intervals to approximate the values of *X*) show that the standard deviation for men is approximately 390 hours, and for women approximately 955 hours.]

Chapter 5, Exercises for optional section

1. **(a)** 8.79 points
   (b) Exactly the same; the two methods are identical
   (c) 77.24 points^2
   (d) Exactly the same; the two methods are identical
2. **(a)** 8.88 points
   (b) Approximately the same, but not identical, because values don't lie exactly on the midpoints
   (c) 78.83 points^2
   (d) Approximately the same, but not identical, because values don't lie exactly on the midpoints

Class interval	X_{mid}	f	fX_{mid}	X^2_{mid}	fX^2_{mid}
96–99	97.5	1	97.5	9,506.25	9,506.25
92–95	93.5	1	93.5	8,742.25	8,742.25
88–91	89.5	7	626.5	8,010.25	56,071.75
84–87	85.5	4	342.0	7,310.25	29,241.00
80–83	81.5	4	326.0	6,642.25	26,569.00
76–79	77.5	0	.0	6,006.25	.00
72–75	73.5	2	147.0	5,402.25	10,804.50
68–71	69.5	2	139.0	4,830.25	9,660.50
64–67	65.5	0	.0	4,290.25	.00
60–63	61.5	1	61.5	3,782.25	3,782.25
		22	1833.0		154,377.50

Chapter 6, Section A: Basic exercises

1. **(a)** 0 **(b)** 1 **(c)** −2 **(d)** 1.5
2. **(a)** 3 minutes **(b)** 80% **(c)** 100%
3. **(a)** −.657 **(b)** 9 minutes
4. 34 plus large half of 14 (about 8), so about 42%
5. **(b)** $z_{\text{by eyeball}} \approx 2$ **(c)** $z = 1.96$ **(e)** 95%
6. $z_{\text{upper 5\%}} = 1.645$
7. **(a)** $z_1 = 0, z_2 = 2.05$, area = .4798
   (b) $z_1 = -2.33, z_2 = 1.6$, area = .935
   (c) $z_1 = -.81, z_2 = .94$, area = .617
   (d) $z_1 = 1.67, z_2 = \infty$, area = .0475
8. **(a)** 2 **(b)** −1 **(c)** .1
   (d) 0 **(e)** −1.05
9. **(a)** 1100 hours **(b)** 800 hours **(c)** 1138 hours
   (d) 980 hours **(e)** 1000 hours
10. **(a)** .13% **(b)** 6.06%
    (c) 6.02% (between .0606 and .0594, but closer to .0606)
11. **(a)** 1232.6 hours **(b)** 23.27%
12. **(a)** −1
    (b) Percentage of area below $z = -1$ is 15.87%.
    (c) $z_{135} = -1.625$, area below mean is .4479 (halfway between .4474 and .4484), $z_{235} = .875$, area above mean is .3092 (halfway between .3078 and .3106), percentage = 75.71%
13. 86.67%
14. **(a)** $z_{79.95} = .84$, $z_{80.23} = .85$, so z_{80} is about $1/5$ (actually $5/28$) of the way from .84 to .85 (much closer to .84), that is, $z_{80} = .842$
    (b) −.841 **(c)** −.385
15. **(a)** 2.28% **(b)** 89.07% **(c)** 66.64%
16. 15.87%
17. **(a)** .65 **(b)** 56.5 **(c)** 109.75 **(d)** 6

Chapter 6, Section C: Extending comprehension

27. .674
28. .3592
29. .0

Chapter 6, Section D: Cumulative review

30. **(a)** 10.10 days **(b)** 9 days **(c)** 8 days
    (d) 13 days **(e)** 3.35 days **(f)** 11.19 days^2
31. **(a)** .56 **(b)** .28 **(c)** .15 **(d)** .27
32. **(a)** −.833 **(b)** 1.67 **(c)** 1.17
33. **(a)** 8.82 hours **(b)** 4.69 hours **(c)** 6.57 hours
34. **(a)** 5.67% **(b)** 17.97% **(c)** 16.99% **(d)** 56.98%

Chapter 6, Section E: From the journals

35. **(a)** Yes **(b)** No **(c)** .5%

Chapter 6, Exercise for optional section

1. **(a)** .2580 **(b)** .2611 **(c)** .2596 **(d)** .2605

Chapter 7, Section A: Basic exercises

1. **(a)** Two **(b)** Too high: no letter
   (c) Too high: no letter **(d)** Too high: no letter
   (e) 15 = "O" **(f)** Too high: no letter
   (g) 06 = "F" **(h)** Too high: no letter
   (i) Too high: no letter **(j)** Too high: no letter
   (k) Too high: no letter **(l)** Too high: no letter
   (m) 15 used already **(n)** 08 = "H"
   (o) Too high: no letter **(p)** 12 = "L"
   (q) 09 = "I"
2. **(a)** Stratified because we wish to have two in each suit
   (b) Queen and 3 **(c)** Queen and 3
   (d) Jack and 7 **(e)** 10 and 9
3. Distribution of means; shape; center; width
4. The means are all near the center of the original distributions.
5. **(a)**

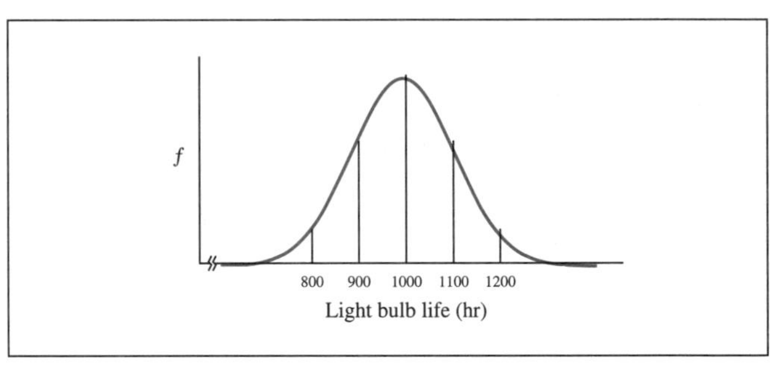

5. **(b)**

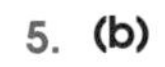

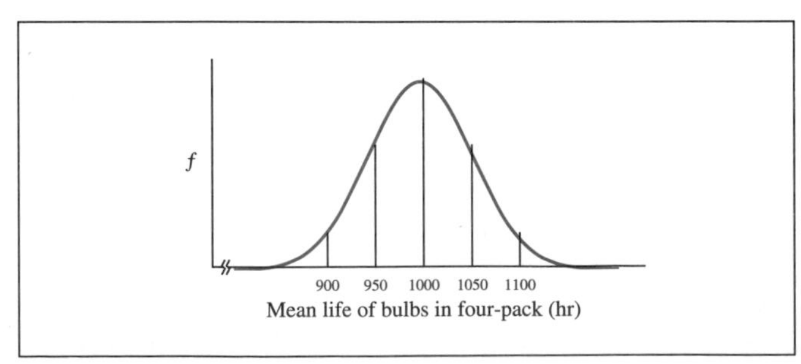

   (c) 1000 hours
   (d) Standard error of the mean
   (e) 50 hours
6. **(a)** 6.68% **(b)** .13%

7. (a)

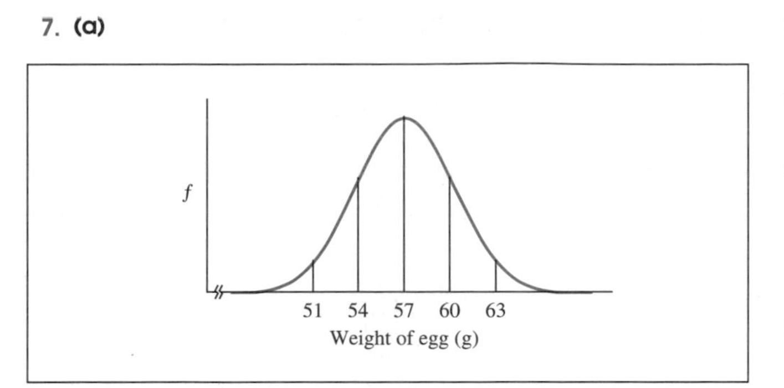

7. (b)

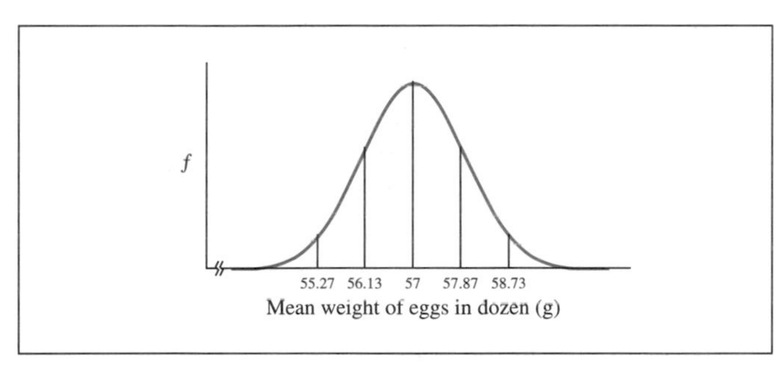

(c) 57 grams (d) .866 gram

8. (a) 15.87% (b) .03%

Chapter 7, Section C: Extending comprehension

15. $n = 1$

16. $n = 4$

17. $n = 1{,}000{,}000$

18. Impossible

Chapter 7, Section D: Cumulative review

19. (a)

Mistakes	f
21	1
18	1
16	2
15	1
14	4
13	2
12	2
11	7
10	7
9	3
8	10
7	5
6	11
5	9
4	5
3	3
2	12
1	8
0	7
	100

(b)

Class interval	f
21–23	1
18–20	1
15–17	3
12–14	8
9–11	17
6–8	26
3–5	17
0–2	27
	100

(c) Yes; positively skewed

20. (a) 6.53 mistakes (b) 6 mistakes
(c) 2 mistakes (d) Yes

21. (a) 21 mistakes (b) $\sigma = 4.59$ mistakes
(c) $\sigma^2 = 21.07$ mistakes^2

22. (a) 6.53 mistakes by the central limit theorem
(b) 1.53 mistakes by the central limit theorem
(c) Slightly skewed, but not as skewed as the distribution of the variable (central limit theorem)

Chapter 8, Section A: Basic exercises

1. $\overline{X} = 5$, $\sigma = 3$, $\sigma_{\overline{X}} = 3/\sqrt{4} = 1.5$, $z_{\text{cv}} = \pm 1.96$, so $2.06 < \mu < 7.94$ with 95% confidence.

2. $\overline{X} = 5$, $\sigma = 3$, $\sigma_{\overline{X}} = 3/\sqrt{9} = 1.0$, $z_{\text{cv}} = \pm 1.96$, so $3.04 < \mu < 6.96$ with 95% confidence.

3. Narrower; the larger sample size makes $\sigma_{\overline{X}}$ smaller.

4. $\overline{X} = 5.000$, $s = 3.3665$, $s_{\overline{X}} = 3.3665/\sqrt{4} = 1.683$, $t_{\text{cv}_{df=3}} = \pm 3.182$, so $-.36 < \mu < 10.36$ with 95% confidence.

5. Wider; primarily because t_{cv} is larger than z_{cv} but also because s is greater than σ

6. (a) 96.2 points
(c) $\sigma = 15$, $\sigma_{\overline{X}} = 4.7434$, $z_{\text{cv}} = \pm 1.96$, "We can say with 95% confidence that the true mean IQ of adults who had low birth weights is greater than 86.9 and less than 105.5 points."

7. (a) No change
(b) We must now compute s and use t instead of z.
(d) $s = 18.72$, $s_{\overline{X}} = 5.92$, $t_{\text{cv}} = \pm 2.262$, "We can say with 95% confidence that the true mean IQ of adults who had low birth weights is greater than 82.81 and less than 109.59 points."
(e) The interval is wider because t_{cv} is larger than z_{cv} and also because s happens to be somewhat larger than 15.

8. (a) 11.6 points
(c) $\sigma = 3$, $\sigma_{\overline{X}} = .6124$, $z_{\text{cv}} = \pm 1.96$, "We can say with 95% confidence that the sixth-grade population mean lies between 10.40 and 12.80 points."

9. (a) No; all measurement has error.
(b) No; because we have no standard error
(c) We have no way of knowing.

10. (a) 30.1 billion meters per second
(c) $s = .2828$, $s_{\overline{X}} = .1155$, $t_{\text{cv}} = 2.571$, "I can say with 95% confidence that the true speed of light is greater than 29.80 and less than 30.40 billion meters per second."
(d) $t_{\text{cv}} = 4.032$, "I can say with 99% confidence that the true speed of light is greater than 29.634 and less than 30.566 billion meters per second."
(e) Wider; the wider the interval, the more likely that the true value lies within it.

11. (a) 16.125 ounces
(c) "I can say with 95% confidence that the true mean volume of beer served at Kelly's Tavern is greater than 15.582 and less than 16.668 ounces."

12. (a) $s_{\overline{X}}$ becomes smaller by a factor of $\sqrt{2}$, and t becomes smaller by a small amount (from 2.365 to 2.131).
(c) "I can say with 95% confidence that the true mean volume of beer served at Kelly's Tavern is greater than 15.78 and less than 16.47 ounces."
(d) It is slightly narrower.

13. "I can say with 95% confidence that the true percentage of the population of voters who support you is greater than 48.1% and less than 61.9%."

Chapter 8, Section C: Extending comprehension

26. $\overline{X} = 12$, $t_{cv_{df=15}} = 2.131$, $s_{\overline{X}_{n=16}} = .704$, $s = 2.82$, $t_{cv_{df=31}} = 2.040$, $s_{\overline{X}_{n=32}} = .498$, so with 95% confidence $10.98 < \mu < 13.02$

27. **(b)** $9.93 < \mu < 14.07$

28. $n = 123$

29. **(a)** Because z_{cv} is smaller than t_{cv}
(b) Yes; if s happens to be smaller than σ

30. $n = 9513$

Chapter 8, Section D: Cumulative review

31. **(a), (f)** Actual mean = 7.78 pounds
(d), (g) Actual standard deviation = .31 pound
(e) Actual variance = .0961 pound^2

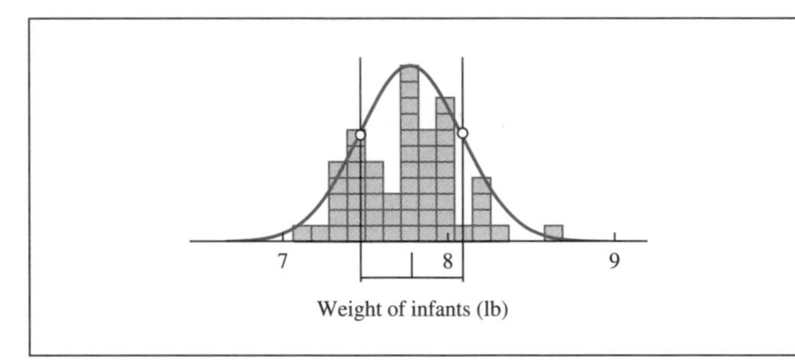

Weights of infants with inflection points indicated

32. **(a)** Actual standard error = .04 pound
(b) Actual $t_{cv} = 2.01$
(c) Actual confidence interval: $7.70 < \mu < 7.86$ pounds

33. **(a), (f)** Actual mean = 110.49 mph
(d), (g) Actual standard deviation = 4.78 mph
(e) Actual variance = 22.85 mph^2

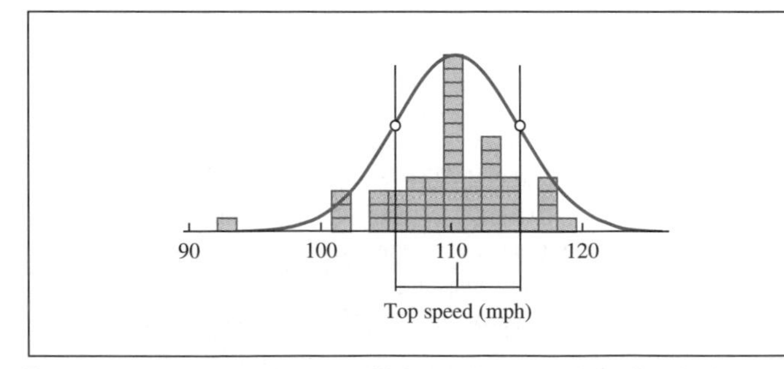

Top speeds of motorboats with inflection points indicated

34. **(a)** Actual standard error = .663 mph
(b) Actual $t_{cv} = 2.01$
(c) Actual confidence interval: $109.10 < \mu < 111.76$ mph

35. **(b)** 48.57% **(c)** 24.87% **(d)** 27.57%

36. **(b)** $s_{\overline{X}} = 1.4$, so 47.15%
(c) 8.74% **(d)** 21.43%

Chapter 8, Section E: From the journals

38. **(a)** 24.8 years **(b)** .97 year
(c) 6.5 years **(d)** 18.4 years
(e) No; it seems likely that the reported confidence interval is too wide because the standard deviation as computed from it is too large.

Chapter 9, Section A: Basic exercises

1. Mean, median, mode, range, standard deviation, variance (also skew, but we did not quantify it)

2. **(a)** H_0: $\mu \leq 81$, H_1: $\mu > 81$
(b) Directional; I expect it to increase students' scores.

3. No; I don't know whether 83 is *significantly* greater than 81.

4. **(a)** Yes; the null hypothesis may have been true but I drew a "lucky" sample.
(b) No; it is impossible to have a Type II error if you reject the null hypothesis.

5. **(a)** α **(b)** β

6. See Table 9.3.

7. Level of significance (α) is the complement of level of confidence ($1 - \alpha$).

8. $$\text{Test statistic} = \frac{\text{(sample statistic} - \text{population parameter)}}{\text{standard error of the sample statistic}}$$

9. I. State null and alternative hypotheses. II. Set criterion for rejecting H_0. III. Collect sample and compute observed values of sample statistic and test statistic. IV. Interpret results.

Chapter 9, Section D: Cumulative review

17. **(b)** 78.81% **(c)** 21.19% **(d)** 28.98%

18. **(a)** 12 ounces **(b)** .5 ounce

19. **(a)** 12 ounces **(b)** .5 ounce

20. **(b)** > 99.999% **(c)** < .001% **(d)** 2.28%

21. We can say with 95% confidence that in the population of voters, the percentage saying yes is greater than 31.3% and less than 40.7%.

22. $12.16 < \mu < 14.49$; with 95% confidence we can say that the average number of patient requests per day in the hospital population is greater than 12.16 and less than 14.49.

Chapter 9, Exercises for optional section

1. H_0 is never strictly true; large samples imply statistical significance; statistical significance is not directly related to effect size; and $P(\text{reject } H_0 \mid H_0 \text{ true}) \neq P(H_0 \text{ true} \mid \text{reject } H_0)$

2. **(a)** Schizophrenia test in textbook
(b) Probability of diagnosing schizophrenia in a normal individual
(c) Probability that a person is normal if he was diagnosed as having schizophrenia

Chapter 10, Section A: Basic exercises

1. **(a)** 5.5 **(b)** -3
(c) ± 1.96 **(d)** 7.06 and 12.94
(e) Yes; $|-3| > |-1.96|$ and (equivalently) 5.5 is farther from 10 than is 7.06.

2. **(a)** H_0: $\mu = 500$, H_1: $\mu \neq 500$
   (b) Nondirectional; "affect performance" does not imply direction.
   (c) SAT score
   (d) Mean SAT score, $\sigma_{\bar{X}} = 25$
   (e) z because we know σ **(f)** .05
   (g) Not applicable with z **(h)** $z_{\text{cv}_{\alpha=.05,\text{ nondirectional}}} = \pm 1.96$
   (i) 451 and 549 **(j)** $\bar{X}_{\text{obs}} = 522.81$
   (k) $z_{\text{obs}} = .913$ **(l)** Yes; always
   (m) No **(n)** Not required
   (o) We do not have reason to believe that meditation affects SAT scores.

2. **(c)**

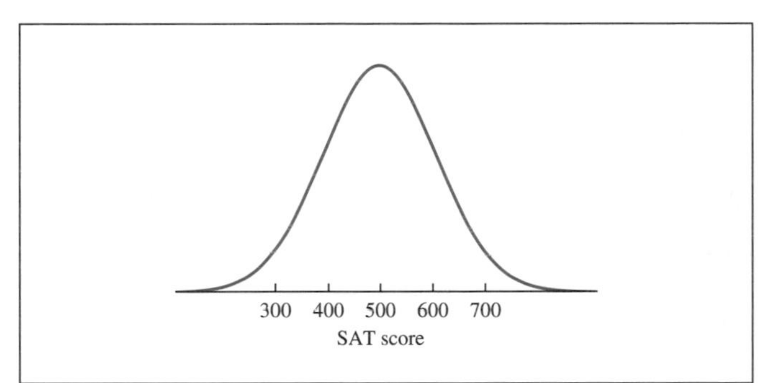

2. **(d) and (e)**

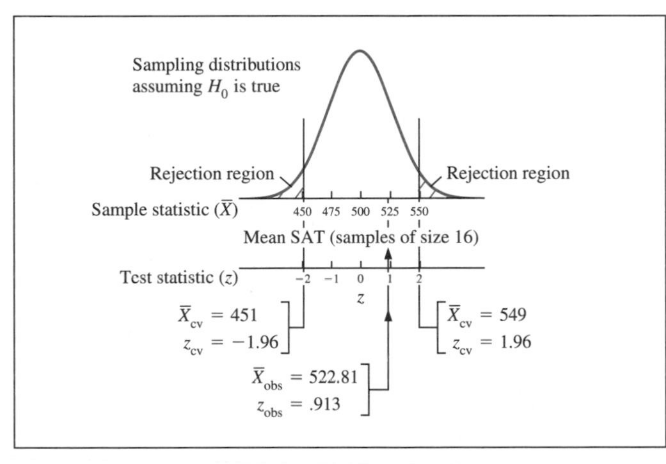

3. **(a)** No change **(b)** No change
   (c) Still SAT score, but $s = 78.00$
   (d) Still mean SAT score, but $s_{\bar{X}} = 19.50$

3. **(c)**

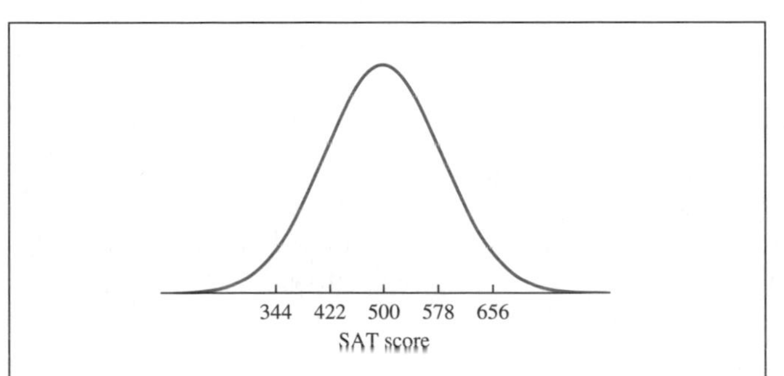

   (e) t instead of z **(f)** No change
   (g) 15
   (h) $t_{\text{cv}_{df=15,\alpha=.05,\text{ nondirectional}}} = \pm 2.131$
   (i) 458.45 and 541.55 **(j)** No change
   (k) $t_{\text{obs}} = 1.17$ **(l)** No change
   (m) No change **(n)** No change
   (o) No change **(p)** $481.3 < \mu < 564.4$
   (q) Yes; yes because we should reject H_0 if 500 is not in the confidence interval.

3. **(d) and (e)**

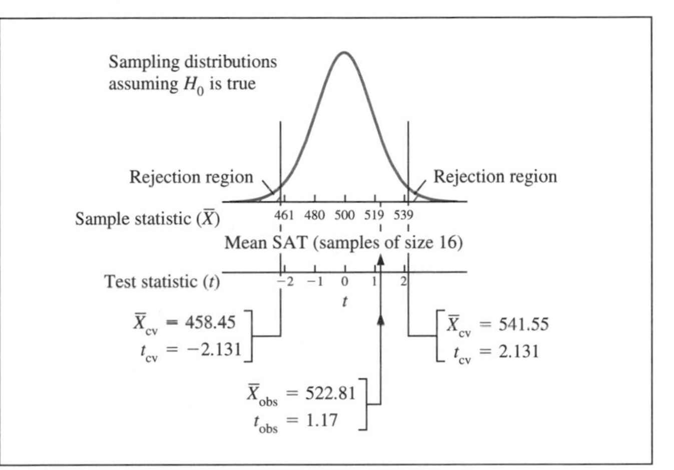

4. There is not enough information; we need the standard deviation of lap times.

5. **(a)** H_0: $\mu \geq 61.7$; H_1: $\mu < 61.7$
   (b) Directional; I expect the injection system to *lower* lap times.
   (c) Lap time
   (d) Mean lap time, $s_{\bar{X}} = .149$
   (e) t because we do not know σ
   (f) .05 **(g)** 20
   (h) $t_{\text{cv}_{df=20,\alpha=.05,\text{ directional}}} = -1.725$
   (i) $\bar{X}_{\text{cv}} = 61.44$ seconds **(j)** $\bar{X}_{\text{obs}} = 61.352$ seconds
   (k) $t_{\text{obs}} = -2.331$ **(l)** Yes; always
   (m) Yes **(n)** $d = .51$; $U = 69.5\%$
   (o) The new injection system has lowered lap times. Our best estimate of the mean lap time using the new injection system is 61.35 seconds, which is .35 second less per lap (faster speeds) than with the original system, amounting to a shift of .51 standard deviation. This new mean lap time is shorter than about 69.5% of the laps run with the original system.

5. **(c)**

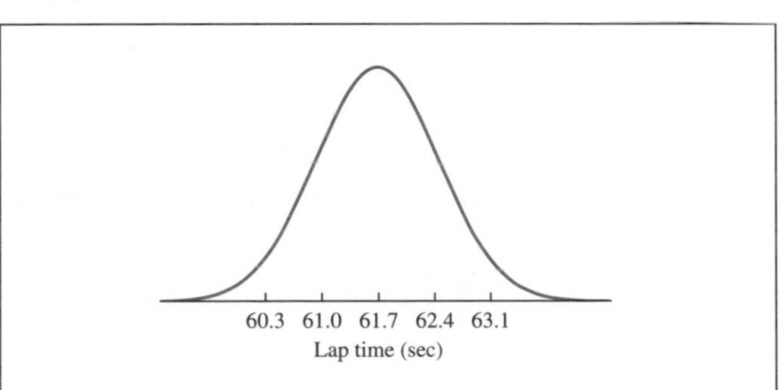

5. **(d) and (e)**

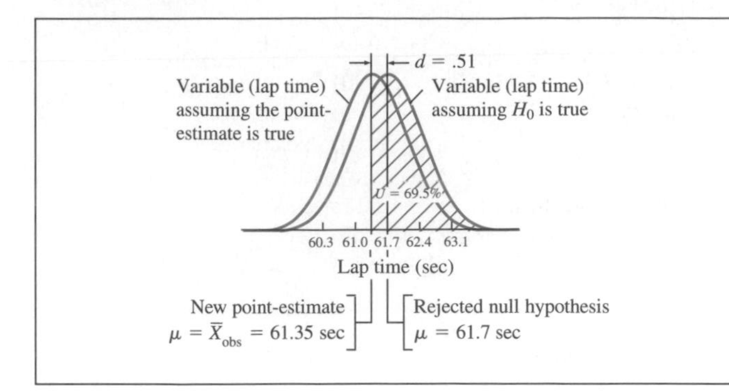

5. **(n)**

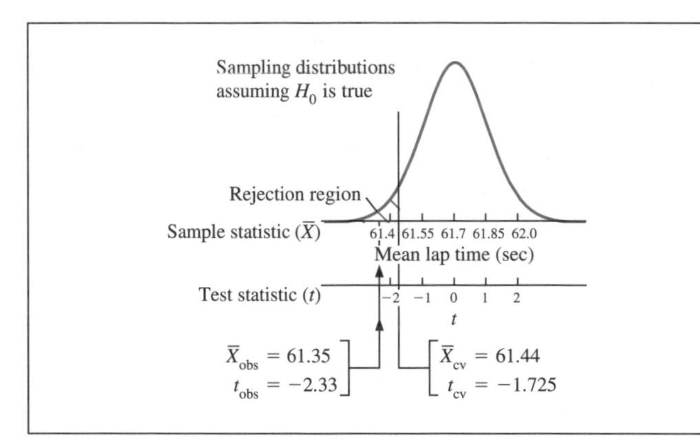

6. **(a)** No change **(b)** No change **(c)** No change
(d) No change **(e)** No change **(f)** .01
(g) No change
(h) $t_{cv_{df=20, \alpha=.01, \text{directional}}} = -2.528$
(i) $\bar{X}_{cv} = 61.32$ seconds (smaller)
(j) No change **(k)** No change **(l)** No change
(m) No; we do not reject H_0 in this situation.
(n) Not required
(o) The new injection system has not lowered lap times.
(p) $\alpha = .01$ is a more stringent test—a "higher hurdle"—so it is possible to reject with $\alpha = .05$ but not to reject with $\alpha = .01$.

7. Yes; he should change; $s_p = .0494$, $t_{obs} = 1.821$, $t_{cv} = 1.666$ (directional)

Chapter 10, Section C: Extending comprehension

15. *Hint:* Choose any five-point data set. How does s change if you add (or subtract) the same constant to each of the five points? How does the numerator of t change under the same alterations? How does s change if you add a constant to the highest value and subtract the same constant from the lowest value? How does the numerator of t change under the same alterations?

16. No; it is impossible. Crossing the higher hurdle implies crossing the lower hurdle.

17. See the hint for Exercise 15.

18. Yes; only if the observed value is in the opposite tail from the critical value of the directional test

Chapter 10, Section D: Cumulative review

19. **(a)** 2 **(b)** 2 **(c)** 2 **(d)** 2
(e) 499 **(f)** 2 **(g)** ±1.96 **(h)** (8.5)

20. **(a)** 3 **(b)** 1 **(c)** 1 **(d)** 2
(e) 99 **(f)** 1 **(g)** 1.666 **(h)** (10.4)

21. **(a)** 3 **(b)** 1 **(c)** 1 **(d)** 2
(e) 24 **(f)** 2 **(g)** 2.064 **(h)** (10.4)

22. **(a)** 1 **(h)** (6.1)

23. **(a)** 2 **(b)** 2
(c) 1 **(d)** 1
(e) Not applicable **(f)** 2
(g) ±1.96 **(h)** (8.1)

24. **(a)** 3 **(b)** 1
(c) 1 **(d)** 1
(e) Not applicable **(f)** 1
(g) 1.645 **(h)** (10.1) or (7.2)

Chapter 11, Section A: Basic exercises

1. **(a)** $s_1 = 2.702$, $s_2 = 1.517$, $s^2_{pooled} = 4.800$, $s_{\bar{X}_1 - \bar{X}_2} = 1.386$
(b) Yes; equal sample sizes; $s_{\bar{X}_1 - \bar{X}_2} = 1.386$
(c) They should be the same.

2. **(a)** H_0: $\mu_1 = \mu_2$; H_1: $\mu_1 \neq \mu_2$ Nondirectional; no expectation that either school is better; two (two different schools)
(b) $s_1 = 2.696$, $s_2 = 2.915$, $s^2_{pooled} = 7.884$, $s_{\bar{X}_1 - \bar{X}_2} = 1.404$
(c) Vocabulary score (note that the figure shows that $\mu_1 = \mu_2$ but does not specify the magnitude of μ_1 or μ_2)

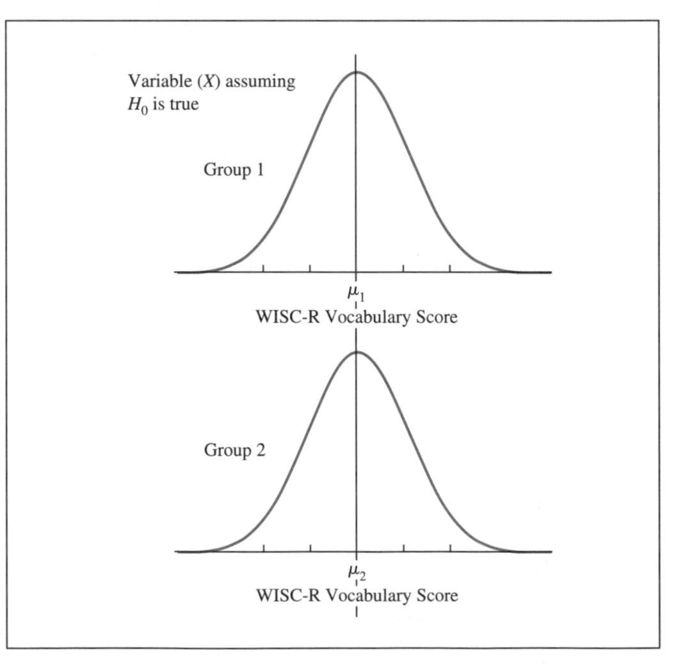

(d) $(\bar{X}_1 - \bar{X}_2)$
(e) t **(f)** .05 **(g)** 14
(h) $t_{cv_{\alpha=.05, df=14, \text{nondirectional}}} = \pm 2.145$
(i) $(\bar{X}_1 - \bar{X}_2)_{cv} = \pm 3.01$ **(j)** $(\bar{X}_1 - \bar{X}_2)_{obs} = -.125$
(k) $t_{obs} = -.089$ **(l)** Yes; always
(m) No **(n)** Not necessary
(o) We cannot say that the two schools are different from each other.

2. (d) and (e)

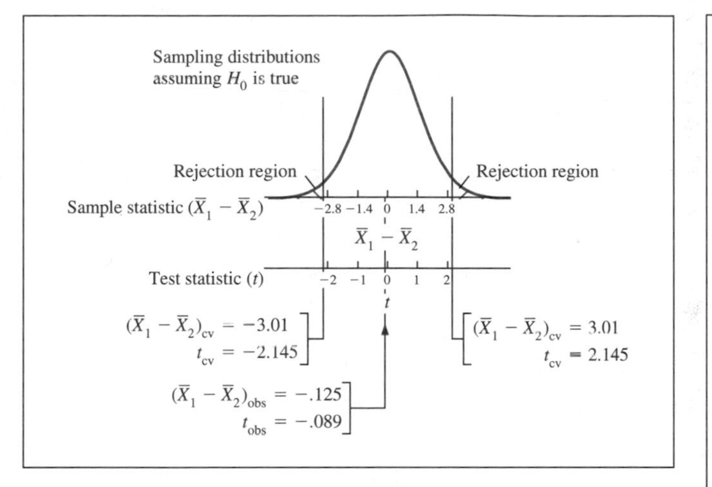

3. (a) H_0: $\mu_{30} \leq \mu_{60}$; H_1: $\mu_{30} > \mu_{60}$ Directional; the younger group is expected to perform better; two (30 year olds and 60 year olds)

(b) $s_1 = 4.111$, $s_2 = 3.568$, $s^2_{\text{pooled}} = 14.518$, $s_{\overline{X}_{30}-\overline{X}_{60}} = 1.603$

(c) Number of words recalled (note that the figure shows that $\mu_1 = \mu_2$ bu does not specify the magnitude of μ_1 or μ_2)

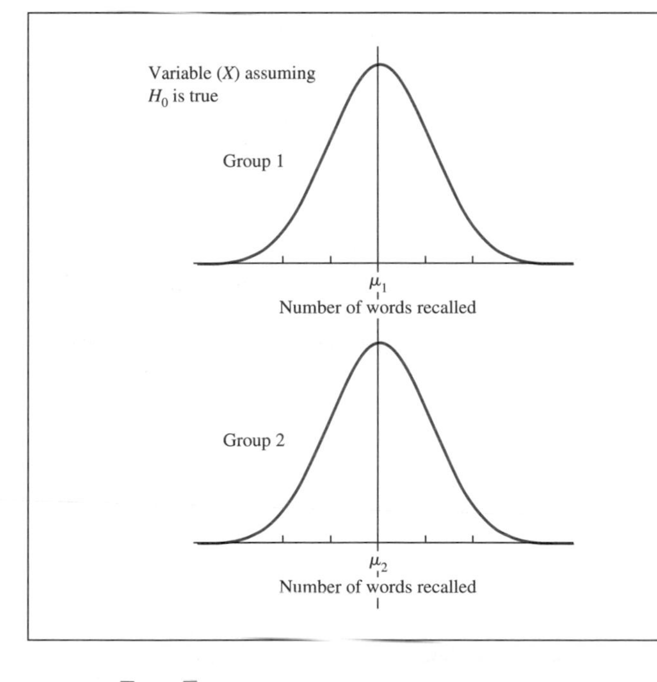

(d) $(\overline{X}_{30} - \overline{X}_{60})$

(e) t **(f)** .05 **(g)** 21

(h) $t_{cv_{\alpha=.05, df=21, \text{directional}}} = 1.721$

(i) $(\overline{X}_{30} - \overline{X}_{60})_{cv} = 2.76$ **(j)** $(\overline{X}_{30} - \overline{X}_{60})_{obs} = 4.61$

(k) $t_{obs} = 2.875$ **(l)** Yes; always

(m) Yes **(n)** $d = 1.21$; $U = 88.7\%$

(o) Memory of 60-year-olds is diminished in comparison with that of 30-year-olds.

3. (d) and (e)

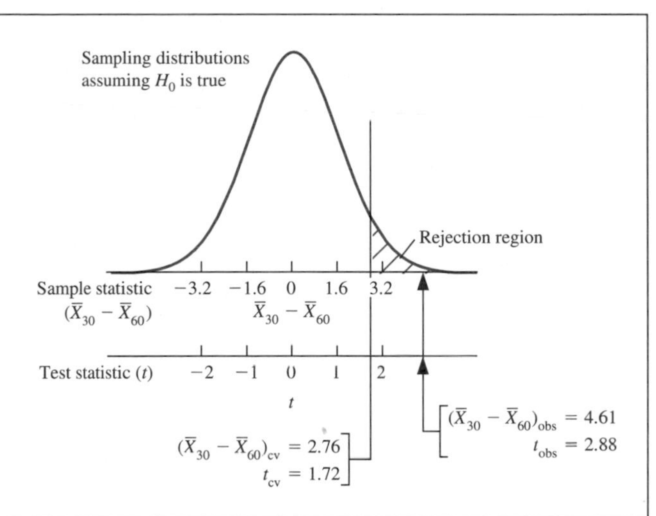

3. (n)

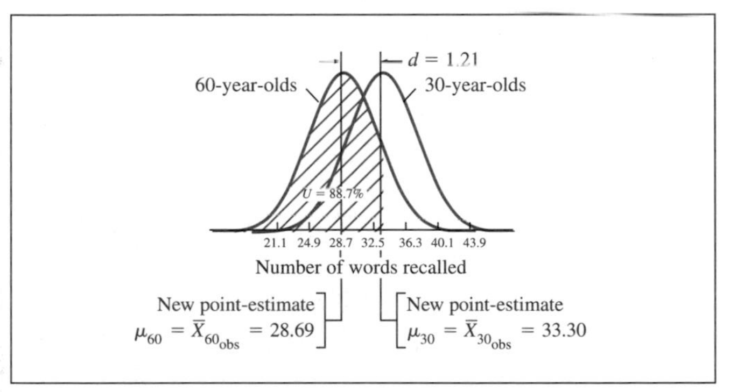

4. (a) H_0: $\mu_{\text{Schindler}} = \mu_{\text{other}}$; H_1: $\mu_{\text{Schindler}} \neq \mu_{\text{other}}$ Nondirectional; no expectation is stated regarding the direction of effect; two (two different groups)

(b) $s_{\text{Schindler}} = 7.782$, $s_{\text{other}} = 7.424$, $s^2_{\text{pooled}} = 57.833$, $s_{\overline{X}_{\text{Schindler}}-\overline{X}_{\text{other}}} = 2.777$

(c) Anti-Semitism score

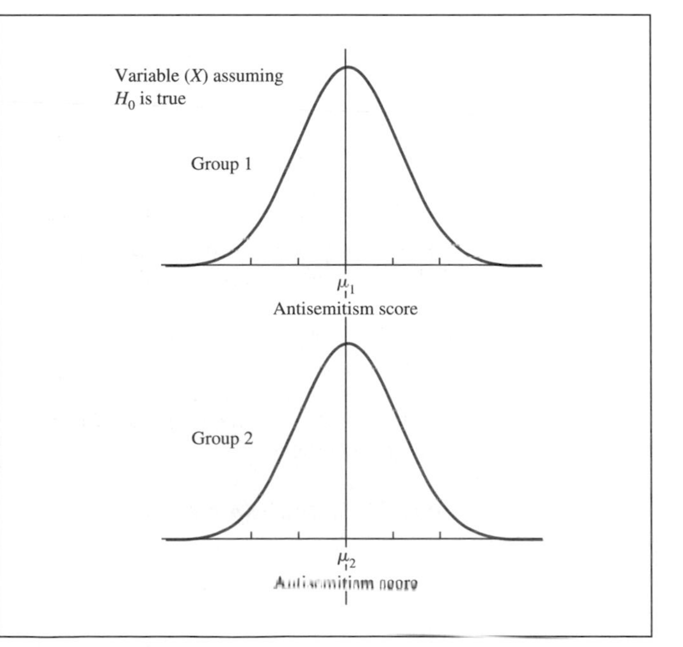

(d) $(\bar{X}_{\text{Schindler}} - \bar{X}_{\text{other}})$
(e) t **(f)** .05 **(g)** 28
(h) $t_{\text{cv}_{\alpha=.05,\,df=28,\,\text{nondirectional}}} = \pm 2.048$
(i) $(\bar{X}_{\text{Schindler}} - \bar{X}_{\text{other}})_{\text{cv}} = \pm 5.69$
(j) $(\bar{X}_{\text{Schindler}} - \bar{X}_{\text{other}})_{\text{obs}} = -3.47$
(k) $t_{\text{obs}} = -1.25$ **(l)** Yes; always **(m)** No
(n) Not necessary
(o) We cannot say that viewing *Schindler's List* changes anti-Semitism.

4. **(d) and (e)**

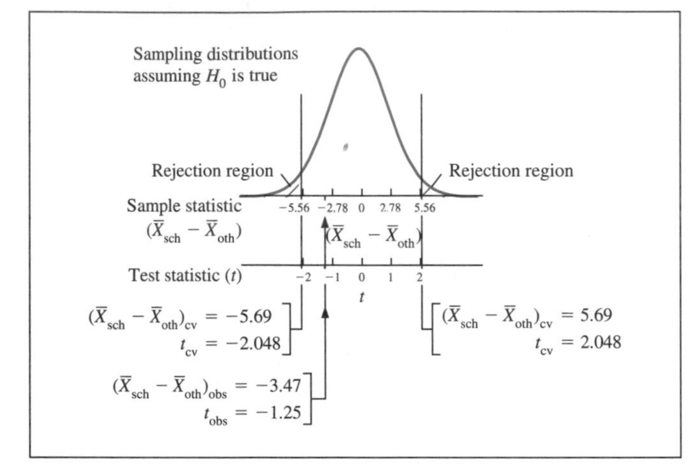

5. **(a)** $s_{\text{pooled}} = 5.101$, $s_{\bar{X}_1-\bar{X}_2} = 2.281$, $t_{\text{obs}} = 2.19$, $t_{\text{cv}} = \pm 2.101$, yes
(b) $s_{\text{pooled}} = 10.198$, $t_{\text{obs}} = 1.10$, $t_{\text{cv}} = \pm 2.101$, no
(c) They are the same; smaller sample standard deviations in experiment I make t larger.

6. **(a)** Same as Exercise 5(a)
(b) $s_{\text{pooled}} = 5.101$, $t_{\text{obs}} = 1.55$, $t_{\text{cv}} = \pm 2.306$; no
(c) Larger n in experiment I makes t larger.

7. **(a)** Yes; $t = 2.10$ **(b)** 1.3 minutes
(c) $d = .074$; $U = 53\%$ **(d)** 53%
(e) I wouldn't; it has low practical significance even though it is statistically significant.

Chapter 11, Section C: Extending comprehension

14. **(b)** No

15. **(b)** Yes; but only if the observed value is in the opposite tail from the critical value

16. **(b)** Depends on your data; for an infinite coin flip, sample size $\sigma = .5$, so $\sigma_{\bar{X}_1-\bar{X}_2}$ for samples of size 5 = .316. Your own $s_{\bar{X}_1-\bar{X}_2}$ should be a point-estimate of that value—that is, close but not exact.
(d) Depends on your data; your answer should be a point-estimate of $\sigma_{\bar{X}_1-\bar{X}_2} = .316$.
(e) Both point-estimate the same value, but they are based on different samples. Thus, they should be expected to be close but not exactly the same.

Chapter 11, Section D: Cumulative review

17. **(a)** 3 **(b)** 1 **(c)** 1 **(d)** 2
(e) 36 **(f)** 1 **(g)** −1.689 **(h)** (10.4)

18. **(a)** 4 **(b)** 2 **(c)** 3 **(d)** 2
(e) 72 **(f)** 1 **(g)** −1.669 **(h)** (11.2)

19. **(a)** 3 **(b)** 1
(c) 1 **(d)** 1
(e) Not applicable **(f)** 1
(g) −1.645 **(h)** (10.1) or (7.2)

20. **(a)** 2 **(b)** 3 **(c)** 1 **(d)** 2
(e) 24 **(f)** 2 **(g)** ±2.064 **(h)** (8.3)

21. **(a)** 4 **(b)** 2 **(c)** 3 **(d)** 2
(e) 28 **(f)** 2 **(g)** ±2.048 **(h)** (11.2)

22. **(a)** 1 **(h)** (7.2)

Chapter 11, Section E: From the journals

23. **(a)** Yes **(b)** Yes; same as Zeidner's
(c) He apparently did not set $\alpha = .05$ in advance, as I recommend. Instead, his logic could be stated, "If we had set α to any value = .03 or larger, then the test would have rejected H_0."

24. **(a)** Similar, but not the same
(b) Larger, 2.73 compared to 1.43
(c) $t_{\text{obs}} = -.64$; smaller than Zeidner's even though the effect size is larger; Zeidner's study has a much larger sample size
(d) No

25. **(a)** Yes
(b) No; there is not enough information.

26. **(a)** Quite close
(b) Slightly smaller; 5.0 compared to 6.67
(c) $t_{\text{obs}} = 3.86$ **(d)** Yes; $t_{\text{cv}} = 2.048$
(e) Our standard deviations must have been smaller than Burger's.

27. **(a)** Yes
(b) Yes; $t_{\text{obs}} = -4.13$ (The difference between −4.13 and −4.03 is probably due to rounding in the reporting of the means and standard deviations.)
(c) Mothers of NOFT infants were significantly less attuned to their children's signals than were mothers of normal infants.

28. **(a)** 28 **(b)** $t_{\text{obs}} = -2.49$
(c) If we use Drotar's level of significance ($\alpha = .01$), then $t_{\text{cv}} = -2.462$ (directional because we can now claim to expect that mothers of normal children are better attuned). Therefore, t is significant, and our replication agrees with Drotar's findings.

Chapter 12, Section A: Basic exercises

1. **(a)**

Subject	*Before*	*After*	*Difference*
1	5	5	0
2	8	10	−2
3	6	8	−2
4	2	6	−4
5	7	9	−2
			$\sum D_i = -10$
			$\sum D_i^2 = 28.00$

(b) −2.00 **(c)** 1.41 **(d)** .63 **(e)** −3.16
(f) Yes; $t_{\text{cv}_{df=4}} = \pm 2.776$

2. (a)

Subject	Before	After	Difference
1	5	10	−5
2	8	5	3
3	6	9	−3
4	2	8	−6
5	7	6	1
		$\sum D_i =$	−10
		$\sum D_i^2 =$	80.00

(b) −2.00 **(c)** 3.87 **(d)** 1.73 **(e)** −1.16

(f) No; $t_{\text{cv}_{df=4}} = \pm 2.776$

(g) In Exercise 1, the relationship between the before and after measures is very strong: Almost all the subjects show an increase, therefore s_D is quite small. In Exercise 2, there is no such strong relationship: Some increase whereas some decrease, so s_D is much larger. Note that the numerators of t are identical in the two cases.

3.

Person	Before	After	Difference
1	190	184	6
2	191	166	25
3	164	172	−8
4	181	192	−11
5	187	174	13
6	178	153	25
7	175	173	2
8	164	158	6
9	198	176	22
10	153	152	1
		$\sum D_i =$	81
		$\sum D_i^2 =$	2165.00
		$\bar{D}_{\text{obs}} =$	8.10
		$s_D =$	12.95
		$s_{\bar{D}} =$	4.09

(a) H_0: $\mu_D \leq 0$; H_1: $\mu_D > 0$

(b) Directional; weight is expected to decrease.

(c) D; $s_{D_{\text{by eyeball}}} \approx [25 - (-11)]/3 \approx 12$; $s_D = 12.95$ pounds

(d) $\bar{D}$; $s_{\bar{D}_{\text{by eyeball}}} \approx 12/\sqrt{10} \approx 4$; $s_{\bar{D}} = 4.10$ pounds

(e) t

(f) $\alpha = .05$ as stated in the problem

(g) $df = 9$

(h) $t_{\text{cv}_{\text{by eyeball}}}$ in absolute value is slightly greater than 1.645—say, 1.80; $t_{\text{cv}_{\alpha=.05,\, df=9,\ \text{directional}}} = 1.833$

(i) $\bar{D}_{\text{cv}_{\text{by eyeball}}} \approx 1.80(4) \approx 7$; $\bar{D}_{\text{cv}} = 7.50$ pounds

(j) $\bar{D}_{\text{by eyeball}} \approx [25 + (-11)]/2 = 7$; $\bar{D}_{\text{obs}} = 8.1$ pounds

(k) $t_{\text{obs}_{\text{by eyeball}}} \sim 7/4 \sim 1.75$; $t_{\text{obs}} = 1.98$

3. (c)

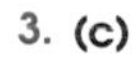

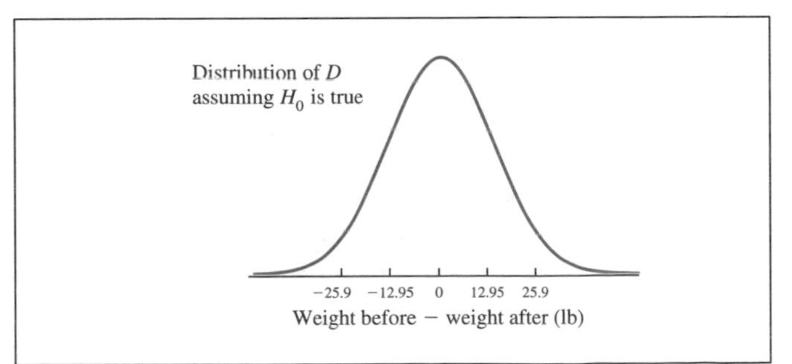

3. (d) and (e)

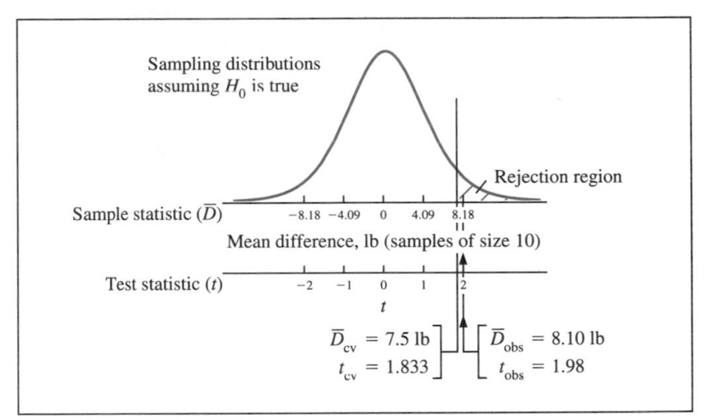

3. (l) Yes; always

(m) Fail to reject H_0 by eyeball but reject H_0 by computation. The discrepancy is because the observed value is quite close to the critical value.

(n) $d = .63$; $U = 73.57\%$

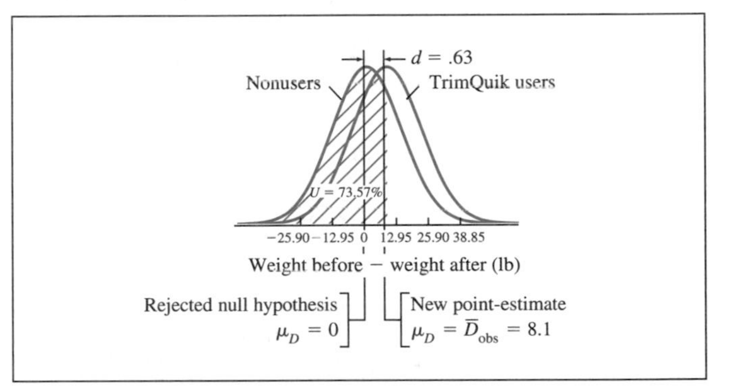

(o) Using TrimQuik does have a statistically significant effect on weight loss. Users on average lose 8.1 pounds (.63 standard deviation of the differences), which indicates that 73.57% of the users of TrimQuik lose more weight than the average nonuser.

4.

Person	Fluores.	Incand.	Difference
1	84	86	−2
2	66	73	−7
3	71	66	5
4	80	74	6
5	78	69	9
6	77	77	0
7	75	70	5
8	84	78	6
9	71	70	1
10	80	74	6
11	83	76	7
12	66	63	3
		$\sum D_i =$	39
		$\sum D_i^2 =$	351.00
		$\bar{D}_{\text{obs}} =$	3.25
		$s_D =$	4.52
		$s_{\bar{D}} =$	1.305
		$t_{\text{obs}} =$	2.49

$t_{\text{cv}_{\alpha=.05,\, df=11,\ \text{nondirectional}}} = 2.201$; reject the null hypothesis.

(a) H_0: $\mu_D = 0$; H_1: $\mu_D \neq 0$
(b) Nondirectional; "affected" direction not specified
(c) D; $s_{D_{\text{by eyeball}}} \approx [9 - (-7)]/3 = 5.33$; $s_D = 4.52$
(d) $\overline{D}$; $s_{\overline{D}_{\text{by eyeball}}} \approx 5.33/\sqrt{12} \approx 1.5$; $s_{\overline{D}} = 1.305$
(e) t **(f)** .05 **(g)** $df = 11$
(h) $t_{\text{cv}_{\text{by eyeball}}}$ slightly greater in absolute value than ± 1.96—say, ± 2.1; $t_{\text{cv}_{\alpha=.05,\, df=11,\ \text{two-tailed}}} = \pm 2.201$
(i) $\overline{D}_{\text{cv}_{\text{by eyeball}}} \approx \pm 2.1(1.5) \approx \pm 3.1$; $\overline{D}_{\text{cv}} = \pm 2.87$
(j) $\overline{D}_{\text{by eyeball}} \approx [9 + (-7)]/2 = 1$; $\overline{D}_{\text{obs}} = 3.25$
(k) $t_{\text{obs}_{\text{by eyeball}}} \approx 1/1.5 = .67$; $t_{\text{obs}} = 2.49$
(l) Yes; always
(m) No by eyeball; yes by computation (the -7 difference score pulls the estimation of $\overline{D}$ down too low)
(n) $d = .72$; $U = 76.42\%$

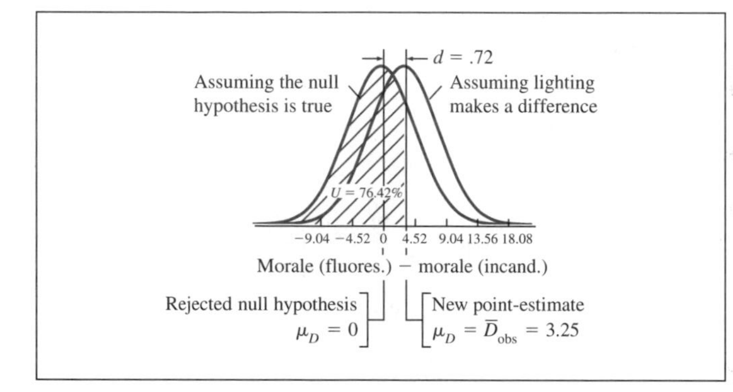

(o) Type of lighting does have a statistically significant effect on morale. Employees' morale is, on average, 3.25 points (.72 standard deviation) higher under fluorescent lighting, which indicates that 76.42% of employees who use fluorescent light will have a higher morale score than the average employee who uses incandescent light.

5.

	Fluores.	*Incand.*
	84	86
	66	73
	71	66
	80	74
	78	69
	77	77
	75	70
	84	78
	71	70
	80	74
	83	76
	66	63
$\sum X_i$	915	876
$\sum X_i^2$	70,233.00	64,352.00
$\overline{X}_j$	76.25	73.00
s_j	6.50	6.06

$s_{\text{pooled}} = 6.28$
$s_{\overline{X}_1 - \overline{X}_2} = 2.56$
t_{obs}(independent samples, $df = 22$) = 1.27
$t_{\text{cv}_{\alpha=.05,\ df=22,\ \text{nondirectional}}} = 2.074$
Fail to reject H_0.

(a) H_0: $\mu_1 = \mu_2$; H_1: $\mu_1 \neq \mu_2$
(b) Nondirectional; "affected" direction not specified
(c) X; $s_{1_{\text{by eyeball}}} \approx (84 - 66)/3 = 6$; $s_{2_{\text{by eyeball}}} \approx (86 - 63)/3 \approx 7.7$; $s_{\text{pooled}_{\text{by eyeball}}} \approx 6.9$; $s_p = 6.28$
(d) $\overline{X}_1 - \overline{X}_2$; $s_{\overline{X}_1 - \overline{X}_{2_{\text{by eyeball}}}} \approx 2.5$; $s_{\overline{X}_1 - \overline{X}_2} = 2.56$
(e) t **(f)** .05 **(g)** $df = 22$
(h) $t_{\text{cv}_{\text{by eyeball}}}$ slightly greater in absolute value than ± 1.96—say, ± 2.05; $t_{\text{cv}_{\alpha=.05,\ df=22,\ \text{nondirectional}}} = \pm 2.074$
(i) $(\overline{X}_1 - \overline{X}_2)_{\text{cv}_{\text{by eyeball}}} \approx \pm 2.05(2.5) \approx \pm 5.1$; $(\overline{X}_1 - \overline{X}_2)_{\text{cv}} = \pm 5.31$
(j) $(\overline{X}_1 - \overline{X}_2)_{\text{by eyeball}} \approx .5$; $(\overline{X}_1 - \overline{X}_2)_{\text{obs}} = 3.25$
(k) $t_{\text{obs}_{\text{by eyeball}}} \approx .2$; $t_{\text{obs}} = 1.27$
(l) Yes; always
(m) No by eyeball; no by computation
(n) Not required
(o) Type of lighting does not affect morale.

6. In Exercise 4, each subject "serves as his own control" and most subjects show a decrease.

7. **(a)** H_0: $\mu_D \geq 0$; H_1: $\mu_D < 0$
(b) Directional; blondes are expected to have *more* fun.
(c) D; $s_{D_{\text{by eyeball}}} \approx [20 - (-23)]/4 \approx 11$; $s_D = 11.78$
(d) $\overline{D}$; $s_{\overline{D}_{\text{by eyeball}}} \approx 11/\sqrt{13} \approx 3$; $s_{\overline{D}} = 3.27$
(e) t
(f) $\alpha = .05$ as stated in the problem
(g) $df = 12$
(h) $t_{\text{cv}_{\text{by eyeball}}} \approx -1.75$, slightly greater in absolute value than 1.645; $t_{\text{cv}_{\alpha=.05,\, df=12,\ \text{directional}}} = -1.782$
(i) $\overline{D}_{\text{cv}_{\text{by eyeball}}} \approx -1.75(3) \approx -6$; $\overline{D}_{\text{cv}} = -5.82$
(j) $\overline{D}_{\text{by eyeball}} \approx [20 + (-23)]/2 = -1.5$; $\overline{D}_{\text{obs}} = -5.54$
(k) $t_{\text{by eyeball}} \approx -1.5/3 \approx -.5$; $t_{\text{obs}} = -1.70$
(l) Yes; always
(m) Fail to reject H_0 both by eyeball and by computation.
(n) Not required
(o) Becoming blonde does not lead to having more fun.

7.

Person	*Brunette*	*Blonde*	*Difference*
1	70	50	20
2	47	56	−9
3	57	63	−6
4	61	73	−12
5	65	51	14
6	57	65	−8
7	58	60	−2
8	58	63	−5
9	62	64	−2
10	56	66	−10
11	52	61	−9
12	60	83	−23
13	51	71	−20
		$\sum D_i =$	−72
		$\sum D_i^2 =$	2064.00
		$\overline{D}_{\text{obs}} =$	−5.54
		$s_D =$	11.78
		$s_{\overline{D}} =$	3.27

8.

	Brunettes	Blondes
	70	50
	47	56
	57	63
	61	73
	65	51
	57	65
	58	60
	58	63
	62	64
	56	66
	52	61
	60	83
	51	71
$\sum X_i$	754	826
$\sum X_i^2$	44,166.00	53,432.00
$\overline{X}_j$	58.00	63.54
s_j	6.01	8.89

$s_{\text{pooled}} = 7.59$
$s_{\overline{X}_1 - \overline{X}_2} = 2.98$
t_{obs}(independent samples, $df = 22$) $= -1.86$
$t_{\text{cv}\,\alpha=.05,\,df=24,\text{ directional}} = -1.711$
Reject H_0.

9. The data in Exercise 7 are not strongly statistically related to one another.

10. **(a) and (b)**

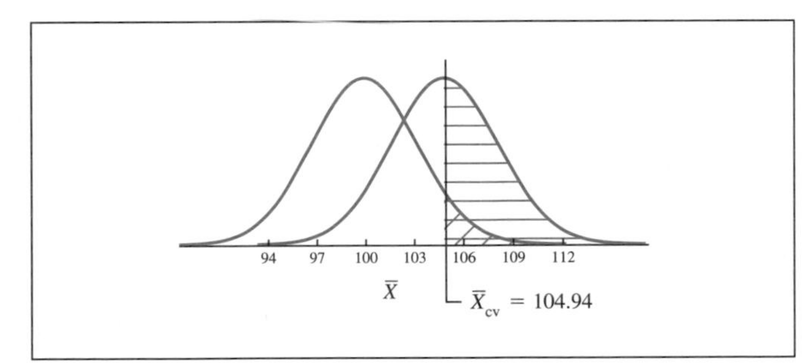

Distribution of sample means assuming H_0: $\mu = 100$ and H_1: $\mu = 105$, $n = 25$, $\alpha = .05$, directional, with rejection region shaded diagonally and power shaded horizontally

(c) Power$_{\text{by eyeball}} \approx .51$ **(d)** Power $= .508$
(e) $\beta = .492$

11. **(a) and (b)**

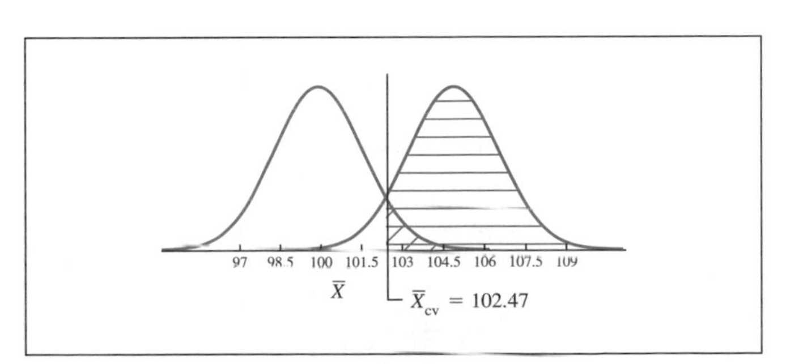

Distribution of sample means assuming H_0: $\mu = 100$ and H_1: $\mu = 105$, $n = 100$, $\alpha = .05$, directional, with rejection region shaded diagonally and power shaded horizontally

(c) Power$_{\text{by eyeball}} \approx .95$ **(d)** Power $= .954$
(e) $\beta = .046$
(f) Increasing sample size increases power.

12. **(a) and (b)**

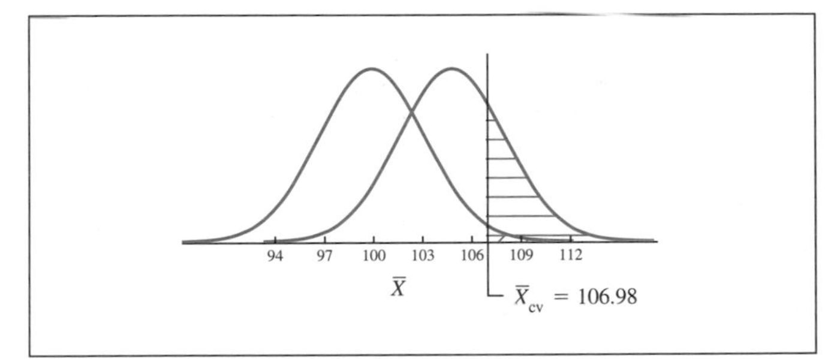

Distribution of sample means assuming H_0: $\mu = 100$ and H_1: $\mu = 105$, $n = 100$, $\alpha = .01$, directional, with rejection region shaded diagonally and power shaded horizontally

(c) Power$_{\text{by eyeball}} \approx .26$ **(d)** Power $= .255$
(e) $\beta = .745$
(f) Decreasing the level of significance (thus making the test more stringent) decreases power.

13. **(a) and (b)**

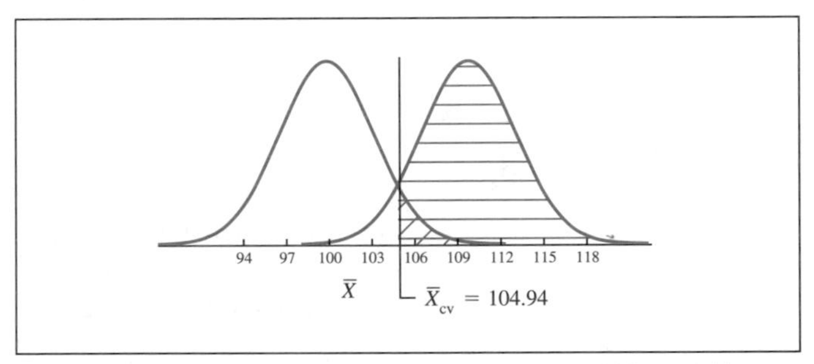

Distribution of sample means assuming H_0: $\mu = 100$ and H_1: $\mu = 110$, $n = 100$, $\alpha = .05$, directional, with rejection region shaded diagonally and power shaded horizontally

(c) Power$_{\text{by eyeball}} \approx .95$ **(d)** Power $= .954$
(e) $\beta = .046$
(f) Increasing effect size increases power.

14. **(a) and (b)**

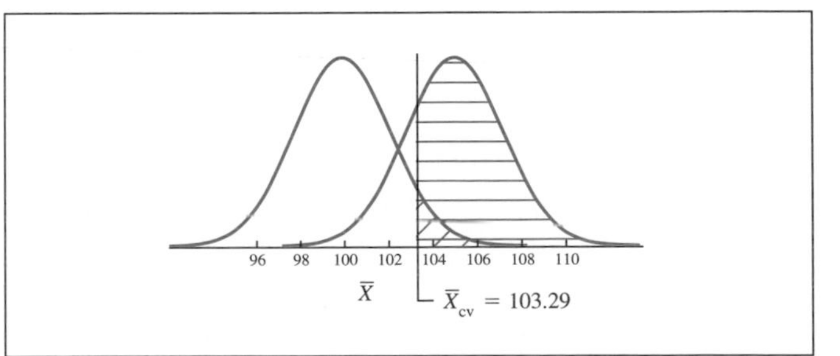

Distribution of sample means assuming H_0: $\mu = 100$ and H_1: $\mu = 105$, $n = 100$, $\alpha = .05$, $\sigma = 10$, directional, with rejection region shaded diagonally and power shaded horizontally

(c) Power$_{\text{by eyeball}} \approx .80$ **(d)** Power $= .804$
(e) $\beta = .196$
(f) Decreasing σ increases power.

15. (a) and (b)

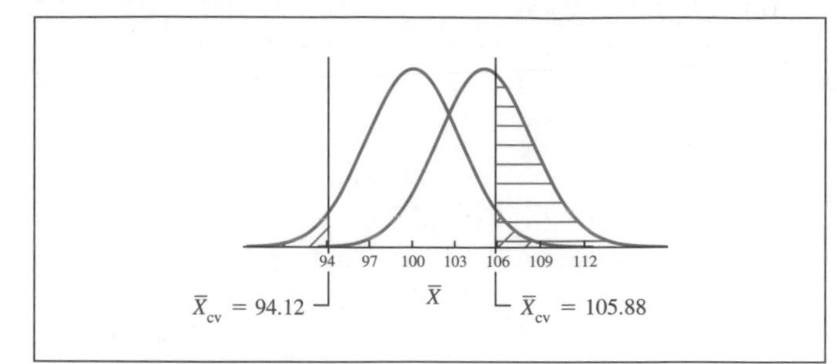

Distribution of sample means assuming H_0: $\mu = 100$ and H_1: $\mu = 105$, $n = 25$, $\alpha = .05$, nondirectional, with rejection region shaded diagonally and power shaded horizontally

(c) $\text{Power}_{\text{by eyeball}} \approx .37$ **(d)** $\text{Power} = .385$
(e) $\beta = .615$
(f) Making a test nondirectional decreases power.

16. 196 subjects

Chapter 12, Section C: Extending comprehension

29. Yes
30. Yes
31. Yes
32. Yes
33. Yes
34. It is 0.
35. It is 1.645 standard errors.
36. Yes; the test must be directional and the parameter as specified by the real distribution must be in the direction of the other tail.
37. Yes; if the parameter as specified by the real distribution is in the opposite tail from the directional rejection region

Chapter 12, Section D: Cumulative review

38. (a) 4 **(b)** 2 **(c)** 3 **(d)** 2
(e) 70 **(f)** 2 **(g)** ± 1.997 **(h)** (11.2)

39. (a) 5 **(b)** 3 **(c)** 4 **(d)** 2
(e) 14 **(f)** 2 **(g)** ± 2.145 **(h)** (12.3)

40. (a) 2 **(b)** 4 **(c)** 1 **(d)** 2
(e) 35 **(f)** 2 **(g)** ± 2.031 **(h)** (8.3)

41. (a) 1 **(h)** (6.1)

42. (a) 4 **(b)** 2 **(c)** 3 **(d)** 2
(e) 18 **(f)** 1 **(g)** 1.734 **(h)** (11.2)

43. 6

44. (a) 3 **(b)** 1
(c) 1 **(d)** 1
(e) Not applicable **(f)** 1
(g) 1.645 **(h)** (10.1)

Chapter 12, Section E: From the journals

45. (a) Yes **(b)** No
(c) $s_D = 1.45$ beats per minute
(d) Heart rate decreased significantly after the exercise program.

46. (a) 13 **(b)** 1.62
(c) Since the study from Exercise 45 gives us reason to believe that the heart rate will decrease, we can use a directional test; thus, $t_{cv} = 1.771$. The replication does not support the idea that exercise decreases heart rate in this population.

47. (a) Yes; 1 second **(b)** No **(c)** 14
(d) 2.144 (otherwise it would be significant)
(e) 1.8 seconds

48. (a) 13 **(b)** -1.42
(c) Given that the test is nondirectional and that $\alpha = .05$, $t_{cv} = 2.160$, there is no significant difference in breath-holding time when subjects smoke potent joints or placebo joints.

49. (a) .63 **(b)** Yes

50. A very small n because then power is very low and the null hypothesis is not likely to be rejected even if Coke is better

Chapter 13, Section A: Basic exercises

1. (a) H_0: $\mu_1 = \mu_2 = \mu_3 = \mu_4$; H_1: The null hypothesis is false.
(b) Nondirectional in the sense that it asks only for *differences* among means with no direction specified (the case for all ANOVAs)
(c) F
(d) $\alpha = .05$ set by problem
(e) 3 and 16
(f) $F_{\text{cv}_{\text{by eyeball}}} \approx 3.5$; $F_{\text{cv}_{\alpha=.05,\, df=3,16}} = 3.24$
(g) $\overline{X}_{j\,\text{by eyeball}} \approx 8, 5, 7, 7$

$s_{\overline{X}_{\text{by eyeball}}} \approx (8-5)/2.5 \approx 1.3$

$s^2_{\overline{X}_{\text{by eyeball}}} \approx 1.7$

$s^2_{\overline{X}_{\text{by eyeball}}}(n) = \text{MS}_{\text{B}_{\text{by eyeball}}} \approx 8.5$

$s_{j\,\text{by eyeball}} \approx .8,\ 2.8,\ 1.6,\ 1.3$

$s^2_{j\,\text{by eyeball}} \approx .64,\ 8,\ 2.6,\ 1.7$

$s^2_{\text{pooled}_{\text{by eyeball}}} = \text{MS}_{\text{W}_{\text{by eyeball}}} \approx 3.5$

$F_{\text{by eyeball}} \approx 2$

	Budweiser	*Miller*	*Coors*	*Strohs*
	8	4	9	6
	8	4	5	7
	7	3	6	6
	8	10	9	7
	9	4	6	4
n_j	5	5	5	5
$\overline{X}_j$	8.00	5.00	7.00	6.00
s^2_j	.500	8.000	3.500	1.500

Source	*SS*	*df*	*MS*	*F*
Between	25.00	3	8.33	2.47
Within	54.00	16	3.38	
Total	79.00	19		

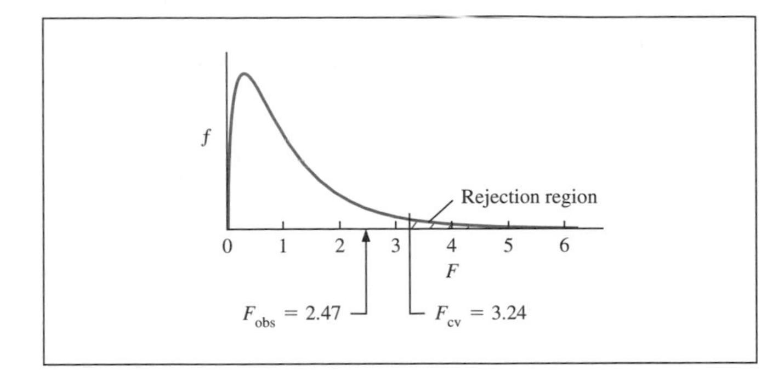

(h) Fail to reject H_0 both by eyeball and by computation.
(i) Not required
(j) There is no significant difference among preference ratings.

2. **(a)** H_0: $\mu_1 = \mu_2 = \mu_3$; H_1: The null hypothesis is false.
(b) Nondirectional in the sense that it asks only for *differences* among means with no direction specified (the case for all ANOVAs)
(c) F **(d)** $\alpha = .05$ set by problem
(e) 2 and 21
(f) $F_{\text{cv}_{\text{by eyeball}}} \approx 3.5$; $F_{\text{cv}_{\alpha=.05, df=2,21}} = 3.47$
(g) $\bar{X}_{j_{\text{by eyeball}}} \approx 214,\ 203,\ 238$ feet
$s_{\bar{X}_{\text{by eyeball}}} \approx (238 - 203)/2.5 = 14$ feet
$s^2_{\bar{X}_{\text{by eyeball}}} \approx 196$ feet^2
$s^2_{\bar{X}_{\text{by eyeball}}}(n) = \text{MS}_{\text{B}_{\text{by eyeball}}} = 1570$ feet^2
$s_{j_{\text{by eyeball}}} \approx 18,\ 14,\ 15$ feet
$s^2_{j_{\text{by eyeball}}} \approx 324,\ 196,\ 225$ feet^2
$s^2_{\text{pooled}_{\text{by eyeball}}} = \text{MS}_{\text{W}_{\text{by eyeball}}} \approx 250$ feet^2
$F_{\text{by eyeball}} \approx 6.3$

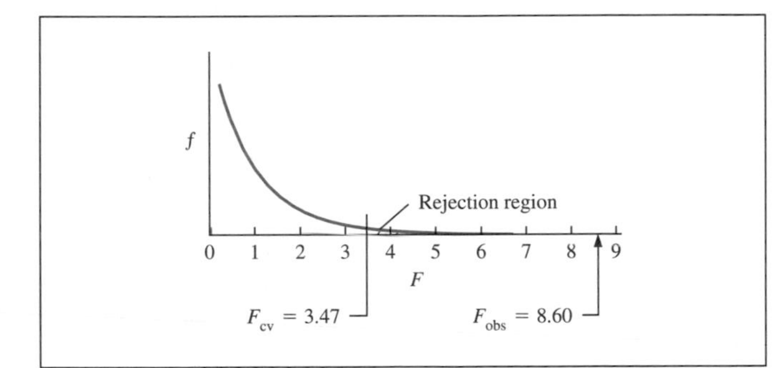

2.

	Process A	Process B	Process C
	213.7	199.8	222.7
	219.3	207.2	245.1
	235.0	191.4	225.0
	217.3	180.1	216.7
	193.8	194.2	223.2
	200.2	209.0	225.1
	222.7	227.1	261.8
	178.8	185.6	236.4
n_j	8	8	8
$\bar{X}_j$	210.100	199.300	232.000
s^2_j	324.50	224.91	224.80

Source	SS	df	MS	F
Between	4441.440	2	2220.72	8.60
Within	5419.580	21	258.08	
Total	9861.020	23		

(h) Reject H_0 both by eyeball and by computation.
(i) $f = 1.04$; $d_M = 2.04$; $U = 98\%$

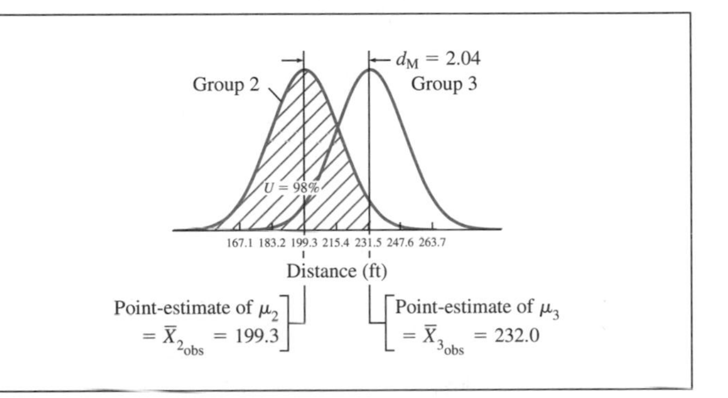

(j) The bat manufacturing process does affect bat power. The mean distance of the longest-hitting bat is greater than 98% of the hits of the shortest-hitting bat.

3. **(i)** It should decrease because the means are now closer together.
(ii) (a) No change **(b)** No change **(c)** No change
(d) No change **(e)** No change **(f)** No change
(g) $\bar{X}_{j_{\text{by eyeball}}} \approx 214,\ 203,\ \underline{208}$ feet (changes underlined)
$s_{\bar{X}_{\text{by eyeball}}} \approx \underline{(214} - 203)/2.5 = \underline{4.4}$ feet
$s^2_{\bar{X}_{\text{by eyeball}}} \approx \underline{19}$ feet^2
$s^2_{\bar{X}\text{ by eyeball}}(n) = \text{MS}_{\text{B}_{\text{by eyeball}}} \approx \underline{152}$ feet^2
$s_{j_{\text{by eyeball}}}$ (no change)
$s^2_{j_{\text{by eyeball}}}$ (no change)
$s^2_{\text{pooled}_{\text{by eyeball}}} = \text{MS}_{\text{W}_{\text{by eyeball}}}$ (no change)
$F_{\text{by eyeball}} \approx \underline{.6}$

3.

	Process A	Process B	Process C
	No change	No change	192.7
			215.1
			195.0
			186.7
			193.2
			195.1
			231.8
			206.4
n_j	No change	No change	No change
$\bar{X}_j$			202.000
s^2_j			224.81

Source	SS	df	MS	F
Between	505.440	No change	252.720	.98
Within	No change	No change	No change	
Total	5925.020	No change		

(h) Fail to reject H_0 both by eyeball and by computation.
(i) Not required
(j) The bat manufacturing process does not affect bat power.

4.

	Sample 1	*Sample 2*
	4	8
	1	6
	9	2
	6	6
	2	3
	5	9
	7	7
	7	1
n_j	8	8
$\bar{X}_j$	5.13	5.25
s_j^2	7.268	8.500

Source	*SS*	*df*	*MS*	*F*
Between	.0625	1	.0625	.0079
Within	110.375	14	7.88	
Total	110.44	15		

(a) $F_{\text{obs}} = .0079 = (t_{\text{obs}})^2 = (-.089)^2$
(b) $F_{\text{cv}_{\alpha=.05, df=1,14}} = 4.60 = (t_{\text{cv}_{df=14}})^2 = (\pm 2.145)^2$
(c) Yes

5. **(a)** H_0: $\mu_1 = \mu_2 = \mu_3 = \mu_4 = \mu_5$; H_1: H_0 is false.
(b) Nondirectional in the sense that it asks only for *differences* among means with no direction specified (the case for all ANOVAs)
(c) F
(d) $\alpha = .05$ set by problem
(e) 4 and 23
(f) $F_{\text{cv}_{\text{by eyeball}}} \approx 3.5$; $F_{\text{cv}_{\alpha=.05, df=4,23}} = 2.80$
(g) $\bar{X}_{j_{\text{by eyeball}}} \approx 23, 25.5, 18.5, 20.5, 24$
$s_{\bar{X}_{\text{by eyeball}}} \approx (25.5 - 18.5)/2.5 \approx 2.8$
$s^2_{\bar{X}_{\text{by eyeball}}} \approx 8$
$s^2_{\bar{X}_{\text{by eyeball}}}(n) = \text{MS}_{\text{B}_{\text{by eyeball}}} \approx 40$
$s_{j_{\text{by eyeball}}} \approx 2.8, 3.3, 3.3, 2, 4$
$s^2_{j_{\text{by eyeball}}} \approx 8, 10, 10, 4, 16$
$s^2_{\text{pooled}_{\text{by eyeball}}} = \text{MS}_{\text{W}_{\text{by eyeball}}} \approx 10$
$F_{\text{by eyeball}} \approx 4$

Table for Chapter 13, Section A, Exercise 5(g)

		Total		*Within groups*		*Between groups*	
	X_{ij}	$X_{ij} - \bar{X}_G$	$(X_{ij} - \bar{X}_G)^2$	$X_{ij} - \bar{X}_j$	$(X_{ij} - \bar{X}_j)^2$	$\bar{X}_j - \bar{X}_G$	$(\bar{X}_j - \bar{X}_G)^2$
Shape 1 ($\bar{X}_1 = 22.5$)	21	−1.5	2.25	−1.5	2.25	.0	.00
	22	−.5	.25	−.5	.25	.0	.00
	27	4.5	20.25	4.5	20.25	.0	.00
	20	−2.5	6.25	−2.5	6.25	.0	.00
Shape 2 ($\bar{X}_2 = 26.0$)	30	7.5	56.25	4.0	16.00	3.5	12.25
	23	.5	.25	−3.0	9.00	3.5	12.25
	26	3.5	12.25	.0	.00	3.5	12.25
	31	8.5	72.25	5.0	25.00	3.5	12.25
	23	.5	.25	−3.0	9.00	3.5	12.25
	23	.5	.25	−3.0	9.00	3.5	12.25
Shape 3 ($\bar{X}_3 = 20.0$)	21	−1.5	2.25	1.0	1.00	−2.5	6.25
	23	.5	.25	3.0	9.00	−2.5	6.25
	20	−2.5	6.25	.0	.00	−2.5	6.25
	20	−2.5	6.25	.0	.00	−2.5	6.25
	15	−7.5	56.25	−5.0	25.00	−2.5	6.25
	23	.5	.25	3.0	9.00	−2.5	6.25
	18	−4.5	20.25	−2.0	4.00	−2.5	6.25
Shape 4 ($\bar{X}_4 = 20.0$)	19	−3.5	12.25	−1.0	1.00	−2.5	6.25
	21	−1.5	2.25	1.0	1.00	−2.5	6.25
	18	−4.5	20.25	−2.0	4.00	−2.5	6.25
	23	.5	.25	3.0	9.00	−2.5	6.25
	19	−3.5	12.25	−1.0	1.00	−2.5	6.25
Shape 5 ($\bar{X}_2 = 24.0$)	26	3.5	12.25	2.0	4.00	1.5	2.25
	20	−2.5	6.25	−4.0	16.00	1.5	2.25
	20	−2.5	6.25	−4.0	16.00	1.5	2.25
	23	.5	.25	−1.0	1.00	1.5	2.25
	31	8.5	72.25	7.0	49.00	1.5	2.25
	24	1.5	2.25	.0	.00	1.5	2.25
			409.00		247.00		162.00

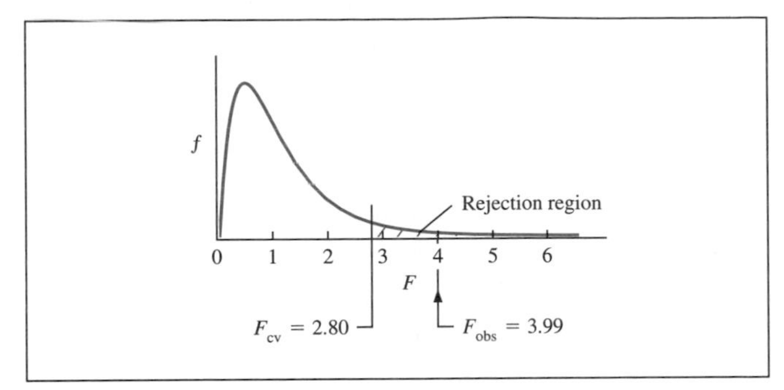

Source	SS	df	MS	F
Between	162.00	4	40.50	3.77
Within	247.00	23	10.74	
Total	409.00	27		

(h) Reject the null hypothesis
(i) $f = 2.60/3.28 = .79$; $d_M = 1.83$; $U = 97\%$

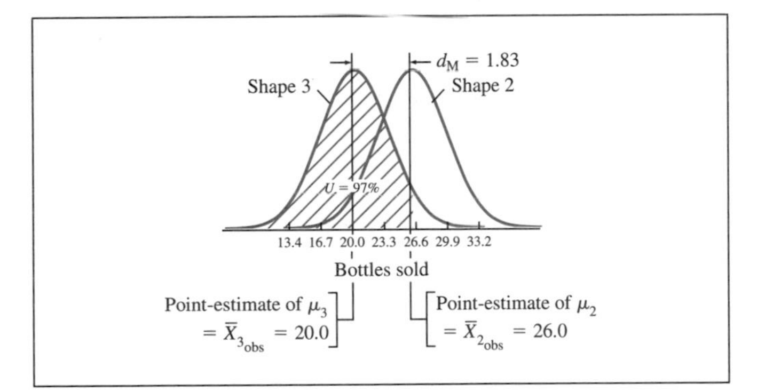

(j) Bottle shape does affect sales; the mean sales of the largest selling bottle is greater than 97% of the sales of the smallest-selling bottle.

5.

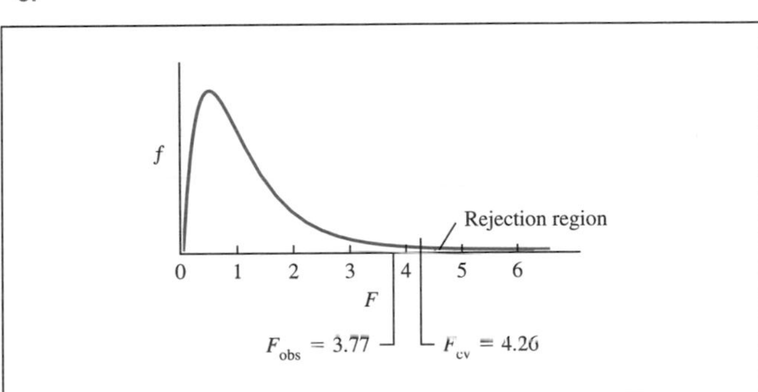

6. **(a)** No change **(b)** No change
(c) No change **(d)** $\alpha = .01$
(e) No change **(f)** $F_{cv_{\alpha=.01, df=4,23}} = 4.26$
(g) No change
(h) Fail to reject the null hypothesis.
(i) No longer required
(j) Bottle shape does not have an effect on sales.

7. (a)

Source	SS	df	MS	F
Between	215.00	5	43.00	2.96
Within	435.55	30	14.52	
Total	650.55	35		

(b) Six groups **(c)** Six subjects in each group

8. (a)

Source	SS	df	MS	F
Between	81.35	3	27.12	9.19
Within	106.22	36	2.95	
Total	187.57	39		

(b) Four groups **(c)** Ten subjects in each group

9.

Source	SS	df	MS	F
Between	60.00	3	20.00	7.0
Within	91.43	32	2.86	
Total	151.43	35		

10. $d_{M_{population}} = (520 - 500)/100 = 20$, which is "small"; the required number of subjects in each group is approximately 274.

Chapter 13, Section C: Extending comprehension

21. .0

22. ∞

23. 0/0, undefined

Chapter 13, Section D: Cumulative review

25. (a) 2 **(b)** 5 **(c)** 1 **(d)** 2
(e) 24 **(f)** 2 **(g)** ± 2.064 **(h)** (8.3)

26. (a) 1 **(h)** (7.2)

27. (a) 6 **(b)** 4 **(c)** 5 **(d)** 3
(e) 2 and 34 **(f)** 2 **(g)** 3.28 **(h)** (13.6)

28. (a) 7

29. (a) 5 **(b)** 3 **(c)** 4 **(d)** 2
(e) 23 **(f)** 2 **(g)** ± 2.069 **(h)** (12.3)

30. (a) 4 **(b)** 2 **(c)** 3 **(d)** 2
(e) 18 **(f)** 1 **(g)** 1.734 **(h)** (11.2)

31. (a) 3 (see Exercise 7 in Chapter 10)
(b) 1 **(c)** 2 **(d)** 2 **(e)** 299
(f) 1 **(g)** 1.645 **(h)** (10.4)

Chapter 13, Section E: From the journals

32. (a) No; there should be 3 and 28 degrees of freedom.
(b) No
(c) Assuming equal n_j's, $MS_B = 117.73$.
(d) $MS_W = 66.89$

(e)

Source	SS	df	MS	F
Between	353.19	3	117.73	1.76
Within	1872.92	28	66.89	
Total	2226.11	31		

(f) We cannot conclude that the four groups have different means to begin with.

33. (a) 3 and 20

(b)

Source	SS	df	MS	F
Between	66.82	3	22.273	.83
Within	536.76	20	26.838	
Total	603.58	23		

(c) There is no significant difference in mean response rates between these groups.

Chapter 14, Section A: Basic exercises

1. (a) Yes; we rejected H_0

(b) Three: $H_{0_1}: \mu_1 = \mu_2; H_{1_1}: \mu_1 \neq \mu_2$
$H_{0_2}: \mu_1 = \mu_3; H_{1_2}: \mu_1 \neq \mu_3$
$H_{0_3}: \mu_2 = \mu_3; H_{1_3}: \mu_2 \neq \mu_3$

(c) Nondirectional because the procedure does not distinguish between positive and negative Q

(d) Three Q statistics

(e) .05 set by original problem

(f) $k = 3, df_W = 21$

(g) $Q_{cv_{k=3;\,df_W=21}} = 3.57$

(h) $\tilde{n} = 8$

(i) $\sqrt{MS_W/n_j} = 5.68$

i	1	2	3	i
$\bar{X}_j$	210.1	199.3	232.0	
$\bar{X}_i - \bar{X}_j$	—	10.8	−21.9	1
	—	—	−32.7	2
Q	—	1.90	3.86*	1
	—	—	5.76*	2

$*p < .05, Q_{cv_{k=3;\,df_W=21}} = 3.57$

(j) Reject H_{0_2} and H_{0_3}.

(k) Manfacturing process C is significantly different from process A or B, but process A does not differ significantly from process B.

2. (a) $\mu_1 = \mu_2 = \mu_3 = \mu_4 = \mu_5$

(b) An omnibus F test; F; 4 and 45

(c) H_0: $\frac{\mu_2 + \mu_4}{2} = \frac{\mu_1 + \mu_3 + \mu_5}{3}$

(d) H_0: $-.333\mu_1 + .5\mu_2 - .333\mu_3 + .5\mu_4 - .333\mu_5 = 0$

(f) Planned comparison; t with 45 degrees of freedom

(g) 3.683 **(h)** 1.245

(i) $t_{obs} = 2.959$ **(j)** 2.015

(k) Yes **(l)** 3.68 LSS points

(m) .85 **(n)** 80.23%

(o) The married individuals (mean 50.45) had statistically significantly higher LSS scores than did the unmarried individuals (mean 46.77), for a difference between groups of 3.68 LSS points or .85 standard deviation. Approximately 80.23% of married individuals would be expected to have higher LSS scores than the average unmarried individual.

3. (a) No; the $\alpha = .05$ critical value F_{cv} with 4 and 45 degrees of freedom is 2.58.

(b) Planned comparisons are substantially more powerful than the omnibus F.

4. (a) Repeated-measures

(b) F with 2 and 8 degrees of freedom

(c) 4.46

(d)

Source	SS	df	MS	F
Between occasions	384.4000	2	192.2000	5.15
Between subjects	7397.7333	4	1849.4333	
Residual	298.2667	8	37.2833	
Total	8080.4000	14		

(e) Yes

(f) $12.4/\sqrt{2(37.2833)} = 1.44$

(g) 92.51%

(h) There is a statistically significant difference between the three times. The largest mean (10:00 P.M.) is 12.4 units or 1.44 standard deviations greater than the smallest mean (5:00 P.M.), which indicates that about 93% of patients would have higher readings at 10:00 P.M. than at 5:00 P.M.

5. (a) No; the $\alpha = .05$ critical value F_{cv} with 2 and 12 degrees of freedom is 3.89.

(b) The repeated-measures ANOVA can be much more powerful if there are substantial differences between the subjects (as there are here).

Chapter 14, Section C: Extending comprehension

11. Yes; post hoc tests are less powerful than ANOVA.

12. No; planned comparisons are always more powerful than ANOVA.

Chapter 14, Section D: Cumulative review

14. (a) 1 **(h)** (6.1)

15. (a) 10

16. (a) 3 **(b)** 1
(c) 1 **(d)** 1
(e) Not applicable **(f)** 2
(g) ±1.96 **(h)** (10.1)

17. (a) 9 **(b)** 4 **(c)** 7 **(d)** 3
(e) 3 and 24 **(f)** 2 **(g)** 3.01 **(h)** (14.14)

18. (a) 8 **(b)** 6 **(c)** 6 **(d)** 2
(e) 28 **(f)** 2 **(g)** 2.048 **(h)** (14.10)

19. (a) 2 **(b)** 7 **(c)** 2 **(d)** 2
(e) 24 **(f)** 2 **(g)** ±2.064 **(h)** (8.3)

20. (a) 6 **(b)** 4 **(c)** 7 **(d)** 3
(e) 2 and 21 **(f)** 2 **(g)** 3.47 **(h)** (13.6)

21. (a) 7 **(b)** 5; ten tests **(c)** 5 **(d)** 4
(e) 5 and 45 **(f)** 2 **(g)** 4.03 **(h)** (14.2)

Chapter 14, Section E: From the journals

22. (a) No; there should be 120 total degrees of freedom (perhaps Erinosho randomly eliminated one subject from the no-treatment group to give the groups equal sizes but failed to report that in her description).

(b) No; SS_T and MS_W do not compute correctly. Perhaps SS_W should actually be 15,440.4, and MS_W should be 132.0.

(c) No; MS_W is the pooled within-cell variance, which should be about $9^2 \approx 81$; 132 is far too high to be possible.
(d) Yes
(e) Apparently so, although they are not computed correctly; the t between the first two groups should be 7.11, and between the first and third groups 7.52 (the remaining t is computed correctly).
(f) No; a post hoc test such as Tukey should be used.
(g) Computation of Q in Tukey's test:

j	1	2	3	i
$\bar{X}_j$	59.31	46.23	44.09	
$\bar{X}_i - \bar{X}_j$		13.08	15.22	1
			2.14	2
Q		7.18*	8.35*	1
			1.17	2

$*p < .05, Q_{\text{cv}_{k=3, df_W=118}} = 4.20$

(h) They are the same.

23. **(a)** ANOVA summary table:

Source	SS	df	MS	F
Between	1476.2	2	738.10	11.60*
Within	1718.1	27	63.63	
Total	3194.3	29		

$*p < .05; F_{\text{cv}} = 3.35$

(b) The groups do not all have the same mean.
(c) Computation of Q in Tukey's test:

j	1	2	3	i
$\bar{X}_j$	58.60	43.20	44.30	
$\bar{X}_i - \bar{X}_j$		15.40	14.30	1
			−1.10	2
Q		6.10*	5.67*	1
			.44	2

$*p < .05, Q_{\text{cv}_{k=3, df_W=27}} = 3.51$

(d) The teacher-directed group is different from both of the other two groups, but the student-directed and no-treatment groups are not significantly different from each other.

Chapter 15, Section A: Basic exercises

1. **(a)**

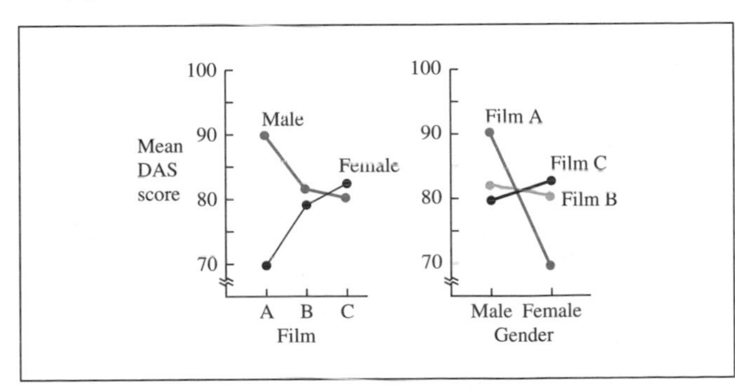

(b) No; it is clearer from the one with Film on the X-axis.
(c) Perhaps; it is hard to tell from these data. Males have somewhat higher DAS scores than females. It is clearer from the one with Gender on the X-axis.
(d) Yes; males have higher DAS scores after watching film A than do females. It is clear on either graph.

2. **(a)** Choice of film has no effect on drug attitude; $\mu_A = \mu_B = \mu_C$.
(b) Gender has no effect on drug attitude; $\mu_{\text{male}} = \mu_{\text{female}}$.
(c) The mean of the male/film A cell will be above (or below) the grand mean by the sum of the distance that the male mean is above the grand mean and the distance that the film A mean is above the grand mean; $\mu_{\text{film A, male}} = \mu + (\mu_{\text{film A}} - \mu) + (\mu_{\text{male}} - \mu)$.
(d) All $\mu_{r,c} - \mu_r - \mu_c + \mu = 0$

3. **(a)** 29 **(b)** 2 **(c)** 1 **(d)** 2 **(e)** 4 **(f)** 24 **(g)** Yes; yes

4. **(a)** Tabulations of DAS data by gender and film: (see p. 526 following)

(b)

Source of variation	Sum of squares (SS)	Degrees of freedom (df)	Mean square (MS)	F
Rows (Film)	61.40	2	30.70	1.20[a]
Columns (Gender)	307.20	1	307.20	12.02[*b]
RC interaction (Film by Gender)	681.80	2	340.90	13.33[*c]
Within cell	613.60	24	25.57	
Total	1664.00	29		

[a] $p > .05, F_{\text{cv}_{2,24}} = 3.40$
[*b] $p < .05, F_{\text{cv}_{1,24}} = 4.26$
[*c] $p < .05, F_{\text{cv}_{2,24}} = 3.40$

(c) The main effect for film was not statistically significant. The main effect for gender was statistically significant: Males ($\bar{X}_M = 83.2$) scored higher on the DAS than did females ($\bar{X}_F = 76.8$). However, this main effect was qualified by the interaction between film and gender.
(d) Residuals:

	Males	Females
Film A	6.5	−6.5
Film B	−1.7	1.7
Film C	−4.8	4.8

Males respond even more favorably to film A than would be predicted by the main effects alone, whereas females respond even less favorably to that film. That interaction effect is reversed for film C.

Table for Chapter 15, Section A, Exercise 4(a)

		Total		*Within cell*		*Row effect*		*Column effect*		*Interaction*	
	X_{irc}	$X_{irc} - \overline{X}_G$	$(X_{irc} - \overline{X}_G)^2$	$X_{irc} - \overline{X}_{rc}$	$(X_{irc} - \overline{X}_{rc})^2$	$\overline{X}_r - \overline{X}_G$	$(\overline{X}_r - \overline{X}_G)^2$	$\overline{X}_c - \overline{X}_G$	$(\overline{X}_c - \overline{X}_G)^2$	$\overline{X}_{rc} - \overline{X}_r - \overline{X}_c + \overline{X}_G$	$(\overline{X}_{rc} - \overline{X}_r - \overline{X}_c + \overline{X}_G)^2$
	80	0	0	−8.0	64.00	−1.7	2.89	3.2	10.24	6.5	42.25
Row 1	86	6	36	−2.0	4.00	−1.7	2.89	3.2	10.24	6.5	42.25
Column 1	90	10	100	2.0	4.00	−1.7	2.89	3.2	10.24	6.5	42.25
$\overline{X}_{11} = 88.0$	93	13	169	5.0	25.00	−1.7	2.89	3.2	10.24	6.5	42.25
	91	11	121	3.0	9.00	−1.7	2.89	3.2	10.24	6.5	42.25
	82	2	4	.6	.36	−.1	.01	3.2	10.24	−1.7	2.89
Row 2	79	−1	1	−2.4	5.76	−.1	.01	3.2	10.24	−1.7	2.89
Column 1	84	4	16	2.6	6.76	−.1	.01	3.2	10.24	−1.7	2.89
$\overline{X}_{21} = 81.4$	76	−4	16	−5.4	29.16	−.1	.01	3.2	10.24	−1.7	2.89
	86	6	36	4.6	21.16	−.1	.01	3.2	10.24	−1.7	2.89
	78	−2	4	−2.2	4.84	1.8	3.24	3.2	10.24	−4.8	23.04
Row 3	91	11	121	10.8	116.64	1.8	3.24	3.2	10.24	−4.8	23.04
Column 1	71	−9	81	−9.2	84.64	1.8	3.24	3.2	10.24	−4.8	23.04
$\overline{X}_{31} = 80.2$	83	3	9	2.8	7.84	1.8	3.24	3.2	10.24	−4.8	23.04
	78	−2	4	−2.2	4.84	1.8	3.24	3.2	10.24	−4.8	23.04
	75	−5	25	6.4	40.96	−1.7	2.89	−3.2	10.24	−6.5	42.25
Row 1	68	−12	144	−.6	.36	−1.7	2.89	−3.2	10.24	−6.5	42.25
Column 2	71	−9	81	2.4	5.76	−1.7	2.89	−3.2	10.24	−6.5	42.25
$\overline{X}_{12} = 68.6$	71	−9	81	2.4	5.76	−1.7	2.89	−3.2	10.24	−6.5	42.25
	58	−22	484	−10.6	112.36	−1.7	2.89	−3.2	10.24	−6.5	42.25
	77	−3	9	−1.4	1.96	−.1	.01	−3.2	10.24	1.7	2.89
Row 2	77	−3	9	−1.4	1.96	−.1	.01	−3.2	10.24	1.7	2.89
Column 2	80	0	0	1.6	2.56	−.1	.01	−3.2	10.24	1.7	2.89
$\overline{X}_{22} = 78.4$	82	2	4	3.6	12.96	−.1	.01	−3.2	10.24	1.7	2.89
	76	−4	16	−2.4	5.76	−.1	.01	−3.2	10.24	1.7	2.89
	86	6	36	2.6	6.76	1.8	3.24	−3.2	10.24	4.8	23.04
Row 3	82	2	4	−1.4	1.96	1.8	3.24	−3.2	10.24	4.8	23.04
Column 2	84	4	16	.6	.36	1.8	3.24	−3.2	10.24	4.8	23.04
$\overline{X}_{32} = 83.4$	79	−1	1	−4.4	19.36	1.8	3.24	−3.2	10.24	4.8	23.04
	86	6	36	2.6	6.76	1.8	3.24	−3.2	10.24	4.8	23.04
Σ	2400	0	$1664 = SS_T$	.0	$613.60 = SS_W$	.0	$61.40 = SS_R$	.0	$307.20 = SS_C$	.0	$681.80 = SS_{RC}$

5. **(a)** 

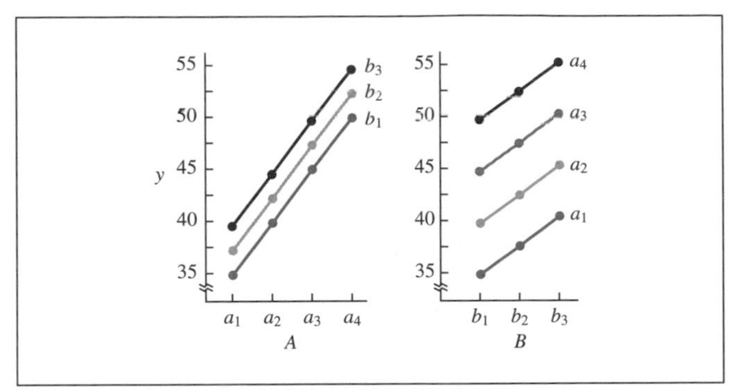

(b) Yes; it seems that all levels of A are different from one another.
(c) Yes; it seems that all levels of B are different from one another.
(d) Interaction appears to be exactly 0 because the lines are exactly parallel; there is no differential effect of any variable in any cell.

6. **(a)** The level of A has no effect on Y; $\mu_{a_1} = \mu_{a_2} = \mu_{a_3} = \mu_{a_4}$.
(b) The level of B has no effect on Y; $\mu_{b_1} = \mu_{b_2} = \mu_{b_3}$.
(c) The effect of b_3 on Y is the same in a_2 as in any other level of A; $\mu_{a_2,b_3} = \mu + (\mu_{a_2} - \mu) + (\mu_{b_3} - \mu)$.
(d) All $\mu_{r,c} - \mu_r - \mu_c + \mu = 0$

7. **(a)** 35 **(b)** 3 **(c)** 2
(d) 6 **(e)** 2 **(f)** 24
(g) Yes; it must.

8. **(a)**

Source of variation	Sum of squares (SS)	Degrees of freedom (df)	Mean square (MS)	F
Rows (B)	150.00	2	75.00	.19[a]
Columns (A)	1125.00	3	375.00	.94[b]
Interaction ($A \times B$)	.00	6	.00	.00[c]
Within cell	9600.00	24	400.00	
Total	10,875.00	35		

[a] $p > .05, F_{cv_{2,24}} = 3.40$
[b] $p > .05, F_{cv_{3,24}} = 3.01$
[c] $p > .05, F_{cv_{6,24}} = 2.51$

(b) The plots do not show any information about the within-cell variance, which is very large in these data. The F ratios are thus all small; we had no way of knowing this from the plots themselves.
(c) Same as part (b).
(d) The lines are in fact exactly parallel.

9. They are identical.

10. **(a)** No change **(b)** No change
(c) No change **(d)** Should be much smaller
(e) No change **(f)** No change
(g) No change **(h)** No change
(i) No change **(j)** No change
(k) No change **(l)** Should be much smaller
(m) Should be much larger **(n)** Should be much larger
(o) Should be many times larger, but many times zero is still zero

Chapter 15, Section C: Extending comprehension

17. No; it is not possible.

18. Yes; the first row within each column must be much different from the second row (for example, the first three elements of each column are much smaller than the second three elements).

19. Yes; the row main effect and interaction must be very small or 0 (for example, if the first three elements in each column are identical to the last three elements in each column), and the one-way ANOVA must be just barely significant so that the loss of degrees of freedom in the two-way test makes it just barely miss being significant.

20. Yes; the mean of all the first-row elements must be identical to the mean of all the second-row elements; the mean of the first column must equal the mean of the second column must equal the mean of the third column; but the individual cell means must differ.

Chapter 15, Section D: Cumulative review

21. **(a)** 3 **(b)** 1
(c) 1 **(d)** 1
(e) Not applicable **(f)** 1
(g) −1.645 **(h)** (10.1)

22. **(a)** 10 **(b)** 7 **(c)** 6 **(d)** 3
(e) Rows: 1 and 24; columns: 2 and 24; interaction: 2 and 24
(f) 2
(g) Rows: 4.26; columns: 3.40; interaction: 3.40
(h) Rows: (15.11); columns: (15.12); interaction: (15.13)

23. **(a)** 9 **(b)** 4 **(c)** 7 **(d)** 3
(e) 2 and 14 **(f)** 2 **(g)** 3.74 **(h)** 14.14

24. **(a)** 7 **(b)** 5; six null hypotheses
(c) 5 **(d)** 4
(e) 4 and 36 **(f)** 2
(g) 3.82 **(h)** (14.2)

25. **(a)** 4 **(b)** 2 **(c)** 3 **(d)** 2
(e) 28 **(f)** 2 **(g)** ±2.048 **(h)** (11.2)

26. **(a)** 2 **(b)** 8 **(c)** 2 **(d)** 2
(e) 299 **(f)** 2 **(g)** ±1.96 **(h)** (8.5)

27. **(a)** 8 **(b)** 6 **(c)** 6 **(d)** 2
(e) 21 **(f)** 2 **(g)** 2.080 **(h)** (14.10)

28. **(a)** 5 **(b)** 3 **(c)** 4 **(d)** 2
(e) 19 **(f)** 1 **(g)** −1.729 **(h)** (12.3)

29. **(a)** 1 **(h)** (6.1)

Chapter 15, Section E: From the journals

30. **(a)** No; rows: 1 and 28; columns: 1 and 28; interaction: 1 and 28
(b) No. Yes; it might be: if F just misses significance, the groups might be considerably (though not significantly) different from each other.
(c) No

31. **(a)** Two-way ANOVA summary table

Source	SS	df	MS	F
Feedback	1,378	1	1,378.00	.24
Irrationality	30,628	1	30,628.00	5.44*
Interaction	22,366	1	22,366.00	3.97
Within	157,730	28	5,633.20	
Total	212,102	31		

$*p < .05; F_{cv} = 4.20$

(b) The low irrationality group ($\bar{X} = 350.00$) scored significantly higher in self-esteem than did the high irrationality group ($\bar{X} = 288.19$) at the beginning of the study. There was no significant main effect for feedback condition and no significant interaction between irrationality and feedback conditions.

32. **(a)** Yes

(b) Yes; but only with difficulty. We recognize that MS_W is the pooled within-cell variance, so $MS_W = [(.58)^2 + (.84)^2 + (.69)^2 + (1.14)^2]/4 = .704$. Therefore, $SS_W = MS_W df_W = .704(120) = 84.53$. The grand mean is $\bar{X}_G = (.26 + .61 + .71 + 1.03)/4 = .6525$. The row (text type) means are $\bar{X}_{unelab} = (.26 + .71)/2 = .485$ and $\bar{X}_{elab} = (.61 + 1.03)/2 = .820$. The column (training) means are $\bar{X}_{untrained} = (.26 + .61)/2 = .435$ and $\bar{X}_{trained} = (.71 + 1.03)/2 = .870$. That allows us to compute the row effects, column effects, and interaction effects for the four cells.
Remembering that each cell has 31 subjects, we see that
$SS_R = 31(4)(.0281) = 3.484$,
$SS_C = 31(4)(.0473) = 5.865$, and
$SS_{RC} = 31(4)(.0000562) = .007$.

Source of variation	Sum of squares (SS)	Degrees of freedom (df)	Mean square (MS)	F
Text structure	3.484	1	3.484	4.95*
Training	5.865	1	5.865	8.33*
Interaction (struct × train)	.007	1	.007	.01
Within cell	84.530	120	.704	
Total	93.886	123		

$^*p < .05, F_{cv_{1,120}} = 3.92$

(c) Their conclusions are correct, but our reconstruction of the main effect for text structure does not quite agree with their report (4.94 instead of 5.04), probably due to roundoff.

(d) Apparently so; at least the pooled variance based on these standard deviations does in fact equal MS_W.

33. **(a)** Two-way ANOVA summary table

Source	SS	df	MS	F
Text structure	.256	1	.256	.789
Training	.004	1	.004	.012
Interaction	.025	1	.025	.077
Within	11.674	36	.324	
Total	11.959	39		

(b) There are no significant differences between any of the means in these data.

Chapter 16, Section A: Basic exercises

1. **(a)**

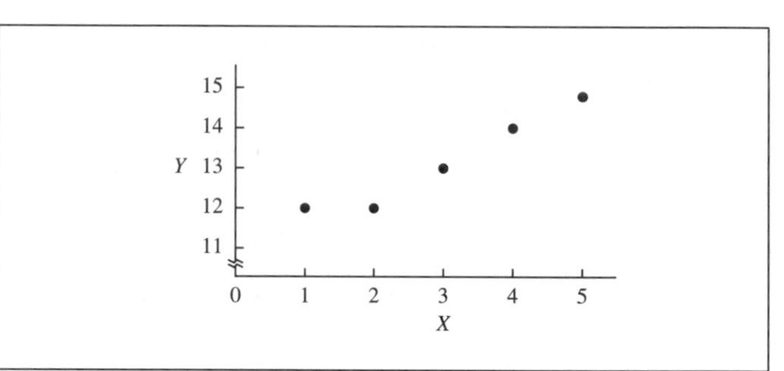

(c)

X	X^2	Y	Y^2	XY
1	1	12	144	12
4	16	14	196	56
2	4	12	144	24
5	25	15	225	75
3	9	13	169	39
15	55	66	878	206

$$r = \frac{\sum XY - \dfrac{\sum X \sum Y}{n}}{\sqrt{\left[\sum X^2 - \dfrac{(\sum X)^2}{n}\right]\left[\sum Y^2 - \dfrac{(\sum Y)^2}{n}\right]}}$$

$$= \frac{206 - \dfrac{(15)(66)}{5}}{\sqrt{\left[55 - \dfrac{(15)^2}{5}\right]\left[878 - \dfrac{(66)^2}{5}\right]}}$$

$$= \frac{8}{\sqrt{[10][6.8]}} = \frac{8}{8.2462} = .97$$

Table for Chapter 15, Section E, Exercise 32(b)

Row/Column	Row effect $\bar{X}_r - \bar{X}_G$	$(\bar{X}_r - \bar{X}_G)^2$	Column effect $\bar{X}_c - \bar{X}_G$	$(\bar{X}_c - \bar{X}_G)^2$	Interaction $\bar{X}_{rc} - \bar{X}_r - \bar{X}_c + \bar{X}_G$	$(\bar{X}_{rc} - \bar{X}_r - \bar{X}_c + \bar{X}_G)^2$
unelab/untr	−.1675	.0281	−.2175	.0473	−.0075	.0000562
elab/untr	.1675	.0281	−.2175	.0473	.0075	.0000562
unelab/tr	−.1675	.0281	.2175	.0473	.0075	.0000562
elab/tr	.1675	.0281	.2175	.0473	−.0075	.0000562

2. **(a)**

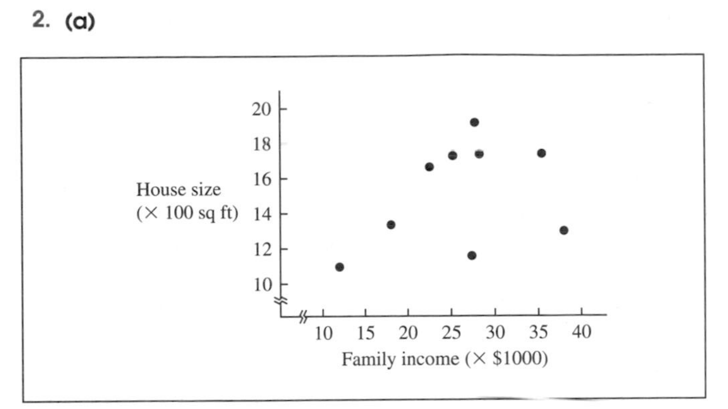

$\sum XY = 3706.89$

(c) $$r = \frac{3706.89 - \dfrac{(238.2)(137.2)}{9}}{\sqrt{\left[6840.420 - \dfrac{(238.2)^2}{9}\right]\left[2156.480 - \dfrac{(137.2)^2}{9}\right]}}$$

$$= \frac{75.6634}{\sqrt{[536.06][64.9423]}} = \frac{75.6634}{186.585} = .41$$

(d) There is a modest positive relationship between income and house size; in general, the higher the income, the larger the house.

3. **(a)** $H_0\!: \rho = 0;\ H_1\!: \rho \neq 0$
(b) Nondirectional; the question asked "significantly different" only.
(c) r
(d) .05 (unless stated otherwise)
(e) $df = 9 - 2 = 7$ **(f)** $r_{cv} = .6664$
(g) $r_{obs} = .41$ **(h)** Fail to reject
(i) There is no significant relationship between income and house size according to these data.
(j) $s_r = .345$; see the figure
(l) $t_{cv} = \pm 2.365$ and $r_{cv} = \pm .816$; r_{cv} depends on s_r, which depends on r_{obs}; if r_{obs} were larger, r_{cv} would be smaller, so that if $r_{obs} = .6664$, r_{cv} would also be .6664.
(n) $t_{obs} = 1.19$
(o) Fail to reject

3. **(j), (k), (l), and (m)**

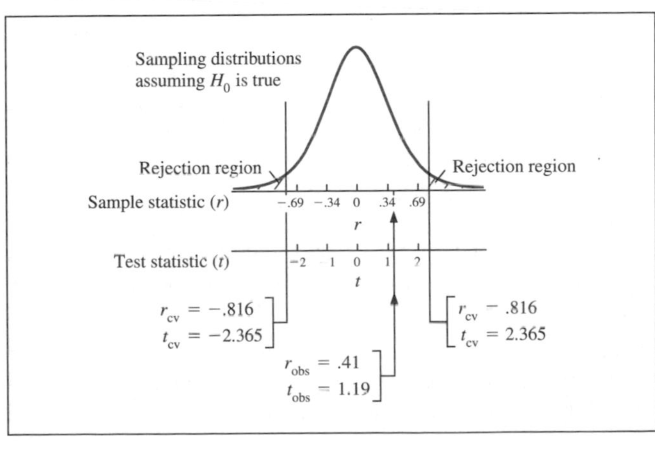

4. **(a)**

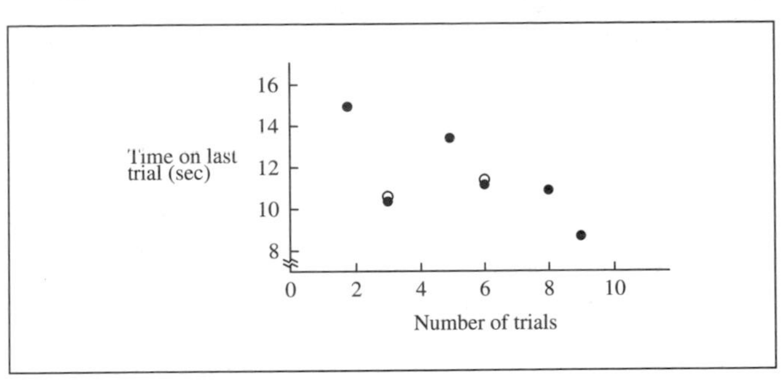

(c)

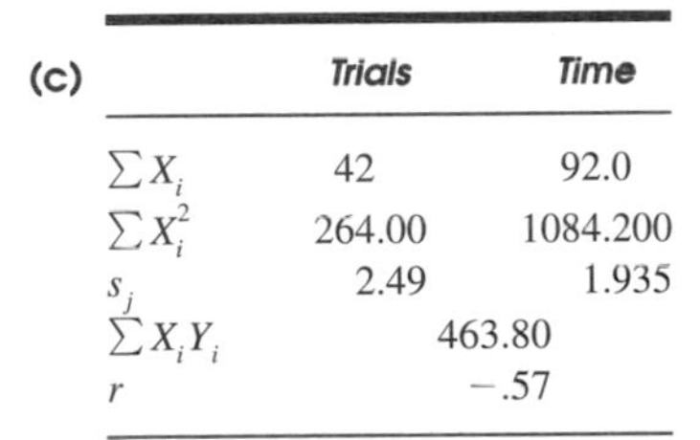

	Trials	*Time*
$\sum X_i$	42	92.0
$\sum X_i^2$	264.00	1084.200
s_j	2.49	1.935
$\sum X_i Y_i$	463.80	
r	−.57	

Pearson r because both variables are interval/ratio.
(d) There is a moderate negative relationship between number of trials and time on the last trial; the more trials, the shorter the time.

5. **(a)** $H_0\!: \rho \geq 0;\ H_1\!: \rho < 0$
(b) Directional; the question asked whether time *decreased* with number of trials.
(c) r
(d) .05 unless stated otherwise
(e) $df = 8 - 2 = 6$ **(f)** $r_{cv} = -.6215$
(g) $r_{obs} = -.57$ **(h)** Fail to reject
(i) There is no significant relationship between time to run a maze and number of trials according to these data.
(j) $s_r = .335$; see the figure
(l) $t_{cv} = -1.943$ and $r_{cv} = -.651$; r_{cv} depends on s_r, which depends on r_{obs}; if r_{obs} were larger (more negative), r_{cv} would be smaller, so if $r_{obs} = -.651$, r_{cv} would also be $-.651$.
(n) $t_{obs} = -1.70$ **(o)** Fail to reject

5. **(j), (k), (l), and (m)**

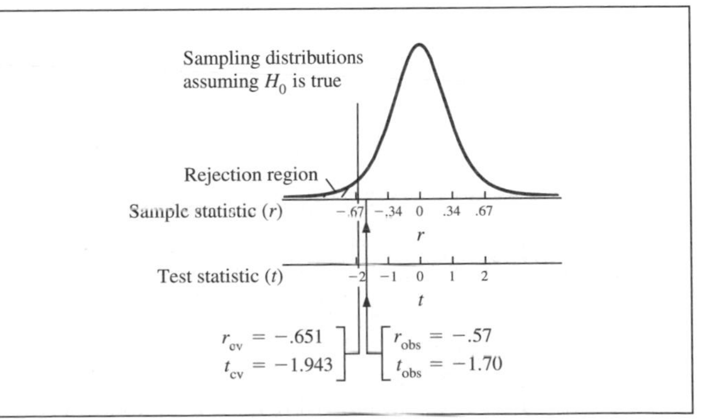

6. **(a)** It shifts all points equally $5000 to the right.
(b) No change **(c)** No change

7. **(a)** It changes the scale of the X-axis, making each pair of points twice as far apart.
(b) No change **(c)** No change

8. (a)

X	Rank of X	Y	Rank of Y	D_i	D_i^2
1	1	12	1.5	−.5	.25
4	4	14	4	0	0
2	2	12	1.5	+.5	.25
5	5	15	5	0	0
3	3	13	3	0	0
				.0	.50

$r_S = .975$

(b) They are quite similar.

9.

Rater	Color Rating	Color Rank	Taste Rating	Taste Rank	D_i	D_i^2
1	5	10.5	7	11	−.5	.25
2	4	5.5	6	8.5	−3	9
3	5	10.5	10	15	−4.5	20.25
4	2	2	1	1	1	1
5	5	10.5	8	12.5	−2	4
6	3	3	6	8.5	−5.5	30.25
7	4	5.5	6	8.5	−3	9
8	6	14.5	5	5	9.5	90.25
9	6	14.5	6	8.5	6	36
10	5	10.5	9	14	−3.5	12.25
11	4	5.5	8	12.5	−7	49
12	5	10.5	5	5	5.5	30.25
13	5	10.5	5	5	5.5	30.25
14	1	1	3	2	−1	1
15	4	5.5	4	3	2.5	6.25
					.0	329.00

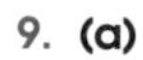
9. (a)

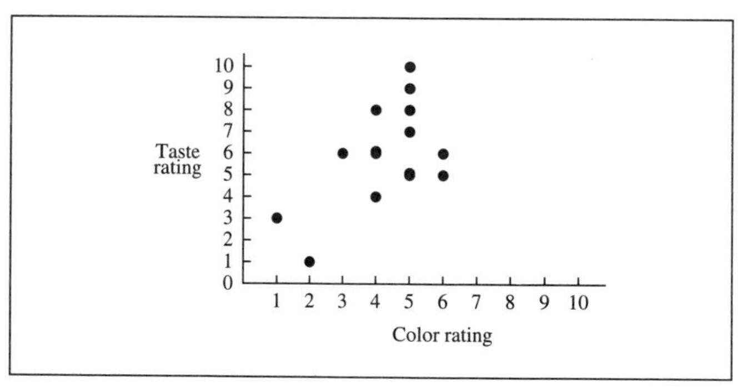

(c) Spearman r_S because ratings are ordinal; $r_S = .412$

(d) Teas with high color ratings also tend to have high taste ratings.

(e) $r_{S_{cv}} = .525$; r_S is not significantly different from 0.

Chapter 16, Section C: Extending comprehension

20. (a) $df = 8$; $t_{cv} = 2.306$; $r = .64$
(b) $df = 28$; $t_{cv} = 2.048$; $r = .37$
(c) $df = 118$; $t_{cv} = 1.98$; $r = .18$
(d) $df = 28$; $t_{cv} = 2.763$; $r = .47$
(e) Yes

22. 0

23. 1

Chapter 16, Section D: Cumulative review

24. (a) 4 **(b)** 2 **(c)** 3 **(d)** 2
(e) 38 **(f)** 1 **(g)** 1.686 **(h)** (11.2)

25. (a) 11 **(b)** 9 **(c)** 8 **(d)** 6
(e) 20 pairs **(f)** 1 **(g)** .377 **(h)** (16.6)

26. (a) 5 **(b)** 3 **(c)** 4 **(d)** 2
(e) 9 **(f)** 2 **(g)** ±2.262 **(h)** (12.3)

27. (a) 10 **(b)** 7 **(c)** 9 **(d)** 3
(e) Rows: 1 and 16; columns: 1 and 16; interaction: 1 and 16
(f) 2
(g) Rows: 4.49; columns: 4.49; interaction: 4.49
(h) Rows: (15.11); columns: (15.12); interaction: (15.13)

28. (a) 2 **(b)** 10 **(c)** 1 **(d)** 2
(e) 9 **(f)** 2 **(g)** ±2.262 **(h)** (8.3)

29. (a) 12

30. (a) 3 **(b)** 1
(c) 1 **(d)** 1
(e) Not applicable **(f)** 1
(g) 1.645 **(h)** (10.1)

31. (a) 8 **(b)** 6 **(c)** 6 **(d)** 2
(e) 24 **(f)** 2 **(g)** ±2.064 **(h)** (14.10)

32. (a) 9 **(b)** 4 **(c)** 9 **(d)** 3
(e) 3 and 27 **(f)** 2 **(g)** 2.96 **(h)** (14.14)

Chapter 16, Section E: From the journals

33. (a) $H_0: \rho = 0$ **(b)** 147
(c) $r_{cv,\, df=147,\, \alpha=.01} < .2540$ (off the scale of our table); yes, the test is significant.

34. (a) 10 **(b)** $r = .49$
(c) Somewhat larger
(d) No; $r_{cv,\, df=10,\, \alpha=.01} = .7079$. Riggio found significance, but we did not even though our observed r was larger. The reason is that our sample size is much smaller.

35. (a) 478 **(b)** Yes
(c) Subjects had smaller dating errors in areas in which they were more knowledgeable.

36. (a) 30 **(b)** $r = -.37$
(c) No; regardless of whether we consider the test directional or nondirectional.
(d) For this subject, we cannot say whether the increase of knowledge is related to a decrease in dating error.

37. (a) Because the size asymmetry classes are not equally wide
(b) $r_S = .97$; the difference may be due to my interpreting data from the scatterplot, or due to rounding.
(c) Yes; $r_{S_{cv,\, \alpha=.01,\ nondirectional}} = .929$

38. (a) $r_S = .930$
(b) Yes; we can use a directional test because we now have reason to believe that increase in size is positively related to non-takeover; $r_{S_{cv,\, \alpha=.01,\ directional}} = .893$

39. (a) 48 **(b)** Yes
(c) Yes; high male dominance is associated with relatively more female-committed homicides.

Chapter 17, Section A: Basic exercises

1. **(a)**

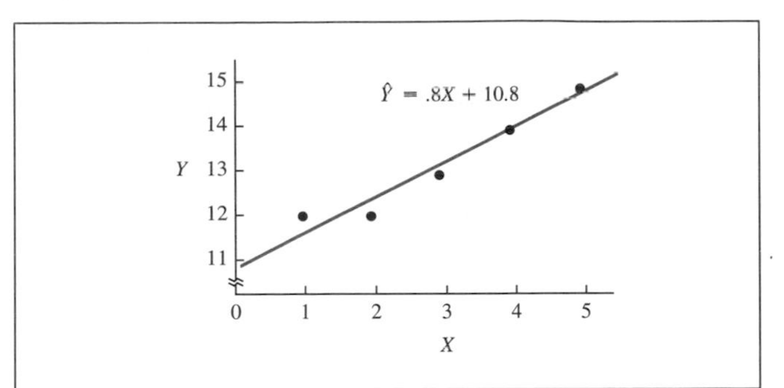

(b) $b = .8$; $a = 10.8$; $\hat{Y} = .8X + 10.8$

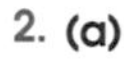

2. **(a)**

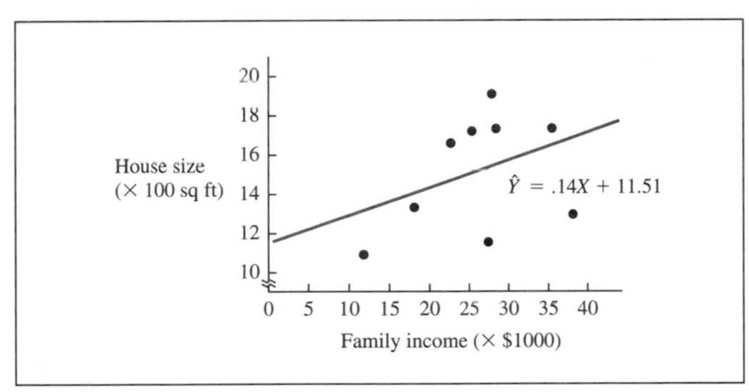

(b) $b = .14$ hundred square feet per thousand dollars
$a = 11.51$ hundred square feet
$\hat{Y} = .14X + 11.51$ hundred square feet

(c) For every $1000 increase in income, house size increases on average by 14 square feet.

(d) Eyeball-estimate: start at $31,000 on the X-axis and go up to the regression line and over to the Y-axis: 1585 square feet

(e) 1332 square feet

3. **(a)** 1585 square feet; we would predict that value for each of the 100 families; the actual house size for each family would (according to our data) be 1585 + some error of prediction; the mean error of prediction would be 0.

(b) The standard deviation of these 100 house sizes would be expected to be equal to the standard error of prediction $s_e = 277.8$; $s_Y = 284.9$ square feet; $r = .41$.

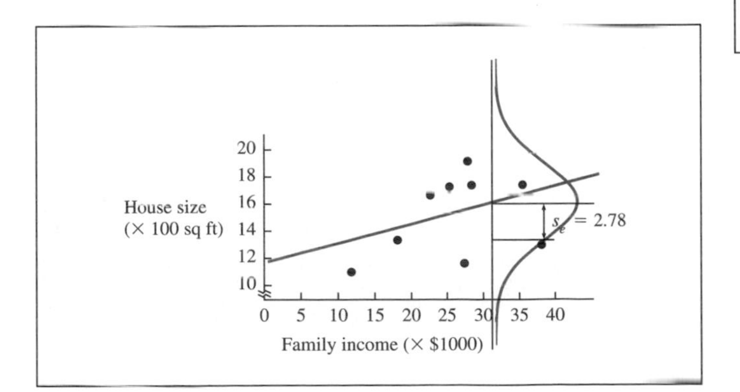

4. **(a)** 1333 square feet [same rationale as in Exercise 3(a)]

(b) Identical to Exercise 3; the number of houses in the sample makes no difference.

(c)

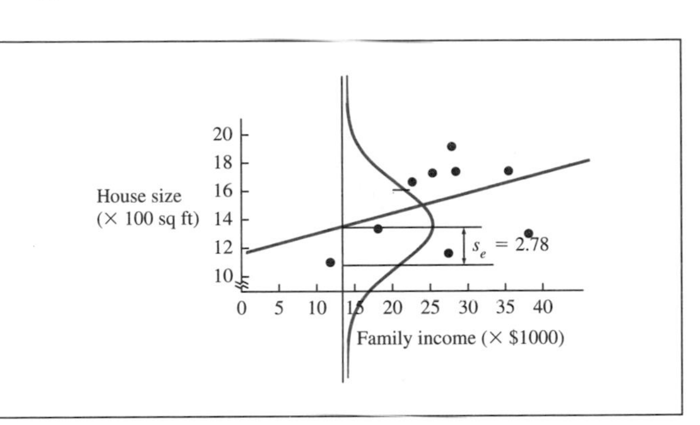

5. Homoscedasticity

6. **(a)** H_0: $b_{\text{population}} = 0$; H_1: $b_{\text{population}} \neq 0$

(b) Nondirectional; the question asked "significantly different" only.

(c) r **(d)** t

(e) .05 unless stated otherwise

(f) $9 - 2 = 7$

(g) $r_{\text{cv}} = \pm .6664$ from Table A.8

(h) $b = .14$ **(i)** $r = .41$

(j) Fail to reject

(k) There is no significant change in house size as a function of income for these data.

(l) This hypothesis test is identical to the one for r in Exercise 2 of Chapter 16.

7. **(b)** $b = -.441$ second per trial; $a = 13.817$ seconds; $\hat{Y} = -.441X + 13.817$ seconds

(c) For every additional trial, the time decreases by .441 second.

(d) 11.18 seconds **(e)** 12.50 seconds

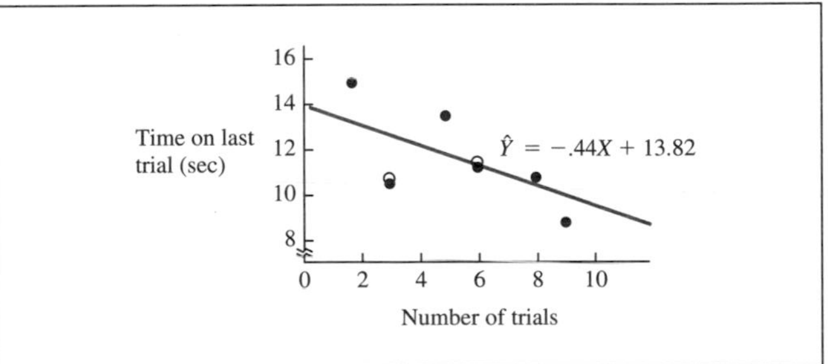

9. See table on page 532

10. **(b)** $b = .110$ vocabulary point per IQ point, $a = -1.19$ vocabulary points

(c) Because the vertical axis is not plotted at $IQ = 0$

Table for Chapter 17, Section A, Exercise 9

X_i	Y_i	$\hat{Y}_i$	$Y_i - \bar{Y}$	$(Y_i - \bar{Y})^2$	$Y_i - \hat{Y}_i$	$(Y_i - \hat{Y}_i)^2$	$\hat{Y}_i - \bar{Y}$	$(\hat{Y}_i - \bar{Y})^2$
8	10.9	10.286	−.6	.36	.614	.3770	−1.214	1.4738
9	8.6	9.845	−2.9	8.41	−1.245	1.5500	−1.655	2.7390
6	11.4	11.169	−.1	.01	.231	.0534	−.331	.1096
5	13.6	11.610	2.1	4.41	1.990	3.9601	.110	.0121
3	10.3	12.493	1.2	1.44	−2.193	4.8092	.993	.9860
6	11.7	11.169	.2	.04	.531	.2820	−.331	.1096
3	10.7	12.493	−.8	.64	−1.793	3.2148	.993	.9860
2	14.8	12.934	3.3	10.89	1.866	3.4820	1.434	2.0564
				$SS_T = 26.20$		$SS_{error} = 17.7285$		$SS_{explained} = 8.4725$

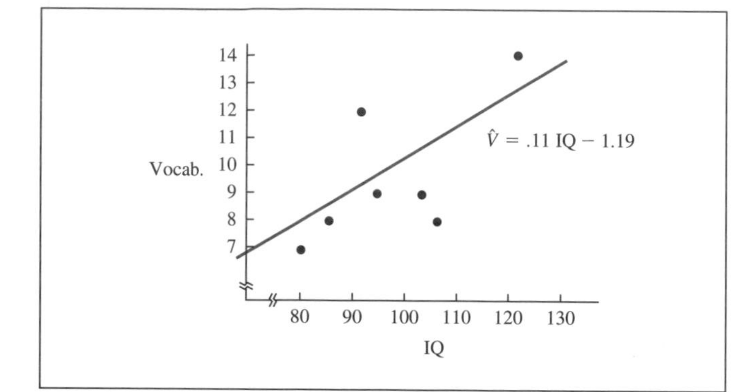

11. (a) .110 vocabulary point (b) 1.10 vocabulary points

Chapter 17, Section C: Extending comprehension

25. $a = \bar{Y}, b = 0$
26. s_X approaches 0.

Chapter 17, Section D: Cumulative review

27. (a) 9 (b) 4 (c) 10 (d) 3
    (e) 2 and 18 (f) 2 (g) 3.55 (h) 14.14
28. (a) 4 (b) 2 (c) 3 (d) 2
    (e) 33 (f) 2 (g) ±2.036 (h) (11.2)
29. (a) 6 (b) 4 (c) 10 (d) 3
    (e) 3 and 36 (f) 2 (g) 2.87 (h) (13.6)
30. (a) 7 (b) 5; six hypotheses
    (c) 5 (d) 4
    (e) 4 and 36 (f) 2
    (g) 3.82 (h) (14.2)
31. (a) 12 (b) 10 (c) 9 (d) 5
    (e) 6 (f) 2
    (g) .7067 (h) (16.1) or (16.5)
32. (a) 5 (b) 3 (c) 4 (d) 2
    (e) 15 (f) 2 (g) ±2.131 (h) (12.3)
33. (a) 2 (b) 11 (c) 1 (d) 2
    (e) 19 (f) 2 (g) ±2.093 (h) (8.3)
34. (a) 10 (b) 7 (c) 10 (d) 3
    (e) Rows: 1 and 32; columns: 3 and 32; interaction: 3 and 32
    (f) 2
    (g) Rows: 4.15; columns: 2.90; interaction: 2.90
    (h) Rows: (15.11); columns: (15.12); interaction: (15.13)
35. (a) 8 (b) 6 (c) 6 (d) 2
    (e) 20 (f) 2 (g) 2.086 (h) (14.10)

Chapter 17, Section E: From the journals

36. (a) Yes
    (b) 1.64 millimeters; 1.22 millimeters
37. (a) $\hat{L} = 1.708 - .016T$ (b) $r = -.81$
    (c) Yes
38. (a) Yes (b) Negative
39. (a) Yes
    (b) Predicted Blessed score = 6(time) + 16.67.
    (c) Because the regression line does not necessarily pass exactly through the data points

Chapter 18, Section A: Basic exercises

1. (a) χ^2 goodness of fit test
   (b) H_0: The population distribution has frequencies [30%, 20%, 20%, 30%] for cashews, pecans, Brazil nuts, and peanuts, respectively. H_1: The population has some other distribution.
   (c) Nondirectional in the sense that a departure from the null hypothesis in any direction will increase χ^2; one-tailed in the sense that the rejection region will be in only the right-hand tail of the χ^2 distribution
   (d) χ^2, $df = 3$ (e) $\alpha = .05$ (f) $\chi^2_{cv} = 7.815$
   (g)

Category	O_i	E_i	$O_i - E_i$	$(O_i - E_i)^2$	$\frac{(O_i - E_i)^2}{E_i}$
Cashews	45	60	−15	225	3.750
Pecans	35	40	−5	25	.625
Brazil nuts	41	40	1	1	.025
Peanuts	79	60	19	361	6.017
	200	200	0		$\chi^2_{obs} = 10.417$

   (h) Reject H_0.
   (i) The distribution is not what the company claims.

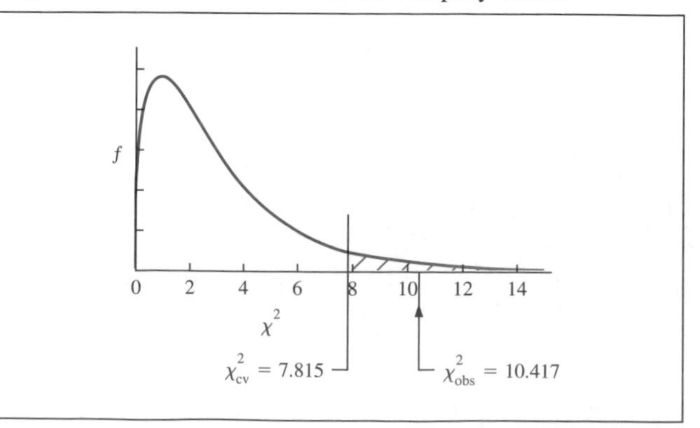

2. **(a)** χ^2 test of independence
(b) H_0: Pattern of wine selection is the same at all three supermarkets. H_1: Pattern is not the same.
(c) Same as Exercise 1(c) **(d)** χ^2, $df = 4$
(e) $\alpha = .05$ **(f)** $\chi^2_{cv} = 9.488$
(g) Contingency table:

Supermarket	*White*	*Wine selection* *Red*	*Blush*	*Row totals*
FoodKing	26 (30.75)	40 (35.51)	42 (41.74)	108
Safeway	29 (25.34)	32 (29.26)	28 (34.39)	89
Lucky	29 (27.91)	25 (32.22)	44 (37.87)	98
Column totals	84	97	114	295

Category	O_i	E_i	$O_i - E_i$	$(O_i - E_i)^2$	$\frac{(O_i - E_i)^2}{E_i}$
FoodKing/White	26	30.75	−4.75	22.562	.734
FoodKing/Red	40	35.51	4.49	20.160	.568
FoodKing/Blush	42	41.74	.26	.068	.002
Safeway/White	29	25.34	3.66	13.396	.529
Safeway/Red	32	29.26	2.74	7.508	.257
Safeway/Blush	28	34.39	−6.39	40.832	1.187
Lucky/White	29	27.91	1.09	1.188	.043
Lucky/Red	25	32.22	−7.22	52.128	1.618
Lucky/Blush	44	37.87	6.13	37.577	.992
	295	295	.01		$\chi^2_{obs} = 5.930$

(h) Fail to reject H_0
(i) The pattern of wine selection does not differ from supermarket to supermarket.

2. **(i)**

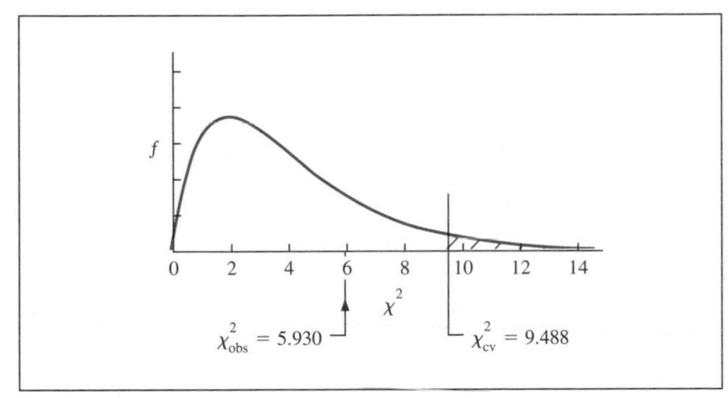

3. The first row of the computation of χ^2 from Exercise 2 is now:

Category	O_i	E_i	$O_i - E_i$	$(O_i - E_i)^2$	$\frac{(O_i - E_i)^2}{E_i}$
FoodKing/White	52	61.50	−9.50	90.250	1.467

and all subsequent rows change similarly.
(a) No change **(b)** No change
(c) No change **(d)** Each entry doubles.
(e) Each entry doubles. **(f)** Each entry doubles.
(g) Each entry is multiplied by 4.
(h) Each entry doubles.
(i) χ^2_{obs} doubles to become 11.860
(j) We now reject H_0 (even though the proportions are the same).

4. **(a)** The McNemar test for significance of change
(b) H_0: The proportion of individuals who say yes in early autumn is the same as the proportion who say yes in late autumn. H_1: The proportion is not the same.
(c) Nondirectional; the proportion could change in either direction.
(d) χ^2 with 1 degree of freedom
(e) .05 **(f)** 3.841
(g) $\chi^2_{obs} = 7.143$ **(h)** Reject H_0.
(i) The confirmation hearings did in fact alter the workers' perceptions of sexual harassment.

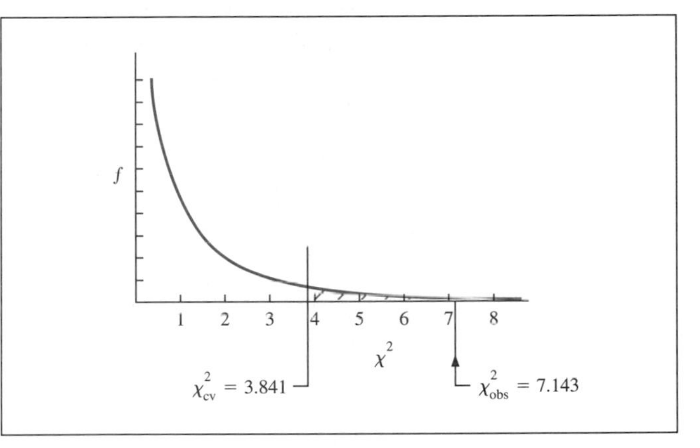

5. **(a)** χ^2 test of independence
(b) H_0: Attitude toward the pledge is independent of when it is measured. H_1: Attitude is not the same.
(c) Same as Exercise 1(c) **(d)** χ^2, $df = 1$
(e) $\alpha = .05$ **(f)** $\chi^2_{cv} = 3.841$
(g) Contingency table:

	Yes	*No*	*Row totals*
Before	42 (47.5)	58 (52.5)	100
After	53 (47.5)	47 (52.5)	100
Column totals	95	105	200

$\chi^2_{obs} = 2.426$

(h) Fail to reject H_0.
(i) The attitude toward the pledge does not differ from time to time. The McNemar test is more powerful than the test of independence.

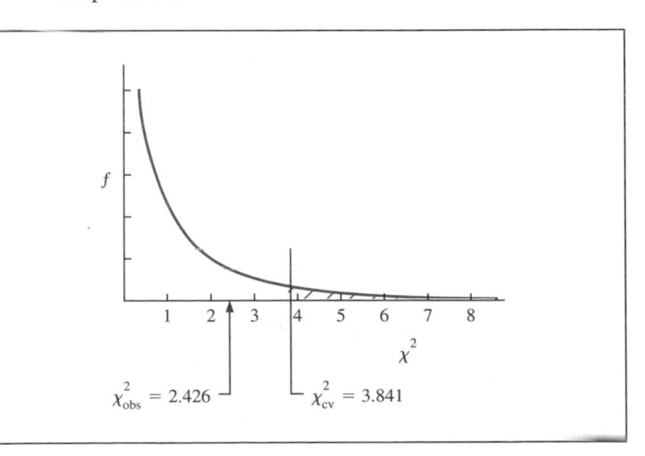

6. (a) Mann–Whitney U test
(b) H_0: The distribution of grades is the same in both groups. H_1: The distribution is not the same.
(c) Directional
(d) U with $n_1 = 11$ and $n_2 = 11$ degrees of freedom
(e) .05 **(f)** 34
(g) $U_1 = 45.5$; $U_2 = 75.5$; $U_{obs} = 45.5$
(h) Fail to reject
(i) We cannot say that the techniques are different.

7. (a) Two-independent-samples t test
(b) H_0: $\mu_1 = \mu_2$
(c) Here the null hypothesis concerns one kind of parameter, the means of the two distributions; the Mann–Whitney null hypothesis is about the shape of the entire distributions.

(d)

	With	Without
$\sum X_i$	63	39
$\sum X_i^2$	435.00	213.00
$\bar{X}_j$	6.30	4.88
s_j	2.06	1.81

$s^2_{pooled} = 3.81$, $s_{\bar{X}_1 - \bar{X}_2} = .93$, $df = 16$, $t_{obs} = 1.54$, $t_{cv} = 2.120$

(e) Fail to reject; same result as Mann–Whitney

8. (a) Wilcoxon matched-pairs signed-rank test
(b) H_0: The distribution of readability is the same in both groups. H_1: The distributions are not the same.
(c) Nondirectional **(d)** T (12 pairs)
(e) .05 **(f)** $T_{cv} = 13$
(g) $T_{obs} = 15.5$ **(h)** Fail to reject
(i) The readabilities are not significantly different.

9. (a) Dependent-sample t test
(b) H_0: $\mu_D = 0$; directional
(c) Here the null hypothesis concerns one kind of parameter, the means of the three distributions; the Wilcoxon null hypothesis is about the shape of the entire distributions.

(d)

Subject	Jug	Fine	D_i
1	71	66	5
2	50	66	−16
3	70	68	2
4	78	75	3
5	60	66	−6
6	86	83	3
7	58	62	−4
8	49	78	−29
9	58	63	−5
10	56	76	−20
11	81	81	0
12	50	58	−8

$\sum D_i = -75$, $\sum D_i^2 = 1685.00$, $\bar{D} = -6.25$, $s_D = 10.52$, $s_{\bar{D}} = 3.04$, $t_{obs}[df = 11] = -2.06$, $t_{cv} = 1.796$; directional

(e) Reject H_0 as in the Wilcoxon test.

10. (a) Kruskal–Wallis test
(b) H_0: The distributions of ratings in the five fraternities are the same. H_1: The distributions are not the same.
(c) Nondirectional
(d) H, or χ^2 with 4 degrees of freedom
(e) .05 **(f)** 9.488
(g) $H = 16.181$ **(h)** Reject H_0.
(i) We do in fact believe that the fraternities are different in political awareness.

11. (a) ANOVA
(b) H_0: $\mu_1 = \mu_2 = \mu_3$
(c) Here the null hypothesis concerns one kind of parameter, the means of the three distributions; the Kruskal–Wallis null hypothesis is about the shape of the entire distributions.

(d)

	Classical	Rock	Standards	Row totals
$\sum X_i$	37.8	48.1	48.7	134.6
$\sum X_i^2$	357.260	462.790	474.430	1294.480
$\bar{X}_j$	9.450	9.620	9.740	
s_j	.129	.130	.152	

(d)

Source	SS	df	MS	F
Between	.187	2	.0936	4.90*
Within	.210	11	.0191	
Total	.397	13		

*$p < .05$, $df = 2, 11$; $F_{cv} = 3.98$

(e) Reject the null hypothesis. Kruskal–Wallis failed to reject; ANOVA is more powerful than the Kruskal–Wallis test.

Chapter 18, Section C: Extending comprehension

21. Yes
22. Yes
23. No
24. Yes
25. Yes
26. Yes

Chapter 18, Section D: Cumulative review

27. (a) 10 **(b)** 7 **(c)** 10 **(d)** 3
(e) Rows: 1 and 32; columns: 3 and 32; interaction: 3 and 32
(f) 2
(g) Rows: 4.15; columns: 2.90; interaction: 2.90
(h) Rows: (15.11); columns: (15.12); interaction: (15.13)

28. (a) 11 **(b)** 9 **(c)** 8 **(d)** 6
(e) 13 pairs **(f)** 2 **(g)** .566 **(h)** (16.6)

29. (a) 14 **(b)** 12 **(c)** 10 **(d)** 7
(e) 6 **(f)** 2 **(g)** 12.592
(h) (18.1) χ^2 test of independence

30. (a) 2 **(b)** 16 **(c)** 2 **(d)** 2
(e) 224 **(f)** 2 **(g)** ±1.96 **(h)** (8.5)

31. (a) 13 **(b)** 11 **(c)** 10 **(d)** 7
(e) 3 **(f)** 2 **(g)** 7.815
(h) (18.1) Goodness of fit test

32. (a) 4 **(b)** 2 **(c)** 3 **(d)** 2
(e) 18 **(f)** 1 **(g)** 1.734 **(h)** (11.2)

33. (a) 19

34. (a) 3 **(b)** 1
(c) 1 **(d)** 1

(e) Not applicable **(f)** 1
(g) 1.645 **(h)** (10.1)

35. (a) 16 **(b)** 14 **(c)** 10 **(d)** 8
(e) $n_1 = 12, n_2 = 12$
(f) 2 **(g)** 37
(h) (18.4) and (18.5) Mann–Whitney *U* test

36. (a) 9 **(b)** 4 **(c)** 10 **(d)** 3
(e) 2 and 10 **(f)** 2 **(g)** 4.10 **(h)** (14.14)

37. (a) 12 **(b)** 10 **(c)** 9 **(d)** 5
(e) 18 **(f)** 2 **(g)** .4438 **(h)** (16.1)

38. (a) 8 **(b)** 6 **(c)** 6 **(d)** 2
(e) 28 **(f)** 2 **(g)** 2.048 **(h)** (14.10)

Chapter 18, Section E: From the journals

39. (a) No; the table shows 1479 subjects, not 1485.
(b) 6
(c) We compute $\chi^2_{obs} = 22.93$, not 24.91 as stated.
(d) Yes; $\chi^2_{cv} = 16.812$
(e) Type of family structure does have a significant effect on percent of victimization.

40. (a) Yes **(b)** Yes
(c) Apparently Goldkamp did not set a level of significance in advance and is reporting the probability provided by some computer package such as SPSS-X. It is a misleading statement because it makes it appear that *p* is negative, which is impossible.

41. (a) 16 and 20 **(b)** 79; yes **(c)** No

42. (a) 2; 5.991; no **(b)** 1; 3.841; no **(c)** Apparently not

43. (a) Yes **(b)** 7.815; 11.345
(c) Yes **(d)** No; yes

List of Symbols and Glossary

Greek Symbols

α: The probability of making a Type I error. Page 172

$\alpha_{\text{experimentwise}}$: Probability of making a Type I error somewhere in the experiment. Page 285

β: The probability of making a Type II error. Page 173

χ^2: A set of distributions underlying nonparametric statistics. Page 443

μ: Mean of a population. Page 59

μ_D: Difference score mean in a population. Page 249

μ_G: Grand mean (mean of all scores) in a population. Page 356

μ_{H_0}: Population mean as specified by the null hypothesis. Page 266

μ_j: Mean of jth population group. Page 287

μ_{real}: Mean of the real (assumed) population. Page 266

Π: Proportion of a population. Page 154

ρ: Pearson correlation in a population. Page 382

ρ_S: Spearman correlation in a population. Page 393

σ: Standard deviation of a population. Page 73

$\sigma_{\bar{X}}$: Standard error of the mean in a population. Page 134

σ^2: Variance of a population. Page 83

$\sum$: Summation operator. Page 18

Roman Symbols

a: Regression line intercept in a sample. Page 411

b: Slope of the regression line in a sample. Page 411

$b_{\text{population}}$: Slope of the regression line in a population. Page 428

C: Number of columns. Page 357

C_j: Coefficients of a comparison. Page 328

cv: Critical value. Page 146

d: Effect size index. Page 192

d_M: Maximum effect size index. Page 299

$d_{\text{population}}$: Effect size index in a population. Page 266

$\bar{D}$: Mean of the sample difference scores. Page 249

D_i: Difference score for the ith individual. Page 246

deviation_i: Distance from the ith point to the mean. Page 72

df: Degrees of freedom. Page 153

e_i: Error of prediction for the ith subject. Page 419

E_i: Expected frequencies of the ith category. Page 443

f: Frequency. Page 27

F: Distribution underlying analysis of variance. Page 290

H: Kruskal–Wallis distribution. Page 456

H_0: Null hypothesis. Page 168

H_1: Alternative hypothesis. Page 168

i: Index variable, indicates the row number and marks position in group. Page 17

j: Group or column number. Page 286

k: Number of groups. Page 286

MS_B: Mean square between groups. Page 289

$\text{MS}_{\text{between occasions}}$: Mean square between occasions. Page 335

$\text{MS}_{\text{between subjects}}$: Mean square between subjects. Page 337

MS_C: Mean square for column variable. Page 357

MS_R: Mean square for row variable. Page 357

MS_{RC}: Mean square for interaction. Page 357

$\text{MS}_{\text{residual}}$: Residual mean square. Page 338

MS_W: Mean square within groups. Page 290

n: Sample size. Page 17

$\tilde{n}$: Harmonic mean of sample size. Page 325

n_j: Sample size of the jth group. Page 286

N: Population size. Page 59

O_i: Observed frequencies of the ith category. Page 443

obs: Observed value. Page 183

p: Proportion in sample. Page 156

P: Probability of. Page 9

q: Proportion of alternative outcome in a sample. Page 156

Q: Studentized range statistic. Page 325

r: Pearson correlation coefficient for a sample. Page 382

r^2: Coefficient of determination. Page 422

r_S: Spearman correlation coefficient for a sample. Page 392

R: Number of rows. Page 357

R_{X_i}: Rank of the ith value of X. Page 393

s: Standard deviation of a sample. Page 73

s^2: Variance of a sample. Page 83

s_D: Standard deviation of the differences. Page 249

$s_{\bar{D}}$: Standard error of the mean differences in a sample. Page 249

s_e: Standard error of estimate. Page 421

s_p: Standard error of a proportion. Page 157

s_{pooled}: Square root of s^2_{pooled}. Page 218

s^2_{pooled}: Pooled estimate of σ^2. Page 218

s_r: Standard error of Pearson r. Page 390

s_X: Standard deviation of X values. Page 411

$s_{\bar{X}}$: Standard error of the mean in a sample. Page 153

$s_{\bar{X}_1-\bar{X}_2}$: Standard error of the difference between two means in a sample. Page 217

s_Y: Standard deviation of Y values. Page 411

SS_B: Sum of squares between groups. Page 295

$SS_{\text{between occasions}}$: Sum of squares between occasions. Page 337

$SS_{\text{between subjects}}$: Sum of squares between subjects. Page 337

SS_C: Sum of squares for column variable. Page 359

SS_{error}: Error sum of squares. Page 424

$SS_{\text{explained}}$: Explained sum of squares. Page 424

SS_R: Sum of squares for row variable. Page 359

SS_{RC}: Sum of squares for interaction. Page 361

SS_{residual}: Residual sum of squares. Page 337

SS_T: Sum of squares total. Page 294

SS_W: Sum of squares within groups. Page 295

t: Student's t distribution. Page 153

t_{obs}: Observed value of the test statistic t in an experiment. Page 183

T: Standardized score with $\mu = 50$ and $\sigma = 10$. Page 113

T: Wilcoxon T distribution. Page 455

U: Cohen's nonoverlap statistic. Page 193

U: Mann–Whitney U distribution. Page 453

X: Symbol for variable. Page 17

X_i: ith value of the variable X. Page 17

X_{ij}: ith value of the variable in the jth group. Page 286

X_{irc}: ith value in the cell in the rth row and cth column. Page 359

X_{mid}: Midpoint of a class interval. Page 93

$\bar{X}$: Mean of the variable X in a sample. Page 56

$\bar{X}_{\text{cv}}$: Critical value of the sample mean. Page 186

$\bar{X}_G$: Grand mean of all individuals in a sample. Page 286

$\bar{X}_j$: Mean of the jth sample. Page 287

$\bar{X}_{\text{obs}}$: Observed value of the sample mean in an experiment. Page 183

Y: Symbol for variable. Page 17

$\hat{Y}$: Predicted value of Y. Page 408

z: Standard score. Page 97

z_{obs}: Observed value of the test statistic z in an experiment. Page 183

Glossary

A priori test: The test of a (pairwise or complex) null hypothesis selected before the data are collected; sometimes called "planned comparison" or "planned contrast." Page 326

Alternative hypothesis: The hypothesis that there is in fact an effect or difference. Page 168

Analysis of variance (ANOVA): The statistical procedure for testing hypotheses about two or more means. Page 283

Asymptotic: Gradually approaching the X-axis. Page 36

Average absolute deviation: The mean of the absolute values of all the deviations in a population or sample. Page 73

Bar graph: A graphical presentation of a frequency distribution of nominal or ordinal data where frequencies are represented by separated bars. Page 40

Between-subjects effect: The portion of the total sum of squares in a repeated-measures design that is due to individual differences between subjects. Page 334

Bimodal distribution: A distribution that has two most-frequently occurring values. Page 35

Blind: Not knowing which subjects are assigned to which experimental condition. Page 7

Cell: The intersection of a row and a column. Page 351

Central limit theorem: The theorem that describes the shape and parameters of the sampling distribution of the means. Page 133

Central tendency: The middle of a distribution. Page 48

Coefficient of determination: The square of the Pearson r. Page 422

Coefficients of a comparison: The constants that multiply the means in a comparison; sometimes called "weights." Page 328

Column effect: Deviation of a column mean from the grand mean. Page 359

Comparison: A linear combination of means; sometimes called a "contrast." Page 328

Complex null hypothesis: A null hypothesis that compares some combination of means to another combination of means. Page 284

Computational formula for the standard deviation: A formula for computing the standard deviation that is somewhat easier to use than the mean squared deviation formula. Page 82

Conditional probability: The probability of an event given that another event has occurred. Page 9

Confidence interval: A range of values that has a specified probability of containing the actual value of the parameter. Page 147

Contingency table: A table presentation of two variables so that their interrelationships can be examined; all combinations of categories of all variables are presented. Page 448

Continuous variable: One with an infinite number of values between adjacent scale values. Page 15

Correlation coefficient: The descriptive statistic that measures the degree of the relationship between two variables. Page 377

Critical value of a statistic: The value of a statistic that marks the boundary of a specified area (such as .05 or .01) in the tail of a distribution. Pages 146, 186

Critical value of the test statistic: A criterion set in advance. If the observed value of the test statistic is greater (in absolute value) than the critical value, we reject the null hypothesis. Page 187

Degrees of freedom: The number of freely varying values in a given data set. Page 153

Dependent samples: Two samples whose subjects are statistically related to each other. Page 245

Dependent variable: The measured outcome of interest. Page 212

Descriptive statistics: The science of describing distributions of samples or populations. Page 166

Deviation: The "distance" any point is from the mean. Page 73

Difference score: A subject's score in one condition minus that same subject's (or related subject pair's) score in the other condition. Page 246

Directional (one-tailed) alternative hypothesis: Where one direction of change (an increase or a decrease) is specified in advance. Page 169

Discrete variable: One with no possible intermediate values between two adjacent points. Page 15

Effect size: A measure of the magnitude of a result. Page 175

Effect size index: The distance between distibutions divided by a measure of the variability of the distributions (such as the pooled standard deviation). Page 192

Error of prediction: The difference between the actual value of Y and the value predicted from X. Page 419

Experimentwise probability of a Type I error: The probability of making one or more Type I errors in an experiment where more than one null hypotheses are being tested simultaneously. Page 285

F ratio: The variance based on the sample means (called the mean square between groups) divided by the pooled within-group variance (called the mean square within groups). Page 290

Factorial design: A two-way (or higher-order) design in which outcomes corresponding to all possible combinations of the treatment variables are observed. Page 351

Frequency: The number of times the particular value of a variable occurs. Page 27

Frequency polygon: A graphical presentation of a grouped frequency distribution with frequencies represented as points. It is appropriate for interval/ratio data. Page 31

Goodness of fit test: A test of the null hypothesis that the sampled data come from a specified distribution. Page 443

Grand mean: The mean of all the observations; the sum of all the observations in all the groups divided by the total number of observations. Page 286

Grouped frequency distribution: A frequency distribution with adjacent values of the variable grouped together into class intervals. Page 28

Histogram: A graphical presentation of a grouped frequency distribution with frequencies represented as vertical bars. It is appropriate for interval/ratio data. Page 30

Homoscedasticity: The assumption in regression that the size of the errors of prediction does not depend on the value of X. Page 422

Hypothesis: An assumption or inference about a parameter or distribution that can be tested. Page 166

Independent (or treatment) variable: The variable whose value defines group membership. Page 212

Inductive statements: Statements whose truth can be assessed by collecting and analyzing data. Page 2

Inferential statistics: The science of using sample statistics to make inferences or decisions about population parameters. Page 166

Inflection point: The point on any curve where the curvature changes from upward to downward or from downward to upward. Page 75

Interaction: The extent to which the outcome associated with one independent variable depends on the level of the other independent variable. Page 355

Interaction effect: The deviation of a cell mean from the grand mean that is not accounted for by either the row effect or the column effect. Page 359

Intercept: The value of Y when $X = 0$. Page 408

Interpolation: Finding a value located proportionately between two values in a table. Page 110

Interval ratio variable: A variable measured at either the interval scale or the ratio scale of measurement. Page 15

Interval scale: A measurement scale for ordered variables that has equal units of measurement. Page 14

Kruskal–Wallis H test: A nonparametric (ordinal data) test for three or more independent samples. Page 456

Least squares criterion: The rule that states that the best regression line is the one that produces the smallest sum of the squared errors of prediction. Page 420

Level of confidence: The probability that a given confidence interval contains the actual value of the parameter. Page 147

Level of significance: The probability (signified by α) of making a Type I error. Page 171

Levels: The possible values of an independent variable. Page 350

Linear regression: Determining the best straight line through a data set and using it to predict Y from X. Page 406

Main effect: The extent to which the dependent (outcome) variable differs from one level of one independent variable to another level of the same independent variable. Page 353

Mann–Whitney U test: A nonparametric (ordinal data) test for two independent samples. Page 452

McNemar test for significance of change: A nonparametric test of dependent-sample proportions. Page 451

Mean: A measure of central tendency appropriate for interval/ratio variables; it is equal to the sum of all values of a variable divided by the number of values. Page 53

Mean deviation: The average of all deviations in a population or sample; it is always equal to zero. Page 72

Mean square: A sum of squares divided by its corresponding degrees of freedom; the ANOVA procedure uses two mean squares: the mean square between groups and the mean square within groups. Page 296

Mean square between groups (MS_B): The between-group point-estimate of σ^2, based on the variation of the sample means; the numerator of the F ratio. Page 289

Mean square within groups (MS_W): The within-group estimator of σ^2, equal to the pooled within-group variance; the denominator of the F ratio. Page 290

Mean squared deviation formula for the standard deviation: A formula for computing the standard deviation that demonstrates that the standard deviation is in fact the square root of the mean of the squared deviations. Page 74

Measurement: The procedure for assigning a value to a variable. Page 13

Median: A measure of central tendency appropriate for ordinal or interval/ratio variables; it is the value that is the midpoint of a data set. Page 51

Mode: A measure of central tendency appropriate for any variable; it is the most frequently occurring value in a distribution. Page 49

Negatively skewed distribution: A distribution whose left tail is longer than its right tail. Page 36

Nominal scale: Classification of unordered variables. Page 13

Nondirectional (two-tailed) alternative hypothesis: Where the hypothesized change can be either an increase or a decrease. Page 169

Nonparametric test: A hypothesis test that does not depend on the parameters of a normal distribution. Page 442

Normal distribution: Any of a family of symmetric, unimodal, asymptotic distributions obtained by specifying values of μ and σ in the equation shown in Box 6.1. Pages 36, 102

Null hypothesis: The hypothesis that there is no effect or no difference. Page 168

Observed value of the statistic: A computed statistic of our sample. Page 186

Observed value of the test statistic: The value of the test statistic computed from our sample. Page 187

Occasions: The columns in a repeated-measures design. Page 335

Omnibus null hypothesis: A null hypothesis that stands for all possible null hypotheses. Page 284

Ordinal scale: Measurement of variables that have an inherent natural order. Page 14

Pairwise null hypothesis: A null hypothesis that compares one mean to another. Page 284

Parameter: Any measured (or assumed) characteristic of a population. Page 8

Parametric test: A hypothesis test about the parameters of a normal distribution. Page 440

Partition: Divide into pieces with no remainder. Page 295

Pearson r: The correlation coefficient appropriate for interval/ratio data Page 382

Percentile: The score below which a specified percentage of scores in the distribution fall. Page 96

Percentile rank: The percentage of scores equal to or less than the given score. Page 96

Personality coefficient: The observation that most correlations between personality measures and behavior are approximately .2 to .3. Page 427

Placebo: In a drug study, a substance that looks like the drug being tested but actually has no effect. Page 7

Point-estimate: A computed statistic that approximates a parameter. Page 145

Pooled variance: The weighted average of two (or more) variances; the weights depend on the respective sample sizes. Page 218

Pooled within-group variance: The weighted average of the variances computed one group at a time; the weights depend on the sample size. Page 289

Population: All the members of the group under consideration. Page 7

Positively skewed distribution: A distribution whose right tail is longer than its left tail. Page 36

Post hoc tests: Hypothesis tests performed after a significant ANOVA to explore which means or combinations of means differ from one another. Page 323

Power of a statistical test: The probability of rejecting H_0 when H_0 is in fact false. Page 257

Practical significance: The degree to which a result is important. Page 174

Pretest–posttest design: An experimental design where each subject is measured twice, once before and again after some treatment. Page 245

Probability of an event: The number of outcomes favorable to that event divided by the total number of possible outcomes. Page 8

Pygmalion effect: People act in accordance with others' expectations. Page 4

Random: Unpredictable given our current knowledge. Page 8

Random sample: A sample chosen so that each member of a population has an equal chance of being included in the sample. Page 126

Range: A measure of the width of a distribution equal to the highest value minus the lowest value. Page 71

Ratio scale: An interval scale of measurement that has a true zero point. Page 14

Raw effect size: The magnitude of an experimental result measured in the scale of its own experiment. Page 192

Raw score: A value of a variable in the orginal scale of measurement. Page 97

Real distribution: The distribution assuming the particular value of the parameter. Page 259

Real limits of a measurement: The points that are half the smallest measuring unit above and below the measured value. Page 15

Regression to the mean: The predicted variable lies closer to its mean than does the predictor variable. Page 418

Rejection region: The portion of the distribution of the statistic or test statistic that lies beyond the critical value(s). If the observed value of the statistic or test statistic lies in the rejection region, we reject the null hypothesis. Page 187

Relative area: The proportional (or fractional) area under a frequency distribution. Page 98

Relative frequency: Frequency divided by the size of the group, expressed as a proportion or percentage. Page 32

Repeated-measures ANOVA: The statistical procedure for testing hypotheses about two or more means when the samples are repeated observations on the same subjects or are otherwise statistically related to each other; also called "dependent-samples" ANOVA. Page 333

Representative sample: A sample of a population that reflects the characteristics of the parent population. Page 126

Residual: What remains after both the subject effect and the occasion effect have been subtracted. Page 336

Robust: Insensitive to the violations of the assumptions. Page 218

Row effect: Deviation of a row mean from the grand mean. Page 359

Sample: Some subset of the group under consideration. Page 8

Sampling distribution of the means: The distribution formed by taking repeated samples from the same population, computing the mean of each sample, and forming the distribution of those sample means. Page 132

Scatter diagram: A graphical method of representing the relationship between two variables where one variable is plotted on the X-axis and the other is plotted on the Y-axis. Page 378

Seed: The point of entry into a random number table, usually chosen in a quasi-random manner. Page 128

Simple random sampling: A random sampling technique where all members of the population are treated equally regardless of their characteristics. Page 127

Slope: The change in Y divided by the change in X. Page 408

Spearman r_S: The correlation coefficient appropriate for ordinal data. Page 392

Standard deviation: A measure of the width of a distribution equal to the square root of the mean of the squared deviations. Page 73

Standard error of estimate: The standard deviation of the errors of prediction. Page 421

Standard error of r: The standard deviation of the distribution of sample correlation coefficients. Page 390

Standard error of the difference between two means: The standard deviation of the distribution that would result from taking two random samples from the same population, computing the sample means, and subtracting one from the other to form the difference between two means; and then repeating the process indefinitely often. Page 217

Standard error of the mean: The standard deviation of the sampling distribution of the means. Page 133

Standard error of the mean of the differences: The standard deviation of the distribution that would result from taking a random sample, measuring each member twice, computing difference scores for each member of the sample, and computing the mean of the difference scores for the sample; and then repeating the process indefinitely often. Page 251

Standard score (or z score): A variable whose value counts the number of standard deviations a score is above or below its mean. Page 97

Standardized variable: A normally distributed variable that has mean μ and standard deviation σ. Page 110

Statistic: Any measurement on a sample. Page 8

Statistically significant: Leading to the rejection of the null hypothesis. Page 172

Stratified random sampling: A sampling procedure where the population is divided into subgroups ("strata") that have the same or similar characteristics, and then simple random samples are taken from each stratum. Page 129

Studentized range statistic (Q): The test statistic used in the Tukey HSD procedure. Page 385

Student's t distribution: A family of distributions that, like z, are unimodal, symmetric, and asymptotic, but the exact shape (unlike z) depends on the degrees of freedom. Page 153

Sum of squares: The sum of the squared deviations. Page 74

Symmetric distribution: A distribution whose left side is a mirror image of its right side. Page 35

Tabular frequency distribution: An ordered listing of all values of a variable and their frequencies. Page 27

Test of independence: A nonparametric test of whether the shape of a distribution depends on the value of another variable (or variables). Page 447

Test statistic: A statistic specifically designed to facilitate the making of inferences. Page 169

True experiment: A study where the experimenter assigns subjects to groups at random. Page 213

Tukey's HSD (honestly significant difference) test: A particular kind of post hoc test appropriate for testing pairwise null hypotheses. Page 325

Two-way design: An experiment in which outcomes corresponding to several levels of two independent variables are observed. Page 350

Type I error: Rejecting the null hypothesis when it is in fact true. Page 170

Type II error: Failing to reject the null hypothesis when it is in fact false. Page 170

Unbiased estimator: A point-estimate such that if we repeated the point-estimating process infinitely often, the same number of point-estimates would be too high as too low. Page 145

Unimodal distribution: A distribution that has one most frequently occurring value. Page 35

Unit normal distribution: A normal distribution with $\mu = 0$ and standard deviation $\sigma = 1$. Page 102

Variable: A characteristic that can take several or many different values. Page 13

Variance: A measure of the width of a distribution equal to the mean of the squared deviations; it is the square of the standard deviation. Page 83

Variation: The width of a distribution; how much values of a variable differ from one another. Page 70

Wilcoxon matched-pairs signed-rank test: A nonparametric (ordinal data) test for two dependent samples. Page 454

I N D E X

CREDITS

Description	Formula	Equation number	Page
ANOVA effect size	$f = \dfrac{s_{\overline{X}}}{s_{\text{pooled}}}$ (as ratio of standard deviation) $\quad d_M = \dfrac{\overline{X}_{\text{max}} - \overline{X}_{\text{min}}}{s_{\text{pooled}}}$ (by analogy to two samples)	(13.16) (13.17)	299 299
Studentized range statistic for Tukey's post hoc tests	$Q = \dfrac{\lvert\overline{X}_i - \overline{X}_j\rvert}{\sqrt{\text{MS}_W/\tilde{n}}}$	(14.2)	325
Harmonic mean of sample sizes	$\tilde{n} = \dfrac{k}{\dfrac{1}{n_1} + \dfrac{1}{n_2} + \dfrac{1}{n_3} + \cdots + \dfrac{1}{n_k}}$	(14.3)	325
Standard error of a comparison	$s_{\text{sample comparison}} = \sqrt{s^2_{\text{pooled}}\left(\dfrac{C_1^2}{n_1} + \dfrac{C_2^2}{n_2} + \cdots + \dfrac{C_k^2}{n_k}\right)}$	(14.8)	329
Partition of the total sum of squares and total *df* in a repeated-measures ANOVA	$\text{SS}_{\text{total}} = \text{SS}_{\text{between occasions}} + \text{SS}_{\text{between subjects}} + \text{SS}_{\text{residual}}$ $df_{\text{total}} = df_{\text{between occasions}} + df_{\text{between subjects}} + df_{\text{residual}} = nk - 1$	(14.11) (14.12)	337 337
Mean square between subjects in repeated-measures ANOVA, definitional formula	$\text{MS}_{\text{between subjects}} = ks^2_{\overline{X}}$	(14.13)	337
Pearson product-moment correlation coefficient for a sample	$r = \dfrac{\sum z_X z_Y}{n-1} = \dfrac{\sum XY - \dfrac{\sum X \sum Y}{n}}{\sqrt{\left[\sum X^2 - \dfrac{(\sum X)^2}{n}\right]\left[\sum Y^2 - \dfrac{(\sum Y)^2}{n}\right]}}$	(16.1) (16.1a)	382 386
Spearman rank correlation coefficient	$r_s = 1 - \dfrac{6\sum D_i^2}{n(n^2-1)}$	(16.6)	392
Regression line	$\hat{Y} = bX + a$ (raw score form) $\quad \hat{z}_Y = rz_X$ (standard form)	(17.1) (17.6)	408 417
Slope of the regression line	$b = \dfrac{\sum XY - \dfrac{\sum X \sum Y}{n}}{\sum X^2 - \dfrac{(\sum X)^2}{n}}$ or $b = r\dfrac{s_Y}{s_X}$	(17.4) (17.3)	411 411
Intercept of the regression line	$a = \overline{Y} - b\overline{X}$	(17.5)	411
Error of prediction for the *i*th individual	$e_i = Y_i - \hat{Y}_i$	(17.7)	419
Standard error of estimate	$s_e = \sqrt{\dfrac{\sum(e_i)^2}{n-2}} = s_Y\sqrt{1-r^2}\sqrt{\dfrac{n-1}{n-2}}$	(17.9–17.10)	421
χ^2 test statistic for goodness of fit test and test of independence	$\chi^2 = \sum \dfrac{(O_i - E_i)^2}{E_i}$ $\quad [df = k - 1]$	(18.1)	444
Expected frequency of cell in contingency table	Expected frequency of cell $rc = \dfrac{f_r f_c}{N}$	(18.3)	449
Mann–Whitney *U* statistic, first	$U_1 = n_1 n_2 + \dfrac{n_1(n_1+1)}{2} - R_1$	(18.4)	453
Mann–Whitney *U* statistic, second	$U_2 = n_1 n_2 + \dfrac{n_2(n_2+1)}{2} - R_2$	(18.5)	453
Kruskal–Wallis *H* statistic	$H = \dfrac{12}{N(N+1)}\sum \dfrac{R_j^2}{n_j} - 3(N+1)$	(18.9)	457